Sinola
Estero de Agiabampo
Bahia Navachiste
Bahia de Topolobampo
a of Cortez
Isla de Carmen
Canal de San Lorenzo
La Paz
San Jose del Cabo
Cabo San Lucas
Boca de Soledad
Bahia Almejas
Bahia Magdalena
A Chart of
THE SOUTHERN RANGE
of the
GRAY WHALE

The Gray Whale
Eschrichtius robustus

The Gray Whale
Eschrichtius robustus

Edited by

Mary Lou Jones
Steven L. Swartz

Cetacean Research Associates
San Diego, California

Stephen Leatherwood

Hubbs–Sea World Research Institute
San Diego, California

with illustrations by

Pieter Arend Folkens

Oceanic Society
San Francisco, California

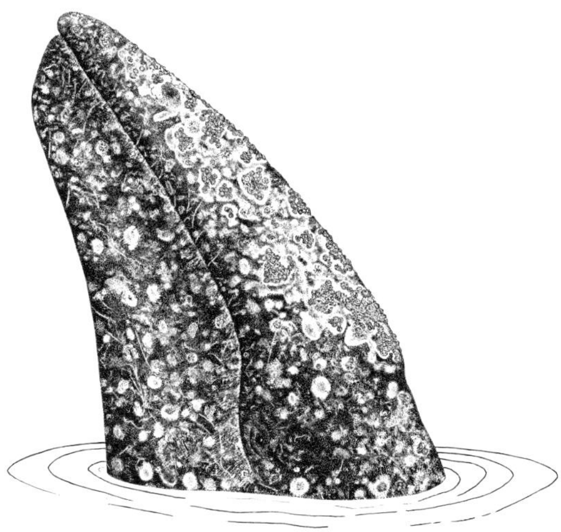

1984

ACADEMIC PRESS, INC.

(Harcourt Brace Jovanovich, Publishers)

Orlando San Diego New York London
Toronto Montreal Sydney Tokyo

ACADEMIC PRESS, INC.
Orlando, Florida 32887

United Kingdom Edition published by
ACADEMIC PRESS, INC. (LONDON) LTD.
24/28 Oval Road, London NW1 7DX

Library of Congress Cataloging in Publication Data

Main entry under title:

The Gray Whale.

Includes index.
1. Pacific gray whale. 2. Whaling--Pacific Coast (North America) II. Jones, Mary Lou. II. Leatherwood, J. Stephen. III. Swartz, Steven L.
QL737.C425G73 1984 599.5'1 84-3113
ISBN 0-12-389180-9 (alk. paper)

PRINTED IN THE UNITED STATES OF AMERICA

84 85 86 87 9 8 7 6 5 4 3 2 1

This volume is dedicated to
Carl L. Hubbs, Laura C. Hubbs (top), and Raymond M. Gilmore (bottom),
three pioneers of modern gray whale research.

Contents

Contributors

Numbers in parentheses indicate the pages on which the authors' contributions begin.

LAWRENCE G. BARNES (3), Section of Vertebrate Paleontology, Natural History Museum of Los Angeles County, Los Angeles, California 90007

ALFRED ANTONOVICH BERZIN (409), Pacific Ocean Science Research Institute of Fisheries and Oceanography, T.I.N.R.O., 690600 Vladivostok, U.S.S.R.

SERGEI ALEKSEEVICH BLOKHIN (487), Pacific Ocean Science Research Institute of Fisheries and Oceanography, T.I.N.R.O., 690600 Vladivostok, U.S.S.R.

LUYDMILA S. BOGOSLOVSKAYA (465), A.N. Severtsoff Institute of Evolutionary Morphology and Animal Ecology, Academy of Sciences of U.S.S.R., 117071 Moscow, U.S.S.R.

HOWARD W. BRAHAM (249), National Marine Mammal Laboratory, Northwest and Alaska Fisheries Center, National Marine Fisheries Service, N.O.A.A., Seattle, Washington 98115

PETER J. BRYANT (375), School of Biological Sciences, University of California, Irvine, Irvine, California 92717

MARILYN E. DAHLHEIM[1] (511), Department of Zoology, University of British Columbia, Vancouver, British Columbia V6T 2A9 Canada

JAMES D. DARLING[2] (267), University of Victoria, Victoria, British Columbia V8W 2Y2, Canada and Vancouver Public Aquarium, Vancouver, British Columbia V6B 3X8, Canada

JEANNETTE DA SILVA CURIEL (451), College of Veterinary Medicine, University of Saskatchewan, Saskatoon, Saskatchewan S7N 0W0, Canada

H. DEAN FISHER (511), Department of Zoology, University of British Columbia, Vancouver, British Columbia V6T 2A9, Canada

PIETER AREND FOLKENS, Oceanic Society, Suite E-225, Fort Mason, San Francisco, California 94123

JAMES T. HARVEY (561, 577), Marine Science Center, Oregon State University, Newport, Oregon 97365

DAVID A. HENDERSON (159), Whaling Museum, New Bedford, Massachusetts 02740

[1]Present address: National Marine Mammal Laboratory, Northwest and Alaska Fisheries Center, National Marine Fisheries Service, N.O.A.A., Seattle Washington 98115.

[2]Present address: West Coast Whale Research Foundation, Vancouver, British Columbia V7X 1L3, Canada.

DENISE L. HERZING (289), Marine Science Center, Oregon State University, Newport, Oregon 97365

MARY LOU JONES[3] (309), Moss Landing Marine Laboratory, Moss Landing, California 95035

IGOR I. KRUPNICK (103), Institute of Ethnography, Academy of Sciences of the U.S.S.R., 117036 Moscow, U.S.S.R.

CHRISTOPHER M. LAFFERTY[4] (375), School of Biological Sciences, University of California, Irvine, Irvine, California 92717

SUSAN K. LAFFERTY[5] (375), School of Biological Sciences, University of California, Irvine, Irvine, California 92717

DONALD K. LJUNGBLAD (543), Naval Ocean Systems Center, San Diego, California 92109

BRUCE R. MATE (289, 561, 577), Marine Science Center, Oregon State University, Newport, Oregon 97365

SAMUEL A. MCLEOD (3), Section of Vertebrate Paleontology, Natural History Museum of Los Angeles County, Los Angeles, California 90007

JAMES G. MEAD (33), Division of Mammals, National Museum of Natural History, Smithsonian Institution, Washington, D.C. 20560

EDWARD D. MITCHELL (33), Arctic Biological Station, Ste. Anne de Bellevue, Quebec H9X 3R4, Canada

SUE E. MOORE (543), SEACO, Inc., San Diego, California 92106

KAREN R. MORIN (451), Department of Zoology, University of British Columbia, Vancouver, British Columbia V6T 2A9, Canada

DEBRA J. MURIE (451), Department of Zoology, University of Guelph, Guelph, Ontario N1G 2W1, Canada

LAURIE D. MURISON (451), Department of Zoology, University of Guelph, Guelph, Ontario N1G 2W1, Canada

MARY NERINI (423), National Marine Mammal Laboratory, Northwest and Alaska Fisheries Center, National Marine Fisheries Service, N.O.A.A., Seattle, Washington 98115

BETH LAURA O'LEARY (79), Department of Anthropology, University of New Mexico, Albuquerque, New Mexico 87107

HIDEO OMURA (57), Whales Research Institute, Tokyo 136, Japan

M. MICHAEL POOLE (389), Division of Biology, Sonoma State University, Rohnert Park, California 94928

RANDALL R. REEVES[6] (187), Division of Mammals, National Museum of Natural History, Smithsonian Institution, Washington, D.C. 20560

[3]Present address: Cetacean Research Associates, San Diego, California 92107.

[4]Present address: 2157 Pattiz, Long Beach, California 90815.

[5]Present address: 2157 Pattiz, Long Beach, California 90815.

[6]Present address: Arctic Biological Station, Ste-Anne-de-Bellevue, Quebec, H9X 3R4 Canada.

STEPHEN B. REILLY (203), Southwest Fisheries Center, National Marine Fisheries Service, N.O.A.A., La Jolla, California 92038

DAVID J. RUGH (225), National Marine Mammal Laboratory, Northwest and Alaska Fisheries Center, National Marine Fisheries Service, N.O.A.A., Seattle, Washington 98115

HAZEL SAYERS (121), American Cetacean Society, San Diego, California 92122

JAMES D. SCHEMPP (511), Ship Silencing Support Division, Puget Sound Naval Shipyard, Bremerton, Washington 98314

STEVEN L. SWARTZ[7] (309), University of California, Santa Cruz, Santa Cruz, California 95064

ALEXEY V. YABLOKOV (465), N. K. Kol'tsoff Institute of Developmental Biology, Academy of Sciences of the U.S.S.R., 117334 Moscow, U.S.S.R.

[7]Present address: Cetacean Research Associates, San Diego, California 92107.

Preface

As student cetologists in the 1970s we learned that scholars had been at work for well over a century trying to piece together the gray whale story. Like many others interested in cetaceans, we often turned to this species to find a story with some sense of depth. The available accounts of other baleen whales were simply less comprehensive. But for the gray whale there existed an extensive body of "facts," mostly the offspring of a marriage of early scientific literature with a recent and rapidly expanding popular literature. Much of what appeared in those earlier publications derived from observations of an exploited whale population during its recovery from a depleted state, and presumed gray whales were increasing in the absence of any density-dependent reduction of their annual reproductive rate. During the 1970s, however, it was generally agreed that gray whales in the eastern Pacific had returned to a stable near preexploitation stock size of between 11,000 and 13,000 animals following their protection from commercial harvest. This created a renewed interest in gray whale population dynamics and reproductive biology as it was hoped that a better understanding of these whales in particular might lead to greater insight into the life processes, reproduction, and growth of large cetacean populations in general.

Much of what we knew or thought we knew about gray whales in the 1970s derived from the writings of Charles M. Scammon, Roy C. Andrews, Carl L. Hubbs, Gordon C. Pike, Raymond M. Gilmore, David A. Henderson, Theodore J. Walker, and others. In 1971 Dale W. Rice and Alan A. Wolman of the National Marine Mammal Laboratory published *The Life History and Ecology of the Gray Whale*, a book that contributed substantially to the understanding of the species. In 1974 the *Marine Fisheries Review* released a special issue devoted to the California gray whale, dealing largely with research on a single gray whale yearling, "Gigi." This animal was collected as a calf in Baja California in 1971, was accessible for study during its year in captivity, and was finally radio tagged and released into the ocean in March 1972.

With the above as background, investigators in the 1970s and early 1980s studied the accessible gray whale in all parts of its range, and research expanded rapidly. Due in part to legal protection afforded this species in Canada, Mexico, and the United States and in part to increased sensitivity to living whales, field researchers recently have accelerated the development and use of "benign nonconsumptive" techniques of study. The study of "live" long-lived mammals requires commitment to long-term programs to provide the basis for population trend analysis and meaningful behavioral studies; such long-term commitments have been beyond the scope of many cetacean field research programs. Therefore, even though the results of some programs completed in the past decades are exciting, most must be viewed as preliminary and not exhaustive treatments. Results of most remain unpublished, and accounts of their progress are scattered in technical and contract volumes difficult for even the most dedicated cetacean biologist to obtain.

Late in the 1970s, at numerous scientific meetings sponsored by the International Whaling Commission, World Wildlife Fund, International Union for Conservation of Nature, Sociedad Mexicana para el Estudio de Mamiferos Marinos, National Marine Fisheries Service–National Marine Mammal Laboratory, Marine Mammal Commission, and other agencies, this new information on gray whales was being discussed. It became increasingly apparent to the participants that a publication featuring the results of these recent programs was highly desirable. With this goal in mind, we announced our intention to collect and edit a volume of research papers on gray whales, and issued a call for contributions at the IV Biennial Conference on the Biology of Marine Mammals in November of 1981 in San Francisco, California. The response was overwhelming and far exceeded our greatest expectations. We and the colleagues who joined us believe this volume will fill a need for better distribution of information. In this sense we see *The Gray Whale* as a transitional vehicle, a summary of current information upon which to build. Scientific research almost always raises as many questions as it answers; it is our sincere hope that this volume will stimulate both more work and publication.

We are particularly pleased that *The Gray Whale* is not provincial. It contains contributions from researchers working in Canada, Japan, Mexico, the United States, and the Soviet Union and information from throughout the gray whale's past and present ranges both in the Atlantic and Pacific. Such diversity points up the fact that neither understanding nor conservation of this species can be effected without truly sincere international commitment and cooperation.

Finally, we wish to extend our thanks to all those individuals, organizations, agencies, translators, and reviewers, anonymous and otherwise, who graciously contributed their time, expertise, and energy to the production of *The Gray Whale.* Special thanks to Pieter Folkens for his original art for the subject sections, cover, and title page and Janice J. Sherer for her maps.

Introduction

The gray whale is clearly an important species from the standpoint of public educational and recreational value, aesthetic appeal, economic significance, and scientific interest. Because of its coastal habits the gray whale is the only large whale that can be regularly observed from shore. Its lengthy annual migration is one of the world's most outstanding wildlife spectacles, viewed by more than two million people each year.

The gray whale is the sole member of the family Eschrichtiidae and is considered by many to be the most primitive surviving baleen whale. There are two extant populations. One, the western Pacific or Korean population, is critically depleted. The other, eastern Pacific or California population, is a well-known conservation success. Whaling reduced the latter population in the 1850s, following the discovery of the breeding lagoons, and again after the turn of the century with the introduction of floating factories. Even so, this population twice made remarkable recoveries. Since it was afforded protection from commercial whaling in 1946 by the International Convention for the Regulation of Whaling, the population has grown to its current estimated level of 16,000 or more animals and continues to grow. Most whaling historians and biologists believe the preexploitation stock size was between 15,000 and 24,000 animals. Presently, a noncommercial quota is set each year. In the Arctic, Alaskan Eskimos from St. Lawrence Island still take a few gray whales annually, and Soviet-government catcher boats take about 140 to 200 gray whales each year for use by Siberian aborigines. In 1978 the International Whaling Commission (IWC) reclassified the California stock, moving it from protected to sustained-management status. Even with this reclassification and continuing annual harvest, however, the Scientific Committee (IWC) has acknowledged that coastal development and industrial activity pose very real threats to the survival of gray whales.

Collectively the chapters in *The Gray Whale* reflect the intimate relationship between gray whales and the coastal habitat in which they have evolved and now live. It is precisely their coastal habits that place gray whales in direct conflict with humans. They cannot avoid exposure to our pollution, vessel traffic, industrial noise, and activities associated with the development of outer continental shelf resources over virtually their entire range. Offshore oil and natural gas exploration and development by the U.S.S.R., the United States, Canada, and Mexico could negatively affect the gray whales on their nothern feeding grounds, during migration, and in their wintering grounds by disturbing these critical habitats. Industrial activities, fishing, commercial whale watching, and private boating along the entire west coast of North America may obstruct the progress of migrating gray whales. Finally, Mexico has jurisdiction over much of their winter range where reproduction is accomplished, and potential disturbances there include local industrial development and foreign recreational pursuits, mainly U.S. tourism.

The balance among aesthetics and scientific desires, local economic needs that contribute to environmental disturbance, and whale well-being is both a complicated and sensitive conservation issue. Gray whales can only be assured of survival if their lives, habitats, and food resources are managed according to an admittedly difficult but responsible ecosystem approach. Concern for the future of gray whales has provided an incentive for private and government support of much recent research within many disciplines, a trend we wish to see continue. Mexico's establishment of gray whale sanctuaries in Laguna Guerrero Negro, Laguna Ojo de Liebre, and Laguna San Ignacio, the first ever for cetaceans, is an encouraging step we hope will be matched by sensitive treatment of gray whale needs in other parts of their range.

The diversity of interest in gray whales was exemplified by the diversity of manuscripts we received. Contributions came from historians, anthropologists, paleontologists, biologists, and naturalists alike. In addition to chapters presenting new findings on the evolution, phylogeny, biology, demography, and behavior of gray whales, we received excellent chapters dealing with the history of their exploitation by both aborigines and industrialized nations. The papers we received fell into four more or less obvious categories which have become the major sections of this volume.

Section I presents two essays dealing with gray whale evolution, fossils, and subfossil remains, systematics, and range in historical times. Barnes and McLeod describe the only known bonafide gray whale fossil and examine the phyletic relationship between gray whales and other mysticetes. Mead and Mitchell provide a detailed review of the gray whale subfossil remains that have been discovered on both sides of the Atlantic Ocean and which confirm that the gray whale was, at sometime, common in the western Atlantic American waters and was taken commercially as recently as American colonial times.

Section II documents historical aspects of gray whale exploitation and the economic importance of these whales to humans, beginning with aboriginal societies in Asia and North America, and leading up to the most recent period of commercial exploitation at the turn of the century. Omura describes two populations of gray whales that migrated to the coastal waters of Japan prior to 1900 and suggests that the Seto Inland Sea of southern Japan served as a calving area. Krupnik traces the origins of aboriginal hunting along the Chukotka Peninsula and, based on excavations of ancient settlements, suggests that the gray whale, particularly young animals, and not the bowhead whale, was the primary object of much of the aboriginal hunting. O'Leary reviews the relevant ethnographic, historical, and archaeological data on aboriginal whaling and challenges previous assumptions about the importance of gray whale subsistence hunting by the Indian groups of the Pacific Northwest. Sayers reconstructs the rise and fall of the colorful and sometimes profitable shore whaling industry of the California coast, from its beginnings in Monterey in 1854 until its final collapse in 1901. Henderson's detailed account of the lagoon whaling period places this episode in the more general perspective of nineteenth century whaling throughout the Pacific. He describes how Yankee and European whalers pursued gray whales during the summer in the Bering, Okhotsk, and

Pacific Arctic Seas, and how the gray whale came to be the focus of "winter whaling" in the coastal waters of Alta (upper) California, Baja (lower) Caifornia, and northwestern Mexico. Reeves completes the section with an account of the modern commercial pelagic whaling that has led to the protective agreements and regulations that protect the species today.

Section III presents chapters on gray whale abundance, migration, and seasonal distribution in the wake of the California population's recovery from depletion. Reilly reviews the history of attempts to assess the gray whale population. He evaluates methods and assumptions used in shore-based censuses during migration and in aerial surveys of gray whales taken on their winter grounds, and addresses the discrepancies generated by the different methods. Rugh presents the results of shore-based censuses of the gray whale fall migration south past Cape Sarichef, Unimak Pass, Alaska during November and December 1977, 1978, and 1979. Braham compiled the National Marine Mammal Laboratory's extensive observations of gray whales in Alaska and has produced a chapter that describes the distribution of gray whales throughout Alaska's coastal waters and their migration through this portion of their range. Darling examines the abundance, distribution, and behavior of gray whales which are summer residents along the west coast of Vancouver Island, British Columbia. His studies of these whales are based on identifications of individual animals from photographs and resightings of these individuals over the 10-year period from 1972 to 1981. Herzing and Mate report the seasonal timing and distribution of gray whale migrations along the Oregon coast from November through May, 1978 through 1981, and demonstrate that these migrations have occurred in three major phases. Swartz and Jones present the findings of 5 years of research on gray whale use of Laguna San Ignacio as a breeding lagoon, including the occupation timetable, the abundance, distribution, and mortality of females with calves and whales without calves, inside the lagoon system and its nearshore waters, and a preliminary evaluation of the impact of whale-watching tourism on the whales wintering in this lagoon. Bryant and colleagues spent three winters studying the gray whales that occupy the smallest of the Baja Lagoons, Laguna Guerrero Negro. Their chapter documents the recent resurgence of gray whales into this lagoon, from which they were displaced during the 1960s.Poole provides an overview of the whales' spring migration out of their southern range based on his shore censuses at Point Piedras Blancas, California. His observations confirm that the departure of gray whales from the southern range is composed of two distinct waves of animals, single whales and female–calf pairs, separated by a month or more and using different migration corridors. Finally, Berzin reviews the recent findings of Soviet studies on the abundance and distribution of gray whales in the Bering and Chukchi Seas from 1968 to 1982.

Section IV contains reports on gray whale biology and behavior that include some very innovative approaches to the study of free-ranging cetaceans. Nerini presents a comprehensive review of gray whale feeding ecology, including a discussion of feeding mechanisms, feeding areas, prey items, rates of food consumption, and the influence of foraging on the prey community based on her research in the northern Bering Sea. She demonstrates that although gray whales rely primarily on shallow benthic invertebrate communities in the northern seas for their annual food supply, they do also frequently exploit other peripheral feeding areas that offer a wide variety of prey. Murison and colleagues describe the behavior of gray whales feeding on the mysid crustaceans that form dense aggregations over the rocky bottom during the summer months off Vancouver Island, British Columbia. Dahlheim and colleagues characterize both the sound produced by gray whales and the ambient noise, both biological and nonbiological, in Laguna San Ignacio, Baja California and propose a model which may account for the way whales and dolphins deal with background noise in their lagoon habitat. Moore and Ljungblad provide data on the distribution and sound production of gray whales in the northernmost portion of their range, the Beaufort, Chukchi, and Bering Seas. Chapters by Mate and Harvey and Harvey and Mate present results of studies of cetacean migration and behavior using radiotelemetry. Within Laguna San Ignacio they obtained continuous measurement of gray whale dive times and surface intervals from whales fitted with radio tags. As the tagged animals left the lagoon, their ocean movements and migration rates were followed from land and aerial receiving stations. Finally, we have two chapters reviewing the most recent findings of Soviet researchers working on gray whales. These include papers by Blokhin, and Yablokov and Bogoslovskaya on the morphology, growth and age, reproduction, feeding, and grouping and social behavior of gray whales on their summer feeding grounds.

As editors we have held ourselves mainly responsible for clarity of content but have left idiosyncracies of style largely to the authors. The contents of each chapter and the opinions expressed therein are the responsibility of the authors. There are some noticeable holes in our coverage. We could not accommodate all recently completed work, the results of some important research in progress were unavailable, and there are major questions about the species which are now identified but not yet addressed. The luxury of time to wait for a more complete picture of the gray whale to emerge was not available; even so, in view of the increasing amount of new findings, we consider it important to bring the record of research on this species up to date. Much of the research presented here is the first of its kind and will undoubtedly be revised and added to in the near future as our understanding of this species continues to grow.

I

Evolution, Fossils, and Subfossil Remains

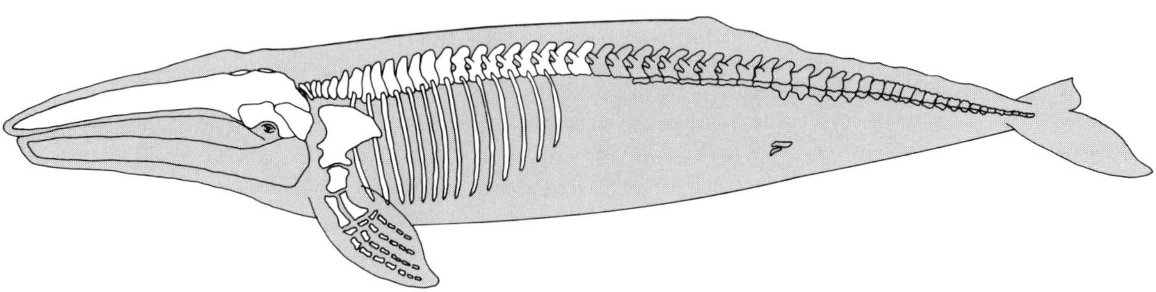

1

The Fossil Record and Phyletic Relationships of Gray Whales

Lawrence G. Barnes and Samuel A. McLeod

Introduction

The systematics of mysticete cetaceans are in serious need of revision, and analyses of their phylogenetic relationships have rarely been attempted. Vertebrate morphologists often rely heavily upon skull anatomy in studies of phyletic relationships and in preparing classifications. Fossil evidence is also commonly used in biogeographic and phyletic studies of animal groups, and in this chapter we review the morphological evidence from both fossil and extant mysticetes to investigate the relationships between gray whales and other mysticetes. The derivation and polarity of characters of mysticetes and of the order Cetacea must be understood within an evolutionary context in order to evaluate whale morphology objectively. We have compared the skulls (which we have illustrated) and other anatomical structures among mysticete families. We review reputed fossil gray whales, present a record of a pre-Holocene fossil gray whale, and suggest some of the reasons for such a poor fossil record of gray whales.

We believe the Cetacea, Mysticeti, and Odontoceti are monophyletic taxa, but there are major gaps in the fossil record which now prevent a comprehensive phyletic

ISBN 0-12-389180-9

analysis. The most primitive mysticetes are known from the Oligocene Epoch, but the fossil record of cetaceans of that age is poor, and we cannot yet clearly define the origin of the Mysticeti or any of the included families. The Cetotheriidae, a family of fossil mysticetes, may contain the ancestry of one or more extant families, but the family is clearly nonmonophyletic and needs revision. Thus we have had to rely on morphological evidence from the other extant families of mysticetes (Balaenidae, Balaenopteridae, and Neobalaenidae) in order to analyze the Eschrichtiidae.

The Fossil Record of Gray Whales

REPUTED FOSSIL GRAY WHALES

Part of a fossil rostrum from Late Pleistocene deposits on the north slope of Alaska has been identified by Repenning (1983) as *Eschrichtius* sp. The specimen, found north of Teshekpuk Lake, was considered by Repenning to have been older than 51,000 years, but according to him (personal communication, 2 February 1984) is no longer available.

We find references to extinct species in the genus *Eschrichtius* Gray, 1864b, in the scientific literature as far back as the middle 1800s (Cope, 1868a). We are initially cautious, because in many cases the generic name was applied to fossils of Miocene age. Owing to the relatively rapid evolutionary rates of most mammalian groups, living mammal genera are rarely identified so far back in geologic time (Simpson, 1953). Upon detailed case-by-case examination of the older names, we find the original type material was usually relatively fragmentary. Many of these fossil species were first described in the voluminous and confusing series of studies by Cope in the latter half of the nineteenth century. Kellogg (1968) reviewed the status of some of the species, and Gillette (1975) summarized the data on the specimens in the Academy of Natural Sciences in Philadelphia (ANSP), which includes much of the type material.

It is well to bear in mind that *Eschrichtius* was initially used for subfossil (Holocene Epoch) bones (Lilljeborg, 1861, 1867; Gray, 1864b, 1865) and was later applied to fossil (Tertiary Period) specimens (Cope, 1868b,d, 1869a, 1870a, 1872). *Rhachianectes* Cope, 1869b, however, was used for the living gray whale (Cope, 1868c,e, 1869b). That the subfossil and living animals were the same species was not demonstrated until Van Deinse and Junge (1937) compared the subfossil bones from Europe with the skeleton of the living gray whale from the North Pacific. They introduced a new problem into the classification, however, by using the new binomen *Eschrichtius gibbosus* (Erxleben, 1777) for the living and subfossil gray whales, by combining *Eschrichtius robustus* (Lilljeborg, 1861) and *Balaena gibbosa* Erxleben, 1777 (which was based on the "scrag whale" of Dudley, 1725). When Cope (1868c) named the living gray whale, he erected the genus *Agaphelus* Cope, 1868c, to contain both the Atlantic "scrag whale", *A. gibbosus* (Erxleben, 1777), and the Pacific gray whale, *A. glaucus* Cope, 1868c. Cope later (1969b) erected the separate genus *Rhachianectes* for *A. glaucus*, and (1884) realized that his *A. gibbosus* was actually a specimen of *Balaenoptera acutorostrata* Lacépède, 1804. Thus

Table I
Previously Described Reputed Fossil Gray Whales

Name	Current status
Eschrichtius cephalus Cope, 1868a	Balaenopteridae, cf. *Balaenoptera*
Eschrichtius davidsonii Cope, 1872	Aff. *Balaenoptera acutorostrata*
Eschrichtius expansus (Cope, 1868d)	Invalid, *nomen vanum*
Eschrichtius leptocentrus Cope, 1868b	Invalid, *nomen dubium*
Eschrichtius mysticetoides (Emmons, 1858)	Invalid, *nomen dubium*
Eschrichtius polyporus Cope, 1870a	Invalid, *nomen vanum*
Eschrichtius priscus (Leidy, 1852)	Invalid, *nomen vanum*
Eschrichtius pusillus (Cope, 1868c)	Invalid, *nomen vanum*

Eschrichtius gibbosus of Van Deinse and Junge (1937) is not a synonym of *Agaphelus gibbosus* as used by Cope (1868c).

Further confusing matters, Cope (1890, 1895a,b, 1896) switched generic names, using the genus *Plesiocetus* Van Beneden, 1859, for *Cetotherium* Brandt, 1843, and *Cetotherium* for *Eschrichtius*. His (Cope, 1890) *Cetotherium* then included *Eschrichtius pusillus* (Cope, 1868c), *E. expansus* (Cope, 1868d), *E. priscus* (Leidy, 1852), *E. polyporus* Cope, 1870a, *E. mysticetoides* (Emmons, 1858), *E. cephalus* Cope, 1868a, *E. leptocentrus* Cope, 1868b, and *E. robustus*, but not *E. davidsonii* Cope, 1872, which he then called *Balaenoptera davidsonii*. Thus one can find as many as three confusing and in part synonymous names used in the same work. For example, Cope (1868d) listed *Agaphelus glaucus*, *A. gibbosus*, and *Eschrichtius? robustus;* Van Beneden and Gervais (1880) listed *Rachianectes* (sic) *glaucus*, *Agalephus* (sic) *gibbosus*, and *Balaenoptera* (*Plesiocetus*) *robustus* (sic), and Trouessart (1898) listed *Plesiocetus robustus*, *Agaphelus gibbosus*, and *Rachianectes* (sic) *glaucus*.

The following is a synopsis of the original and subsequent references for the nominal species of fossil gray whales, listed alphabetically, and their current status. Table I summarizes this information.

Eschrichtius cephalus

This species was established on the basis of mandibles, parts of a skull, limb bones, and vertebrae of one individual of Middle Miocene age from the Calvert Formation in Maryland (Cope, 1868a,b, 1890; Gillette, 1975, p. 64). The holotype is Academy of Natural Sciences of Philadelphia (ANSP) number 12691–12692. The original describer, Cope, later (1890, pp. 612, 616; 1895a, p. 141; 1895b) changed its generic assignment to *Cetotherium*.

Kellogg (1968, pp. 105–113) reviewed the holotype and referred specimens, illustrated the critical parts of the skeleton, and came to the different conclusion that they represent a balaenopterid whale that compares well with the modern genus *Balaenoptera* Lacépède, 1804. We agree that the specimen is not related to *Eschrichtius*.

Eschrichtius davidsonii

Cope (1872) based this name on a fragment of fossil baleen whale mandible (ANSP 12922) found in the Pliocene age San Diego Formation at San Diego, California (Gillette, 1975, p. 64). We believe that Bowers (1889) incorrectly referred bones of Pleistocene age from Ventura, California, to this same species.

Cope's original (1872) generic assignment was incorrect, if only because the holotype fossil mandible fragment has a large coronoid process and *Eschrichtius* does not (Barnes, 1976, p. 333). Cope later (1890, p. 616) changed his mind and put this species in *Balaenoptera.*

Eschrichtius expansus

Originally Cope (1868d) placed this species in the genus *Megaptera* Gray, 1846, but later transferred it first (1869a) to *Eschrichtius,* then (1890) to *Cetotherium,* and finally (1895a,b) to *Siphonocetus* Cope, 1895a (see also *Eschrichtius priscus*). The original basis of the species is a collection of vertebrae (ANSP 12769) labeled as types and cotypes (Kellogg, 1968, p. 116). Unfortunately, these were not only from different animals but were also found in different states (Maryland and Virginia), probably in the Middle Miocene Calvert Formation. Kellogg (1968, p. 118) correctly declared the species invalid because it was based on nondiagnostic specimens.

Eschrichtius leptocentrus

This species was based on a cervical vertebra (ANSP 12693) of Miocene age from Virginia (Cope, 1868b,d, 1869a; Kellogg, 1968, p. 105; Gillette, 1975, p. 65), which Cope believed resembled *Eschrichtius* by having a large neural canal. Later (1890, 1895a, 1896) he put it in *Cetotherium.* We see no basis for naming this species based on a vertebra nor for placing it in the genus *Eschrichtius,* and declare it a *nomen dubium.*

Eschrichtius mysticetoides

Emmons (1858, p. 205, Fig. 26) described *Balaena mysticetoides* based upon an earbone of Miocene age from North Carolina. Cope (1875, p. 42) listed the species in *Eschrichtius,* but in 1890 (p. 616) he assigned it to *Cetotherium.* Trouessart (1898) listed it as both *Cetotherium mysticetoides* (p. 1071) and *Balaena mysticetoides* (p. 1092). The original specimen, an incomplete auditory bulla, was not given an institutional catalog number, and its present location is not known. We agree with Dr. Frank C. Whitmore (personal communication) that the figure presented by Emmons (1858, Fig. 26; not Fig. 27 as cited by Cope, 1875, p. 42) is a nondiagnostic auditory bulla of an indeterminate mysticete other than the genus *Eschrichtius.*

Eschrichtius polyporus

This species (see Cope, 1870a, p. 192; 1870b, p. 285; 1872, p. 30; 1875, pp. 42–43) was based on two mandible fragments (ANSP 12694) of Miocene age from North

Carolina (Gillette, 1975, p. 65). Cope later 1890, 1895a,b) transferred the species to *Cetotherium*. Because of the fragmentary nature of the specimens, we regard the species as a *nomen vanum*.

Eschrichtius priscus

Leidy originally (1852) called this species *Balaena prisca*. It was based on a jaw fragment (ANSP 12915) of Miocene age from Virginia (see Gillette, 1975, p. 65). Cope (1868b, p. 144; 1868d, p. 193) first cited it as *Balaenoptera prisca*. He then (1869a, p. 11; 1872, p. 30) placed it in *Eschrichtius* and was followed in doing so by Leidy (1869). In 1890 (p. 616) Cope used the combination *Cetotherium priscum* Leidy, but this combination was preoccupied by *Cetotherium priscum* (Eichwald, 1840) (see Cope, 1895a, p. 147). Later Cope (1895a, p. 140; 1895b) made *Balaena prisca* the type species of his new genus *Siphonocetus*, and Kellogg (1928, p. 188, Fig. 18) and Barwick (1938, 1940) referred skulls and another mandible to the species. Subsequently, however, Kellogg (1968, p. 127) declared the species invalid because the holotype mandible fragment is not diagnostic.

Eschrichtius pusillus

This species was first described by Cope (1868c, p. 159) as *Balaenoptera pusilla*, based on a mandible fragment and vertebrae (ANSP 12917) of Miocene age from Maryland (see Gillette, 1975, p. 65). Cope later listed it as a species of *Eschrichtius* (1868d, p. 193) and then (1890, pp. 612, 616; 1895a, pp. 141, 145) called it *Cetotherium pusillum*. Trouessart (1898, p. 1071) found the second combination, however, to be preoccupied by *Cetotherium pusillum* Nordmann, 1860 and so proposed the new binomen *Cetotherium parvum* Trouessart, 1898, to replace Cope's name. Kellogg (1968, p. 128) declared Cope's species invalid because it was a composite and based on nondiagnostic bones.

THE SAN PEDRO GRAY WHALE

In 1971, a fossil gray whale skeleton was collected in California (Barnes, 1976, pp. 334–335; McLeod and Barnes, 1983). To our knowledge it is the only currently available fossil referable to the family. It is described as follows.

Systematics

Family Eschrichtiidae Ellerman and Morrison-Scott, 1951

Agaphelidae Gray, 1871, p. 47; type genus: *Agaphelus* Cope, 1868c.
Rhachianectidae Weber, 1904, p. 575; type genus: *Rhachianectes* Cope, 1869b.
Eschrichtiidae Ellerman and Morrison-Scott, 1951, p. 713; type genus: *Eschrichtius* Gray, 1864b.

Genus *Eschrichtius* Gray, 1864

Balaenoptera Lacépède, 1804 (part); Lilljeborg, 1861, p. 602; used for *Balaenoptera robusta* Lilljeborg, 1861.
Megaptera Gray, 1846 (part); Gray 1864b, p. 350; to include *Megaptera* ? (*Eschrichtius*) *robusta* (Lilljeborg, 1861).
Eschrichtius Gray, 1864b, p. 350; type species: *Balaenoptera robusta* Lilljeborg, 1861; used as a subgenus.
Eschrichtius Gray, 1864b; Gray 1865, p. 494; first use at generic rank.
Agaphelus Cope, 1868c, p. 159 (part); type species: *Balaena gibbosa* Erxleben, 1777.
Rhachianectes Cope, 1869b, p. 15; type species: *Agaphelus glaucus* Cope, 1868c.
Eschrichitus Cope, 1872, p. 30; misspelling.
Rachianectes Cope, 1879, p. 655; Van Beneden and Gervais, 1880, p. 235; Trouessart, 1898, p. 1087; Elliot, 1901, p. 9; incorrect spelling.
Balaenoptera Lacépède, 1804 (part); Van Beneden and Gervais, 1880, p. 290; to include *Balaenoptera* (*Plesiocetus*) *robustus* (sic) Lilljeborg, 1861.
Cetotherium Brandt, 1843 (part); Cope, 1890, p. 614; replacement of generic name for *E. robustus* (Lilljeborg, 1861).
Plesiocetus Van Beneden, 1859 (part); Trouessart, 1898, pp. 1072, 1074; generic assignment for *Eschrichtius robustus* (Lilljeborg, 1861).
Rhachionectes Winge, 1921, pp. 20, 22; incorrect spelling.

Type species: *Balaenoptera robusta* Lilljeborg, 1861, a subfossil skeleton from Gräsön, Roslagan, Upland, Sweden, in the University Museum of Upsala; type by original designation by Gray, 1864.

Note: *Balaena gibbosa* Erxleben, 1777, is considered by some authors (e.g., Van Deinse and Junge, 1937; Hall and Kelson, 1959; Hershkovitz, 1966) to be the type species of *Eschrichtius* but the species was based on the "scrag whale" (Dudley, 1725, p. 258) of the New England coast, for which there is no type specimen (Schevill, 1952). Mead and Mitchell (this volume) consider its taxonomic use to be inappropriate.

Eschrichtius cf. *E. robustus* (Lilljeborg, 1861)
(Figs. 1 and 2)

cf.*Eschrichtius*. Barnes, 1976, pp. 334–335.

Specimen. Natural History Museum of Los Angeles County (LACM) 122322, a partial skeleton, including the skull, vertebral column from the cervicals to the anterior lumbars, ribs, and limb bones, found by Paul R. Kirkland in December, 1970. Collected in late January and early February, 1971 by volunteers, students, and staff of Carson High School, Harbor College, California State University at Northridge, Cabrillo Beach Museum, and the Natural History Museum of Los Angeles County.

Locality. LACM locality 3270, between Gaffey Street and Harbor Freeway (California Highway 11) in a cut slope between property owned by Westoil Terminals Co. and the City of Los Angeles, in the district of San Pedro, Los Angeles County, California, 33°45′51″ North latitude, 118°17′12″ West longitude, at an elevation of approximately 110 ft (33.5 m).

Formation and Age. San Pedro Sand, late Pleistocene, older than approximately 50,000 years and possibly somewhat older than 100,000–120,000 years.

Morphology

We see no osteological basis for separating the fossil skull (Figs. 1 and 2) from modern *Eschrichtius robustus* (see Fig. 5a, and Andrews, 1914, Plate 25). The fossil is typical of gray whales in that it has a narrow and moderately arched rostrum whose bones interdigitate with the cranial bones and are markedly telescoped into the cranial vertex, a relatively small and steeply sloping occipital shield, a large dorsal narial opening, and temporal fossae widely open dorsally. The nasal bones on the fossil skull are more tapered posteriorly than in the modern skull we illustrate in Fig. 5a, because of median compression between the posterior ends of the premaxillae. This is a variable feature, however, among Recent gray whales (cf. Andrews, 1914, Plate 25, Figs. 1 and 2).

The fossil skull resembles modern gray whale skulls in such other details as large nasal bones, presence of paired tuberosities on the occipital shield, massive and short zygomatic processes of the squamosals, large and posteriorly directed paroccipital processes, large occipital condyles, lateral walls of the cranium bulging into the temporal fossae, several large maxillary foramina scattered lateral to the narial region, premaxillae and maxillae rising nearly vertically on either side of the elevated narial opening, a thick lateral process of the maxilla anterior to the supraorbital process of the frontal, and premaxillae peaked medially on the rostrum over the mesorostral groove anterior to the narial opening.

Age and Paleoecology

The gray whale skeleton was found in a slightly bedded, fine-grained sand. This sand, which we identify as the San Pedro Sand, is widespread in the area and during a Pleistocene interglacial period was deposited around an island created by the present Palos Verdes Hills (Woodring *et al.*, 1946; Marincovich, 1976). Several feet above the gray whale was an unconformity, and above that, a cobble-bearing quartz sand, which we identify as the Palos Verdes Sand. This overlying deposit represents a marked change, after deposition of the skeleton, to a shallow-water and/or nearshore environment, and contains a diverse assemblage of fossil mollusks and vertebrates, including the harbor seal, *Phoca* cf. *P. vitulina*, reported by Barnes and Mitchell (1975), and the tooth of a horse, *Equus* sp. At other localities the Palos Verdes Sand has produced fossils of *Bison* sp. (Miller, 1971), indicating that in the North American land mammal chronology (Savage, 1951) this formation overlying the gray whale skeleton is Rancholabrean in age. While the Rancholabrean Age is usually considered to have extended from 10,000 to 50,000 years BP (Evernden *et al.* 1964), Fanale and Schaeffer (1965) have reported dates for the Palos Verdes Sand that are as old as 100,000 to 120,000 years,

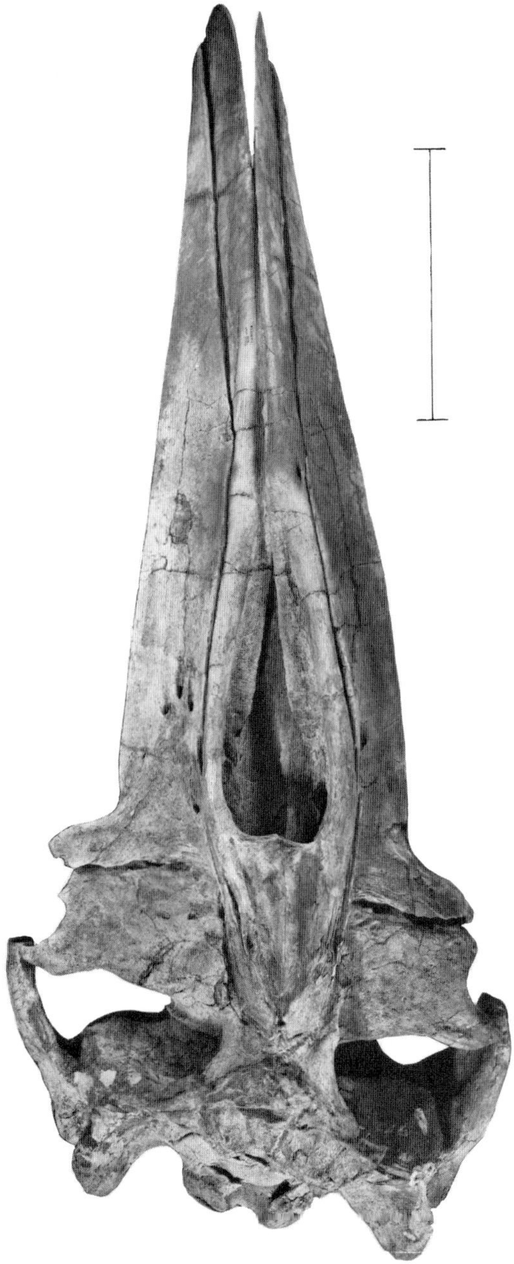

Fig. 1. *Dorsal view of fossil skull of* Eschrichtius *cf.* E. robustus *(Lilljeborg, 1861), LACM 122322, from San Pedro, California, Pleistocene, San Pedro Sand. Scale line = 0.5 m.*

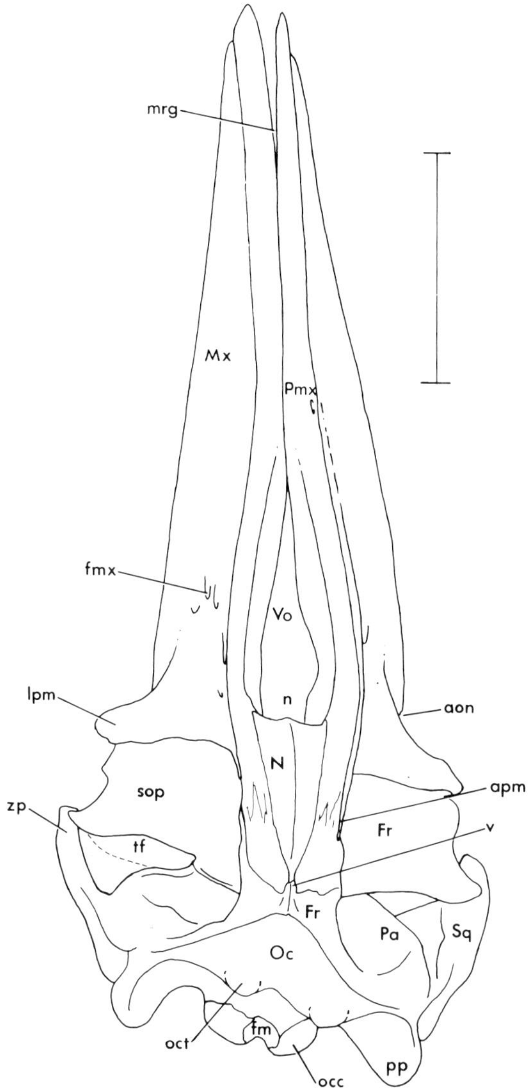

Fig. 2. *Dorsal view of fossil skull of* Eschrichtius *cf.* E. robustus *(Lilljeborg, 1861), LACM 122322, Pleistocene, San Pedro California, with anatomical structures labeled. Abbreviations: aon, antorbital notch; apm, ascending process of the maxilla; fm, foramen magnum; fmx, maxillary foramina (there are several on each side); Fr, frontal; lpm, lateral process of the maxilla; mrg, mesorostral groove; Mx, maxilla; N, nasal; n, narial opening; Oc, occipital; occ, occipital condyle; oct, occipital tuberosity; Pa, parietal; Pmx, premaxilla; pp, paroccipital process; sop, supraorbital process of the frontal; Sq, squamosal; tf, temporal fossa; v, cranial vertex; Vo, vomer; zp, zygomatic process of the squamosal. Scale line = 0.5 m.*

based on U–He and U–Th ratios. Thus the fossil gray whale, being in the San Pedro Sand, was deposited prior to the advent of the Rancholabrean North American Land Mammal Age at least 50,000 years ago and possibly more than 100,000–120,000 years ago.

Either while the whale's carcass was still floating in the ocean and being scavenged by sharks, or during the time it lay on the ocean bottom, the lower jaws became separated from the body and were transported elsewhere. The rest of the skeleton remained virtually articulated, lying ventral side up, and with its anterior end pointing approximately due north. Numerous cuts on the bone of the snout prove that sharks were able to bite through the flesh to the premaxillae and maxillae. One tooth of the great white shark, *Carcharodon* sp., lay on its inverted palatal surface, but we cannot prove whether this was lost by one of its original killers or by a postmortem scavenger. The carcass must have been buried relatively quickly because the ribs were virtually connected to the vertebral column, and the forelimb bones were only slightly disarticulated from the body. Decomposition of the flesh, however, was quick enough for bones to be exposed prior to burial in sediment, because the articulated skeleton of a fish was found lying directly on the upper surface of the left scapula.

Much of the Los Angeles basin was inundated by the sea during parts of the Pleistocene, and the present Palos Verdes Peninsula was then an island measuring up to 7 miles long by 3 miles across but varying in size depending on glacial or interglacial stages. During deposition of the San Pedro Sand when the gray whale skeleton was deposited, this island was 8 to 10 miles offshore, and the skeleton came to rest on the sandy ocean bottom off the landward side of the island. Later in the Pleistocene Epoch, the water became relatively shallower at the site, and during the Rancholabrean Land Mammal Age the shore was close enough that while the coarse Palos Verdes Sand was being deposited, parts of land animals were being washed in by streams flowing from the land. Further uplift and/or eustatic sea level drop in comparatively very recent (geological) time caused the Palos Verdes Hills and surrounding area to become a peninsula, and the landward side of the former island now stands above sea level.

Phyletic Relationships of the Gray Whale

THE ANCESTRY OF MYSTICETES

In our opinion the order Cetacea is a monophyletic evolutionary group, in which the two monophyletic modern suborders Odontoceti (toothed whales) and Mysticeti (baleen whales) both have their ultimate ancestries within the primitive suborder Archaeoceti (see Barnes and Mitchell, 1978). This belief was not held, however, by such earlier authors as Thompson (1889), Beddard (1900), True (1908), Gidley (1913), Miller (1923), Kleinenberg (1958, 1959), and Yablokov (Kleinenberg and Yablokov, 1958; Yablokov, 1964) who based their opinions upon some items of information which today are considered incorrect or invalid. For example, Van Valen (1968), responding to papers by

Yablokov (1964) and Kleinenberg (1958, 1959), rejected a list of such commonly cited misconceptions.

Dorudontine archaeocetes (see Fig. 3a) probably gave rise to both primitive mysticetes and odontocetes in the Oligocene (Barnes and Mitchell, 1978, pp. 594–595; Fordyce, 1980, Fig. 1). Fordyce (1977) has reaffirmed that middle Oligocene species of *Mauicetus* Benham, 1939 are the earliest known mysticetes and have cranial features very much like those of archaeocetes (Marples, 1956; Mchedlidze, 1970; Whitmore and Sanders, 1976, p. 317; Fordyce, 1977, p. 266; 1980). *Mauicetus*, classified in the Cetotheriidae, was edentulous and probably had baleen but had a primitive degree of telescoping. *Aetiocetus* Emlong, 1966 (Fig. 3b), however, which is latest Oliogcene in age, still retained teeth and had a skull that generally resembled that of *Mauicetus* (Fig. 3c) but was more telescoped. It is classified in the monotypic family Aetiocetidae.

Obviously, different lineages were evolving at different rates. The important point is that among the earliest mysticetes in Oligocene time are animals having primitive characters shared with archaeocetes, derived mysticete cranial features, and in some taxa, vestigial teeth.

OVERVIEW OF MYSTICETE FAMILIES

At present, there are five families of fossil and modern baleen whales commonly recognized. Listed roughly from most primitive to most derived these are Aetiocetidae, Cetotheriidae, Eschrichtiidae, Balaenopteridae, and Balaenidae. We have resurrected a sixth, the extant family Neobalaenidae (see Gray, 1874; Miller, 1923) for *Caperea* Gray, 1864a, recognizing the differences between it and the balaenids.

Aetiocetidae

We consider the family to be represented solely by the latest Oligocene *Aetiocetus cotylalveus* Emlong, 1966 (see Fig. 3b). The family was originally assigned to the suborder Archaeoceti by Emlong (1966, p. 3), a decision based primarily on the presence of teeth in the species. Van Valen (1968) transferred it to the Mysticeti based on relationships of cranial bones and considered its retention of teeth to be merely primitive (see also Barnes, 1976, p. 324; Barnes and Mitchell, 1978, p. 595). Other authors have considered it to be intermediate between archaeocetes and mysticetes (Thenius, 1969, p. 489; Whitmore and Sanders, 1976, p. 317).

Aetiocetus is definitely a mysticete because it has a long, flat rostrum, a premaxilla that is not dense and lacks a foramen like that of odontocetes, a single bony dorsal naris, elongate nasal bones, broad, flat supraorbital processes of the frontals, a lateral maxillary process sweeping along the anterolateral corner of the supraorbital process, a thick jugal, a triangular, forward-thrusting occipital shield, an elongate zygomatic process of the squamosal with an indistinct glenoid fossa for articulation of the mandible, a wide basioccipital with large lateral tuberosities, a periotic firmly ankylosed to the skull and bearing a long posterior process, and a middle ear sinus system restricted mainly to a peribullary sinus.

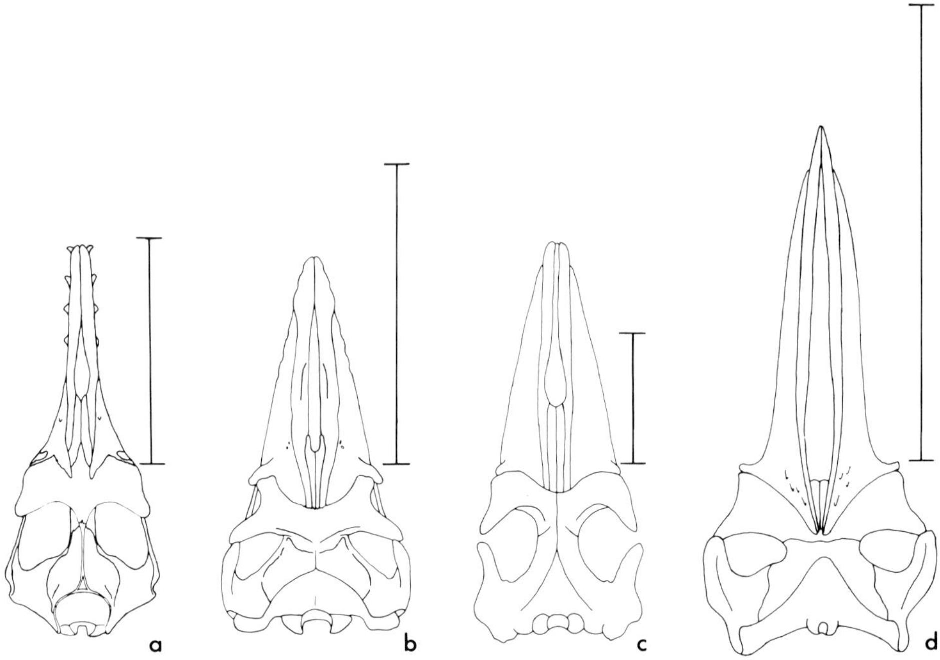

Fig. 3. *Dorsal views of skulls of an archaeocete and some primitive fossil mysticetes: (a)* Zygorhiza kochii *(after Kellogg, 1936, Fig. 29); (b)* Aetiocetus cotylalveus *(after Emlong, 1966; Fig. 3); (c)* Mauicetus lophocephalus *(after Marples, 1956; Fig. 1c); (d)* Cetotherium rathkii *(after Van Beneden and Gervais, 1880, Plate 17, Fig. 6; Brandt, 1873, Plate 2, Fig. 1). For this and all subsequent figures skulls are reduced to the same cranium length, measured from the occipital condyles to the antorbital notch. Scale line = 0.5 m.*

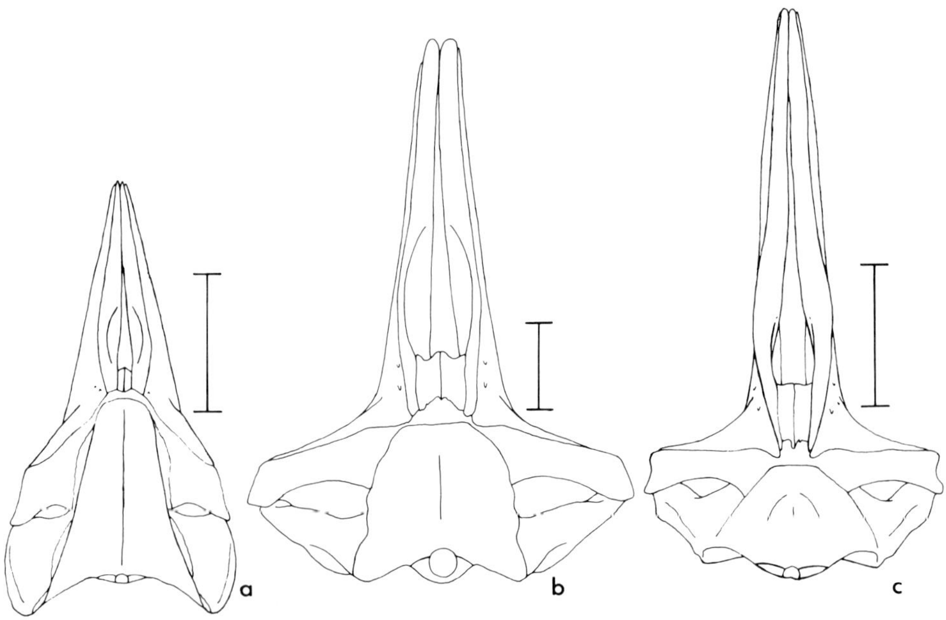

Fig. 4. *Dorsal views of skulls of living mysticetes: (a)* Caperea marginata *(after Oliver, 1922, Plate 1, Fig. 2); (b)* Eubalaena glacialis *(after True, 1904, Plate 42, Fig. 1); (c)* Balaena mysticetus *After Nishiwaki and Kasuya, 1970b, Plate 2, Fig. 1). Scale line = 0.5 m.*

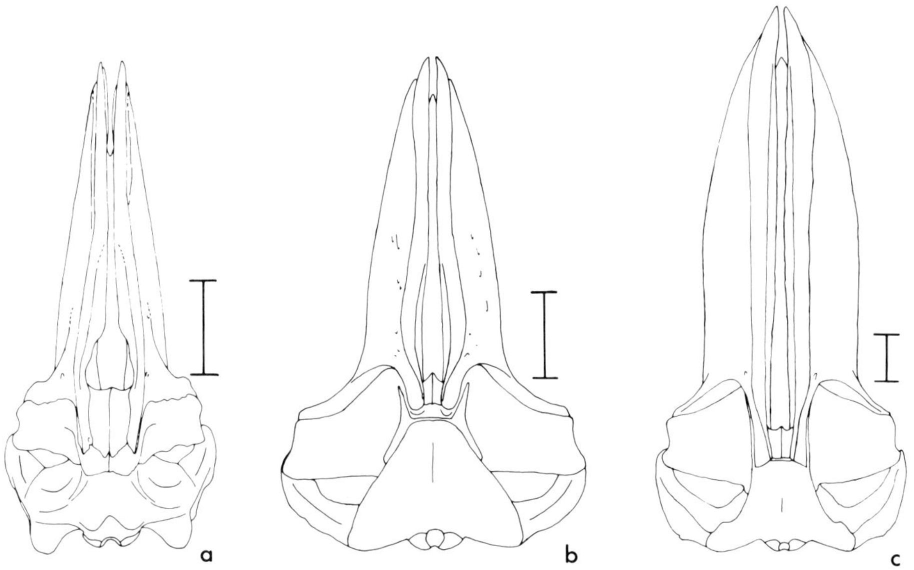

Fig. 5. *Dorsal views of skulls of living mysticetes: (a)* Eschrichtius robustus *(after Andrews, 1914, Plate 25, Fig. 2); (b)* Megaptera novaeangliae *(after True, 1904, Plate 29, Fig. 2); (c)* Sibbaldius musculus *(after Miller, 1924a, Plate 1). Scale line = 0.5 m.*

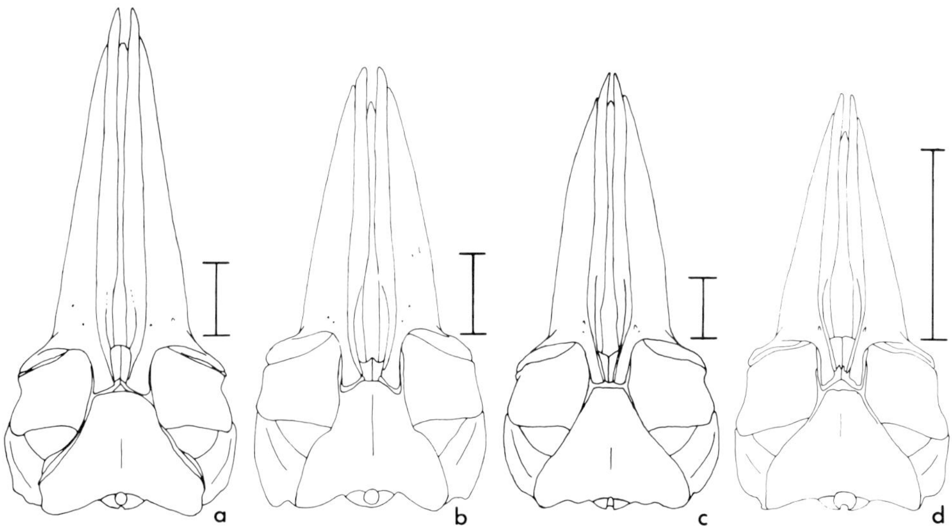

Fig. 6. *Dorsal views of skulls of living mysticetes: (a)* Balaenoptera borealis *(after Miller, 1924b, Plate 1); (b)* Balaenoptera physalus *(after True, 1904, Plate 1, Fig. 3); (c)* Balaenoptera edeni *(after Omura, 1959, Plate 1, Fig. 1); (d)* Balaenoptera acutorostrata *(after True, 1904, Plate 22, Fig. 1). Scale line = 0.5 m.*

Cetotheriidae

Contrary to Simpson (1945, p. 104) the authorship of this family name is attributable not to Cabrera (1926) but to Brandt (1872). Cetotheriinae was first proposed as a subfamily of Balaenopteridae by Brandt (1872, p. 120). His earlier (1871, p. 196) use of Cetotherien is an improperly formed name. At present, there are no less than 34 nominal genera assigned to the family.

Miller (1923, pp. 21, 41) defined the family, and Cabrera (1926, pp. 369–370), trying to put some order into it, recognized three "groups." Kellogg (1928, p. 186, Fig. 18; 1934, pp. 77–78) recognized two "groups" in the family, combining Cabrera's second and third group. The first "group" of both authors included only animals on whose skulls the posterior extensions of the rostral bones (the maxillae, premaxillae, and nasals) wedge between the frontals at the cranial vertex in a deep V-shaped pattern. This condition is seen in the type genus of the family, *Cetotherium* (see Fig. 3d) and should be considered the characteristic and probably primitive condition for the family. This condition is also present in *Aetiocetus* (Fig. 3b). A very different situation exists, however, for animals in the second group of Kellogg, in which the juncture between the rostral bones and the frontals is more or less a straight transverse line or only slightly indented posteriorly. This condition exists in Middle and Late Oligocene *Mauicetus* (see Fig. 3c), which is usually considered to be related to, or to actually be, a cetothere (Kellogg, 1942, pp. 450–451; Marples, 1956, p. 565; Whitmore and Sanders, 1976, p. 317). Another reputed cetothere, *Aglaocetus* spp. (see Kellogg, 1934) shows similar structure of the cranial vertex. Whether species in this second group are true cetotheres or represent another family group taxon awaits further study. They are not "typical" cetotheres, yet they are among the earliest of baleen whales.

Unifying features that are shared by all supposed cetotheres (*sensu lato*) are a wide, flat, elongate snout, devoid of teeth, with grooves on the palatal surface marking the course of nutrient arteries for the baleen, a single dorsal bony narial opening on the proximal part of the rostrum, a relatively simple sutural relationship between the rostral and cranial bones, a large, triangular, occipital shield, elongate zygomatic arches of the squamosals with indistinct glenoid fossae for articulation with the mandible, and middle ear air sinus system limited largely to the peribullary sinus.

The family ranges from Middle Oligocene, circa 30 million years ago, to Late Pliocene, circa 3 million years ago.

Balaenidae

This was the first named family of mysticetes, and in many early classifications it included all living baleen whales, which were sometimes ranked in separate subfamilies. Currently it includes living right whales, *Eubalaena* Gray, 1864a, and bowhead whales, *Balaena* Linnaeus, 1758, and in most classifications, the pygmy right whale *Caperea* Gray, 1864a (*Neobalaena* Gray, 1870) (see Fig. 4a), but we follow Gray (1874) and Miller (1923) by putting it in a separate family, the Neobalaenidae Gray, 1874.

There have been described in the literature many extinct species of balaenids of Late Miocene and Pliocene age from European and North American deposits which show close relationships to either the modern bowhead or right whales. It is a big jump both morphologically and chronologically, however, to the oldest balaenid, *Morenocetus parvus* Cabrera, 1926, from Argentina, which is about 23 million years old.

Typical balaenids (Figs. 4b,c), excluding *Caperea*, are distinguished from other baleen whales by having a narrow, dorsally arched rostrum to accommodate greatly elongated baleen plates. Other concommitant cranial characters are a highly arched cranial vertex, elongate and steeply sloping supraorbital processes of the frontals (literally to "reach down" to the eyes), steeply sloping occipital shield, and unusually descending zygomatic processes of the squamosals. The rostral bones are not extensively interdigitated with the cranial bones, and the suture between them is almost straight across. The glenoid fossa is nearly flat and relatively wide to accommodate the large, upwardly directed condyle on the lower jaw. The mandible has a very reduced coronoid process and is greatly arched upward at mid-length, somewhat twisted and narrow at the front end.

Neobalaenidae

Caperea marginata (Gray, 1846) differs from the balaenids, not only by being smaller but also in many morphological characters (Gray, 1874, p. 93; Miller, 1923). It differs from balaenids by having a dorsal fin, incipient throat grooves, a different type of baleen, a smaller head size compared to the body, a proportionally shorter humerus, and four instead of five digits in the manus. Its skull is of dramatically different architecture (Fig. 4a), having a larger, more anteriorly thrust occipital shield, a shorter, wider, and less arched rostrum, smaller nasal bones, shorter supraorbital processes of the frontals, and orbits and glenoid fossae that are not located so far ventrally on the skull.

These unique characters set *Caperea* off from the balaenids with which it is commonly classified. Since there are no fossils of *Caperea*, we have no evolutionary data on familial interrelationships.

Eschrichtiidae

Gray whales stand apart morphologically from other baleen whales, and the absence of a long fossil record clouds any clear hint at their evolutionary history. In overall appearance, the skull (Fig. 5a) resembles that of balaenids with the narrow and arched snout. The rostral bones meet the supraorbital processes in a curved suture differing from the balaenids, and they interdigitate with the cranial bones in a complex interfingering manner which is unlike that of balaenopterids. The narial opening is unusually large, as are the nasal bones. The supraorbital processes are not elongate as in Balaenidae, but are broad. The occipital shield is not thrust as far anteriorly as in balaenopterids, and the frontals are widely exposed on the vertex. A unique feature is the presence of paired occipital tuberosities for insertion of muscles which originate in the neck region. The baleen plates differ from those of balaenids by being fewer, thicker, and white. In balaenids, they are more numerous, longer, and dark in color. In balaenopterids they are also dark and numerous, but shorter.

The mandible of gray whales resembles that of balaenids by having only a rudimentary coronoid process and by being arched upward at midlength, but it is heavier and not so narrow at the front. Unlike balaenids, gray whales have two to four grooves on the throat, but this is insignificant in comparison with the numerous grooves on balaenopterids. Gray whales have no dorsal fin like the one in balaenopterids but do have a series of dorsal median bumps, which make them different from balaenids which have neither bumps nor fins.

Balaenopteridae

The fossil record of balaenopterids extends back about 10–12 million years ago to the Late Miocene. The earliest fossil members of the family are not particularly primitive but the cranial proportions, degree of forward thrusting of the occipital shield, and flatness of the rostrum cause the balaenopterid skull (Figs. 5b,c and 6a–d) to resemble those of cetotheres *sensu lato* more closely than either balaenids or gray whales. Kellogg (1942, p. 450) considered cetotheres to be ancestors of balaenopterids. Compared with cetotheres and balaenids, the rostral bones of balaenopterids have a more complex interdigitation with the cranial bones, but this complexity is based on a different pattern than in gray whales. Compared with gray whales, the narial opening and nasal bones are smaller, the supraorbital process, although equally short and broad, slopes more dramatically from the vertex, and the frontals are not exposed on the vertex because the occipital shield has pushed farther forward. The baleen plates are short as in gray whales but are more numerous, not as heavy, and are darker [except in *Balaenoptera acutorostrata* and on the right side of *Balaenoptera physalus* (Linnaeus, 1758)].

The mandible of balaenopterids is more generalized, and more like those of cetotheres by being elongate, of nearly equal height throughout its length, and by having a well-developed coronoid process and a smaller mandibular condyle.

In external morphology, balaenopterids have a dorsal fin, unlike gray whales and balaenids, and have numerous throat grooves.

PREVIOUS INTERPRETATIONS OF GRAY WHALE RELATIONSHIPS

Since Cope (1868c,e, 1869b) brought the gray whale to the attention of the scientific community, its taxonomic position within the Mysticeti has been problematical. From the time of Linnaeus (1758), all of the mysticetes had been placed in only one family, the Balaenidae. This classification was retained by several authors (Flower and Lydekker, 1891; Elliot, 1901; Winge, 1921), despite the fact that separate mysticete families had been recognized by Gray as early as 1864. Gray's (1864b) paper is the one in which he named the subgenus *Eschrichtius* for *Balaenoptera robusta* and questionably referred it to the genus *Megaptera* within his family Balaenopteridae. When Gray (1865) raised *Eschrichtius* to generic status, he still allied it with *Megaptera,* noting that both had some characters similar to balaenids. This arrangement was retained by Gray (1868, 1871) when he placed *Megaptera* and *Eschrichtius,* still known only by bones (along with a synonymous genus of *Megaptera, Poescopia* Gray, 1864a) in a separate family, the

Megapteridae; but at the same time, he also classified Cope's *Agaphelus gibbosus*, the scrag whale, and *Rhachianectes glaucus*, the gray whale, in a separate family, the Agaphelidae. He placed both families, along with the Balaenopteridae and the Physalidae (Balaenopteridae), in the Balaenopteroidea. Weber (1904) placed the gray whale in its own family, Rhachianectidae, and this was followed by most recent authors, with the subsequent change of name to Eschrichtiidae (Ellerman and Morrison-Scott, 1951). Gaskin (1982, p. 200) mentioned the possibility of classifying gray whales as a subfamily of cetotheres.

The arrangement of four modern families of baleen whales, Balaenidae, Neobalaenidae, Balaenopteridae, and Eschrichtiidae, however, does not elucidate the possible relationships between them. Nearly all students of extant mysticetes have considered the gray whale to be more closely related to the rorquals (Balaenopteridae) than to the right whales (Balaenidae). They either proposed this relationship directly or implied it in their classifications (Gray, 1864b, 1865; Lilljeborg, 1867; Cope, 1868b, 1869a; Van Beneden and Gervais, 1880; Winge, 1921; Slijper, 1936). Others have stressed the primitive characters of the gray whale and given it an intermediate position between the balaenids and balaenopterids (Andrews, 1914; Miller, 1923; Fraser and Purves, 1960). When considering the phyletic relationships of the gray whale, Andrews (1914) titled the section "Primitive characters of *Rhachianectes*."

Including fossil mysticetes in their considerations, Andrews (1914), Winge (1921), Miller (1923), and Kellogg (1928) indicated relationships between the gray whale and cetotheres. Kellogg (1928, p. 190) briefly stated, ". . . that *Rhachianectes* (Andrews, 1914) stands apart from other whalebone whales, and is the least modified, and furthermore that the conditions observed are readily derivable from the cetothere skull." Apparently, this statement has led subsequent authors to declare a specific relationship between the gray whale and cetotheres. For example, Mitchell (1966, Fig. 7 legend) wrote, "Some cetotheres were probably the ancestors of the modern gray whale . . . ," and Rice (1967, p. 303) says, "The living gray whale is closely related to the extinct family Cetotheriidae." We consider such statements to be without basis partly because the family Cetotheriidae, as currently recognized, is almost certainly not a monophyletic group but rather an evolutionary grade. It contains diverse evolutionary clades and has come to be used as a taxonomic unit to contain primitive baleen whales that have baleen but no teeth and which cannot be placed in an extant family. We believe that the origin of any living mysticete family has yet to be objectively postulated from sound paleontologic or morphologic evidence.

MORPHOLOGIC AND PHYLETIC EVIDENCE FOR CHARACTER ANALYSIS

Several, and differing, characters have been used by authors (Gray, 1864b, 1865; Cope, 1868d, 1869b; Weber, 1904; Winge, 1921; Kellogg, 1928) to align gray whales with other mysticetes. In the following sections we discuss some of these and other characters in terms of their primitive states and the various, and different, derived states within the mysticetes.

Dorsal Fin

The majority of living cetaceans have some sort of dorsal fin. The lack of a dorsal fin occurs universally in the families Balaenidae and Monodontidae and in a minority of genera in the Delphinidae (*Lissodelphis*) and the Phocoenidae (*Neophocaena*). Some other types of living cetaceans possess a series of knobs along the dorsal ridge (*Physeter*, *Eschrichtius*) instead of a dorsal fin. Having neither skeletal nor muscular support, the presence or absence of a dorsal fin cannot be determined from the skeleton alone of either modern or fossil species. Based on its widespread occurrence in modern cetaceans, however, we believe the dorsal fin evolved very early as a primitive trait for cetaceans (Barnes and Mitchell, 1978, pp. 596–597), and that its secondary and independent loss in some lineages is a derived character (Mitchell, 1970).

Ventral Throat Grooves

Several cetaceans possess longitudinal grooves in the skin of the throat region. These function as pleats that allow the throat region to expand during feeding. They appear in great length and numbers in the balaenopterids, whose feeding behavior includes gulping large quantities of food-laden water. They are absent in balaenids. *Caperea* possesses two parallel depressions in the throat region which may be incipient ventral grooves (Ross *et al.*, 1975). Among the odontocetes, ziphiids have two or more ventral grooves that meet posteriorly to form a V-shaped pattern in the throat region. *Physeter* has numerous short ventral grooves in the throat region. Most cetaceans do not possess such grooves, however, and we consider their presence to be a derived and convergent character. Gray whales have only two to four grooves (Andrews, 1914) but these do not extend past the throat region as they do in balaenopterids.

Baleen

All living mysticetes possess baleen and it is one of the diagnostic characters of the group. A few specimens of fossil baleen exist, and we would expect its presence in the earliest and most primitive fossil mysticetes that have edentulous, flattened rostra. The longest, thinnest, and greatest numbers of baleen plates are found in the mysticetes with the most highly arched rostra (Balaenidae), and we believe this is the derived condition because the earliest mysticetes had flat rostra. *Eschrichtius* has the thickest and fewest number of baleen plates among the living mysticetes and is primitive in this respect.

Skull–Skeleton Length Ratio

Early cetaceans had relatively short skulls, comprising only 20% or less of the skeletal length. Archaeocetes had relatively short skulls, ranging from 8% of the skeletal length in *Basilosaurus* to 17% in *Zygorhiza*. This condition was inherited with their terrestrial mammalian ancestry. Modern mysticetes have values ranging from 15% in *Sibbalidius* Gray, 1864a, to 41% in some specimens of *Balaena* Linnaeus, 1758, with *Es-*

chrichtius having a moderate value of about 20%. The greater skull length is therefore a derived character.

Cervical Vertebrae

The seven cervical vertebrae are shortened anteroposteriorly in all archaeocetes, mysticetes, and odontocetes, but to varying degrees. In addition, many cetaceans have two or more of the cervical vertebrae fused. Such diverse cetaceans as the balaenids, neobalaenids, and the odontocetes *Hyperoodon* and *Kogia* have all seven cervicals fused, whereas the balaenopterids, eschrichtiids, platanistids, iniids, and pontoporiids retain the primitive free condition of the cervicals. Fusion and shortening of the vertebrae is the derived condition.

Manus

All cetaceans exhibit hyperphalangy. Among the living mysticetes, only the Balaenidae retain the primitive condition of having five digits. All other living mysticetes, including *Eschrichtius*, have only four digits. Kükenthal (1893) reported that at least in *Sibbaldius* the lost digit may be the third rather than the first. This is based on embryological evidence. Hyperphalangy and reduction of the number of digits to four is the derived character state.

Humerus–Radius Length Ratio. Most cetaceans have a relatively short humerus compared to the radius and ulna (which are approximately equal in length). About one half of the flipper length is due to hyperphalangy. In the archaeocetes, however, the humerus is considerably longer than the radius, and we consider that to be the primitive condition. Living balaenids have a humerus nearly equal in length to the radius and *Megaptera* has a humerus less than 40% the length of the radius. Other living mysticetes have a value between 50 and 75%, including *Eschrichtius* which has a value of about 70%.

Scapula

In archaeocetes, the acromion and coracoid processes of the scapula both are well developed, with the acromion being about twice the length of the coracoid. Many odontocetes retain the primitive condition of both a large acromion and a large coracoid process, but among the living mysticetes only *Eschrichtius* retains this condition. *Megaptera*, which has a peculiarly elongated flipper, completely lacks both processes. In other cetaceans, the coracoid process tends to be reduced.

Mandible

Coronoid Process. In the early mysticete *Aetiocetus*, as well as all archaeocetes, the mandible has a large coronoid process for the attachment of the temporal musculature. This process is nearly uniformly reduced among living odontocetes (only

Platanista retains a distinct process). Among mysticetes, it is of moderate size in cetotheriids and balaenopterids but is represented by only a slightly upraised rugose area in *Eschrichtius*, *Caperea*, and the balaenids. This reduction of the process is derived and convergent.

Arch. The mandible in most cetaceans presents a straight horizontal profile when viewed laterally. In *Eschrichtius*, however, the mandible arches upward to follow the arch of the rostrum, thereby acting as a cheek flap. This is true also in *Caperea*, and to a lesser extent in balaenids. This is a derived character.

Curvature. Primitive cetaceans, including aetiocetids and cetotheriids, like most mammals have a mandible that appears straight when viewed dorsally. This is true of most odontocetes, but several odontocete taxa have medially bowed mandibles due to a long, fused symphysis (physeterids, pontoporiids, etc.). In all living mysticetes the mandible curves laterally from the condyle to the symphysis, and in balaenopterids the widest part of this curvature actually sweeps out farther laterally than the condyle. In *Eschrichtius*, however, the mandible is more or less straight for the first two-thirds of its length and we consider this state to be more primitive.

Skull

Rostral Arch. Early and generalized mysticetes have flat rostra. Among the living mysticetes the rostrum arches only slightly in the balaenopterids (less than 5° between the basicranium and the base of the rostrum), moderately in *Eschrichtius* and *Caperea* (10 and 17°, respectively), and greatly in the balaenids (over 20°). In the balaenids the high rostral arch accomodates the long baleen plates, whereas in *Eschrichtius* and *Caperea*, which have shorter baleen plates, the arch of the mandible roughly conforms to the rostral arching. Rostral arching is derived and evolved differently in different groups.

Rostral Telescoping. In archaeocetes and early mysticetes like *Mauicetus*, the narial opening had not migrated very far posteriorly on the rostrum (Figs. 3a,c). This opening was retracted posteriorly (to varying degrees) in later mysticetes and all odontocetes. As the narial opening retracted, the nasals and the ascending processes of the premaxillae and the maxillae retracted with it. In the archaeocetes, the ascending process of the maxilla is well demarcated from the supraorbital process and tapers to a posterior projection where it contacts the posterior portion of the nasals (Fig. 3a). In mysticetes, the premaxillae are interposed between the nasals and the posterior ends of the maxillae. In the balaenids and neobalaenids, the maxillae and the nasals terminate entirely anterior to the orbits and the ascending process of the maxilla merges smoothly with the supraorbital process (Figs. 4a–c). In the balaenopterids and eschrichtiids, as well as in many cetotheres, the maxillae are separated from the supraorbital process by a fissure and the nasals, premaxillae, and ascending processes of the maxillae insert between the frontals where the rostral and cranial bones interdigitate to form a complex suture. In balaenopterids, the ascending process of the maxilla is narrow near the anterior margin of the nasals but flares posteriorly (Figs. 5b,c and 6a–d). In *Eschrichtius*

Table II
Derived Characters Used in Phyletic Analysis and Classification of Mysticetes

				Balaenopteridae	
Family name	Eschrichtiidae	Balaenidae	Neobalaenidae	Megapterinae	Balaenopterinae
Dorsal fin	+[a]	++	—	—	—
Ventral throat grooves	+	—	+?	++	+++
Baleen	—	+++	++	++	+
Skull/skeleton	—	+	—	—	—
Cervical vertebrae	—	++	++	—	—
Manus	+	—	+	+	+
Humerus/radius	+	—	+	++	+
Scapula	—	++	+	+++	+
Mandible					
Coronoid process	++	++	++	+	—
Arch	+	—	+	—	—
Curvature	—	+	+	+	+
Skull					
Rostral arch	+	+++	++	—	—
Rostral telescoping	++	+	—	++	++
Occipital telescoping	—	++	+++	+	+
Number of derived characters	8	10	11	9	8
Degree of derived states	10	19	17	15	11

[a]Plus signs indicate derived states of characters as described in the text. The number of plus signs indicates the degree of derivation. Blank spaces (—) indicate the presence of a relatively primitive character state.

the ascending process of the maxilla is merely a thin vertical plate lateral to the wide proximal portion of the premaxilla (Fig. 5a).

Occipital Telescoping. In archaeocetes, the occipital shield is relatively small, is oriented almost vertically, and in some specimens even projects posteriorly leaving an elongated cranial roof composed of the frontals and parietals (Fig. 3a). In all mysticetes the occipital shield thrusts anteriorly. In derived species this forces the parietals to either side of the braincase, overspreading the frontals to obscure them almost completely from exposure at the vertex. In modern balaenopterids the anterior edge of the occipital shield projects anterior to the posterior border of the supraorbital processes of the frontals. In the balaenids, it projects anterior to the orbits (Figs. 4b,c), and in *Caperea* even further to the antorbital notch of the maxilla (Fig. 4a). In *Eschrichtius*, however, the occipital shield does not even extend anteriorly as far as the temporal fossae, leaving a broad strip of the frontal exposed on the vertex (Fig. 5a).

The foregoing characters are summarized in Table II, where the derived states are compared relative to one another among the extant mysticetes. Only five of these characters are shared in a similarly derived state by both the gray whale and some other extant family of mysticetes (number of digits in the manus, humerus/radius ratio, reduction of the coronoid process, mandibular arch, and rostral telescoping). The manner of rostral telescoping may be the only unique, derived (i.e., nonconvergent) character state that the gray whale shares with another family of extant mysticetes, the Balaenopteridae. Even so, the relative widths of the proximal portions of the maxillae and premaxillae, and the proportions of the nasals are quite different in the two families, suggesting this may also be a convergent character. The size of the nasals has often been cited as a primitive and key character of the gray whale. As Miller (1923) pointed out, however, the nasals of *Eschrichtius* are relatively larger than in any other cetacean, living or extinct, and this may therefore be a derived rather than a primitive character for the gray whale.

Based on all of these characters, *Eschrichtius* appears to be the least derived of the living mysticete genera. Based on the manner of rostral telescoping, it may be more closely related to the Balaenopteridae than to either the Balaenidae or Neobalaenidae, but we do not consider this relationship to be clearly demonstrable at present. We know of no unique derived character that *Eschrichtius* shares with any cetothere. Thus the gray whale is not demonstrably more closely related to cetotheres than it is to any other mysticete family. It is between the Balaenopteridae and Cetotheriidae that we see the closest osteological resemblances.

CLASSIFICATION

We present here a classification, down to subfamilies where they have been proposed, which places the gray whales in context with other mysticetes. The position of the Eschrichtiidae follows our findings that the family is no closer to cetotheres than is any of the other three living baleen whale families and that it is the least derived of the living mysticetes. The extinct families Aetiocetidae and Cetotheriidae are more primitive than any living mysticete family.

We decided that the family Balaenidae should not include *Caperea marginata*, which we place in the family Neobalaenidae. In so doing we have introduced no new names because the family name Neobalaenidae is available from the date that Gray (1874) proposed it, based on the junior synonym of *Caperea*, *Neobalaena*. A family group name has never been proposed based upon the name *Caperea*.

The family Rhachianectidae Weber, 1904, based on Cope's genus *Rhachianectes* for the gray whale, predates the name Eschrichtiidae Ellerman and Morrison-Scott, 1951. However, in this case, although it does not have priority, Eschrichtiidae should be retained as valid because it came into general acceptance (e.g., Miller and Kellogg, 1955, p. 665; Hall and Kelson, 1959, p. 832) prior to 1960.

We have used the two available subfamily names to divide the Balaenopteridae based on the substantial morphological differences between the humpback whale and the diverse species in the *Sibbaldius–Balaenoptera* complex. We recognize the separate genus *Sibbaldius* for the blue whales, which have substantially different cranial morphology from the other four living species of balaenopterines, which are all placed in the genus *Balaenoptera* (see Figs. 5c and 6a–d; also Miller, 1923; Miller and Kellogg, 1955). We use the spelling *Sibbaldius* (Flower, 1864, p. 391) in preference to *Sibbaldus* (Gray, 1864a, p. 223), an emendation with the proper patronymic suffix later accepted by Gray (1866, p. 169). Compared with the *Balaenoptera* species, the skull of blue whales has a wider rostrum, the occipital shield is shaped differently, the zygomatic process of the squamosal is more slender and diverges more laterally from the braincase, the posterior end of the maxilla extends into the cranial vertex differently, the supraorbital process of the frontal is shaped differently, and the palatine bones extend farther posteriorly beneath the braincase. The superfamily names Balaenoidea and Balaenopteroidea exist in the literature (cf. Gray, 1871), but in the absence of any established superfamily breakdown of all the mysticetes, we refrain from using them here.

Class MAMMALIA Linnaeus, 1758
- Order CETACEA Brisson, 1762
  - Suborder Mysticeti Flower, 1864
    - Family Aetiocetidae Emlong, 1966
    - Family Cetotheriidae Brandt, 1872
    - Family Eschrichtiidae Ellerman and Morrison-Scott, 1951
    - Family Balaenopteridae Gray, 1864a
      - Subfamily Balaenopterinae Gray, 1864a
      - Subfamily Megapterinae Gray, 1868
    - Family Neobalaenidae Gray, 1874
    - Family Balaenidae Gray, 1825

Discussion and Conclusions

The family Eschrichtiidae is monotypic with *Eschrichtius* being the sole fossil or living genus. None of the pre-Pleistocene fossils attributed to the family or the genus have been correctly identified. The Pleistocene fossil record is limited to an occurrence in California that appears to represent the living species, *Eschrichtius robustus*, and a less certain one from the North Slope of Alaska. Within historic Holocene time there were two allopatric populations of this species, one with a circum-North Atlantic distribution and the other with a circum-North Pacific distribution (Andrews, 1914; Junge, 1936, p. 78; Van Deinse and Junge, 1937; Fraser, 1970; Nishiwaki and Kasuya, 1970a; Brownell and Chun, 1977; Mead and Mitchell, this volume). The Atlantic population is now extinct. Monotypic families in mammals, including cetaceans, are not common, but among the Mysticeti, Aetiocetidae, Neobalaenidae, and Eschrichtiidae are currently monotypic. We presume that there are undiscovered taxa in the fossil record which should belong in these groups. Based on the known evidence, one could argue either that the eschrichtiid lineage is of very recent origin or that its evolutionary rate is very slow. Some mammalian genera exist for as long as several million years (Simpson, 1953, p. 32), and based upon its unique morphology and lack of derived characters we would expect that the family Eschrichtiidae should extend back much earlier in time than the Pleistocene.

The absence of a good fossil record for gray whales may be a result of their life habits. Modern gray whales spend their lives close to shore, and when they die their carcasses wash ashore in environments not conducive to fossilization. The only known fossil gray whale, which we described above, was deposited in a shallow-water, nearshore environment consistent with these generalizations. The shoreline environment is one of high energy in which organic material, particularly the light, porous bone of cetaceans, is rapidly destroyed. Beaches are also not areas of thick sedimentation, despite the presence of much sand, because they are reworked seasonally and eroded away during transgressive or regressive sedimentary cycles. Very commonly, it is the offshore sediments that we see in the preserved geologic record. Of the Pleistocene sediments in most parts of the world, it is usually only the emergent beach and nearshore deposits which are now accessible for collecting. Nearly all deep-water Pleistocene sediments simply are not accessible because they have not yet been uplifted from the ocean by tectonic activity. Therefore, as a general rule the Pleistocene fossil record of whales is poor in most parts of the world (cf. Barnes, 1976, p. 335).

This does not explain, however, why such an apparently unique group as gray whales does not have an antecedent fossil record in the numerous accessible shallow-water deposits of Pliocene age or older. For example, the Pliocene San Diego Formation of southern California and northern Baja California is probably only about 3 to 5 million years old; it was laid down in shallow water in a broad bay (Hertlein and Grant, 1944; 1954) and has yielded abundant shallow-water marine vertebrate fossils (Howard, 1949; Barnes, 1973; Deméré, 1981a,b). Fossil balaenids and balaenopterids are relatively abundant in this formation, but there are no specimens representing Eschrichtiidae.

Even the genus of whale barnacle, *Cryptolepas*, which is an obligate parasite found only on gray whales (Tomilin, 1967), has a fossil record extending back only to the late Pleistocene. It occurs in a warm water invertebrate fauna at San Quintín Bay, Baja California (Zullo, 1961), which Jordan (1926, pp. 247–248) has correlated with the upper part of the San Pedro Formation in southern California. Fossil whale barnacles have also been found in the older San Diego Formation (Zullo, 1969), but these all represent taxa which today are found on balaenids and balaenopterids, not on gray whales. Thus neither gray whales nor their specific barnacles are found prior to the late Pleistocene, but the more generalized genus *Coronula*, which is known on various modern balaenids and balaenopterids, occurs as far back as the late Miocene. Zullo (1961) suggested that whale barnacles diversified rapidly after the late Miocene, and this may be concomitant with evolution of modern mysticetes.

We know that gray whales in historic times occupied the North Atlantic as well as the North Pacific Ocean. Their distribution is antitropical and unipolar. Important questions exist regarding their place of origin, the degree of past interocean interchange, and whether or not they ever lived in the southern hemisphere. Modern gray whales migrate farther than any other mammals, although their habitat is restricted to a narrow band within a few kilometers of shore. They are tolerant of climatic extremes since they travel between their northern summer feeding grounds in the ice-free parts of the Arctic Ocean and their southern calving grounds in warm lagoons of Mexico. Colder climates during extreme Pleistocene glacial maxima undoubtedly pushed gray whale distributions farther south than at present.

If the extinct North Atlantic populations of gray whales were as mobile as the present North Pacific populations, one would expect some sort of interocean interchange. The Panamanian seaway was closed far back in time, however, and might have been too far south to have allowed the two populations any interchange. The only alternative link would have been via the Arctic Ocean during interglacial periods. During periods of glacial maxima, the polar ice pack and emergent Bering Isthmus would have excluded them entirely from the Arctic Basin. During mild interglacials (such as at the present time), the polar ice still would have blocked dispersal. Apparently only during warm interglacial periods would the polar ice cap have melted enough for a complete connection throughout the Arctic Ocean.

Although gray whales apparently are not part of a diverse or rapidly evolving lineage, they do not seem to be a dying or poorly adapted group. They are highly mobile, spend their lives between extreme latitudes, are not especially stenophagous, are gregarious, seem individually adaptable and tenacious, and have recovered well from a very reduced population after massive human exploitation.

Summary

The fossil record of gray whales is meager, being limited to a single Pleistocene occurrence of an animal which appears to belong to the modern species, *Eschrichtius robustus*. All previous published references to extinct species of *Eschrichtius* in the fossil record are spurious, either representing some other family of cetacean or being unidentifiable. Based on their unique morphology, gray whales belong in the monotypic family Eschrichtiidae. In some characters they are similar to both modern Balaenidae and Balaenopteridae, but they are not particularly closely related to either family. Gray whales are the least derived of the living mysticetes, but should not be considered the closest living relatives of the extinct Cetotheriidae. Based on osteological features,

Balaenopteridae are probably the most like Cetotheriidae (*sensu lato*) of all living mysticetes. Based upon an overview of modern mysticetes families and the position of Eschrichtiidae, we propose that the pygmy right whale, *Caperea marginata*, should be classified not in the Balaenidae but as the sole representative of a fourth living family, the Neobalaenidae.

Acknowledgments

We thank Mr. Paul R. Kirkland, the original discoverer of the fossil gray whale from San Pedro, for investing much time and energy in its eventual collection. Mr. William Samaras (Carson High School), Mr. John Olguin (Director, Cabrillo Beach Museum), and Dr. Eugene Fritche (California State University at Northridge) were especially helpful in calling attention to the discovery and in directing the excavation along with Dr. Shelton P. Applegate, then of the LACM.

We thank Westoil Terminals Company and specifically Mr. Graham Polk, Mr. C. Edward Miller, Jr., and Mr. Fred Jonas of that company for the generous donation of the fossil to the Museum. Also instrumental in enhancing public awareness of the find and in negotiating with Dr. David P. Whistler (LACM) for its deposition in the Museum were Mr. John S. Gibson (Los Angeles City Council), the late Mr. Vincent Thomas (State Assemblyman), and Mr. C. L. "Jack" Allen (San Pedro Chamber of Commerce).

We thank Ms. Angela Bennett, Mrs. Joan C. Brown, Ms. Hilary J. Hicks, Mrs. Diana Weir, and especially Ms. Pat Tyus for their incredible patience and valuable service rendered in preparing the fossil for study. Mr. Dick Meier (LACM) took the photograph in Fig. 1. Mr. Ron Sabados (LACM) helped with the production of artwork prepared by McLeod.

For their help and comments during preparation of the manuscript we thank Drs. Robert L. Brownell, Jr., James G. Mead, Edward D. Mitchell, Albert C. Myrick, Jr., Frank C. Whitmore, Jr., and Edward C. Wilson.

References

Andrews, R. C. (1914). Monographs of the Pacific Cetacea. I. The California gray whale (*Rhachianectes glaucus* Cope). *Mem. Am. Mus. Nat. Hist.* [N.S.] **1**(5), 227–287.

Barnes, L. G. (1973). Pliocene cetaceans of the San Diego Formation, San Diego, California. *In* "Studies on the Geology and Geologic Hazards of the Greater San Diego Area, California" (A. Ross and R. J. Dowlen, eds.), pp. 37–42. San Diego Association of Geologists, San Diego, California.

Barnes, L. G. (1976). Outline of eastern North Pacific fossil cetacean assemblages. *Syst. Zool.* **25**(4), 321–343.

Barnes, L. G., and Mitchell, E. D. (1975). Late Cenozoic Northeast Pacific Phocidae. *Rapp. P.-V. Reun., Cons. Int. Explor. Mer* **169,** 34–42.

Barnes, L. G., and Mitchell, E. D. (1978). Cetacea. *In* "Evolution of African Mammals" (V. J. Maglio and H. B. S. Cooke, eds.), pp. 582–602. Harvard Univ. Press, Cambridge, Massachusetts.

Barwick, A. R. (1938). Jaw of fossil whalebone whale: *Siphonocetus priscus. Proc. Biol. Soc. Wash.* **51,** 121–122.

Barwick, A. R. (1940). Skull of fossil cetotherean whale, *Siphonocetus priscus*, from the Miocene of the Chesapeake Bay region. *Am. Midl. Nat.* **23**(3), 746–750.

Beddard, F. E. (1900). "A Book of Whales." G. P. Putnam's Sons, New York.

Benham, W. B. (1939). *Mauicetus:* A fossil whale. *Nature (London)* **143,** 765.

Bowers, S. (1889). An abstract of a letter regarding the finding of the remains of fossil vertebrates in Ventura, California. *Am. Geol.* **4,** 391–392.

Brandt, J. F. (1843). De Cetotherio, novo balaenarum familiae genere, in Rossia meridionali anti aliquot annos effoso. *Bull. Acad. Imp. Sci. St. Petersbourg* [2] **1,** 145–148.

Brandt, J. F. (1871). Bericht über den Fortgang meiner Studien über die Cetaceen, welche das grosse zur

Tertiärzeit von Mitteleuropa bis Centralasien hinein ausgedehnte Meeresbecken bevölkerten. *Bull. Acad. Imp. Sci. St. Petersbourg* [3] **16,** 563–566.

Brandt, J. F. (1872). Ueber eine neue Classification der Bartenwale (Balaenoidea) mit berücksichtigung der untergegangenen Gattungen derselben. *Bull. Acad. Imp. Sci. St. Petersbourg* [3] **17,** 113–124.

Brandt, J. F. (1873). Untersuchungen über die fossilen und subfossilen Cetaceen Europa's. *Mem. Acad. Imp. Sci. St. Petersbourg* [7] **20**(1), 1–372.

Brisson, M. J. (1762). "Regnum animale in classes IX distributum sive synopsis methodica. Editio altero auctior." Leiden, Theodorum Haak.

Brownell, R. L., Jr., and Chun, C.-I. (1977). Probable existence of the Korean stock of the gray whale (*Eschrichtius robustus*). *J. Mammal.* **58**(2), 237–239.

Cabrera, A. (1926). Cétáceos fosiles del Museo de la Plata. *Rev. Mus. La Plata* **29,** 363–411.

Cope, E. D. (1868a). (for 1867). (Description of *Eschrichtius cephalus, Rhabdosteus latiradix, Squalodon atlanticus,* and *Squalodon mento.*) *Proc. Acad. Nat. Sci. Philadelphia* **19,** 131–132.

Cope, E. D. (1868b). (for 1867). An addition to the vertebrate fauna of the Miocene period, with a synopsis of the extinct Cetacea of the United States. *Proc. Acad. Nat. Sci. Philadelphia* **19,** 138–156.

Cope, E. D. (1868c). (Description of *Ixacanthus coelospondylus, Balaenoptera pusilla, Agaphelus gibbosus,* and *Agaphelus glaucus.*) *Proc. Acad. Nat. Sci. Philadelphia* **20,** 159–160.

Cope, E. D. (1868d). Second contribution to the history of the Vertebrata of the Miocene period of the United States. *Proc. Acad. Nat. Sci. Philadelphia* **20,** 184–194.

Cope, E. D. (1868e). On *Agaphelus,* a genus of toothless Cetacea. *Proc. Acad. Nat. Sci. Philadelphia* **20,** 221–227.

Cope, E. D. (1869a). Third contribution to the fauna of the Miocene period of the United States. *Proc. Acad. Nat. Sci. Philadelphia* **21,** 6–12.

Cope, E. D. (1869b). (In C. M. Scammon) On the cetaceans of the western coast of the North America. Part 1. Systematic synopsis of the species of the cetaceans of the west coast of North America. *Proc. Acad. Nat. Sci. Philadelphia* **21,** 14–32.

Cope, E. D. (1870a). (for 1869). (Remarks on *Hyperaodon bidens* and descriptions of *Eschrichtius polyporus, Hypsibema crassicauda, Hadrosaurus tripos,* and *Polydectes biturgidus.*) *Proc. Acad. Nat. Sci. Philadelphia* **21,** 191–192.

Cope, E. D. (1870b). Fourth contribution to the history of the fauna of the Miocene and Eocene periods of the United States. *Proc. Am. Philos. Soc.* **11,** 285–294.

Cope, E. D. (1872). On an extinct whale from California. *Proc. Acad. Nat. Sci. Philadelphia* **24,** 29–30.

Cope, E. D. (1875). Synopsis of the Vertebrata whose remains have been preserved in the formations of North Carolina. *In* "Report of the Geological Survey of North Carolina" (W. C. Kerr, ed.), Vol. 1, App. B, pp. 1–51. Raleigh, North Carolina.

Cope, E. D. (1879). The California gray whale. *Am. Nat.* **13,** 655.

Cope, E. D. (1884). Catalogue of aquatic mammals of the United States, by F. W. True (Review). *Am. Nat.* **18,** 1123–1124.

Cope, E. D. (1890). The Cetacea. *Am. Nat.* **24,** 599–616.

Cope, E. D. (1895a). Fourth contribution to the marine fauna of the Miocene period of the United States. *Proc. Am. Philos. Soc.* **34,** 135–155.

Cope, E. D. (1895b). The phylogeny of the whalebone whales. *Am. Nat.* **29,** 572–573.

Cope, E. D. (1896). Sixth contribution to the knowledge of the marine Miocene fauna of North America. *Proc. Am. Philos. Soc.* **35,** 139–146.

Deméré, T. A. (1981a). Fossil whales of San Diego. *Environ. Southwest* **492,** 17–20.

Deméré, T. A. (1981b). Fossil whales of San Diego. Part II. *Environ. Southwest* **493,** 22–25.

Dudley, P. (1725). An essay upon the natural history of whales, with a particular account of the ambergris found in the Sperma Ceti whale. *Philos. Trans R. Soc. London* **33**(387), 256–269.

Eichwald, E. (1840). "Die Urwelt Russlands, durch Abbildungen erlaeatert," Part 1. St. Petersburg.

Ellerman, J. R., and Morrison-Scott, J. C. S. (1951). "Checklist of Palaearctic and Indian mammals, 1758–1946." British Museum (Natural History), London.

Elliot, D. G. (1901). A synopsis of the mammals of North America and the adjacent seas. *Field Columbian Mus. Publ., Zool. Ser.* **2,** 1–471.

Emlong, D. (1966). A new archaic cetacean from the Oligocene of Northwest Oregon. *Bull. Mus. Nat. Hist., Univ. Oreg.* **3,** 1–51.

Emmons, E. (1858). "Report of the North Carolina Geological Survey. Agriculture of the eastern counties, together with descriptions of the fossils of the marl beds." Raleigh, North Carolina.

Erxleben, J. C. P. (1777). "Systema Regni Animalis." Weygandianis, Lipsiae.

Evernden, J. F., Savage, D. E., Curtis, G. H., and James, G. T. (1964). Potassium-argon dates and the Cenozoic mammalian chronology of North America. *Am. J. Sci.* **262,** 145–198.

Fanale, F. P., and Schaeffer, O. A. (1965). Helium-uranium ratios for Pleistocene and Tertiary fossil aragonites. *Science* **149,** 312–316.

Flower, W. H. (1864). Notes on the skeletons of whales in the principal museums of Holland and Belgium, with descriptions of two species apparently new to science. *Proc. Zool. Soc. London* **1864,** 384–420.

Flower, W. H., and Lydekker, R. (1891). "An Introduction to the Study of Mammals Living and Extinct." Adam & Charles Black, London.

Fordyce, R. E. (1977). The development of the circum-Antarctic current and the evolution of the Mysticeti (Mammalia: Cetacea). *Palaeogeogr., Palaeoclimatol., Paleoecol.* **21,** 265–271.

Fordyce, R. E. (1980). Whale evolution and Oligocene southern ocean environments. *Palaeogeogr., Palaeoclimatol., Palaeoecol.* **31,** 319–336.

Fraser, F. C. (1970). An early 17th century record of the California grey whale in Icelandic waters. *Invest. Cetacea* **2,** 13–20.

Fraser, F. C., and Purves, P. E. (1960). Hearing in cetaceans. *Bull. Br. Mus. (Nat. Hist.),* Zool. **7**(1), 1–140.

Gaskin, D. E. (1982). "Ecology of Whales and Dolphins." Heinemann Educational Books, Ltd., London.

Gidley, J. W. (1913). A recently mounted *Zeuglodon* skeleton in the United States National Museum. *Proc. U.S. Natl. Mus.* **44,** 649–654.

Gillette, D. D. (1975). Catalogue of type specimens of fossil vertebrates, Academy of Natural Sciences, Philadelphia. Introduction and Part 1: Marine mammals. *Proc. Acad. Nat. Sci. Philadelphia.* **127**(8), 63–66.

Gray, J. E. (1825). Outline of an attempt at the disposition of the Mammalia into tribes and families with a list of the genera apparently appertaining to each tribe. *Ann. Philos.* **26,** 337–344.

Gray, J. E. (1846). Mammalia. III. On the cetaceous animals. *In* "The Zoology of the Voyage of H. M. S. *Erebus* and *Terror* under the Command of Captain Sir James Clark Ross, R.N., F.R.S., during the years 1839 to 1843" (J. Richardson and J. E. Gray, eds.), Vol. 1, pp. 13–53. Longman and Co., London.

Gray, J. E. (1864a). On the Cetacea which have been observed in the seas surrounding the British islands. *Proc. Zool. Soc. London* **1864,** 195–248.

Gray, J. E. (1864b). Notes on the whalebone-whales; with a synopsis of the species. *Ann. Mag. Nat. Hist.* [3] **14,** 345–353.

Gray, J. E. (1865). Notice of a new whalebone whale from the coast of Devonshire, proposed to be called *Eschrictius robustus. Ann. Mag. Nat. Hist.* [3] **15,** 492–495.

Gray, J. E. (1866). "Catalogue of Seals and Whales in the British Museum." British Museum, London.

Gray, J. E. (1868). "Synopsis of the Species of Whales and Dolphins in the Collection of the British Museum." Bernard Quaritch, London.

Gray, J. E. (1870). Notes on the skull of *Balaena marginata,* the type of a new genus, *Neobalaena. Ann. Mag. Nat. Hist.* [4] **6**(32), 154–157.

Gray, J. E. (1871). "Supplement to the Catalogue of Seals and Whales in the British Museum." British Museum, London.

Gray, J. E. (1874). Notes on Dr. Hector's paper on the whales and dolphins of the New Zealand seas. *Trans. Proc. N. Z. Inst.* **6**(18), 93–97.

Hall, E. R., and Kelson, K. R. (1959). "The Mammals of North America," Vol. II. Ronald Press, New York.

Hershkovitz, P. (1966). Catalog of living whales. *Bull. U.S. Natl. Mus.* **246,** 1–259.

Hertlein, L. G., and Grant, U. S., IV (1944). The geology and paleontology of the marine Pliocene of San Diego, California. Part 1. Geology, *Mem. San Diego Soc. Nat. Hist.* **2,** 1–72.

Hertlein, L. G., and Grant, U. S., IV (1954). Geology of the Oceanside-San Diego coastal area, southern California. *Bull. Calif. Div. Mines Geol.* **170,** 53–63.

Howard, H. (1949). New avian records for the Pliocene of California. *Carnegie Inst. Washington Publ.* **584,** 177–199.

Jordan, E. K. (1926). Expedition to Guadalupe Island, Mexico, in 1922. No. 4. Molluscan fauna of the Pleistocene of San Quentín Bay, Lower California. *Proc. Calif. Acad. Sci.* [4] **15**(7), 241–255.

Junge, G. C. A. (1936). Bones of a whale from the Wieringermeer, Zuider Zee. *Nature (London)* **138,** 78.

Kellogg, A. R. (1928). The history of whales—their adaptation to life in the water (concluded). *Q. Rev. Biol.* **3**(2), 174–208.

Kellogg, A. R. (1934). The Patagonian fossil whalebone whale, *Cetotherium moreni* (Lydekker). *Carnegie Inst. Washington Publ.* **447,** 63–81.

Kellogg, A. R. (1936). A review of the Archaeoceti. *Carnegie Inst. Washington Publ.* **482,** 1–366.

Kellogg, A. R. (1942). Tertiary, Quaternary, and Recent marine mammals of South America and the West Indies. *Proc. Am. Sci. Congr., 8th, 1940* Vol. 3, pp. 445–473.

Kellogg, A. R. (1968). Fossil marine mammals from the Miocene Calvert Formation of Maryland and Virginia. 5. Miocene Calvert mysticetes described by Cope. *Bull. U.S. Natl. Mus.* **247**(5), 103–132.

Kleinenberg, S. E. (1958). K voprosu o proiskhozhdenii kitoobraznykh. (The origin of the Cetacea.) *Dokl. Akad. Nauk SSSR* **122**(5), 950–952 (in Russian).

Kleinenberg, S. E. (1959). On the origin of Cetacea. *Proc. Int. Congr. Zool., 15th, 1958* pp. 445–447.

Kleinenberg, S. E., and Yablokov, A. V. (1958). O morfologii verkhnikh dykhatel'nykn putei kitoobraznykh. (The morphology of the upper respiratory passages in the cetaceans.) *Zool. Zh.* **37,** 1091–1099.

Kükenthal, W. (1893). Vergleichend-anatomische und entwickelungsgeschichtliche Untersuchungen an Walthieren. *Denkschr. Med. Naturwiss. Ges. Jena* **3**(2), 221–448.

Lacépède, B. G. E. (1804). "Histoire naturelle des Cétacés. Chez Plassan, Paris.

Leidy, J. (1852). (1851). (Descriptions of two fossil species of *Balaena, B. palaeatlantica* and *B. prisca*). *Proc. Acad. Nat. Sci. Philadelphia* **5,** 308–309.

Leidy, J. (1869). The extinct mammalian fauna of Dakota and Nebraska, including an account of some allied forms from other localities, together with a synopsis of the mammalian remains of North America. *J. Acad. Nat. Sci. Philadelphia* [2] **7,** 1–472.

Lilljeborg, W. (1861). Hvalben funna i jorden på Gräsön i Roslagen i Sverige. *Forh. Skand. Naturf., 8th Mote, 1860* pp. 599–616.

Lilljeborg, W. (1867). On two subfossil whales discovered in Sweden. *Nova Acta Regiae Soc. Sci. Ups.* [3] **6**(2), 1–48.

Linnaeus, C. (1758). "Systema naturae," 10th ed., Vol. 1. Laurentii Salvii, Stockholm.

Marincovich, L. (1976). Late Pleistocene molluscan faunas from upper terraces of the Palos Verdes Hills, California. *Nat. Hist. Mus. Los Angeles County, Contrib. Sci.* **281,** 1–28.

Marples, B. J. (1956). Cetotheres (Cetacea) from the Oligocene of New Zealand. *Proc. Zool Soc. London* **126,** 565–580.

Mchedlidze, G. A. (1970). "Nekotorye obshchie cherty istorii kitoobraznykh. Chast' I. (Some features of the historical development of the Cetacea. Part 1.)." Akad. Nauk Gruz. S.S.R., Inst. Paleobiol., Metsniereba Press, Tbilisi.

McLeod, S. A., and Barnes, L. G. (1983). Gray whales. *Terra* (The Natural History Museum of Los Angeles County) **22**(2), 19–21.

Miller, G. S., Jr. (1923). The telescoping of the cetacean skull. *Smithson. Misc. Collect.* **76**(5), 1–71.

Miller, G. S., Jr. (1924a). Some hitherto unpublished photographs and measurements of the blue whale. *Proc. U.S. Natl. Mus.* **66**(7), 1–4.

Miller, G. S., Jr. (1924b). A pollack whale from Florida presented to the National Museum by the Miami Aquarium Association. *Proc. U.S. Natl. Mus.* **66**(9), 1–15.

Miller, G. S., Jr., and Kellogg, R. (1955). List of North American Recent mammals. *Bull. U.S. Natl. Mus.* **205,** 1–954.

Miller, W. E. (1971). Pleistocene vertebrates of the Los Angeles basin and vicinity (exclusive of Rancho La Brea). *Bull. Los Angeles Co. Mus. Nat. Hist.* **10,** 1–124.

Mitchell, E. D. (1966). Faunal succession of extinct North Pacific marine mammals. *Nor. Hvalfangst Tid.* **1966**(3), 47–60.

Mitchell, E. D. (1970). Pigmentation pattern evolution in delphinid cetaceans: An essay in adaptive coloration. *Can. J. Zool.* **48**(4), 717–740.

Nishiwaki, M., and Kasuya, T. (1970a). Recent record of gray whale in the adjacent waters of Japan and a consideration on its migration. *Sci. Rep. Whales Res. Inst.* **22,** 29–37.

Nishwaki, M., and Kasuya, T. (1970b). A Greenland right whale caught at Osaka Bay. *Sci. Rep. Whales Res. Inst.* **22,** 45–62.

Nordmann, A. (1860). "Palaeontologie Südrusslands," Part IV, pp. 271–360. H. C. Friis, Helsingfors.

Oliver, W. R. B. (1922). A review of the Cetacea of the New Zealand Seas. I. *Proc. Zool. Soc. London* **1922,** 557–585.

Omura, H. (1959). Bryde's whale from the Coast of Japan. *Sci. Rep. Whales Res. Inst.* **14,** 1–33.

Repenning, C. A. (1983). New evidence for the age of the Gubik Formation, Alaskan North Slope. *Quat. Res. (N.Y.)* **19,** 356–372.

Rice, D. W. (1967). Cetaceans. *In* "Recent Mammals of the World" (S. Anderson and J. K. Jones, eds.), pp. 291–324. Ronald Press, New York.

Ross, G. J. B., Best, P. B., and Donnelly, B. G. (1975). New records of the pygmy right whale (*Caperea marginata*) from South Africa, with comments on distribution, migration, appearance, and behavior. *J. Fish. Res. Board Can.* **32**(7), 1005–1017.

Savage, D. E. (1951). Late Cenozoic vertebrates of the San Francisco Bay region. *Univ. Calif. Publ. Dept. Geol. Sci.* **28**(10), 215–314.

Schevill, W. E. (1952). On the nomenclature of the Pacific gray whale. *Breviora* **7,** 1–3.

Simpson, G. G. (1945). The principles of classification and a classification of mammals. *Bull. Am. Mus. Nat. Hist.* **85,** 1–350.

Simpson, G. G. (1953). "The Major Features of Evolution." Simon & Schuster, New York.

Slijper, E. J. (1936). Die Cetaceen, vergleichend-anatomisch und systematisch. *Capita Zool.* **7,** 1–590.

Thenius, E. (1969). "Phylogenie der Mammalia." de Gruyter, Berlin.

Thompson, d'Arcy W. (1889). On the systematic position of *Zeuglodon. Stud. Mus. Zool. Univ. Coll., Dundee* **1**(9), 1–8.

Tomilin, A. G. (1967). Cetacea. *In* "Mammals of the U.S.S.R. and Adjacent Countries" (V. G. Heptner, ed., O. Ronen, transl.), Vol. 9. Israel Program Sci. Transl., Jerusalem.

Trouessart, E. L. (1898). "Catalogus Mammalium," Vol. II. pp. 665–1469. R. Friedlander & Son, Berlin.

True, F. W. (1904). The whalebone whales of the western North Atlantic. *Smithson. Contrib. Knowl.* **33,** 1–332.

True, F. W. (1908). On the classification of the Cetacea. *Proc. Am. Philos. Soc.* **47,** 385–391.

Van Beneden, P. J. (1859). Sur la découverte d'ossements fossiles faite à Saint-Nicolas. *Bull. Acad. Sci. Belg.* [2] **8,** 123–146.

Van Beneden, P. J., and Gervais, P. (1880). "Ostéographie des Cétacés" (Text and Atlas). Arthus Bertrand, Paris.

Van Deinse, A. B., and Junge, G. C. A. (1937). Recent and older finds of the California gray whale in the Atlantic. *Temminckia* **2,** 161–188.

Van Valen, L. (1968). Monophyly or diphyly in the origin of whales. *Evolution* **22**(1), 37–41.

Weber, M. C. W. (1904). "Die Säugetiere." Fischer, Jena.

Whitmore, F. C., Jr., and Sanders, A. E. (1976). Review of the Oligocene Cetacea. *Syst. Zool.* **25,** 304–320.

Winge, H. (1921). A review of the interrelationships of the Cetacea. (Translated by Gerrit S. Miller, Jr.) *Smithson. Misc. Collect.* **72**(8), 1–97.

Woodring, W. P., Bramlette, M. N., and Kew, W. S. W. (1946). Geology and Paleontology of the Palos Verdes Hills, California. *Geol. Surv. Prof. Pap. (U.S.)* **207,** 1–145.

Yablokov, A. V. (1964). Konvergentsiya ili parallelizm v razvitii Kitoobraznykh. (Convergence or parallelism in the evolution of cetaceans.) *Paleontol. Zh.* **1964**(1), pp. 97–106; transl. in *Int. Geol. Rev.* **7**(8), 1461–1468 (1965).

Zullo, V. A. (1961). A new whale barnacle from late Pleistocene deposits at San Quintín Bay, Baja California. *Veliger* **4**(1), 13–14.

Zullo, V. A. (1969). Thoracic Cirripedia of the San Diego Formation, San Diego County, California. *Los Angeles County Mus. Contrib. Sci.* **159,** 1–25.

2

Atlantic Gray Whales

James G. Mead and Edward D. Mitchell

Introduction

The gray whale (*Eschrichtius robustus*), common in the north Pacific at the present time, was first known to science on the basis of subfossil remains from Europe. There are historical references to gray whale in the North Atlantic, such as the "sandloegja" of Iceland (Fraser, 1970) and the "scrag whale" (Dudley, 1725) of early New England whalers. In the early literature pertaining to the North Atlantic, there are a few reports of whales that do not seem to correspond to Atlantic species. Some of these are, no doubt, erroneous and garbled accounts of species that are known from that ocean, but some may represent unrecognized accounts of the Atlantic gray whale. We will go through the historical accounts and add newly found subfossil specimens that demonstrate the existence of a population of gray whales in the western North Atlantic until relatively recent times.

To make sense of these reports, one must have criteria to differentiate accounts of gray whales from those of other whales. The main external characters that set gray whales apart from other large cetaceans are

1. Appearance of the dorsal region. Most species of whale have a well-formed dorsal fin. Gray whales have a series of low knob-like crenulations on the back starting where the dorsal fin would be. There are three species of whale that could be mistaken for a gray whale; the sperm (*Physeter catodon*), humpback (*Megaptera novaeangliae*), and right whale (*Balaena glacialis*). The sperm whale has a low dorsal fin or hump followed by a series of crenulations. On closer examination, it may be distinguished from a gray whale by the shape of its head and the presence of teeth. Some humpback whales have a low triangular dorsal fin that could be interpreted by persons not familiar with whales as not being a dorsal fin. In the North Atlantic Ocean, the humpback's

ISBN 0-12-389180-9

extremely long white flippers are markedly different from gray whales' short gray flippers. Right whales have no dorsal fin at all. Their smooth black back sets them apart from gray whales, which have mottled gray backs.

2. Ectoparasites and pigmentation. Gray whales are heavily infested with barnacles (*Cryptolepas rhachianecti*) and whale lice (*Cyamus scammoni*, *C. ceti*, and *C. kessleri*). This, coupled with their coarsely mottled pigmentation, is one of the species' most distinctive traits. The only other whales with ectoparasite infestations sufficiently dense to cause them to be confused with grays are humpbacks and right whales. However, ectoparasites are generally confined to the head in right whales and to the flukes, flippers, chin, and genital slit in humpbacks, whereas the ectoparasites may be distributed over the entire body of gray whales. Other species of large whales are usually dark on the back and lighter below. Gray whales are coarsely mottled with varying shades of gray both above and below.

3. Color of the baleen. Gray whale baleen is "cream to pale yellow," similar in color to the baleen of finbacks (on part of their right sides) and most minke whales, but it is much thicker and coarser in grays.

Balaenopterids can be mistaken for gray whales. While working at a whaling station in Newfoundland in 1971, one of us (Mead) was approached by the skipper (Clarence George) of one of the small vessels used to take minke whales and asked what would happen if he took a gray whale. Several years before he had seen an animal with white markings over its body and no dorsal fin but rather a series of bumps in its place. He had looked it up in a book and come to the conclusion that it had to be a gray whale, and besides he had killed at least one of everything else in the temperate North Atlantic (including the right whale) and "knew" it was none of those. He knew that the gray whale was protected in the Pacific but wondered if that protection extended to the Atlantic. A few days later Mead happened to be on board the regular catcher vessel which worked out of that station (M. V. *West Whale 8*) and asked the skipper (Arne Borgen), who had whaled out of a station on Vancouver Island and was familiar with gray whales, if he had ever seen anything that looked like a gray. He at first said he had not and wondered why Mead asked. When Mead told him Clarence George's story, his face lit up and he said that he remembered that whale. He had seen it a few miles out of the station and had initially been surprised because it looked for all the world like a gray whale. He followed it and finally came to the conclusion that it was an injured balaenopterid. He harpooned it and when it was hauled up on the station slipway it turned out to be an emaciated finback that had lost the dorsal fin and was scarred over large portions of the body.

To clarify past accounts, we shall review the pertinent literature in chronological order.

Literature Accounts without Specimens

The earliest account of cetaceans in the North Atlantic containing information relevant to this problem of the existence of gray whales in the North Atlantic is the "Speculum Regale" (kongespiel; konungs-skuggsja; kings' mirror), an Icelandic manuscript of the mid-thirteenth century. This work has been commented upon in some detail by Eschricht and Reinhardt (1866, pp. 23–34). The manuscript contains a list, with descriptions, of the cetaceans of Iceland which number some 20 species. It apparently also once contained illustrations that have since been lost. We have examined an English translation of the Speculum Regale (Larson, 1917) and found no descriptions that are obviously of gray whales. There are one or two that could potentially be gray whales but there is insufficient information given to allow verification.

Eschricht and Reinhardt (1866, p. 31) mention a list of the cetaceans of Iceland that was written in the seventeenth century and sent by an Icelandic cleric to Olaus Worm. This list appears to have formed the basis for the chapter on cetaceans in Worm's *Museum Wormianum*, published in 1655 (see Allen, 1882, p. 417 for comments). As abridged and amended by Worm the information presented consists merely of a list of 22 species, many of which have no description and none of which can be related to the gray whale.

In 1611, The Muscovy Company of England sent an expedition consisting of two ships under the command of Thomas Edge to Spitzbergen to investigate the possibility of whaling there. The instructions for this expedition included a list of the various whales that they might encounter and their relative commercial values (Lubbock, 1937, pp. 60–63; Allen, 1882, p. 412). Allen indicated that these instructions appeared in "Purchas His Pilgrimes" in 1625, bearing the date of issue of 31 March 1611. Lubbock quoted portions of this list which clearly describe the Greenland right whale (*Balaena mysticetus:* bearded whale), the right whale (sarda), the sperm whale (trumpa), one of the finner whales (*Balaenoptera* sp.: gibarta), another finner whale, possibly the blue whale (*Balaenoptera musculus:* sedeva), and what was apparently the humpback (sedeva negro). In addition to these six there was another sort of whale described, the otta sotta, which Lubbock was unable to identify. We repeat the account quoted by Lubbock (1937, p. 63) from the instructions given to Edge:

> The fourth sort of whale is called Otta Sotta, and is of the same colour as the Trumpa having finnes in his mouth all white but not above halfe a yard long, being thicker than the Trumpa but not so long. He yeeldes the best oyle but not above 30 hogs' heads.

The color and size of the baleen and oil yield are more consistent with an identification of a gray whale than with any other known species. The only other whale with all-white baleen is the minke whale, which is shorter, of a uniform darker color dorsally and white ventrally, and has a much lower oil yield. Of particular note is the statement about the quality of the oil, which indicates that it was not only taken commercially but that it was particularly sought after.

The commission from the Muscovy merchants to Thomas Edge bears further examination in the of the state of knowledge about Spitzbergen at that time. Eschricht and Reinhardt (1866, p. 26) stated that news of the abundance of whales in Spitzbergen was brought by Jonas Poole who visited there in 1610. Thus it seems unlikely that the commission of 1611 was based on information derived from activities at Spitzbergen. Edge took with him to Spitzbergen a number of Basques known for their skill in pursuit of whales, and it is likely that the information of the commission was obtained from Basque sources.

Also of interest is an engraving (Fig. 1), unfortunately of somewhat dubious origin, which has appeared in a variety of publications (e.g., Walton Advertising and Printing Company, 1915, p. 15; Dow, 1925, Plate 44). It appears to have come from Churchill's Voyages, a collection of narratives of travel and exploration published in various editions from 1704 onward. In the edition that we have examined, this engraving is associated with an account of the voyage of John Monck (Jens Munk) to Hudson Bay in 1619 and 1620 (Churchill and Churchill, 1704, pp. 541–569). Appended to the narrative of the voyage is an account of uncertain origin of whaling at Spitzbergen. The title page bears the comment "translated from the High-Dutch original, printed at Frankfort upon the Maine, 1650." The compilers of this edition have associated it with Monck, but neither the authorship nor the source of the engraving is clear. The engraving shows a whale on the beach surrounded by what appear to be whalers and merchants. The whale pictured clearly represents none of the balaenopterids or the black right whale. This leaves only the Greenland right whale and the scrag whale. If we assume for the moment that the scrag whale is the gray whale and use this species as the basis for comparison with the engraving, some interesting features become apparent. In the engraving the head is relatively less bowed than in the Greenland right whale and in general is more similar to the gray whale, particularly in the relative depths of the lower jaw and rostrum. The pigmentation pattern, generally light along the whole ventral surface, is also more similar to the gray whale than to the Greenland right whale, in which the underside of the tip of the lower jaw and sometimes the ventral surface of the tailstock are usually

Fig. 1. *Dutch engraving.*

much lighter than the rest of the ventral surface, which is usually black. The shape of the flipper is more similar to that of the gray whale than to the broad flippers of the Greenland right whale. The pigmentation shown on the flippers and flukes resembles the mottling seen on the appendages of the gray whale.

Unfortunately the engraving turns out to be a copy of a portion of a painting of a Danish whaling station at Spitzbergen done in 1634 (Dalgård, 1962; Plate 4). There are numerous resemblances between the original painting and the engraving in Churchill, from the arrangement of the people and their dress to the equipment shown scattered about. The one thing that has changed is the whale. In the original painting it is much more like a Greenland right whale than a gray whale. The resemblance of the image in the copy to a gray whale was apparently solely the result of the engravers' inability to copy accurately the whale in the painting.

Fraser (1970, p. 17) gave the translation of Gudmundsson's seventeenth century (about 1640) description of a whale that the Icelanders referred to as the "sandloegja":

> Sandloegja (fig. 5). Good eating. It has white baleen plates, which project from the upper jaw instead of teeth, as in all other baleen whales, which will be discussed later. It is very tenacious of life and can come on land to lie as seal like to rest the whole day. But in sand it never breaks up.

It is sufficient to say that we agree with Fraser's conclusion that the sandloegja of Gudmundsson represents the gray whale, indicating that this animal was known to the Icelanders at this time by the palatability of its flesh, its habits, and its physical characteristics. We initially thought this information might have derived from earlier Icelandic works, specifically the "Speculum Regale," but as stated earlier we could find no accounts in that work from which this information could have come.

De Vries (1853) gave an account of whaling in Delaware Bay (a large embayment formed at the mouth of the Delaware River and bounded by the states of New Jersey and Delaware) in the early part of the seventeenth century. Whales were apparently very numerous there, but as de Vries points out in several places, yielded very little oil. At one point he mentions the capture of seven whales which yielded only 32 "cartels" (65 barrels) of oil. De Vries paid considerable attention to documenting economical aspects of whaling in Delaware Bay, with much comment on the low oil yield of the whales. From the fact that baleen, which was extremely valuable at that time, was not mentioned, we can conclude that the whales taken may not have been right whales.

Van der Donck (1841) presented considerable information on whaling in "New Netherlands" (New York state) in the mid-seventeenth century. He spoke of a good whale fishery in Delaware Bay and described the whales there in the following terms: "Whales are numerous in the winter on the coast; and in the bay where they frequently ground on the shoals and bars; but they are not as fat as the Greenland whales." Van der Donck also mentioned an instance in March of 1647 in which two whales "of common size" swam 43 miles up the Hudson River. One of these stranded there; the other returned to within 12 miles of the sea before it also stranded ("near which place four others also stranded the same year"). As from de Vries' account (1853), one gathers from van der Donck that whaling took place only in the winter. It is notable that there is no mention of baleen as a product of this industry, although considerable stress is placed on the economics of the venture. This leads to the conclusion that they also may not have been right whales.

Dudley (1725), in his famous letter relating to the whales of the coasts of New England, provided the first mention of an enigmatic whale known as the scrag whale. Dudley gives accounts of recognizable right, sperm, finback, and humpback whales. In addition he describes another kind of whale in these words (1725, p. 258):

> The Scrag whale is near a-kin to the Fin-back, but instead of a Fin on his Back, the Ridge of the After-part of his Back is scragged with a half Dozen Knobs or Knuckles; he is nearest the right Whale in Figure and for Quantity of Oil; his Bone is white, but won't split.

There has been much written about the identity of the "scrag whale of Dudley." Authors fall into two groups, those who thought the scrag whale was the gray whale and those who did not believe one could determine its identity. We intend to demonstrate that there is enough evidence that gray whales existed along the American coast to ascertain that what Dudley meant by scrag whale is in fact the gray whale.

Douglass (1760, p. 58), in a discussion of the British territories in North America, gave an account of the whales and the whaling industry in which he mentions the "bunch" of the humpback and merely states "the scrag whale has several of these bumps."

St. John (1782, p. 169) presented a list of the whales known at Nantucket that included right, humpback, finback, sulphur-bottom, and sperm whales, but nothing which could pertain to the gray whale. As True (1904, p. 43) noted, St. John seems to have had first-hand knowledge of whaling in this area, which may indicate that the gray whale was no longer present in any numbers at this time.

In his history of the colony of Nantucket, Macy (1835, p. 28) mentioned that in the early days of the colony (prior to 1672) a whale of the kind called "scragg" entered the harbor and was pursued and killed by the settlers. He remarks that this instance provided the start of the Nantucket whaling enterprise. It is unclear, in the reference of Macy (1835, p. 28) to scrag whales at Nantucket, Massachusetts prior to 1672, whether the term was in use in 1672 or was simply supplied by Macy at the time of his writing.

Cuvier (1836, p. 309) dismissed Dudley's scrag whale as probably an erroneous account of a balaenopterid. Cuvier had interpreted Dudley's remark about the protuberances on the back of the scrag whale to mean they were bony, leading him to discredit the rest of Dudley's account as improbable.

Gray (1850) commented on the validity of *Balaena gibbosa* (the scientific name based on Dudley's account of the scrag whale), which he felt was "a *Balaena*, probably well known formerly." To support his case he cited Beale (1839) as saying that the scrag whale was known to contemporary whalers. We have read Beale (1835, 1839) and were unable to locate this information. Gray himself seems to have had no original information on this subject.

Eschricht and Reinhardt (1866, p. 42) commented upon the usage of the term "scrag whale" by whalers in the Pacific and concluded that most of these could be simply references to young right whales. They stated (1866, p. 43) that

> Scrag and Scrag-whale have been common appellations during at least 150 years, and, like the name of the right-whale itself, have at different times and in different places been applied to cetaceans quite different from one another.

It seems from their account that some of the whales termed "scrag" may have been gray whales. Brown (1868, p. 547), in a discussion of the whales of Davis Strait and Baffin Bay, said that he could not imagine what Dudley's scrag whale was, as the whalers did not know of such an animal.

The comparisons which Cope (1868c) (see p. 44) made between the *Agaphelus gibbosus* specimens and the Pacific gray whale material to which he had access are sufficient to indicate that *A. gibbosus* was quite different. In our opinion the shape of the base of the baleen plates, the nature of the coronoid process, the shape of scapula, and the pattern of pigmentation of Cope's *A. gibbosus* are definitely balaenopterid. The supposed lack of dorsal fin and ventral grooves is most certainly due to errors arising from the description which Cope received of the badly decomposed and mutilated carcass of the Long Beach specimen.

It is not entirely clear, however, which species of balaenopterid was represented by the Long Beach stranding. Based on the white color of the baleen there are two possibilities; the finback and the minke whale. Cope's size estimate of 43 feet (13 m), plus the information that the vertebral epiphyses were unfused, strongly suggests that the specimen was a young finback. The size of the auditory bulla, the pigmentation pattern, and the size of the baleen, however, indicate that it was a minke. If it were the latter, it would have been an extremely large animal (adults of this species rarely excede 10 m in length); this is contrary to the other evidence that this was a very young animal.

All evidence argues that the Long Beach stranding was probably a minke, the details of which were perhaps confused with defective information or remains of an independent stranding of a finback. This conclusion agrees with Cope's initial (1867) and final (1884) diagnoses, with the detailed account of True (1904, p. 105), and the discussion of Hershkovitz (1961).

Captain Atwood of Provincetown, Massachusetts wrote (in Allen, 1869, p. 203) of a scrag whale sometimes taken near that locality. He said that some of the whalers were of the opinion that it was a right whale. However, he stated that its most prominent feature was a small number of projections on the dorsal ridge near the tail, "having some resemblance to the teeth of a saw." Allen had sent the account of Atwood to Cope, who attached the name *Agaphelus gibbosus* to this animal.

Allen (1869, p. 203) placed *Eschrichtius robustus* on the list of mammals of Massachusetts on the basis of the jaw which Cope mentioned from the New Jersey coast (not the "*Agaphelus*" stranding at Long Beach). In 1869 Cope published further descriptions of fossil material which he assigned to *Eschrichtius*, still not recognizing its relationship to *Agaphelus* (Cope, 1869a; see Chapter 1, this volume).

Gray (1870) stated that the scrag whale (*A. gibbosus*) was abundant in America and was still caught now and then by whalers. This was probably derived from the account of Atwood in Allen (1869). Gray (1871) included *Eschrichtius robustus* as a part of the recent Atlantic cetacean fauna based on Cope (1868b).

Flower (1872) described subfossil material of *Eschrichtius* from Cornwall and cited the jaw from the New Jersey coast (Cope, 1868b) as evidence of the occurrence of this species on the western coasts of the Atlantic. He did not, however, make the connection between the species and the Pacific gray whale.

Ten years later Cope (1869b), having examined more material of the Pacific gray whale, recognized its distinctness from the specimen which he had assigned to *Agaphelus gibbosus* and formed the new genus *Rhachianectes* for it. In so doing he lost the one fortuitous correlation he had made, that of the scrag whale and the Pacific gray whale.

Van Beneden and Gervais (1880, p. 236) considered *Agaphelus gibbosus* to be a young right whale. This conclusion is certainly based on the earlier idea of "scrag" whales as young or emaciated right whales rather than on any evaluation of the basis of Cope's taxon. They cited (1880, p. 236) the jaw from New Jersey which Cope (1868b) referred to *Eschrichtius robustus*. References to this specimen have probably been confused with the variety of fossil cetaceans from the Miocene which Cope assigned to *Eschrichtius* (see Barnes and McLeod, this volume). Fischer (1881, p. 35) noticed the account of *Agaphelus* from Cope's works and included it in his list of cetaceans from the southwest of France, without adding anything new to the account.

Finally, in 1884, Cope recognized the errors in the account of the Long Beach stranding and referred the specimen to *Balaenoptera rostrata* (minke), as he had originally done in 1867. He noted that "the species may, however, be the *Balaena gibbosa* of the old authors," indicating that the scrag whale question remained unsettled in his mind.

True (1884, p. 628) provided a good indication of the state of knowledge of the problem in a tabular presentation of the cetaceans of North America and their distribution. In this table he listed *Rhachianectes glaucus* (Pacific gray whale) as occurring only in the Pacific, *Agaphelus gibbosus* (the scrag whale of Dudley and Cope) only in the western Atlantic, and *Balaenoptera robusta* (the subfossil *Eschrichtius* material) as occurring on both sides of the Atlantic but not in the Pacific. Clearly he had noted all of the specimens in the literature but, like his predecessors, had made none of the connections.

Lydekker (1894, p. 1185) was apparently the first to make the connection between *Eschrichtius robustus* and the Pacific gray whale. He said, of the geographic distribution of gray whale:

> from the evidence of certain bones found in the superficial deposits of the British Islands, and described under the name of *Eschrichtius*, it is, however, not improbable that it formerly frequented the Atlantic.

True (1904) commented extensively on both the scrag whale and *Agaphelus gibbosus*. He was of the opinion that the scrag whale was an aberrant right whale. He presented a detailed analysis (1904, p. 105) of the history of *Agaphelus gibbosus*, concluding that it was a mutilated specimen of a minke. On p. 48 of this work he presented a list of the whales that Gray (1871) considered to inhabit American waters, a list which contained *Eschrichtius robustus*. True commented that the inclusion of this species must rest on the statement of Cope that there was a jaw of this form in the

Rutgers museum. Although True presented measurements of some of the material at Rutgers, he apparently never examined the collection himself as he made no further mention of this jaw and did not include *E. robustus* in his account of the Atlantic fauna. In addition, True's monograph contains numerous useful references to early whaling along the Atlantic coast of North America, many of which have information in them on the scrag whale question.

Allen (1916) was of the opinion that "scrag whale" referred to small or emaciated individuals of the right whale, and went so far as to illustrate (Plate 9) a "scrag whale" (an immature female of a right whale). He also made note of the confused state of *Balaena gibbosa* Erxleben (1777), which included some accounts of what were probably the humpback. He included Gray's (1865) specimen of *Eschrichtius robustus* in his synonomy of *Megaptera nodosa*, but gave no reasons for this decision. He also assigned *Balaenoptera robusta* Lilljeborg (the holotype of *Eschrichtius robustus*) to *Balaenoptera physalus*, again with no discussion of the obvious differences between the two.

The question of the relationship of the subfossil occurrence of *Eschrichtius robustus* to the Pacific gray whale was finally examined by van Deinse and Junge (Junge, 1936; van Deinse and Junge, 1937). They reviewed all of the known occurrences of *E. robustus* in the eastern Atlantic and concluded that this material could not be differentiated from the Pacific gray whale. In conclusion they state:

> Provided with these facts we can state, therefore, that *Eschrichtius gibbosus* [= *robustus*] must have occured still along the European coasts in the first centuries A.D. Along the American coast of the Atlantic this species still must have been present in the beginning of the 18th century, for at that time it was still caught by the whalers. And we think it not at all improbable that along the Atlantic coasts of North America in future *skeleton fragments of this species will be found.* (emphasis ours)

Cederlund (1938) made a detailed comparison of the *Eschrichtius robustus* remains described by Lilljeborg with Andrews' (1914) description of the gray whale and with two European skeletons of the gray whale. He concluded that the material of the holotype, although lacking any skull remains, was remarkably similar to comparable elements of gray whale. He also discussed the question of the scrag whale and concluded that this term had been applied to both the Atlantic gray whale and to aberrant individuals of the right whale.

Hubbs (1959), in a review of the natural history of gray whale, stated that gray whales were present in the Atlantic until early whaling and might still be there, but he did not document his conclusions. In his account of the description of the sandloegja, Fraser (1970) equated it with both the scrag whale and gray whale. He suggests, on the basis of the account of the sandloegja, that the North Atlantic gray whale may have been exterminated at an early date by human predation.

Eastern Atlantic Specimens

The first scientific mention of gray whale came in 1861 (Table I), when Lilljeborg described remains found on the coast of the Gulf of Bothnia at Gräsö, Roslagen,

Table I
Subfossil Specimens of Atlantic Gray Whales

Year found	Locality	Material	Age (years BP)	Authority
European material				
1829	Pentuan, England	Partial skeleton		Flower (1872)
1859	Gräsö, Sweden	Partial skeleton	4000–6000	Lilljeborg (1861)
1861	Babbacombe Bay, England	Vertebra		Gray (1864)
1865	Babbacombe Bay, England	Vertebra		Gray (1866)
1879	Ijmuiden, Netherlands	Partial skull		Deinse and Junge (1937)
1916	Ijmuiden, Netherlands	Partial skull	1400	Deinse and Junge (1937)
1935	Wieringermeer-Polder, Netherlands	Partial skeleton	4000–6000	Deinse and Junge (1937)
North American material[a]				
1850s	Toms River, New Jersey	Jaw	455 ± 90	This chapter
1959	Myrtle Beach, South Carolina	Partial skeleton	865 ± 165	This chapter
1969	Chesapeake Bay, Virginia	Partial skull	10140 ± 125	This chapter
1970s	Nags Head, North Carolina	Jaw	865 ± 50	This chapter
1976	Corolla, North Carolina	Squamosal	2415 ± 90	This chapter
1977	Southampton, New York	Jaw	275 ± 35	This chapter
1977	Corolla, North Carolina	Partial skull		This chapter
1978	Rehobeth, Delaware	Squamosal		This chapter
1979	Cape Lookout, North Carolina	Partial skull	1190 ± 245	This chapter

[a]Carbon 14 dates on the North American specimens determined by the Radiocarbon Laboratory, Smithsonian Institution. Samples were run on collagen fraction remaining in the bone.

Sweden (Fig. 2) under the name *Balaenoptera robusta*, in the belief that it was a new species of finner whale (Lilljeborg, 1861). The remains included the mandibles and most of a postcranial skeleton but no skull fragments. All seem to have belonged to a single physically mature individual, inasmuch as all vertebral epiphyses remained attached to the centra. The description was subsequently translated into English (Lilljeborg, 1866) with some additional figures added. A complete description of all the elements was prepared and published in English (Lilljeborg, 1867).

Lilljeborg's material was found in a field, 10–15 ft (3–5 m) above sea level and 840 ft (260 m) from the current shoreline. The remains were in a stratum of sand and clay 2–4 ft (60–120 cm) beneath the surface, associated with shells of *Mytilus edulis* and *Macoma balthica*. Lilljeborg considered this deposit to be postglacial and estimated its age at 4000–6000 years. Van Deinse and Junge (1937) reconsidered the age of this find and agreed that Lilljeborg's estimate was essentially correct.

Although it was not described until 1872, the first specimen of gray whale to be found along the northeast coast of the Atlantic was a partial skeleton excavated in 1829 from the Happy Union Tin Stream Works at Pentuan, Cornwall, England. The specimen consisted of the right lower jaw, a lumbar vertebra, a humerus, radius, and two metacar-

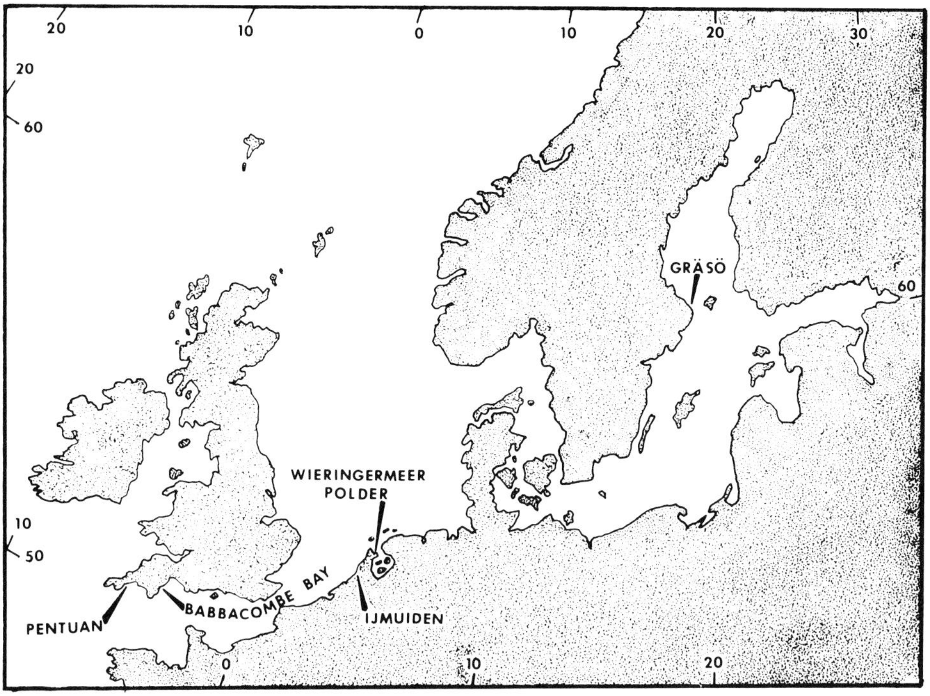

Fig. 2. *Location of eastern Atlantic specimens.*

pals, all probably from the same individual. According to Flower (1872), who described this material, it was slightly smaller than Lilljeborg's specimen from Gräsö and was a somewhat younger individual, as the vertebral epiphyses were not fused to the centra. These remains came from a depth of about 7 m in a stratum of marine sands, about 850 m from the present shoreline (Colenso, 1832, p. 36). The overlying sediments consisted of river sands and gravels, and the immediately underlying sediments were sands containing the remains of a number of terrestrial mammals, including man.

The third specimen to be described of this species from the northeast Atlantic was a cervical vertebra washed ashore at Babbacombe Bay, England and found by a Mr. Pengelly on November 24th, 1861. This vertebra was sent to Gray who immediately associated it with Lilljeborg's material on the basis of the extremely wide neural canal. Gray (1864, p. 350) initially thought that it was from a species of humpback whale and accordingly designated it "*Megaptera? robusta.*" He felt it was sufficiently different to warrant a subgeneric distinction and accordingly proposed the subgenus *Eschrichtius*, which he subsequently (1865, p. 41) elevated to full generic rank. In June of 1865, Mr. Pengelly recovered from the same locality another cervical vertebra which Gray (1866, p. 373) also identified as *Eschrichtius*. No estimates have been made of the age of this specimen.

No further remains of Atlantic gray whales came to light until 1937, when van Deinse and Junge published on a series of finds from the Netherlands. Stimulated by a recently found specimen, they searched museum collections and uncovered some earlier material. The first was a partial skull of a juvenile, found in 1879 at Ijmuiden on the North Sea coast of the Netherlands, just west of Amsterdam. Unfortunately there was no information on the circumstances under which this specimen was found on which to base an estimate of its age. The second was also from Ijmuiden and was a partial skull of an adult found in 1916 at a depth of about 1 m during excavation for a new harbor. Van Deinse and Junge (1937, p. 184) suggested an age of about 500 AD, based upon the amount of accumulated silt overlying it. The final specimen was a partial skeleton, consisting of a broken skull, mandible, hyoid, miscellaneous vertebrae, ribs, and limb elements, found by a farmer in 1935 while plowing a field in the Wieringermeer-Polder, a portion of the Zuiderzee about 40 miles (60 km) north of Amsterdam which had been drained in 1930. Van Deinse and Junge estimated the length of this animal at 8 m. Based on a pollen analysis of associated sediment, they estimated an age of 4000–6000 years (van Deinse and Junge, 1937, p. 184).

Fredén (1975) examined a number of subfossil remains of seals and whales from the west coast of Sweden. Although he did not present information on any specimens identified as gray whales and specifically excluded any material found along the current coastline, he did list a number of finds of unidentified whales, some of which might prove to be this species.

Western Atlantic Specimens

Edward Drinker Cope (1867) assigned a series of Miocene fossil cetaceans to the subgenus *Eschrichtius*, which Gray had formed for subfossil remains from Europe. These have been determined to be unrelated to the gray whale (see Chapter 1, this volume). Gray and Cope were deliberating on the strength of certain resemblances of these remains to the humpback. Thus when Cope first encountered recent material of *Eschrichtius* in the form of the Pacific gray whale, he did not make the connection between the two. It is in this same paper (Cope, 1867, p. 147) that Cope first mentioned a whale stranded at Long Beach, New Jersey. He referred this specimen to the minke whale.

In the next year Cope (1868a) made an informal announcement of a new genus *Agaphelus*, based upon the Long Beach stranding, the scrag whale of Dudley, and the Pacific gray whale. Subsequently another short notice appeared (Cope, 1868b) mentioning *Agaphelus* and giving a brief reference to a whale jaw, which Cope assigned to *Eschrichtius robustus* (Lilljeborg). In a third paper of the same year, Cope (1868c) formally defined *Agaphelus* and described the two species, *A. gibbosus* (based upon the Long Beach stranding and the scrag whale of Dudley) and *A. glaucus* (based upon the Pacific gray whale). It must be borne in mind that much of Cope's concept of *Agaphelus* was based upon the scrag whale of Dudley, and although he was familiar with remains of *Eschrichtius* described by Lilljeborg, he never made the connection between it and *Agaphelus*.

The specimen of *Eschrichtius robustus* that Cope (1868b) mentioned was a jaw sent

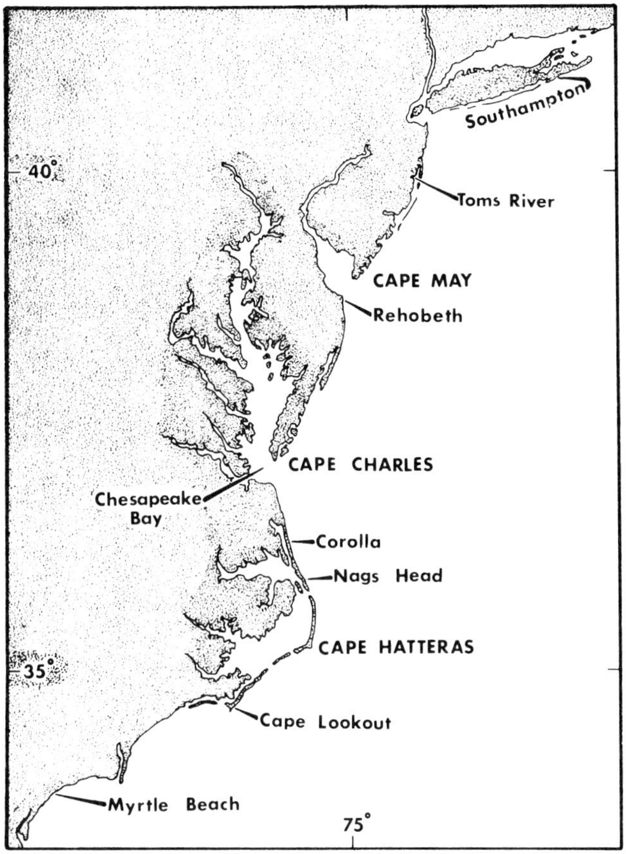

Fig. 3. *Location of western Atlantic specimens.*

to the Rutgers College museum by a Mrs. Dunham sometime about 1855. The data associated with this specimen are limited to those found on the paper label glued to the specimen and include only the name of the donor and the locality, Toms River, New Jersey (Fig. 3). There are a number of other bones in the museum collection which were contributed by the same donor, all apparently from Toms River. The jaw was examined by Cope (1868b, p. 194) who commented on it in a list of the recent mysticetes of the Atlantic coast of the United States:

> The *Eschrichtius robustus* is admitted on the evidence of a ramus of the under jaw in the Museum Rutger's College, which is of peculiar form, and closely resembles the figure given by Lilljeborg of that portion of this rare species.

Although this reference was cited in numerous subsequent works (Gray, 1871, p. 52; Flower, 1872, p. 442; van Beneden and Gervais, 1880, p. 236; True, 1904, p. 48) the jaw was apparently never reexamined and the connection between it and the Pacific gray whale was never made.

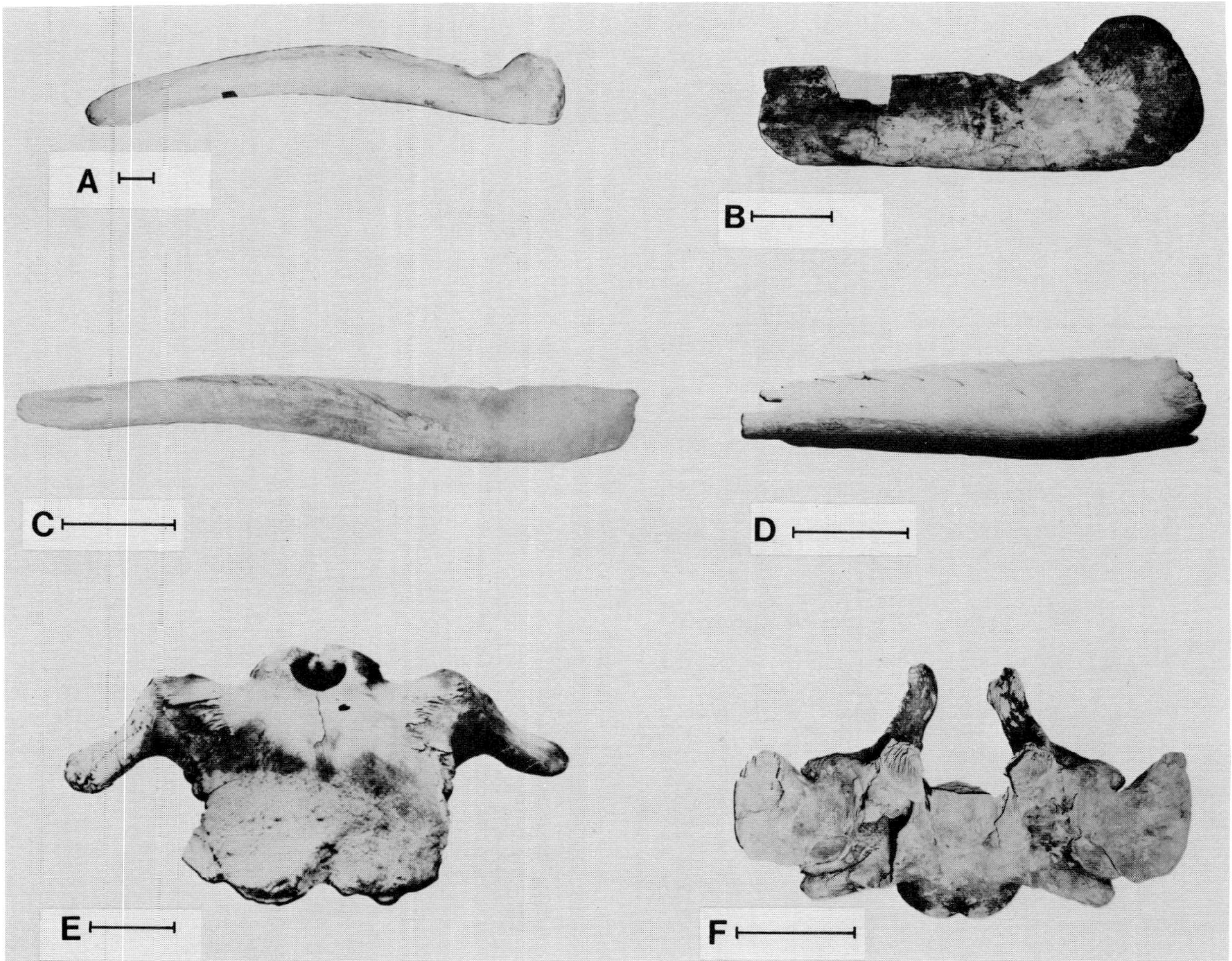

Fig. 4. *Photographs of North American specimens of the Atlantic gray whale. Scale bars = 20 cm. (A)–(D) are lateral views of left mandibles. (A) USNM 187448, Toms River, New Jersey; (B) USNM 23260, Myrtle Beach, South Carolina; (C) USNM 244465, Nags Head, North Carolina; (D) USNM 244307, Southampton, New York; (E) posterodorsal view of partial skull of USNM 299838, Corolla, North Carolina; (F) ventral view of partial skull of USNM 187449, Chesapeake Bay, Virginia.*

The specimen is a left mandible (Fig. 4A), 272 cm long and in excellent condition. Comparing its length with the mandible of the Korean specimen described by Andrews (1914, p. 264), a 1250-cm male with a mandibular length of 243 cm, the total length of the Rutgers specimen is estimated as 1400 cm. This specimen has been transferred to the United States National Museum (USNM 187448)[1] (Table I).

The second specimen of a gray whale from the western north Atlantic was found September 7, 1959 by amateur divers at Myrtle Beach, South Carolina. This specimen, apparently including most of a skeleton, was discovered about 50 ft (15 m) offshore in about 5 ft (1.5 m) of water. The divers recovered the posterior portion of the mandible (Fig. 4B) and brought it to the Smithsonian Institution where Charles Handley identified it as from a gray whale. As encouraged by Handley, the divers returned to attempt to recover more of the specimen but discovered that a storm had disrupted the locality; they were unable to locate any further remains. This specimen (USNM 23260) is from a large adult, comparable in size to the Rutgers specimen.

In 1969 Captain Floyd Durette recovered a partial skull of a juvenile gray whale which he sent to the United States National Museum (USNM). Exact locality data are unavailable, but it is presumably from near the mouth of Chesapeake Bay, Virginia, as this was the general area of Captain Durette's work at the time. The specimen (USNM 187449; Fig. 4F) consists of the posterolateral portion of a cranium with most of the dorsal and ventral elements missing (i.e., it consists of squamosals, exoccipitals, parietals, and the basioccipital). This is the oldest specimen of the Atlantic gray whale that we have radiocarbon dated. The age of the specimen is 10,140 ± 125 years BP.

A fragment of a right squamosal of a young gray whale was found on the beach at Corolla, North Carolina on 26 March 1976 by Avean Meekins. He took it to the North Carolina Marine Resource Center in Manteo, North Carolina, where one of us (Mead) examined it. We borrowed the specimen and compared it to juvenile gray whale material in the USNM, ascertaining that it was from an animal about 25 feet long. A cast was made of the specimen (USNM 244308), and a sample was taken for radiocarbon dating. The original was returned to the North Carolina Marine Resource Center in Manteo, North Carolina.

A midportion of a badly weathered left mandible (Fig. 4D) of a juvenile was found by Mae St. Germaine on the beach at Southampton, Long Island, New York, on May 31, 1977. A cast was prepared (USNM 244307), a sample taken for radiocarbon dating, and the original specimen returned to Mrs. St. Germaine. The height at the coronoid process was 18.2 cm (plus approximately 2–3 mm for abrasion).

A complete left mandible (USNM 244465; Fig. 4C) of a juvenile was donated to the USNM by Betty Clark of Nags Head, North Carolina, who said it had been found by a friend in the vicinity of Nags Head in the 1970s. The length of the mandible was 120 ± 5 cm.

A cranium of an adult (Fig. 4E) was found on the beach at Corolla, North Carolina in the fall of 1977 by Virgil Lanning. The width across the zygomatic processes was about

[1]USNM numbers refer to specimens that are in the collection of the Division of Paleobiology, United States National Museum, Washington, D.C.

141 cm (131 cm measured, plus about 10 cm estimated to have been lost by abrasion). The width across the articular condyles was 31 cm. This specimen was donated to the USNM (299838). We tried to radiocarbon date this specimen but it had been contaminated by current atmospheric radioactive carbon to such an extent that dates were useless.

A fragment of the squamosal of an adult was found on the beach at Rehobeth, Delaware in August of 1978. This was donated to the USNM (256749) by Gloria Waggener.

The cranium of a juvenile was discovered in April 1979 on the beach at Cape Lookout, North Carolina. The width across the zygomatic processes was 80 cm. This specimen is currently housed at the Hampton Mariners Museum in Beaufort, North Carolina.

We have just learned of an additional find of an Atlantic gray whale. It was found on Jupiter Island, Florida (approximate geographic coordinates 27° 3′ north, 80° 6′ west). Dr. Daniel Odell (University of Miami) sent some photographs of the specimen and the data to us in January 1983. The specimen consists of most of the cranium of what looks like an adult which was found on the beach after a storm. This extends the known distribution of the Atlantic population nearly 500 miles to the south. The specimen is currently in the possession of the Hobe Sound National Wildlife Refuge.

The radiocarbon ages for the samples from the western North Atlantic are shown in Table I. The most recent date (1675 AD) is from a period when colonists had settled in America and the next youngest is potentially so (1405–1585 AD). The oldest specimen that we radiocarbon dated was USNM 187449, which gave an estimated age of 10,140 ± 125 years. No sample was run on USNM 256749 because it was felt that the removal of so much bone would essentially destroy the specimen. The importance of these radiocarbon ages is not to define exactly the period of existence of gray whales in the north Atlantic but merely to give an idea of how recently they became extinct.

Taxonomy of the Atlantic Gray Whale

Eschrichtius

Eschrichtius Gray 1864, p. 350. [Subgenus of *Megaptera*. In the following year he elevated it to full generic rank (Gray, 1865)]. Type species *Balaenoptera robusta* Lilljeborg 1861, p. 602.

Agaphelus Cope 1868a, p. 159. Type species *Balaena gibbosa* Erxleben 1777, p. 610.

Rhachianectes Cope 1869a, p. 15. Type species *Agaphelus glaucus* Cope 1868a, p. 160.

Eschrichtius robustus

Balaenoptera robusta Lilljeborg 1861, p. 602. Type specimen the Gräsö whale preserved in the University Museum of Upsalla, Sweden.

Megaptera? robusta Gray 1864, p. 350.

Eschrichtius robustus Gray 1865, p. 42.

Agaphelus glaucus Cope 1868a, p. 160. Lectotype a side of baleen in the museum of the Essex Institute, Salem, Massachusetts (True, 1904, p. 81), which has subsequently been lost.
Rhachianectes glaucus Cope 1869a, p. 15.
Eschrichtius glaucus Maher 1961, p. 257.

One of the taxonomic problems concerning the Atlantic gray whale has been the disposition of the name *Balaena gibbosa* Erxleben (1777, p. 610). This has usually been associated with the scrag whale, which formed the basis for the species name *gibbosus* being applied to the gray whale. Allen (1882, p. 467) stated that Erxleben had included the scrag whale in the *species obscurae* part of his compilation and that *Balaena gibbosa* Erxleben was a *nomen nudum.* However, as True (1904, p. 42) pointed out, this was an error on the part of Allen, and indeed Erxleben included the scrag whale within the synonomy of *Balaena gibbosa,* although it is the last of nine items in the synonomy. Although Erxleben's compilation does not refer specifically to Dudley's work, he does give white baleen plates in his diagnosis of *Balaena gibbosa* which is one of the more striking characters in Dudley's (1725) description of the scrag whale. While there is good reason to believe the Dudley's scrag whale formed part of the basis of *Balaena gibbosa* Erxleben, the humpback whale also formed part of the basis of *B. gibbosa.* The confusion that has centered around the name *B. gibbosa* renders it highly undesirable. Should its validity arise again as a possibility, an appeal should be made to the International Commission on Zoological Nomenclature to suppress it in favor of *Eschrichtius robustus,* which has a type specimen and a long history of unambiguous usage.

We agree with Cederlund (1938, p. 282) that the holotype of *Eschrichtius robustus* is sufficient evidence to demonstrate the conspecificity of the extinct Atlantic and extant Pacific populations of the gray whale. This is contrary to the opinion expressed by Schevill (1952). In that paper he quotes van Deinse and Junge (1937, p. 181) as follows:

> We must emphasize that the bones we have now at hand we cannot give any argument that the Pacific and Atlantic representative of *Eschrichtius* should be specifically different.

Schevill (1952, p. 1) goes on to say that "But by the same token these bones cannot prove that these whales are specifically identical." We agree that proof of specific identity is not conclusive when only the bones of an extinct animal are at hand, but is commonly the usage in paleontology to judge animals specifically identical if differences judged by competent taxonomists to be specifically distinctive cannot be found between the bones of the animals. Geographic separation, in the case of sharks, large fish, marine birds, and whales, cannot in itself form the basis for a judgement of specific identity. Rice and Wolman, who did the last major work on the gray whale (1971), recognized the Pacific population as *Eschrichtius robustus.* Any further questions along this line can only be resolved by a critical comparison of all of the Atlantic gray whale material with a broad suite of specimens of the Pacific gray whale in order to take such variables as age and sex into account.

In the event that the Atlantic and Pacific populations of gray whales are demonstrated to warrant specific or subspecific distinction, the specific name *Eschrichtius glaucus* (Cope) will be available for the Pacific form. The lectotype of the species, as designated by True (1904, p. 81), is a set of baleen in the museum of the Essex Institute, Salem, Massachusetts. It is probable that a neotype will have to be designated because the lectotype material cannot be found (W. Schevill, personal communication).

Conclusion

There are three accounts in the literature that we interpret to be reliable records of gray whales in the North Atlantic. One of these was an Icelandic record of the sixteenth century (Fraser, 1970), another was the account of Dudley in 1725, the third was a commission from the Muscovy Merchants to Thomas Edge in 1611 which dealt with a whale called the "otta sotta," which fits the description of a gray whale. There are several other reports which, although they do not give enough information to allow positive identification of the whales concerned, appear to describe gray whales.

The species was first known to science on the basis of subfossil specimens from Europe. It is now known in the Atlantic from a total of seven European specimens and nine American specimens. Six of the American specimens were found in the last 10 years, despite the fact that there has been no special effort to find them. Based upon these finds we conclude that the gray whale was, at some time, common in American waters and was sometimes known to whalers as the scrag whale. Radiocarbon dates show the most recent specimen to have existed in colonial times, around 1675 AD.

Summary

Eschrichtius robustus was first known to science on the basis of a subfossil skeleton from Gräsö, Sweden described by Lilljeborg in 1861. In subsequent years six more subfossil specimens were described from the coasts of England and the Netherlands. Van Deinse and Junge (1937) compared these remains to the Pacific gray whale and concluded that they could not be separated on a specific basis.

Dudley (1725), in a short paper enumerating the whales then known off New England, described a whale known as the "scrag" whale. This species was discussed at length by biologists, most of whom came to the conclusion that it represented young or emaciated individuals of the right whale. This was to a large part based upon the supposed lack of evidence of a western Atlantic population of *Eschrichtius*, but in fact Cope (1868a) had published on a jaw of this species that had been found on the New Jersey coast. Since that time there have been eight more discoveries of remains of *Eschrichtius* on the Atlantic coast of North America, upon which we report in this paper. Radiocarbon dates reveal that the youngest specimen died during colonial times (1675 AD).

Fraser (1970) published on a sixteenth century Icelandic account which he interpreted as evidence of the persistence of a gray whale (sandloegja) population in the North Atlantic at that time. Lubbock (1937) published a seventeenth century account of a whale called the "otta sotta," whose description fits the gray whale. Several other more anecdotal accounts also fit the gray whale.

The conclusion is that a population of *Eschrichtius robustus* existed on both sides of the Atlantic and was present on the coasts of North America up to the seventeenth century AD.

Acknowledgments

This investigation was prompted by Clarence George's inquiry in 1971 about Atlantic gray whales. Since that time it has involved help from the following people: Doug Oliver, a student at Rutger's, who took photographs of the whale remains in the attic of their museum in response to our telephone requests; Dr. R. C. Murray of that same institution, who suggested that the remains be transferred to the United States National Museum; Dr. Robert Brownell, who kindly transported it to Washington; Dr. Charles Handley, who recognized the second specimen as a gray whale, asked the collectors, Mr. Maklion and Mr. Aleksandrov to try to obtain more of the specimen, and kindly made the specimen available to us; Dr. Lawrence Barnes, who brought an unidentified specimen in the United States National Museum to our attention; Charles Potter, who deserves special thanks for recognizing two of the specimens as being gray whales. We thank Floyd Durette, Betty Clark, Avean Meekins, Steve Daniels, Mae St. Germaine, Virgil Lanning, Gloria Waggener, and Karen Davis, who were involved in the collecting of specimens.

References

Allen, G. M. (1916). The whalebone whales of New England. *Mem. Boston Soc. Nat. Hist.* **8**(2), 105–322.

Allen, J. A. (1869). Catalog of the mammals of Massachusetts. *Bull. Mus. Comp. Zool.* **1**(8), 143–252.

Allen, J. A. (1882). Preliminary list of works and papers relating to the mammalian orders cete and sirenia. *U.S. Geol. Geogr. Surv. Territ. Bull.* **6**(3), 399–562.

Andrews, R. C. (1914). Monographs of the Pacific cetacea. I. The California gray whale (*Rhachianectes glaucus* Cope). *Mem. Am. Mus. Nat. Hist.* [N.S.] **1**(5), 227–287.

Beale, T. (1835). "A Few Observations on the Natural History of the Sperm Whale." Effingham Wilson, London.

Beale, T. (1839). "The Natural History of the Sperm Whale." J. Van Voorst, London.

Brown, R. (1868). Notes on the history and geographical relations of the Cetacea frequenting Davis Strait and Baffin's Bay. *Proc. Zool. Soc. London* **35,** 533–556.

Cederlund, B. A. (1938). A subfossil gray whale discovered in Sweden in 1859. *Zool. Bidr. Uppsala* **18,** 269–286.

Churchill, A., and Churchill, J. (1704). "A Collection of Voyages and Travels," 6 vols. Churchill, London.

Colenso, J. W. (1832). A description of the Happy Union Tin Stream-work at Pentuan. *Trans. R. Geol. Soc. Corn.* **4,** 29–39.

Cope, E. D. (1867). An addition to the vertebrate fauna of the Miocene period, with a synopsis of the extinct Cetacea of the United States. *Proc. Acad. Nat. Sci. Philadelphia* **19,** 138–156.

Cope, E. D. (1868a). Minutes of the meeting of June 23. *Proc. Acad. Nat. Sci. Philadelphia* **20,** 159–160.

Cope, E. D. (1868b). Second contribution to the history of the vertebrata of the Miocene period of the United States. *Proc. Acad. Nat. Sci. Philadelphia* **20,** 184–194.

Cope, E. D. (1868c). On *Agaphelus*, a genus of toothless Cetacea. *Proc. Acad. Nat. Sci. Philadelphia* **20,** 221–227.

Cope, E. D. (1869a). Third contribution to the fauna of the Miocene period of the United States. *Proc. Acad. Nat. Sci. Philadelphia* **21,** 6–12.

Cope, E. D. (1869b). Introductory note by the editor, pages 13 to 32. On the cetaceans of the western coast of North America (by C. M. Scammon, edited by E. D. Cope). *Proc. Acad. Nat. Sci. Philadelphia* **21,** 13–63.

Cope, E. D. (1884). Catalogue of aquatic mammals of the United States, by F. W. True (Review). *Am. Nat.* **18,** 1123–1124.

Cuvier, F. (1836). "De l'histoire naturelle des cétacées." Libraire Encyclopédique de Roret, Paris.

Dalgård, S. (1962). "Dansk-Norsk Hvalfangst 1615–1660." G. E. C. Gads, Kopenhagen.

de Vries, D. P. (1853). "Voyages from Holland to America, A.D. 1632 to 1644." Billin & Brothers, New York (transl. from the Dutch by H. C. Murphy).

Douglass, W. (1760). "A Summary, Historical and Political, of the First Planting, Progressive Improvements, and Present State of the British Settlements in North America," 2 vols. R. and J. Dodsley, London.

Dow, G. F. (1925). "Whale Ships and Whaling," Publ. No. 10. Marine Research Society, Salem, Massachusetts.

Dudley, P. (1725). An essay upon the natural history of whales. *Philos. Trans. R. Soc. London* **33,** 256–269.

Erxleben, J. C. P. (1777). "Systema regni animalis." Weygandianis, Lipsiae.

Eschricht, D. F., and Reinhardt, J. (1866). On the Greenland right whale. *In* "Recent Memoirs on the Cetacea by Professors Eschricht, Reinhardt and Lilljeborg" (W. H. Flower, ed.), pp. 1–143. Ray Society, London (transl. of the Danish original: Om Nordhvalen, Copenhagen, 1861).

Fischer, M. P. (1881). Cétacés du Sud-ouest de la France. *Soc. Linn. Bordeaux, Actes* **35,** 1–219.

Flower, W. H. (1872). On a Subfossil Whale (*Eschrichtius robustus*) discovered in Cornwall. *Ann. Mag. Nat. Hist.* [4] **9,** 440–442.

Fraser, F. C. (1970). An early 17th century record of the California grey whale in Icelandic waters. *Invest. Cetacea* **2,** 13–20.

Fredén, C. (1975). Subfossil finds of Arctic whales and seals in Sweden. *Sver. Geol. Unders., Ser. C,* No. 710, Arsb. 69, No. 2, 1–62.

Gray, J. E. (1850). "Catalogue of the Specimens of Mammalia in the Collections of the British Museum," Part 1, Cetacea. British Museum, London.

Gray, J. E. (1864). Notes on the whalebone-whales; with a synopsis of the species. *Ann. Mag. Nat. Hist.* [3] **14,** 345–353.

Gray, J. E. (1865). Notice of a new whalebone whale from the coast of Devonshire, proposed to be called *Eschrichtius robustus. Proc. Zool. Soc. London* pp. 40–43.

Gray, J. E. (1866). "Catalogue of the Seals and Whales in the British Museum," 2 ed. British Museum, London.

Gray, J. E. (1870). Observations on the whales described in the "Ostéographie des Cétacés" of Mm. Van Beneden and Gervais. *Ann. Mag. Nat. Hist.* [4] **6**(33), 193–204.

Gray, J. E. (1871). "Supplement to the Catalogue of Seals and Whales in the British Museum." British Museum, London.

Hershkovitz, P. (1961). On the nomenclature of certain whales. *Fieldiana, Zool.* **39**(49), 547–565.

Hubbs, C. L. (1959). Natural history of the gray whale. *Proc. Int. Cong. Zool. 15th, 1958,* pp. 313–316.

Junge, G. C. A. (1936). Bones of a whale from the Wieringermeer, Zuider Zee. *Nature (London)* **138,** 78.

Larson, L. M. (1917). "The King's Mirror (Speculum Regale—Konungs Skuggsja)," Scand. Monogr., Am. Scand. Found.

Lilljeborg, W. (1861). Hvalben, Funna i jorden paa Gräsön i Roslagen i Sverige. *Forh. Skand. Naturf., 8th Mote, 1860,* pp. 599–616.

Lilljeborg, W. (1866). Synopsis of the cetaceous mammalia of Scandinavia (Sweden and Norway). *In* "Recent Memoirs on the Cetacea by Professors Eschricht, Reinhardt and Lilljeborg" (W. H. Flower, ed.), pp. 219–309. Ray Society, London (translation of the Swedish original: "Öfversigt af de inom Skandinavien (Sverige och Norrige) anträffade Hvalartade Däggdjur (cetacea)." Uppsala Universitets, Årsskrift, 1862).

Lilljeborg, W. (1867). On two subfossil whales discovered in Sweden. *Nova Acta Regiae Soc. Sci. Ups.* [3] **6**(2), 1–48.

Lubbock, B. (1937). "The Arctic Whalers." Brown, Son & Fergusson, Glasgow.

Lydekker, R. (1894). "The Royal Natural History," Vol. 3. Frederick Warne & Co., London and New York.

Macy, O. (1835). "The History of Nantucket." Hilliard, Gray & Co., Boston.

Maher, W. J. (1961). Record of the California grey whale. *Arctic* **13**(4), 257–265.

Rice, D. W., and Wolman, A. A. (1971). The life history and ecology of the gray whale (*Eschrichtius robustus*). *Spec. Publ. Am. Soc. Mammal.* **3,** 1–142.

St. John, J. H. (1782). "Letters from an American Farmer." Thomas Davies, London.
Schevill, W. (1952). On the nomenclature of the Pacific gray whale. *Mus. Comp. Zool. Breviora*, **7,** 1–3.
True, F. W. (1884). Catalogue of the aquatic mammals exhibited by the United States National Museum. *Bull. U.S. Nat. Mus.* **27,** 623–644.
True, F. W. (1904). The whalebone whales of the western north Atlantic. *Smithson. Contrib. Knowl.* **33,** 1–332.
van Beneden, P. J., and Gervais, P. (1880). "Ostéographie des cétacés vivants et fossiles." A. Bertrand, Paris.
van Deinse, A. B., and Junge, G. C. A. (1937). Recent and older finds of the California Gray Whale in the Atlantic. *Temminckia* **2,** 161–188.
van der Donck, A. (1841). "Description of the New Netherlands," Collect., 2nd ser., Vol. 1, pp. 125–242. New York Historical Society, New York (transl. by Jeremiah Johnson from the Dutch original of 1655).
Walton Advertising and Printing Company, Boston (1915). "The Whale Fishery of New England." Printed for the State Street Bank and Trust Company (reprint by State Street Bank and Trust Company and the Old Dartmouth Historical Society of New Bedford, 1968).
Worm, O. (1655). "Museum Wormianum." Johannus Elsevirius, Leyden.

II

Historical Relationships and Exploitation

C.M.SCAMMON.DEL. ZINCOT.X.VAN DE CASTELLE.S.F

WHALEBOAT WITH GREENER'S GUN MOUNTED.

3

History of Gray Whales in Japan

Hideo Omura

Introduction

The occurrence of gray whales in the coastal waters of Japan prior to the turn of the century was described by Omura (1974). He referred to a population occurring in the south of Wakayama and Kochi prefectures and very briefly to the presence of another population occurring off the north coast of Yamaguchi prefecture, in the Sea of Japan, and off the west coast of Kyushu, in the East China Sea. Catches were made from both populations in the days of net whaling (1675–1890).

The Wakayama/Kochi population is thought to have followed a migration route along the east coast of Japan (Nishiwaki and Kasuya, 1970; Omura, 1974). The Yamaguchi/Kyushu population is thought to have been a portion of the Korean stock. This stock migrated down the east coast of Korea in winter. The majority turned west along the south coast of Korea (Andrews, 1914) but a fraction is assumed to have continued southward until they reached the north coast of Yamaguchi prefecture (near Kayoi, a present Nagato city) and then to have turned along the west coast of Kyushu.

Subfossil mandibles excavated some years ago from the city of Ichikawa were recently identified as belonging to a gray whale. This further supports the contention that this species occurred in coastal waters of Japan.

ISBN 0-12-389180-9

Brief History of Old Whaling in Japan

Whaling off Japan appears to have a very long history. A poem describing the catching of whales by a small boat is dated prior to the tenth century. Nothing is known, however, about the method and gear used or about the species of whales taken.

In the years of Genki (1570–1573), whaling in Mikawa and Qwari districts (now Aichi prefecture) facing Ise Bay was conducted by villagers who formed teams of 7–8 vessels and used hand harpoons (Otsuki, 1808; Fukumoto, 1960; Hashiura, 1969; Nagasaki, 1981). It can be assumed this whaling began as bay whaling and its center was Morosaki (see Fig. 7). Nothing is known about the species of whales taken, but there is a possibility that gray whales were the target. In the Edo era (1603–1867), gray whales en route to the Seto Inland Sea were thought to enter Ise Bay, passing sufficiently close to shore to become accessible to fishermen.

In years of the Bunroku (1592–1595), a skilled harpooner from Morosaki named Sukebei Mase came to Misaki (now in Kanagawa prefecture) and began whaling (Hashiura, 1969). The species of whales taken are not known, but a considerable number of whales was taken and local fishermen were stimulated to begin whaling with the same technique. It was reported this period did not last more than 20 years, because of a scarcity of whales.

There are grounds to believe that the harpoon method of whaling was transferred to Katsuyama, the village opposite Misaki across Tokyo Bay in Chiba prefecture. According to Yoshihara (1976a), new whaling started there in the years of the Keicho (1596–1603). This was the origin of the small cetacean whaling now in operation in Chiba prefecture, which takes mainly Baird's beaked whales (*Berardius bairdii*) and some other small toothed whales but no gray whales. Hand harpoons, and after 1907 small harpoon guns, were used for killing whales. Nets were not used at all in this fishery, contrary to the practice in other whaling sites in western Japan. The whaling grounds were later shifted to the east coast of Chiba prefecture.

Hand-harpoon whaling originated in Morosaki and spread to the south along the coast of Ise and Kumano districts (present Mie and Wakayama prefecture). In the eleventh year of Keicho (1606), Yorimoto, head of the then powerful Wada clan, established five whaling groups at Taiji, Kumano. He took command of the operation in the Taiji area, consulting with two fishermen, Denji from Morosaki and Jiemon from Sakai near Osaka (Hashiura, 1969). It is thought Denji, possibly a harpooner, assisted Yorimoto with the technical aspects and that Jiemon handled the commercial aspects of whaling. In 1618 a skillful harpooner named Yoheiji, from Onoura, near Morosaki, was employed at Taiji and given the name of Hazashi (chief harpooner), showing superior status to that of any other whaling crew.

In 1675, Yoriharu Wada, a grandson of Yorimoto later renamed Kakuemon Taiji, invented a new method of whaling using nets. Whales trapped and entangled by many folds of nets were easily harpooned. Straw nets were used at first, but they were too weak and were replaced in following years by hemp.

The use of nets was a revolutionary event in the history of whaling in Japan, allowing takes of humpback whales, other balaenopterid whales, and right whales, none of which could have been taken by hand harpoon previously. This method, however, required more fishing boats, more manpower, and a coordinated cooperative operation. At Taiji, five whaling groups were reorganized into one and Yoriharu Wada took the command.

The basic organization of the net-whaling fleet consisted of the following groups:

1. Seko-bune or beater boats (for driving and killing whales), one group of about 15–20 boats, 13 m long and 2.3 m wide, with eight oars. The crew of each boat consisted of 15 men, namely 1 captain, 13 sailors, and 1 apprentice.
2. Ami-bune or netting boats, one group of 6 boats, 13 m long and 3.5 m wide, with eight oars and a crew of 10 men.
3. Mosso-bune or tug boats, a group of 4 boats, nearly the same as beater boats but with a broader beam, and a crew of 1 captain and 12 sailors.

From a hillside hut commanding a wide view, watchmen scanned the sea surface for whale blows. When a whale was sighted within range, the watchmen sent signals with flags or rockets informing the boat crews of the species, position, and swimming direction of the whale. The boats described above waited in scattered positions and moved into action in an orderly fashion when the commander gave the order.

The beater boats surrounded the whale from a distance and drove it toward the netting boats which moved into proper position to set their nets. Nets were sometimes set two or three deep to prevent the whale's escape and to reduce its struggle. The netted whale was harpooned from the beater boats. When the whale was sufficiently weakened by a number of harpoons, a sailor jumped into the water and climbed onto the head of the whale with his knife to make a hole on the septum of the blowholes through which a rope was passed. Another hole was made in a similar way near the dorsal hump. In both cases the sailor stayed on the slippery surface of the whale body by grasping the handle of a struck harpoon. These harpoons also prevented the nets from slipping off.

The invention of net whaling was a major event in the economy of Japanese fishing villages. Whaling supported many hundreds of people, including workers processing the whale carcass, making nets, or engaged in other related activities, as well as sailors at sea.

Net whaling was soon introduced at Koza, a town close to Taiji, and at other places in Kumano district (part of the present Mie and Wakayama prefectures). In 1683 this method was transferred to Tosa (in the present Kochi prefecture), where the two whaling groups of Ukitsu and Tsuro were already in operation. In the following year a whaling boss named Gidayu Fukazawa from Omura, west Kyushu, came to Taiji and learned this method from Yoriharu (Hashiura, 1969). When Fukazawa returned to Kyushu, net whaling soon spread over the west coast of Kyushu and the north coast of Yamaguchi prefecture.

Descriptions of Gray Whales in Old Literature

Old Japanese whaling flourished in the Edo or Tokugawa era (1603–1867). Several books and picture scrolls of whales and whaling published during this period are useful because they contain such information on the whales migrating to Japanese waters as their species composition, migratory pattern, and catch statistics.

"Geishi" (the Treatise of the Whale), written in 1758 and printed in 1760 by Jiemon Kandoriya, the castellany of Wakayama, is the earliest printed monograph of the cetacea of Japan. It contains a drawing and brief description of the gray whale (Fig. 1), named Kokujira from *ko* meaning small and *kujira* meaning whale. The gray whale was so called because it was thought to be the smallest species among whales. There is evidence to suggest that at that time the minke whale was not identified as an independent species of baleen whale.

Kokujira had no dorsal fin but was illustrated as having several knobs in the posterior portion of the back. On both upper and lower jaws there were many hairs that spread over the entire head. This agrees with the description by Andrews (1914) who states "they (hairs) are more widely and more uniformly spread over the entire head, than in any other baleen whales."

The body color of Kokujira is described as pale blue, and there are many circular or

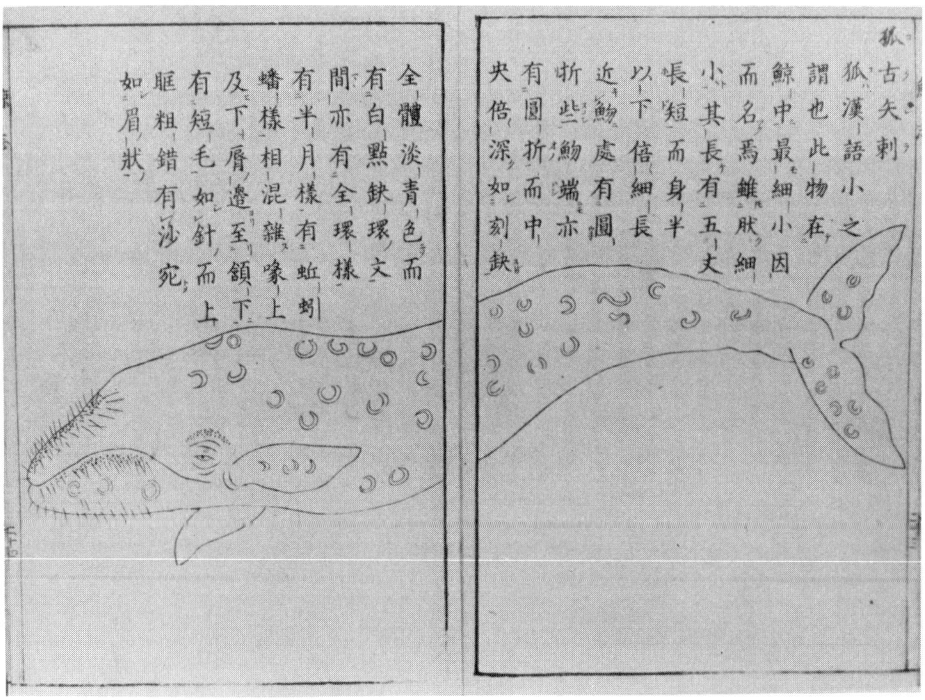

Fig. 1. *Kokujira in Geishi, 1760 (from Hawley, 1958).*

various stages of semicircular markings on the body surface. These are thought to be scars left by detached barnacles (*Cryptolepas* sp.). These barnacles embed themselves deeply in all parts of the body as well as the flukes and pectoral fins. When a barnacle detaches, it creates a circular grayish wound which in time becomes white as it heals (Andrews, 1914).

Another drawing of Kokujira (Fig. 2) appears in a scroll appended to "A History of Whaling at Taijiura, Kumano," by the Committee for the Compilation of the History of Whaling at Taijiura (Hashiura, 1969). It is assumed this scroll was made during the early years of Kambun (1661–1673), 100 years before "Geishi" was printed. These two drawings are quite similar, except for the shape of markings on the surface which is shown as pale blue in color. In both there are needle-like white hairs approximately 6 cm long. In the second (Fig. 2), barnacles and lice are shown attached to the body. The positions of blowholes, eye, ear hole, mouth, flippers, navel, genital aperture, anus, and tail flukes are all correctly indicated by arrows.

Kiyonori Otsuki (1773–1850), a famous scholar in the Tokugawa era, left an undated manuscript on whales and whaling entitled "Geishiko (A Draft of a History of the Whale)." It was written perhaps in 1808, but remained unpublished until 1925 when it was printed in a series of various collections in Sendai. In March, 1951 this book was reprinted by the Japan Whaling Association. It is by far the most systematic survey of whales and whaling written in the Tokugawa period (Hawley, 1958).

The author of the "Geishiko" refers to various books on whales and whaling which

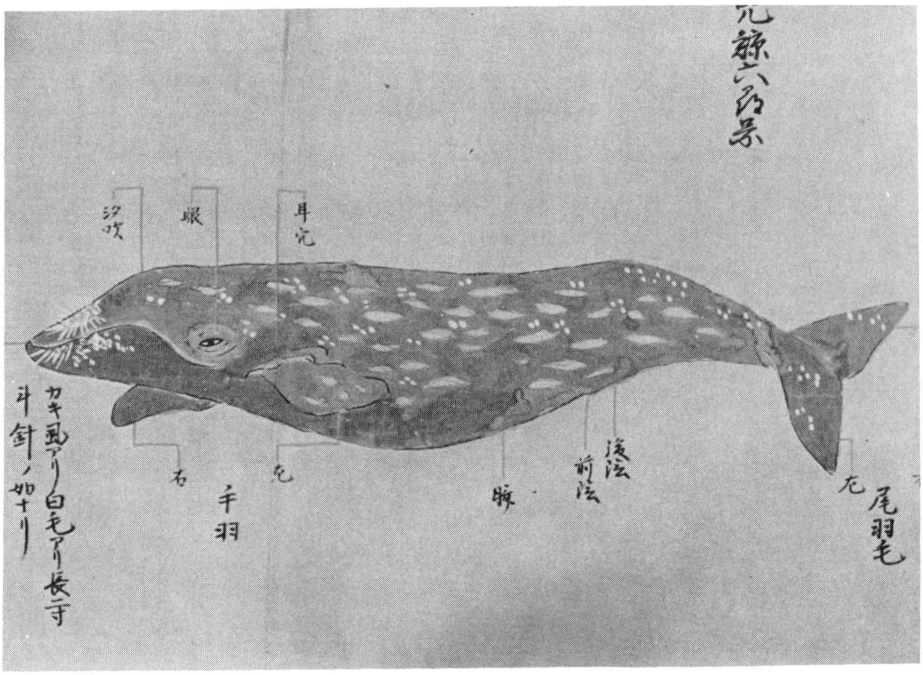

Fig. 2. *Kokujira drawn in 1661–73 (from Hashiura, 1969).*

were then available to him. Kokujira was also called Chikokujira, *chiko* also meaning small. The baleen plates of Chikokujira were described as being white. Two kinds of Kokujira were reported, one called *Aosagi* and the other *Share*. Aosagi was preferred to Share because its oil was white. However, interpretation of this old description became somewhat confused in later years. Andrews (1914) states

> The blubber is thick and fat, and varies in color from red to flesh pink. Because of this the Japanese recognized two kinds of gray whale, the 'Aosaki' (red blubber) and the 'Shirasaki' (white blubber). Although specimens with blubber strongly red, almost white, and of every intermediate shade, were taken during my stay at Ulsan, I could detect no difference, other than those purely individual, between them.

In Andrews' description, *Aosaki* is the same as Aosagi. There is only a slight difference in pronounciation of the two words, and Aosagi means blue heron.

I have been unable to find the names of *Shirasaki* or *Shirasagi* in any of the old books on whales and whaling. The only two kinds of Kokujira named are "Aosagi" and "Share." And, contrary to the above description, the color of blubber in Aosagi is said to be white. Hattori (1887–1888) thought that Share were young animals and Aosagi full-grown adults. This may be the correct interpretation. The individual differences in color of the blubber may be due to the different foods they took, as suggested by Andrews (1914). There is a description (Anonymous, 1890) in which the bigger whales which come from the west are called Aosagi (at Kawajiri). If this statement is correct, Aosagi referred to whales that were migrating from the calving ground to feeding grounds.

Biological Evidence of Occurrences

SUBFOSSIL EVIDENCE

In 1966, both mandibles, some vertebrae, and ribs of a whale were excavated from a depth of 4 m in a sand stratum in Ichikawa city, a neighboring city of Tokyo, during construction of the Ichikawa Telegram and Telephone Office. This stratum contained fossil shells, which were radiocarbon dated to 6000 years BP, or the time of the Jomon Sea Regression. The whale was erroneously identified as the sei whale, *Balaenoptera borealis*. The skeleton was then mounted with an artificial skull, vertebrae, and ribs, and exhibited in the main hall of the Ichikawa Municipal Museum.

In March, 1982 I visited the museum, with Dr. T. Kasuya of the Ocean Research Institute, University of Tokyo, and Mr. H. Kato of the Whales Research Institute and positively identified the specimen to be a gray whale. In gray whales the "lower jaws are massive, wide, and without coronoid processes: their inner surface somewhat concave at the proximal part" (Tomilin, 1967). There was no difficulty with the identification, because these features are very prominent even at a glance (Fig. 3).

Both mandibles of the Ichikawa whale have been broken anteriorly. The straight-line length of the right bone from broken tip to posterior end is 2.22 m. The height is 35

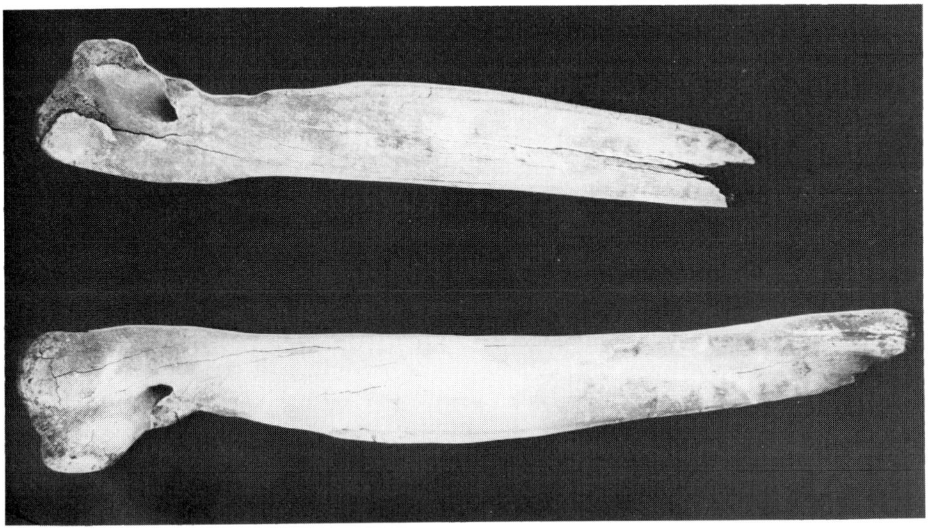

Fig. 3. *Mandibles of the Ichikawa whale (courtesy of the Ichikawa Municipal Museum).*

cm at the articulation, 21.5 cm at the lowest part just in front of articulation, and 29.7 cm at the top of the remaining part. Total length of this mandible is estimated to have been about 2.5 m and the body length of the whale approximately 12 m.

Ichikawa city is situated at the innermost part of the Tokyo Bay. The positive identification of this specimen is direct evidence that gray whales once occurred in the eastern coastal waters of Japan.

RECENT HISTORICAL EVIDENCE

In 1864 two gray whales were taken in the Seto Inland Sea, at Kawanoe, Ehime prefecture, the first on February 20 and the second on February 26. These whales were reported first by Shindo (1968) and then by Omura (1974). It is not necessary to describe them here in detail, but I wish to mention briefly their relationships to Aosagi and Share. Drawings of these whales are kept at the Kawanoe City Library (Figs. 4 and 5).

The first whale is smaller than the second and possibly would have been referred to as Share, although nothing was noted on the drawing (Fig. 4); the second whale (Fig. 5) was labeled as Aosagi. The body surface of the first whale was slightly infested with barnacles. The second whale bears heavy white markings of barnacle infestation, suggesting it was older than the first. A left scapula, possibly from the second whale, has been preserved at Hachiman shrine in the city, presented as Ema in memory of the whaling event (Fig. 6). The scapula of *Rhachianectes* (*Eschrichtius*) is distinctive in being intermediate between the wide, low scapula of *Balaenoptera* and the high, narrower, and more symmetrically fan-shaped scapula of *Eubalaena* (Andrews, 1914). Measurements of

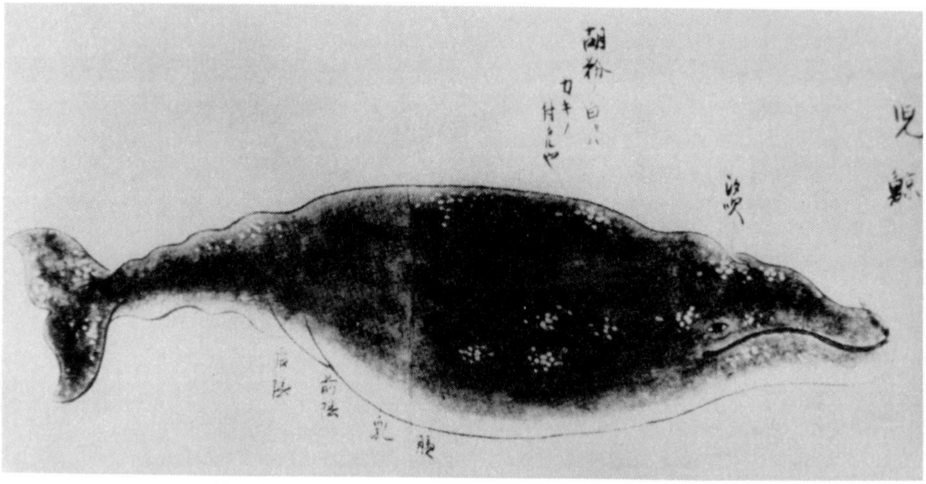

Fig. 4. *Kawanoe whale—first whale (from Omura, 1974).*

the at Hachiman shrine scapula were given by Omura (1974). Greatest breadth and greatest height are 99 and 74 cm, respectively, and the ratio of breadth to height is 1.34.

Omura (1974) thought the nearby waters of Iwaijima in the Seto Inland Sea was a calving ground of the gray whales, which occurred on the south coast of Japan. According to Saito (1977), Engelbert Kaempher, a German traveler and physician, sighted small whales in the waters near Mitajiri on April 30, 1691, while he was returning from a visit to the Shogun's capital of Edo (Tokyo). Mitajiri is close to Iwaijima and not far from Shimonoseki, the western entrance to the Inland Sea. Kaempher referred to these

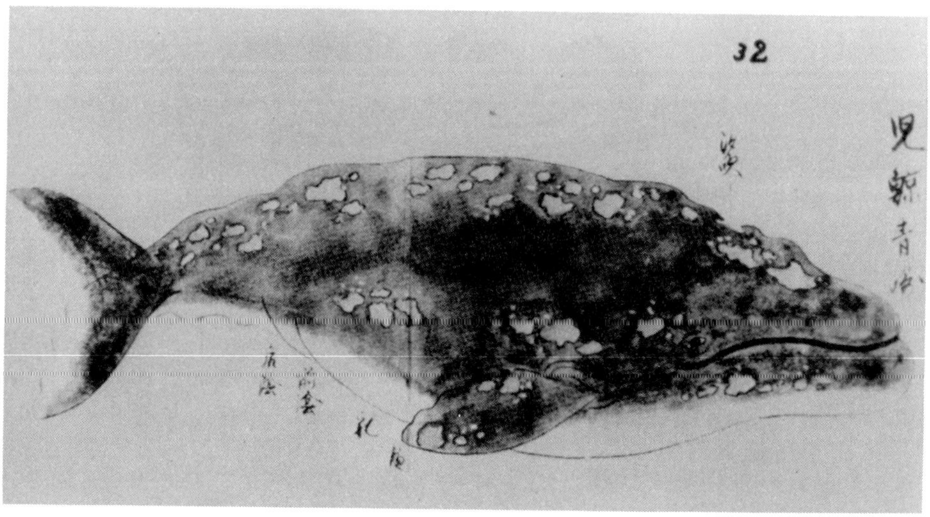

Fig. 5. *Kawanoe whale—second whale (from Omura, 1974).*

Fig. 6. *Left scapula of the Kawanoe whale (from Omura, 1974).*

whales as "Nordcaper," which means right whale, a species not reported from the Inland Sea; he further describes them as small whales. It is possible the whales he sighted were in fact gray whales. Gray whales and right whales have often been confused, even by gunners of modern whale catcher boats.

Nishiwaki and Kasuya (1970) report the accidental catch of a young female gray whale in 1968 at Shingu, Wakayama prefecture (the skeleton of this whale is maintained at Taiji Whale Museum). They also reported the sighting, by a skillful gunner of long experience, of a gray whale in nearby waters around 1959.

Catches of Gray Whales

Net whaling flourished nearly 200 years in the Edo era (1603–1867), during which time gray, right, humpback, and other species of whales were taken. The gray whales were usually taken without nets, because of their relatively smaller size (Taiji, 1937).

Practically no catch statistics exist for Wakayama and Mie prefectures. At Taiji, whaling continued after the Meiji Revolution (1868) until a tragedy occurred in 1878. In that year more than 100 people were killed by a heavy storm while they pursued a right whale accompanied by a calf. This incident was practically the end of old whaling at Taiji.

There remain at Taiji some fragmentary records of the catch. For example, from December 24, 1799 to January 15, 1800, a total of 9 whales was taken, including 1 right, 6 humpback, and 2 gray whales. However, because most of the records were lost by fire or during several floods, it is not clear whether this was an exceptionally good season. This seems to have been a good catch, but further comparison of catch records is not possible.

Whaling on the coast of the Sea of Japan, at Ine, Kyoto prefecture, had been conducted since very ancient times, probably as early as Tenmon (1532–1554). The methods were somewhat different from those used in other places. The village of Ine is located on a small inlet, called Inewan, into which whales occasionally swam. When this happened the fishermen of Ine blocked the entrance of the inlet with nets and then caught the whale using hand harpoons and nets. Catch records of the Ine whaling operation during a period from 1656 to 1913 are available by species (Yoshihara, 1976b). During this time a total of 357 whales, including 167 humpback, 149 "fin" (including minke), and 41 right whales, was taken. No gray whales were taken, evidence that gray whales did not occur on the east side of the Sea of Japan at that time.

Whaling was conducted, however, in several villages on the north coast of Yamaguchi prefecture, including Kayoi in the east and Kawajiri in the west (Tokumi, 1957; Tada, 1978). At Kayoi, at present Nagato city, there is a temple named Koganji where notes are preserved which contain the Buddhist names of each whale taken and from which catches of species have been summarized (Kimura, 1956). Japanese people were given special names from a Buddhist priest when they died, and in this respect the whales were treated somewhat like human beings. The first volume of the Koganji notes is missing, but from remaining volumes catch figures are available for about 50 years, from 1802 to 1850. During this period a total of 308 whales, including 116 fin, 105 humpback, 59 right, and 28 gray whales, was taken. The average catch per year was 6.3 animals; gray whales comprised 9% of the catch.

For Kawajiri, catch statistics arranged by species and by 10-year increments are available from 1699, when whaling was started, until 1888 (Anonymous, 1890). Yearly figures for the 8 years from 1894 to 1901 were given by Tada (1978) (Table I). For a period of 50 years, from 1769 to 1818, only the total number of catches was reported, but catches by species can be compared for the periods 1699 to 1768 and 1819 to 1888. Catches of gray whales increased from 12% in the former period to 16% in the latter. Both figures are greater than proportions of gray whales in catches at Kayoi. Catches of right and humpback whales decreased as those of "fin" whales (again possibly including minke whales) increased, possibly due to decrease of the former two species. Tada (1978) states that no right whales were taken after 1884. Catch statistics at Kawajiri in the later years are given by Tada (1978) for a period from 1894 to 1901. The decline of gray whale catches and the increase of "fin" whale catches during this period are remarkable. Whaling at Kawajiri was virtually terminated in 1902.

In addition to Kayoi and Kawajiri, whaling was conducted at several villages including Mishima (Tada, 1968), a small island about 45 km northwest of Hagi city. Whaling in these villages began around 1680 and lasted until near the end of the nineteenth century. Unfortunately, details of the catches are not available.

On the west coast of Kyushu, whaling was conducted at various places. Otsuki (1808), for example, listed 68 localities. This does not mean, however, that 68 whaling groups existed in Kyushu; rather, it represents the total number of places where whales,

Table IA
Catches of Whales at Kawajiri, Yamaguchi Prefecture: 1699–1888[a]

Years	Gray	Right	Humpback	Fin	Others[b]	Year total
1699–1708	11	29	94	2	7	143
1709–1718	17	17	103	14	9	160
1719–1728	21	29	103	3	9	165
1729–1738	12	31	92	3	2	140
1739–1748	21	26	93	—	—	140
1749–1758	13	20	48	—	—	81
1759–1768	15	14	58	—	1	88
Species totals	110	166	591	22	28	917
Average per year	1.6	2.4	8.4	0.3	0.4	13.1
Percentage	12.0	18.1	64.4	2.4	3.1	100
1769–1778						140
1779–1788						136
1789–1798						117
1799–1808						84
1809–1818						105
Species totals						582
Average per year						11.6
1819–1828	8	19	55	—	37	119
1829–1838	10	29	27	—	72	138
1839–1848	18	48	59	4	—	129
1849–1858	37	23	88	24	—	172
1859–1868	34	2	55	32	7	130
1869–1878	44	4	34	65	—	147
1879–1888	19	3	49	130	—	201
Species total	170	128	367	255	116	1036
Average per year	2.4	1.8	5.2	3.6	1.7	14.8
Percentage	16.4	12.4	35.4	24.6	11.2	100

[a]Anonymous (1890).
[b]Species were not noted.

Table IB
Catches of Whales at Kawajiri, Yamaguchi Prefecture: 1894–1901[a]

Year	Gray	Humpback	Fin	Others	Total
1894	0	3	10	1	14
1895	0	0	4	1	5
1896	4	1	7	1	13
1897	0	4	9	2	15
1898	1	4	10	1	16
1899	0	2	8	0	10
1900	2	1	4	1	8
1901	0	13	3	2	18
Species totals	7	28	55	9	99
Average per year	0.9	3.5	6.9	1.1	12.4
Percentage	7.0	28.3	55.6	9.1	100

[a]Tada (1978).

especially right whales, were taken. These numbers were reduced to only 3 a few years before the turn of the century, due to heavy reduction in numbers of right whales migrating there (Takahashi, 1899).

Whaling in Kyushu also started as a small enterprise using hand harpoons. After the invention of net whaling, this method spread over the west coast of Kyushu, and whaling flourished there more than in any other place (Anonymous, 1980). Whaling in Kyushu was operated by several groups, such as the Nakao group of Yobuko and Ogawajima (Saga prefecture), the Toi group of Iki Island (Nagasaki prefecture), and the Masutomi group of Hirado and Ikitsuki (Nagasaki prefecture). Among them the Masutomi group was the largest. From 1725, the year when the group was formed by Matazaemon Masutomi, to 1874, a total of 21,790 whales was taken (Yoshihara, 1977), a yearly average of 150 whales for all locations combined. The group operated at several locations in Kyushu, and sometimes at Mishima and Kayoi (both in Yamaguchi prefecture) as well. It was said this whaling group employed 3,000 people and about 200 fishing vessels when in operation.

Whaling at Ogawajima survived long after the introduction of modern whaling (Anonymous, 1980). A hut called Yamami, or lookout, was built on the top of a hill. When the watchmen sighted a whale he telephoned to a modern catcher waiting in the port of Yobuko. The catcher then pursued and killed the whale and returned it to the village, where it was processed by the villagers using traditional methods. This type of whaling lasted until 1948.

It is possible that gray whales were also taken in the waters west of Kyushu, but there are no catch statistics by species. Catch figures of whales at Iki Island in the 16 years from 1845 to 1860 are shown in Table II. Whaling from Iki Island was conducted by two whaling groups, Masutomi and Kuramitsu, operating at Katsumoto one season and

Table II
Catches of Whales at Iki, 1845–1860[a]

Year	Katsumoto	Maeme	Year totals
1845	60	78	138
1846	40	45	85
1847	42	32	74
1848	34	40	74
1849	14	11	25
1850	19	19	38
1851	22	19	41
1852	7	14	21
1853	12	4	16
1854	9	20	29
1855	10	14	24
1856	7	7	14
1857	19	Unknown	19
1858	Unknown	7	7
1859	Unknown	7	7
1860	4	7	11
Species totals	299+	324+	623+
Average per year	21.4	21.6	38.9+

[a]Anonymous (1890).

Maeme the next. The catches of whales at Iki decreased after 1849, probably because of the decrease of right whales.

In Tosa (present Kochi prefecture) whaling with hand harpoons dates from the years of Kanei (1624–1643) (Izukawa, 1943), and net whaling from 1683 when it was introduced from Taiji. Two groups of whalers, Ukitsu and Tsuro, operated on the coast of Tosa, splitting their activities between an east and a west whaling ground. There are two peninsulas in Kochi, the Muroto Peninsula in the east and the Ashizuri Peninsula in the west. The east whaling grounds were on the east side (in winter), and west side (in spring) of the Muroto Peninsula. The west ground was on the east side of the Ashizuri Peninsula in both winter and spring.

Each year the Ukitsu whaling group operated in one ground and the Tsuro whaling group on the other, and the two alternated grounds each year. Both groups left good catch records by year, from which general trends of the whaling in Kochi can be detected (Table III).

Table IIIA shows the total catches of whales by the Tsuro group in the years 1693–1712, inclusive. The average catch per year was 20.6 whales, but no species breakdown was available. For the 35 years from 1800 to 1835 (Table IIIB), the Ukitsu group took 959 whales, or 27.4 whales per year. Humpback whales comprise more than one-half of the total catch. In the years 1849 and thereafter (Tables IIIC, D, and E), right whales decreased considerably, while catches of Bryde's, blue, and fin or balaenopterid whales increased. Catches of gray whales appear to have been rather stable.

Table IIIA
Catches of Whales in Kochi Prefecture: Tsuro Group, 1693–1712[a]

Year	Catch	Year	Catch
1693	26	1704	11
1694	14	1705	32
1695	30	1706	16
1696	5	1707	26
1697	23	1708	26
1698	11	1709	26
1699	25	1710	21
1700	21	1711	22
1701	22	1712	22
1702	12		
1703	21		
		Total	412
		Average per year	20.6

[a]Anonymous (1931).

Table IIIB
Catches of Whales in Kochi Prefecture: Ukitsu Group, 1800–1835

Year	Gray	Right	Humpback	Bryde's	Others	Year total
1800	5	2	4	—	1	12
1801	0	1	5	—	2	8
1802	5	2	6	1	3	17
1803	1	4	17	—	—	22
1804	2	4	29	—	1	36
1805	0	11	17	—	—	28
1806	3	3	23	—	1	30
1807	2	5	15	—	—	22
1808	5	5	29	—	1	40
1809	4	11	11	—	—	26
1810	3	2	11	—	—	16
1811	1	7	20	—	—	28
1812	4	7	14	—	—	25
1813	6	5	10	—	—	21
1814	5	8	5	2	2	22
1815	6	3	41	—	—	50
1816	2	4	8	—	—	14
1817	4	5	13	—	—	22
1818	4	8	12	—	1	25
1819	4	3	10	—	6	23
1820	6	7	6	—	1	20
1821	4	5	9	—	5	23
1822	5	10	3	1	1	20

(continued)

Table IIIB (*Continued*)

Year	Gray	Right	Humpback	Bryde's	Others	Year total
1823	5	7	18	—	1	31
1824	3	8	20	—	—	31
1825	3	11	27	—	—	41
1826	1	16	8	—	—	25
1827	4	10	17	—	—	31
1828	2	5	12	—	—	19
1829	4	9	15	—	—	28
1830	7	9	25	—	2	43
1831	6	10	21	—	—	37
1832	7	12	16	1	—	36
1833	6	11	10	—	—	27
1834	7	15	4	—	2	28
1835	8	14	10	—	—	32
Species total	144	259	521	5	30	959
Average per year	4.0	7.2	14.5	0.1	0.8	26.6
Percentage	15.0	27.0	54.3	0.5	3.1	100

[b]Prepared from Yoshihara (1974).

Table IIIC
Catches of Whales in Kochi Prefecture: Tsuro Group, 1849–1865[c]

Year	Gray	Right	Humpback	Bryde's	Others	Year total
1849	4	—	10	3	—	17
1850	8	4	14	4	—	30
1851	6	1	25	—	—	32
1852	5	—	11	1	—	17
1853	5	1	22	—	—	28
1854	5	1	9	5	—	20
1855	8	2	10	2	—	22
1856	5	—	7	3	—	15
1857	2	—	14	1	—	17
1858	9	—	9	3	2	23
1859	6	1	8	1	1	17
1860	5	1	21	—	2	29
1861	9	1	3	1	1	15
1862	10	—	14	2	—	26
1863	6	3	7	5	—	21
1864	7	4	18	1	—	30
1865	1	—	7	3	—	11
Species total	101	19	209	35	6	370
Average per year	5.9	1.1	12.3	2.1	0.4	21.8
Percentage	27.3	5.1	56.5	9.5	1.6	100

[c]Prepared from Anonymous (1937).

Table IIID
Catches of Whales in Kochi Prefecture: Tsuro Group, 1874–1896[d]

Year	Gray	Right	Humpback	Bryde's	Others	Year total
1874	9	2	4	4	2	21
1875	5	—	5	2	—	12
1876	4	1	6	2	4	17
1877	5	3	6	—	2	16
1878	5	—	14	1	1	21
1879	7	—	5	2	—	14
1880	9	2	13	3	1	28
1881	4	2	—	1	3	10
1882	8	7	9	4	—	28
1883	3	—	1	3	3	10
1884	9	—	10	3	2	24
1885	2	1	2	1	1	7
1886	2	1	11	7	1	22
1887	3	1	5	2	5	16
1888	5	1	8	4	3	21
1889	1	—	2	—	3	6
1890	1	—	7	2	2	12
1891	2	2	3	4	3	14
1892	3	—	3	8	2	16
1893	2	—	—	1	4	7
1894	4	—	7	2	8	21
1895	4	—	4	3	4	15
1896	2	—	9	13	2	26
Species total	99	23	134	72	56	384
Average per year	4.3	1.0	5.8	3.1	2.4	16.7
Percentage	25.8	6.0	34.9	18.7	14.6	100

[d]Anonymous (1937).

Table IIIE
Catches of Whales in Koch: Prefecture: Ukitsu Group, 1875–1896[e]

Year	Gray	Right	Humpback	Bryde's	Others	Year total
1875	1	1	2	4	3	11
1876	2	—	6	5	4	17
1877	11	—	19	3	—	33
1878	1	6	4	4	3	18
1879	7	3	9	2	1	22
1880	6	1	2	8	6	23
1881	5	2	10	5	1	23
1882	3	—	4	2	3	12
1883	3	1	7	3	—	14
1884	1	2	1	2	6	12
1885	3	2	5	4	5	19

Table IIIE (*Continued*)

Year	Gray	Right	Humpback	Bryde's	Others	Year total
1886	1	—	6	3	10	20
1887	4	—	13	4	5	26
1888	1	—	6	4	4	15
1889	2	—	4	7	2	15
1890	1	—	—	4	3	8
1891	1	—	9	—	5	15
1892	1	1	1	3	2	8
1893	4	1	3	1	4	13
1894	1	1	3	4	1	10
1895	3	—	6	5	3	17
1896	2	—	6	4	7	19
Species total	64	21	126	81	78	370
Average per year	2.9	1.0	5.7	3.7	3.5	16.8
Percentage	17.3	5.7	34.0	21.9	21.1	100

[e]Shibusawa (1939).

Discussion and Conclusions

Except for a few stragglers, gray whales do not occur in the coastal waters of Japan at present. Prior to the turn of the century, however, they were found in two regions, one on the Pacific side of Japan, from Tokyo Bay to Kochi prefecture, and the other on the west side of Japan off Kyushu, including the north coast of Yamaguchi prefecture. This conclusion is supported by subfossil mandibles excavated from Ichikawa city, old books on whales and whaling, and catch statistics of old whaling operations in the Edo era (1603–1867). The whales in these two regions may represent different populations of gray whales. The latter (the Kyushu group) is clearly a portion of the Korean stock. Distribution of the two populations in the waters off Japan and Korea is shown in Fig. 7. The two populations may have intermingled in the Seto Inland Sea. If so, then both populations belong to the Korean stock.

In summer, the Korean stock of gray whales occupies, or at least formerly occupied, the northern Okhotsk Sea (Rice and Wolman, 1971). It is not clear whether or not the gray whale population occurring off the south coast of Japan also spends the summer in Okhotsk Sea mingling with the Korean stock. Bowen (1974) states that the young female gray whale accidentally taken in 1968 at Shingu (Nishiwaki and Kasuya, 1970) resembled members of the California stock in morphological features and should be regarded as a stray from the Bering Sea. There still exists the possibility, however, that the two populations occurring in the coastal waters of Japan intermingle with each other in the Seto Inland Sea.

The annual catch of gray whales in Kochi prefecture in the days of net whaling was rather small (Table III). The largest catch was 11 in 1877 by the Ukitsu group; the yearly

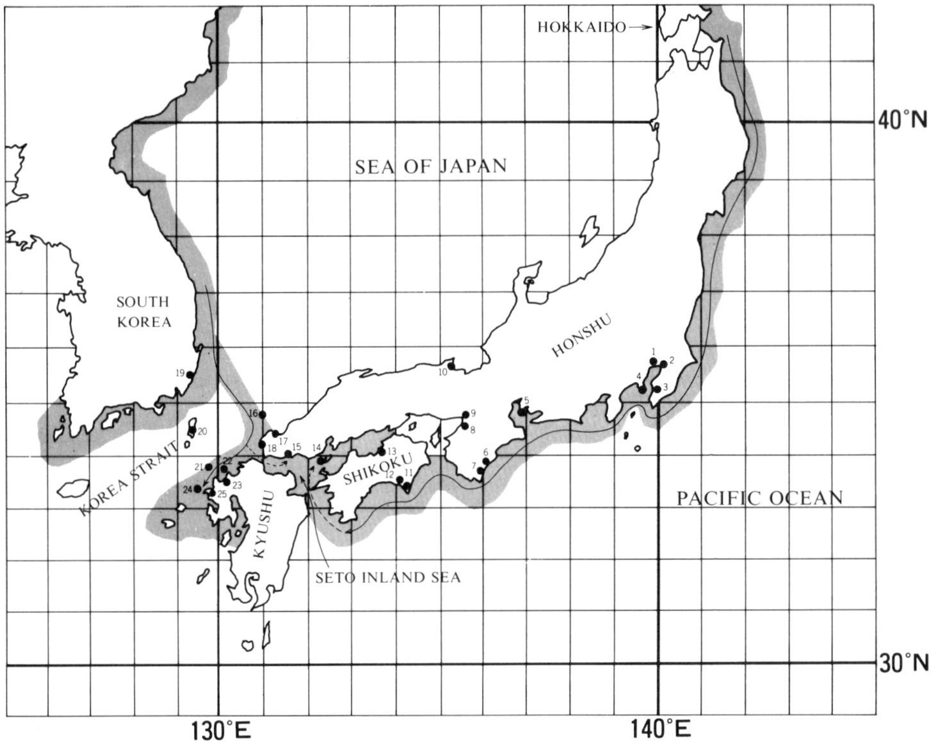

Fig. 7. *Map showing the distribution of gray whales in the coastal waters of Japan and Korea. Place names appearing in the text are indicated by numerals as follows: (1) Tokyo, (2) Ichikawa, (3) Katsuyama, (4) Misaki, (5) Morosaki, (6) Taiji, (7) Koza, (8) Sakai, (9) Osaka, (10) Ine, (11) Tsuro, (12) Ukitsu, (13) Kawanoe, (14) Iwaijima, (15) Mitajiri, (16) Mishima, (17) Kayoi, (18) Kawajiri, (19) Ulsan, (20) Tsushima, (21) Iki, (22) Ogawajima, (23) Yobuko, (24) Ikitsuki, (25) Hirado. Solid arrow, migration routes and area of distribution of Gray whales; broken arrow, movements uncertain.*

average was 2.9–5.9, and there were no remarkable differences between the two whaling groups or among different years. The total yearly catch of gray whales by the two groups in Kochi prefecture is estimated to have been around 10 whales. This suggests gray whales occupied a less important position than right and humpback whales.

Catch figures of gray whales in Wakayama and Mie prefectures are not known. Presumably gray whales were taken in numbers such as in the Kochi prefecture. If one assumes that equal numbers were taken in each of these prefectures, then the total catch of gray whales from the population would have been 30 per year. The catch of gray whales from this population appears to have been rather stable (Table III), suggesting that the catch rate from this population was not high.

The catch of gray whales at Kawajiri, Yamaguchi prefecture during the period of 1699–1768 totaled 110 whales, or an average of 1.6 whales per year (Table I). In the third period, 1819–1888, the corresponding figures are 170 total and 2.4 whales per year, indicating an average increase in take of 0.8 whales per year. These catches are very low compared with the corresponding figures in Kochi prefecture. For the period 1894–1901, the average catch per year was 0.9, less than in previous years. Kimura (1956) also shows a reduction in gray whale catch at Kayoi after 1831.

It is not known how many gray whales were caught on the west coast of Kyushu. It is assumed, however, these catches were far more dependent on availability of right, humpback, and fin whales, as in Kawajiri and Kayoi, and that gray whales were caught only secondarily to the whaling there. It is assumed as many as 20–30 gray whales were taken per year but the catch of gray whales decreased towards the end of net whaling around 1900 and these stocks were finally extirpated.

This reduction does not necessarily mean a reduction of the Korean stock. Mizue (1951) reports heavy catches of gray whales at Ulsan, Korea after 1910. Therefore, the decrease of gray whales migrating to the coast of Yamaguchi and to the west coast of Kyushu may have resulted in a change in migration route.

Modern-type whaling in Japan commenced in 1898, when the first catcher Hokamaru, built of wood, caught three whales (Akashi, 1910). After that year, whaling operated first on the east coast of Korea, the west coast of Kyushu, and the south coast of Japan, and then shifted to the northeast coasts of Japan and Hokkaido. As pointed out by Mizue (1951), only a few gray whales were taken during modern whaling, except in Korea and Tsushima. Omura (1974) speculated the gray whale population on the Pacific side of Japan was driven from its calving ground in the Seto Inland Sea sometime prior to this century by the increase in boat traffic and industrial development on the coast. It now appears this matter is more complicated and that additional material is needed to support firm conclusions.

Summary

Prior to the turn of the century, two populations of gray whales migrated to the coastal waters of Japan, one to the southern coast of Hokkaido and Honshu and the other to the west coast of Kyushu via the north coast of Yamaguchi prefecture. The former migrated along the east coast of Japan from the north and entered the Seto Inland Sea, the supposed calving ground for this population. The latter, considered to be a portion of the Korean stock, migrated along the east coast of Korea and then to southwest Honshu and northwest Kyushu. Intermingling of the two populations in the Seto Inland Sea cannot be denied.

The annual catch of gray whales in the days of net whaling (1675–1890) was approximately 30 whales from the former population and 20–30 from the latter. Virtually no gray whales, however, have been observed in the waters around Japan since about 1898 when modern-type whaling was introduced.

The southern coast population was probably driven from the Inland Sea sometime prior to this century due to increased industrial development and boat traffic on the coast. The matter is complicated, however, and additional material is needed before final conclusions can be drawn.

Acknowledgments

I wish to acknowledge the help of Dr. Fukuzo Nagasaki of the Japan Whaling Association and Dr. Toshio Kasuya of the Ocean Research Institute, University of Tokyo and thank them for their constructive criticisms on the draft of this manuscript.

References

Akashi, K. (1910). "Hompo no Norway-shiki Hogeishi" (History of Norwegian-type whaling of Japan)." Toyo Hogei Kabushikigaisha, Osaka (in Japanese).

Andrews, R. C. (1914). Monographs of the Pacific cetacea. I. The California gray whale (*Rhachianectes glaucus* Cope). *Mem. Am. Mus. Nat. Hist.* [N.S.] **1,** 227–287.

Anonymous (1890). Hogei Tokei. (Whale catch statistics.) *Dainihon Suisankai Hokoku* **104,** 142–145 (in Japanese).

Anonymous (1937). "Tsurohogeishi (Whaling History of Tsuro)." Tosataro Yamaji, Tokyo (in Japanese).

Anonymous (1980). "Genkai no Kujiratori (Whaling in the Western Sea)." Sagakenritsu Hakubutsukan, Saga (in Japanese).

Bowen, S. L. (1974). Possible extinction of the Korean stock of the gray whale (*Eschrichtius robustus*). *J. Mammal.* **55,** 208–209.

Fukomoto, K. (1960). "Nihon Hogeishiwa (Whaling History of Japan)." Hoseidaigaku Shippankyoku, Tokyo (in Japanese).

Hashiura, Y. (1969). "Kumano Taijiura Hogeishi (A History of Whaling at Taijiura, Kumano)." Heibonsha, Tokyo (in Japanese).

Hattori, T. (1887–1888). "Nihon Hogei Iko (Studies of Whaling of Japan)," Vols. I and II. Dainihon Suisankai, Tokyo (in Japanese).

Hawley, F. (1958). "Whales and Whaling in Japan," Misc. Jpn. II. Vol. 1, Part 1. MCMLVIII–MCMLX. Kyoto. 354pp.

Izukawa, A. (1943). "Tosa Hogeishi (Whaling History of Tosa)." Nihon Jomin Bunka Kenkyujo, Tokyo (in Japanese).

Kandoriya, J. (1760). "Geishi (The Treatise of the Whale)." Kobunkan, Kyoto (in Japanese).

Kimura, S. (1956). Kujira no Kakocho. (Buddhistic records of dead whale.) *Whales Res. Inst. Geiken Tsushin, Tokyo* **63,** 144–150. (in Japanese).

Mizue, K. (1951). Gray whales in the east sea of Korea. *Whales Res. Inst. Sci. Rep. Tokyo* **3,** 71–79.

Nagasaki, F. (1981). "Gyoshoku no Tami (Japanese Nation and Fish)." Hokuto Shobo, Tokyo (in Japanese).

Nishiwaki, M., and Kasuya, T. (1970). Recent records of gray whale in the adjacent waters of Japan and a consideration on its migration. *Sci. Rep. Whales Res. Inst.* **22,** 29–37.

Omura, H. (1974). Possible migration route of the gray whale on the coast of Japan. *Sci. Rep. Whales Res. Inst.* **26,** 1–14.

Otsuki, K. (1808). "Geishiko (A Draft of a History of the Whale)." 1951 ed.: Japan Whaling Association, Tokyo. (in Japanese).

Rice, D. W., and Wolman, A. A., (1971). The life history and ecology of the gray whale (*Eschrichtius robustus*). *Spec. Publ. Am. Soc. Mammal.* **3,** 1–142.

Saito, M. (1977). "Edo Sampu Ryoko Nikki (Diary of the Journey to and from Edo)." A translation of

"Geschichte und Beschreibung von Japan" von Engelbert Kaempher, aus den Originalhandschriften des Verfassers 1777–79. Heibonsha, Tokyo (in Japanese).
Shibusawa, K. (1939). "Tosa Muroto Ukitsugumi Hogei Shiryo (Records of the Ukitsu Whaling Group, Tosa Muroto)," Vol. 36. Attique Museum Iho, Tokyo (in Japanese).
Shindo, N. (1968). "Setonaikai no Kujira no Kenkyu (Studies of Whales in Seto Inland Sea)." Kobeshi Ishi Kyodo Kumiai, Kobe (in Japanese).
Tada, H. (1968). "Mishima to Kujira (Mishima Island and Whales)." Mishima to Kujira Hensankai, Toyokita-machi, Yamaguchi Prefecture (in Japanese).
Tada, H. (1978). "Meijiki Yamaguchiken Hogeishi no Kenkyu (Studies of Whaling History in Yamaguchi Prefecture in Meiji Period)." Matsuno Shoten, Tokuyama, Yamaguchi Prefecture (in Japanese).
Taiji, G. (1937). "Kumano Taijiura Hogei no Hanashi (Story of Whaling at Taiji, Kumano)." Kishujinsha, Osaka (in Japanese).
Takahashi, S. (1899). Kyushu no Hogeigyo. (Whaling industry in Kyushu.) *Dainihon Suisankai Hokoku* **201,** 85–95 (in Japanese).
Tokumi, K. (1957). "Choshu Hogeiko (Studies of Whaling in Yamaguchi Prefecture)." Kammon Mingeikai, Shimonoseki, Yamaguchi Prefecture (in Japanese).
Tomilin, A. G. (1967). "Mammals of the U.S.S.R. and Adjacent Countries" (V. G. Heptner, O. Ronen, transl.), Vol. 9. Israel Program Sci. Transl., Jerusalem.
Yoshihara, T. (1974). Kujira Gounjo Hikae niyoru Bakumatsu no Tosa niokeru Hogei nitsuite (On whaling in Kochi prefecture in the late Edo era, Known from whale tax records). *Whales Res. Inst. Geiken Tsushin, Tokyo* **278,** 77–84 (in Japanese).
Yoshihara, T. (1976a). History of the whale fishery off Kanto district. *Tokyo Suisan Daigaku Ronshu* **11,** 17–144 (in Japanese).
Yoshihara, T. (1976b). On the whaling at Ine, Kyoto prefecture. *Tokyo Suisan Daigaku Ronshu* **11,** 145–184 (in Japanese).
Yoshihara, T. (1977). Tomb of whale. *Tokyo Suisan Daigaku Ronshu* **12,** 15–101 (in Japanese).

4

Aboriginal Whaling from the Aleutian Islands to Washington State

Beth Laura O'Leary

Whale, I want you to come near me, so that I will get hold of your heart and deceive it. . .

Whale . . . when you come ashore there, young men will cover your great body with bluebill duck feathers and with the down of the great eagle, the chief of all birds; for this is what you are wishing, and this is what you are trying to find from one end of the world to the other, everyday you are traveling and spouting.

Nootkan prayer (Curtis, 1916)

Introduction

Long before man settled on the Northwest coast of North America, whales cruised the waters of the Pacific. The earliest known coastal archaeological sites with cetacean remains date to 3120 ± 59 years BP (before the present) at Chaluka in the Aleutian Islands (Laughlin, 1963; Denniston, 1966), to 9140 ± 200 BP at Namu on the coast of British Columbia (Borden, 1975), and to 3000 BP at Yuquot on Vancouver Island (Dewhirst, 1975). When the area was visited by Europeans in the eighteenth century, aborig-

ISBN 0-12-389180-9

inal whaling cultures existed in several places in the Aleutians and on the Northwest coast of the United States and Canada (Cook, 1785; Meares, 1790). Some traditional whaling practices continued until the early twentieth century (Waterman, 1920). Although today no traditional whaling takes place on the Northwest coast of North America, such whaling does continue among some Eskimo groups located on the Arctic Ocean and the Bering and Beaufort Seas (Spencer, 1976; Marquette and Braham, 1982).

The importance of the whale to aboriginal cultures has been assumed without being well defined. This chapter reviews the relevant ethnographic, historical, and archaeological data on aboriginal whaling groups from the Aleutian Islands to Washington state, information upon which previous assumptions about the whale's importance have been based. The chapter focuses on evaluating a set of criteria for defining the importance of the whale to the cultural and subsistence systems of native groups. It also examines the insufficiency of available data, especially for use in extrapolating whale population sizes prior to commercial whaling and suggests a potential source of new data from archaeological research.

The discussion covers the period from the late eighteenth to the early twentieth century. Prior to the first European arrivals, the native people lived lives based on hunting and gathering of wild resources. But their lives were changed by trade, war, and missionary contact with European, Canadian, and American society. Some of the data on native groups were collected in the early twentieth century by ethnographers interested in the "remembered past," the memories of traditional lifestyles recounted by their informants. Additional data were gathered by archaeologists at prehistoric and historic sites located in the territories of these groups. I use the past tense throughout the descriptions of these groups, but it does not necessarily mean that the group itself is extinct. Many continue to live in the same areas and practice traditional lifestyles, while others participate fully in a modern national economy.

THE STUDY AREA

The study area extends from the Aleutian Islands to Washington state. Within that area, examination focuses on areas having aboriginal whaling populations in the Aleutians, and along the coasts of southern and southeastern Alaska, British Columbia, and Washington. It also includes groups using stranded whales within that area and in southern California.

The Aleutian Islands are a partly submerged continuation of the Alaska Peninsula. They extend approximately 1200 miles southwest from the Peninsula. Samalga Pass, at the southern end of Umnak Island, formerly marked the southwestern extension of the Bering Land Bridge which connected Asia and America some 13,000 years ago (Laughlin, 1974).

East of the Alaska Peninsula, there is a series of islands, the largest of which, Kodiak Island, marks the northwestern edge of the Gulf of Alaska. Northeast of Kodiak Island, the coastline of southcentral Alaska is marked by large fjords and sounds, the two most important of which are Cook Inlet and Prince William Sound (Fig. 1).

Glaciers once covered most of the Northwest coast of North America, creating the fjords and sounds characteristic of southeastern Alaska, British Columbia, and Puget

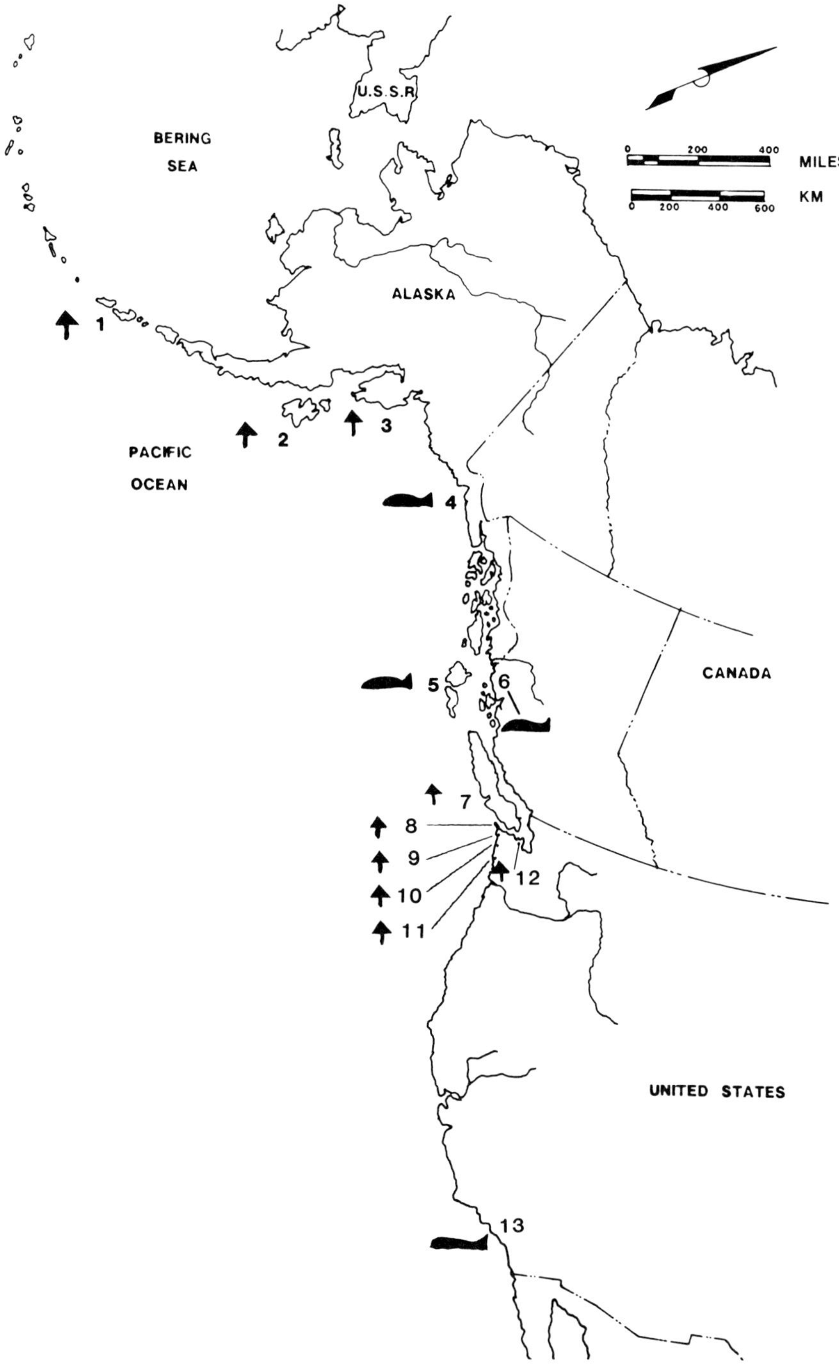

Fig. 1. *Map of study area showing aboriginal whaling groups (arrows) and use of stranded whales (outlines of whales). (1) Aleuts; (2) Koniag; (3) Chugash; (4) Tlingit; (5) Haida; (6) Tsimshian; (7) Nootka; (8) Makah; (9) Ozette; (10) Quileute; (11) Quinault; (12) Klallam; (13) Chumash.*

Sound, Washington. There are also many islands, one of the largest of which is Vancouver Island. South of the Olympic Peninsula, Washington, the coastline is relatively straight.

The climate of this area is maritime, known for its relatively mild winters, cool summers, and heavy precipitation. Such weather is caused by the interaction of physiographic factors, wind patterns, and ocean currents, especially the warm Japanese current. In general, from north to south, there is an increase in temperature and a decrease in precipitation. Precipitation tends to be seasonal in the south (usually occuring in winter) and more year round in the north (Schalk, 1978).

Native Peoples

GROUPS

According to Murdock (1967), there were 44 ethnographic groups that lived on, or had access to, the coast of the study area. Well-documented evidence of aboriginal whaling was available for only 8 groups: Aleut, Koniag, Chugash, Nootka, Makah, Klallam, Quileute, and Quinault. Four additional groups, the Tlingit, Haida, Tsimshian, and Chumash, utilized only stranded whales. Other groups occurring within the area have been excluded from the present study if they took only an occasional small cetacean or if their use of stranded whales is not well documented. Following are brief descriptions of the 12 groups named above, including information on location, population, social organization, and subsistence systems.

Aleuts

The Aleuts lived on the Aleutian Islands. The total population at the time of discovery by the Russians in 1741 was 16,000, with most Aleuts living on the eastern islands (Laughlin, 1975a). This represents a density of 64.70 persons per 100 km^2 (Osborne, 1977).

Aleutian society was stratified. Veniaminov (1840), an early Russian observer, divided the Aleuts into three classes: honorables or chiefs, common people, and slaves. As will be discussed later, this stratification was reflected in the whaling done by these people.

Although the islands did not support many land mammals, they had abundant marine resources including seals, principally harbor seals (*Phoca vitulina*), northern sea lions (*Eumetopias jubata*), sea otters (*Enhydra lutris*), fish [cod (*Gadus* sp.) and in some areas salmon (*Oncorhynchus* sp.)], and a variety of mollusks, crabs, sea urchins, and cetaceans (whales, dolphins, and porpoises) (Laughlin, 1975a). Whales have been sufficiently abundant in the area to support modern whaling activities, undertaken principally by Japan and Russia (Rice, 1974). Leatherwood *et al.* (1982) reported that at least six species of baleen whale [blue (*Balaenoptera musculus*), fin (*B. physalus*), sei (*B. borealis*), minke (*B. aculorostrata*), humpback (*Megaptera novaeangliae*), and gray (*Es-*

chrichtius robustus)]; five species of larger toothed whales [sperm (*Physeter macrocephalus*), Baird's beaked (*Berardius bairdii*), Stejneger's beaked (*Mesoplodon stejnegeri*), Cuvier's beaked (*Ziphius cavirostris*), and killer whale (*Orcinus orca*)], and a variety of smaller toothed whales currently inhabit waters around the Aleutians.

The gray whale migrates through Unimak Pass, the easternmost pass between the islands (Chapter 10, this volume). Historically, the Pacific right whale (*Eubalaena glacialis*) was also present (Murie, 1959; True, 1904), and the bowhead whale (*Balaena mysticetus*) may have been an occasional visitor (Murie, 1959; Rice and Scheffer, 1968).

It is interesting to note that Veniaminov (1840) reported that most of the Aleut villages were on the north side of the islands because it was good for driftwood and whales. W. F. Laughlin (personal communication, 1977), in his archaeological work, located many villages on the southern side also.

Koniag

The Koniag, a Pacific Eskimo group, lived on Kodiak Island and the adjacent mainland. According to Kroeber (1939), they had the relatively high population density of 30.60 persons per 100 km^2. The Koniag had a stratified society similar to that of the Aleuts (Lantis, 1938a). Villages were found near the mouths of major rivers and at certain strategic coastal points and were inhabited about 9 months of the year (Graburn and Strong, 1973).

The area is rich in resources, similar to the Aleutians with sea otters, seals, sea lions, halibut (*Hippoglossus stenolepis*), cod, and a large variety of shellfish. There are abundant salmon runs. Land mammals include bear (*Ursus* spp.), and mountain goat (*Oreamnos americanus*), and generally there are the same species of cetaceans that are found in the Aleutians (Leatherwood *et al.*, 1982).

Chugash

The Chugash were also a subarctic Eskimo group, living along Cook Inlet and Prince William Sound in Alaska. The population at the turn of the eighteenth century was estimated to be about 500–600 people (Meares, 1790). In contrast to most Eskimo groups, which are egalitarian, the Chugash had hereditary chiefs and a bottom class of slaves (Birket-Smith, 1953). The area surrounding their territory has high mountain ranges close to the coast. Unlike most areas of Eskimo occupation, it has a thick coastal forest with a wet, mild climate which supports mountain goats, Dall sheep (*Ovis dalli*), and moose (*Alces alces*) (Birket-Smith, 1953). Aquatic resources include seals, sea lions, sea otters, and various shellfish. Very important are five species of salmon which ascend the rivers to spawn. Birket-Smith (1953) reported that fin whales[1] and humpback whales were hunted there and that gray whales, sperm whales, and right whales were present from time to time.

[1]Contrary to ethnographies, routine successful hunting of fin whales by aboriginal groups was highly unlikely. It was only after the invention of the Foyn harpoon gun in 1868 that commercial whaling of these swift rorquals was successful (Slipjer, 1976).

Nootka

The Nootka lived on the west side of Vancouver Island. Koppert (1930) distinguished 19 subtribes or bands and estimated a population of 3698 in 1881. This estimate represents a population reduced by European diseases in the early nineteenth century. An early report by Cook (1785) estimated the population of Nootka Sound, where only 1 subtribe of Nootka dwelt, of 2000 people. Jewitt (1815) estimated the population in the same sound as 1500.

The Nootka migrated between two villages seasonally. Most winter villages were located in the upper reaches of inlets or rivers. Summer villages were on the lower portion of the inlets or outside beaches (Drucker, 1951). The society was stratified by wealth and heredity into chiefs, commoners, and slaves.

The seasonal round of food getting included exploiting the various rich marine resources; a winter herring run, the spring migration of waterfowl and seals, summer clamming, and fishing for halibut and cod. In fall, dog salmon (*Oncorhynchus keta*) and coho salmon (*O. kisutch*) were taken, dried, and stored. Whales available included gray whales (which are present in spring as they move north to Arctic feeding grounds and in fall when heading south to calving grounds), humpback, minke, blue, right, fin, and killer whales, and various species of dolphins and porpoises (Leatherwood *et al.*, 1982).

Makah

The Makah have been variously identified as a subtribe of the Nootka (Gibbs, 1877) or as a separate multivillage tribe of their own (Riley, 1968). Their territory extended from the tip of Cape Flattery, on the Olympic Peninsula of Washington, east into the interior and south along the coast to the area inhabited by the Quileute. One of the earliest estimates of their population is that of Quimper, who in 1790 reported 1000 inhabitants (Riley, 1968).

As with the Nootka, Makah society was stratified. The fundamental principles of social organization were kinship, hereditary transmission of status and privilege, and an emphasis on material wealth (Drucker, 1965).

According to Swan (1870) "the principal subsistency of the Makahs is drawn from the ocean and is formed nearly of all its products . . ." There was an abundance of fur seals (*Callorhinus ursinus*), sea lions, and sea otters, as well as halibut and other fish. There were a few salmon streams in the southern part of the territory (Huelsbeck, 1981). The Makah also had access to the same species of whale as did the Nootka.

Makah/Ozette. Ozette is an archaeological site located on the Olympic Peninsula which was occupied by members of the Makah until the early 1900s (Huelsbeck, 1981). It is included in the present report because of the information it provided on the protohistoric period of the Makah. A portion of the site containing at least four houses was destroyed by a mudslide, which preserved many of the artifacts. This portion of the site was radiocarbon dated to 440 $\pm$ 90 years BP (Huelsbeck, 1981). Although the data from the site have not been completely analyzed or published, preliminary reports and unpublished manuscripts significantly advance our understanding of the subsistence of the

Makah. Analysis of faunal remains, especially those of cetaceans, provides rare quantitative data on the catch of various species.

Quileute

The Quileute territory extended 30 miles south from Cape Flattery, Washington along the coast as well as 30 miles inland (Pettitt, 1950). In aboriginal times (pre-nineteenth century), the population was probably close to 1000 (Pettitt, 1950). Major settlements were at the mouths of rivers and up to 20 to 30 miles inland. The society was stratified, with headmen or chiefs in each village who were succeeded by their sons (Murdock, 1975). Their resources were similar to those of the Makah, with salmon being the most important. The whale species present were essentially the same as for the Nootka and Makah.

Klallam

The Klallam formerly occupied the southern shore of the Strait of Juan de Fuca, Washington between the mouth of the Hoko River and Port Discovery Bay (Gunther, 1927). Most villages were situated on the shore, though one was about 20 miles inland. In 1845, the Hudson Bay Company counted 1500 people. The Klallam were divided into two groups, an upper and lower class, with membership depending partly on birth but principally on wealth.

Fishing was the most important activity, especially for salmon, which run from late April or early May until December. Fishing was carried on either in the straits or in the rivers. The Klallam whaled infrequently (Gunther, 1927). Because of their location, on the inland side of the Strait, the Klallam had access to fewer large whales than did the Pacific Coast dwellers (Scheffer and Slipp, 1948).

Quinault

The Quinault, located directly south of the Quileute, were the southernmost people to pursue whales along the Northwest coast of North America. Their territory extended north of Gray's Harbor, Washington to the mouth of the Queets River and inland considerable distances from the coast; therefore, whaling was impractical except for residents of villages near the mouth of the river (Olson, 1936). The population in precontact times is estimated to have been 900 (Olson, 1936) to 1500 (Kroeber, 1939). There were village chiefs, and the village community was the fundamental social unit (Olson, 1936). The Quinault had resources similar to those of the Quileute and depended heavily on their huge salmon runs. Salmon was bartered in trade with other groups. The Quinault also hunted elk (*Cervus canadensis*) (Olson, 1936).

NATIVE PEOPLES' USE OF STRANDED WHALES: TLINGIT, HAIDA, TSIMSHIAN, AND CHUMASH

Clark (1952) has demonstrated that whales stranded on European shores of the Atlantic Ocean and the North and Baltic Seas were used during times when farming productivity was low. Such strandings on the coast of Britain were frequent, for example, 407 between 1913 and 1926 (Clark, 1952). We do not know the frequency of strandings in the study area, but the apparent benefit of stranded whales was recognized by at least four aboriginal groups including some which did not whale.

The Tlingit inhabited the coast of southeastern Alaska. There is evidence that the Yakutat Tlingit utilized stranded whales. Delaguna (1972) mentioned a stranding that occurred at Situk in winter in the mid-1880s and at Lost River in the 1850s after a smallpox epidemic at the village. Krause (1956) cited one report that the Tlingit hunted whales with harpoons but also claims that Veniaminov and Holmberg insisted that the Tlingit disdained whale meat.

The Haida, who lived on the Queen Charlotte Islands off the coast of British Columbia, may have occasionally hunted whales and utilized whales that drifted ashore (Blackman, 1981). However, Heizer (1943) stated that the Haida did not whale and made little or no use of stranded whales. He believes that such drift whales were poisoned by aconite in the Aleutian and Kodiak Islands and drifted to Haida territory. The meat may have poisoned the Haida, who Heizer feels feared to use it.

The Tsimshian lived on the coast and interior of British Columbia east of the Queen Charlotte Islands. Gunther (1927) reported that they used stranded whales.

The Chumash lived on the Santa Barbara Islands and nearby mainland of California. The one mention that the Chumash hunted whales is disqualified by Heizer (1974) on the grounds that the information is not reliable and that there is no mention by other observers of whaling by this group. It is known from both ethnographic and archaeological sources that the Chumash, like a few other coastal Californian groups, welcomed a stranded whale (Kroeber and Barrett, 1960). Baja California groups may have occasionally used a stranded whale as well, but Hubbs and Roden (1964) commented on the rarity of cetacean bones in middens there.

TECHNIQUES OF WHALING

Among the groups that actively whaled, there were two basic methods (Fig. 2). One was the use of the poison, aconite, on lances or spears thrown by single whalers or groups of whalers, each in a skin-covered bidarka (kayak). This method was used by the Aleuts, Koniag, and perhaps the Chugash. The other method was the use of harpoons with line and float attached, delivered from canoes containing large crews. This method was used by the Nootka, Makah, Quileute, Klallam, Quinault, and perhaps the Chugash.

Poisoned-Lance Whaling

Poisoned-lance whaling has been described by various observers (Veniaminov, 1840; von Wrangell, 1839; Markoff, 1856). Basically, it consisted of the following: one or two men in a bidarka approached a whale close enough to throw a stone-tipped lance from a throwing board into the whale (Fig. 3). The tip of the lance was coated with a concentrate of aconite rendered from the plant monkshood (*Aconitum* sp.). The struck whale was left to die. If the hunters were lucky, in 3 to 6 days it would drift ashore to their island.

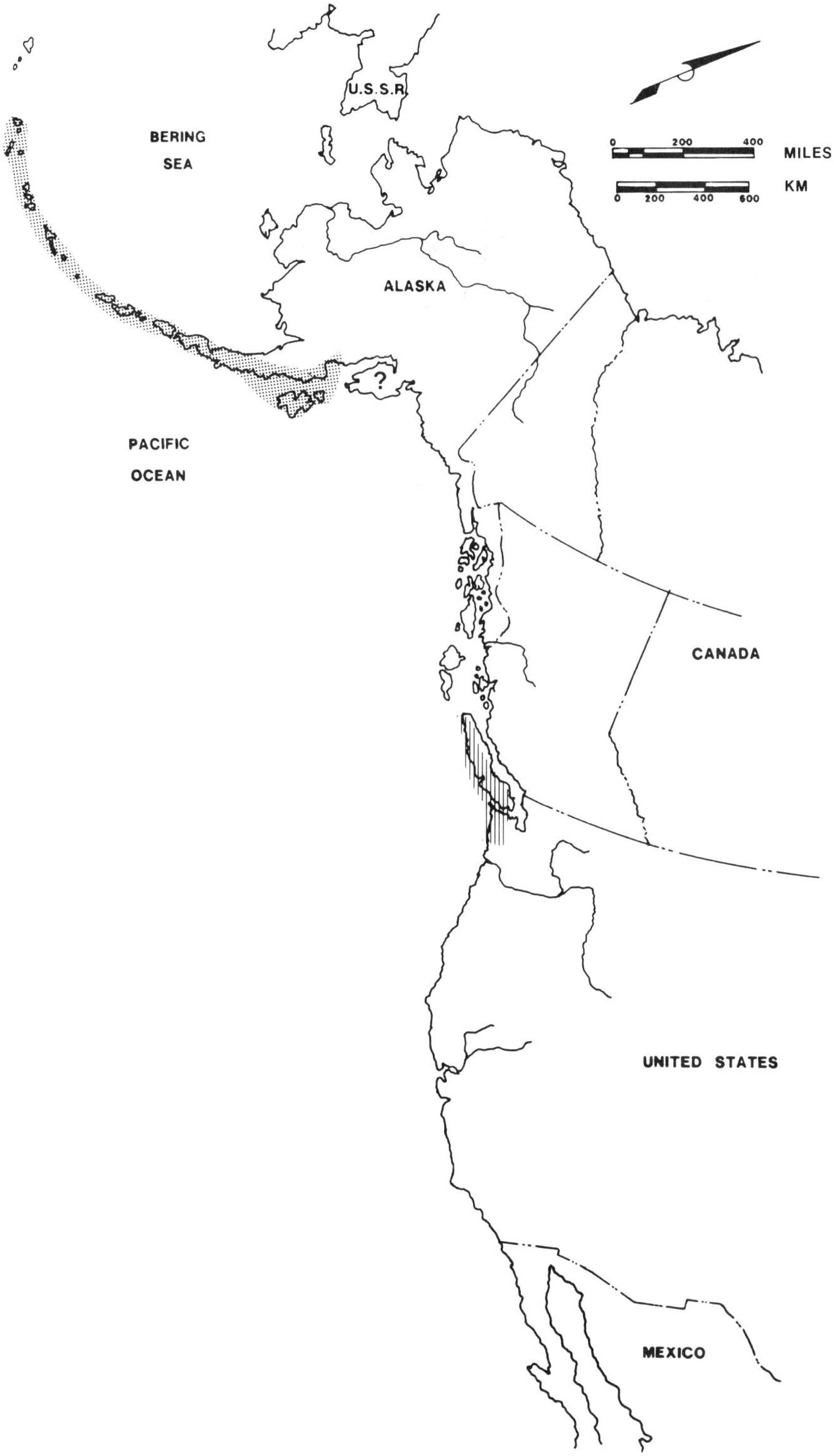

Fig. 2. *Map of study area showing techniques of whaling. Stippled areas, poisoned-lance whaling; hatched areas, harpoon–float whaling.*

CANOES of OONALASHKA.

Fig. 3. *Aleuts and bidarkas.*

There are two main problems in reconstructing the use of aconite in whaling. First, there is evidence that aconite was not used on lances at all. Secondly, it has been suggested that even if aconite had been used it could not have been delivered in sufficient quantities on the lance to constitute a lethal dose.

Albert von Chamisso, a naturalist who lived on Unalaska Island in 1815, reported on whales and on several species of aconites but did not mention any connection between the two (Laughlin, 1975b). Turner (1886), an Aleutian observer, failed to mention the use of aconite at all. Laughlin (1975b) concluded from his interviews with Aleuts on Akutan, Nikolski, and Atka that aconite, along with such things as bumblebee legs, may have been one of the various items used as a magical poison. Dall (1877) reported that the Aleuts used slate-tip points which they felt to be poisonous in and of themselves. By looking at the archaeological distribution of slate points in the north, Osborne (1976) concluded that slate was the preferred material for use with aconite in marine mammal hunting. He further states that the spread of slate assemblages correlates with the initial use of alkaloid poison and is indicative of early marine mammal hunting of virgin populations. However, other authors (Pinart, 1872, von Longsdorff, 1813) reported that the Aleuts also used obsidian points with aconite for whaling.

The question of dosage is more problematical. A research pharmacologist recently calculated, based on the effects of aconite on other animals, that the lethal dose of aconite for a large (40,000–50,000 kg) whale was over 100 g, a quantity he felt too large to put on a lance. However, the lethal dose of aconite for whales is unknown and it is possible, given some known facts about the poison and whale physiology, that a smaller dosage might have serious effects.[2] First, death by the poison results from respiratory and cardiac failure. Whales have a massive vascular network (*retia mirabilia*) in several locations in their bodies; thus, poison, even a sublethal dose, might spread rapidly (Bisset, 1976). Further, a wounded whale, laboring with its breathing, would be an easier target for a second lance. Laughlin (1975b) states the Aleuts were capable of achieving good penetration with their lances. This penetration, together with the poison, may have hastened the death. The hunter did not depend on killing the whale at the hunting site but on retrieving the dead whale drifting to shore.

Harpoon Whaling with Line and Floats

The method of taking whales by means of a harpoon with line and floats attached was employed by the groups mentioned previously and by traditional Eskimo of the Arctic Ocean, Bering, and Beaufort Seas, although under very different environmental conditions (Spencer, 1976). Basically, large boats, one or several, were each paddled by a crew of six to eight men. When a whale surfaced, the harpooner struck it with a harpoon on a detachable pole. Attached to the harpoon was a long line with appended sealskin buoys. The equipment for this kind of whaling was very specialized. Prior to the direct introduction of metal through European contact,[3] a mussel shell harpoon blade was manufactured by the Nootka and other groups (Waterman, 1920). The line was made of cedar bark and sinew and could be as long as 750 feet (Fig. 4).

The harpooner waited until the boat was about three feet from the whale, then aimed for the area just behind the flipper. Fisken (1980) discovered mussel shell blades embedded in three caudal vertebrae, as well as in the lateral surfaces, of five humpback whale scapulae at the Makah site of Ozette. When the wounded animal sounded its progress was impeded by the floats and it could be harpooned again. When it was exhausted, it could be approached and dispatched with a lance.

After the whale died, it was necessary to tie its mouth shut to prevent the carcass from filling with water and sinking. One of the crew members dived overboard and cut a hole through the whale's jaw and another through the upper lip and then passed a rope through both. Floats were attached to the carcass to keep it as high in the water as possible and a tow line was passed underneath the canoe to tow it back. As one Nootka told it, "When I looked back at the whale my heart felt sick. Every time a swell went over

[2]Dr. Sam Ridgway (personal communication) commented that the body-weight ratios for a lethal dose of aconite may be much less than Bisset suggests. A strike to the retia mirabilia would allow the poison to work more efficiently on respiration and cardiac function. Also, the spear itself, with or without aconite, would introduce bacteria into the wound, which could cause injury or death.

[3]Gleeson (1981) reported on excavation at Ozette. He found in a protohistoric deposit "dating to 440 ± 90 years BP . . . metal artifacts, not of any obvious European design. This was the first incontrovertible evidence for the use of iron and steel on the coast before known European contact."

Fig. 4. *Nootkan whaler with harpoon, line, and floats. (From Philadelphia Museum of Art: Purchased with funds from the American Museum of Photography.)*

it, the whale seemed to go backward: I couldn't see how we would ever get it to shore" (Drucker, 1951).

It was not unknown for a whaling crew to be out 4 days before returning to shore with a whale (Drucker, 1951). Although the whale was the "chief's fish" and he was the one to throw the first harpoon, he may not have been the most skillful harpooner (Jewitt, 1815;

Koppert, 1930). For example, Jewitt (1931) reports that Chief Maquinna, at Nootka Sound, was at sea 53 days, struck and lost eight whales, and got one.

The Nootka whale fishery, for one example, involved heavy investment in technology and many man-hours of pursuit, with sometimes only a small rate of return.

THE WHALING RITUAL

Although groups differed in their techniques of whaling, common to all groups was the ritual preparation of whalers prior to the hunt.[4] The material, customs, and supernatural interaction were as important to the aboriginal mind as the actual whaling equipment. Studies by Dr. Lewis Binford (personal communication, 1982) have suggested ritual played an important part in preliterate societies as a means of transferring information. He believes that the more complex and highly involved the ritual, the less frequent the activity to which it is linked. The highly complex whaling ritual on the Northwest coast may thus indicate the relative infrequency of the whale hunt among these groups. Lantis (1938b) has documented the presence of a whale "cult" in many whaling groups from Alaska to the Northwest coast. A general patterning of ritual can be discerned for the groups discussed here.

In general, only men were whalers. They formed an elite group, tied together by mutual participation in secret activities. Whaling was also tied to economic status in society. Among the Nootka, the chief was the central figure in the hunt as he was the one to throw the first harpoon and distribute the whale (Jewitt, 1815). Among the Aleuts, both the whaling boats and whale fishery belonged only to chiefs and particular families, which handed them down to their children (Heizer, 1941).

For a man to become a whaler he was forced to undergo certain restrictions, such as abstinence from sex prior to the hunt. Among the Nootka, a man was required to bathe nightly during the waxing of eight moons (Drucker, 1951). For the first four moons he bathed in fresh water and rubbed his body with nettles:

> For the second four moons he bathed in the sea. He swam about slowly, submerged, and came up making a noise like a whale "blowing" four times, then floated quietly in the water. This was all direct dramatization of the behavior he wished of whales (Drucker, 1951)

Among the Aleuts it was different:

> . . . the spear heads used in hunting the whale were greased with human fat, or portions of human bodies were tied to them, obtained from corpses found in burials, or portions of a widow's garments, or some poisoned roots or weed. All such objects had their own special properties and influence and the whalers always kept them in their bidarkas (Veniaminov, 1840)

The use of magic and ritual in the hunt date back at least to the Paleolithic (45,000 years BP). Magical items and activities were not thought to be effective against the whale itself. They could, however, affect the outcome of the hunt in that they gave confidence to the hunter, essential when facing a prey as large as a whale. Failure to

[4]Except for the Klallam, who infrequently went out (Gunther, 1927).

take a whale can be blamed on the hunter not having approached close enough or having made a ceremonial mistake (O'Leary, 1977). In addition to procuring food, whalers were affecting their own status and increasing their own prestige.

COMPARISON OF TECHNIQUES

The only similarities between the two radically different techniques described above are that both involved whaling in the open ocean and both contained ritual or magical components. For the Aleut, success depended on accuracy, depth of penetration, and efficacy of the aconite poison. The Aleut method of taking whales involved low investment in technology and fewer man-hours in pursuit time. It was basically a method of increasing the probability of obtaining a whale, although the ability to predict where a whale might be obtained when stranded was low. After the whale had been struck, the hunt was over and the whalers depended on the ocean currents and on their own ability to predict where the carcass would strand. Within the complex of islands where the North Pacific meets the Bering Sea, ocean currents and wind patterns are complex and incompletely understood. It was just as likely for a whale killed on the open sea to drift away as it was for the whale to be cast upon the beach of the hunter or on the beach of a neighboring group. Von Kittlitz (1858), describing the whaling in the Fox Islands, remarked

> [The Aleuts] . . . hope that the monstrous beast, after it has died of the wound, will finally be grounded by the wind and waves on one of their islands. The sea connection of the inhabitants of the islands with one another, but particularly the geographical situation of the islands themselves, make it possible that this practice can be accomplished by some success. But the aforementioned hunting of the Aleuts is so frightfully wasteful. [This passage is cited by Heizer (1943).]

Von Langsdoff (1813) and others reported that the weapon found in the dead whale was distinguished by various marks which identified its owner, who could rightfully claim his share. He might not, however, live where the whale had been stranded; therefore he must have depended on others to tell him of the beaching. Given the extensive kin relations among aboriginal groups, relatives of the hunter would certainly have been found on neighboring islands, and he would have been informed. However, whales without identifying marks or those washing up on more distant islands would have been claimed by the inhabitants.

Harpoon whaling with line and float involved large whaling crews, intensive preparation, and lengthy gear-up time. It was conducted from dugout canoes, made of large cedar logs, with an attached square sail of cedar-bark matting (Swan, 1870). Curtis (1916) noted nine items of whaling tackle alone, not counting other boat equipment. Waterman (1920) reported the whalers put to sea at sunset so as to be on the whaling grounds at daybreak. Judging from the meager quantitative information on whales actually killed, the success rate was low.

WHALING SUCCESS BY SPECIES: PROBLEMS WITH THE DATA

To understand the nature of aboriginal whaling and evaluate its importance in aboriginal subsistence systems, it is preferable to know the species and number of individuals taken. Most ethnographies and historical accounts deal with the general seasonal round of activities of native groups, rarely including specific or quantitative information. Careful reading of the record, however, reveals some quantifiable data.

Most early observers (with such notable exceptions as Charles M. Scammon) were not trained in natural history or biology and usually referred only to "the whale," making no accurate distinctions as to species. For example, Pinart (1875) suggested that the Aleuts hunted the "smaller whales," which he interpreted to be the humpback and fin whales. Veniaminov (1840) reported "they ate all varieties of whales . . . except one," but added "but the whales here are generally of a smaller species, so that the meat of a whole whale is easily packed into a single baidar." Although this suggests reliance on smaller whales, such as minke or perhaps dolphins and porpoises, good quantitative data on cetacean remains in Aleut sites are scarce. Humpback whale remains have been reported in some Aleut sites (Denniston, 1973).

Other authors mentioned the Aleuts attacking the "giants of the sea" (von Wrangell, 1839) or referred to a "monstrous beast" (von Kittlitz, 1858) or simply "the monster" (von Langsdoff, 1813), as cited in Heizer (1943). Others collected native names for whales by species but did not report which were most frequently hunted (e.g., Waterman, 1920).

One of the most neglected sources of information on aboriginal whaling is the analysis of faunal remains from archaeological sites. In well-stratified sites with good preservation, one can quantify cetacean remains to the minimum number of individuals of each species. However, the study of faunal remains is relatively new and still being developed, and most published reports are frustratingly incomplete (Binford, 1978). Even in major, early sites on the Northwest coast faunal remains are listed as a percentage of the total and identified as "cetacean" or "delphinid." More recent work is encouraging. One example is that of Fisken (1980), whose analysis of whale bones from Ozette revealed that the humpback whale, despite its infrequent mention in the literature, was present in this Makah site. Fisken's analysis of frequencies of the occurence of whale bones of various species from different time periods should provide data on the changes of whale use through time.

In some cases the species caught or used might reflect an interaction with commercial whaling during the nineteenth century. It is possible that strandings of gray whales on the Northwest coast increased with commercial whaling, as a certain percentage of whales struck by whaling ship crews were lost and stranded on native beaches. According to Pettitt (1950), in 1874 the Quileute were having trouble catching gray whales, which they felt had decreased in number. As early as 1839, the Northwest Coast was already a favorite hunting ground for commercial whalers. Bancroft (1890) claimed that the Makah sold $8000 worth of whale oil in 1856.

UTILIZATION OF THE WHALE

> *Whale . . . you will be proud to see the young men come down on the fine sandy beach of my village of Yahksis to see you; and the young men will say to one another: "What a great whale he is! What a fine whale he is! and you, whale will be proud of all that you will hear of your greatness"* (Curtis, 1916).

When a whale was caught by the Makah at Neah Bay the whole village came out to receive it (Waterman, 1920). On Unalaska, "The whale had been lately caught, was washed from the shore by the high water, in the late storm, and ran aground . . . and as they has cut much flesh and blubber from the head, the back part and the preponderance, and sunk to the bottom" (von Kotzebue, 1821). Their partial dependence on stranded whales placed the Aleuts and others in competition with carnivorous marine and land predators (bears, foxes, gulls, etc.).

Certain procedures were followed in butchering the whale. These procedures are better known for the Nootka, Makah, Quileute, and Quinault than for the Aleut. The harpooner instructed the whale carcass be "measured" from blowhole to dorsal "fin." The whaler's special piece was the "saddle" (a large square on the back around the "fin") which was ritually decorated and sung over. The blubber was cut up into strips for a feast, and the whaler gave it away to his tribe in order of rank. Oil was rendered and stored. The whole process was accompanied by dance and song (Drucker, 1951). Certainly, the problem of dealing with an animal of such a large size was common to all.

There was also the critical problem of efficiently processing whale products for storage, because not all of a whale could be eaten in a short period of time. As all groups in the study area lived within a subarctic or maritime region with relatively mild climate, the processes of decay were relatively fast. This is in direct contrast to arctic regions inhabited by whaling Eskimo groups. Arctic Eskimo groups had the relatively easy option of freezing whale products, evidenced by the ample storage facilities (ice cellars) dating from the Thule period (AD 1000–3000) until the present historic time in Alaska and the Canadian Arctic (McCartney, 1980a).

The problem of spoilage was recognized by Petroff (1884), who wrote the Koniag would not touch a whale which they considered dangerously decayed, as indicated by the absence of a certain species of bird. Thus, some whales were worthless as food.

It is clear that all groups, to a greater or lesser extent, used the blubber and oil from the whale and other marine mammals as food. The predominant view is that whale meat may have been eaten fresh, but I feel that because of the enormous amounts of meat and the small populations an entire kill could not be completely consumed as fresh meat. The eating of fresh whale meat is also in question (Waterman, 1920). There is little mention in the literature of processing whale meat for storage, although Petroff (1884) reports that Holmberg describes Koniags butchering a whale, stripping the whale meat and then boiling it by an unknown method. It was then put in the ground to putrify and was used when "set." This seems more of a culinary process than a process for long-term storage. I believe that this is because the only other alternative for processing was smoking and drying. This process as described for salmon and other fish is not described for whale. It involves a great deal of labor, time, and attention in both the initial butchering and maintenance of the right conditions (i.e., smoking fires, wind, protection from rain). This, coupled with the rendering of the blubber, is a substantial undertaking. Many other resources, especially salmon, would have been more predictably and easily obtained at certain locations where facilities for the processing had been built and maintained.

The three most important products from whales were blubber/oil, meat, and bone. Other products used as raw material were of minor importance by comparison.

Blubber/Oil

The oil of cetaceans resides in the outer covering of fat and in the bones. Although edible itself, the bulk of the blubber was rendered by boiling, yielding a great quantity of oil. The Nootka rendered oil by boiling in long wooden tubs heated by hot stones placed inside. Whale oil was the most prized commodity among these Northwest groups. It was stored fresh, or used as a preservative for other foods. It was ubiquitously used as a gravy on foods and considered necessary to the diet (Swan, 1870).

It is reported throughout the literature on the Northwest coast that whale oil indicated prestige, the amount of oil a chief had being an indication of his wealth (Pettitt, 1950). Other accounts reveal that whale oil was stored by the chief in preparation for large gatherings where much oil was consumed by neighbors and local groups, and some was spewed into the wood fire to impress spectators with the flame (Pettitt, 1950; Sapir, 1920).

Whale oil was a prized commodity for trade as well (Olson, 1936; Swan, 1870). Even early observers such as Jewitt (1815) reported that extensive trade and gift-giving among groups involved oil or blubber. Thus, whale oil may have functioned as a type of currency on the Northwest coast (Colson, 1953; Swan, 1870; Gibbs, 1877). Whaling and whale products may have reinforced a stratified society.

Ethnographic evidence indicating the importance of oil encouraged Fisken (1980) to evaluate the relative numbers of humpback as compared to gray whales indicated by the faunal remains at Ozette. Despite the fact that humpbacks would have been less available to the people at Ozette, Fisken found gray and humpbacks almost equally represented. Since a humpback contains, on the average, twice as much oil as a gray whale (Scammon, 1968), Fisken reasoned that the people of Ozette were selecting the less available but more oil-rich humpbacks over the gray, this despite the proximity of the gray whale's migratory route.

In sharp contrast to the Eskimo of the Arctic, for groups south of the Chugash, whale oil/blubber was not critical as a fuel for lighting, cooking and heating. Most whaling groups from the Tlingit to the Quinault lived in large plank houses and had reserves of timber which they used for fuel (Drucker, 1951). The Aleuts and Koniag did have stone lamps which were used for holding blubber/oil for domestic purposes (Denniston, 1966; Clark, 1975). But other animal oils (e.g., seals) were also used as fuel, and the maritime subarctic climate was less severe than the cold arctic.

Meat

There are conflicting reports concerning the processing and consumption of whale meat, which is very lean and must be dressed or cured to prevent spoilage. Some ethnographies reported that the meat of the whale was not often used. Some included accounts of fresh meat being eaten but not stored.

It has been tacitly understood by modern anthropologists that the whale was totally utilized by native groups. It has also been assumed that the whale meat was stored. These assumptions may be misleading. There are few references to the actual techniques of processing meat for storage. Veniaminov (1840) noted there was rarely a steady supply of food among the Aleuts; morever, the facilities for preserving food were limited. Waterman (1920) reported the Makah did not generally utilize the meat because decay proceeds more rapidly in the flesh than in the blubber. After the time spent towing the dead whale to shore, coupled with the long time taken butchering the whale, the meat was considered rancid. Sapir (1920) wrote that the Makah used blubber and flesh for food. There is one brief mention in Curtis (1916) that the cooked blubber and flesh of the whale was dried in the sun and smoke. While not explicit, it is implied that the meat was stored. However, Drucker (1951) reported the flesh of various mammals and birds, including the whale, was never dried. Thus whale meat was most likely not processed for storage but utilized only fresh, if at all; therefore a great deal may have spoiled and been unavailable for later use. Due to their locations, none of the groups had the option of simply freezing the meat.

Bone

Bone and baleen were easily worked and were selected by native peoples for a variety of reasons. Whale bone artifacts have been identified, ranging from combs (Kent, 1975) to clubs (Fisken, 1980). Bone was also used in the building of structures. Because of its density and size, whale bone is slow to decay and some bones may be reused over many years.

Hrdlicka (1945) reported the use of whale bone as roof supports on the Aleutian Islands, although wooden roof supports were also used. Even with the abundance of trees near the site of Ozette, large whale bones (ribs, skulls, and mandibles) were used to retain mud banks and to channel high runoff after the rains.

Although whale bones occur in archaeological sites and are present in historic collections, a large percentage of whale bones lie on the beach where the whales were butchered. Any sampling of a coastal site should include areas of beach that may be contemporaneous with site deposits. There should also be a theoretical consideration of faunal remains. Butchering procedures, selection of certain bones for culinary, tool, or construction purposes is usually described in the existing ethnographies and indirectly by the bone artifacts found at sites. By understanding how the site was formed at a whaling location as reflected in the structure of the faunal remains (i.e., differential selection of osteological elements), a more accurate accounting of the numbers of whales taken and their importance becomes possible.

THE IMPORTANCE OF WHALING IN THE STUDY AREA

Whaling among aboriginal groups has captured the imagination of anthropologists and laymen. Its importance in the cultural and political systems of native peoples cannot be denied. Even today, the subject of whaling remains controversial, especially in regard to the North Alaskan Eskimo (Braham *et al.*, 1980).

Within the study area, anthropologists have recognized the need to evaluate the importance of the whale relative to other resources. Drucker (1951) felt that the prestige value of whaling among the Nootka outweighed its economic importance, and that in proportion to the expenditure of time and energy the economic reward was slight.

The dramatic art of whaling was passed down through families and intermarriage among the Nootka and other groups to the south, and came to symbolize the greatest individual, or in most cases, the chief. Among the Aleuts, whalers represented the elite group who knew the "secrets" about whaling. One of these secrets was the use of the poison aconite. The amount of ritual and prestige involved in whaling may have overshadowed the economic and subsistence gains. The Northwest coast peoples' emphasis on fats and oils was cultural as well as dietary. Other groups in the same area that did not whale acquired sufficient amounts of fat and oil from other animals (e.g., seals, salmon, etc.).

The annual round of subsistence activities included fishing for and processing salmon. All along the Northwest coast, salmon was a primary resource (Schalk, 1978). The salmon runs differed in their periodicity and abundance. The runs father north occurred in a shorter time period than those to the south (Schalk, 1977). Salmon was smoked, dried, and put up for winter use.

Among the Nootka, all salmon streams were owned by local groups who controlled access to them (Dewhirst, 1975). Drucker (1951) felt the early Nootkans inhabiting areas without salmon streams invented whaling to compensate for the absence of salmon. He based this idea on Nootka legend locating the first whalers at Bajo Point, which was completely exposed to the "outside" or outer coastline and had no significant salmon streams (Dewhirst, 1975). Eventually, this group allied itself to a group with "inside" properties, or settings of inlets and river mouths. If whaling in itself represented a sufficient economy for Nootkan groups, then we would expect to find all sites located in the most advantageous positions to whale (i.e., on the "outside"). Early and historic site patterning reveals sites both on the "outside" and the "inside," an adaptation to two major environmental settings on a seasonal basis (Dewhirst, 1975). Examples of historic northern and central Nootkan groups attempting to obtain and hold properties in both locations are well documented (Swadesh, 1948).

The structure of resources available to aboriginal groups determined, to a large extent, their seasonal location and activities. Resources, especially in northern areas, are not available year round. In many cases the availability of resources overlaps in time, and this overlap presents problems in scheduling to allow each resource to be best utilized. For example, for the Makah at Neah Bay, halibut was a staple and there were only limited salmon streams (Swan, 1870). Halibut would be exploited during the spring and summer when they congregated at depths of 20 to 30 fathoms. Then they were accessible to the traditional hook-and-line technology (Huelsbeck, 1981). Halibut was dried and stored for future use and trade (Swan, 1870). Fur seals were also available. Ethnohistoric information on the Makah does not indicate they hunted fur seals aboriginally, but archaeological finds do. At Ozette, more than 80% of the faunal remains were fur seals (Daugherty and Kirk, 1974). Each spring (April–May), large groups of seals passed by the village on their way north to breed. Also on its way north during this time was the gray whale.[5] Thus, during spring there was an overlap of several potential food resources. Whaling was not pursued exclusively by any group. The whale was one of a number of resources, many of which were more easily obtained and stored.

[5]The gray whale was not pursued in the fall (October–November) on its southern migration due to stormy and risky weather at that time. M. Fisken (personal communication) also reports the presence of six young gray whale calves in the faunal remains at Ozette, evidencing whaling when the young were going north for the first time.

South of the Makah, the Quileute and Quinault seemed to have been less involved in whaling. Their principal resource was fish, mainly salmon. The Quinault were famous for their dried salmon, sea otter skins, and dried elk meat (Olson, 1936). Only a handful of men took to whaling. The whaling canoe was obtained in trade from the Makah, and usually only two crews went out (Olson, 1936). Farther south, Gunther (1927) reported the Klallam "do not go out to sea hunting for whales as do the Makah and Nootka, but wait until one is in sight."

The Aleuts lived in an area rich in marine and coastal resources. Although the Pacific Ocean in this area has one of the world's most productive whaling grounds, there were many other resources available to the Aleuts. Laughlin (1974) has examined the hunting and collecting habits of Aleuts, which he feels closely parallel those of the sea otters which are indigenous to the area. Most of their food was secured within 3 miles of the coastline (including strandflats, streams, and inside bays) roughly inside the 30-fathom isobath. By implication, if one can obtain all he needs close to home there is more hesitation about going to sea to whale. Hence, two factors influenced the importance of whaling to the Aleut. The first was the availability of abundant resources close to and on shore. The second was the unpredictability of their whaling strategy. Even if we accept as demonstrable the effectiveness of the aconite and lance, the fact is that the whalers returned to land and depended on the winds and currents to strand the whale in a place both known and accessible.

ESTIMATES OF ABORIGINAL WHALE KILLS

Recent controversy over aboriginal harvesting of whales, most notably the bowhead and gray, has sent many biologists to the anthropological and ethnohistorical literature in search of the aboriginal "take," the numbers and species of whales harvested by native groups (Ohsumi, 1976; Mitchell, 1979; Mitchell and Reeves, 1980; Braham *et al.*, 1980). One attempt has been made to reconstruct the number of gray whales in "initial" populations (populations prior to the advent of commercial whaling around 1840) in order to make recommendations for the modern management of present populations (Mitchell, 1979). The problem is that sufficient quantitative information in the anthropological and ethnohistorical literature does not exist, nor is it always possible to determine which species are being described.

Ohsumi (1976) bases his estimate for aboriginal catches on the number of gray whales presently provided to the Chukchi by the U.S.S.R. whale fishery. As Mitchell (1979) states ". . . extrapolating past aboriginal catches from current takes by a modern catcher boat may not be valid." This is an understatement. The actual number of gray whales taken prior to commercial fishing may have been well above or below Ohsumi's estimates. The U.S.S.R. currently supplies an average of 165 gray whales per annum to the Chukchi, hence Ohsumi's estimates may be inflated if a significant percentage of whales taken aboriginally were not grays. Mitchell (1979) reported only that the aboriginal catch was "mainly gray whales."

By taking a different approach, we raise questions as to whether Ohsumi's numbers may be too low. If the Chukchi needed 165 whales per annum to subsist before the Soviet fishery acquired whales for them, surely many more than this must have been struck and lost. Depending on what rate of loss we assume, the number of whales destroyed could be extremely high. Also, Ohsumi's estimates may be too low because aboriginal populations may have been higher. It is impossible to draw a conclusion based on such speculative premises.

Mitchell (1979) tries to use the available ethnographies to imply that aboriginal takes of gray whales were far in excess of what most biologists have assumed. He assumes that the aboriginal take was all gray whales. As reported in this chapter, this was definitely not the case. In examining his catalog of North Pacific coastal groups who took whales it is clear that other key factors in the equation are missing. We need more quantitative information on species of whales killed, numbers of whales struck and lost and the size of aboriginal populations doing the whaling. To his credit, Mitchell (1979) is cautious about drawing fast conclusions from the available data.

With the paucity of quantitative data, the key to looking at aboriginal whaling is to view the whale as part of the total subsistence system by using the ethnographic and ethnohistorical record. Through the use of faunal analysis, the archaeological record may provide us with the best qualified quantitative data on whale use by aboriginals. We should use archaeology to clarify and extend the other information. Preliminary findings, such as those at Ozette, indicate the potential of these data. Recently in the Canadian arctic, McCartney (1980b) has analyzed cetacean faunal remains at Thule sites in order to reconstruct Bowhead whale stocks and use by prehistoric Inuit. The archaeological record is available, and new research on coastal sites continues to provide another way to assess our present estimates of the aboriginal take of whales, and the importance of the whale in aboriginal cultural systems.

Summary

Popular opinion has attributed great importance to whaling as a subsistence activity among aboriginal Northwest coast and Aleut groups. Actual criteria to define such importance are lacking. This chapter reviews the relevant ethnographic, historical, and archaeological data upon which such assumptions have been based. It evaluates the importance of whales in aboriginal cultural and subsistence systems by examining (1) strategies and techniques of hunting whales, (2) success at whaling, (3) the position of whaling in the social system, (4) availability of other resources, (5) problems of scheduling conflicts between whaling and other subsistence activities, (6) use, processing, and storage of whale products, and (7) trade of whale products. The chapter also examines the data available, evaluates its use in current studies of whale populations, and suggests the potential importance of archaeological research in such studies.

Acknowledgments

For comments on drafts of this chapter and providing stimulating ideas I would like to express appreciation to Lewis R. Binford and John M. Campbell. For editing and consistent support throughout the writing of the manuscript I thank Thomas P. Keyes. I am grateful to Marian (Fisken) Byse, William S. Laughlin, Edward Mitchell, Stephen Reilly, and Sam Ridgway for specific information and source material. Finally, thanks are due to Dorrit O'Leary for assistance in final editing, to Lewis Binford for the Aleut illustration, and to Marc H. Hudson for the map illustrations.

References

Bancroft, H. H. (1890). "History of Washington, Idaho and Montana, 1845–1889." Works of H. H. Bancroft, Vol. XXXI. San Francisco.

Binford, L. R. (1978). "Nunamiut Ethnoarchaeology." Academic Press, New York.

Birket-Smith, K. (1953). "The Chugash Eskimo," Nationalmus. Skr., Ethnogr. Række 6. København.

Bisset, N. G. (1976). Hunting poisons of the North Pacific Region. *Lloydia* **39** (2 and 3), 87–124.

Blackman, M. (1981). "Window on the Past: The Photographic Ethnohistory of the Northern and Kaigani Haida," Nat. Mus. Man Mercury Ser., Can. Ethnol. Serv., Pap. No. 74. National Museum of Man Ottawa.

Borden, C. E. (1975). "Origins and Development of Early Northwest Coast Culture to about 3000 B.C.," Nat. Mus. Man Mercury Ser., Archaeol. Surv. Can., Pap. No. 45. National Museum of Man Ottawa.

Braham, H. W., Marquette, W. M., Bray, T. W., and Leatherwood, J. S., eds. (1980). The bowhead whale: Whaling and biological research. *Mar. Fish. Rev.* **42,** 9–10.

Clark, D. W. (1975). "Koniag-Pacific Eskimo Bibliography," Nat. Mus. Man Mercury Ser., Archaeol. Surv. Can., Pap. No. 35. Museum of Man Ottawa.

Clark, J. G. D. (1952). "Prehistoric Europe: The Economic Base." London.

Colson, E. (1953). "The Makah Indians." Univ. of Minnesota Press, Minneapolis.

Cook, Capt. J. (1785). "A Voyage to the Pacific Ocean for Making Discoveries in the Northern Hemisphere," Vol. II. H. Hugh, London.

Curtis, E. S. (1916). "The North American Indian," Vol. XI. Nootka, Haida, Norwood, Illinois.

Dall, W. H. (1877). "On Succession in the Shell-heaps of the Aleutian Islands," Contrib. Am. Ethnol., Vol. I. Washington, U.S. Govt. Print. Off. Dept. Int. U.S. Geogr. and Geol. Surv., Rocky Mt. Region.

Daugherty, R. D., and Kirk, R. (1974). "Hunters of the Whale, An Adventure in Northwest Coast Archaeology." W. Morrow, New York.

Delaguna, F. (1972). Under Mount Saint Elias: The History and Culture of the Yakutat Tlingit. *Smithson. Contrib. Anthropol.* **7.**

Denniston, G. (1966). Cultural change at Chaluka, Umnak Island: Stone artifacts and features. *Arct. Anthropol.* **3**(2), 84–125.

Denniston, G. (1973). Ashishik Point: An economic analysis of a prehistoric Aleutian community. Ph.D. Thesis, University of Wisconsin, Madison (unpublished).

Dewhirst, J. (1975). Nootka Sound: A 4000 Year Perspective. *Sound Heritage.* Vol. VII, No. 2. Provincial Archives of B.C.

Drucker, P. (1951). The Northern and Central Nootkan Tribes. *Bull. Bur. Am. Ethnol.* No. 144.

Drucker, P. (1965). "Cultures of the North Pacific Coast." Chandler Publ. Co., New York.

Fisken, M. (1980). Whale bones from the Ozette site. Paper presented at the Northwest Coast Studies Conference, March.

Gibbs, G. (1877). Tribes of Western Washington and Northwestern Oregon. *Contrib. North Am. Ethnol.* **1,** pt. 2.

Gleason, P. (1981). "Ozette Wood Technology," Proj. Rep. No. 3. Washington State University, Pullman.

Graburn, N. H. H., and Strong, B. S. (1973). "Circumpolar Peoples: An Anthropological Perspective." Goodyear Publ. Co., Inc., California.

Gunther, E. (1927). Klallam ethnography. *Univ. Wash. Publ. Anthropol.* **1**(5), 171–314.

Heizer, R. F. (1941). Aboriginal whaling in the Old and New Worlds. Ph.D. Dissertation, University of California, Berkeley (unpublished).

Heizer, R. F. (1943). Aconite poison whaling in Asia and America: An Aleutian transfer to the New World. *Bull.* 133(24) *Bur. Am. Ethnol.* 415–468.

Heizer, R. F. (1974). "Were the Chumash whale hunters? Implications for ethnography." *J. Calif. Anthropol.* **1**(1), 26–32.

Hrdlicka, A. (1945). "The Aleutian and Commander Islands and their Inhabitants." Wistar Institute, Philadelphia, Pennsylvania.

Hubbs, C. L., and Roden, G. L. (1964). Oceanography and marine life along the Pacific coast of Middle America. *In* "Handbook of Middle American Indians," (R. Wauchope, ed.) Vol. 1, pp. 143–186. Univ. of Texas Press, Austin.

Huelsbeck, D. R. (1981). "Utilization of Fish at the Ozette Site, Proj. Rep. No. 11. Laboratory of Archaeology and History, Washington State University, Pullman.

Jewitt, J. R. (1815). "Narrative of the Adventures and Suffering of J. R. Jewitt; Only Survivor of the Crew of the Ship Boston, During a Captivity of Nearly Three Years among the Savages of Nootka Sound: with an Account of Manners, Mode of Living etc." S. Richards, Middletown, Connecticut.

Jewitt, J. R. (1931). "A Journal kept at Nootka Sound." Boston, Massachusetts.

Kent, S. (1975). An analysis of Northwest coast combs with special emphasis on those from Ozette. M. A. Thesis, Washington State University, Pullman (unpublished).

Koppert, V. A. (1930). "Contributions to Clayoquot Ethnology," Anthropol. Ser. No. 1. Catholic University of America.

Krause, A. (1956). "The Tlingit Indians" (E. Gunther, transl.). Univ. of Washington Press, Seattle.

Kroeber, A. L. (1939). Cultural and natural areas of native North America. *Univ. Calif., Berkeley, Publ. Am. Archaeol. Ethnol.* No. **38.**

Kroeber, A. L., and Barrett, S. A. (1960). Fishing among the Indians of Northwestern California. *Anthropol. Rec.* **21,** 1–210.

Lantis, M. (1938a). The Mythology of Kodiak Island, Alaska. *J. Amer. Folklore* **51,** 123–172.

Lantis, M. (1938b). The Alaskan whale cult and its affinities. *Am. Anthropol.* **40,** 438–464.

Laughlin, W. F. (1963). Eskimos and Aleuts: Their origins and evolution. *Science* **142,** 633–645.

Laughlin, W. F. (1974). Holocene history of Nikolski Bay, Alaska and Aleut evolution. *Folk* **16–17,** 96–115.

Laughlin, W. F. (1975a). Aleuts, holocene history and Siberian origin. *Science* **189,** 507–515.

Laughlin, W. F. (1975b). Review of N. G. Bisset's unpublished manuscript. "Hunting Poisons of the North Pacific Region."

Leatherwood, S., Reeves, R. R., Perrin, W. F., and Evans, W. E. (1982). Whales, dolphins and porpoises of the Eastern North Pacific and adjacent Arctic waters; a guide to their identification. *NOAA Tech. Rep., NMFS Circ.* **444.**

McCartney, A. P. (1980a). The nature of Thule Eskimo whale use. *Arctic* **33**(3), 517–541.

McCartney, A. P. (1980b). Study of Archaeological Whale Bones for the Reconstruction of Canadian Arctic Bowhead Whale Stocks and Whale Use by Prehistoric Inuit. Northern Environmental Protection Branch, D.I.N.A. Ottawa. Canada.

Markoff, A. (1856). "The Russians in the Eastern Ocean, St. Petersburg" (transl. by Ivan Petroff). MS in Bancroft Library, University of California, Berkeley.

Marquette, W., and Braham, H. W. (1982). "Gray Whale distribution and catch by Alaskan Eskimos: A replacement for the Bowhead Whale?" *Arctic* **35**(3), 386–394.

Meares, J. (1790). "Voyages Made in the Years 1788 and 1789 from China to the North West Coast of America." J. Walter, London.

Mitchell, E. (1979). Comments on Magnitude of Early Catch of East Pacific Gray Whale (Eschrichtius robustus). *Rep. Int. Whal. Commn.* **29,** SC/30/Doc. 41, 307–314.

Mitchell, E., and Reeves, R. R. (1980). The Alaska Bowhead problem: A commentary. *Arctic* **33**(4), 686–723.

Murdock, G. P. (1967). "Ethnographic Atlas." Yale HRAF Press, New Haven, Connecticut.

Murdock, G. P. (1975). "Ethnographic Bibliography of North America." Yale HRAF Press, New Haven, Connecticut.

Murie, O. (1959). "Fauna of the Aleutian Islands and Alaskan Peninsula." U.S. Govt. Printing Office, Washington, D.C.

Ohsumi, S. (1976). Population assessment of the California whale. *Rep. Int. Whal. Commn., Sci. Rep.* **26,** 350–359.

O'Leary, B. L. (1977). Magic and poison: The whaling technologies of three northern cultures. Paper presented at the Kendall Whaling Museum, October.

Olson, R. L. (1936). The Quinault Indians. *Univ. Wash. Publ. Anthropol.* **6,** No. 1.

Osborne, A. J. (1976). Aconite and old slates: An explanatory argument for change in lithic assemblages and raw materials. Paper presented at the meeting of the Society for American Archaeology, April.

Osborne, A. J. (1977). Aboriginal exploitation of marine food resources. Ph.D. Thesis, University Microfilms, Ann Arbor, Michigan (unpublished).

Petroff, I. (1884). "Report on the Population, Industries and Resources of Alaska." Department of the International Census Office, Washington Printing Office.

Pettitt, G. A. (1950). Quileute of LaPush. *Univ. Calif. Anthropol. Rec.* **14.**

Pinart, A. (1875). "La chasse aux animaux marins et les pêcheries chez les indigènes de la côte nord-ouest d'Amérique." Charles Aigne, Boulogne-sur-mer.

Rice, D. W. (1974). Whales and whale research in the eastern North Pacific. *In* "The Whale Problem" (W. E. Schevill, ed.), pp. 170–195. Harvard Univ. Press, Cambridge, Massachusetts.

Rice, D. W., and Scheffer, V. B. (1968). A History of Marine Mammals of the World. *U.S. Fish. Wildl. Serv., Spec. Sci. Rep.-Fish.* **579,** i–iv, 1–16.

Riley, C. L. (1968). The Makah Indians: A study of political and economic organization. *Ethnohistory* **15,** 57–95.

Sapir, E. (1920). The life of a Nootka Indian. *Queens Q.* **27.**

Scammon, C. M. (1968). "The Marine Mammals of the Northwest Coast Of North America." Dover, New York.

Schalk, R. F. (1977). Structure of an anadromous fish resource. *In* "For Theory Building in Archaeology" (L. R. Binford, ed.), pp. 207–249. Academic Press, New York.

Schalk, R. F. (1978). Foragers of the northwest coast of North America: The ecology of aboriginal land use systems. Ph.D. Thesis, University of New Mexico, Albuquerque (unpublished).

Scheffer, V. B., and Slipp, J. W. (1948). The whale and dolphin of Washington state with a key to the cetacea of the west coast of North America. *Am. Midl. Nat.* **39,** 257–337.

Slipjer, E. J. (1976). "Whales and Dolphins." Univ. of Michigan Press, Ann Arbor.

Spencer, R. F. (1976). "The North Alaskan Eskimo, A Study in Ecology and Society." Dover, New York.

Swadesh, M. (1948). Motivations in Nootka warfare. *Southwest. J. Anthropol.* **4**(1), 76–93.

Swan, J. G. (1870). The Indians of Cape Flattery at the entrance to the Strait of Fuca, Washington Territory. *Smithson. Contrib. Knowl.* **16,** No. 220.

True, F. W. (1904). The Whalebone whales of the western North Atlantic compared with those occurring in European waters with some observations on the species of the North Pacific. *Smithson. Contrib. Knowl.* **33.**

Turner, L. M. (1886). "Contributions to the Natural History of Alaska, Results of Investigations Made Chiefly in the Yukon District and the Aleutian Islands, Conducted Under the Auspices of the Signal Service, U.S.A. extending from May 1874 to August 1881." Washington.

Veniaminov, I. (1840). "Notes on the Unalaska District," 2 vols. St. Petersburg. (MS in Bancroft Library, University of California, Berkeley.)

von Kittlitz, F. H. (1858). "Denwürdigkeiten einer Reise nach dem russischen Amerika," 2 vols. Gotha.

von Kotzbue, O. (1821). "Entdeckungs-Reise in die Sud-See und nach der Berings-Strasse . . . Unternommen in den Jahren 1815, 1816, 1817 und 1818," Vols. 1–3. Weimar.

von Langsdorff, G. H. (1813). "Voyages and Travels in Various Parts of the World, during the years 1803 . . . 1807." H. Colburn, London.

von Wrangell, F. (1839). Statistsche und Ethnografiske Nachrichten über die Russischen Besitzaungen an der Nordwestküste von Amerika. *Betr. Kent. Russ. Reiches* **1.**

Waterman, T. T. (1920). The whaling equipment of the Makah Indians. *Univ. Wash. Publ. Anthropol.* **1**(1), 1–67.

5

Gray Whales and the Aborigines of the Pacific Northwest: The History of Aboriginal Whaling

Igor I. Krupnik

Introduction

The appearance of whaling by the aborigines of the northwestern Pacific Ocean can be traced with certainty along the number of archaeological sites dating to the first centuries AD (Rudenko, 1947; Arutyunov and Sergeev, 1969, 1975; Vasil'yevskii, 1971; Dikov, 1971, 1974, 1979). Its significance and importance in the aboriginal economy extending into the following centuries can be substantiated by the numerous discoveries of hunting and flensing implements, petroglyphs depicting whaling scenes, and abundant whale bones, baleen, and handicrafts from the ancient sites which suggest the wide

ISBN 0-12-389180-9

use of the whale bones for construction of dwellings, ritual, and memorial household objects and graves. Additional authentic information comes from local oral and folklore traditions, and narratives dating from the eighteenth to twentieth centuries.

However, to the best of our knowledge, until recent times the primary objective of the aboriginal hunting along the northeastern shore of Asia was unquestioned and rarely even considered. The predominant opinion was that whaling was directed primarily toward the bowhead whale (*Balaena mysticetus*). The problem of differentiation of areas of aboriginal whaling according to the species taken surfaced only recently following the discovery on the east coast of Chukotka Peninsula, in the islands of the Senyavin Strait and in Mechigmen Bay, of the ancient settlements and memorials built from the bones of the gray (*Eschrichtius robustus*) and bowhead whales (Arutyunov *et al.*, 1979, 1982; Bogoslovskaya *et al.*, 1979).

Materials collected in the last few years obviously show that the role of whaling varied among the different groups of the aborigines of the Northwestern Pacific. The ancient whaling stations or shore camps indicate aboriginal cultures specializing in the hunting of different species of cetacean within their particular cultural or ecological traditions. To explain these new facts, it was necessary to summarize the existing ethnohistorical data on the aboriginal whaling in this region. This was done in part in our previous publications (Krupnik, 1979, 1980; Arutyunov *et al.*, 1982; Bogoslovskaya *et al.*, 1982; Krupnik *et al.*, 1983).

For this chapter, data was first collected on aboriginal whaling in the entire northwest part of the Pacific, or more precisely in northeastern Asia, inasmuch as aboriginal whaling continued beyond the Bering Strait and into the Arctic shores of the Chukotka Peninsula. In the entire adjacent area the distributions of the gray, bowhead, and other large whales overlap or were overlapped in the recent past. Therefore, the preferred catch of the different whale species by the aboriginal hunters was strictly determined by local ecological peculiarities or specific cultural traditions.

General Characteristics of Research

Reviewing the area populated by the gray whale along the northwest coast of Asia, we came across five groups of aborigines: Asiatic (Siberian) Eskimos, Chukchi, Koryaks, Kereks, and Itle'mens (Kamchadals). This vast region of aboriginal whaling may be divided into northern and southern subregions or areas. First was the territory inhabited by Eskimos and Chukchi along the shores of Chukotka Peninsula, from the Kresta Gulf to the Bering Strait and further westward along the shores of the Arctic Ocean, where the gray whale is currently very common up to the western Chukchi Séa (Tomilin, 1957; Votrogov and Bogoslovskaya, 1980). The existence of ancient aboriginal whaling can be traced even further west, at least to the Chaun Inlet or probably to the mouth of the Kolyma River (Okladnikov and Beregovaya, 1971).

The second area includes the coasts populated by Kereks, Koryaks, and Itle'mens, and stretches from the southern shores of Anadyr Gulf to the Tauyskaya Inlet of the Okhotsk Sea, encompassing the western part of the Bering Sea and northeastern part of the Okhotsk Sea but apparently excluding Kamchatka Peninsula and its adjacent is-

Fig. 1. *Cultural variations of aboriginal gray whaling in northeastern Asia. I, Northern/Chukchi; II, Mechigmen; III, Kerek; IV, Olyutora; V, Northern/Okhotsk; ?, whaling probably existed; however, data are absent or rather obscure.* ⸬ *Aboriginal whaling was oriented mainly to bowhead whales; there is no evidence for gray whale hunting prior to the twentieth century.* ☷ *Aboriginal whaling; utilization of strandings only.*

lands. A distinct border between the two areas probably does not exist, although they are separated by a "gap" of several hundred kilometers along the west coast of Anadyr´ Gulf (Fig. 1).

The necessity of separating the history of the aboriginal whaling of these two areas is dictated not only by the differences in the ethnic and cultural traditions, but also by the different amounts of information available. In the northern area aboriginal whaling was maintained until the 1960s; ship whaling for the gray whales on behalf of the native population is still carried out at the present time. A vast amount of data on aboriginal whaling including statistics of the catch by native settlements is thus available. A special investigation of the local types of whaling according to the species of whales taken was carried out by the author and his colleagues through inquiries of the local people and from the study of archeological sites between 1975 and 1981 (Krupnik, 1979, 1980; Arutyunov *et al.*, 1982; Krupnik *et al.*, 1983). This information allows us to follow chronologically the fundamental changes in the history of aboriginal whaling off Chukotka Peninsula.

For the southern area, where aboriginal whaling for large whales probably ceased in the second half of the nineteenth century, similar data are practically unavailable. Therefore, the information presented here is a preliminary attempt to summarize the literature on the role of the gray whale hunt in the aboriginal economy of this area. In contrast to the northern area, I chose the regional presentation of data, collecting facts from the main parts of the seacoast with regard to the ethnic composition and cultural traditions of the aborigines.

Northern Region

Until recently, the existing opinion in the literature was that ancient aboriginal whaling around the Chukotka Peninsula (similar to that in Alaska) was oriented exclusively towards bowhead whales. Fully recognizing the lack of information from ancient times and the widespread extent of the past hunt of the bowhead whale off Chukotka, we can present new reliable data about fairly active aboriginal whaling of the gray whales in several regions of the peninsula (Krupnik, 1979; Krupnik *et al.*, 1983). The aboriginal whaling history off Chukotka may be conveniently divided into four periods: (1) the aboriginal (from ancient times to the last third of the nineteenth century); (2) the traditional (from the end of the nineteenth century up to 1930); (3) the transitional (from the end of 1930 to the beginning of the 1960s); and (4) the modern or "contemporary" (from the beginning of the 1960s to the present). Each period differs not only by the methods adapted, but also through the correlation of the different species of whales hunted and their importance within the structure of the aboriginal economy (Krupnik, 1982; Krupnik *et al.*, 1983).

ABORIGINAL PERIOD

Numerous indications about the existence of whaling among aboriginals of Chukotka Peninsula (ancestors of the Asiatic Eskimos) are known from the first centuries AD in the archaeological sites of the Old Bering Sea culture (OBS) (Rudenko, 1947; Arutyunov and Sergeev, 1969, 1975; Dikov, 1977). The earliest evidence suggesting the possible catch of gray whales (together with the catch of bowhead whales) was received during the excavation of the Ekven cemetery on the northeastern tip of the Peninsula. In the published data from this excavation there are drawings depicting bones of the small whales: scapula 35–45 cm in length, mandibles up to 1.5 m, and small vertebrae and intervertebral discs (Arutyunov and Sergeev, 1975). Although special osteological analyses of these bones have not been conducted, they clearly suggested belonging to juvenile or calf gray whales (S. A. Arutyunov, personal communication).

In the Ekven old Bering Sea graves (first centuries AD), the presence of the bones of these "small" whales is rare and can be considered as a product of random (episodical) whaling, or even as occasional findings on the seashore (S. A. Arutyunov, personal communication). Much later, in Punuk graves dating from the eighth to fourteenth centuries AD, these small bones were practically absent, but the bones of the bowheads were used exclusively and in enormous quantities as a main building material. Thus, even more remarkable is the presence of the large numbers of the remains of the small gray whales in the graves of the Birnirk culture (of the fifth and seventh centuries AD), where bones of large bowheads were absent. In the Birnirk graves, some specific toggle-head harpoons were found which were larger than those used for the walrus (*Odobenus rosmarus*) hunt, yet smaller than the harpoons used for the hunt of the bowhead whales. Methods of whaling by the ancient aborigines from Chukotka Peninsula are well depicted in whaling scenes in Pegtymel petroglyphs (Dikov, 1971). Nine whale scapulae ranging in size from 40 to 45 cm, which were found in one Birnirk tomb, only indicated that such a hunt was fairly intensive (Arutyunov and Sergeev, 1975).

A new area of ancient aboriginal whaling of the gray whale was discovered recently in the central part of the eastern coast of the Chukotka Peninsula. In 1981, during the shore inventory from Arakamchechen Island to the Bay of Lawrence, six large ancient settlements were discovered and investigated. The former inhabitants of these settlements apparently specialized in the hunt of the gray whales, predominantly nursing calves and yearlings (M. A. Chlenov and I. I. Krupnik, field notes). Only in the Masik settlement, located at the entry of Mechigmen Inlet, were we able to count more than 1000 skulls of the small gray whales with a skull width from 70 to 90 cm (Arutyunov *et al.*, 1982). The bones of more than 20 bowhead whales and more than 10 other baleen whales of undetermined species were also discovered (Krupnik *et al.*, 1983).

The skulls, and to a lesser extent other bones of gray whales (mandibles, vertebrae, ribs, and scapula), were used as main building materials in the construction of underground dwellings and meat caches. A considerable number of gray whales skulls are also preserved in the other ancient settlements of the Mechigmen Bay: Kukun (about 130), Nykhsirak (about 120), and Raupelyan (about 70). According to the local inhabitants, a number of the same skulls were formerly present at the ancient sites of Lyuren (Lorino) and Ilyan, which no longer exist (M. A. Chlenov and I. I. Krupnik, field notes). Some 50 skulls were also found in the ancient settlements of Kiginin on Arakamchechen Island (Bogoslovskaya *et al.*, 1979; Arutyunov *et al.*, 1982); more than 10 others were discovered on the other ancient sites on the southeastern shore of Arakamchechen Island (Fig. 2). Practically all of the skulls seen have a width of 70 to 80 cm and thus

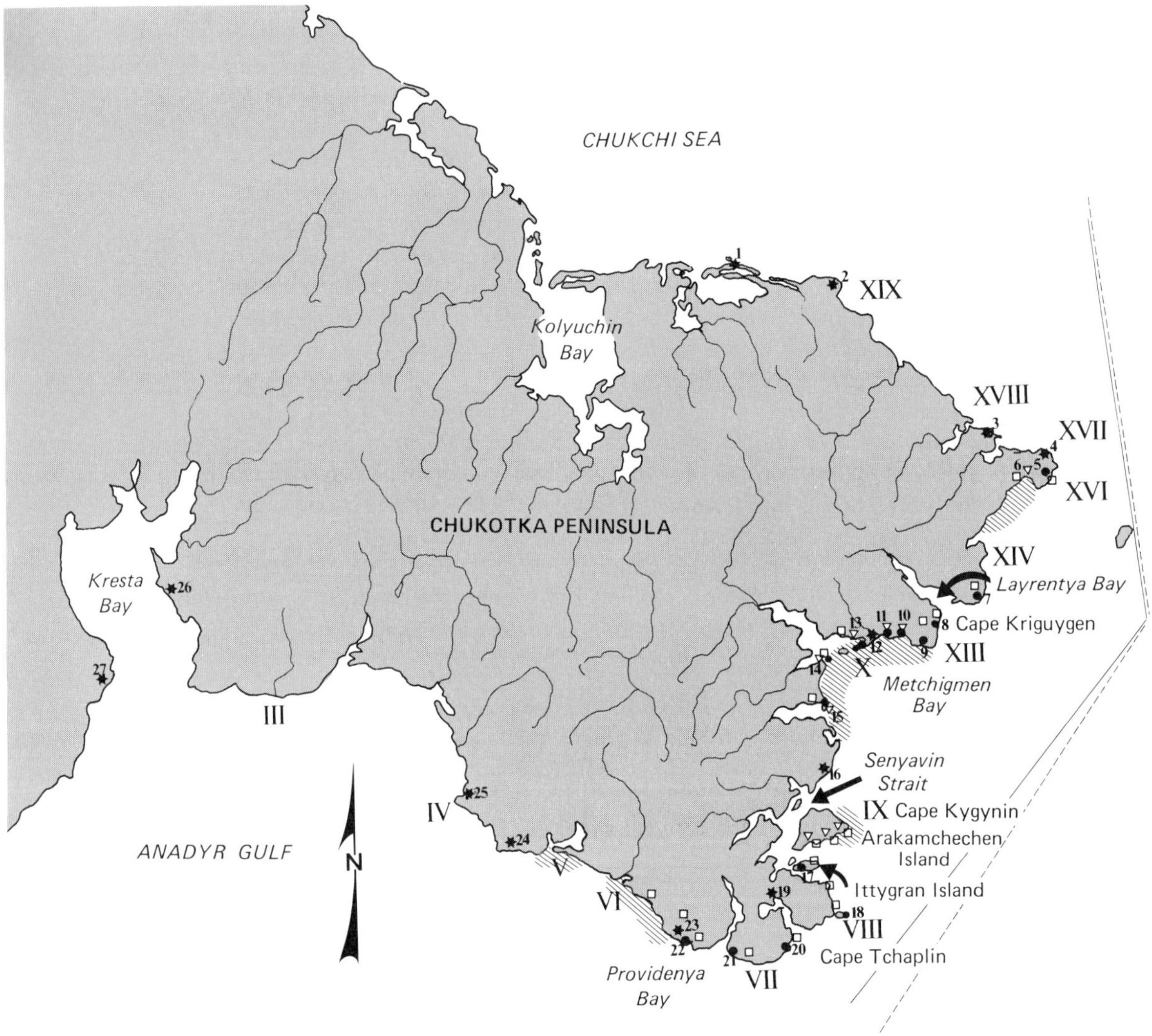

Fig. 2. *Areas of gray whale distribution; ancient and modern gray whaling off Chukotka.* ★, *Existing native communities;* ●, *native villages mentioned in the text which existed until 1940–1970;* △, *ancient constructions of gray whale bones and skulls;* □, *ancient constructions of bowhead skulls and mandibles; hatched areas, observed areas of gray whale calves coming very close to shore; IX, areas of modern processing and utilization of gray whales. The roman numerals are given according to the map in Bogoslovskaya and Votrogov, 1981, pp. 436–437. (1) Neshkan, (2) Enurmino, (3) Inchoun, (4) Uelen, (5) Naukan, (6) Ekven cemetery, (7) Nunyamo, (8) Yandogai, (9) Akkanı, (10) Kukoon, (11) Ilyan, (12) Lorino, (13) Raupelyan, (14) Masik (Metchigmen site), (15) Nikhsirak, (16) Yanrakkynnot, (17) Napakutak, (18) Ungazik, (19) Novoe Tchaplino, (20) Kiwak, (21) Avan, (22) Imtuk, (23) Sireniki, (24) Nunligran, (25) Enmelyn, (26) Konergino, (27) Uelkal.*

belonged to calves 6–8 months old. Considering that the majority of the bone remnants are presently buried underground, it can be assumed that in all of the settlements of this area there are as many as 2000 skulls of the small gray whales. Noticeably in all the settlements studied the skulls of the adult gray whales are absent, but the skulls, jaws, and other bones of the bowhead whales were present in small numbers.

Such peculiar uniformity in the type of the bone remnants is a possible indication of a highly specialized whaling industry or peculiar culture, which was oriented toward the hunt of the yearling and nursing calves of the gray whales and to a lesser degree to some other marine mammals such as small seals, walruses, and bowhead whales. The northern limit of this culture area was Kukun site, to the west from Cape Kriguygun, and the southern limit, the small stations on the southeastern coast of Arakamchechen Island (M. A. Chlenov and I. I. Krupnik, field notes). According to some data, the ancient settlements of Mechigmen Bay and Arakamchechen Island apparently existed from the middle of the second millennia AD up to the eighteenth to nineteenth centuries (Chlenov, 1982), and on some of these sites the cultural continuity was maintained up until the middle of the twentieth century. In all the other areas of Chukotka Peninsula, the main object of aboriginal whaling during the entire precontact period was evidently the bowhead whale. In any case, in all the numerous ancient sites investigated along the southeast coast of Chukotka Peninsula and near the Bay of Lawrence, we did not come across any kind of ancient structures built of gray whale bones or any remnants of these animals.

TRADITIONAL PERIOD

The beginning of the traditional period falls into the last third of the nineteenth century, which is considered a period of severe population reduction of all principal cetaceans off Chukotka Peninsula as the result of Yankee commercial whaling and the vicious depletion of the California gray whale stock in the lagoons of Baja California. For the aborigines of the Chukotka Peninsula, this situation was realized as a severe reduction of aboriginal whaling, massive disasters and starvation, particularly between the 1870s and the 1890s. In the beginning of the twentieth century, when the situation was somewhat stabilized, the number of whales taken in the aboriginal settlements was probably just half of that in precontact period (Krupnik, 1980). In many of the coastal regions, native whaling almost ceased or completely disappeared (Gondatti, 1898; Bogoras, 1907).

The native whalers at that time had acquired the American (Yankee) whaling gear (darting and shoulder guns, wooden whale boats, etc.); however, these progressive changes did not at first lead to any significant modifications in the distribution of the aboriginal whaling. The new equipment was utilized almost exclusively for the bowhead whale hunt, which still maintained its significance and cultural domination in the northern and southeastern parts of the Chukotka Peninsula (Bogoslovskaya *et al.*, 1982).

The traditional methods of whaling with the iron toggle-head harpoon and killing lances were preserved, at the beginning of the twentieth century, in only one region of the Chukotka Peninsula: the coast of Mechigmen Bay. Four aboriginal Chukchi settlements existed here: Mechigmen (Masik), Lorino, Raupelyan, and Ilyan, with a total population of some 170–200 people which continued as before with the hunt of gray whale calves and yearlings. Hunting practices were described earlier by N. F. Kallinnikov (1912) and N. B. Schnakenburg (1933); additional information was collected by the author from the local Chukchi hunters of the elder generation (I. I. Krupnik, field notes; Krupnik *et al.*, 1983).

According to the estimate of the local inhabitants, during a "good" year in the 1920s the Mechigmen and Raupelyan hunters would take, in combined effort, from 2 to 3 gray whales; the same number would be taken respectively in Lorino and Ilyan, so the total catch would be 6–10 whales each year. According to the elders, the object of the hunt would always be the calves immediately after they were weaned. The hunters never attempted to take the adult whales as they were considered too dangerous and too heavy for further transportation and butchering. The local statistics from 1925 (although incomplete) indicated that 5 gray whales were taken from this area, and a total of 28 animals were taken within 10 years (1923–1932) in the two native settlements of Lorino and Ilyan.

It is well known that gray whaling predominated in Mechigmen Bay and has remained unchanged on the Chukotka Peninsula during the whole of the "traditional" period. In the remaining parts of the coast in all of the large settlements whether Eskimo or Chukchi, aboriginal whaling was still oriented to the catch of the bowhead whale (Krupnik, 1980). Although it is true that from 1910 to the 1920s the first evidence of episodical gray whaling appeared in the native settlements of the northeastern villages of Uelen and Naukan and southeasten Ungazik and Avan around 1922, and in Napakutak in the middle 1920s. The hunters in the tips of the Peninsula, Uelen and Naukan, took primarily medium to small whales using rifles and lances. In the southern areas, the large adult whales were killed with darting and shoulder guns (I. I. Krupnik, field notes). According to some information, the catch of the gray whales in Uelen can be traced even earlier, at least from the second half of the nineteenth century using aboriginal lances and harpoons with floats as described by Scammon (1874) (Gondatti, 1898; V. V. Leont´yev, personal communication).

TRANSITIONAL PERIOD

The transitional period is characterized by the gradual expansion of the area and size of the gray whale catch in the native settlements of Chukotka Peninsula. During all this time, the reduction of aboriginal bowhead whaling continued because of the absence of new equipment and the limitations of the outdated Yankee whaling gear available. In all the coastal settlements the catch of the last bowhead whale is still well remembered; in Avan in 1933, Kivak in 1936, Enurmino in 1938, Uelen about 1937 or 1938, Chaplina in 1941, Naukan in 1946 or 1952, and Sireniki in 1960 (I. I. Krupnik, field notes). From this moment on, as a rule, local hunters were forced to change to the hunt of the gray whales.

The hunt of the gray whales from the 1930s to the 1940s was conducted at first from wooden motorboats and traditional skin or hide boats with the aid of rifles and iron harpoons with floats; from the 1950s on, the hunt continued to expand on a much larger

Table I
Catch of Gray Whales in Aboriginal Settlements of the Chukotka Peninsula (1945–1960)[a,b]

Year	Lorino	Uelen	Sireniki	Akkani	Inchoun	Yarrakkynnot	Yandogai	Enmelyn	Nunligran	Chaplino	Nunyamo	Naukan[c]
1945	8	—	—	1	—	—	1	—	—	—	—	1
1946	—	—	—	—	—	—	—	—	—	—	—	—
1947	7	—		2	—	—	—	—	—		—	0
1948	—	—		5	—	—	—	—	—		—	0
1949	—	—	0	—	—	—	—	—	—	0	—	3
1950	—	—		—	—	—	—	—	—		—	2
1951	—	—		—	—	—	—	—	—		—	—
1952	—	—	—	—	—	—	—	—	—	—	—	—
1953	26	—	—	—	—	—	—	—	—	1	—	7
1954[d]	25	1	2	10	—	—	5	—	—	2	5	5
1955	—	—	—	—	—	—	—	—	—	3	—	—
1956	—	—	0	—	—	—	—	—	—	4	—	—
1957	43	4	5	21	6	6	—	—	0	1	6	5
1958	68	10	0	28	11	14	—	4	—	1	5	—
1959	60	19	2	29	1	26	—	2	2	3	32	—
1960	71	14	4	22	10	25	—	1	2	6	16	—[e]
			2							8		

[a]A more complete version of the table with sources was presented in Krupnik *et al.* (1983).

[b]Data by years; the catch reported by local sources.

[c]Sireniki, Chaplino, and Naukan are Eskimo villages; Lorino, Akkani, Uelen, Inchoun, Yandogai, Enmelyn, Nunyamo, Nunligran, and Yanrakkynnot are predominantly Chukchi.

[d]The data for 1954, except Chaplino, were planned; the actual catch of the gray whales for the entire year of 1954 on Chukotka Peninsula was just 36 whales.

[e]In addition, in Sireniki, two bowheads were killed in 1945, three in 1947, and one each year in 1949, 1950, 1953, 1957, and 1960. In Naukan, in 1946 and 1952 one bowhead was taken each year.

scale and included the utilization of the small schooners and cutter boats equipped with large caliber guns and cannons. Both of these methods of hunting are described by Bezumov (1960), Leont'yev (1973), Ivashin and Mineev (1978), Smolyak (1957), etc. and are well preserved in the memory of the older generations of local inhabitants. The hunt of the gray whales from the whale boats included some of the elements of the traditional hunt. The toggle-head harpoons, and in some instances the iron killing lances, were used to take young and immature gray whales. In the 1940s and the early 1950s, the average weight of the whales was not over 5–8 tons. The take of large whales of 15–20 tons was a rare exception. The struck-and-lost ratio of whales during this period was very high, and some estimates suggest that of the whales killed up to 30% sank and 30% of those struck escaped (Ivashin and Mineev, 1960). For this reason, mainly, whaling from whale boats and medium-sized vessels was curtailed in the 1960s.

By the end of the 1950s, gray whaling was conducted in 15 aboriginal settlements of the Chukotka Peninsula along the coast from the Kresta Gulf up to the Koluchin Inlet (the most extreme points respectively documented are settlements of Konergino and Vankarem). The total figures of whales taken since 1948 were published (Ivashin and Mineev, 1978); however, more detailed information on the of the catch by native settlements between 1945 and 1960 is now available from local sources and native informants of older generations (Krupnik *et al.*, 1983) (Table I).

By the end of the 1930s, the Chukotka aboriginals were apparently taking not more than 8–12 gray whales in a "good" year, mainly in the vicinity of the Mechigmen Bay (the villages of Lorino, Mechigmen, Raupelyan, and Akkani) and around Cape Dezhnev (the villages of Uelen and Naukan). In 1945 the catch was approximately 14 whales, and by the middle of the 1950s it increased to 40–60 whales per year (Krupnik *et al.*, 1983). The main centers of the whaling remained, as before, the area of Mechigmen Bay, the waters around Cape Dezhnev, and along the southern coasts of the Peninsula from Cape Chaplina to Providenya Bay and further westward. Transition to massive whaling from whaling cutters and motor schooners boosted the number of whales taken to 100–150 per year by the end of the 1950s, and also extended the regions of active whaling.

MODERN (CONTEMPORARY) PERIOD

Whaling for all large cetaceans by local aboriginal hunters from small whale boats ceased in the 1960s and was replaced by a specialized vessel industry for only enough gray whales to meet the needs of the local aboriginal people. The modern ship whaling is conducted now along the entire coast of Chukotka Peninsula (Votrogov and Bogoslovskaya, 1980) according to International Whaling Commission quota, which fluctuated in the last 20 years from 150 to 200 whales per year (Ivashin and Mineev, 1978). Whales are taken now for 10 native villages where they are towed ashore and butchered by native hunters. The main centers of butchering and consumption of the gray whales are the villages of Lorino, Uelen, Novoye Chaplina, Sireniki, Yanrakkynot, and Uel´kal (Krupnik *et al.*, 1983).

Whaling for the gray whales by aboriginals continued longest in the settlements of Uelen and Lorino up to the beginning of the 1970s (V. V. Leont´yev, personal communication). By the end of the 1960s, from the total quota of 120–140 whales some 20 animals were still taken by local hunters. In all, within the last 20 years almost 3500 gray whales were taken off the Chukotka Peninsula (Ivashin and Mineev, 1978; Zimushko and Ivashin, 1980) and about 30 whales of other species, including 7 bowheads which were delivered to the native communities and distributed for the direct consumption of the aboriginal population. Three additional bowheads were killed in the 1960s and 1970s by native hunters from their whale boats. The problem of nutritional value and importance of the gray whale meat in the aboriginal diet has been reported (Krupnik *et al.*, 1983).

Southern Region

WESTERN SHORE OF ANADYR´ GULF

This coastal region was historically apparently an area of contacts and a tempory habitat for marginal groups of Eskimos, Kereks, and maritime Chukchis. Although this area has not been sufficiently studied archeologically, the existing evidence indicates the absence of traces of aboriginal whaling. In the descriptions of the ancient coastal settlements of Uel´kal´ (Y. M. Rodnyy, personal communication) it is specially noted that a minimum of whale bones were discovered, compared with the usual amount of whale bones in ancient settlements of the Chukotka Peninsula. The whale remnants are not even mentioned in the descriptions of the ancient sites on Cape Nizkyi, Zemlya Geka, and Sedmoi Prichal (Leont´yev, 1976; Dikov, 1977; Dikov, 1979) where the remains of the old semisubterranean dwellings and meat cellars were constructed of logs. We assume that due to the former absence or scarcity of whales in the coastal waters of this region it was not suitable for aboriginal whaling. Nonetheless, when in 1918–1920 Eskimos from Cape Chaplin broke ground for their new settlement on Uel´kal´, they made an attempt to conduct regular whaling but never encountered any whales close to shore between 1920 and the 1940s. Only one standed whale was found about 1935 (Krupnik, 1980).

SOUTHERN SHORE OF ANADYR GULF AND KORYAK HIGHLAND COAST

The bones of whales appeared in large quantities only in the ruins of the settlements of the southern coast of the Anadyr´ Gulf, from Lakhta site on Cape Barykov to Cape Navarin and further to the southwest along the shore of Koryak highland. Large congregations of the whale bones were found throughout this entire territory (the ancient settlements in Ugol´nyy Bay, Gabriel Inlet, the Meynypyl´gino, Khatyrka, Opukha lagoons, and a number of ancient sacrificial sites) up to Cape Olyutora (Leont´yev, 1976, 1983; Orekhov, 1979). Although biological identification of the remains in these excavations was not specifically conducted, V. V. Leont´yev (personal communication) determined that the majority of the bones belonged to the small gray whales, and a small number to the bowhead whale. The abundance of young gray whales in the lagoons and bays of these areas is noted by Tomilin (1957) and has been observed recently (V. V. Leont´yev, personal communication; Votrogov and Bogoslovskaya, 1980). All these facts, in combination with the ancient whaling gear found and oral information from the local inhabitants (the Kereks), allow the conclusion that there existed in the past an aboriginal whaling industry in this region. The Kereks substantiate that their ancestors formerly hunted the whales with the use of a special large arrow or lances possibly tipped with poison (Leont´yev, 1976, 1983; Orekhov, 1979). The young whales could have also been driven by the whalers into shallow waters and then killed by lances from skin boats. The whaling industry of the Kereks totally ceased in the second half of the nineteenth century due to the extermination of whales by Yankee whalers and the impoverishment of the aboriginal population (Leont´yev, 1983).

SHORES OF OLYUTORA AND KORF BAY

The gray whale hunt by aborigines of this part of the coast (Olyutora-Koryak or Olyutortsy) by means of nets made of walrus-skin lines with the "width of the arm of a full-grown man" was once described in detail in the middle of the eighteenth century by S. Krasheninnikov (1949). A. G. Tomilin and I. S. Volovin (personal communication) believe that the Olyutora used these nets primarily to take mainly the young gray whales, although direct evidence is not available. Sources of the eighteenth and nineteenth centuries noted the great importance of whaling and the significant role of whale products in the economy of the Olyutora people (whale meat, blubber, skin, and baleen), and they described whaling festivals at the beginning of the hunt and after the catch of the whales.

According to the memories of the Olyutora elders in 1954, the nets for the whales were formerly placed at Cape Olyutora and Cape Goven where these animals would come close to the shore (Vdovin, 1973). However, in the nineteenth century the whale hunt took on more of an incidental character while hunting seals by net became more important. When a whale was entangled in the net, the natives would approach it in their skin boats and attempt to kill it by hitting it in the abdomen with stone-tipped lances. This style of fishery ceased completely at the end of the nineteenth century, following the decimation of whales in the coastal waters by Yankee whalers (Vdovin, 1973).

There are only a few fragmented accounts of the whaling industry of the Apuka–Koryaks or Apukintsy, which existed in the eighteenth-century Russian literature. These people inhabited the coast to the north from Olyutortsy (Vdovin, 1973). The lower jaws of the whales were frequently placed at the ancient sacrificial sites as poles, such as, for example, near the mouth of the Apuka River, where until the present the large whale mandible (probably of the bowhead, according to the photographs) was well preserved (Vdovin, 1971).

EASTERN SHORE OF KAMCHATKA PENINSULA

The natives of the eastern shore of the Kamchatke Peninsula, Karaga-Koryaks and Itel´mens (Kamchadals), did not occupy themselves with whaling. This is asserted by I. S. Vdovin (1973) for Karaga-Koryaks and by S. P. Krasheninnikov (1949) for Itel´mens. Nevertheless, the meat and especially blubber of stranded whales were widely consumed by these people (Krasheninnikov, 1949; Starkova, 1976), and the whale bones and baleen were used in boat and dog-sled construction (Krasheninnkov, 1949; Vdovin,

1973). However, the biological species of whales utilized are never specifically described, but with great degree of probability it can be assumed that the gray whales were among them (I. S. Vdovin, personal communication). Their bones and carcasses that washed ashore were reported along the entire eastern coast of Kamchatka Peninsula up to Kronotskiy and Avacha Bays (Ditmar, 1901; Tomilin, 1957) and possibly even to the Cape Lopatka (Krasheninnikov, 1949). Tomilin (1957) reported from Cape Lopatka that whaling was conducted by "Blizhnyi" (neighbor) Kuriles, a mixed Ainu–Itelmen population, with the aid of poison arrows. Neither the fact of whaling by the Lopatka Kuriles in the early eighteenth century, nor the species of whales hunted, was further described in later sources.

WESTERN SHORE OF KAMCHATKA PENINSULA

Although the close approaches and strandings of the gray whales were described at many points along the western coast of the Kamchatka Peninsula (Ditmar, 1901; Tomilin, 1957), the local inhabitants of its central and southern part (the Itel´mens) and old Russian settlers were not involved in whaling. It was documented in the eighteenth and nineteenth centuries that only the standard whales were utilized as food for dogs and consumed by the people (Krasheninnikov, 1949; Ditmar, 1901).

However, at least during the eighteenth century, the aborigines of the northwestern shore of the peninsula, maritime Palana Koryaks (Palana) took whales as well as utilizing stranded carcasses (Vdovin, 1973). A description of the Palana whale festival in the late eighteenth century is preserved, but it is impossible to determine the species of the whales hunted from the information available. At any rate, the aboriginal whale hunting in this part of the coast ceased in the first half of the nineteenth century (i.e., prior to the appearance of the Yankee whaling industry), primarily as a result of diminishing of the aboriginal population as a result of starvation and epidemic diseases.

THE SHORE OF PENZHINA BAY

There is a great deal of information about the maritime Koryaks (Kamentsy, Parentsy, and Itkantsy). According to Russian sources, they conducted active whale hunting during the entire eighteenth and the first half of the nineteenth century and undoubtedly from earlier times. It is known that in total the aboriginal hunters in Penzhina Bay took 22 whales in 1778 and 25 in 1779 (Vdovin, 1973).

The aboriginal methods of whaling prior to and during the eighteenth century were not recorded in the literature. Ditmar (1855) reported that in the mid-nineteenth century Penzhina Koryaks hunted whales with hand harpoons and firearms. In the early twentieth century Koryak whaling in Penzhina Bay was described in detail by W. Jochelson (1905); the whaling itself however ceased one to two generations earlier because of the depletion of the whale stocks by Yankee commercial Whalers. Koryaks utilized only the carcasses of stranded whales or those brought ashore by the Yankee whalers; however, by 1925 a new description of whale hunting by the Penzhira Koryaks was published (Arsen´yev, 1925). It seems that this author described the reviving of the aboriginal whaling industry from large multiseat skin boats and the use of toggle-head harpoons. It was reported that aboriginals took one or two whales every whaling season (Arsen´yev, 1925).

Which whale species were formerly hunted by the Penzhina Koryaks is now difficult to determine. Jochelson (1905) specified that Koryaks celebrated the whale festivals when the large black whales with the long baleen (i.e., bowheads ?) were taken, but they never hunted the small whales with the white baleen (probably the gray whales) although they willingly consumed the meat and blubber of these whales when stranded. It is doubtful that all 20–25 whales taken yearly in Penzhina Bay in the late eighteenth century were bowhead whales, particularly when considering the widespread hunting of the gray whales by the neighboring groups of Koryaks.

GIZHIGA BAY AND NORTHWESTERN SHORE OF OKHOTSK SEA

A number of Russian sources of the eighteenth century described the existence of active whaling among the sedentary maritime Koryaks in this part of the coast. The Koryaks hunted three species of whales: large ones (judging by the local names it was probably the bowhead or right whale), "medium," and "small". In the latest sources of the eighteenth and nineteenth centuries it mentions that Koryaks during their hunt selected small- and medium-sized whales not more than 10–12 m in length (Vdovin, 1973). Therefore, we can make an assumption that some portion of the catch consisted of the gray whale, which were abundant in these waters until the middle of the nineteenth century. This is substantiated by the presence of the special nets used for whaling among the Gizhiga Koryaks in the eighteenth century (Vdovin, 1973). It is reported that in 1777 Gizhiga Koryaks took three whales, in 1778 five whales (including three with the assistance of Russian old-settlers), in 1779 three whales (one with the help of Russian old-settlers) and in 1780 eight whales (two with the help of Russians). The aboriginal whaling industry, however, declined during the first part of the nineteenth century and completely ceased with the appearance of Yankee whalers.

Large quantities of whale bones were discovered by archeologists on the ancient Koryak sites throughout the entire territory from Gizhiga Bay up to Tauyi Bay, as well as on neighboring islands. In one ancient settlement at Srednyy Bay (Kony Peninsula) the remains of more than 200 whales were found (Vasil´yevskii, 1971). Although the biological identification of the bones was not conducted, R. S. Vasil´yevskiy (personal communication) reported that the majority of the whale bones on the ancient Koryak sites on the Okhotsk coast belonged to the medium- and even small-sized whales with mandibles of not more than 1.2–1.5 m, skull width not more than 1 m, and vertebrae of about 35 cm in diameter. Judging from these measurements and oral descriptions of the bone structure, the majority of these bones probably belonged to young gray whales. It is considered that whaling by aboriginals of the Okhotsk Sea probably could have begun in the first millennium BP, reaching its peak by the thirteenth or fourteenth centuries A.D. and being preserved up to the appearance of the Russians in the seventeenth and eighteenth centuries (Vasil´yevskii, 1971). According to archeological information available, the western limit of this whaling may be fixed now at Khmitevsky Peninsula to the west of Tauyi Bay (A. I. Lebedintsev, personal communication).

Conclusions

The information presented here paints a relatively complex history of the evolution of the aboriginal whaling industry in Northeast Asia. It was one of the most ancient areas of active aboriginal whaling, spread along the coast of three seas and settled by a number of aboriginal populations. In the eighteenth and first half of the nineteenth centuries the following variants (types) of aboriginal whaling could be identified here:

1. The northern Chukchi (maritime Chukchi, formerly Eskimos). The gray whale hunt occupied a subordinate position compared to the catch of the bowhead whale. The hunt was conducted from large multiseat skin boats with the help of the toggle-head harpoons and killing lances. The harpoon line was not fastened to the boat, but was connected to several skin floats.
2. Mechigmen (Eskimos and later maritime Chukchis). Prefered to hunt the young gray whales and nursing calves, for which specialized harpoon points and highly effective killing lances were developed.
3. Kerek (Kereks). Hunted mainly juveniles and calves in the lagoons and shallow waters where the whales were killed by arrows and darts, possibly with poison.
4. Olyutora (Koryaks–Olyutortsy). Caught the young whales in shallow waters with nets.
5. Northern Okhotsk (maritime Koryaks of northeastern Okhotsk Sea). Hunted gray whale and bowhead from multiseat skin boats with toggle-head harpoons and lances with the harpoon line fastened to the boat.

The reduction of the aboriginal whaling area in northeast Asia began in the early contact period in the mid-nineteenth century as a result of a decline in the aboriginal population and changes in its cultural traditions. An additional blow to aboriginal whaling was brought in the mid-nineteenth century by the appearance of the Yankee commercial whaling fleet in the northwest Pacific. This brought a complete cessation in the catch of the gray whale by aboriginals in the entire southern area and in many settlements along the coast of the Chukotka Peninsula.

The catch of gray whales on the Chukotka Peninsula increased again in the 1930s and 1940s as the catch of the bowhead whale continued to decline and the lack of whaling gear made gray whaling the only active industry. In the 1950s aboriginal gray whaling flourished while the bowhead whaling practically ceased throughout the northern area. The state of ship gray whaling continues here up to the present for the needs of the native populations of Chukotka Peninusla, and thus preserves the importance of the whale in aboriginal consumption. Nonetheless, such complete substitution of all species of whales by the gray whale in nutritional and cultural traditions of the aboriginal populations has its own negative consequences. These might be partially restituted by rehabilitation of the hunt of other whales (and first of all, of the bowhead whale) by aboriginals in those settlements where they are attempting to preserve their aboriginal maritime hunting traditions (Bogoslovskaya *et al.*, 1982; Krupnik *et al.*, 1983).

Summary

The origin of aboriginal whaling in the northwest Pacific can be definitely traced to the early centuries AD; at the end of the precontact[1] period, whale hunting had become the major form of aboriginal economy in most of the coastal areas of the Chukchi, Bering, and Okhotsk Seas. Until recently, however, the exact areas of aboriginal whaling, especially for the gray whale (*Eschrichtius robustus*) and the bowhead whale (*Palaena mysticetus*) were not precisely delineated. It was presumed by most scientists that the bowhead whale was the main object of aboriginal whaling. Compilation of narratives, local documentation, and field data indicated that aboriginal whaling for gray whales existed in northwest Asia in two areas: northern and southern. Several cultural variations were observed in both areas. Among them we determined the following: (1) northern Chukchi, (2) Mechigmen, (3) Kerek, (4) Olyutora, and (5) northern Okhotsk types. The differences between them were based on the role patterns in aboriginal cultural traditions of different species of Cetacea. In the southern area, aboriginal gray whaling almost completely ceased in the second half of the nineteenth century. In the northern area, aboriginal whaling was very active up to the 1960s, and at the present it is replaced by state ship whaling for the needs of the aboriginal population.

Acknowledgments

Field materials for this paper were collected in Chukotka between 1975 and 1981 as part of the scientific program sponsored by the Institute of Ethnography, Academy of Science of the U.S.S.R. Assistance in this work of my colleagues Dr. M. A. Chlenov, Dr. I. S. Bogoslovskaya, and the late Dr. L. M. Votrogov must be stressed with most gratitude. I am particularly grateful to my local informants, the people and hunters of the Chukotka Peninsula, members of native communities of Sireniki, Novoye Chaplino, Lorino, and Providenya. The understanding of the aboriginal whaling industry in other regions was possible through the literature and the help of Drs. S. A. Arutyunov, V. V. Leont´yev, R. S. Vasil´yevskiy, I. S. Vdovin, A. A. Orekhov, A. I. lebedintsev, Yu. M. Rodnyy, and also the kind assistance with field materials by Dr. A. V. Smolyak (1954).

The editors wish to express their thanks to Elena Baskin and Vladimir Gurevich for their assistance with the translation of this chapter.

References

Arsen´yev, V. K. (1925). Gizhiga commercial region. *In* "Economic Life of the Far East," Vol. 6, pp. 17–34 (in Russian).

Arutyunov, S. A., and Sergeev D. A. (1969). "Ancient Cultures of the Asiatic Eskimos. Uelen Cemetery," "Nauka," Moscow.

Arutyunov, S. A., and Sergeev, D. A. (1975). "Problems of Ethnic History of the Bering Sea region. Ekven Cemetery." "Nauka," Moscow (in Russian).

[1]The new era refers to the first contact with the western world, in this case Yankee whalers.

Arutyunov, S. A., Krupnik, I. I., and Chlenov, M. A. (1979). "Whale alley"; ancient Eskimo memorial on Itygran Island. *Sov. Ethnogr.* **4**:12–29 (in Russian).

Arutyunov, S. A., Krupnik, I. I., and Chlenov, M. A. (1982). "Whale Bone Alley. Antiquities of Senyavin Strait Islands." "Nauka," Moscow (in Russian).

Bezumov, R. Ya. (1960). "Hunting for the Sea Mammals." Magadan Publishing House, Magadan (in Russian).

Bogoras, V. (1907). The Chukchee. Material culture. *Mem. Am. Mus. Nat. Hist.* **7** (1), 276.

Bogoslovskaya, L. S., Chlenov, M. A., and Yurtsev, P. A. (1979). Islands of the Senyavin Strait; a unique natural history complex. *Priroda (Moscow)* **9,** 91–97.

Bogoslovskaya, L. S., Votrogov, L. M., and Krupnik, I. I. (1982). Bowhead whale off Chukotka: Migrations and Aboriginal whaling. *Rep. Int. Whal. Commn.* **32,** SC/33/PS 16, 1–45.

Chlenov, M. A. (1982). Expeditions in brief. *Sov. Ethnogr.* **1,** 158–159 (in Russian).

Dikov, N. N. (1971). "Rock-mysteries of the Ancient Chukotka. Petroglyphs of Pegtymel." "Nauka," Moscow (in Russian).

Dikov, N. N. (1974). "Chini Cemetery (To the History of Sea Hunters of the Bering Strait). "Nauka," Novosibirsk (in Russian).

Dikov, N. N. (1977). "Archeological Monuments of Chukotka, Kamchatka and Upper Kolyma River. Asia at the Junction with America in Antiquity." "Nauka," Moscow (in Russian).

Dikov, N. N. (1979). "Ancient Cultures of the Northeastern Asia. Asia at the Junction with America in Antiquity." "Nauka," Moscow (in Russian).

Ditmar, K. (1855). On the Koryaks and their close relatives the Chukchi. *In* "Herald of Russian Geographical Society," Vol. XV, p. 6 (in Russian).

Ditmar, K. (1901). "Journey through the Life in Kamchatka in 1851–1855," Part 1. St. Petersberg (in Russian).

Gondatti, N. L. (1898). Journey from the village Markovo on Anadyr´ River to the Bay of Providenya, Berling Strait. In "Notes of Priamurskiy Section of Russian Geographical Society," Vol. IV(I), pp. 1–42 (in Russian).

Ivashin, M. V., and Mineev, V. I. (1978). On the status of gray whale stocks. *Rybn. Khoz. (Moscow)* **3,** 15–17.

Jochelson, W. (1905). The Koryak. Material culture and social organisation. *Mem. Am. Mus. Nat. Hist.* **6**(2), 383–873.

Kallinikov, N. F. (1912). "Our Far East." St. Petersburg (in Russian).

Krasheninnikov, S. P. (1949). "Description of the land of Kamchatka Peninsula." "Glavsevmorput," Moscow-Leningrad (in Russian).

Krupnik, I. I. (1979). Aboriginal Asiatic Eskimo whaling in the 1920–1930's. *Proc. Pac. Sci. Congr., 14th*, Comm. F. Sect. F III, pp. 145–148.

Krupnik, I. I. (1980). Asiatic Eskimo sea-mammal hunting in the 1920s–1930s. *In* "Marine Mammals," pp. 66–79 (in Russian).

Krupnik, I. I. (1982). Gray whale in the economy and aboriginal whaling of Chukotka Peninsula. *In* "Study, Conservation and Rational Exploitation of the Sea-mammals," pp. 179–180 (in Russian).

Krupnik, I. I., Bogoslovskaya, L. S., and Votrogov, L. M. (1983). Gray whaling off Chukotka Peninsula: Past and present status. *Rep. Int. Whal. Commn.* SC/34/PS 25, 1–25.

Leont´yev, V. V. (1973). "Economy and Culture of the People of Chukotka Peninsula (1958–1970)." "Nauka," Novosibirsk (in Russian).

Leont´yev, V. V. (1976). "In the Land of the Ancient Kereks." Magadan Publishing House, Magadan (in Russian).

Leont´yev, V. V. (1983). "Ethnography and folklore of the Kereks." "Nauka," Moscow (in Russian).

Okladinkov, A. P., and Peregovaya, N. A. (1971). "Ancient Settlements of the Cape Baranov." "Nauka," Novosibirsk (in Russian).

Orekhov, A. A. (1979). Ancient cultures of the northwestern region of the Bering Sea. *In* "New Archeological Monuments of the Far-North East," pp. 64–81. Magadan Publishing House, Magadan (in Russian).

Rudenko, S. I. (1947). "Ancient Culture of the Bering Sea and the Eskimo Problem." "Glavsevmorput," Moscow-Leningrad (in Russian).

Scamman, C. M. 1874. The Marine Mammals of the Northwestern Coast of North America. John Carmany and Co., San Francisco, Calif. Reprinted 1968 by DOVER, New York.

Schnakenburg, N. P. (1933). Whaling on Chukotka Peninsula. *Pac. Star (Khabarovsk)* **259:**3 (in Russian).
Smolyak, A. V. (1957). Materials to the characteristics of the socialist culture and the way of life of the aboriginal population at Chukotka Peninsula. *Sib. Ethnogr. Proc.* **2,** 3–37.
Starkova, N. K. (1976). "Itel´mens´ Traditional culture. From 18th century until 1960's." "Nauka," Moscow (in Russian).
Tomilin, A. G. (1957). "Mammals of the U.S.S.R. and Adjacent Countries," Vol. 9. Publ. House U.S.S.R. Acad. Sci., Moscow (in Russian).
Vasil´yevskii, R. S. (1971). "Origin and Ancient Culture of Koryaks." "Nauka," Novosibirsk (in Russian).
Vdovin, I. S. (1971). Sacrificial sites of Koryaks and their historical and ethnographical value. *Proc. Mus. Anthropol. Ethnogr.* **27,** 275–300 (in Russian).
Vdovin, I. S. (1973). "Essay of Ethnic History of Koryaks." "Nauka," Leningrad (in Russian).
Votrogov, L. M., and Bogoslovskaya, L. S. (1980). Gray whales off the Chukotka Peninsula. *Rep. Int. Whal. Commn.* SC/31/Doc. 55, 435–437.
Zimushko, V. V., and Ivashin, M. V. (1980). Some results of Soviet investigation and whaling of gray whales. *Rep. Int. Whal. Commn.* SC/31/Doc.19, 237–246.

6

Shore Whaling for Gray Whales along the Coast of the Californias

Hazel Sayers

Introduction

Shore whaling for gray whales along the coast of the Californias took place from the mid-1850s to the early 1900s. This industry as practiced off California was defined by Scammon (1874) as follows: Shore whaling was the pursuit of a whale from a boat launched from the shore. When the whale was captured, it was towed back to the shore where it was flensed and its oil and other by-products were processed for market.

There has been very little authoritative information published to describe this era in whaling history. The sources most often quoted include "The Marine Mammals of the Northwestern Coast of North America" by Charles M. Scammon, 1874, "A History of California Shore Whaling" by Edwin C. Starks, 1922, and the Federal Fish Commission Reports by Collins (1892), Goode (1887), Jordan (1887), and Townsend (1886). Historians have invariably quoted from this same limited material.

ISBN 0-12-389180-9

In an effort to supplement this usual published information, the following sources were examined. Newspapers published from 1852 to 1900 were read in the library of the *San Diego Union–Tribune*, the San Diego Public Library, and the Monterey Public Library. All articles found were extracted in their entirety. References were checked at the libraries of Scripps Institution of Oceanography, University of California at San Diego, and the city libraries of San Diego, Monterey, Pacific Grove, and Los Angeles. Photographs and charts from the Title Insurance Historical Collection were viewed at San Diego and Los Angeles. Unpublished manuscripts were read. Papers dealing with related subjects were consulted. Letters explaining the nature of the search were sent to newspapers in selected California coastal cities appealing to descendants and friends of whaling families to share their personal recollections and family records. The information collected from all these sources was put together in an attempt to create an orderly chronology of this era, which has been poorly documented.

Data obtained pertaining to profits, oil yield, and catch numbers are fragmentary. As a result, few conclusions can be drawn concerning the financial significance of this industry and its impact on the populations of the hunted species. However, considerable fascinating anecdotal material is available, lending color and excitement which helps to make up for the minimal statistical data.

Some of the information from early books, magazine articles, and newspapers is quoted fully, as it conveys the feelings of the whalemen and the ambiance of the stations and the hunt. Other material is paraphrased with an effort to retain its essence. This introduction of historical material into a reference book that is primarily intended for scientists and scholars requires that the historian record some disclaimer as to the precise accuracy of the information. Conclusions are necessarily based on weighing the available data and making a judgment on the credibility of the source.

The Whalemen

The California shore whaling industry was manned and operated, with a few exceptions, by Portuguese immigrants from the Azores and Cape Verde (Hittel, 1882). These Portuguese islanders had been active in the Yankee whaling industry for several years. The first one officially recorded was a seaman named Dios, who served aboard the *Sharon* in 1842. The islanders sought berths aboard whalers to escape burdensome taxes and long compulsory military service in their homeland (Dias, 1979).

Lured by the discovery of gold at Sutter's Mill in California, many crewmen signed on in New England simply to secure passage to California and its promise of instant prosperity. Desertion by whalemen when their boats called in at California ports was a well-documented phenomenon that reached its peak in the late 1840s and early 1850s (Dias, 1979). When the gold fever subsided, these erstwhile whalers sought employment at their former occupation. Shore whaling was a seasonal pursuit that allowed its men to have small farms and a family life as well as the excitement of the whale hunt. The Azoreans and Cape Verdians were attracted by this combination life style, as their own culture also embraced farming and fishing.

Indirect evidence indicating participation of the Portuguese in California industry is supported by a recorded increase in their population from 109 in 1850 to 1459 in 1860 (Dias, 1979). The majority of the Azoreans and Cape Verdians who immigrated to the west coast were illiterate, and their movements were not recorded anywhere in detail. They found it expedient to anglicize their surnames, which ships' officers, not familiar with Portuguese, would record phonetically. Thus Bernardo became Bennett, Machado became Marshall, Periera became Perry or Peters, etc. This practice only served to add confusion to the meager records of California's shore whaling history.

Species Harvested by the Shore Whalers

The California shore whalers were limited in their hunting range to about 10 miles, and therefore they preyed only on whales frequenting nearshore waters. With this limitation, the two species killed in the greatest numbers were the gray (*Eschrichtius robustus*) and the humpback (*Megaptera novaengliae*), both coastal swimmers (Scammon, 1874).

The first California shore station was established at Monterey in 1854 specifically to take humpbacks, which were then found in large numbers within the bay. However, a catch record from 1855 records that 24 whales were killed at the Monterey station between April and September: 6 grays, 14 humpbacks, and 4 killer whales (*Orcinus orca*), indicating that grays were preyed upon from the beginning (*Monterey Sentinel;* November 10, 1855).

The two species migrated annually at different times, grays from December to February and humpbacks from August to December. By taking advantage of both migrations, the northern California shore whalers were able to extend their hunting season to as much as 8 months each year (Watkins, 1925).

Stations along the southern coast of California confined their whaling to winter months, concentrating on the grays which passed during "going down" season from December to February and the "coming up" season from February to April (Watkins, 1925). The whales killed on the "down season" were fat, well nourished, and rendered a fine quality of oil. Whales harvested on the "up season" were smaller, having lost up to one-third of their body weight during the long migration (Rice and Wolman, 1971; Scammon, 1874). The yield of oil was commensurately less and not of the same fine quality as from the "down season" whales.

Gray whales measured 35–45 ft and yielded 25–35 barrels of oil. The exceptional animal reached 45–47 ft and gave 60+ barrels of oil. Humpbacks were somewhat larger, averaging 40–50 ft and 35–45 barrels of oil. An occasional cow would measure up to 75 ft, yielding 73 barrels of oil (Scammon, 1874).

The baleen of the gray was short and had no commercial value. The humpback's baleen was a little longer and although not of premium quality could be harvested for marketing (*San Diego Daily Union;* March 2, 1872).

The impact of the shore whalers on gray and humpback populations would seem to be significant, but the lack of solid information makes it impossible to calculate. Other species of whales were seen only infrequently by the shore whalers, but if a right

(*Balaena glacialis*), blue (*Balaenoptera musculus*), or finback (*Balaenoptera physalus*) wandered in range of their boats, it was pursued with great enthusiasm, as these great whales yielded commensurately larger amounts of oil and baleen than either the gray or humpback (Scammon, 1874).

Operation of the Shore Whaling Stations

The whale company, boats, weaponry, and processing techniques of the shore whaling fishery undoubtedly varied from station to station. The following descriptions are paraphrased from articles written by authors of the times and give examples of methods used at the larger stations. The other stations probably used these basic techniques with variations adapted to their locations and available equipment.

THE WHALING COMPANY

The organization of the personnel of the shore station was similar to that of a whale ship's officers and crew. Each member was paid a portion of the profit in amounts corresponding to their position or services rendered. This portion was called a "lay."

The whaling company consisted of 16 men: a captain, a mate, a cooper, 2 boat steerers, and 11 men. Their chores were divided as follows: two whaleboats[1] were crewed by 6 men each and 4 men were left on shore to cover the lookout station and to attend the try-pots. The stock of the company included the boats, whaling implements, and whaling gear. The value of the stock was divided into 16 parts and each member of the company received an equal share, 1/16 lay. The captain and the mate received an annual bonus of $200 to $300 each in addition to their lay (Scammon, 1874).

WHALEBOATS AND WEAPONRY

The boats used by the shore whalers were the same type as used in the pelagic fishery. They were built in New Bedford, Connecticut at a cost of approximately $100 each, ranged from 15 to 31 ft long, were up to 6 ft wide, and depending on size, were manned by a crew of two to six. Both bow and stern were sharp ended, so the boat could be maneuvered rapidly in either direction. Each boat was equipped with a sail and five oars: (1) the 22-ft stroke oar, aftermost starboard; (2) the 18-ft tub oar, next forward; (3) the 18-ft midships oar, next; (4) the $17\frac{1}{2}$-ft bow, second from the bow; and (5) the 17-ft harpoon, forwardmost (Bradford, 1952). Scammon (1874), in his description of whale-boat crews, included a sixth oarsman. This position was called the after-oarsman, who in addition to pulling the lightest oar was responsible for tending the lines as they were hauled in or paid out and for bailing the boat (Fig. 1).

[1]Because it was not an uncommon happening for a whaleboat to be stove in or swamped by an enraged whale, boats from the shore whaling stations always hunted in pairs and never alone. The men refused to go out unless accompanied by another boat (Starks, 1922).

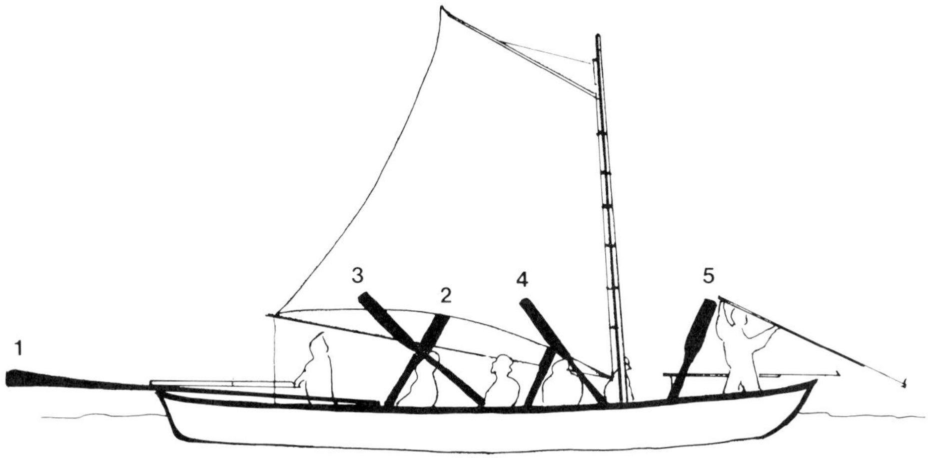

Fig. 1. *Whale boat with oarsmen in position. (1) stroke oar; (2) tub oar; (3) midships oar; (4) bow oar; (5) harpoon oar. Unmarked oar was indicated by Scammon (1874) as after-oarsman (sketch by B. Girard).*

In addition to oars and sail, the boats carried paddles, hand harpoons, one or two line tubs (to hold the 225–300-fathom length of $\frac{3}{4}$-in. manila whale line), hand-lances, bomb-lances, and possibly a Greener's harpoon gun, swivel mounted on the bow.

The placement of the crew on the whaleboat was as follows: The officer in charge (boat-header) steered the boat from the stern with the stroke oar, the tub oarsman saw that the line ran clear from the tub, the midship oarsman provided power, and forward of him the bowsman attended the line when "bowing on." The boat-steerer pulled the harpooner oar and also darted the harpoon. When the boat was fast to the whale, the boat-steerer changed ends with the boat-header and proceeded to steer the boat while the latter killed the whale (Scammon, 1874).

THE KILL

When all the crew, equipment, and weaponry were in place, the whaleboat set out for the whaling ground. As soon as a spout was sighted, everyone pulled his oar with a will, heading for the whale. The guns and/or harpoons were readied to fire. As the boat neared the whale, the oars were "peaked" and the crew used paddles to propel the boat quietly. At about 20 m from the prey, the harpoon was thrown or fired into whatever portion of the animal was visible. At the impact of the harpoon, the whale usually sounded. The line was paid out for several hundred meters and then held fast. The whale generally made a direct course toward the open ocean, dragging the whaleboat along at high speed. When the animal wearied and surfaced to breathe, the whalers fired a bomb-lance. If the lance hit a vital part, death was quick. This however, occurred infrequently and most times several bombs were necessary for the kill. Usually the dead whale sank immediately. When this happened, a buoy was attached to the line and the animal left until the gases of decomposition caused it to rise, a process that took from 3 to 9 days. If the carcass did not sink, it was towed immediately to the shore station where it was tried-out (Watkins, 1925).

PROCESSING THE WHALE FOR MARKET

The trying-out process readied the whale products for market. The carcass was made fast by a chain and cable attached to a flipper. The shore end of the cable was attached to a wooden capstan on the bluff where the try-pots stood. At the next high tide, the "ship-keepers" and "blubber-carriers" (names given the shore crew) manned this capstan and floated the carcass as high on the beach as possible. The stripping of the blubber then commenced, using an impressive collection of implements which included long-handled blubber spades, huge hooks and forks, large knives (some with two handles for mincing), and stretchers to transfer the blubber from the beach.

The first cut or "scarf" was made under the flipper, followed by parallel cuts a foot or more apart. The strips were then divided transversely and separated from the underlying tissue (Fig. 2). These strips were picked up with the long-handled forks or hooks and tossed on the stretchers. The stretchers, manned by two stout bearers, were carried up the carefully sanded incline to the mincing vats.

Here the blubber was minced and tossed into the four-barrel capacity try-pots set in masonry. The furnace fire was kept burning by using the remains of the blubber that was rendered by boiling. The smell of this burning fuel has been described as a "villainous

Fig. 2. *Flensing a whale, probably at Zuniga Point, San Diego (1890) (photo courtesy Historical Collection, Title Insurance and Trust).*

stench," or more delicately as emitting a dense black smoke "not over grateful to other nostrils than those of born whalers."

In addition to the try-pots, other cauldrons, called fat–lean pots, were used to render the tongue and other oleaginous parts. The yield of oil was reckoned by the barrel, which was supposed to measure 31.5 gallons. These measurements were approximate, at best, because the cooperage varied downward from huge tuns and hogsheads (Berwick, 1900).

The Stations

The shore stations are considered in geographical sequence starting at the northernmost boundary of Upper California and proceeding south to include stations in Lower California (Figs. 3 and 4). The years of operation are depicted in Table I. Captains and owners are listed whenever information was available. Catch records are rare and incomplete, but the data retrieved is shown in the Appendix.

UPPER CALIFORNIA STATIONS

Crescent City

This northernmost of the California shore stations was probably in operation at or before the establishment of the Monterey station. In September 1854, the *Crescent City Herald* stated:

> Almost daily we have the opportunity of witnessing the ponderous whale in our bay. A few days since two whale boats being out and within sight of the city, one of them made fast to a large humpback. For the space of a quarter of an hour the scene was most exciting, the wounded animal now lashed the water in a perfect fury, then darted off with the speed of an arrow. The boat had paid out nearly the whole of the line when the harpoon parted its hold much to the disappointment of the spectators as well as the boat's crew who we hope, may soon meet with better luck.

From fragmentary reports in newspapers, there appear to have been two stations, one called the Crescent Whaling Company and the other headed by a Captain McCoy. One was located on an island in Crescent Bay (*Sacramento Daily Union*, May 31, 1855), and the other on the south shore just inside the mouth of Lake Earl (personal correspondence with M. L. Van Meter, Gasquet, California).

Further mention of the station(s) was made in the *Monterey Sentinel* in June, 1855

> . . . Already we notice that a whaling company has been formed in Crescent City to carry on the whaling business on this coast during this season. Of course, their business for the present will not be very extensive, but still, we believe it will be profitable. All that is necessary is to make a beginning, and establish the fact that it can be carried on successfully.

The *Sacramento Daily Union* on October 2, 1856 comments that "The Crescent Whaling Company on Wednesday, September 17th took two whales, and on the following Friday Captain McCoy's company killed one, the largest of the season at Crescent

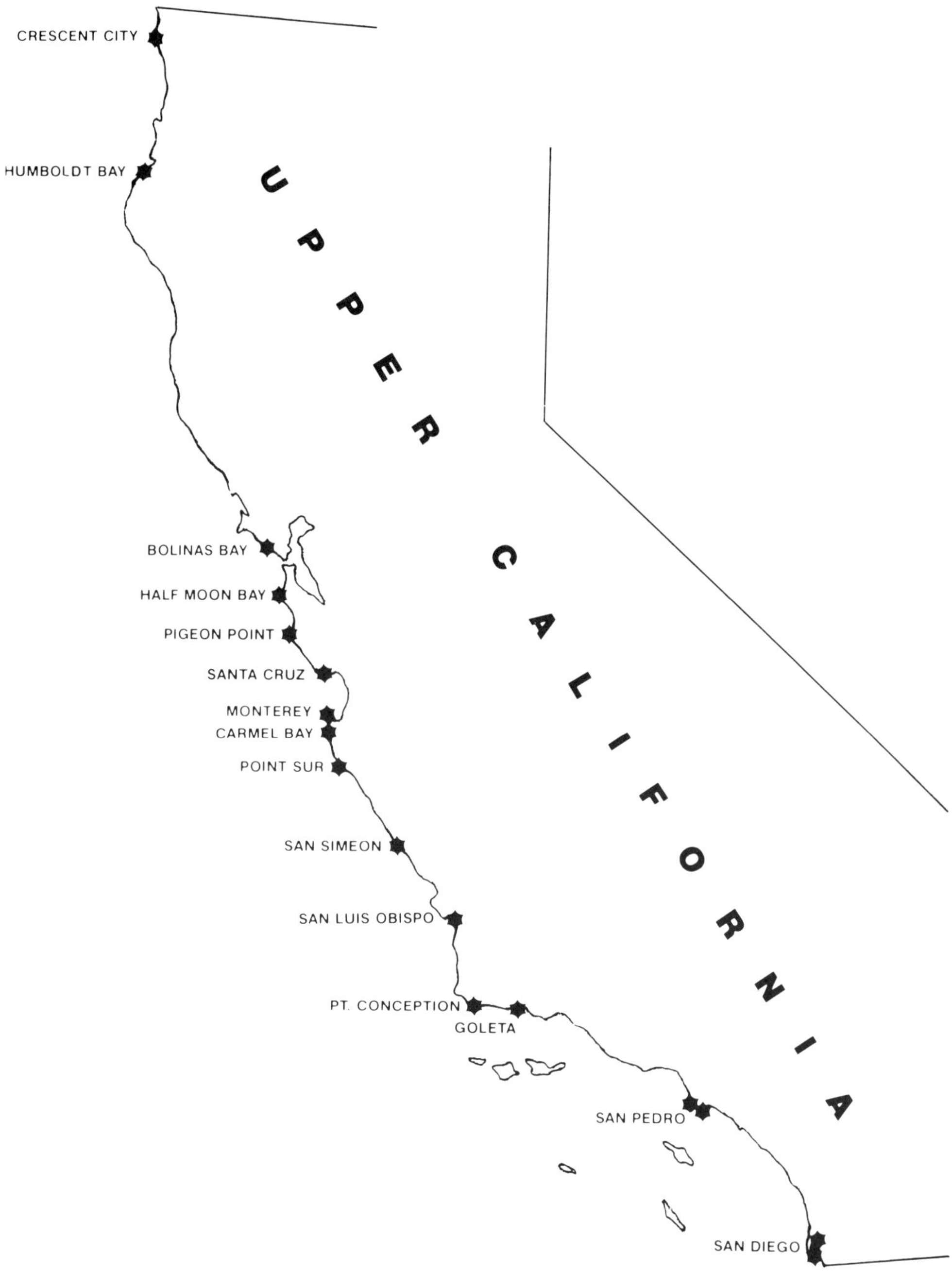

Fig. 3. *Upper California shore whaling stations, 1854–1900 (drawing by B. Girard).*

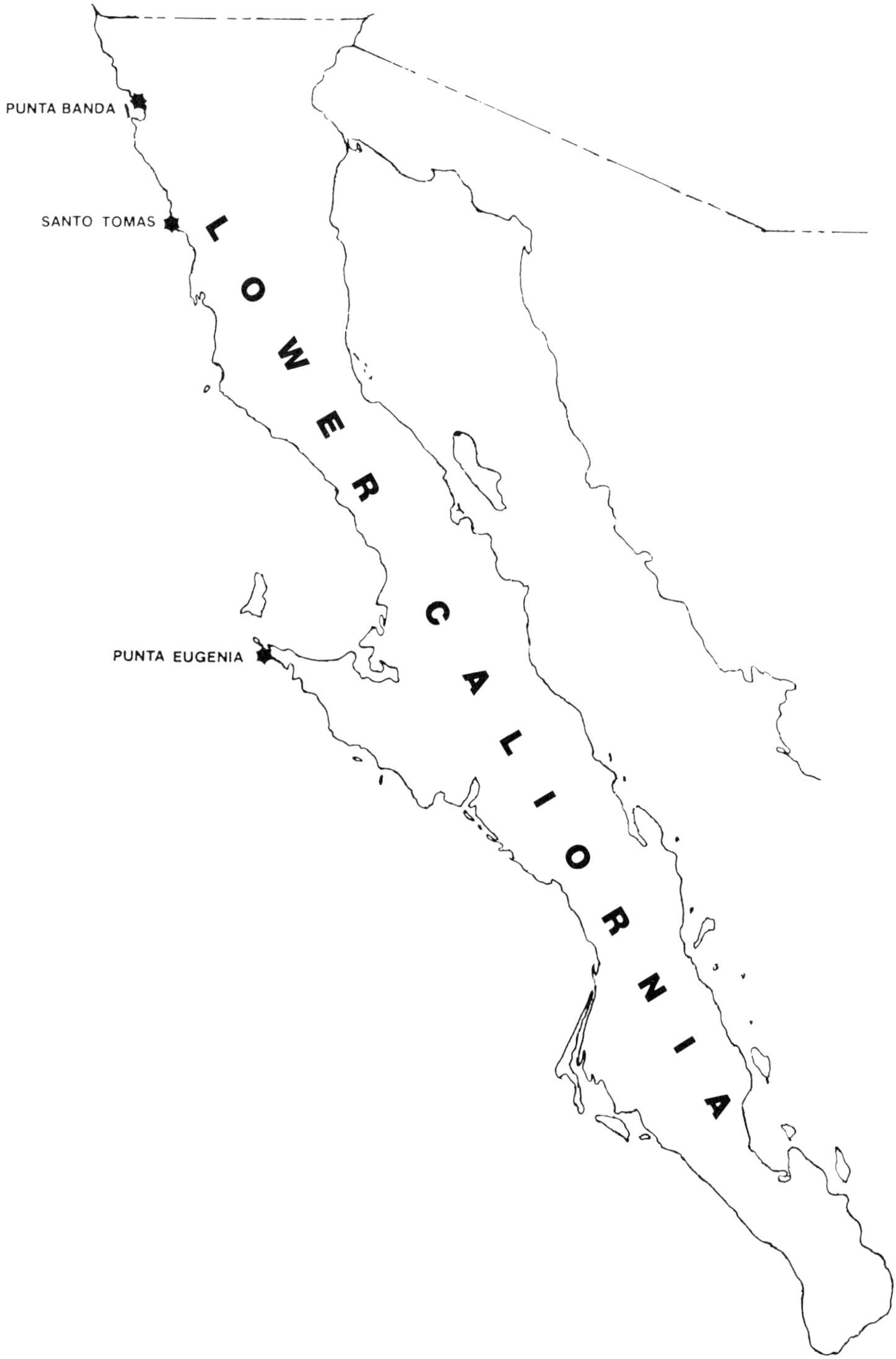

Fig. 4. *Lower California shore whaling stations, 1868–1885 (drawing by B. Girard).*

Table I
Years of Operation of California's Shore Whaling Stations[a]

Location	Years of operation (1854–1900)
UPPER CALIFORNIA	
1 CRESCENT CITY	1854 - 1856-57
2 HUMBOLDT BAY	1855 - ?
3 HALF MOON BAY	1860 or 61 - 1880
4 BOLINAS BAY	1857 - ?
5 PIGEON POINT	1862 - 1895
6 SANTA CRUZ	Mid 1850's - 1870
7 MONTEREY	1854 - 1900
8 CARMEL BAY (Point Lobos)	1861 - 1884
9 POINT SUR	1877 - 1879
10 SAN SIMEON	1865 - 1893
11 SAN LUIS OBISPO (Port Harford)	1868 - 1887
12 POINT CONCEPTION (Cojo Viejo)	1879 - 1886
13 GOLETA (More's Landing)	1867 - 1880
14 SAN PEDRO (Dead Man's Is. & Portugese Bend)	1861 - 1885
15 SAN DIEGO	1858 - 1886
LOWER CALIFORNIA	
16 PUNTA BANDA & SANTO TOMAS	1868 - 1885
17 PUNTA EUGENIA	Late 1870's - 1885

[a]Drawn by B. Girard.

City." References from the newspapers describe whaling incidents occurring from May through September. One can assume from these dates that the primary species taken was the humpback, since the grays do not ordinarily start their migration until October, even in this northernmost area of the California coast. The Crescent City stations are not mentioned by any of the historians usually quoted, and I have been unable to determine their closing dates.

Humboldt Bay

The only reference to this station stated that "Whales were taken in the bay in 1855 by the crew of the steam tug *Mary Ann*. They tried out at Humboldt Point." (*Eureka Humboldt Times;* January 20, 1855).

Bolinas Bay

This company, the Bolinas Bomb-Lance Whaling Association, did not operate as a shore station according to Scammon's definition cited earlier in this chapter. It had a fleet of small vessels that cruised about searching for whales. The catch was flensed alongside the ships, which docked biweekly and unloaded their cargo of blubber to be tried-out and refined. This method of operation was similar to that used in the so-called modern shore stations, which were active in the early 1900s.

The financial backing of this station was reportedly $100,000, a figure that far exceeded the recorded capitalization of any other shore station of the time (*Daily Alta California*, San Francisco; November 13, 1857). I was unable to determine how long this station operated. Scammon did not list it in 1874, so it probably closed before that year. Starks (1922) states that it did not last long, but he made no estimate of its closing date.

Half Moon Bay

This company operated for about 2 years at Pillar Point on Half Moon Bay, then moved to Pigeon Point (Starks, 1922). A second company, organized to replace it, was still operating in 1874 (Scammon, 1874). The station was finally abandoned sometime in the 1880s.

Pigeon Point

The first company that opened on Half Moon Bay in 1861 operated there for about a year and then moved south to Pigeon Point in 1862. It was under the command of Captain J. F. Bennett for many years (Starks, 1922).

Captain Frank Anderson, who had run a company at Portuguese Bend, came to Pigeon Point in 1877. He may have taken over Bennett's company or brought his own men with him. He operated there for 2 years and then moved the entire company to Cojo Viejo near Point Conception (Goode, 1887). After Anderson left, another company under Captain Perry opened in 1879 and operated until 1895 (Starks, 1922).

Both humpbacks and grays were taken at Pigeon Point. Jordan (1887) noted that there were many humpbacks but that the water was too rough in winter to make it a good station.

Santa Cruz

The operation of the Santa Cruz station is summarized only from interviews with former whalemen many years after their actual whaling experience. I have found no references except that in Starks (1922), which I quote in its entirety.

> . . . I have found no printed account of the existence of a whaling station at this point, but the son of Captain Bennett assured me that whaling was carried on here for some time, and Mr. Noon of Monterey Bay, told me that Captain Davenport[3] of the New Company of Monterey whaled here for a season or two. I was unable to get even approximate dates.

Monterey Bay

The Monterey Whaling Company, established by charter in 1854 (Townsend, 1886), was the first of the shore whaling stations on the California coast. The company was owned and managed by J. P. Davenport, a New England whaling captain from Tiverton, Rhode Island. Davenport was half-owner of the 180-ton schooner *Alfred* that plied regularly between the California coast and Hawaii from 1845 to 1852, engaged in trading and whaling (Orlando, 1980).

On his regular calls into Monterey, Davenport took note of the large numbers of humpback whales that congregated in the Bay. In 1852, after 7 years in the Pacific, he sailed back to New England with the intention of returning to Monterey to start a coastal or shore whaling station. While in New England, he married Ellen Tabor of Fairhaven, Massachusetts and shortly thereafter set out to return to Monterey. The couple arrived in Monterey and moved into the house now known as "the first brick house in California" to set up housekeeping.

Davenport attempted to collect enough capable men for a whaling crew, but encountered considerable difficulty as the gold fields were infinitely more appealing than life on a whaling station (Orlando, 1980). He managed to recruit 12 men for his company. Only 3 or 4 of these crewmen were experienced whalers. The remainder were green hands, who needed to be trained. The company had two boats and whaled with hand harpoons and hand-lances. (Davenport had brought some bomb-lances with him from New England, but they were defective and proved useless.) Their first year in business was reasonably successful, as the whales were plentiful and easy to approach.

However, despite this modest success, the company disbanded at the end of the first season, reportedly because the price of whale oil dropped to only \$0.25/gallon (Watkins, 1925). However, Scammon (1874), in his chart showing oil prices, recorded the following figures: 1853, \$0.58/gallon; 1854, \$0.58¼/gallon; and 1855, \$0.71/gallon. Additionally, in November, 1855 the *Monterey Sentinel* stated that oil was worth \$0.55/gallon,

[3]J. P. Davenport moved with his family to Soquel (near Santa Cruz) in 1865. It is known that he ran a shore station from Point Ano Nuevo for 3 years before moving to what is now the town of Davenport (see p. 133). It is possible that this is the Santa Cruz station referred to by the whalers interviewed by Starks (1922).

and in April of that year, gray oil was sold for $0.85/gallon in San Francisco (*Monterey Sentinel;* November 10, 1855). From these figures, one wonders if in fact the company disbanded because the novice whaling crew was unable to successfully pursue and capture whales in sufficient numbers to show a substantial profit.

In early 1855, another company called the Portuguese Whaling Company was formed in the Monterey Bay area. All 17 members of this company, including its captain, Joseph Clark (born Machado) were Portuguese from the Cape Verde Islands. They were seasoned whalemen, having previously shipped on American whalers and consequently had a marked advantage over Davenport's inexperienced crew (*Monterey Sentinel;* November 10, 1855).

For the next 3 years the Portuguese Whaling Company operated successfully, taking about 800 barrels of humpback oil annually using only hand harpoons and hand-lances (Watkins, 1925).

The winter of 1858 brought Captain Davenport back into the whaling business. This time he started with two boats manned by experienced Portuguese whalers and equipped with bomb-lances and harpoon guns. This group was known as the New Company of Portuguese Whalers (Watkins, 1925). Davenport was involved with this company at the beginning and then intermittently until 1865, when he moved his family (now 7 children) to Soquel. In that year, he established a whaling station at Point Ano Nuevo. In 1868 he finally moved to an area just 14 miles north of Santa Cruz where he built a large wharf known as Davenport's Landing (Orlando, 1980). By the late 1870s, whaling had become less profitable and Davenport's Landing primarily handled shipping for several shingle companies and lumber mills in the area. There is still a small community called Davenport, although few remnants of its whaling days remain.

Mr. Tony Silveira, one of Monterey's venerable whalemen, told Starks (1922) that in 1861 there were four stations on Monterey Bay. These probably included the "Old" and "New" Portuguese Whaling Companies and a station that operated on Monterey Bay for only a year, then moved to Point Lobos becoming the Carmel Bay station. The fourth station may have been the one that moved to Pigeon Point in 1862. This station operated for a year from Macabee's Beach on Monterey Bay before moving to Pigeon Point (Starks, 1922).

In 1865 the "Old" and "New" Portuguese Whaling Companies, both controlled by joint stockholders, consolidated under Captain Thomas G. Lambert, a whaling master from Martha's Vineyard. The newly consolidated company was called the Monterey Whaling Company (Barrows, 1893).

The number of whales harvested gradually decreased and Townsend (1886) reported that in 1885–1886 there was an especially "dull" season, due to rough weather and gun-shy whales. Whalers told Townsend that the decline along the coast was due to general use of the bomb gun by novice whalers. Many whales were wounded but not killed, causing them to avoid swimming too close inshore (Townsend, 1886). This problem was anticipated as early as April 5, 1856 in an article in the *Monterey Sentinel* reading "It should be stated also that the whales are every year getting more shy from the use of the bomb lance. It is said that they hear the bomb explode in the water, even though ten or twenty miles off."

After 1886, whaling from the Monterey station was sporadic. In 1896, Captains H. Schaufele and Jose Pedro opened a station (*Monterey New Era*, March 7, 1896) which operated for 2–3 years. In 1901 a U.S. Fish Commission Report stated: "The whale fishery was followed by eight Portuguese and eight Japanese, who used whaleboats to pursue passing whales, which on being killed are towed to shore stations" (Starks, 1922).

Of all California shore sites, the city of Monterey has the best record of its whaling history. The First Brick House (Davenport's home) and the Old Whaling Station were purchased by the State of California in 1979 and became a part of the Monterey State Historic Park. The Brick House has been scheduled for rehabilitation. Renovation of the Old Whaling Station (Fig. 5A) at 391 Decatur Street has begun. This building formerly served as a boarding house for the Portuguese whalers who worked for Captain Davenport. The passing whales could easily be spotted from windows on the second floor. The pavement in front of this structure is set with whale vertebrae, which are readily identifiable to this day (Fig. 5). On the second floor verandah sits a rocking chair constructed from whale ribs and vertebrae (Fig. 5F). Down the street at California's First Theatre (sometimes called the Jenny Lind Theatre), an arch of whale ribs is in place over the entry and whale vertebrae form the newel posts for the steps leading to the entry (Fig. 6). This building was purported to be the working and equipment storage area for Davenport's company. (Restoration and preservation of these artifacts of the whaling era has been undertaken by the Junior League of Monterey County).

Carmel Bay, Point Lobos

This company was organized in 1861 and its station operated on Monterey Bay for only 1 year. In 1862 it was moved to Point Lobos (Fig. 7) on Carmel Bay (Walton and Curtis, 1875). The station was owned by Captain Manuel Mariano and four partners. Mariano ran the establishment with a crew of 17, most of whom were Azoreans. Company operations extended from October to March, at which time the crew was disbanded to be rehired in October for the new season (Jordan, 1887).

David Starr Jordan, who visited the station in 1880, stated that it was one of the best stations on the coast, second only to Monterey. At the time of his visit there were two boats in operation. The station was apparently abandoned by 1884 (Starks, 1922).

The physical surroundings of this station were fully described by Scammon (1874) in his account of the American whale fishery.

> . . . The localities of several of the stations are quite picturesque. Some of them are nearly concealed from seaward view, being inside some rocky reef, or behind a jagged point, with its outlying rocks, upon which each successive wave dashes its foam, as if forbidding the approach of ship or boat. The one which most interested us is half-hidden in a little nook, on the southern border of the Bay of Carmel, just south of Point Pinos. Scattered around the foot-hills, which come to the water's edge, are the neatly whitewashed cabins of the whalers, nearly all of whom are Portuguese, from the Azores or Western Islands of the Atlantic. They have their families with them, and keep a pig, sheep, goat or cow, prowling around the premises; these, with a small garden-patch, yielding principally corn and pumpkins, make up the general picture of the hamlet, which is a paradise to the thrifty clan in comparison with the homes of their childhood. It is a pleasant retreat from the rough voyages experienced on board the whale-ship. The surrounding natural

Fig. 5. *(A) Old Whaling Station, Monterey, California, 1902 (photo courtesy Pat Hathaway Collection, Pacific Grove, California). (B) Old Whaling Station, Monterey, 1924 (photo courtesy Pat Hathaway Collection, Pacific Grove, California). (C) Old Whaling Station, Monterey, 1982. (D) Sidewalk constructed of whale vertebrae in front of station, Monterey, 1982. (E) Close-up of whale vertebra. (F) Rocking chair constructed of whale ribs and vertebrae. (Photographs by author unless otherwise indicated.)*

Fig. 5. *(Continued)*

Fig. 6. *(A) California's First Theater, whale rib archway, Monterey, 1982. (B) Steps made from vertebrae at entrance to theater, Monterey, 1982. (Photographs by author.)*

A

B

Fig. 7. *(A) Whaling station at Carmel Bay, Point Lobos (from Scammon, 1874). (B) Carmel Bay, Point Lobos, 1982 (photo by author).*

scenery is broken into majestic spurs and peaks, like their own native isles, with the valley of the Rio Carmel a little beyond, expanded into landscape loveliness.

Under a precipitous bluff, close to the water's edge, is the station; where, upon a stone-laid quay, is erected the whole establishment for cutting-in and trying-out the blubber of the whales. Instead of rolling them upon the beach, as is usually done, the cutting-tackles are suspended from an elevated beam, whereby the carcass is rolled over in the water—when undergoing the process of flensing—in a manner similar to that alongside a ship. Near by are the try-works, sending forth volumes of thick, black smoke from the scrap-fire under the steaming cauldrons of boiling oil. A little to one side is the primitive storehouse, covered with cypress boughs. Boats are hanging from davits, some resting on the quay, while others, fully equipped, swing at their moorings in the bay. Seaward, on the crest of a cone-shaped hill, stands the signal-pole of the lookout station. Add to this the cutting at the shapeless and half-putrid mass of a mutilated whale, together with the men shouting and heaving on the capstans, the screaming of gulls and other sea-fowl, mingled with the noise of the surf about the shores, and we have a picture of the general life at a California coast-whaling station. (Fig. 7A,B)

Point Sur

The Point Sur station was maintained by the Carmel Company from 1877 to 1879 and was run by Captain Manuel Mariano. There were more whales at Point Sur than at Carmel, but water conditions were hazardous during the winter and the men could not go out regularly (Jordan, 1887).

San Simeon

The San Simeon station was owned outright by Captain Joseph Clark, who had started his whaling career in Monterey in 1854 (Starks, 1922). He purchased land on the inshore side of San Simeon Point and set up his operation. It proved a prosperous undertaking and Clark headed the station until his death at age 56 in 1891. Some whaling continued until 1893 under the direction of Hipolite Marshall, a cousin of Clark. Mr. Marshall was not particularly interested in the whaling business. This fact, combined with the diminishing numbers of whales, prompted Marshall to sell the property to Phoebe Apperson Hearst, the widow of George Hearst. This transaction was completed in 1894, ending the whaling industry in San Simeon. The property was used by fishermen intermittently after that time (Hamilton, 1974).

Complete catch records are available from this station (see Appendix), a fact attributable to the stability of the company, operated by an experienced whaleman, who supervised its entire 27 years of operation. Clark's company consisted entirely of Portuguese whalers who were hired each season. The crew ranged from 10 to 20 men. The catch was primarily gray whales, but occasionally a humpback was taken, and 1884–1885 recorded three right whales.

San Luis Obispo, Port Harford

Captain Marshall (Marsiali) established the station in 1868 and was still there when Jordan visited in 1880. Marshall and three or four others owned the outfit, which consisted of three six-oar boats and other standard equipment (Jordan, 1887).

It was this station that Captain Frank Anderson presumably ran in 1873–1874.

References consulted do not say if Captain Marshall simply interrupted his command for that season. Starks (1922) stated that Marshall stayed on at the station at least until 1880, and perhaps until it closed in 1887 due to the scarcity of whales and the low price of oil (Collins, 1892).

The station's crew was 21 men, including 20 Azoreans and 1 American, Michael Noon; Noon retired in Monterey and was interviewed by Starks in 1922. The crew was hired seasonally, and as was customary, paid by lays of 1/17 to 1/20 for harpooners to as little as 1/65 for boat pullers.

Both grays and humpbacks were taken at this station, but Jordan, who reported partial catch records, did not indicate relative numbers (1887).

Point Conception, Cojo Viejo

The Cojo Viejo station set up operation on April 25, 1879 under the command of Captain Frank Anderson. The company had operated at Pigeon Point for the 2 years prior to this move (Starks, 1922).

It was a joint stock arrangement with land and equipment owned by 15 members of the company who shared the profits equally. Captain Anderson, employed as the station manager, received a 1/17 lay and $100 annual bonus. There were 4 more members of the company (nonowners). Two received lays of 1/35 and 1/40 respectively. The other 2 were Chinese laborers who received the sinews of the whales as payment for their work. The sinews from each whale weighed 20–25 lb. and sold on the San Francisco market for about $0.25/lb. They were then exported to China, where they brought a high price. The soup made from whale sinews was highly prized by the Chinese (Dias, 1979).

Three of the whalers were from the Madeira Islands. All the rest, except the two Chinese, were from the Azores. The Azoreans included Captain Anderson who had considerable experience running a shore station, having headed the camps at San Luis Obispo in 1873–1874, Portuguese Bend in 1874–1877, and Pigeon Point in 1878–1879.

The camp was set up with the try-pots on the beach and the living quarters on a bluff overlooking the work place. The company men occupied a large house. Captain and Mrs. Anderson had a separate house, as did the Chinese men.

The investment in the company, aside from the houses and a symbolic $1.00 a year for the rental of the land, consisted of $2000 worth of equipment and annual operating expenditures of $4000–5000 (Dias, 1979).

Both grays and humpbacks were taken regularly, and during the 1884–1885 season, a right whale capture was recorded (Townsend, 1886).

Goleta, More's Landing

There is contradictory information about this station, with several historians giving different dates of opening and closing. The confusion may be a result of at least three different companies having operated between 1867 and 1880.

Samuel Hackett of San Diego whaled from the Goleta station for two seasons, 1867 to 1869. He sold his whaling interests in 1869, because petroleum floating on the ocean "frightened the whales away" and gummed up the whale lines (McGrew, 1922). Starks

(1922) apparently did not know about Hackett's activities as he wrote, "This station was established about 1870 and suspended about 1880. During this time, there were three different companies here, the first being composed of Jamaican negroes."

This was apparently a small station which operated casually at irregular intervals. Dias (1979) quotes an unidentified 1873 reference as saying the Goleta station was comprised of only two houses where six whalemen lived together with a Chinese cook. Their food consisted primarily of whale meat, which they consumed in the form of steaks or meatballs, and seal liver. Collins (1892) reported that by 1880 the station had been abandoned.

San Pedro, Dead Man's Island and Portuguese Bend

"A whaling party from San Diego has located on Dead Man's Island at San Pedro, and has succeeded in capturing two whales from which forty-five barrels of oil were extracted" (*Los Angeles Star;* January 12, 1861). This party of whalers was the first to operate in the area and probably came from the Johnson Company, located at Ballast Point in San Diego. Captain Johnson was reported by Hayes (1861) to be "also interested in the company at Dead Man's Island" (Starks, 1922).

In March, 1861, a right whale was taken, along with five other whales in a 2-week period. The whales were reckoned to net an average of $300 each for the whalers (*Los Angeles Star;* March 9, 1861).

This year, 1861, also found Captain W. W. Clark, skipper of the *Ocean*, whaling in San Pedro Bay. Captain Clark, a New Englander, was the first whaler in San Diego Bay. Henderson (1972) records that he (Clark) killed two gray whales in that bay as early as 1842. In San Pedro Bay, he was able to take enough whales to yield an average of 450–500 barrels of oil over the 6 weeks he was in the area (Newmark, 1930).

In the exceptionally active season of 1861–1862, Captain Hart was in charge of the Dead Man's Island station (Starks, 1922). Twenty-five whales had been taken by March 15, 1862. Newmark (1930) remarked that "For six days in March 1862, San Pedro whalers harpooned a whale a day, bringing to the landing over 200 barrels of oil as a result of the week's labor."

Jordan (1887) reported that the San Pedro station closed in 1866, but in January 1867 several vessels were operating in and out of the bay. One of them, the schooner *Ned Beale*, was reported to have had its "try-works" on Dead Man's Island (K. Nisheim, unpublished manuscript).

Dead Man's Island no longer exists. It was removed from the bay in 1929 during extensive harbor renovations (Fig. 8). The Portuguese Bend station was a seasonal operation from the outset and probably operated only intermittently during the years from 1864 to 1885. Captain Joseph Clark[4] an Azorean and whaling captain from Mon-

[4]In reviewing the history of shore whaling, the names of two Captain Clarks invariably come up. Usually the two identities are not explained and there is considerable confusion about the two men. Captain W. W. Clark was a native of New Haven, Connecticut. He was master of the *Ocean*, a craft that had been engaged in whaling just off shore in California as early as 1831. He spent many years whaling out of San Diego Bay. He was reported to be whaling off San Pedro in 1861, but was never associated with a shore station. The *Ocean*

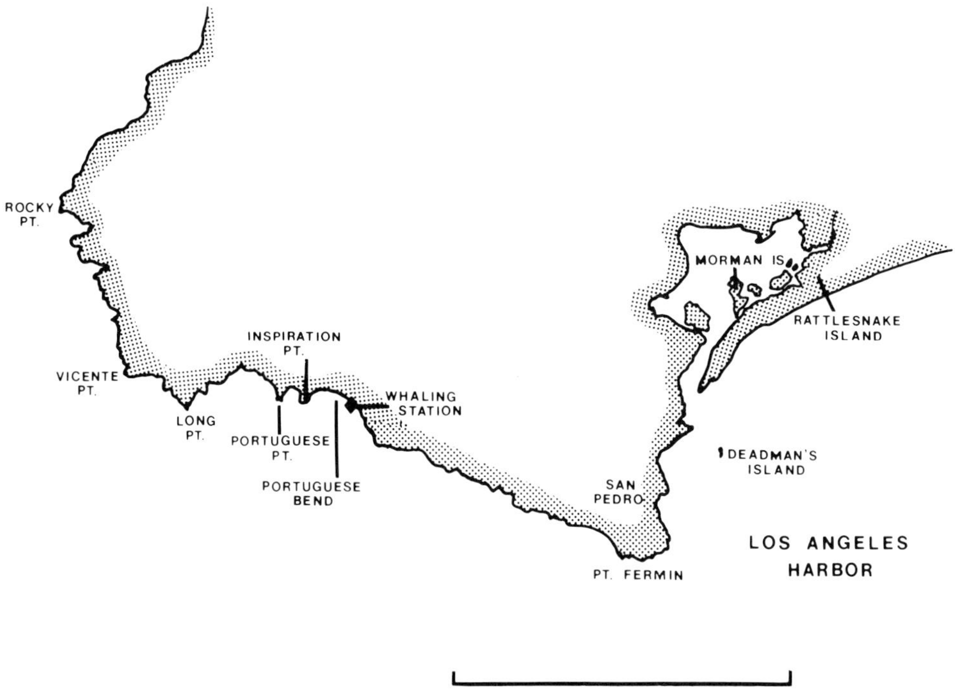

Fig. 8. *Map of San Pedro, Deadman's Island, and Portuguese Bend. Scale bar = 5 miles.*

terey, reportedly came to the San Pedro area in 1864 and whaled from Portuguese Bend. He left after a year and started a station at San Simeon (Starks, 1922).

In 1869, the Portuguese Bend station was run by the John Brown Whaling Company (*Los Angeles Star*; March 13, 1869). From 1874 to 1877 the company was run by Captain Frank Anderson, who moved on to Pigeon Point and finally to Cojo Viejo. The kill from this station was made up entirely of grays and it operated only from December to April. Anderson's tenure marked its most active period, during which 2166 barrels of oil were tried-out.

Some whaling went on after 1877, following Anderson's departure, but it did not appear to be on an organized basis. Townsend (1886) reported that 13 whales were taken at San Pedro in the season of 1884–1885. The station was abandoned after this season due to lack of firewood for the try-pots and inadequate fresh water, rather than to the scarcity of whales. The whalers had never established a permanent settlement at this site (Dias, 1979). Ray Wallace, the well-known San Pedro marine artist, painted a watercolor of the old whaling station at Portuguese Bend from an old photograph that

operated as far north as San Francisco. Captain Joseph Clark, born Machado, was an Azorean who came to Monterey in 1852. He headed the Old Company of Portuguese Whalers from 1855 to 1860. I have seen no record of his whereabouts from 1860 to 1864. Although other historians have said he came to San Pedro from San Diego, it appears he may be confused with W. W. Clark.

Fig. 9. *Portuguese Bend Whaling Station, 1885 (painting by Ray Wallace).*

depicted the station as it was in 1885. It shows a small cabin with a tiny porch, an even smaller shed with try pots, barrels, and three whale boats pulled up on the sand (Fig. 9). Nothing remains of this station, as the area has been the site of numerous landslides. However on January 15, 1983, the Rancho de los Palos Verdes Historical Society mounted a bronze plaque at the site of the station to commemorate this romantic era. The plaque is a replica of Mr. Wallace's watercolor.

San Diego

The station (Fig. 10) which opened at La Playa in 1858 marked the beginning of the shore whaling era in San Diego. The owners, Prince and Alpheus Packard, were from a New England family with an English mother and Portuguese father.

In February 1859 the whalers recorded a kill of about a dozen whales, only five of which they were able to return to port. The *San Diego Herald* on February 5, 1858 stated that . . . "If some means could be devised to prevent the whales from sinking, 3 or 4 parties could do a good business during the season catching whales within 10 miles of the entrance to the harbor."

From the beginning, San Diego whalers used a hand-held bomb-lance to pursue their "bloody business" (*San Diego Weekly Herald;* February 5, 1859). Judging from the record of 5 whales brought to shore of 12 killed, this weapon did not make them more efficient than their northern colleagues who used hand-lances.

The second company, located at Ballast Point, was run by Captain Miles Johnson

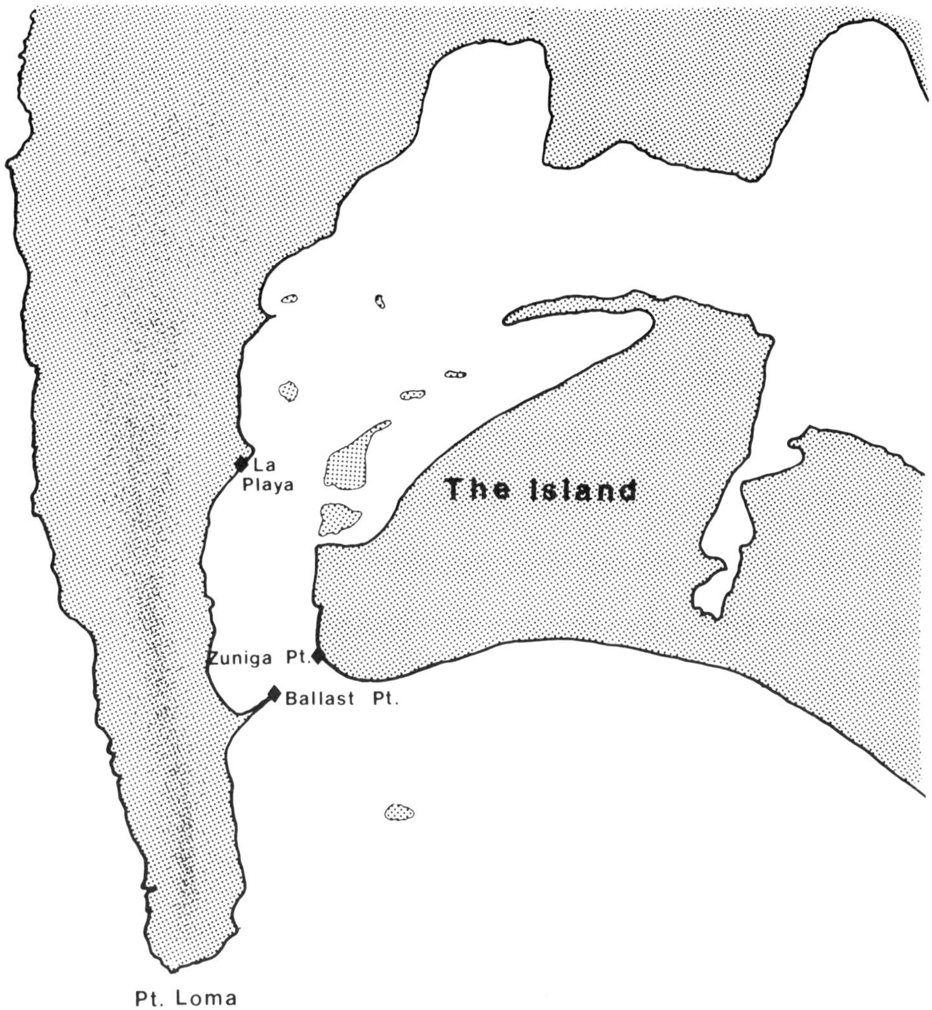

Fig. 10. *San Diego's whaling stations. San Diego Bay. California (from a Trigonometrical Survey of the Coast of the United States, 1859).*

and his cousins, Henry and James Johnson. This station opened about 1860. It is mentioned in the 1860 census along with another company called the Tilton Company.

There was no newspaper published in San Diego from 1860 to 1868. There are no references during this period in the public library or the *San Diego Union* archives, so information is minimal. It appears, however, that Packard and Johnson operated throughout these years, because when a newspaper did begin publication in 1868, both companies were mentioned regularly.

Captain Samuel Warren Hackett, an important participant in San Diego's shore whaling history, recorded details of the whaling industry in diaries he kept from 1864 to

1869. Unfortunately, these diaries, used and quoted by Davidson (1929), have been lost and with them the only details of whaling activities for this site.

When Hackett arrived in San Diego from his home in Massachusetts, he was 23 years old. He whaled from San Diego until 1863, working in the off season aboard the Pacific Mail Steamship Company's store ship *Clarissa Andrews*. He left San Diego in the spring of 1865 to go to San Francisco, and returned in November of that year to run a whaling company owned by a San Franciscan named Watson. After 2 years with this company he married a San Diego girl and moved to Goleta to open his own shore company. The remainder of his association with whaling is detailed in the Goleta station history.

In 1868, when the new *San Diego Weekly Union* newspaper started, the December 12 issue reports that Captains Packard and Johnson were engaged in whaling off Ballast Point. The two companies, when working at full complement, kept 20 men working outside the harbor and about a dozen at the Point tending the try-pots.

On December 26 the paper reported that there were four companies at work. The names of the other two companies are nowhere mentioned. Captains Price and Saddler are mentioned as having captured whales, but it is not immediately obvious if at that time they each headed a company (to account for the four mentioned earlier) or if they worked for Packard or Johnson. A later reference (*San Diego Union Weekly;* January 2, 1869) mentions Saddler and Price as separate companies.

In 1868 the first references to shore whaling operations in Lower California appear. Captain Saddler captured a whale at Punta Banda and Price was operating from Santo Tomas. This was the start of their shift of effort to the south (*San Diego Union Weekly;* January 16, 1869).

In the summer of 1870, the government decided to occupy Ballast Point and ordered the whalers to vacate the premises. This land had been claimed in 1852 by the Federal Government for use as a military reservation (*San Diego Union;* June 11, 1870). However, Packard and Johnson were permitted to continue their operations until 1873, at which time the military took over both Ballast Point and Point Loma and the whalers moved across the bay to Zuniga Point on North Island.

From 1868 to 1874, reports are available on catches from Lower California. Two of the stations there, Punta Banda and Santo Tomas, were operated by the joint efforts of San Diego Companies: Packards, Johnson, Saddler, and Price. These stations will be more fully covered later in this chapter.

The whalers in San Diego continued to capture a sufficient number of animals to make their business profitable through 1872, but the move southward was, most certainly, the result of a diminishing number of whales. By 1873, the last year at Ballast Point, Johnson took 5 whales, and Packard, 2. This same season Packard took 11 at Santo Tomas and 6 at Punta Banda, giving a total yield of 980 barrels of oil, a very light catch for these two companies (New Bedford, Massachusetts, *Republican Standard;* May 23, 1873).

From 1874 to 1883, most of the action of the San Diego companies occurred in Lower California, with only one recorded kill off San Diego (on December 22, 1876). On that day, Flanders and Company took a "big spouter" off the Point. At least a 60-barrel

Fig. 11. *Tourists viewing whale carcass at Ballast Point, San Diego, 1890 (photo courtesy of Historical Collection, Title Insurance and Trust).*

oil yield was expected. The whale was displayed at Ballast Point, and the *Cuyamaca* scheduled regular trips for tourists to see it, at 10:30 AM and 1:30 PM for $1.00 (*San Diego Union;* December 22–23, 1876). This show (Fig. 11) may have launched the million-dollar whalewatching industry which still prospers along the coast of the Californias.

In the autumn of 1883, Captains Plummer and Wall (Wall was the skipper of the whaling schooner *Sierra* of San Francisco) joined forces to find a suitable trying-out location on San Diego Bay. Their plan was to cruise in the *Sierra* off the coast and bring their captured whales into San Diego for trying-out. Subsequently, they built a camp on the Peninsula opposite Ballast Point (Whaler's Bight) where there was a freshwater spring. On December 23, 1883, Captain Wall captured the first whale taken off this harbor for several years, with an anticipated yield of 50 barrels of oil (*San Diego Daily Union;* November 18, 1883). The 1883–1884 season was moderately successful for Wall and his company and by April 1, the close of the season, they had captured 13 whales from the *Sierra* which were tried-out on shore.

The success of Wall's company encouraged two other San Diego men, T. J. Higgins and J. R. Scranton, to investigate sites for a station. They purchased equipment from San Francisco and took a crew to Punta Banda, but made no kills during their first season (*San Diego Daily Union;* January, 1884).

At the start of the 1884–1885 season, in November, 1884, Higgins built his whaling company in the same general area as Wall, on the peninsula opposite Ballast Point (*San Diego Union;* November 8, 1884). Captain Wall died in 1885, and it appears that his company ceased operation (Stewart, 1965). Reports from the 1884–1885 season mention only the whaling camp of T. J. Higgins and Son (*San Diego Daily Union;* April 28, 1885).

The *San Diego Union* reported on March 22, 1885

> . . . T. J. Higgins whaling camp on the Peninsula will probably be broken tomorrow, the run of whales is nearly over and as three Portuguese belonging to the crews are anxious to return to their home in the Cape Verde Islands before Easter, Mr. Higgins has consented to release them before the contract time: all the heavier equipment of the camp will remain where they are for use in the fall. . .

On November 5, 1885, an article appeared in the *San Diego Daily Union* stating that

> . . . On the Pacific Coast there are 5 or 6 whaling stations as follows: One, but little operated and that by natives, 300 miles down the coast at a point known as Eugenia, and the San Diego, Point Conception, Port Harford, San Simeon, and Monterey Bay Whaling Stations.[5] The San Diego station is the most completely equipped of any on the Coast and its crew the most skillful to be had. The San Diego Station is owned and controlled by men of this city. A warehouse will be constructed on the waterfront at the foot of G Street to make storage room for the product of the station. . . .

In March of 1886, the Higgins station closed, recording a take of 300 barrels of oil and 1000 lb. of baleen. Half the oil was from a rare right whale taken on March 9. (This whale was called by the reporter "a rare *Arctic* right whale," but present-day cetologists doubt that the Arctic right whale ever came this far south.) The right whale yielded 150 barrels of prime oil, estimated to be worth $1500, and baleen worth $2500. The remainder of the harvest was from seven grays taken over the course of the season (*San Diego Daily Union;* March 5, 1886).

No evidence of San Diego's shore whaling industry remains at the station sites. All of Ballast Point and Point Loma was declared a historical landmark in 1940, and a marker was put in place denoting it officially as the Site of the San Diego Whaling Station. In 1975 I walked the Ballast Point area with Commander Franke of the Coast Guard in search of this marker, but could not find it. I also had an opportunity to speak to Retired Chief and Mrs. Radford Franke, parents of the Commander, who were the lighthouse keepers at Ballast Point from 1930 to 1957. They recalled whale bones and oil remains, but on the inner side of the Point. At that time the tide came in over the neck at the Point. This probably aided the whalers in floating their catch onto land for processing. A few photographs in the Title Insurance Company Historical Collection and some news clippings preserved on microfilm are essentially all that remain of this most interesting era in San Diego's history.

LOWER CALIFORNIA WHALING STATIONS

Three shore whaling stations operated in Lower California from 1868 to 1885. These included Punta Banda and Punta Santo Tomas (Fig. 12), which operated from 1868 to 1885, and Punta Eugenia, which operated from the late 1870s to 1885.

Punta Banda and Santo Tomas

These two stations were run by the various San Diego whaling companies and reports of their catches, openings, closings, and commanding officers are inextricably entwined with the records of these companies.

[5]The report omitted the Pigeon Point station which operated until 1895. See p. 131, this chapter.

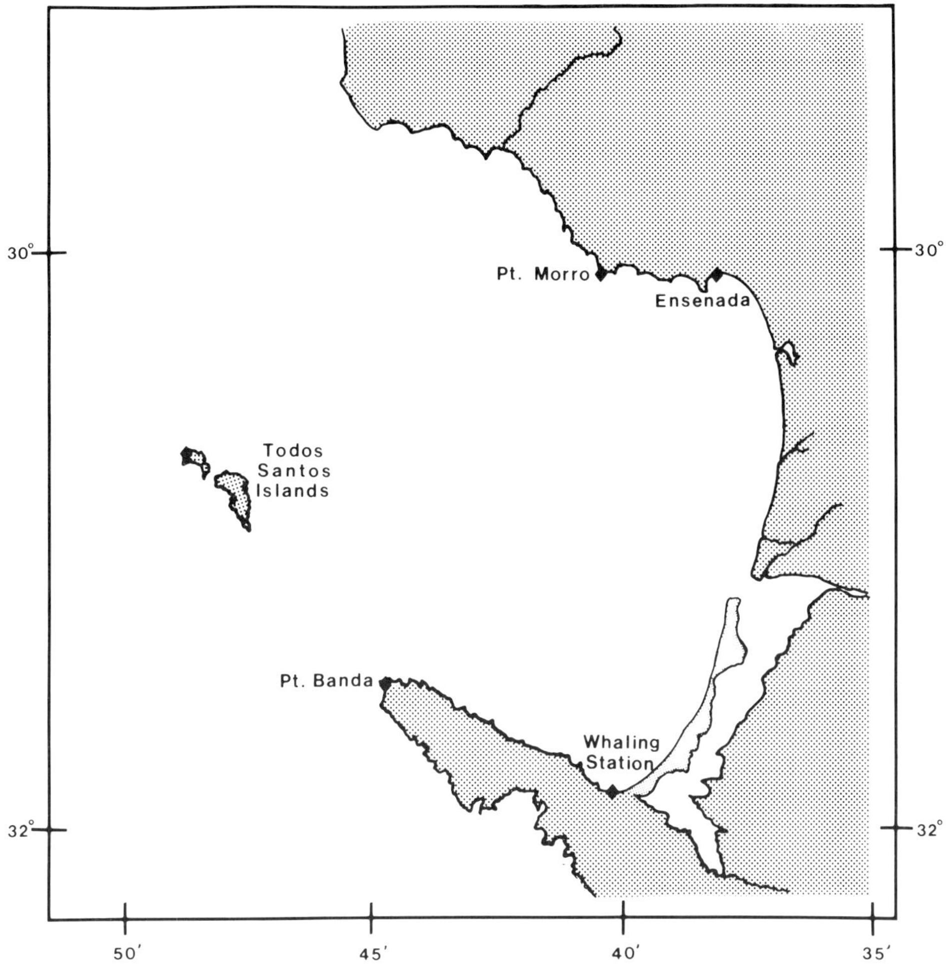

Fig. 12. *Site of whaling station, Punta Banda, Lower California (from the Todos Santos Harbor Survey of the* U.S.S. Ranger, *1886).*

The year 1868 found the Price Company at Santo Tomas and the Saddlers at Punta Banda (*San Diego Union;* January 16 and March 6, 1869).

There is evidence that some of the San Diego companies combined their efforts and worked cooperatively at these two sites. An article in the *San Diego Union* on March 2, 1872 gives the following account:

> The Whaling Season. Mr. I. Matthius (agent for the San Diego Companies) arrived last evening from Santo Tomas and Punta Banda. He gives us some interesting items concerning the whaling season of 1871. Mr. M. left at Punta Banda two large whale boats with a Captain and a crew of 18 men. The "up" season at Punta Banda commences the 15th of February as on all whaling points on this coast. Two days after the arrival of the party at Punta Banda, a right whale appeared within a quarter mile of the camp; immediately all the boats were set in motion and chase given: the whale received two harpoons but got away. The next day the weather being

> rough, the boats could not leave their moorings, but the day after a large cow came in sight and was captured after an exciting chase. She showed considerable pluck but had to succumb to the deadly bomb gun. This was one of the largest whales taken this season and will fill from 50–60 barrels—the yield would be at least 10 barrels more had the whale been taken at Point Loma where there are better facilities for getting out the blubber.
>
> . . . The species of whale caught during the present season is known as the California gray whale or devil fish. Further down the coast and after the gray whale season (12/10 to 4/10) chase is given to the humpback, a much larger whale—50–100 barrels of oil. The bone of the humpback is of some value.

The crews operating these two stations included the names of all the men prominent in San Diego's whaling community. Only one company, the Johnsons, remained in San Diego. During the 1872–1873 season, the last years that the whalers worked from Ballast Point, only a few whales were taken and nearly all the men and equipment were shifted to the two Lower California stations.

Whaling continued at these stations until 1878. That year, Captain Packard, who had become skipper of the *New Hope*, a vessel that plied regularly between Lower California and San Diego, lost a whaleboat and other equipment when returning the Santo Tomas crew to San Diego at the end of the season (*San Diego Union;* February 26, 1878).

Nothing more is recorded until 1884 when a new San Diego company, Higgins and Scranton, started operations on Whaler's Bight in San Diego and established a camp in Punta Banda as well. Their first year no whales were taken from the Lower California station, but confidence remained high that this situation would soon turn itself around (*San Diego Union;* March 21, 1884). Meanwhile, whales had returned to the San Diego area and the Higgins company enjoyed considerable success and profit there through 1886, when it closed.

No further mention was made of the Punta Banda station. A U.S. Hydrographic Office survey, in notes taken by the crew of the *U.S.S. Ranger* in 1886, makes this observation: "The whaling station should not be mentioned as the building has been removed or destroyed."

Punta Eugenia

Punta Eugenia was one of the last shore stations to be established. It was operated by "native whalers," probably immigrant foreigners and Mexicans. It operated until at least 1885 (Henderson, 1972).

The *San Diego Union* reported on February 19, 1885, that the schooner *City of San Diego*, skippered by Captain Behen, arrived yesterday afternoon with 80 barrels of oil from Point Eugene, Lower California. Dr. Raymond Gilmore of San Diego mentioned that he saw the remains of a whaling station at Punta Eugenia when he flew over the area in 1954 (personal communication).

Summary

Shore whaling along the California coast was a colorful and sometimes profitable industry that lasted from 1854 until 1901. Documentation of the industry is sketchy, and accurate data are difficult to obtain. Much of the published information is contradictory, as it was dependent on the recollections of aged whalemen interviewed many years after their active whaling careers. After reviewing the frequently cited references, reading the newspapers published from 1855 to 1900, and seeking the aid and generosity of published and unpublished researchers, a fuller picture of this little-known era has been developed.

The first station was established on Monterey Bay in 1854, and over the next 45 years, 15 stations were operated at various times from Crescent City on California's northernmost border to San Diego in the south. Additionally, three were active in Lower California. The shore whalers worked daily during the short season and returned to their homes each evening. In the off season, many of them worked their own small farms.

The daily logs kept by pelagic whalers, who spent long months or years at sea, were not required by the shore whaling stations, so this excellent source of documentation is not available to historians. Companies were formed, disbanded, and reformed, often in different locations at irregular intervals. The captains moved freely between stations, sometimes buying the business and other times employed as managers. The shore whaling stations, particularly the northern ones, were established to hunt the humpback (*Megaptera novaengliae*), but the gray (*Eschrichtius robustus*) was included almost immediately. The southern stations were established to take advantage of the regular habits of the gray on its annual southward migration. The industry was sporadically profitable for about 40 years, declining when the whale population became scarce and the demand for whale oil products diminished. By the early 1900s, shore whaling in California was obsolete.

Epilogue

The definitive history of the California shore whaling fishery has not yet been and may never be written. The official records are not complete, and most of the efforts to document the fishery have introduced or perpetuated inaccuracies.

In each local area there are descendants of these hardy men, who may have personal records, photographs, and other historical treasures that would aid us in completing this study with some degree of accuracy. These records need to be unearthed so that a sympathetic recording of the accomplishments and contributions of the shore whalers should be made. Only then can we truly appreciate the impact this pursuit had on the populations of grays and humpbacks as they swam along the shore of the coast of the Californias.

Acknowledgments

The author gratefully acknowledges assistance from Barbara Girard, who composed, drew, and laid out the maps and charts; Kenneth Nesheim, who permitted review of his unpublished manuscript on California shore whaling; and K. L. Seavey, who shared his collection of information on Monterey's shore whaling era.

Glossary

Baleen ("Whalebone"). The cornified fibrous plates attached to the upper jaw of certain whales for filtering food organisms from the seawater.

Blubber. The fat of whales and other large marine mammals from which oil is obtained.

Bomb-lance. Steel tube filled with gunpowder and fitted with a fuse causing it to explode inside the whale, after being fired from a heavy shoulder gun. It carried no line and was used solely for killing the harpooned whale.

Bone. Whaler's term for baleen. Short for whalebone.

Capstan. A kind of winch with a vertical drum, used for weighing anchor and lifting heavy weights.

Cooper. The man who made and maintained water casks and barrels for oil.

Cutting-In. Flensing and dismembering a whale.

Davit. Small crane that projects over the ship's sides for hoisting boats.

Flense. The process of removing the blubber blanket from a whale.

Greener's Harpoon Gun. Small swivel-mounted gun, $1\frac{1}{2}$-in. bore, 3-ft barrel, weighing 75 lb. The harpoon is $4\frac{1}{2}$ ft long with an 84-yd range.

Hand-lance. Preceded the bomb-lance as a weapon for killing the whale after it had been harpooned. It was hand thrown and had a range of about 20 yd.

Harpoon. Barbed spear used to faston on to the whale, thus enabling the whalers to get close enough to use a hand- or bomb-lance to effect the kill.

Hogshead. A barrel-like container with a 63-gal capacity (238.5 liters).

Lay. A share of a whaling company. 1/16 lay meant 1 barrel of oil in 16.

Peak. To raise to a position perpendicular or more nearly so.

Pelagic. Of, or performed on, the open sea.

Sound. Term signifying the descent of the whale after being fastened to (harpooned).

Stove. To break or make a hole in, by crushing inward.

Swivel-mount. A revolvable base that can be rotated horizontally and vertically.

Trying-out. To extract the oil from the blubber by boiling.

Tun. A large cask for oil or wine with a capacity of 252 gallons or four hogsheads.

Vertebra. Individual bone or segment of spine or backbone.

Appendix. Fragmentary Catch Record, California Shore Whaling Stations

Year	Number of whales	Barrels	Price ($) Barrels/gallon	Value	Source[a]	Comments
Crescent City						
1854	1				1	Humpback—harpooned and lost
1856	3				2	Probably humpbacks
Pigeon Point						
1877–1878		1000			24	Both humpbacks and grays
1878–1879		561			24	
Monterey Bay						
1854	18[b]				3	5 grays, 9 humpbacks, 4 killers Whalers netted $438 each for season
1855	24[b]	300	20.44/0.65	6,132.50	5	Lost 5 humpbacks and 1 gray after kill so yield was from 18 whales.
1856		509	23.58/0.75	12,000.00	4	Included 1 right whale.
1857	23	1016	22.15/0.70	22,500.00	26	
1858		1500	17.01/0.54	25,515.00	27	
1859		1800	9.14/0.29	16,443.00	28–34	These figures are given as net amounts, the • only ones so indicated. One-third from grays remainder from humpbacks and 1 right whale.
1862		2500	18.44/0.59	46,105.00	35	Mostly humpbacks.
1863	64	1930	30.96/0.98	59,755.25	24	
1865		679	45.66/1.45	31,000.00	22	Includes catch of two Monterey stations and one Carmel station
1867		800	23.36/0.74	18,689.00	30	Includes Monterey and Carmel stations
1869		1260	21.93/0.70	27,631.80	31	Includes Monterey and Carmel stations
1873		175	11.43/0.36	2,000.00	33	Right whale which also yielded 1500 lbs. of baleen worth over $2000
1877–1878		500			24	
1883–1884	11				35	
1884–1885	12				35	
1885–1886	5				35	

(*continued*)

Appendix (Continued)

Year	Number of whales	Barrels	Price ($) Barrels/gallon	Value	Source[a]	Comments
Carmel Bay, Point Lobos						
1879–80	7	200			24	
Point Sur						
1877–1878					24	1 gray and 1 blue
1878–1879					24	1 humpback, 3 grays, 1 right
San Simeon						
1865	25				34	
1866	23				34	
1867	24				34	
1868	25				34	
1869	20				34	
1870	23				34	
1871	22				34	
1872	21				34	
1873	22				34	
1874	16				34	
1875	12				34	
1876	7				34	
1877	13				34	
1878	3				34	
1879	14				34	
1880	13				34	
1883–1884	11				35	
1884–1885	15				35	
1885–1886	14				35	
1889	5				36	
1890	7				36	
1891	7				36	
1892	5				36	
San Luis Obispo–Port Harford						
1878–1879	11				24	Grays and humpbacks
1879–1880	9				24	
1886–1887	5				23	

Point Conception–Cojo Viejo						
1879–1880	20	544	14.18/0.45	7,711.30	24	
1883–1884	25				35	
1884–1885	18				35	Includes one right whale
1886	11				35	
San Pedro–Deadman's Island, Portuguese Bend						
1862	25	600			34	
1874–1875		722			24	Total take for three seasons from December–April 1874, 1875, and 1876 was 2166 barrels
1875–1876		722			24	
1876–1877		722			24	
1884–1885	13				35	
San Diego						
1858–1859	20	775	18.90/0.60	14,647.00	8	
1859–1860		900	10.00/0.32	9,000.00	32	
1861		650			32	
1868–1869	22				9	
1870–1871	23	695			33	
1871–1872	21	1750	21.71/0.69	38,000.00	11	Peak profit year for San Diego, includes two Lower California stations.
1872–1873	7	280			12	
1873–1874	17	645			33	
1874–1875		400			33	
1876–1877	1	60			14	
1883–1884	13 (5)				18	These 13 whales were captured by the vessel *Sierra* and were brought to shore for tryingout.
1884–1885	11 (6)	273			19	The figures in ( ) for 1884, 1885, and 1886 were reported by Townsend
1885–1886	10 (8)	300			21	Included one right whale

(*continued*)

Appendix (Continued)

Year	Number of whales	Barrels	Price ($) Barrels/gallon	Value	Source[a]	Comments
Lower California–Santo Tomas and Punta Banda						
1860	5				10	
1870–1871		684	19.20/0.60	13,132.80	11	
1871–1872	8	240			11	
1872–1873	17	700			12	
1873–1874	17	400			13	
1875–1876	12	432			15	
1876–1877		292			16	
1877–1878	10	286			17	
Punta Santa Eugenia						
1885		80			21	

[a](1) *Crescent City Herald*, September 1854; (2) *Sacramento Union*, October 1852; (3) *Sacramento Union;* March 11, 1855, (4) *Sacramento Union;* August 22, 1855, (5) *Monterey Sentinel;* November 10, 1855, (6) *Los Angeles Star;* January 12, 1861, (7) *Los Angeles Star;* March 9, 1861, (8) *San Diego Weekly Herald;* March 22, 1859, (9) *San Diego Weekly Union;* February 2, 1869, (10) *San Diego Weekly Union;* March 6, 1869, (11) *San Diego Union;* January 1, 1872; February 11, 1872; March 9, 1872, (12) *San Diego Union;* March 20, 1873; April 10, 1873; July 4, 1873; November 18, 1873, (13) *San Diego Union;* April 10, 1874, (14) *San Diego Union;* December 22, 1876, (15) *San Diego Union;* March 20, 1875; May 9, 1875; January 25, 1876, (16) *San Diego Union;* January 5, 1877; February 18, 1877; April 17, 1877, (17) *San Diego Union;* February 20, 1878; May 4, 1878, (18) *San Diego Union;* December 24, 1883; March 14, 1884, (19) *San Diego Union;* March 10, 1885; March 13, 1885, (20) *San Diegc Union;* March 21, 1886, (21) *San Diego Union;* February 19, 1885, (22) Barrows, 1893, (23) Collins, 1892, (24) Jordan, 1887, (25) K. Nesheim, unpublished manuscript; note that sources 27–32 were cited by Nesheim in his unpublished manuscript, (26) *State Register and Year Book of Facts for the Year 1857*, Langley and Mathews, p. 229. San Francisco, (27) State Register and Year Book of Facts for the year 1859, Langley and Mathews, p. 286. San Francisco, (28) California Surveyer General, Reports of the County Assessors, in Appendix to Journals of Assembly, 11 Session, 1860 III p. 72, (29) Annual report of the Surveyor General, 1861 and 1865, Sacramento. O. M. Clayes, State Printer, 1866, (30) Transactions of the California State Agricultural Society During the Years 1866 and 1867, Sacramento; D. J. Gelwicks. State Printer, 1868. p. 579, (31) Report of the Surveyor General of California from November 1, 1867. Sacramento: D. J. Gelwicks, State Printer, 1868. p. 85, (32) U. S. Bureau of the Census, Eighth Census of the United States: 1860, (33) Smythe, W. E., History of San Diego: San Diego Printing Co., 1892, (34) Starks, 1922, (35) Townsend, 1886, (36) Wilcox, 1895

[b]The figures quoted for these two years are exactly the same. Quotations come from two different papers on different dates. I believe the figures should be considered for 1855, since a third paper (*San Francisco Daily Bulletin*, 12 November 1855) gives 24 whales killed from April to November 1855.

References

Barrows, H. D. (1893). "Memorial and Biographical History of Coast Counties of Central California." Lewis Publ. Co., Chicago, Illinois.

Berwick, E. (1900). Offshore whaling in the Bay of Monterey. *Cosmopolitan* **29**(6), 631–637.

Bradford, G. (1952). "The Mariner's Dictionary." Weathervane Books, New York.

Collins, J. W. (1892). "Report on the Fisheries of the Pacific Coast of the United States in 1888," Report of the Commissioner, U.S. Commission of Fish and Fisheries. U.S. Govt. Printing Office, Washington, D.C.

Davidson, W. (1929). "Where California Began." McIntyre Publ. Co., San Diego, California.

Dias, E. M. (1979). A pesca costeira na California. "Baleeiros Portugueses na America," Sep. do Bol. No. 35, pp. 21–25. Composta e Impresso nas Oficinas Graficas da Tipographia Angrense.

Goode, G. B. (1887). "Fishery Industries of the United States." Sect. 5, Vol. II. U.S. Govt. Printing Office, Washington, D.C.

Hamilton, G. (1974). "Where the Highway Ends," Chapter 12, pp. 141–149. Padre Productions, San Luis Obispo, California.

Hayes, B. I. (1861). "Scrapbooks, Southern California, Los Angeles County, Monterey, Santa Barbara and San Luis Obispo Counties, 1851–61." Bancroft Library, Berkeley, California.

Henderson, D. A. (1972). "Men and Whales at Scammon's Lagoon," pp. 91–200. Dawson's Bookshop, Los Angeles, California.

Hittel, J. S. (1882). "The Commerce and Industries of the Pacific Coast of North America," p. 349. Bancroft, San Francisco, California.

Jordan, D. S. (1887). "Boat Whaling in the Present Century. The Fisheries and Fishing Industries of the United States." Sect. V, Vol. II, pp. 55–56. U.S. Govt. Printing Office, Washington, D.C.

McGrew, C. A. (1922). "City of San Diego and San Diego County, the Birthplace of California," Vol. I, pp. 318–320. The American Historical Society, Chicago, Illinois.

Newmark, H. (1930). *In* "Sixty Years in Southern California" M. R. Newmark and M. H. Newmark, (eds.), 3rd ed., pp. 1–744. Houghton Mifflin, Boston, Massachusetts.

Orlando, A. (1980). "A Day on the Bay. CIAO Study Group." University of California, Santa Cruz.

Rice, D. W., and Wolman, A. A. (1971). Food and feeding. The life history and ecology of the gray whale (Eschrichtius robustus). *Spec. Publ.—Am. Soc. Mammol.* **3,** 32.

Scammon, C. M. (1874). "The Marine Mammals of the Northwestern Coast of North America," Chapter V, pp. 247–251. John H. Carmany & Co., San Francisco, California (Dover Edition, New York, 1968).

Starks, E. C. (1922). "A History of California Shore Whaling," Fish Bull. No. 6. State of California Fish and Game Commission, California State Printing Office, Sacramento.

Stewart, D. M. (1965). "Frontier Port, a Chapter in San Diego's History," p. 97. Ward Ritchie Press, Los Angeles, California.

Townsend, C. H. (1886). Present condition of the California gray whale fishery. *Fish. Bull.* **6,** 346–350.

Walton, V., and Curtis, A. (1875). "The Handbook of Monterey and Vicinity." Bacon & Co., San Francisco, California.

Watkins, R. C. (1925). "History of Monterey and Santa Cruz Counties, California," pp. 217–222. S. J. Clarke Publ. Co., Chicago, Illinois.

Wilcox, W. A. (1895). "Fisheries of the Pacific Coast in 1893," Report of the Commissioner, U.S. Commission of Fish and Fisheries. U.S. Govt. Printing Office, Washington, D.C.

7

Nineteenth Century Gray Whaling: Grounds, Catches and Kills, Practices and Depletion of the Whale Population

David A. Henderson

Introduction

Catching gray whales was not easy, but because it could be profitable catches in the nineteenth century greatly depleted the number of gray whales. Yankee whalers who dared the dangerous coastal and lagoon waters of Baja California in winter and the treacherous ice-filled seas of the North Pacific and Pacific Arctic in summer to chase the

ISBN 0-12-389180-9

wily and destructive devil fish, or gray whale, have much to tell us about their prey. Experiences of the chase, catch reports, and observations that would increase catches reveal much about the populations, behavior, migrations and their timing, the ranges and thus whaling grounds, food, predators, whale size, and other data related to the gray

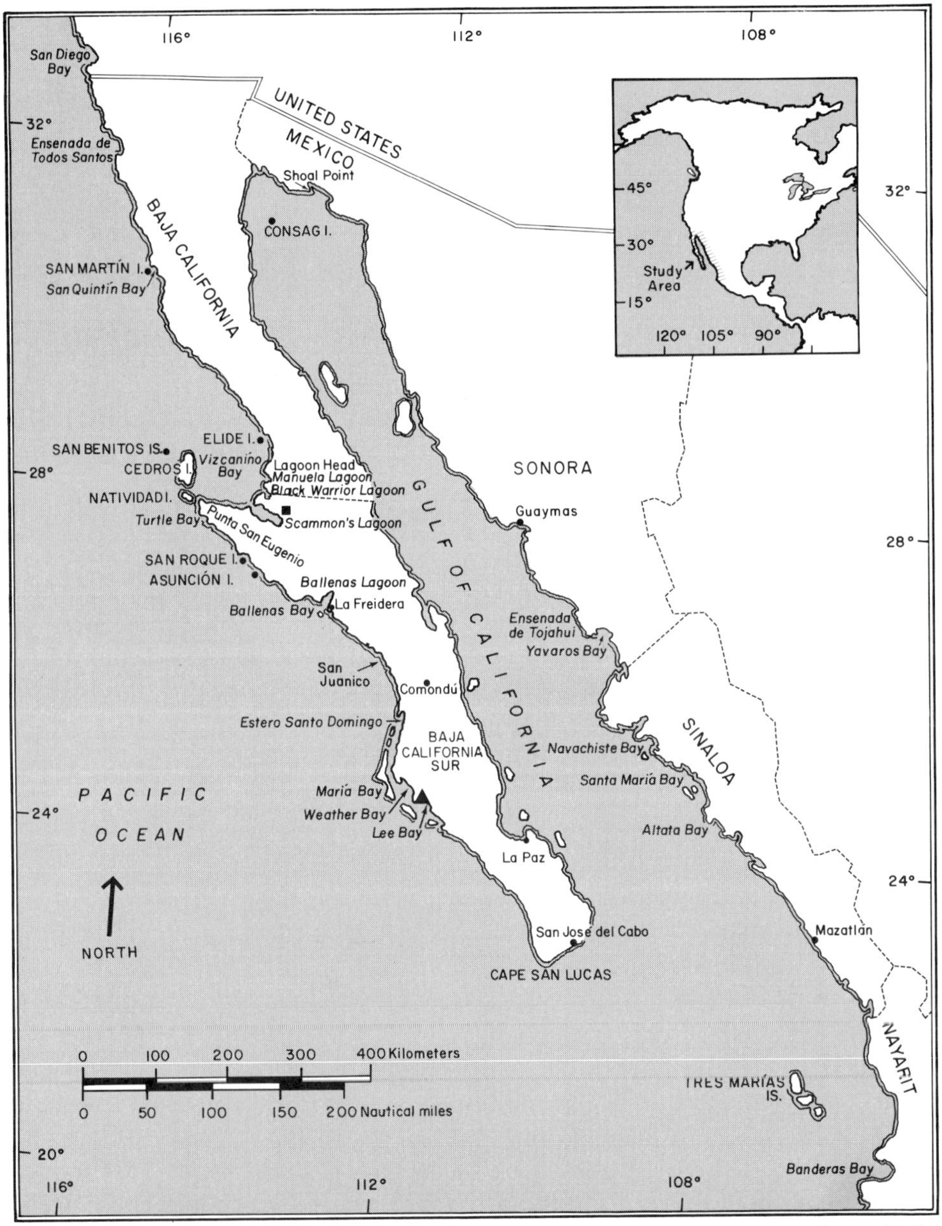

Fig. 1. *Gray whaling grounds of Northwest Mexico.* ■, *Ojo de Liebre Waterhole or Spring;* ▲, *mud hole.*

whales. The history and geography of nineteenth century gray whaling, which shed light on the preceding subjects, are important concerns of this chapter. First, however, we should follow whalemen through history to the gray whaling waters which were confined to the North Pacific.

It was a long sail, historically, from the Atlantic shores from which high-seas whalers had set forth in their vessels since the thirteenth century (Sanderson, 1956, p. 136), to distant Pacific coasts along which Yankee and European whalers of the nineteenth century found an unfamiliar prey awaiting their harpoons and lances, the gray whale. These whalemen found and pursued gray whales in coastal waters of Alta (Upper) California, Baja (Lower) California, Northwest Mexico, and in the Bering, Okhotsk, and Pacific Arctic seas. Once in the Pacific Ocean, whalers searched out that ocean's whaling grounds rapidly.

The first whaler to enter the Pacific was the ship *Amelia* (*Emelia*) of London which rounded Cape Horn in 1789 (Henderson, 1972, p. 16). Whalers then pushed their ships up the west coast of South America, and by 1793 the Englishman Colnett had conducted a survey for whales on the ship *Rattler* as far north as Cape San Lucas (Fig. 1), (Henderson, 1972, p. 15). By 1795, Captain John Locke on the English ship *Resolution* had carried out the first genuine whaling venture in the eastern North Pacific Ocean, during which he reached Baja California. Locke sailed around the Cape of Good Hope with supplies for Australia; he probably sailed next to India and/or China, and thence across the broad North Pacific without a landing until he reached Baja California in October–November 1795 (Cumpston, 1964, p. 29; Henderson, 1972, pp. 17–21). His landfall was near the mouth of the Guadalupe River, and he put ashore just south of there for badly needed "recruits" (supplies) at Ensenada de Todos Santos and later at the end of the peninsula at San José del Cabo. At these places, the Spaniards on the weak mission frontier met his needs and learned that he had 500 barrels of whale oil on board, undoubtedly the sperm whale oil which whalers of those years most desired.

Following Locke's visit, armed English whaleships came to Baja California and Alta California waters, at the end of the eighteenth century and in the early years of the nineteenth century, to prey on sperm whales and the small and isolated Spanish-governed communities alike (Henderson, 1972, p. 21). These whalers followed closely in the wake of English and Yankee vessels that had been guided to the sea otter and fur trade of the Northwest Coast of America by the discoveries of Captain James Cook in 1778 (Henderson, 1972, p. 16). Yankee whalers soon followed the English, and after 1815 American whaleships in the Pacific Ocean outnumbered those of other nations. It was from the coast of Baja California that the first two whaleships arrived in the Hawaiian Islands in 1819—the American ships *Balaena* (New Bedford, Captain Gardner) and *Equator* (Nantucket, Captain Folger) (Stackpole, 1964, pp. 267–268). From that island mecca of Pacific whalers, ships subsequently cruised northward in summer to Japanese seas, the northwest coast of America, and the northeast coast of Asia (both known as the Northwest Coast to whalers), to the Bering and Okhotsk seas, and finally in 1848 to the Pacific Arctic Ocean. The coasts of the Californias became important sperm whale grounds where American, English, and European whalers made abundant catches (Henderson, 1972, pp. 21–22).

North Pacific Whaling

When whalemen began to depart the Hawaiian or Sandwich Islands in spring to cruise for sperm, right, and bowhead whales, they also left ice and cold at summer's end to return to the Hawaiian Islands in autumn. Thence in winter the whalemen could steer for a variety of warmer grounds: New Zealand or even Chilean waters, the line or equator, and the Marianas; most important for this history was the Coast of California ground which to the whalers meant the ground on the west coast of Baja California. Among the various winter cruises, the sail to California was considered a short and easy one during which southern westerlies blew ships to the coast and trade winds carried them back to the Islands. The crossing in each direction took only about 3 weeks.[1] Another recommendation for the cruise to California was that ships could lie at anchor all season in a bay or lagoon or make short runs up and down the coast instead of ranging where wind and wave could be damaging and costly.

A whaling pattern evolved in the North Pacific which involved visits to the Hawaiian Islands in spring and autumn and whaling in northern parts of the Pacific in summer and in warmer parts of that ocean in winter.[2–4] Hundreds of vessels called each spring and fall at the Hawaiian Islands for recruits and to get rid of some whaling hands and sign on others, many of whom were Kanakas. To the Yankee, a Kanaka was any Pacific islander, but most were from Polynesian islands and of those most were from the Hawaiian Islands.[1] Because the northern season when right and bowhead whales were captured was the most productive season for most whalers, the number of "seasons out" from home meant the number of summer seasons spent in the Pacific. Winter whaling on the coast of California or elsewhere was known as "between the seasons" whaling (Henderson, 1972, pp. 85–86). In addition to the cruise patterns which sent most gray whalers from the Hawaiian Islands to Baja California and back to the Islands in the spring, some whalers from New England or San Francisco sailed out of that port directly down the coast to the gray whaling grounds. Some left San Francisco in the spring and summer in search of sperm whales or elephant seals before going after the grays. Other whalers out of New England ports and San Francisco remained in peninsular waters or went south from there to search for sperm whales after the winter gray whaling season had ended (Henderson, 1972, pp. 85, 91–94, 188, 268–270).[1]

Gray Whaling Beginnings

During the several decades preceding gray whaling when whalers worked the sperm whale fishery of Baja California, they recognized gray whales as "scrag" whales

[1]See also logbooks in Whaling Archives (see Appendix).

[2]See also reports in *Seaman's Friend and Temperance Advocate*, Honolulu.

[3]See also reports in *Pacific Commercial Advertiser*, Honolulu.

[4]See also reports in *Whalemen's Shipping List and Merchant's Transcript*, New Bedford, Massachusetts.

and even chased them, but so far as the record shows they never caught any. Of course, native peoples of Baja California and the shores of the northern Pacific Ocean, Okhotsk Sea, and Arctic Ocean, as well as Spaniards, Mexicans, and Siberian Russians, also knew and recognized this whale (Henderson, 1972, pp. 35, 59, 71, 73; Mitchell, 1979, pp. 309–311).[5] Eventually two Connecticut captains who came to the Coast of California decided to go after the grays. Magdalena Bay, known to the whalers as Marguerita Bay, an immense complex of bays and lagoons on the southwest coast of Baja California, was a well-known "between the season" protected haven which sperm whalers from the coast knew well, and it was also a calving and mating ground to which gray whales migrated each winter (Fig. 1).

Thus, whalers had long known that gray whales frequented the bay when Captain James Smith of the New London ship *Hibernia* and Captain Josiah Stevens of the Stonington bark *United States* decided to chance pursuit of grays there in the winter of 1845–1846. That winter in Magdalena Bay, in the first gray whaling by western man on record, the whalers of *Hibernia* captured 22 grays and those of *United States* took 10 (Henderson, 1972, p. 82). They found that the gray whales had no bone worth saving. More importantly, they learned, as did all whalers who chased grays, that they were in a dangerous business. The grays killed the second mate and badly injured the first mate of *United States*. Thus was initiated the nineteenth century gray whaling which would eventually disastrously deplete the population of gray whales in the eastern Pacific California herd and make a significant reduction in the Okhotsk Sea or Korean herd of the western Pacific.

Whalers' Names for Grays

Some of the characteristics of the gray whale are revealed by the various names that the nineteenth century gray whalers applied to this whale which was new to the chase. Even before gray whaling began on the Coast of California, whalemen there hunted sperm whales and also recognized this different whale, which they called a scrag whale because they identified it with a whale of that name reportedly taken off New England coasts in the seventeenth and eighteenth centuries (Dudley, 1725, p. 258). However, early in the history of the fishery of this whale the whalers began to call it California gray, gray whale, gray back, and California whale. Devil fish and hard head were names plainly derived from the reaction of the whales to their hunters, who recognized the gray as a dangerous prey knowing that in the lagoons of Baja California the whales commonly crashed into and stove boats with their heads. The names mussel digger, mud digger, or digger referred to the whales' habit of bottom dredging, which had been observed by the whalers especially in the lagoons. Whalers used two names more on the northern summer grounds than on the Coast of California; one was ripsack, which apparently referred to the manner in which blubber came off the whale in cutting-

[5]See also logbooks in Whaling Archives recording cruises to the Okhotsk Sea (see Appendix).

in (flensing), and scamper down, also the name of another whale but unmistakably used to refer to grays, was the other. Curiously, some whalers referred to the whales by one name on the summer grounds and another on the winter grounds—for instance, ripsack in the Arctic or Okhotsk seas and devil fish or gray on the Coast of California (Henderson, 1972, pp. 26–27, 35).[1]

Initial Period of Gray Whaling

The first years of gray whaling, including the winters of 1845–1846 through 1853–1854, have characteristics that distinguish them from later years. At that time, Magdalena Bay was the only significant gray whaling ground being visited and thus supplied nearly the entire catch (Table I). A few vessels did whale along shore on the coast in addition to entering the bay; the first was the ship *Dromo* of New London in 1851–1852.[6a] Moreover, the captains of some vessels which whaled in the Bering Sea and Arctic Ocean made small incidental gray whale catches when their main quarries, bowhead and right whales, were not available to be chased. Thus, the catch per vessel was small as was the total catch in northern waters.[1,2,4,6]

Connecticut vessels, perhaps from venturesomeness born of competition with the New Bedford and other southern Massachusetts ports, sought out the new fishery and dominated it. Perhaps commercial connections and information drew Connecticut ships together in winter at Magdalena Bay. At any rate, in the second winter of whaling at Magdalena Bay, 1846–1847, three New London, two Stonington, and two Mystic whalers met with one New Bedford ship on that ground.[1,2,4] In the early period, half of the perhaps 50–60 cruises to Magdalena Bay were made by vessels from Connecticut ports. Southern Massachusetts ports, including New Bedford, sent their ships to the new ground only 10 or 11 times. Two vessels each came from Sag Harbor, Long Island, and Warren, Rhode Island, and one from Cold Spring Harbor, Long Island. In 1852–1853, one of the first whalers out of San Francisco, the ship *Aquetnet* (Captain Curry) fished at Magdalena Bay. Foreigners also were willing to try their luck in the new fishery, and from Le Havre came five French vessels and from Bremen and Amsterdam one each.[1,2,4,6] Thus, the representation of hailing ports by vessels at the new ground was distinctive; it was not dominated by ships from southern Massachusetts ports, most of the captains and owners of which probably had not yet the need or desire to chance the pursuit of gray whales.

In this earliest period many of the vessels on the coast or that entered Magdalena Bay to escape weather or cooper oil were sperm whalers, while some captains pursued humpback whales in and outside the bay; however, on at least 50–60 cruises, the whalers concentrated on taking gray whales. The daring and crafty crews of a few vessels captured many; for example, the men of the New London ship *Brooklyn* (Captain Jeffrey) took 29

[6]See also "Dennis Wood Abstracts," manuscript records of whaling voyages 1831–1873. New Bedford Free Public Library, New Bedford, Massachusetts.

[6a]Whalemen's Shipping List and Merchant's Transcript; March 30, 1852, p. 18.

whales in 1846–1847.[1,2,4,6] Most, however, took far fewer. Captain Curry of the *Aquetnet* of San Francisco got only four grays in 1852–1853, although he was at Magdalena Bay from late December to mid-February.[1,2,4,6,7] Some crews were "skunked," that is, they left the bay without taking any oil and were undoubtedly glad to quit such a dangerous business. During the earliest gray whaling period, few captains brought their ships and men back for a second season of the new and hazardous whaling. One who did was Captain James Smith of the New London ship *Hibernia* who had originated gray whaling in 1845–1846 with an excellent catch of 22 whales; he returned the following winter to make another catch at least equal to his first.

A Mexican government report from the mid-nineteenth century archives at La Paz tells us that during the first 3 years of gray whaling at Magdalena Bay, 1845–1846 through 1847–1848, 38 whaleships captured 338 whales in Magdalena Bay (Espinosa, 1854, p. 125). If this report is added to other reports of whalers at the bay later, then the total number of cruises to Magdalena Bay in the early period may have been nearly 70, and the total capture may have been nearly 500 whales. However, if estimates are based on whalemen's records in logbooks and newspapers, the estimate for the early period is closer to 50–60 cruises and 400–450 captures. The capture discrepancy could then amount to as many as 100 whales. This is probably explained by the Mexican report having included some vessels whose crews were not gray whaling, and thus vessel and catch totals were increased by including sperm and humpback whalers. If there were 50–60 cruises for grays and 400–450 captures, then the average catch per vessel was 7–9 whales, figures that seem reasonable in a new fishery in which the men of some vessels caught many whales and those of others caught few or none.

Scammon's statement that in the winter of 1848 (1847–1848) 50 whaleships went to Magdalena Bay is an obvious exaggeration (Scammon, 1874, p. 268). Scammon did not arrive in California until 1850 and he did not whale at Magdalena Bay until the winter of 1855–1856, when he began his gray whaling career at the bay on the San Francisco ship *Leonore;* this was 2 years before he first whaled in Scammon's Lagoon on the brig *Boston* (Henderson, 1972, pp. 82, 267–271). In the winter of 1847–1848, records show that only 20 gray whaling vessels were at Magdalena Bay and that their crews captured at most some 140 gray whales, or an average of 7 each.[1,2,4,6]

A pattern that would persist throughout the history of the fishery had been established; some crews, experts at gray whaling, made exceptional catches, many took few whales, and some who had no knowledge of or taste for chasing the tough grays came away from the Coast of California without a drop of oil.

Right in the middle of the earliest gray whaling period there were three seasons, 1848–1849, 1849–1850, and 1850–1851, during which pursuit of the gray backs stopped. Scammon attributed the abandonment of gray whaling to the disasterous destruction of boats, loss of life, and frequent injuries sustained by those who pursued the grays or devil fish and to the concentrated whaling and associated noise and try-pot fires, which drove the whales outside Magdalena Bay (Scammon, 1874, pp. 268–269). Moreover, the low price of the dark-colored gray whale oil and the fact that the whale's bone

[7]See also reports in *San Francisco Alta California*, San Francisco, California.

(baleen) was next to worthless contributed to the temporary abandonment of the fishery (Henderson, 1972, p. 84).[1,2,4,6] However, the proximity of the California ground to the Hawaiian Islands and San Francisco and probably also overfishing of sperm whales on the Coast of California soon brought revival of the gray whale fishery at Magdalena Bay in the winter of 1851–1852. During that season and the two seasons following it, between 4 and 12 vessels visited Magdalena Bay each winter, and the catch of 1852–1853 was between 65 and 75 whales.[1,2,4,6]

The number of whales which were mortally wounded but escaped the whalers in the early period was probably about 1 for every 10 captured. Although the bomb-lance was not much used in the early period and more whales therefore probably escaped after being attacked, fewer of those which escaped would have been mortally wounded than when the bomb-lance came into wide use in the fishery. Thus, unrecovered dead whales probably averaged about the same relative to those caught both before and after the use of the bomb-lance, which mortally wounded more whales but which also brought death before the wounded animal could swim beyond the inside waters in which its carcass could be recovered. Thus, in the early period of gray whaling at Magdalena Bay, whalemen probably captured 400–450 grays and killed about 450 to 500.

The few cruises alongshore in the early period probably garnered whalemen only about 25 grays, and the number of grays captured on the northern summer grounds probably amounted to only about 52 whales. By the late 1840s whalers in the northern Bering Sea and Arctic Ocean, as well as in the Okhotsk Sea, had observed and commented upon gray whales in their logbooks. They often referred to them as scrag whales, but they chased and captured few as these early whalers were still impressed with the abundance of right and bowhead whales in the northern waters. The loss of grays in the northern waters was high as the wounded whales could escape in fog and under ice to die unretrieved; thus the size of the kill was nearly one-half again that of the catch (see Table I, footnote *g*). As shore whaling had only just begun in this period (at Monterey in 1854), probably only some 10 gray whales were captured there in the initial year of whaling. Whalers made some 75–85 cruises for grays in the early period, and one shore station had been opened along their migration route. The total catch on all grounds in the early period was only some 487–537 grays, and the total kill was between 555 and 610 (Table I). These figures are lower than the estimates I made for this period in 1972 because I have since reviewed more and better records. Nevertheless, my estimate of the total catch and kill of gray whales in the nineteenth century remains about the same because of increases corresponding to other times and places.

Bonanza Period of Gray Whaling

By far the most important gray whaling period was that which lasted the 11 winters from 1854–1855 through 1864–1865. In that period whalers made about two-thirds of all the captures and kills of grays in the California herd that took place prior to the near abandonment of gray whaling from ships in the mid-1870s. Whaling increased at the old ground of Magdalena Bay; in the winter of 1856–1857, whalers captured 250 grays there and for six seasons (1855–1856 through 1860–1861, gray whaling remained intensive at the bay (Henderson, 1972, p. 151).

At Magdalena Bay, whalers frequented the northern bay of the complex known as Main or Weather Bay as in past seasons, but they also learned of new highly productive parts of the complex. Many vessels anchored in Lee (Almejas) Bay, south of Weather Bay, where they chased the whales especially at the "Mud Hole" at the northeast corner of Lee Bay and along the eastern shore of this bay. More daring captains with large ships, but especially those with small brigs and schooners, had their boat crews tow and kedge their vessels northward into the tortuous, shallow channels of Estero Santo Domingo, north of Weather Bay and behind the long barrier bar, where at low tide vessels rested on their beam ends (Fig. 1). In the bonanza period, 1854–1855 through 1864–1865, whalers on some 125 cruises to Magdalena Bay captured about 1400 grays but killed about 1540 (Table I).

As important as the increase in whaling at Magdalena Bay were the discoveries of new grounds. At the Mexican *Laguna Ojo de Liebre* in 1857–1858, Captain Charles Melville Scammon with the San Francisco brig *Boston* and her schooner–tender *Marin* opened a short-lived bonanza of gray whaling at this calving and mating ground of the gray whale (Henderson, 1872, pp. 113–116, 261). The lagoon became known among whalers as Scammon's Lagoon. During the bonanza period at this lagoon, whalers on 36 cruises took some 553 devil fish and caused the death of about 608 (Table I). Just north of Scammon's Lagoon lie Black Warrior (*Guerrero Negro*) Lagoon and Manuela lagoons. They lie behind the same barrier bar as does Scammon's Lagoon, but are separated from it and one another by transverse bars. Both are mostly tidal flats and have little channel area for gray whale occupance. Few captains poked their vessels' bows into these lagoons, especially after the Honolulu bark *Black Warrior* had entered the lagoon that acquired her name and with her tender, the schooner *J. D. Carr*, had cruised in the lagoon for 2 weeks without success in December 1858 and then was lost just outside while being towed out by the boats of several vessels when swept against the shore by a current (Henderson, 1972, pp. 29–30, 119–121; Scammon, 1874, p. 123). Laguna Manuela, north of Black Warrior Lagoon, has other names, but probably acquired that name from the San Francisco whaling brig *Manuella* which was wrecked in 1871, apparently along its outer coast (Henderson, 1972, p. 48). Whaling vessels did not enter that lagoon. What few gray whales were caught in the two lagoons were taken by boats from whaleships lying outside along the coast.

The same sagacious whaler who first entered the lagoon now bearing his name, Captain Scammon, was also first into another calving and mating area, Ballenas or San Ignacio Lagoon. The husband of Scammon's wife's sister, Captain Jared F. Poole, had seen the lagoon in 1857 and in 1858 he tried to find a channel into it. In the winter of 1859–1860, Scammon took a small fleet of San Francisco vessels into the lagoon (Henderson, 1972, pp. 122–123). That year his schooner–tender *Kate* led his bark *Ocean Bird* and two other barks, Poole's *Sarah Warren* and Captain Easton's *Carib* with two other tenders, between pounding breakers and across the shallow bar into Ballenas Lagoon (Fig. 1). This was a feat with which Scammon had become familiar at Scammon's Lagoon. Boats of two New Bedford ships joined whaling at the lagoon that winter, but only *Henry Kneeland* went in; *John Howland* remained outside. In subsequent winters many of the grays captured at Ballenas Lagoon were taken in shallow waters behind bars lying southeast of the lagoon mouth (Henderson, 1972, p. 154). Moreover, whalers who chased grays inside Ballenas Lagoon and its southeastern lagoons (*esteros*) also attacked the whales in Ballenas Bay outside the lagoon (Henderson, 1972, pp. 154–155). Thus, it is sometimes difficult to determine if a particular ship captured a whale inside or outside the lagoon itself; only if one has logbook records at hand, rather than newspaper accounts, can he determine how many whales were taken inside or outside the lagoon. The best estimate I can make is that during the bonanza period whalers on some 20 cruises captured about 350 whales, but killed some 385 at the main lagoon and in the smaller lagoons (*esteros*) to the southeast of Ballenas Lagoon (Table I).

Other new whaling grounds and practices were quickly discovered and exploited. At San Diego Bay, San Quintín Bay, and at Turtle Bay (Saint Bartholomew's Bay), ships anchored out of the weather and sent their boats into the kelp beds outside where they lay in wait for grays traveling their migration routes to and from northern waters (Henderson, 1972, pp. 28–29). At San Quintín and Turtle bays, whaleships sometimes waited outside the harbors; at San Quintín, in the lee of tiny San Martín Island (Henderson, 1972, p. 29). At María Bay, south of Cape San Lázaro on the sand bar coast outside northern Magdalena Bay, many whalers also awaited migrating whales. Alongshore whaling, in which vessels waited along the migration route, sometimes anchored and sending their boats off for whales, became a common practice especially when lagoon whale populations began to decline. Thus, whalers along the coast north of several lagoons had a chance to waylay whales en route to several calving grounds.

The stretch of coast where alongshore whaling was most concentrated was between the south shore of Vizcaíno Bay, around Natividad Island and Cape San Eugenio, and on south to San Roque and Asunción islands (Fig. 1). The lee sides of islands and headlands here afforded protection for the waiting ships. From the ships, boats set out down the migration route propelled by sails and steady northern winds to capture grays by the method that Scammon called "sailing them down" (Scammon, 1874, p. 27). Boats with captured whales could row (pull) the whales back to the ships or the ships could sail south after their boats. A practice employed by some captains was that of buoying their anchors so that they could slip anchor quickly to sail after boats fast to rapidly swimming wounded whales.[1] Off the San Benitos Islands and Cedros Island, off San Juanico, the last point between Ballenas and Magdalena bays, and off Cape San Lucas, whalers also practiced alongshore whaling (Fig. 1). In Ballenas Bay, whalemen also took advantage of whaling along the migration route by pursuing whales entering and leaving Ballenas Lagoon. During the peak period of gray whaling, whalers on some 80 alongshore cruises captured some 900 whales, but killed about 990 (Table I).

Some masters took their vessels to places in the Gulf of California, but especially to grounds along the mainland coast of the southern Gulf and south from there (Henderson, 1972, pp. 30–31, 96–97, 165–166). Most important of these grounds was Banderas Bay, which was the southern limit of gray whaling at about 20°N latitude. Others

Table I

Catch and Kill of Gray Whales in the California Herd, 1845–1846 through 1873–1874[a]

Whaling grounds	Cruises	1845–1846 through 1853–1854: captured[b]	Cruises	1854–1855 through 1864–1865: captured[b]	Cruises	1865–1866 through 1873–1874: captured[b]	Cruises	Totals: 1845–1846 through 1873–1874: captured[b]
Magdalena Bay	50–60	400–450[c] (440–495)[d]	125	1400 (1540)	25	150 (165)	200–210	1950–2000 (2145–2200)
Scammon's lagoon			36	553 (608)	5	38 (42)	41	591 (650)
Ballenas Lagoon (San Ignacio Lagoon) and adjacent waters			20	350 (385)	7	50 (55)	27	400 (440)
Baja California lagoons and bays: Totals	50–60	400–450 (440–495)	181	2303 (2533)	37	238 (262)	268–278	2941–2991 (3235–3290)
Alongshore whaling on the coasts of the Californias	5	25 (28)	80	900 (990)	45	400 (440)	130	1325 (1458)
Shore Whaling Stations and Camps		10 (12)[e]	—	1190 (1428)	—	960 (1152)	—	2160 (2592)
Gulf of California and mainland Mexico	?	?	15	150 (165)	10	50 (55)	25	200 (220)
Northern Summer[f] Grounds	20	52 (75)[g]	80	175 (232)	40	175 (232)	140	402 (539)
Totals	75–85	487–537 (555–610)	356	4718 (5348)	132	1823 (2141)	563–573	7028–7078 (8044–8099)

[a]Information from reports in *Seaman's Friend and Temperance Advocate*, Honolulu; *Pacific Commercial Advertiser*, Honolulu; *San Francisco Alta California*; *San Francisco Bulletin*; *San Francisco Chronicle*; "Dennis Wood Abstracts," manuscript records of whaling voyages 1831–1873, New Bedford Free Public Library; and *Whalemen's Shipping List and Merchants Transcript*, New Bedford, Massachusetts. Sources: Summer catch and kill data based on information collected by Dr. John R. Bockstoce, Old Dartmouth Historical Society, New Bedford, Massachusetts, when making his studies of bowhead whaling in the Western Arctic and Bering Sea. Henderson, 1972, 258. Logbooks in Whaling Archives listed in Appendix. Scammon, 1874, pp. 250–251, 270. Starks, 1922, pp. 17–18, 20, 29. Townsend, 1935, pp. 19–20.

[b]Numbers in parentheses are the whales killed (not all were retrieved or captured).

[c]Average yield per whale 35 barrels.

[d]Includes whales captured plus 1 lost mortally wounded for every 10 captured. Ratios the same throughout Table I except for shore and northern gray whaling.

[e]Includes whales captured plus 1 lost mortally wounded for every 5 captured.

[f]Includes summers following winters designated at head of column.

[g]Catch and kill information based on data collected by Dr. John R. Bockstoce. See footnote *a*.

included inlets and alongshore waters at Yavaros and Tojahui in southern Sonora, and Navachiste, Santa María, and Altata on the coast of Sinaloa (Fig. 1). A few captains also chased grays at the Tres Marías Islands where the primary quarry was the humpback whale. Scammon once made a notation that grays also went to the head of the Gulf. He wrote "Around Consag Island and off Shoal Point near the head of the Gulf of California large numbers of California Gray whales have been seen in March (Henderson, 1972, pp. 31–32). In the bonanza period whalers in perhaps 15 cruises captured about 150 gray whales, but killed some 165 in the Gulf and along the Mexican mainland coast (Table I).

During this peak period of gray whaling, 1854–1855 through 1864–1865, shore whaling, using shore-based whaleboats and try-works on the beach, was initiated and expanded at a series of places on the Alta California shore and in Baja California. This low-cost kind of whaling without ships was never important in Mexican waters where the gray whalers on ships made almost all of their catch, and some of the shore stations there were not started until the final, declining period of gray whaling. However, in the peak period at Magdalena Bay, local Mexicans, as well as Americans and English whalers (probably deserters) who lived northeast of the bay at Comondú, came to the shores in winter to practice "carcassing." This operation involved the retrieval of "stinkers" or dead whales and also carcasses which still had tongues and organs in them, to try-out the blubber and meat in order to sell oil to ships in the bay (Henderson, 1972, p. 127). These shore whalers also caught a few live whales. Moreover, in 1860–1861 at least there was a shore station on the southeast side of Ballenas Lagoon, at a place that was designated as Carcass Town on Scammon's map of the lagoon and is still called *La Freidera* (Try-pot or Tryworks) by local Mexicans (Henderson, 1972, p. 100). Those vessels anchored for the season at San Diego and San Quintín bays, which sent their boats outside to whale, became in essence floating shore stations. Shore whalers at all stations probably captured some 1190 grays, but caused the death of 1428 during the peak period of gray whaling (Table I). Not many of these whales were captured or killed in Baja California waters.

Beginning in the 1840s, as whalers in summer cruised farther and farther northward up the coast of Asia to enter the Arctic Ocean in 1848, gray whales on their northern summer grounds were attacked when right and bowhead whales were nowhere to be found; the whalers had really come north for those two whales. By the peak period of gray whaling, whalers in many vessels pursued grays in the north, but with rare exceptions they captured few on any one northern cruise. In 1858, when ice conditions were bad in the Bering Straits, whalers caught gray whales early in the season, as they often did, before they could sail on north in search of bowheads.[8] In that summer Captain Fisher of the Edgartown bark *Rose Pool* reported on the catches of 30 vessels in the Bering and Arctic seas as of July 18. Of the 30, 5 had gray whale catches amounting to a total of 13; none had captured more than 4. However, that year another report tells us that the New Bedford bark *Emerald* (Captain Peirce) as of August 25 had a catch of 2 bowheads, but 9 devil fish.[9] As mentioned earlier, many were killed but lost in fog and

[8] *Seaman's Friend and Temperance Advocate*, Honolulu; October, 1858, p. 80; November, 1858, p. 86; December, 1858, p. 94.

[9] *Seaman's Friend and Temperance Advocate*, Honolulu; November, 1858, p. 86.

under ice; so that more whales were lost, relative to those caught, than in any other sector of the gray whale fishery. Grounds of the California herd where grays were most pursued and captured included the Koryak Coast and Anadir Sea, waters in and off Plover and Saint Lawrence bays, the coast of Saint Lawrence and King islands, the Bering Straits (especially off East Cape), the Chukchi Sea, and all along the north coast of Alaska eastward nearly to Point Barrow. Even during the later period of steam whaling, none appears to have been sighted in the open northern Arctic Ocean on such bowhead grounds as those around Herald Island.[1–4] In the north, it appears that there was more of a mix of large and small whales, for many of the grays captured in the north yielded less oil than those taken on the Coast of California. In all, whalers on about 80 northern cruises probably captured 175 grays, but killed some 232 during the bonanza period (Table I).

As the waters in which gray whales could be caught became widely known and as techniques and equipment improved, the number of vessels that came to the Coast of California each winter increased. The years of maximum cruises and catches were between 1857–1858 and 1861–1862. After rapid depletion, the grounds sustained significant whaling until the mid-1860s. Some captains sought not only grays but also sperm whales or humpbacks or elephant seals. Some cruises by small vessels out of San Francisco were called "pick-up" cruises, and on such a cruise in 1856–1857 the schooner *Henry* (Captain Reynolds) was out 10 months, excluding only summer, and returned with gray and sperm oil, green turtles, and the dried meat and shells of abalone (Henderson, 1972, pp. 94–95).

Catches were low for crews that knew not the ways of the gray whales and gray whaling but large for the experts. In 1856–1857, the few men on the schooner *Henry* filled their casks in just 4 weeks in Estero Santo Domingo where they obtained 500 barrels of oil from 19 whales. Scammon, on the bark *Ocean Bird* in 1858–1859 at Scammon's Lagoon, captured 47 whales in 3 months of whaling and returned to San Francisco with 1700 barrels of oil. On 9 different days from December 9 through February 27, his boat crews took 2 whales per day, and between January 13 and 17, 1859, they captured 6 whales (Henderson, 1970, pp. 39, 45–48, 52, 58; 1972, p. 91). In a catch which may have been made partially at Scammon's Lagoon but probably mostly in Ballenas Bay and Lagoon, Captain Whelden of the New Bedford ship *John Howland* took 1700 barrels of gray whale oil in the winter of 1860–1861.[10]

There are a number of reasons why gray whaling revived after its initiation and decline at Magdalena Bay in the late 1840s. Some reasons included the discovery of new grounds, and others were the new practices, techniques, and equipment that aided gray whaling. Among the most significant contributions to the improvement of gray whaling was the use of the bomb-lance. This device was a large shoulder-held gun which fired a pointed bomb so that a tough gray could be wounded or killed while a whale boat and crew were already attached to it by harpoon and line (a "fast" boat) although out of the range of danger (Henderson, 1972, p. 98). However, sometimes the bombs were fired while a fast (attached) boat was right next to a whale, in order to take the fight out of the whale before hand-lancing began, and sometimes bombs were fired

[10] *Whaleman's Shipping List and Merchants Transcript;* records of her catches, January–June, 1861.

into whales even before the harpoons had been thrown (darted).[1] Nevertheless, because the bombs were expensive, many gray whales were killed without the use of bomb guns even after the instrument became common.[1] Scammon said that the first winter's whaling at Scammon's Lagoon would have been a failure had it not been for the bomb-lances he had on board (Scammon, 1874, pp. 263–264). To shore whalers who had to catch grays swimming rapidly past the station along the migration routes, bomb-lances were especially useful in waylaying whales that were some distance from the boats.

Greener's guns, which were swivel-mounted in the bows of whaleboats, fired harpoons greater distances than they could be thrown, and these contrivances became especially useful in shore whaling or whaling outside lagoons in the breakers, where the migrating whales or rough water precluded close approach and use of the hand-thrown harpoon (Henderson, 1972, p. 98). Scammon termed this practice "whaling in the breakers" (Scammon, 1874, p. 28). Those whalers who employed the Greener's gun also used bomb lances and other kinds of guns which attached boats to whales or killed whales.

During the bonanza period of gray whaling on the Coast of California, vessels from a wide range of home ports entered the fishery. The Connecticut whalers continued to come, but so now did many from the ports of southern Massachusetts such as New Bedford, Fairhaven, and Nantucket. Others came from the Long Island ports. Moreover, vessels from San Francisco and Honolulu increasingly joined in the gray whaling. Some of these vessels were small brigs and schooners, especially well suited for reaching shallow grounds such as the long lagoon, Estero Santo Domingo, which leads northward from Magdalena Bay. German whalers, some of Hawaiian registry, and French whalers continued in the fishery; one Russian vessel, the brig *Constantine*, came to Scammon's Lagoon in 1859–1860 (Henderson, 1972, p. 261).[1–4] Her master Captain Lindholm, a Swede from Russian Finland, whaled in the Okhotsk Sea with the Russian whalers, and undoubtedly had been induced to come to the Coast of California by gray whalers who sought bowheads and rights in the Okhotsk Sea in summer (Beane, 1905, pp. 208–209).

The total catch and kill of gray whales in the California herd in the bonanza period were probably about 4718 and 5348 respectively (Table I). For whaling vessels, Magdalena Bay remained the most important ground; the alongshore ground was second in importance, followed by Scammon's Lagoon and Ballenas Lagoon. Grounds of lesser significance for the whaleships in the winter season were those in the Gulf of California and along the mainland shore of Mexico. However, whalers were catching a fair number of grays now on their summer grounds in the Bering Sea, Arctic Ocean, and Okhotsk Sea. Very significantly, shore whalers on the coasts of both Alta and Baja California, but especially in Alta California, probably had caught and killed more grays than whalers in any other single sector of the fishery except at Magdalena Bay.

Because the number of whales in the California herd was not large, gray whaling grounds were rapidly depleted during the 11 seasons in the bonanza period. Whalers essentially abandoned Scammon's Lagoon by the mid-1860s, and by that time Ballenas Lagoon was little more productive than was Scammon's. Magdalena Bay still had more whales left than did its northern counterparts but had declined rapidly as a gray whaling ground. The Coast of California was entering its final period as a gray whaling ground for sailing ships.

Period of Declining Gray Whaling

The last period of significant whaling from ships on the Coast of California was from the mid-1860s into the early 1870s. It was a period of such declining rewards that at its end masters of whaleships had abandoned most of the gray whaling grounds. During this last period, the formerly important grounds at Magdalena Bay, Scammon's Lagoon, and Ballenas Lagoon were reduced to minor whale grounds. For example, only 5 vessels entered Scammon's Lagoon. However, because so little whaling had occurred there since 1861–1862, Captain James McKenzie made fair catches of 14 whales each season in the winters of 1868–1869 and 1869–1870. He and his crew visited the lagoon alone on the New Bedford barks *Eagle* and *Hercules*, successively (Henderson, 1972, pp. 263, 266). At Magdalena Bay in late January of 1867, J. Ross Browne visited the shores during his explorations for the Lower California Company, and at that time he found only 2 whaleships in the bay. Although the whalers had "scoured" the waters of the bay for 2 months, they had captured only 2 whales, and the whalemen told Browne that it was then an achievement to capture 2 or 3 whales per month (Henderson, 1972, p. 172). Actually some 16 vessels did visit Magdalena Bay in 1866–1867, but most were whaling alongshore outside and had come in for wood and water and to look for grays.[1] The New Bedford ship *Milo* (Captain Hawes) spent only 2 weeks in the bay and saw a few whales, but captured none.[11]

Whalers on vessels engaged in alongshore whaling and shore whalers, however, continued to do relatively well (Henderson, 1972, pp. 174, 178). Both exploited the grays along their migration route and so had the advantage of waylaying whales headed to or from the several calving and mating grounds. At San Quintín, masters of vessels anchored inside the bay made moderate catches as late as the winter of 1866–1867 by whaling outside the bay (Henderson, 1972, p. 39, n. 24). Moreover, shore whalers, with their low investment, could make a living from small annual catches. This opportunity thereby attracted enough men into shore whaling to bring about the establishment of numerous shore stations and the largest catch of any sector of the gray whale fishery. As whaling declined on the west coast of Baja California, a few whalers still sought out grays in the Gulf of California and along the mainland coast of Mexico where they made small catches (Henderson, 1972, pp. 175–176). However, the northern grounds remained important because although few vessels now cruised the Coast of California, a good many went north into the Arctic Ocean and Okhotsk Sea in search of bowhead whales. In the northern Bering Sea and the Arctic Ocean, the California herd probably suffered as great a catch and kill as it did during the peak period of gray whaling (Table I). More whaleships caught grays than they did earlier, but usually only a few grays were taken by

[11]Ship *Milo* 1863–1868; New Bedford Free Public Library.

any one vessel. In 1871, however, Captain Valentine Lewis of the New Bedford bark *Thomas Dickason* captured 1 at Plover Bay and 10 off King Island, which lies in the northeastern Bering Sea.[12]

In the final period of significant gray whaling from sailing ships, whalers on some 25 cruises probably took about 150 whales at Magdalena Bay, but killed 165. At Scammon's Lagoon, some 38 whales were captured and about 42 killed on 5 cruises, whereas at Ballenas Lagoon and in adjacent waters whalers on about 7 cruises probably killed some 55 whales and retrieved perhaps 50. Alongshore whaling from about 45 cruises accounted for the death of approximately 440 grays and the capture of 400 of them. In the Gulf of California and on the coast of mainland Mexico, whalemen probably made only 10 cruises and captured 50 grays from a kill of 55. On the northern summer grounds, whalers took more grays per cruise as bowhead and right whales became more scarce. On some 40 cruises, 175 grays were probably captured out of 232 killed. The most important part of the gray whale fishery had become shore whaling, because of the low costs involved and the opportunity to prey on all of the much reduced California herd as it passed along the coast. Shore whalers, almost all of whom were from stations in Upper California, probably captured 960 grays from among the 1152 they killed (Table I). During the period of declining gray whaling from ships, considerably fewer than one-half as many gray whales were captured or killed as during the preceding bonanza period. Two-thirds of the whales caught and killed fell victim to shore and alongshore whalers. Whaling at the formerly productive lagoons and Magdalena Bay was essentially abandoned by the end of the period.

Throughout the entire period of significant gray whaling from whaleships on the Coast of California (1845–1846 through 1873–1874), whalers had made a total of about 563–573 cruises to California and northern waters on which they captured and killed gray whales. Moreover, shore stations, easily established and moved, had become involved in the gray whale fishery on the California coasts (Chapter 6, this volume). In all, the gray whalers captured approximately 7028–7078 whales of a total kill of about 8044–8090 (Table I).

Whaling and Whale Population

From its inception, nineteenth century gray whaling had a devastating effect on the population of the California herd. The kill was, to a large extent, concentrated on the cows and calves that were easily killed in the crowded lagoons and bays. Calves, however, were not counted in the recorded catch or kill. Because most of the cows carried fetuses, or would have been impregnated had they not died, or had calves which either were killed or died of starvation, the reproductive capacity of the population was greatly reduced through this excessive slaughter of cows and calves relative to the kill within the total population (Henderson, 1972, pp. 132, 149).

[12]Nicholson Whaling Collection, C798, 1866; Providence Public Library.

The number of grays killed during the initial period of 9 winters of concentrated whaling and the summers following those winters averaged only 62–68 whales per year. This was undoubtedly a sustainable kill for the existing population. During the peak period of gray whaling (1854–1855 through 1864–1865), in 11 winters and succeeding summers the number killed averaged 486, which must have far exceeded a sustainable kill. And in the 10-year period of whaling decline (1865–1866 through 1873–1874), although the kills declined each year, the average kill of 214 per year was still well in excess of that sustainable by a rapidly declining population.

The population decrease is clearly demonstrated by the precipitious decline in the ability of the major calving grounds to sustain whaling. At Scammon's Lagoon, after only 4 years of intensive whaling (1858–1859 through 1861–1862), so few whales remained that few whaling vessels continued to visit the lagoon. With a total kill at Scammon's Lagoon of 650 whales, the lagoon was essentially emptied of whales. Captain Nye of the New Bedford bark *Louisa* reported on February 20, 1873 that whales were scarce and that he had taken only 2 whales at the lagoon during the entire season (Henderson, 1972, p. 170).

At Ballenas Lagoon, a total catch of some 400 whales eliminated most of the population on that whaling ground. Scammon and a few other masters had initiated whaling there in 1859–1860. That winter and the next were seasons of big catches. However, only seven seasons later, in 1866–1867, whalers apparently had essentially abandoned Ballenas Lagoon as an exploitable ground.

The huge complex of waters at Magdalena (Marguerita) Bay sustained more whaling and for a longer time than did the two lagoons to the north. However, after a total kill there of a maximum of 2200 whales over a period of some 25 to 30 years, Magdalena Bay was ignored as a gray whaling ground.

Of course, many whales headed for the calving and mating grounds at the lagoons and bays were killed by alongshore whalers on ships and by other whalers from shore stations. These same whalers killed many of the grays as they made their way back to the northern summer grounds. Whether the grays were killed at the lagoons and bays, en route to or from those waters, or on their northern grounds, a total kill of only about 8100 eliminated pursuit of California gray whales as a truly profitable sector of the nineteenth century whaling business.

So few gray whales remained after 1874 that only whalers at a few shore stations or those on ships seeking right or bowhead whales in northern seas continued to kill a small number of graybacks. Nevertheless, those which survived may have numbered as many as 4000–5000 (Storro-Patterson, 1980, p. 9; Ohsumi, 1976), although they were no longer pursued intensively because of the inefficiencies of nineteenth century whalemen in small boats compared to those in modern powered catcher boats. Many nineteenth century whalng grounds appear to have been left behind for other places as soon as whales ceased to be easily found and captured.

Whalers captured or killed most of the whales at the known and originally crowded grounds within a period of two decades. At some individual grounds, such as Scammon's Lagoon, they eliminated most of the whales in only a few years. Thus, the whale kill must have represented most of the population of California gray whales alive at the time gray whaling began. However, some gray whales survived at the well-known grounds, and a fair number must have spent winters sparsely distributed at various places along their migration route. Considering the kill of some 8100, unreplaced natural mortality, and those perhaps 4000 that may have survived in 1874, the population of the California herd of gray whales probably was not over 15,000 when gray whaling began at Magdalena Bay in 1845–1846 (Henderson, 1972, p. 185; Rice and Wolman, 1971, pp. 118, 130; Storro-Patterson, 1977, p. 46).

Scammon, estimating the number of gray whales killed by 1874 as 10,800, and the number of the California herd between 1853 and 1856 as 30,000–40,000, was undoubtedly high on both estimates (Henderson, 1972, p. 179). His estimate of 8,000–10,000 whales that may have survived in 1874 is also too high (Henderson, 1972, p. 183). Captain Packard (of a San Diego shore station) was probably low in his estimate that the population was only about 1,900 in 1874 (Henderson, 1972, p. 183). The size of catches and kills in the twentieth century, on the other hand, make the extremely low appraisals of population size in the late nineteenth century and early twentieth century obviously wrong (Henderson, 1972, pp. 183, 229).

Gray whales of the California herd did survive the onslaught of the nineteenth century Yankee and other whalers on ships and at shore stations in considerable numbers, as is illustrated by both sightings and catches in the late nineteenth and early twentieth centuries. Whalers on vessels from San Francisco and some other ports continued to exploit the Arctic bowhead fishery in those times, and they also continued to catch a few grays and sight many more (Henderson, 1972, p. 232).[1] As late as 1922, the San Francisco motor ship *Herman* arrived at Nome on July 10 with a catch of four devil fish which had an average yield of 28 barrels, only slightly lower than the average yield of grays on the coast of California during the boom years of gray whaling.[13] Moreover, in the twentieth century, on floating factories, at shore stations where grays have been taken commercially and for scientific study, and from or for aboriginal communities in northeast Siberia and Alaska whalers have probably killed at least 4000 California gray whales (Rice and Wolman, 1971, p. 123; Storro-Patterson, 1977, p. 47; 1980, p. 8).

Nineteenth Century Whaling for Gray Whales in the Okhotsk Sea

Perhaps this is the place to insert that nineteenth century whalers when in northern waters in summer captured many gray whales in the Sea of Okhotsk, that is, grays of the so-called Korean herd.[14] They did not capture grays until the late 1840s. By that time

[13]An "account book of P. James Laflin, shipping agent, San Francisco," in "San Francisco Crew Lists 1906–1928." Original at San Francisco Maritime Museum. Transcript read at Old Dartmouth Historical Society, New Bedford, Massachusetts.

[14]The information upon which the following observations are based is now being utilized by the author in a study of whaling in the Okhotsk Sea, and has been derived from newspaper accounts, books on the Okhotsk Sea, and manuscript whaling records and logbooks.

whalers had pushed their vessels far enough into the Sea to be on the northern and southwestern bowhead grounds of the Upper Sea, that part of the Sea north of 53°N where they found gray whales in abundance. By the early 1850s some captains chased and captured grays and soon some good-sized catches had been made, that is, catches of six or seven grays per vessel.[15] The primary quarry of the whalers in the Okhotsk Sea was the bowhead; so whalers, as in the Bering Sea and Arctic Ocean, took grays only when no bowheads were around. Sometimes this meant a concentrated chase of grays for several weeks. As on other northern grounds, many of the grays captured were smaller than those taken on the Coast of California. Whales of only 10–15 barrels yield, less than half the average yield on the coast of California, were not uncommon in the Okhotsk catches, an indication that many young whales born the previous winter must have been captured. As in the Bering Sea and Arctic Ocean, whalers continued to pursue gray whales in the Okhotsk Sea into the late nineteenth century, that is, at least into the mid-1880s. The total number of grays taken and killed in the Okhotsk Sea by nineteenth century whalers probably approached that of those captured and killed in the Bering Sea and Arctic Ocean; and as the Okhotsk (Korean) herd must have been smaller than the California herd, greater inroads into that population must have been made in summers than were made in the California herd in that season.

Gray Whaling: Problems and Dangers, Practices, Equipment, Resources of the Shores and Bays, Trade, Catches and Yields

Shipwrecks on the Coast of California were surprisingly few considering the currents which swept the coast and the difficulties of working whaleships into, out of, and within the shallow lagoons with only very shallow water over bars across their mouths. Only about 10 vessels were lost during the nineteenth century whaling (Henderson, 1972, pp. 110, 117, 144–145, 170).[1–4,6] Probably the most remembered of the wrecks is that of the Honolulu bark *Black Warrior* (Captain Brown) which, as previously described, was swept against the coast in December 1858 while being towed out of the lagoon that bears her name (*Guerrero Negro* in Spanish) (Henderson, 1972, pp. 109–110, 119–121).

The absence of hard bottoms throughout much of the lagoon area and on the bars saved a good many vessels from disaster, although damage from grounding hulls of heavily laden whalers was not uncommon. Once inside the lagoons and bays, whalers usually anchored their ships for long periods of time and sent their boats on what often turned out to be long chases. To facilitate cutting-in whales far from the ships and save a long boat tow of the whale or his rafted blubber back to the ship, schooner tenders accompanied some larger vessels into the lagoons (Henderson, 1970, p. 14). These schooners could work into the very shallow calving grounds, like those at the Mud Hole in Lee (Almejas) Bay, at Magdalena Bay, and the Nursery in Main Lagoon at the inner reaches of Scammon's Lagoon. It was in such places that cows and calves were concen-

[15] *Seaman's Friend and Temperance Advocate*, Honolulu; December, 1858, p. 84.

trated, whereas bulls when taken were in the outer lagoons, outside in the breakers, and along the migration routes. In the winter of 1861–1862 at Ballenas Lagoon, the 299-ton bark *Delaware* served as a tender to the larger ship (377 tons) *John Howland* so that the ship could remain completely outside the lagoon in Ballenas Bay (Henderson, 1972, p. 158; Starbuck, 1878, pp. 556, 560). Whalers staked out the channels of the lagoons to guide whaleships, tenders, and boats away from grounding and to prevent a 3- or 4-ft falling tide from leaving them high and dry (Henderson, 1970, p. 27). In Estero Santo Domingo, low tide at some places exposed a mile of sandy bottom on either side of the channel (Henderson, 1970, p. 27).

Dangers other than those of managing the ship awaited the gray whalers when they set out in their boats after gray whales in lagoons, bays, or alongshore.[16] Whalers with long experience gave precedence to the grays as the most dangerous of whales. Many men were killed or badly injured in the business of gray whaling. In the bays and lagoons, equipment (whalecraft) and practices were geared to dangers as well as other peculiarities of "lagoon whaling." Of prime importance in all kinds of gray whaling was the use of the bomb lance, which has already been described. To make sure that the wily grays did not escape, often men in four or more boats would attack one whale which meant that if some got fast with their harpoons the whale would have less chance to escape. However, it meant also that lines could become fouled and boats overturned, so special rules of attack were normally observed. One of these was that the men in the first boat to locate and pull toward a whale would set a waif, which was supposed to give that boat and boats cooperating with it by ownership or agreement first chance at the whale.

In order to bring as many boats as possible to bear down upon a whale or several whales, captains often mated their vessels, that is, worked together and shared the catch. Sometimes as many as three vessels in a lagoon were mated. Lagoon whaling could be a crowded affair in which one whale could stave several boats and endanger many men. Unfortunately, so many boats often entered the small areas of calving waters where boats and whales were packed together that great confusion during the chase, disputes over claims to whales, and many accidents occurred. Such waters included the Mud Hole in Almejas Bay and the inner Main Lagoon at Scammon's Lagoon. Also in lagoon calving waters, whalers often had problems with calves fouling their lines and with confusing a rising calf with its mother so that they attacked the calf while the cow escaped. Whalers handled attacks on calves in two ways: some preferred to harpoon the calf first so that the cow would stay close by; others left calves alone out of fear that wounded and dying calves provoked the cows into more destructive behavior.

In the lagoons, boat crews sometimes bombed whales repeatedly so that if they escaped they would die and eventually rise to the surface to be captured. Whalers sometimes marked dead whales with a waif (flag) so that they could be located later, while boat crews went in search of more whales (Henderson, 1970, p. 72). Bombing and harpooning could sometimes be done while the boat was stationed where the whale could not reach it, as in shallow water on a tidal flat next to a channel. Moreover, in the

[16]The following observations about gray whaling have been abstracted from Whaling Archives (see Appendix).

channels currents could be swift and sweep a boat onto its quarry, to the dismay of the crew.

Some other kinds of equipment, in addition to the advantageous bomb lances and Greener's guns, were very useful to gray whalers. Because whalers often had to cut lines to keep their boats from being pulled a long way from their ships and even outside the lagoons or to keep from becoming foul of other boats' lines, a large supply of line was necessary. If ships were anchored in channels where currents were swift, a supply of extra anchors helped hold them in place. Kedges were needed in ships and boats alike to keep from running with the tide, to pull into shallow water, and to warp ships off the mud when grounded (Henderson, 1970, p. 32).[1] Gray whalers carried grapnels to haul sunken whales up from the shallow bottom. Some whalers carried cutting spades in their boats in order to stab into the "small" of the backs of the tough grays just ahead of the flukes, to slow the escape of a whale towing a boat about the lagoon.

Whalers at the lagoons spent much time ashore. They often beached their boats when away from the ships on long searches for whales, at times overnight (Henderson, 1970, p. 14).[1] Sometimes a dead whale beached itself and was then cut-in ashore and the blubber rafted back to the ship (Henderson, 1970, p. 27).[1] To make more room aboard the ships during the busy whaling season, carpenters and coopers sometimes worked ashore; sails, rigging, and casks also were sometimes stored ashore (Henderson, 1970, pp. 21–22, 59).[1] Moreover, whalers often went ashore to cut wood, obtain water, and shoot birds, rabbits, and deer; they also traded with Mexicans who traveled to the shores with cattle, cheese, fruit, game, and other trade items from near or distant ranches and towns (Henderson, 1972, pp. 125; 1975, pp. 51–52). Even some "ladies of the evening" trekked across the desert from La Paz to Magdalena Bay to come aboard ships waiting to whale (Henderson, 1975, p. 53). Some whalers with time on their hands, when whales were scarce both before and after the whaling season, walked inland to visit Mexican and American ranchers who lived near the shores of Magdalena Bay (Henderson, 1975, p. 53).[1] Deserters, who did not find berths on other vessels and thus remain undiscovered, were usually unsuccessful except for those few who found refuge on ranches and in towns near Magdalena Bay. However, most deserters were caught by Mexicans who returned them for a reward, or the desperate men themselves returned from the desolate desert. Some died in their attempted escapes (Henderson, 1970, p. 26; 1972, p. 128; 1975, p. 53).[1]

The waters of the bays and lagoons or those just outside were especially attractive to the whalers in the food supply they afforded. During their stays the whalemen were well provided with green turtles, spiny lobsters, abalone, oysters, and clams. Whalers salted fish and caught live turtles in quantity at the end of the season to provide free food on board after they left the Coast of California (Henderson, 1970, p. 71).[1]

The many inducements to whalers on the Coast of California included the possibility of doing little if any sailing for the season; the provision of cheap food from the shore, especially the favorites of New Englanders (i.e., beef, cheese, and fresh milk); the abundance of free food in the sea; the availability of cheap Mexican laborers aboard ships and ashore as collectors and processors of "stinkers;" the low number of successful desertions; the ready supply of wood and water ashore; and the mild weather.

Even if many catches were low, the cruise to the Coast of California was a cheap and a pleasant way to spend the "between the seasons" winter in anticipation of a good northern summer catch in "the season."

Because the bays and lagoons were quiet places where whalers could repair ships and sails and paint their vessels, rather than remain in such places as Honolulu and San Francisco where expenditures were necessary and desertions easy, many whaleships arrived at the Coast of California grounds early in the winter. But then, so did the whales. Although January and February were the months in which most whales were caught, whalers captured and killed many before and after those months. In the winter of 1858–1859 at Scammon's Lagoon, Captain Scammon on the San Francisco bark *Ocean Bird* was 20 miles into the lagoon in The Narrows when he saw the first whales on November 24.[17] By November 26, he had anchored for the season off Cooper's Lagoon (Nursery Inlet), and from December 11 on he and his crew saw whales regularly from The Narrows to the inner lagoon. His men chased their first whale on December 18 and captured their first on December 22. Nevertheless, on December 20 Scammon and Captain Poole, on the San Francisco bark *Sarah Warren*, had made the kind of agreement that became common among bay and lagoon whalers on the Coast of California; they agreed to let a good number of whales come into the lagoon before beginning to whale. Scammon's men killed their last whale on March 13; of 47 grays captured that season, they took 7 in late December and 6 in early March. As did whalers at other lagoons and bays, Scammon's crew spent the last days or weeks on the ground at such tasks as catching turtles and fish and cutting wood. That season *Ocean Bird* went over the bar and out of Scammon's Lagoon on March 27.

Unfortunately, whalers made essentially no reports on the measurements of grays but were interested only in the oil yield of their catches, either per whale or for the season. However, oil yields reveal relative sizes of whales from one time or place to another. Most gray whales yielded 35 barrels (31.5 gal per barrel). In fact, newspaper reports from whale grounds almost always equated 1 gray whale to 35 barrels of oil. If a ship was reported to have taken 10 grays, her catch would be reported as 350 barrels.[2–4,7] During the entire period of gray whaling up to the early 1870s, Scammon also equated 1 gray to 35 barrels of oil (Henderson, 1972, p. 138). In some reports he reported that the yield averaged between 20 and 40 barrels (Henderson, 1972, p. 138). In the early years of gray whaling whalers caught some whales that were much larger. In 1858–1859 at Scammon's Lagoon on *Ocean Bird*, Scammon caught a whale on December 25 that yielded 55 barrels (Henderson, 1970, p. 37), and he said that very large grays might yield 60–70 barrels (Henderson, 1972, p. 137). The records of huge gray whales, which could have yielded the 101 barrels reported by a New London whaler and the 90 barrels reported taken from 1 whale caught by San Diego shore whalers, appear to be mistakes in catch size or species taken (Henderson, 1972, p. 137). Both could have been right whales. Perhaps there was some decline in the size of whales captured; various reports of the oil taken in 1846–1847 at Magdalena Bay by the New London ship *Hibernia* and the Stonington bark *United States*

[17]This and the following information about Scammon's whaling in 1858–1859 is from Henderson, 1970, pp. 17, 19–20, 21, 28, 31, 33, 35, 67, 68, 71; and from logbooks in Whaling Archives (see Appendix).

in the first year of gray whaling indicate an average yield of 45 to 55 barrels per whale.[2,4,6] Nevertheless, except for that indication, average yields on the Coast of California do not seem to have declined with total catch declines, perhaps because some large bulls were taken outside the bays and lagoon in the late years of gray whaling (Henderson, 1972, pp. 140, 148–149). Catches on the northern grounds in the Bering Sea and Arctic Ocean as well as in the Okhotsk Sea did seem to produce lower yields per whale, however, probably because whalers on those grounds caught more young whales born the previous winter (Henderson, 1972, pp. 137–138).[1]

Revisions of Gray Whaling History and Geography

The historical record supports a number of revisions in conceptions about gray whales and gray whaling which I should like to emphasize before ending this chapter. First, the idea that before gray whaling began there was a large population of grays in the California herd numbering 30, 40, or even 50 thousand is not borne out by the historical record of catches and the number of whales that survived in the twentieth century. The population of the California herd when nineteenth century whaling began was probably around 15,000 and certainly not more than 20,000 (Henderson, 1972, pp. 132, 154, 169–191).

Another misconception is that Scammon's Lagoon contained more gray whales than any other lagoon or bay on the Coast of California. Records of vessels on the grounds and reported catches demonstrate that in Magdalena Bay whalers caught and killed over three times as many whales as they did in Scammon's Lagoon (Table I). Moreover, in 4 years of intensive whaling at Scammon's Lagoon, it was effectively ruined as an important whaling ground, although Magdalena Bay endured at least 12 years of intensive whaling. Of course, the Magdalena Bay complex is much larger than Scammon's Lagoon and contains three distinct parts; these are, from south to north, Lee (Almejas) Bay; Weather, Main, or Magdalena Bay; and Estero Santo Domingo, the long lagoon behind the barrier bar north of Weather Bay. Nevertheless, in comparing the total size of Scammon's Lagoon to that of the Magdalena Bay complex, it would appear that whales were just as concentrated in Magdalena Bay as in Scammon's Lagoon.

Among the persistent errors concerning gray whale migration and range is that which places the most northern bays of winter habitation at San Diego and San Quintín bays. The report that Captain W. W. Clark killed two whales of some kind in the bay in 1842 may be true; on the other hand, it may simply have meant that he had his ship in the bay and captured the whales outside the bay as he did some 15 years later (Henderson, 1972, p. 28). This may be a good place to point out that Captain W. W. Clark was a Yankee who conducted shore whaling outside the bay in boats of the ship *Ocean* which was anchored in the bay (Henderson, 1972, p. 28). This Clark has been confused with another Captain Clark who was a shore whaler in California but who came from the Azores and changed his name from Machado to Clark (Henderson, 1972, p. 210, footnote 357). A few grays may have visited San Diego Bay; but if the account, in a 1908 history of San Diego County (Smythe, 1908, p. 110), of many calving whales passing in and out of the bay in the early 1840s were correct, it would seem that such a well-known and well-stocked potential whaling ground would have attracted many whalers to those waters after gray whaling was initiated at Magdalena Bay in the winter of 1845–1846. Such was not the case, and it appears unlikely that the harbor traffic, which is credited with driving the grays from the bay, increased that much in the early years of this quiet port between 1840 and 1846 (Henderson, 1972, pp. 28, 82–83, 133–134).

The maritime traffic, both at Scammon's Lagoon, since salt works were re-established there in 1857, and at Magdalena Bay, where fishing, agricultural export, and other marine traffic has been increasing since World War II, has not driven gray whale populations from those places. Thus, how could the few merchant ships visiting San Diego Bay in the early 1840s ruin the bay as a whaling ground for whalers who knew it well and had been taking large numbers of grays at Magdalena Bay between 1845–1846 and 1847–1848? Certainly by the time shore whaling from stations just inside San Diego Bay began in the late 1850s, the whalers were sailing their boats outside the bay to attack whales along the migration route (Henderson, 1972, p. 28); whaling vessels reported sighting shore boats whaling outside the bay.[1]

Like San Diego Bay, the much smaller and shallower San Quintín Bay approximately 190 miles south of San Diego in Baja California was a protected place from which masters of whaleships could send out their boats. Some vessels reported at San Quintín were actually anchored outside, south of the small island of San Martín off Point San Quintín. From vessels anchored inside or outside the bay, whalemen in their boats chased the grays along the migration route outside the bay (Henderson, 1972, pp. 28–29).

An historical myth about the discovery of Scammon's Lagoon has Captain Scammon following grays through the surf to find himself inside a lagoon full of whales (Henderson, 1972, p. 114). In fact, Indians and Spaniards had passed knowledge of the lagoon and its whales down to nineteenth century Mexicans (Henderson, 1972, pp. 53–54; 60–77). Sea otter hunters knew the lagoon prior to Scammon's entry (Henderson, 1972, pp. 102–103), and early sperm whalers on the coast of California, including those of the English ship *Toward Castle* (misidentified by Scammon as *Tower Castle*) which was wrecked in 1836 on the west side of the mouth of the lagoon, surely knew that the lagoon was there.[18] Early hunters of Guadalupe fur seals and elephant seals at Cedros Island and the small island rookeries in Vizcaíno Bay also must have known the lagoon (Henderson, 1972, pp. 102–104). At Elide Island almost on the coast and just north of Scammon's Lagoon, guano collectors had contacts with Mexican traders who must have told them about Scammon's Lagoon at least a year before Scammon went into it (Henderson, 1972, pp. 104–105). A most believable account of whaling on the Coast of California says that Scammon learned of the lagoon from a Mexican at Magdalena Bay, where Scammon whaled in 1855–1856 on the San Francisco ship *Leonore* and in 1856–1857 on the San Francisco brig *Boston* (Henderson, 1972, pp. 113, 115, 268). And in fact,

[18]Scammon's misidentification is in Scammon, 1869, p. 124 (see Henderson, 1972, pp. 109, 202, footnote 153). The correct identity of the vessel as *Toward Castle* is found in *Lloyd's Register*, July 1, 1838–June 30, 1839, T301; *Sandwich Island Gazette*, October 21, 1837, p. 3.

after a disappointing cruise on *Boston* whaling, sealing, and sea elephant hunting off the coasts of the Californias in the summer and autumn of 1857, Scammon decided to attempt to find a way into the gray whale lagoon of which he had heard at Magdalena Bay (Henderson, 1972, p. 115). Scammon had taken his brig from Cedros Island to the peninsular coast north of Vizcaíno Bay and had worked his way southward to Lagoon Head, just north of the three lagoons on Vizcaíno Bay (Henderson, 1972, pp. 115, 268). From the protected anchorage on the lee side of this headland, which other whalers must have visited by 1857, Scammon could easily have seen the mouths of the two northern lagoons (later named Manuela and Black Warrior lagoons). Later on the voyage, Scammon had arranged to have the small schooner-tender *Marin* sent south from San Francisco, and the two vessels met at Catalina Island to sail southward. After they arrived at Lagoon Head, Scammon sent his tender and three whale boats to explore southward along the coast. This expedition partially explored the two lagoons north of *Ojo de Liebre* (Scammon's) Lagoon. Then, after 2 days, news was sent back to Scammon that the tender had entered the large southernmost lagoon. Scammon then followed *Marin* and took his brig *Boston* into the lagoon which was to acquire his name (Henderson, 1972, pp. 115–116, 268).

A clarification of the translation of the Spanish name for Scammon's Lagoon, *Laguna Ojo de Liebre*, is needed at this point. Many authors have translated the name as Eye of the Jack Rabbit Lagoon or Jack Rabbit Eye Lagoon, rather meaningless names. The correct translation is Jack Rabbit Spring Lagoon. At the eastern end of the Lagoon is a freshwater hole or spring called *Ojo de Liebre* which takes its name from the *Llanos de Liebre* (Jack Rabbit Plains) on whose western edge the spring is located. These are old Spanish names, which probably go back to the mission period prior to the nineteenth century. *Ojo*, however, has other meanings than eye, one of the most common of which is spring, its obvious meaning at Scammon's Lagoon. The lagoon was named after the spring because it undoubtedly had more significance to Spaniards on the trail, which passed the spring, than did the lagoon. Thus, they named the Lagoon after the spring which in turn got its name from the plains to the east.

Another myth which has been perpetuated about Captain Scammon and the lagoon is that he kept his "discovery" concealed for several seasons, and that only by smelling the odors emanating from Scammon's trypots did other whalers later discover his secret (Henderson, 1972, p. 118). In fact, Scammon himself made arrangements in his second winter at the lagoon, 1858–1859 on the San Francisco bark *Ocean Bird*, to join his tender *A. M. Simpson* and the San Francisco bark *Sarah Warren*, commanded by the husband of Scammon's wife's sister, Jared F. Poole, at Lagoon Head from which the vessels entered the lagoon together on November 20, 1858 (Henderson, 1970, p. 17). Although Scammon's journal makes it clear that *Sarah Warren* awaited Scammon at Lagoon Head and sailed into the lagoon with him, in his book "The Marine Mammals" Scammon does not acknowledge Poole's company but indicates that Poole was among the captains who "succeeded in finding their way in" after Scammon had entered the lagoon (Henderson, 1970, pp. 17–18; 1972, p. 204, footnote 197; Scammon, 1874, pp. 265–266). Furthermore, captains on whaleships recruiting in the Hawaiian Islands had learned of the lagoon by the second year of whaling there, through reports from Scammon's mate on *Boston* during the first year's whaling at the lagoon. Thus, by the second year of whaling at the lagoon a fleet of 18 vessels came from the Islands to attempt whaling at Scammon's Lagoon and those lagoons adjacent to it. Most of the masters found entry difficult, and after the loss of *Black Warrior* at Middle Lagoon (now Black Warrior or *Guerrero Negro* Lagoon) they went elsewhere on the coast. In the second winter of whaling at Scammon's Lagoon, however, seven vessels excluding tenders whaled at the lagoon, including the barks *Ocean Bird* and *Sarah Warren* of San Francisco; four from the Islands, and one, flying a New Granada flag of convenience, which was a merchant vessel-become-whaler and probably came south from San Francisco, the bark *Cleopatra* (Henderson, 1970, pp. 44–45, footnote 62; 1972, pp. 119, 161).

Summary

After whalers had worked the sperm whale fishery of the Coast of California and had largely turned their attention northward in the Pacific Ocean, they returned to the Coast of California in many vessels to attack gray whales in winter when they had fled the cold and icy northern grounds. The new ground was easy to reach from the whalers' recruiting ports in the Hawaiian Islands and San Francisco, and it afforded mild weather, little sailing, food supplies, and good catches for those who understood the ways of gray whales and gray whaling. Some whalers avoided this peculiar and dangerous kind of whaling. The gray whale fishery lasted only some three decades, but other nineteenth century whale fisheries were also short lived because whalers attacked whales without thought of conservation, moving on when whales were hard to find but often far from extinct on the grounds. Nevertheless, nineteenth century gray whaling from ships and shore stations on the coasts of the Californias caused a devastating decline in the gray whale population, and probably only the early whalers' inability to pursue a species to its end saved the California grays from near extinction or complete extirpation.

Appendix: Whaling Archives in Which Source Manuscripts, Chiefly Logbooks, Are Located

Bancroft Library
University of California
Berkeley, California

Dukes County Historical Society
Edgartown, Massachusetts

East Hampton Free Library
East Hampton, Long Island, New York

Falmouth Public Library
Falmouth, Massachusetts

Kendall Whaling Museum
Sharon, Massachusetts

The Mariners Museum
Newport News, Virginia

Mystic Seaport, Inc.
G. W. Blunt White Library
Mystic, Connecticut

Nantucket Historical Association
Peter Foulger Museum
Nantucket, Massachusetts

New Bedford Free Public Library
Melville Whaling Room Collection
New Bedford, Massachusetts

New London County Historical Society
New London, Connecticut

Old Dartmouth Historical Society
Whaling Museum
New Bedford, Massachusetts[19]

Peabody Museum of Salem
Salem, Massachusetts

Providence Public Library
Nicholson Whaling Collection
Providence, Rhode Island

San Francisco Maritime Museum
San Francisco, California

Stonington Historical Society
Stonington, Connecticut

Suffolk County Whaling Museum of Sag Harbor, L. I.
Sag Harbor, New York

Whaling Museum Society, Inc.
Cold Spring Harbor, New York

References

Beane, J. F. (1905). "From Forecastle to Cabin." The Editor Publ. Co., New York.

Cumpston, J. S. (1964). "Shipping Arrivals and Departures. Sydney, 1788–1825," Parts I, II, and III. Aranda, Australia (privately published, 1964).

Dudley, P. (1725). An essay on the natural history of whales. *Philos. Trans. R. Soc. London* **33,** 256–269.

Espinosa, R. (1854). Reseña estadística sobre la antigua o Baja California. *Bol. Soc. Mex. Geogr.* **4,** 121–127.

Henderson, D. A. (1970). "Journal Aboard the Bark Ocean Bird on a Whaling Voyage, Winter of 1858–59. Chas. M. Scammon's Voyage Edited and Annotated." Dawson's Book Shop, Los Angeles, California.

Henderson, D. A. (1972). "Men and Whales at Scammon's Lagoon." Dawson's Book Shop, Los Angeles, California.

Henderson, D. A. (1975). Whalers on the coasts of Baja California: Opening the peninsula to the outside world. *Geosci. Man* **12,** 49–56.

Lloyd's Register (1838–1839). "Lloyd's Register of British and Foreign Shipping from 1st July 1838 to 30th June 1839." J. L. Cox & Sons, London.

Mitchell, E. (1979). Comments on magnitude of early catch of East Pacific Gray Whale (*Eschrichtius robustus*). *Rep. Int. Whal. Commn.* **29,** SC/30/Doc. 41, 307–314.

Ohsumi, S. (1976). Population assment of the California gray whale. *Rep. Int. Whal. Commn.* **25,** 350–359.

Rice, D. W., and Wolman, A. A. (1971). The life history and ecology of the Gray Whale (*Eschrichtius robustus*). *Spec. Publ. Am. Soc. Mammal.* **3,** pp. 1–142.

Sanderson, I. T. (1956). "Follow the Whale." Little, Brown, Boston, Massachusetts.

Scammon, C. M. (1869). Report of Capt. C. M. Scammon, of the U.S. Revenue Service, on the west coast of Lower California. *In* "Resources of the Pacific Slope," (J. Ross Browne, ed.), Appendix, pp. 123–31. Appleton, New York.

Scammon, C. M. (1874). "The Marine Mammals of the Northwestern Coast of North America." John H. Carmany & Co., San Francisco, California.

Smythe, W. E. (1908). "History of San Diego 1842–1908." The History Co., San Diego, California.

Stackpole, E. A. (1964). "The Sea Hunters. The New England Whalemen During Two Centuries 1635–1835." Lippincott, Philadelphia, Pennsylvania.

Starbuck, A. (1878). "History of the American Whale Fishery from its Earliest Inception to the Year 1876." Washington, D.C.: (facsimile ed., preface by Stuart C. Sherman, 2 vols. Argosy-Antiquarian, Ltd., New York, 1964).

[19]Logbooks and Dr. John R. Bockstoce's information on summer gray whale catch and kill in the Arctic Ocean and Bering Sea, collected in making his bowhead whaling studies.

Starks, E. C. (1922). "A History of California Shore Whaling," Fish Bull. No. 6. State of California Fish and Game Commission, California State Printing Office, Sacramento.

Storro-Patterson, R. (1977). Gray whale protection. *Oceans* **10**(4), 45–49.

Storro-Patterson, R. (1980). Biological aspects of the eastern Pacific stock of gray whale *Eschrichtius robustus*. *In* "World Wildlife Fund, U.S. and I.U.C.N.," Contract Rep., pp. 1–165. Whale Center, Oakland, California.

Townsend, C. H. (1935). The distribution of certain whales as shown by logbook records of American whaleships. *Zoologica* **19**(1), 1–50.

8

Modern Commercial Pelagic Whaling for Gray Whales

Randall R. Reeves

Introduction

The history of modern commercial pelagic whaling for gray whales, *Eschrichtius robustus*, is short and reasonably well documented. Only four nations were involved: the United States, Japan, Norway, and the Soviet Union. The number of gray whales killed, compared both to the numbers of other large baleen whales taken during the modern phase of commercial pelagic whaling and to the number of gray whales taken by nineteenth century (premodern) pelagic whalers and shore whalers, was modest. In their Table 18, called "Recorded Catch of Gray Whales by Modern Style Whaling from 1910 to 1946," Rice and Wolman (1971) list an aggregate catch of 933 for the California stock and 1474 for the Korean stock. These totals include catches by shore stations.

An indication of its relative lack of importance to the modern whaling industry is that the gray whale, along with the right whales, *Eubalaena* spp., and the bowhead, *Balaena mysticetus*, has never merited a separate column in the statistical tables prepared by the Committee for Whaling Statistics (*International Whaling Statistics*, 1930 *et seq.*). Rather, the gray whale and the balaenids are listed under the column headed "Others," sometimes with footnotes giving details (e.g., Ruud, 1942).

An important distinction can be made between nineteenth century Yankee-type whaling and modern Norwegian-type whaling. Whaling historians have always recognized the differences between the two eras (e.g., Tønnessen, 1967; Risting, 1931). For purposes of this chapter, the definitions developed by Mitchell and Reeves (1980, Table 1) are used to identify modern commercial pelagic whaling and to distinguish it from

ISBN 0-12-389180-9

other types. Modern whaling refers to "techniques and implements centered on the use of explosives and engines, involving deck-mounted cannons, explosive grenades, direct fastening to the whale, and diesel, gas or steam-powered boats and ships" (see definition of "mechanical whaling" in Mitchell and Reeves). Commercial whaling is done "for the main or ancillary purpose of selling one or more products in a cash economy." Pelagic whaling is "conducted by one or more vessels that catch the whales and process them on board, or deliver the carcass to a floating processing plant." Thus, the type of whaling considered here is narrowly defined to exclude nineteenth century Yankee pelagic whaling (Chapter 7, this volume), subsistence whaling by or on behalf of native people (Wolman and Rice, 1979; Marquette and Braham, 1982), and shore-based operations of any kind (Chapter 6, this volume).

This chapter is organized as follows: First, the history of protective agreements and regulation of the whaling industry is reviewed as it pertains to the gray whale. Second, the involvement of each of the four nations is discussed in turn, with statistical tables giving as many details about the catches as could be derived from available published and unpublished sources. Finally, a short essay is presented concerning the completeness and accuracy of the data. The main purpose of the chapter is to provide investigators attempting to model the population dynamics of the gray whale with a complete and accurate profile of recent catch history.

History of Protection and Regulation

The first serious steps to control the international whaling industry were taken from 1932–1933 to 1936–1937 (Jahn *et al.*, 1942, p. 6). These consisted of voluntary agreements between Norwegian and British whaling companies concerning quotas on whale oil production and opening and closing dates for whaling in the Southern Ocean. Japanese, German, Soviet, American, and South African expeditions did not adhere to these agreements.

The International Agreement for the Regulation of Whaling was concluded in 1937 (for text, see *IWS*, Vol. 16, pp. 58–66; Vol. 17, pp. 26*–34*). Signatories included the Union of South Africa, the United States of America, the Argentine Republic, the Commonwealth of Australia, Germany, the United Kingdom of Great Britain and Northern Ireland, the Irish Free State, New Zealand, and Norway. Article 4 of the agreement stated "It is forbidden to take or kill grey-whales and/or right-whales." "Grey-whale" was defined in Article 18 to mean "any whale known by the name of grey-whale, California grey, devil fish, hard head, mussel digger, grey back, rip sack." Thus, the gray whale was protected from hunting after 1937 by Norwegians, Americans, and nationals of other signatory states. It was not, however, protected from continued whaling by Japan and the Soviet Union. Also, a precedent was set in Article 10 of the International Agreement for the Regulation of Whaling which allowed contracting governments to grant scientific research permits to their own citizens. Such permits could be for killing, taking, and treating whales according to the conditions set by the government involved, regardless of the terms of the international agreement. This precedent cleared the way for "scientific collections" of gray whales by the United States (316 from 1959 to 1970; Rice and Wolman, 1971) and Canada (10 in 1953; Pike, 1962).

Canada acceded to the International Agreement for the Regulation of Whaling at the 1938 International Whaling Conference in London (for text, see *IWS*, Vol. 17, pp. 35*–45*). A Japanese delegation participated in this conference and indicated its willingness to accede. However, Japan insisted that its right to hunt whales in the Pacific Ocean between 66° N latitude and 72° N latitude be entrenched in the protocol. The protocol was amended accordingly, and at the Informal Conference in London in 1939 the Japanese delegation reported that its government "had definitely decided to complete all the necessary legislation to enable them to accede to the Agreement and the Protocol by the opening of the whaling season 1939/40" (for text, see *IWS*, Vol. 17, pp. 46*–51*). The outbreak of hostilities, however, precluded final Japanese accession to this agreement. There was no Japanese (or Soviet) participation in the International Whaling Conferences in London in 1944 (*IWS*, Vol. 17, pp. 52*–57*) and 1945 (*IWS*, Vol. 17, pp. 58*–66*).

In summary, gray whales were protected from commercial whaling by Norway and the United States after 1937, and from commercial whaling by Canada after 1938. Protection from Japanese and Soviet whaling did not come until after World War II.

Another International Whaling Conference took place in Washington, D.C., November 20–December 2, 1946 (*IWS*, Vol. 18, pp. 26–53). The United States, Soviet Union, Norway, and Canada were among the 14 nations that sent plenipotentiary delegations. The Final Act of the conference included as Article X the following statement:

> *The International Whaling Conference*
> Supports:
> and considers justified the request of the Delegation of the Union of Soviet Socialist Republics that the taking of gray whales in the Bering and Chukotsk seas should be permitted when the meat and products of such whales are to be used exclusively for local consumption by the aborigines of the Chukotsk and Korjaksk areas.

Under the Schedule adopted within the framework of the International Convention for the Regulation of Whaling established at the conference, Section 2 made it "forbidden to take or kill gray whales or right whales, except when the meat and products of such whales are to be used exclusively for local consumption by the aborigines." Thus, except for the killing by or on behalf of aborigines for "subsistence" and the killing authorized under special scientific permits (Article VIII of the 1946 convention was almost identical to Article 10 of the 1937 International Agreement for the Regulation of Whaling; see above), gray whales were fully protected from whaling by signatory states, which this time included the U.S.S.R.

The 1946 convention was ratified by the required number of states and became operational with the first meeting of the International Whaling Commission in London in 1949 (IWC, 1950). Japan's whaling interests apparently were represented at this meeting by the observer from the Supreme Commander of Allied Powers. In April 1951 Japan formally adhered to the convention (IWC, 1951). Consequently, since 1951 all nations with factory ships operating in the North Pacific have been subject to the provisions protecting gray whales from commercial whaling.

It is important to note that the whaling commission agreed in 1978 to reclassify the "Eastern [California] Stock" of gray whales from a "Protected Stock" to a "Sustained Management Stock" with a catch limit of 178 per annum (Bollen, 1979, p. 26). This catch was reserved "to be taken by aborigines or a Contracting Government on behalf of aborigines" for noncommercial purposes. It was stated that "No member nation is proposing commercial operations on gray whales."

Norwegian Whaling for Gray Whales

The first factory ship to operate in the North Pacific was the Norwegian steamer *Sommerstad* (Chamberlain and Bower, 1913, pp. 68–69; Tønnessen, 1973 [1967], p. 28). She operated with American registry, and her three catcher boats flew American flags. The enterprise was backed by Norwegian capital, and the crew and gunners were Norwegian. The *Sommerstad* did not actually work on the high seas during her first season (1912); rather, she was based at Port Armstrong on Baranof Island, Alaska. Between May 2, when the first whale was taken, and mid July, 71 whales were delivered to the *Sommerstad*, producing 2500 barrels of oil. The operation (including two of the three catchers) relocated on July 14 to an area west of the Shumagin Islands along the Alaskan Peninsula. Between July 19 and September 17, 184 whales were caught; they produced only 3000 barrels. These were all listed as blue whales ("sulphur bottoms," *Balaenoptera musculus*), humpback whales (*Megaptera novaeangliae*), and fin whales (*Balaenoptera physalus*). I assume, judging by the localities and dates as well as the whalers' expressed preference for blue whales and sperm whales (*Physeter macrocephalus*), that few of the whales taken by the *Sommerstad*'s catcher boats in 1912 were gray whales. The *Sommerstad* went back to Norway after this season and never returned to Alaska.

A second Norwegian factory ship, the *Admiralen*, was dispatched to Alaska in 1912, equipped with an important innovation, wireless telegraph (Tønnessen, 1973 [1967], p. 34). The *Admiralen* operated in Akutan Harbor, on the north shore of Akutan Island, from June 3 to October 21, taking a total of 310 whales (Chamberlain and Bower, 1913, p. 68). There is no evidence that the *Admiralen* took any gray whales. Nor did the *Kit*, a combined Norwegian catcher boat/factory ship that cruised in the Sea of Okhotsk in 1911 and in the Arctic Ocean north of Bering Strait in 1912, 1913, and 1914 (Tønnessen, 1973 [1967], p. 39).

The first gray whale catches by a Norwegian factory ship were made in spring of 1914, when the *Capella I* took 19 off Baja California (Rice and Wolman, 1971, p. 122). The most substantial catches, however, were made from 1925 through 1929. During these years two whaling companies, A/S *Vega* and A/S *Mexico*, sent expeditions to the North Pacific (Anonymous, 1926; Risting, 1928, pp. 83–84; E. Vangstein, in letter to D. W. Rice,

May 11, 1965). Some data from these expeditions are tabulated in Tables I and II. For the most part, Norwegian pelagic whaling activities off Baja California centered on the capture of blue and humpback whales. During the years 1925–1929, gray whales (n = 182) represented only about 7% of the catch (n = 7245). Blue whales (n = 870) and humpback whales (n = 1568) comprised approximately 32 and 57% of the catch, respectively.

In 1925, when 100 gray whales were caught by the two Norwegian factory ships operating off Baja California, one of these, the *Kommandoren I*, went to Natal'inskiy Bay off northeastern Kamchatka for the summer and took an additional 33 gray whales (Rice and Wolman, 1971, p. 124). Most of the catch was made between July 27 and August 22.

Table I
Data on the Catch of Gray Whales by Norwegian Floating Factories, 1924–1929[a]

Period of operation	Locality	Expedition	Whales
Dec. 12, 1924–Jan. 16, 1925	Magdalena Bay	*Mexico* (3 catchers)	11
Jan. 19–Apr. 4, 1925	Cabo San Lucas	*Mexico* (3 catchers)	7
Apr. 9–May 31, 1925	Magdalena Bay	*Mexico* (3 catchers)	0
Nov. 14, 1924–Apr. 7, 1925	Magdalena Bay	*Kommandoren I* (4 catchers)	82
Oct. 1, 1925–Apr. 23, 1926 (gray whales taken only in Dec.–Feb.)	Magdalena Bay and Cabo San Lucas	*Mexico* (3 catchers)	5
Nov. 30, 1925–Apr. 15, 1926 (gray whales taken only in Jan.–Feb.)	Magdalena Bay	*Kommandoren I* (5 catchers)	36
Dec. 28, 1926–May 12, 1927 (gray whales taken only in January)	"Coast of Mexico"	*Kommandoren I* (5 catchers)	16
Nov. 15, 1926–June 10, 1927 (gray whales taken only in Dec.–Jan.; one in April)	Baja California	*Esperanza* (4 catchers)	13
Nov. 1927–June 1928 (gray whales taken only in Dec.–March)	"Mexico's West Coast"	*Esperanza* (4 catchers)	9
March 7–July 4, 1929 (gray whales taken only in March)	"Mexico's West Coast"	*Esperanza* (4 catchers)	2
		Total	181[b]

[a]Source: Committee for Whaling Statistics, courtesy E. Vangstein via D. W. Rice; also see Risting (1928) and Anonymous (1926).

[b]In addition, 19 were taken off Baja California by the *Capella I* in 1914, and 33 off Kamchatka by the *Kommandoren I* in 1925 (see text).

Table II
Biological Data on 150 Gray Whales Caught Off Baja California by Norwegian Floating Factories, 1925–1929[a]

Date		Total length in feet	Sex	Pregnant	Length of fetus in feet
January	1	41	F	X	14
	2	34	F	X	12
	3	38	F	X	10
	4	38	F	X	6
	4	38	F		
	7	44	F	X	12
	7	39	F	X	$15\frac{1}{2}$(F)
	7	42	F		
	7	43	F	X	16(M)
	7	40	F		
	9	41	F	X	16
	9	37	F	X	14
	9	34	F		
	9	34	F		
	9	42	F	X	8
	10	42	F	X	13
	10	35	M		
	11	35	F	X	
	12	40	F		
	14	41	F	X	$16\frac{1}{6}$
	14	43	F	X	17
	15	38	F		
	15	44	F		
	15	36	M		
	17	35	F		
	17	44	F		Accompanied by a calf
	19	50	F		
	20	50	F		
	22	34	M		
	22	35	F		
	23	36	M		
	23	35	M		
	23	32	M		
	23	35	F		
	24	50	F		
	24	36	M		
	24	38	M		
	24	38	M		
	25	34	F		
	25	35	F		
	25	37	F		
	25	32	M		
	25	34	M		
	25	38	F		

Table II (*Continued*)

Date		Total length in feet	Sex	Pregnant	Length of fetus in feet
January	26	38	M		
	26	38	F		
	26	36	F		
	26	30	M		
	27	36	F		
	28	60 (sic)	M		
	28	34	F		
	28	33	M		
	28	36	M		
	28	36	M		
	28	34	M		
	28	35	F	X	13
	28	38	M		
	28	34	M		
	28	32	M		
	28	32	M		
	29	34	M		
	29	35	F		
	29	34	F		
	29	45	F		
	29	33	M		
	29	36	F		
	29	35	M		
	29	34	M		
	30	34	M		
	30	34	M		
	30	35	F		
	30	32	M		
	30	33	M		
	30	32	M		
	30	35	M		
	30	34	M		
	30	33	M		
	30	35	M		
	31	34	M		
	31	36	M		
	31	32	M		
	31	32	M		
	31	32	M		
February	1	32	M		
	1	34	M		
	1	35	M		
	1	36	M		
	1	37	M		
	3	34	M		

(*continued*)

Table II (*Continued*)

Date		Total length in feet	Sex	Pregnant	Length of fetus in feet
February	3	34	M		
	3	36	M		
	3	33	M		
	4	36	F		
	4	35	M		
	4	32	M		
	5	35	F		
	5	36	M		
	5	36	M		
	6	34	M		
	8	30	M		
	9	34	M		
	9	34	M		
	10	34	M		
	11	34	M		
	13	36	M		
	14	30	M		
	16	30	M		
	16	44	F	X	9
	19	45	M		
	24	45	M		
March	19	40	F		
	25	30	F		
	27	30	F		
December	21	40	F		
	27	42	F		
	29	50	F	X	18
January	4	41	F	X	16
	5	40	F	X	15
	6	41	F	X	17
	8	32	M		
	10	39	F	X	16
	11	38	F	X	17
	11	40	F	X	
	11	40	M		
	11	35	F	X	
	12	40	F	X	
	12	35	F	X	
	12	35	M		
	13	35	M		
	14	35	F		
	14	35	F		
	15	36	M		
	15	36	M		
	16	35	F	X	
	16	36	F	X	
	17	38	M		

Table II (*Continued*)

Date	Total length in feet	Sex	Pregnant	Length of fetus in feet
17	38	F	X	
17	36	F		
17	38	M		
19	36	M		
19	35	M		
19	35	F		
19	37	F		
19	36	F		
20	40	M		
20	38	F	X	
20	39	F		
21	35	M		
21	37	F		
22	34	F		

[a]This material was analyzed by Risting (1928), but Rice and Wolman (1971, p. 3) cautioned that "body lengths were estimated rather than measured . . . and therefore the conclusions concerning fetal growth and size at sexual maturity are not valid." The main usefulness of presenting the data here is to demonstrate that, overall, sex composition was almost even, with 76 males and 74 females reported. Source: same as Table I.

Of the 31 gray whales taken during this period, 19 (61%) were males and 12 (39%) females (Risting, 1928, p. 84). The year's catch, totaling 133 animals, was the largest made in any 1 year by the Norwegian fleet.

American Whaling for Gray Whales

Although gray whales were taken occasionally by modern catcher boats working from shore stations in California (Chapter 6, this volume), Washington (Scheffer and Slipp, 1948), and Alaska (Kellogg, 1931), there is no evidence of substantial catches by any American factory ship. The floating factory *Lansing* began operating off California in 1927, taking 3 gray whales that year and 1 the next (Rice and Wolman, 1971, pp. 123–124). The *Lansing* whaled off southern California between December 20, 1927 and April 26, 1928, taking 156 whales; off Kodiak until the beginning of October, taking 237 whales; and off southern California again from October 29 to December 31, taking 175 whales (Tønnessen, 1973 [1967], p. 53). Thus, less than 1% of the catch consisted of gray whales. The *Lansing* continued to operate off California through 1930, employing six catcher boats the last year. Although Rice and Wolman (1971, p. 125) were unable to locate catch data by species for the 1930 season, it is reasonable to assume few if any gray whales were taken by the *Lansing* after 1928.

The *Lansing* was replaced in 1932 by the *California*, which operated until 1937. According to Tønnessen (1973 [1967], pp. 53–54), one of the reasons the *California* stopped whaling in 1937 was "because the international whaling convention prohibited catching of gray whales." Apparently the *California* operated off Mexico and California, so it is curious that no gray whale catches have been attributed to her (Rice and Wolman, 1971, Table 18).

I reviewed reports on whales inspected on the factory ship *California* during the period 1939–1941 (available on microfilm from the Old Dartmouth Historical Society, New Bedford, Massachusetts). Catches were made in the general area of 37–40°N latitude and 122–124° W longitude during May through July. Primarily humpback whales were taken, with a few fin whales and sperm whales. No mention of gray whales appears in these reports.

I am assured by D. W. Rice (in letter, March 24, 1982) that he has been unable to locate more than the "bare facts" published in 1971 concerning gray whale catches by American vessels. J. N. Tønnessen (in letter, 1982) was unable to provide further information on the catch by the *California* from 1932 to 1937. E. Vangstein (in letter, January 28, 1982) also stated that the International Whaling Statistics office in Sandefjord, Norway, has no further information on this subject.

Soviet Whaling for Gray Whales

The floating factory *Aleut*, a converted cargo steamer, began whaling in the western Bering Sea in late October 1932 (Zenkovich, 1935, 1955). Gray whales were first taken by the *Aleut* in 1933. The *Aleut* almost never encountered gray whales along the Kamchatka coast south of Mys Olyutorskiy, so nearly all the hunting took place north of this cape (Zenkovich, 1934, 1937). In 1934, 2 gray whales were taken in August in the region of Bukhta Dezhneva and Bukhta Natalii; the expedition then proceeded through Bering Strait and into the Chukchi Sea, where the principal gray whale catch (more than 50) was made (Zenkovich, 1935).

During the following 11 seasons (except 1944), gray whales were caught by the *Aleut* without restriction (Table III). This occurred in spite of concerns raised by Zenkovich (1937), who warned after noting the high proportion of immature females in the 1936 catch (~34%): "This figure cannot but arouse concern regarding the correctness of whaling practice, especially since we are dealing with a rapidly disappearing species, whose destruction in view of its biological characteristics could be accomplished very easily." Zenkovich recommended that gray whales be protected by international agreement. Until an internationally agreed moratorium could be achieved, he recommended that a unilateral Soviet ban be instituted on gray whaling in the Mys Navarin and Mys Rubicon areas, where young gray whales were mainly caught. As stated above, the U.S.S.R. participated fully in the International Whaling Conference of 1946 and agreed to complete protection of gray whales from commercial whaling.

Table III
Data on the Catch of Gray Whales by the Soviet Floating Factory *Aleut*, 1933–1946

Date	Number of whales	Mean total length (m)	Range of total length (m)	Source
1933	2 (2 M, 0 F)	12.23	12.12–12.35	Zenkovich, 1934, 1937
Aug.–Sept. 1934	54 (30 M, 24 F)[a]	12.0 M, 12.8 F	10.1–13.5 M, 9.1–15.0 F	Zenkovich, 1934, 1935, 1937; Tomilin, 1937
Aug.–Sept. 1935	34 (11 M, 23 F)	11.9 M, 12.7 F	11.2–13.2 M, 9.2–14.1 F	Zenkovich, 1937
Aug.–Oct. 1936	102 (45 M, 57 F)[b]	10.7 M, 11.1 F	8.6–14.6 M, 8.5–14.3 F	Zenkovich, 1937
1937	14	10.4		Sleptsov, 1955
1938	54	11.3		Sleptsov, 1955
1939	29	11.9		Sleptsov, 1955
1940	47	10.8		Sleptsov, 1955
1941	57	11.1		Sleptsov, 1955
1942	101	12.1		Sleptsov, 1955
1943	99[c]	11.2		Sleptsov, 1955
1944	—	—		—
1945	30	9.2		Sleptsov, 1955
1946	22[d]	11.0		Sleptsov, 1955

[a]Three females were pregnant; overall, most whales in the catch were immature.
[b]Twenty of the females were less than 10 m long and sexually immature.
[c]Sleptsov (1955) indicates 77; this figure was corrected by A. Yablokov (in letter, April 7, 1982).
[d]Zenkovich (1955) indicates no catch for this year and a catch of one gray whale for 1947.

I attempted unsuccessfully to obtain more detailed information on the gray whale catch from 1937 to 1946 by the *Aleut*. A. Yablokov (in letter, April 7, 1982) assured me that the statistics tabularized by Sleptsov (1955) are the most complete available (Table III).

Japanese Whaling for Gray Whales

Japanese factory ship whaling in the North Pacific began with the *Tonan Maru* expedition in 1940. After two seasons, 1940 and 1941, this fishery was suspended and did not resume until 1952, by which time Japan belonged to the International Whaling Commission. A Japanese pelagic fleet operating near the Bonin Islands from 1945 to 1949 (Rice, 1978) took no gray whales (H. Omura, in letter, February 17, 1982).

During 1940 the *Tonan Maru* caught 58 gray whales, most of them in the Arctic Basin (Sakiura *et al.*, 1953; Table IV). This represented less than 9% of the expedition's whale catch for this year (Japan Whaling Association, 1943). Thirty-five (60.3%) of the 58

Table IV
Data on the Catch of Gray Whales by the Japanese Floating Factory *Tonan Maru*, 1940[a]

Date	Noon position (Lat. N, Long. W)	Water temperature (°C)	Whales
August 4	64°23′, 171°00′	5.5	2
5	65°30′, 169°40′	5.0	4
6	67°04′, 171°00′	7.2	3
10	67°17′, 171°47′	5.0	1
21	67°02′, 171°14′	7.2	1
22		8.8	16
23	67°, 171°39′	7.0	18
24	67°08′, 170°38′	10.5	3
25	67°20′, 170°42′	7.5	4
26	67°09′, 170°42′	6.8	6
			58

[a]From Japan Whaling Association (1943).

gray whales were female; 23 (39.7%), male. Mean length for females was 41.3 m; for males, 37.9 m. These are the only documented removals by the Japanese pelagic fleet from the California stock (H. Omura, in letter, February 17, 1982). The *Tonan Maru* operated southeast of Kamchatka in 1941 and took no gray whales.

Discussion and Conclusion

Although documentation is incomplete, it seems clear that modern commercial pelagic whaling for gray whales represents a comparatively short phase in the species' history of exploitation. The catches documented in this paper total 940 gray whales taken by four countries over a period of 32 years, from 1914 to 1946.

I have not been able to locate information on the struck-but-lost rate in this fishery. Presumably, most struck whales were eventually secured. In a study of humpback whaling, Mitchell and Reeves (1983) estimated a loss-rate factor of 1.06 as necessary for correcting catch data from a modern Newfoundland shore station. Using a similar figure for the modern commercial pelagic fishery for gray whales, a corrected estimate of the total catch is 996 or about 1000 whales. Especially considering the fact that documentation is incomplete, such an estimate is probably reasonable for the total kill.

In comparison to most other phases of whaling history, this phase is well documented. The emergence of international regulation of the industry (the IWC) and centralized record keeping (*IWS*) has resulted in more detailed reporting. Also, because factory ship expeditions are financially, socially, and politically more conspicuous than many small-scale, shore-based enterprises, it is likely that their activities would be better documented.

Acknowledgments

I am indebted to the following individuals who responded generously and promptly to my inquiries concerning gray whale catches: H. Omura, A. Yablokov, E. Vangstein, D. W. Rice, and J. N. Tønnessen. The manuscript was reviewed by D. W. Rice.

References

Anonymous (1926). Graahvalen. *Nor. Hvalfangst-Tid.* **15**(3), 35–36, 39, 42.

Bollen, A. G., chairman (1979). Chairman's report of the thirtieth annual meeting. *Rep. Int. Whal. Commn.* **29,** 21–37.

Chamberlain, F. M., and Bower, W. T. (1913). Fishery industries. *In* "Fishery and Fur Industries of Alaska in 1912," (B. W. Evermann, ed.), p. 18 *et seq.* U.S. Dept. of Commerce, Washington, D.C. (Bur. Fish. Doc. No. 780).

International Whaling Commission (IWC) (1950). International Commission on Whaling (Constituted under the International Whaling Convention signed at Washington on 2nd December, 1946). "First Report of the Commission (covering the first fiscal year 1st June, 1949 to 31st May, 1950)," pp. 1–32. Issued from the Office of the Commission, London.

International Whaling Commission (IWC) (1951). International Commission on Whaling (Constituted under the International Whaling Convention signed at Washington on 2nd December, 1946). "Second Report of the Commission (covering the second fiscal year 1st June, 1950 to 31st May, 1951)," pp. 1–16. Issued from the Office of the Commission, London.

International Whaling Statistics (IWS) (1930 *et seq.*) Published annually by the Committee for Whaling Statistics, Oslo.

Jahn, G., Bergerson, B., and Paulsen, H. B. (1942). Introduction. *Int. Whal. Stat.* (Oslo) **16,** 3–57.

Japan Whaling Association (1943). "Hogei Binran (Handbook of Whaling)," Vol. I (mimeographed in Japanese) (Reference and excerpts received by author from H. Omura, in letter, February 17, 1982.)

Kellogg, R. (1931). Whaling statistics for the Pacific coast of North America. *J. Mammal.* **12**(1), 73–77.

Marquette, W. M., and Braham, H. W. (1982). Gray whale distribution and catch by Alaskan Eskimos: A replacement for the bowhead whale. *Arctic* **35**(3), 386–394.

Mitchell, E., and Reeves, R. R. (1980). The Alaska bowhead problem: A commentary. *Arctic* **33**(4), 686–723.

Mitchell, E., and Reeves, R. R. (1983). Catch history, abundance, and present status of Northwest Atlantic humpback whales. *Rep. Int. Whal. Commn. Spec. Issue* **5,** 153–212.

Pike, G. C. (1962). Migration and feeding of the gray whale (*Eschrichtius gibbosus*). *J. Fish Res. Board Can.* **19,** 815–838.

Rice, D. W. (1978). The humpback whale in the North Pacific: Distribution, exploitation, and numbers. *In* "Report on a Workshop on Problems Related to Humpback Whales (*Megaptera novaeangliae*) in Hawaii" (K. S. Norris and R. R. Reeves, eds.), Contract Rep., pp. 29–44. Marine Mammal Commission (available from Natl. Tech. Inf. Serv., PB-280 794).

Rice, D. W., and Wolman, A. A. (1971). The life history and ecology of the gray whale (*Eschrichtius robustus*). *Spec. Publ.—Am. Soc. Mammal.* **3,** 1–142.

Risting, S. (1928). Whales and whale fetuses. *Rapp. P.-V. Reun, Cons. Int. Explor. Mer* **50,** 1–122.

Risting, S. (1931). The development of modern whaling. *Int. Whal. Stat.* (Oslo) **2,** 4–15.

Ruud, J. T. (1942). A review of the investigations on whales and whaling in recent years. *Int. Whal. Stat.* (Oslo), **16,** 67–77.

Sakiura, H., Ozaki, K., and Fujino, K. (1953). "Biological Investigation of the Northern Pacific Baleen Whales Caught by the Japanese Whaling Fleet in 1952," pp. 1–64. Fisheries Agency of Japanese Government, Tokyo (issued by the Japan Whaling Association).

Scheffer, V. B., and Slipp, J. W. (1948). The whales and dolphins of Washington State with a key to the cetaceans of the west coast of North America. *Am. Midl. Nat.* **39**(2), 257–337.

Sleptsov, M. M.(1955). "Kitoobraznye dal'nevostochnykh morei (Cetaceans of Far Eastern Seas)," 2nd ed. Primorskoe Knizhnoe Izdatel'stvo, Vladivostok (citation from Rice and Wolman, 1971, p. 139).

Tomilin, A. G. (1937). Kity Dal'nego Vostoka. (The whales of the Far east.) *Uch. Zap. Mosk. Gos. Univ., Ser. Biol. Nauk.* **13,** 119–167.

Tønnessen, J. N. (1967). "Den moderne hvalfangst historie," Vol. 2: Norges Hvalfangstforbund, Sandefjord.

Tønnessen, J. N.(1973) [1967]. "History of Modern Whaling. Foundation and Development," 2nd vol. Part I, pp. 129–173 (transl. from Norwegian by Randi Mathisen).

Wolman, A. A., and Rice, D. W. (1979). Current status of the gray whale. *Rep. Int. Whal. Commn.* **29,** 275–279.

Zenkovich, B. A. (1934). Materialy k poznaniyu kitoobraznykh dal' nevostochnykh morei. (Seryi Kaliforniiskii kit-*Rhachianectes glaucus* Cope.) [Research data on Cetacea of far eastern seas (the gray California whale—*Rhachianectes glaucus* Cope)]. *Vestn. Akad. Nauk SSSR, Dal'nevostochnyi Fil.* **10,** 9–25 (citation from Rice and Wolman, 1971, p. 140).

Zenkovich, B. A. (1935). (Whaling in the USSR during 1934.) "Rybnoe Khoziaistvo Dalnego Vostoka (Pacific Scientific Institute of Fisheries)," Issue 8, 1st Quarter, pp. 65–66. Vladivostok (Typescript translation by L. A. Hutchinson, UCSD, La Jolla, California).

Zenkovich, B. A. (1937). Esche o serom kaliforniiskom kite (*Rhachianectes glaucus* Cope, 1864). [More on the gray California whale (*Rhachianectes glaucus* Cope, 1864).] *Vestn. Akad. Nauk SSSR, Dal'nevostochnyi Fil.* **23,** 91–103 (citation from Rice and Wolman, 1971, p. 140).

Zenkovich, B. A. (1955). Kratkaia istoriia kitoboinogo promysla i sovremennoe ego sostoianie v SSSR. (A brief history of whaling in prerevolutionary Russia and an account of present-day whaling in the USSR.) *In* "Kitoboinyi Promysel Sovetskogo Soyuza (The Whaling Industry of the Soviet Union)" (S. E. Kleinenberg and T. I. Makarova, eds., Part I, Chapter 1. Rybnoe Khozyaistvo, Moskva. (trans. from Russian by Israel Program Sci. Transl., Jerusalem, 1968).

III

Demography, Distribution, and Migration

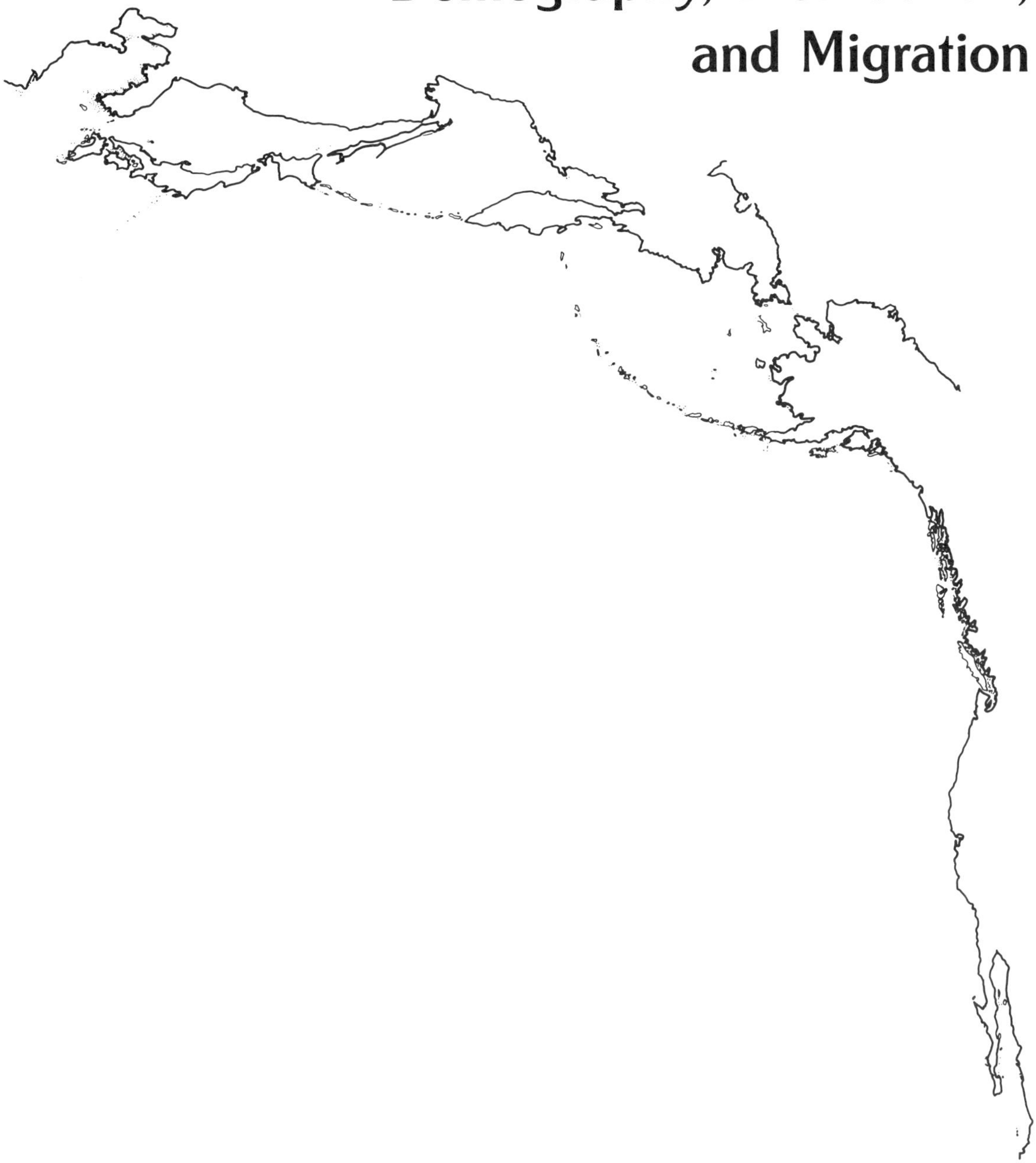

9

Assessing Gray Whale Abundance: A Review

Stephen B. Reilly

Introduction

Recent estimates of gray whale abundance from two parts of the population's range are quite different in magnitude. Shore station censuses conducted during the north and south migrations range from about 15,000 to 20,000 for 1978 through 1980 (Chapters 10 and 13, this volume; Poole, 1984; Reilly *et al.*, 1983). The most recent estimate from an aerial census of the winter grounds is 7,600 for 1981 (Rice *et al.*, 1983). This is close to winter range estimates from the early 1970s (approximately 7,000 from Gard, 1974, 1978). The twofold disparity between estimates from different areas is an apparent dilemma. The two census types use different methods for data collection, different assumptions to infer total abundance, and have different problems.

This chapter reviews the history of gray whale population assessment and the methods and assumptions used by the two primary techniques. Estimating gray whale abundance is a difficult problem throughout their range. Each geographic–seasonal area presents specific difficulties, necessitating field and analytic techniques tailored to deal with differences in local abundance, distribution, habitat, and temporality of occupation. Neither assessment method is fully verified in its present state. (My personal bias favors the shore counts as being nearer to verification.) The winter range estimates

ISBN 0-12-389180-9

are probably too low, but it is uncertain by how much. The shore station estimates are too high if gray whales in migration slow their rate of travel at night. There are reasonable scenarios under which the data from both locations can indicate similar population sizes.

In the following pages I first review the winter range assessments, then those from along the migratory route. Finally, I suggest how alterations in some of our assumptions for which little data exist can result in comparable population estimates from the data on hand.

Winter Range Assessment

On their winter grounds, gray whales occupy a vast area. The known range includes a number of physiographic settings within major and minor lagoons, near and offshore from lagoon entrances, and offshore zones throughout much of the west coast of the Baja Penninsula (Scammon, 1874; Gilmore, 1960; Rice *et al.*, 1981, 1983), parts of southern California (Chapter 14, this volume) and mainland Mexico (Gilmore *et al.*, 1967). Each subsection of each lagoon and alongshore area potentially requires individual and intense study for the design and execution of surveys of abundance.

Most winter range surveys have been made in small aircraft. Whales observed in the nearshore zone and within the lagoons were recorded (Gilmore, 1960; Hubbs and Hubbs, 1967; Gard, 1978; Rice *et al.*, 1981, 1983). The resulting raw data have consisted of counts per lagoon or per alongshore stretch. For abundance estimation, the problem is inferring total population size from these counts. This is essentially the same problem in any wildlife sighting survey, but the Baja situation presents some unique problems.

HISTORY AND METHODS

The history of population assessment on the winter grounds has been briefly outlined by Gard (1978), Rice *et al.* (1981), and Storro-Patterson (1981). Table I and the discussion which follows were drawn largely from those sources and from the original papers for some details.

There have been at least 21 separate aerial surveys of some portion of the winter range between 1952, when Carl and Laura Hubbs (1967) began the enterprise, and one in 1981 conducted by Rice *et al.* (1981, 1983) (Table I). Rice *et al.* (1981) stated that the winter range survey data are not comparable over time because of unstandardized field and analytic techniques, and I agree. The surveys were flown over varying portions of the winter range, along different track lines, at different times, and at different altitudes in varying visibility conditions, and so forth. Data treatment in analysis has been similarly nonstandardized.

This is not meant to depreciate past winter range aerial surveys. In the context of its time, each series contributed significantly to our knowledge. Our present understanding of the vastness of the area inhabited, and the complexity of the task of assessing abundance there, has emerged slowly. The logistic constraints (e.g., fuel limits, lack of

Table I
Aerial Censuses of Gray Whales on the Winter Range

Dates	Observers	Areas surveyed[a]	Verification[b]	Source
1952 (Feb. 16–20)	Hubbs and Hubbs	1,2		Gilmore (1960)
1953 (Feb. 25–27)	Gilmore	1,2		Gilmore (1960)
1953 (Jan. 31–Feb. 4)	Rechnitzer	1,2		Gilmore (1960)
1954 (Feb. 14–21)	Hubbs and Hubbs	1,2		Gilmore (1960)
1954 (Feb. 1–7)	Gilmore	1,2,4		Gilmore (1960) Gilmore *et al.* (1967)
1955 (Feb. 26–Mar. 3)	Gilmore	1,2,4		Gilmore *et al.* (1967)
1956 (Feb. 14–17)	Gilmore	1,2,4		Gilmore *et al.* (1967)
1957 (Feb. 27–Mar. 3)	Gilmore	1,2,4		Gilmore *et al.* (1967)
1959 (Feb. 20–26)	Hubbs and Hubbs	1,2		Hubbs and Hubbs (1967)
1960 (Feb. 18–21)	Hubbs and Hubbs	1,2		Hubbs and Hubbs (1967)
1961 (Feb. 25–27)	Hubbs and Hubbs	1,2		Hubbs and Hubbs (1967)
1961 (Feb. 23)	Harrison	4		Gilmore *et al.* (1967)
1962 (Mar. 8–22)	Harrison	4		Gilmore *et al.* (1967)
1962 (Feb. 18–21)	Hubbs and Hubbs	1,2		Hubbs and Hubbs (1967)
1963 (Feb.–Mar. ?)	Harrison	4		Gilmore *et al.* (1967)
1964 (Feb. 20–24)	Hubbs and Hubbs	1,2		Hubbs and Hubbs (1967)
1965 (Jan. 3–Apr. 20)	Harrison	4		Gilmore *et al.* (1967)
1966 (Mar. 6–27)	Harrison	4		Gilmore *et al.* (1967)
1970 (Jan. 31–Feb. 1 and Mar. 4)	Gard	1		Gard (1974)
1973 (Feb. 27–28, Mar. 4)	Gard	1		Gard (1974)
1974 (Feb. 14–22)	Gard	1,2		Gard (1978)
1975 (Feb. 15–27)	Gard	1,2	1	Gard (1978)
1976 (Feb. 11–19)	Gard	1,2	2	Gard (1978)
1980 (Jan. 17–Mar. 28)	Swartz and Jones	5	3	Swartz and Jones (1980b)
1980 (Feb. 4–15)	Rice *et al.*	1,2	1,3	Rice *et al.* (1981)
1981 (Feb. 5–15)	Rice *et al.*	1,2,3	4	Rice *et al.* (1983)
1981 (Feb. 23–25)	Swartz and Jones	5	3	Swartz and Jones (1981)

[a]Areas: (1) nearshore Baja west coast, (2) calving lagoons, (3) offshore zones, (4) mainland calving areas, (5) Laguna San Ignacio only.

[b]Verification: (1) air versus lagoon tower, (2) ground-based behavioral observations for aerial correction factor, (3) air versus lagoon boat, (4) air versus offshore boat.

radio contact in case of emergency) imposed on anyone venturing to fly down Baja in a small plane in the 1950s and early 1960s effectively precluded offshore ventures and repetitious patterns over any one area (R. M. Gilmore, personal communication).

The early surveys conducted by Hubbs and Hubbs (during 1952, 1954, 1959, 1960, 1961, 1962, and 1964) and Gilmore (during 1953, 1954, 1955, 1956, and 1957) were pioneering efforts. However, the details of their efforts and methods are not made entirely clear in the published literature (e.g., Hubbs and Hubbs, 1967; Gilmore, 1960). They reportedly flew the entire coast from San Diego to the Cabo San Lucas area and over the major calving lagoons, tallying the whales seen. Hubbs and Hubbs (1967) discussed many potential sources of error and inconsistency in the methods used, including different aircraft, varying visibility conditions, dates surveyed, and number and experience of persons involved.

Gilmore (1960) gave a year-by-year summary of some further details of the early aerial censuses, enough to lend the reader a sense of the variation in methods used. Gilmore (1960, p. 29) also stated that early aerial surveys of the winter grounds were unreliable for estimating population size: "Because the air-survey method introduced many variable factors, and information needed to apply corrections was not available, it is unwise to estimate the size of the total population on air-survey data."

Although the early efforts conducted no ground-based verification experiments, it was acknowledged that many whales were not seen from the aircraft. Gilmore (1960) speculated that "About one-fourth of the whales beneath the plane could perhaps be seen by the observers at any one time." Hubbs and Hubbs (1967) assumed that "about half of the total population was observed in the area covered, and that almost all the individuals were concentrated in that area at the time of the counts . . ."

After a lapse of 5 years following the 1964 Hubbs survey, Gard (1974, 1978) conducted a series of six surveys in 1970 and in 1973 through 1976. In 1970 and 1973, only the major lagoons were surveyed; in later years coverage included the "entire western shore of Baja California." The distance offshore was not reported nor were full details on data-recording protocol. For the first time, separate censuses were made outside the entrances to the major lagoons to 3 km (1.8 mi) offshore. Gard's surveys were conducted during varying time periods in February and March. As with the earlier aerial surveys, track lines flown at each location, dates, and times of day were not reported. Consequently, it is not possible to determine to what extent his surveys were replicates of each other or of earlier attempts.

Gard was the first to address the problem of extrapolating from recorded counts to the total numbers present by directed data collecting during the winter ground surveys. He estimated the proportion of time gray whales spend at the surface, to determine what proportion might be missed by passing aircraft. Gard also compared his aerial estimates for Laguna Ojo de Liebre with ground-based counts made by White (1975). These verification experiments are discussed further in the section Assumptions and Summary (Winter Range).

In 1980 and 1981, Rice *et al.* (1981, 1983) conducted winter range aerial censuses, including the nearshore area between San Diego and Boca del Colorado (24° 20′N) and the known Baja calving areas. They also surveyed offshore areas in the large bays. To verify the overall surveys, tower, small boat, and aircraft studies were made in and near Laguna Ojo de Liebre in 1980. In 1981 they conducted a vessel survey of Vizcaino Bay during the same period as the aerial survey of that area. In both years replicate surveys were made of Lagunas Ojo de Liebre, Guerrero Negro, and San Ignacio. Dates, locations, and other details of survey segments were reported.

No estimate of population total was made by Rice *et al.* (1981) for the 1980 census, as the survey was considered exploratory for determination of total range and relative importance of areas within the range. A limitation of the 1980 (and 1981) surveys by Rice *et al.* is their southern termination at Boca del Colorado near Magdelena Bay. Both Norris *et al.* (1983) and Storro-Patterson (1981) report sighting congregations of gray whales in the Cabo Falso–Cabo San Lucas Region [174 were actually sighted by Norris *et al.* (1983)].

Rice *et al.* conducted the 1981 survey using three methodologies and related analysis schemes: "systematic transects" of open areas followed predefined straight parallel lines. Data were analysed as strip censuses for density estimates and variances; "coastline transects" were made along most of the outer coast covered. Only raw counts were used, and no extrapolations were made from these data; "channel transects" were made along narrow lagoon channels by flying along the channel sides so that its entire width was visible from one side of the plane. Again only raw counts were used for the population estimates, with no extrapolations for whales missed.

Total population estimates were made by Rice *et al.* (1983) for both adults and calves. For adults, statistical estimates from "systematic" transects were summed with raw counts from coastline and channel transects to produce an estimate of 7601.

Estimation of the number of calves was yet more complex, including ratio correction schemes for some areas and raw counts for others. The total estimate, summed from the combination of raw counts and various types of estimates, was 1439 calves. In relation to the estimate of 7601 "adults" by Rice *et al.* (1983), this indicates a crude birth rate near 19%. Other information, discussed below, suggests that either the "adult" estimate is too low or the calf estimate is too high, or both.

ASSUMPTIONS AND SUMMARY OF CURRENT STATUS

A list of factors affecting or involved in estimating total abundance from winter range aerial surveys is presented in Appendix A. This list is admittedly armchair in nature (and in extent), but I believe that all of the points mentioned require at least minimal attention before we can have a great deal of confidence in population estimates from the winter range. Fortunately, many of these points have been or are being addressed by past or present research. A quantitative synthesis of the winter range assessment situation is the most outstanding current need. Some of the verification work done to date is discussed in the following sections.

Behavior

A considerable amount of information on behavior has been accumulated by ground-based research projects in the calving lagoons: in Laguna San Ignacio (LSI) by

Swartz and Jones (1978, 1980a,b, 1981, this volume); in Laguna Guerrero Negro (LGN) by Bryant *et al.* (this volume); in Laguna Ojo de Liebre (LOL) by Rice *et al.* (1981, 1983), Fleischer (1984), and White (1975); and by Norris *et al.* (1977, 1983) in and near Magdelena Bay.

In conjunction with his overall surveys, Gard (1978) timed the above- and below-surface occurrence for 25 groups of whales for a minimum of 3 min each, in and outside of LSI and LOL. Pooling the data by location and type of social group (cow–calf versus others), Gard estimated that gray whales on average spent 29.7% of their time at the surface. Consequently he multiplied observed aerial counts by a factor of 1/0.297 or 3.37.

Recent attempts by S. L. Swartz and M. L. Jones (personal communication) to replicate this study in LSI met with severe logistic problems. They found it extremely difficult to identify and follow individuals or groups for periods sufficiently long to record meaningful data. According to Swartz and Jones, a wide variety of types of behavior can be noted from any one vantage point, ranging from rapid traveling to quiescent lolling. They found surface times in general to be markedly variable in the lagoons. Consequently, the use of the mean surface time as a correction factor without considering its variance is questionable.

Much less detailed observation has been made outside the lagoons, where Rice *et al.* (1981, 1983) recorded nearly 70% of the whales observed in their survey during the peak occurrence period in February. Rice *et al.* (1983) performed a ship-based strip transect survey in Vizcaino Bay (see below), and Norris *et al.* (1983) recorded occurrences and some behavioral observations at the entrance to the Magdalena Bay complex. We now need additional detailed, systematic observations on behavior outside the lagoons in open coast and other areas of whale concentration which emphasize age, sex, temporal, and physiographic differences in activity. These data are necessary to determine if whales in along shore regions are differentially "available" to our observation from passing survey aircraft, and how much variability should be expected from replicate flights over each area.

Geographic Distribution

As recent surveys have expanded their efforts, new, and in some cases large, concentrations of whales have been seen in previously under- or unsurveyed areas: Vizcaino Bay is one example. Even the latest surveys probably did not cover the entire winter range.

It is apparently appropriate to consider the winter range as extending from about Point Conception in the north to Cabo San Lucas, including the Gulf of California and parts of the Mexican mainland (Chapter 14, this volume). In February when the Baja occupation is at its peak, there are still a few individuals passing Monterey (California) both north and southbound. There is probably no one time when the entire population is between San Diego and Magdalena Bay, the termination points of the latest surveys. There has been considerable coverage of smaller scale distribution in and near the major calving areas.

Interplatform Comparisons

Aerial to ground comparisons have been made for LOL and Vizcaino Bay (Gard, 1978; Rice *et al.*, 1981, 1983), and LSI (Swartz and Jones, 1981). Gard (1978) first compared counts obtained from the air and on the ground, for LOL. For a comparison period in mid-February 1975, he estimated 2013 adult whales inhabited LOL. For approximately the same period, White (1975) estimated 1963 as the adult population within LOL. This is compared to a February 1980 estimate of 895 for LOL by Rice *et al.* (1981).

That the aerial and land estimates were within 3% suggests either that both Gard (1978) and White (1975) had developed accurate assessment procedures or that the procedures perhaps were similarly biased. Even with the benefit of hindsight, it is not yet possible to distinguish between these possibilities, but some points regarding their methods are worth considering.

First, White's (1975) method consisted of monitoring the lagoon entrance from a tower for varying periods per day and week, subtracting weekly average exits per hour from average entrances per hour and multiplying the difference by 168 hr/week. Recent, similar observations of all daylight hours for 7 days per week (Rice *et al.*, 1981) have shown a great deal of variation in turnover rates. This and other lagoon studies (Swartz and Jones, 1978, 1980a,b, 1981; Norris *et al.*, 1977, 1983) suggested that turnover rates are probably more reflective of tidal fluxes and visibility conditions than actual abundance of whales in the lagoons. Second, Gard's (1978) aerial data also showed a great deal of variation between replicate surveys of the same areas. Third, as discussed above there is some question regarding estimation of the surface time statistic and its use as a correction factor.

In the lagoon comparisons, small single-engined aircraft were the principal platform. Because of this, results may not be equally applicable to the larger twin-engined aircraft (which were used for the comprehensive surveys) due to possible differences in visual perspective from the two types of planes. If air speed and altitude are not duplicated, this would exacerbate any such lack of comparability.

Visibility Conditions

Rice *et al.* (1983) found that weather and visibility conditions had a major effect on sighting efficiency in their repeated vessel and aerial censuses of Vizcaino Bay. Their results implied that other censuses made in less than adequate visibility conditions, such as nearly every afternoon on the Baja west coast when the strong prevailing wind blows, are probably negatively biased. I am aware of no other winter range data which represent visibility effects on census results.

In summary, the recent surveys have made notable progress toward an effective, standardized field methodology for estimating gray whale abundance on the winter range. We now need (1) a survey effort that covers the entire range and somewhat beyond to effectively define its limits. Further refinement of area- and time-specific field methods would also contribute to a reliable data base. (2) A more rigorous analytic methodology is also needed. All counts from an airplane are in fact estimates, with inherent variance. The basic properties, at least, of each type of estimate must be defined before they can be combined to give crudely comprehensive estimates of total and variance.

There are many nontrivial problems involved in the development of a more rigorous analytic methodology for the Baja survey situation, and much difficult work remains. Some important points which have emerged from winter range surveys to date are (1) aircraft are necessary to cover the entire area in a short time, but censuses from them systematically underestimate whale presence; and (2) interreplicate variation is high and is greatly affected by visibility conditions.

Migratory Route Assessment

Gray whales travel very close to shore, especially during their southward trek from feeding to calving grounds (Pike, 1962; Rice and Wolman, 1971). It is consequently feasible to census, from strategic points, whales passing during daylight hours. As with winter range aerial surveys, the essence of the resulting estimation problem is inferring total population size from recorded counts. There are still a number of problems to confront when inferring population size from these data, but the magnitude and scope of these problems is considerably smaller and more tractable than those resulting from winter range aerial surveys. The relative simplicity results largely from having only one site's complexities to deal with and not having to rely on instantaneous glimpses from airplanes for measuring absolute abundance. Also, the vast majority of the population is likely to pass a location that is well chosen.

HISTORY AND METHODS

An outline summary of in-migration censuses is presented in Table II. Townsend (1887) reported the first shore census of migrating gray whales from a sighting effort that was limited to December, 1885. Townsend identified three of the basic problems common to all shore counts made to estimate total abundance: How many whales pass unseen at night? on days before and after the census? too far offshore to be seen? Regarding his observed total, Townsend stated "What proportion this number bears to the number passing offshore would be hard to say, but it is certainly less than half. . . ."

Carl Hubbs, of the Scripps Institution of Oceanography in La Jolla, organized the next attempts at censusing southbound migrating gray whales 61 years later, in 1946–1947 (Walker, 1949). These counts, made by students from a campus rooftop, were continued through 1951–1952 (skipping 1950–1951). The details of effort expended are not clear in the published accounts (Walker, 1949; Gilmore 1960). The recorded annual counts were 250, 500, 600, 600, and 880, respectively.

During 1952–1953, the U.S. Fish and Wildlife Service conducted censuses of southbound migrating gray whales from San Diego, supervised by Dr. Raymond Gilmore (1960). Point Loma was the principal census site, although some concurrent censusing was done at La Jolla. These early San Diego censuses are of questionable utility as measures of population abundance for a number of reasons. By the mid-1960s, it became clear that a large (but as yet undefined) proportion of southward migrating whales passed far offshore through the Southern California Bight (Rice, 1965). This

Table II
Shore Station Censuses of Gray Whales

Dates	Migration direction	Observers	Location	Source
1885–1886 (December)	S	C. H. Townsend	San Simeon, California	Townsend (1887)
1946–1947 (Dec.–Jan.?)	S	Hubbs and students	La Jolla, California	Gilmore (1960); Walker (1949)
1947–1948 (Dec.–Jan.?)	S	Hubbs and students	La Jolla, California	Gilmore (1960); Walker (1949)
1948–1949 (Dec.–Jan.?)	S	Hubbs and students	La Jolla, California	Gilmore (1960); Walker (1949)
1949–1950 (Dec.–Jan.?)	S	Hubbs and students	La Jolla, California	Gilmore (1960); Walker (1949)
1951–1952 (Dec.–Jan.?)	S	Hubbs and students	La Jolla, California	Gilmore (1960); Walker (1949)
1952–1953 (Dec. 15–Feb. 15)	S	Gilmore	Pt. Loma and La Jolla	Gilmore (1960)
1953–1954 (Dec. 15–Feb. 15)	S	Gilmore	Pt. Loma and La Jolla	Gilmore (1960)
1954–1955 (Dec. 15–Feb. 15)	S	Gilmore	La Jolla	Gilmore (1960)
1955–1956 (Dec. 15–Feb. 15)	S	Gilmore	La Jolla	Gilmore (1960)
1956–1957 (Nov. 2–Mar. 25)	S	Gilmore	La Jolla	Gilmore (1960)
1959–1960 (Dec. 14–Feb. 22)	S	Gilmore	La Jolla	Rice (1961)
1966–1967 (Nov. 27–Feb. 17)	S	Adams	Point Lobos, California	Adams (1968)
1967–1968 (Dec. 18–Feb. 3)	S	Rice and Wolman	Yankee Point, California	Reilly *et al.* (1983)
1967–1968 (Dec. 18–Feb. 5)	S	Rice and Wolman	Point Loma	D. W. Rice and A. A. Wolman (unpublished)
1968–1969 (Dec. 10–Feb. 6)	S	Rice and Wolman	Yankee Point	Reilly *et al.* (1983)
1968–1969 (Dec. 20–Feb. 14)	S	Rice and Wolman	Point Loma	D. W. Rice and A. A. Wolman (unpublished)
1969–1970 (Dec. 8–Feb. 8)	S	Rice and Wolman	Yankee Point	Reilly *et al.* (1983)
1970–1971 (Dec. 9–Feb. 12)	S	Rice and Wolman	Yankee Point	Reilly *et al.* (1983)
1971–1972 (Dec. 18–Feb. 7)	S	Rice and Wolman	Yankee Point	Reilly *et al.* (1983)
1972–1973 (Dec. 16–Feb. 16)	S	Rice and Wolman	Yankee Point	Reilly *et al.* (1983)
1973–1974 (Dec. 14–Feb. 8)	S	Rice and Wolman	Yankee Point	Reilly *et al.* (1983)
1974–1975 (Dec. 10–Feb. 7)	S	Rice and Wolman	Yankee Point	Reilly *et al.* (1983)
1975–1976 (Dec. 10–Feb. 3)	S	Rice and Wolman	Granite Canyon, California	Reilly *et al.* (1983)
1975–1976 (Dec. 15–Feb. 11)	S	Rice and Wolman	Point Loma	D. W. Rice and A. A. Wolman (unpublished)

(*continued*)

Table II (*Continued*)

Dates	Migration direction	Observers	Location	Source
1976–1977 (Dec. 10–Feb. 6)	S	Rice and Wolman	Granite Canyon	Reilly *et al.* (1983)
1976–1977 (Dec. 15–Feb.?)	S	Rice and Wolman	Point Loma	D. W. Rice and A. A. Wolman (unpublished)
1977 (?)	N	Hall *et al.*	Unimak Pass, Alaska	Hall *et al.* (1977)
1977 (Nov. 20–Dec. 9)	S	Rugh	Unimak Pass, Alaska	Rugh and Braham (1979)
1977–1978 (Dec. 15–Feb.)	S	Rice and Wolman	Point Loma	D. W. Rice and A. A. Wolman (unpublished)
1978 (Nov. 7–Dec. 20)	S	Rugh	Unimak Pass	Rugh (Chapter 10, this volume)
1978–1979 (Dec. 2–Feb. 1)	S	Herzing and Mate	Newport, Oregon	Herzing and Mate (Chapter 12, this volume)
1978–1979 (Dec. 10–Feb. 8)	S	Rice and Wolman	Granite Canyon	Reilly *et al.* (1983)
1979–1980 (Oct. 31–Jan. 3)	S	Rugh	Unimak Pass	Rugh (Chapter 10, this volume)
1979 (Feb. 27–?)	N	Herzing and Mate	Newport, Oregon	Herzing and Mate (Chapter 13, this volume)
1979–1980 (Dec. 11–Feb. 4)	S	Herzing and Mate	Newport, Oregon	Herzing and Mate (Chapter 13, this volume)
1979–1980 (Dec. 10–Feb. 6)	S	Rice and Wolman	Granite Canyon	Reilly *et al.* (1983)
1980 (Feb. 21–?)	N	Herzing and Mate	Newport, Oregon	Herzing and Mate (Chapter 13, this volume)
1980 (Mar. 10–May 31)	N	Poole	Piedras Blancas	Poole (Chapter 16,this volume)
1980–1981 (Dec. 10–Feb. 9)	S	Herzing and Mate	Newport, Oregon	Herzing and Mate (Chapter 13, this volume)
1981 (Feb. 22–?)	N	Herzing and Mate	Newport, Oregon	Herzing and Mate (Chapter 13, this volume)
1981 (Feb. 9–May 30)	N	Poole	Piedras Blancas	Poole (Chapter 16, this volume)
1981 (Mar. 23–Jun. 17)	N	Hessing	Unimak Pass	Hessing (1983)

proportion may have changed throughout the years, in response to increasing levels of boating and other human activities nearshore (Gilmore, 1978a). The method used to correct for whales missed due to poor visibility was rather arbitrary for the first 3 years, being based on a subjective appraisal of the percentage reduction in visibility each day. Effort expended varied between years. For these and other reasons (discussed in Reilly,

1984) it is inadvisable to look to the early Point Loma censuses for accurate measures of abundance. However, through trial and error these early shore censuses did establish much of the basic field methodology that was used by subsequent censuses through the 1979–1980 Monterey study.

Five more censuses were conducted at Point Loma between 1967–1968 and 1977–1978. Although field and data recording procedures were well systemized in those years, including detailed observations on visibility conditions, these data are also of questionable accuracy because of the unknown proportion passing offshore.

In 1966–1967, Adams (1968) conducted a fall census from Point Lobos, near Monterey, California. Rather than censusing all day during each day of the migration, he counted only during periods of good visibility and when convenient (his was a privately conducted and funded venture). The period of November 27 through February 17 reported includes the full duration of the southern migration near Monterey (Reilly *et al.*, 1980). Actual effort expended and raw counts are not reported. From average counts during hours watched, Adams extrapolated over the remainder of the day. The method sounds reasonable, but the results are questionable. A histogram (his Fig. 1) shows a maximum of about 55 whales per hour counted, with a number of days above 40/hr on average. This is almost certainly too high. In the highest days during the 13-year NMFS Monterey series, the average observed daily count was never more than 25 whales/hr. In no single hour did more than 30 whales pass the station (Reilly, 1981). Using arbitrary extrapolations of 70% daylight count for whales passing at night and 5% for whales passing offshore, Adams estimated the total population in 1966–1967 to be 18,300. As with the average hourly count, this is probably too high. The upper 95% confidence limit for the 1967–1968 census from Monterey was only 15,597 (Reilly *et al.*, 1983). This comparison assumes that the Monterey census was unbiased.

In 1967–1968, Dale Rice and Allen Wolman of the Marine Mammal Laboratory (Seattle) began a 13-year series of consecutive annual censuses near Monterey, California. These censuses were made from Yankee Point until 1973–1974, and from Granite Canyon (3.7 km south of Yankee Point) in the remaining years. Boat traffic is at a minimum in the vicinity of these sites along the beginning of the "Big Sur" coast. Further, there are no nearshore islands which might divert some whales offshore, as suggested for the southern California Bight south of Point Conception (Rice, 1965; Gilmore, 1978b).

At Monterey, counts were conducted for 10 hours per day (the duration of daylight) during approximately 2 months each year. Estimates were recorded of the number of whales present in each passing group and the distances of the groups offshore. Details of visibility conditions were also recorded. Areas considered in abundance estimation included the effects of varying visibility conditions, observer accuracy in estimating the number within and distance to passing pods, whales missed as a function of their distance offshore, those passing at night, and those passing before the first and after the last days of censusing. Raw counts recorded during 10 daylight hours ranged from a low of 2,667 in 1971–1972 to a high of 4,924 in 1979–1980. For this final year the "best estimate" of total abundance was 15,647, with 95% confidence limits of 13,450 and 19,201. (Confidence limits refer to precision here and assume an unbiased estimate.)

In November and December of 1977, 1978, and 1979, David Rugh conducted censuses from Cape Sarichef in Unimak Pass, Alaska (Rugh and Braham, 1979; Chapter 10, this volume). This is the point at which the population filters into and out of the Bering Sea. The first season comprised only 20 days during which 2,055 whales were observed in 82.6 hr of systematic watch. During 1978 and 1979, the duration of the census was extended in an attempt to cover the entire migration. Mean counts recorded those years were 5,050 and 4,061. Rugh also conducted experiments to verify some of the assumptions necessary to extrapolate to total population size. As with the California censuses, the most important estimation problem regarded the number of whales passing at night (approximately 16 of 24 hr at Unimak Pass are in darkness during November and December). Other areas considered but not applied to the abundance estimation included the "tails" of the migration, whales missed due to varying visibility and storm conditions, offshore distribution, differences among observers, and accuracy in counts even in ideal conditions. A "best estimate" was given for 1978 only of approximately 17,000.

The first attempt to census northbound gray whales during their spring migration was made by Hall *et al.* (1977) at Unimak Pass in 1976. The duration of the spring migration, nearly 4 months, makes it more difficult to census than the 2-month (or less) fall migration. The duration of the Hall *et al.* census was less than 2 months, and they estimated that only 9000 whales passed into the Bering Sea during the census.

During 1978, 1979, and 1980, counts were conducted by Herzing and Mate (Chapter 13, this volume) (of Oregon State University) from Yaquina Head in Newport, Oregon. They surveyed both north- and southbound whales, observing only on a "sampling" basis during good conditions in daylight hours. Population estimates were made from southbound counts only. Raw counts were 253 in 1979–1980 and 653 in 1980–1981 (actual effort was not reported). Estimates for 1980 ranged from 13,627 to 21,854, depending upon factors corrected for. The minimal figure considered whales missed due to lack of effort during all night and some daylight hours. The maximal figure included corrections for underestimating pod size and whales missed offshore.

During 1980 and 1981, Poole (1984; Chapter 16, this volume) conducted censuses of northbound migrants from Point Piedras Blancas along the central California coast, just north of San Simeon. In 1980, 1,496 whales were sighted during 659 hours of effort; during 1981, 3,296 were seen during 755 hours. Population estimates of 15,725 and 16,140 included corrections for whales passing at night, bias in estimating pod size, and whales passing too far offshore to be seen.

In 1981, Hessing (1983) conducted the second spring census at Unimak Pass. Between March 23 and June 17, 3,851 whales were sighted. From 2 to 10 of ~18 daylight hours were censused each day. Hessing estimated that 14,146 whales passed into the Bering Sea in 1981. A constant night travel rate was assumed, and no corrections were made for poor visibility conditions or for whales passing before and after the census.

ASSUMPTIONS AND SUMMARY OF CURRENT STATUS

Five major topics must be quantitatively addressed in order to estimate population size from shore station counts of migrating whales [Appendix B; see Reilly *et al.* (1983) for a more complete discussion].

Offshore Distribution

Most recent shore censuses have included verification of the offshore distribution of passing whales via aerial surveys [see "History and Methods" (Migratory Route)]. The offshore distribution appears to change during the course of the migration, both southbound (Reilly *et al.*, 1980) and northbound (Chapter 16, this volume). Consequently it is important that aerial observations be compared only to shore observations made at roughly the same time. Observers to date have assumed that all pods passing within some fixed distances are seen. The proportion assumed missed outside of this zone is corrected for by comparing the inshore proportions seen from aircraft and land. At Monterey the results indicated about 20% were missed for this reason (Reilly *et al.*, 1983); for the Newport, Oregon counts the estimate was 38% (Chapter 13, this volume); for Piedras Blancas, 28% during "phase a" only (Poole, 1984). For Unimak Pass in autumn, it was shown that few if any whales were missed solely as a function of their distance offshore (Chapter 10, this volume).

Migratory Timing

To quantify migratory timing, we fit a probability density function to the observed daily Monterey counts, for each year individually. This was useful to compare annual timing between years at one location and between locations for any one year. In assessment, the timing models were used to estimate population proportions passing on days with poor visibility (next section) and on days before and after the censuses.

For the Monterey censuses, extrapolations for whales passing before and after the field efforts increased the total estimates only by ~5% per year. For other censuses which cover a smaller proportion of the migration, estimating the tails is of greater importance. Rugh (Chapter 10, this volume) compared a core period of days between two years, estimating the tails of the earlier, less well-covered year by the ratio of the core periods (but did not estimate total abundance for the earlier year). Poole (Chapter 16, this volume) used a similar procedure for 2 years' data from Point Piedras Blancas. Herzing and Mate (Chapter 13, this volume) did not account for missed "tails" but their censuses may have been long enough for this not to matter.

Visibility Conditions

Lack of fit of the Monterey migratory timing models for any 1 day (after finding model parameters that produced the best overall fit) was highly correlated with visibility conditions, recorded in an ordinal scale (Reilly *et al.*, 1983). For days with conditions worse than "fair" on average, the predicted values from the probability function were used instead of the raw data to estimate the number of whales passing. If in fact there are daily pulses of whales passing the site, independent of weather, this method may bias the resulting estimate. The direction such a bias would take is difficult to predict.

Rugh (Chapter 10, this volume) used an empirical approximation of this method for the Unimak Pass data but worked on a finer time scale of hours rather than days. He noted drops in hourly rate by visibility code and corrected data from hours of "poor" and "unacceptable" conditions by interpolating between periods before and after. Others have used variations of this same technique (Rice, 1961; Poole, 1984; Chapter 13, this volume).

There should be no additional problems in resulting estimates of total abundance from the above visibility corrections, unless gray whales actually slow down during periods of poor visibility (as suggested by R. M. Gilmore, personal communication). If this is so, we are all overestimating population size by some amount which would vary by the method used and the amount of poor weather for which correction was made. The variance of the Monterey estimates increased proportionately with the number of poor visibility days, so this is at least partially reflected in the results for that site. Rugh (Chapter 10, this volume) used deviations from migratory course to suggest that gray whales do not slow down as a function of elevated Beaufort conditions. Perhaps some radio tracking or other electronic-based observation could help clear up the remaining ambiguity on this point.

Observer Accuracy and Precision

When an observer at a shore-based counting station records the passing of a whale or pod, at least two estimates are usually made: the number of individuals present and their distance offshore. To test a sample of "whale counters" for systematic biases in the above estimates, experiments were conducted at the Granite Canyon site during the 1978–1979 migration, involving independent estimates from a mixture of experienced and naive observers for 50 events of passing whales. Results from these experiments included (1) People vary widely in their accuracy and precision in estimating both number and distance; (2) even considering this variability, there are consistent (i.e., significant) biases for both estimates; (3) there is no benefit from experience in accuracy of distance estimation, although there is some benefit in precision. That is, over time a person is likely to become consistent in his inaccuracy if it exists (Reilly, 1981; Reilly *et al.*, 1980, 1983).

Correction for whales missed offshore (Section III,C,1) is confounded by any systematic inaccuracy in estimating distance to passing whales. For the Monterey data, such inaccuracy was detected and corrected for by redistributing the observed data according to results of the observer bias experiments. The overall change in the Monterey population estimates as a result of this correction was quite small. There are potential problems even with this correction before correction, relating to the range of between observer variability. This is discussed further in the following section.

Some results from the Monterey observer bias experiments have been applied to other surveys (Poole, 1984; Chapter 13, this volume). There are two possible problems with this, the first applicable to our original use as well. First, the "sample" of people was relatively small (12, 3 experienced and 9 naive). Second, the sample size of observed pods (50) was not large enough to allow partitioning of pod size estimates by distance to test for interaction. If such interaction exists, the results may not be applicable to sites with migratory corridors significantly different than those at the Monterey sites.

Night Travel Rate

The very limited data applicable to the determination of night travel rate were reviewed in Reilly *et al.* (1980) and Reilly (1981). Only indirect evidence exists: There are no systematically collected data of sufficient sample size which unequivocally measure night travel rate during migration. The indirect information available is more suggestive of a constant 24-hr rate than not, and most recent studies have assumed this (Reilly *et al.*, 1983; Poole, 1984; Chapter 13, this volume; Rugh and Braham, 1979). Rugh (Chapter 10, this volume) collected data using night-vision goggles at Unimak Pass. (These were not useful at Monterey because the whales were farther offshore there.) Two slightly different experiments were run during dusk, comparing sightings made with and without the goggles. These gave two different results. The whales either maintained their daytime rate or slowed to about 73% of that rate. Rugh chose the latter result in computing total abundance. Gilmore (1960, and personal communication) feels that gray whales slow at night in response to limited visual cues for orientation.

Because it is necessary to interpolate more than one-third to two-thirds of the total migration due to darkness (depending on season and latitude), estimation of whales passing at night is the single most important aspect of inferring abundance from shore counts. To date, it is not clear what migrating gray whales do at night or if this behavior changes between locations or between seasons. Considering the relative importance of this, and the relative paucity of our knowledge, it is instructive to investigate how our total estimates would change if gray whales slowed their travel rate at night by various hypothetical amounts. This topic is pursued in the following section.

Discussion

Both the winter range and in-migration censuses require further research before we can be reasonably assured of the accuracy of our population estimates. As stated, my personal bias favors the present estimates from shore counts over those from the winter range. Aside from this, there are reasonable scenarios in which the data from the two sites can indicate similar population estimates.

The following is an exercise in the nature of "what if"? It is not intended as a postulation that the true population size is at any particular value intermediate between shore and winter range estimates. Rather, I wish to demonstrate that within the regions of our uncertainty about night migration rate and the overall accuracy of winter range estimates, there are population sizes that are consistent with data from both sources.

All of the recent shore count estimates are relatively close, about 15,000 to 20,000 for 1980. This is not surprising in view of the essentially common assumptions and corrections used. If gray whales do slow down at night, our data would indicate a smaller population. I have used the 1980 Monterey estimate of 15,647 as an example to indicate how the population estimate would be reduced as a function of a reduced night travel rate (Table III). For example, if the animals traveled at 75% of the day rate, the estimate would be about 13,365. It is logically possible that gray whales speed up their

Table III
Total Population Size Estimates for 1980 from the Monterey Shore Census, as a Function of Hypothetical Night Travel Rates

If night rate is this proportion of day rate	The 1980 population estimate is
1.00	15,647
0.75	13,647
0.67	12,635
0.50	11,083
0.33	9,532
0.25	8,801
0.10	7,432[a]

[a]The 1980 winter range estimate from Rice *et al.* (1983) is 7,601.

rate of travel at night, rather than slowing or maintaining the daylight rate. I can think of no reason why this would be probable and therefore have not pursued an increase in rate in Table III.

In spite of the assumption that night travel is reduced to 73% of the day rate at Unimak Pass in the fall, Rugh (Chapter 10, this volume) produced a 1978 estimate of 17,648 from a maximum raw count of 5,050. This may be increased evidence in support of the magnitude of other shore census estimates but does not clarify the issue of night travel rate. There are other possible ways in which his methods could result in an overestimate, but the amount would be minimal. Previously discussed ways include the possibly inappropriate application of the Monterey pod-size adjustment (because whales pass closer to shore at Unimak, accurate estimation of pod size may be less of a problem there), and Rugh's method of correcting for whales missed in periods of poor visibility. But, if these two corrections are removed entirely, his estimate would still be more than 15,000. It is perhaps relevant to note that the average daylight period is almost 8 hr at Unimak Pass during the fall migration, whereas it is about 10 hr at Monterey. Rugh consequently had a larger proportion of the day over which to extrapolate.

The present status of population estimation from the winter range suggests that 7,600 adults (from the 1981 effort) is an underestimate. These efforts are substantially limited by the physiographic vastness and complexity of the study area. This is exacerbated by the protracted migration into and out of the area, such that the entire population is not present at any one time. Other factors pointing to an underestimate from the winter range are that airplane censuses systematically miss whales (this has not been consistently accounted for), and that the observed ratio of calves to adults from the 1981 survey (Rice *et al.*, 1983) implies a reproductive rate higher than that which gray whales are likely to achieve under any conditions.

The ratio of calves to adults from the 1981 count was 0.19 (Rice *et al.*, 1983). If our current interpretation of gray whale vital rates and age structure is reasonably accurate (Rice and Wolman, 1971; Reilly, 1984) a crude birth rate greater than 0.14 is quite unlikely. This is calculated from a pregnancy rate of 0.467, a sex ratio of 0.5, and a mortality schedule that results in 60% mature ($0.467 \times 0.5 \times 0.6 = 0.14$). If we assume that Rice *et al.* (1983) gave an accurate calf estimate, then 1,439 calves and a maximum crude birth rate of 0.14 indicate a minimum adult population of 10,279. Summing adults and calves gives 11,718. Removing ~7% for 1 year mortality (from 0.10 net for whales under the age of sexual maturity and ~0.05 for those over that age; Reilly, 1984) gives a minimum figure of about 10,900 passing central California the next autumn. If in fact the calf estimate is low, for example, 30% (as suggested by the data of Swartz and Jones, Chapter 14, this volume), then the number passing Monterey would be approximately 14,300. This is within the 95% confidence limits for the 1980 Monterey census (Reilly *et al.*, 1983).

Again, I am not postulating that the population size is exactly 14,300, or that gray whales do in fact slow down at night. Lacking new evidence, it still seems most probable that the night travel rate is on average not different from the daytime rate, and therefore that the population size in 1980 was approximately 15,600. The most important point here is that the Baja estimate is probably an underestimate of total population size. Also, if *not* accurate, the shore station estimates are probably too high. Perhaps research in some key areas will help us reinterpret our census data and result in a resolution to the present dilemma. Unequivocal documentation of night travel rates during migration is needed. Further field work on the winter grounds, directed at verification of specific problem areas and coupled with a comprehensive, quantitative appraisal of the entire estimation problem there, are prerequisite to reliable population estimation.

Appendix A. Factors Affecting or Involved in Estimating Total Abundance from Winter Range Aerial Surveys

I. Specific research topics
  A. Behavior of animals
    1. Time spent at the surface
      a. Age/sex group differences
      b. Group size differences
      c. Location differences
      d. Diel differences
      e. Intraseason changes
      f. Response to platform
    2. Immigration–emigration
      Same factor list as in 1. above
  B. Geographic distribution
    1. Lagoon channel
    2. Lagoon mouth
    3. Lagoon open area
    4. Alongshore bay

        5. Alongshore exposed, sandy beach
        6. Alongshore exposed, rocky beach
    C. Interplatform comparisons
        1. Lagoons
            a. Air to boat
            b. Air to tower
            c. Boat to tower
        2. Alongshore
            a. Air to boat
    D. Visibility conditions
        1. Quantity (scale effects)
        2. Confounding with behavior?
        3. Confounding with field methods?
    E. Observer accuracy and precision
        1. Experience
        2. Fatigue
        3. Intraobserver variation
        4. Interobserver variation
II. Methods
    A. Field methods
        1. Altitude
        2. Air speed
        3. Location
        4. Track line repetition
        5. Data recording
        6. Intraseason timing
        7. Boat, land-based verification experiments
    B. Analytic methods
        1. Post factor stratification, pooling
        2. Sighting model definition
        3. Parameter estimation
        4. Parameter comparison
            i. Between platforms
            ii. Between locations
            iii. Between observers
        5. Data, parameter "adjustment" (i.e., application of correction factors)
        6. Estimation of total abundance and variance
            i. Model derivation
            ii. Computation conventions

Appendix B. Factors Affecting or Involved in Estimating Total Abundance from Shore Station Censuses for Migrating Whales

I. Specific research topics
  A. Offshore distribution
    1. Site characteristics
    2. Season
      a. Between (fall versus spring) changes
      b. Within-season changes
    3. Group size differences
  B. Migratory timing
    1. Model estimation
    2. Comparisons between years
    3. Comparisons between sites
    4. Estimation of "tails"
    5. Estimation of poor visibility periods
  C. Visibility conditions
    1. Establish scale of effect
    2. Confounding with A., B.
  D. Observer accuracy and precision
    1. Distance estimation
      a. Change with distance?
      b. Change with pod size?
    2. Pod size estimation
      a. Change with distance?
      b. Change with pod size?
    3. Interobserver differences
    4. Intraobserver differences
  E. Diel migration rate
    1. Night versus day rates
    2. Change during daytime?
II. Methods
  A. Field methods
    1. Site choice
    2. Data collection conventions (i.e., variables chosen, frequency recorded, etc.)
    3. Verification experiments (for distance, pod size)
  B. Analytic methods
    1. Post facto stratification, pooling (distance only)
    2. Parameter estimation
      a. Distance offshore
      b. Pod size
      c. Number per time interval

3. Parameter adjustment (re: verification experiments)
   a. Offshore distribution
   b. Pod size
4. Estimation of total abundance and variance
   a. Model derivation
   b. Computation conventions

Acknowledgments

I thank the editors for their encouragement, advice, and help in assembling this review. The idea for the chapter, and part of its content, resulted from a series of lively discussions with the late Raymond Gilmore. He, Izadore Barrett, Douglas Chapman, Dale Rice, David Rugh, and an anonymous reviewer provided helpful comments on an earlier draft.

References

Adams, L. (1968). Census of the gray whale, 1966–67. *Nor. Hvalfangst-Tid.* **57**(2), 40–43.

Fleischer, L. (1984). Studies of gray whales in Laguna Ojo de Liebre. (in progress)

Gard, R. (1974). Aerial censuses of gray whales in Baja, California lagoons, 1970 and 1973, with notes on behavior, mortality and conservation. *Calif. Fish Game* **60**(3), 132–144.

Gard, R. (1978). Aerial census and a population dynamics study of gray whales in Baja California during the 1976 calving and mating season. *NTIS Rep.* PB 275 297, 1–20.

Gilmore, R. M. (1960). A census of the California gray whale. *U.S. Fish Wildl. Serv., Spec. Sci. Rep. Fish.* **342,** 1–30.

Gilmore, R. M. (1978a). Some news and views of the gray whale. Part 1. Migration into and out of the Bering sea. *Whale Watcher* **12**(1), 4–8.

Gilmore, R. M. (1978b). Some news and views of the gray whale, 1977. Part 2. Migration south and north along and between the islands of southern California. *Whale Watcher* **12**(2), 9–13.

Gilmore, R. M., Brownell, R. C., Mills, J. G., and Harrison, A. (1967). Gray whales near Yavaros Southern Sonora, Golfo de California, Mexico. *Trans. San Diego Nat. Hist. Soc.* **14**(16), 197–204.

Hall, J. D., Harrison, C. S., Nelson, J., and Taber, A. (1977). The migration of California gray whales into the Bering Sea. *Proc. Conf. Biol. Mar. Mamm., 2nd, 1977* Abstract, p. 8.

Hessing, P. (1983). "Gray Whale (*Eschrichtius robustus*) Migration into the Bering Sea, Spring 1981," Final Rep. Contract NA 81SRGA 00080. Outer Continental Shelf Environ. Assess. Program, NOAA Juneau Project Office.

Hubbs, C. L., and Hubbs, L. C. (1967). Gray whale censuses by airplane in Mexico. *Calif. Fish Game* **53**(1), 23–27.

Norris, K. S., Goodman, R. M., Villa-Ramirez, B., and Hobbs, L. (1977). Behavior of California gray whale, *Eschrichtius robustus*, in southern Baja California. *Fish. Bull.* **75**(1), 159–172.

Norris, K. S., Villa-Ramirez, B., Nichols, G., Wursig, B., and Miller, K. (1983). Lagoon entrance and other aggregations of gray whales *Eschrichtius robustus*. *In* "Behavior and Communication of Whales" (R. Payne, ed.), AAAS Sel. Symp. Westview Press, Boulder, Colorado.

Pike, G. C. (1962). Migration and feeding of the gray whale *Eschrichtius robustus*. *J. Fish. Res. Board Can.* **19,** 315–838.

Poole, M. (1984). Preliminary assessment of annual calf production of the California gray whale, *Eschrichtius robustus* from Pt. Piedras Blancas, California. *Rep. Int. Whal. Commn., Spec. Issue* **6** (in press).

Reilly, S. B. (1981). Population assessment and population dynamics of the California gray whale *Eschrichtius robustus*. Ph.D. Dissertation, University of Washington, Seattle.

Reilly, S. B. (1984). Observed and maximum rates of increase in gray whales. *Rep. Int. Whal. Commn., Spec. Issue* **6** (in press).
Reilly, S. B., Rice, D. W., and Wolman, A. A. (1980). Preliminary population estimate for the California gray whale based upon Monterey shore censuses, 1967/68 to 1978/79. *Rep. Int. Whal. Commn.* **30,** 359–368.
Reilly, S. B., Rice, D. W., and Wolman, A. A. (1983). Population assessment of the gray whale, *Eschrichtius robustus,* from California shore censuses, 1967–1980. *Fish. Bull.* **81**(2), 267–281.
Rice, D. W. (1961). Census of the California gray whale. *Nor. Hvalfangst-Tid.* **50,** 219–225.
Rice, D. W. (1965). Offshore southward migration of gray whales off southern California. *J. Mammal.* **46**(3), 500–501.
Rice, D. W., and Wolman, A. A. (1971). The life history and ecology of the gray whale *Eschrichtius robustus. Spec. Publ.—Am. Soc. Mammal.* **3,** 1–142.
Rice, D. W., Wolman, A. A., Withrow, D. E., and Fleischer, L. A. (1981). Gray whales on the winter grounds in Baja, California. *Rep. Int. Whal. Commn.* **31,** 477–493.
Rice, D. W., Wolman, A. A., and Withrow, D. E. (1983). Distribution and numbers of gray whales on their winter grounds. NMML/NWAFC, NMFS, NOAA, Seattle, Washington. (unpublished manuscript)
Rugh, D. J., and Braham, H. W. (1979). California gray whale *Eschrichtius robustus* Fall migration through Unimak Pass, Alaska, 1977: A preliminary report. *Rep. Int. Whal. Commn.* **29,** 315–320.
Scammon, C. M. (1874). "The Marine Mammals of the Northwestern Coast of North America." John H. Carmany & Co., San Francisco, California.
Storro-Patterson, R. (1981). Biological aspects of the eastern Pacific stock of gray whale *Eschrichtius robustus. In* "World Wildlife Fund, U.S. and I.U.C.N.," Contract rep., pp. 1–165. Whale Center, Oakland, California.
Swartz, S. L., and Jones, M. L. (1978). The evaluation of human activities on gray whales, *Eschrichtius robustus,* in Laguna San Ignacio, Baja California, Mexico. *NTIS* PB 289737.
Swartz, S. L., and Jones, M. L. (1980a). Gray whales, *Eschrichtius robustus* during the 1977–1978 and 1978–1979 winter seasons in Laguna San Ignacio, Baja California Sur, Mexico. *U.S. NTIS PB Rep.* **PB80-202989,** 1–34.
Swartz, S. L., and Jones, M. L. (1980b). "Gray Whales in Laguna San Ignacio and its Nearshore Waters during the 1979–1980 Winter Season," Final Tech. Rept. to World Wildlife Fund, U.S. Washington, D.C.
Swartz, S. L., and Jones, M. L. (1981). Demographic studies and habitat assessment of gray whales, *Eschrichtius robustus,* in Laguna San Ignacio, Baja California Sur, Mexico. *U.S. NTIS, PB Rep.* **PB82-123373,** 1–56.
Townsend, C. H. (1887). Present condition of the California gray whale fishery. *Bull. U.S. Fish. Comm.* **6,** 340–350.
Walker, L. W. (1949). Nursery of the gray whales. *Nat. Hist. (N.Y.)* **58**(6), 248–256.
White, S. B. (1975). Population and behavioral patterns of gray whales wintering in Scammon's Lagoon. MS Thesis, Colorado State University Ft. Collins.

10

Census of Gray Whales at Unimak Pass, Alaska, November–December 1977–1979

David J. Rugh

Introduction

Unimak Pass, the first prominent waterway west of the Alaska Peninsula (Fig. 1), has long been considered a potential route for gray whales migrating between calving lagoons in Baja California and feeding grounds in the Bering and Chukchi Seas. According to Scammon's (1874, p. 32) description, this species "makes regular migrations from the hot southern latitudes to beyond the Arctic Circle; and in its passages between the extremes of climate it follows the general trend of an irregular coast. . . ." Such a coastal route would most probably take these whales through Unimak Pass. Pike (1962) cited evidence of a coastal route between Baja California and British Columbia and included a scattering of coastal sightings from southern Alaska. Because of the paucity of sightings along the Alaskan coast, however, other authors have assumed that north of Vancouver Island whales traveled directly across the Gulf of Alaska and passed through

ISBN 0-12-389180-9

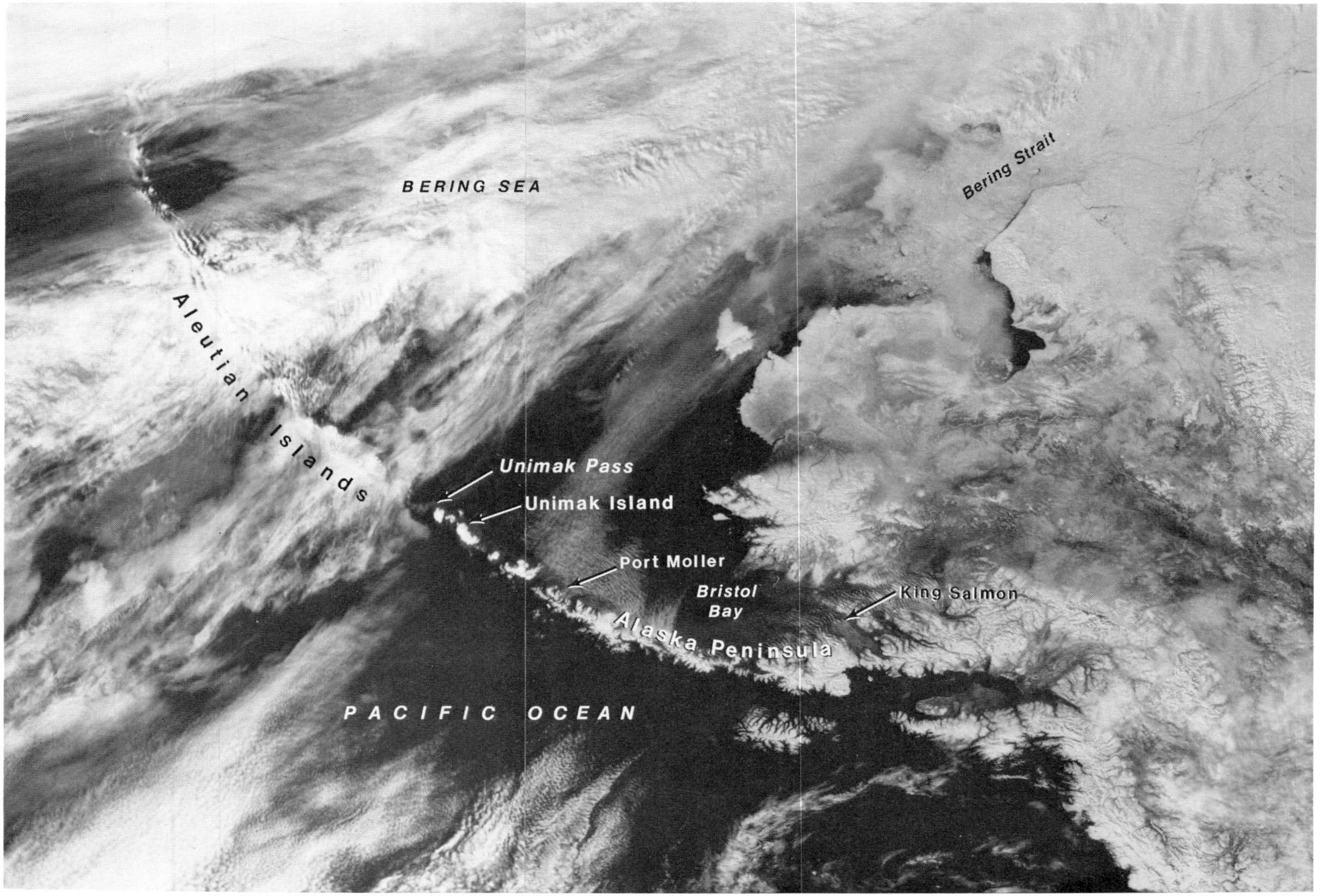

Fig. 1. *Satellite view of Alaska, April 17, 1978 (courtesy of U.S. Geological Survey, EROS Data Center, South Dakota). The study site, located on the west end of Unimak Island, was at the edge of the easternmost principal waterway in the Aleutian Islands.*

the Aleutian Island chain well west of Unimak Pass (Gilmore, 1955, 1960, 1968; Hubbs, 1959; Ichihara, 1958; Maher, 1960). On the other hand, no gray whales have ever been reported west of Unimak Pass (Ichihara, 1958; Pike, 1962) in spite of many aerial surveys in that area (Braham *et al.*, 1977; Chapter 11, this volume) and numerous shipboard surveys, especially during the 1962–1965 pelagic fur seal studies (C. Fiscus, personal communication). Furthermore, there are some early records of gray whales in Unimak Pass (Turner, 1886; Ichihara, 1958), and G. T. Bush (of the U.S. Coast Guard) at Cape Sarichef on Unimak Island described hundreds of gray whales passing the station between late October and mid-December of 1961.[1] The above evidence, coupled with spring sightings along the north shore of the Alaska Peninsula and Unimak Island (Braham *et al.*, 1977; J. Hall, personal communication), prompted the decision to attempt a census of the gray whale population from the west side of Unimak Island.

The southward migration was selected for this census because it is of shorter duration than the northward migration (Pike, 1962). Records show that gray whales remain near Bering Strait from June through October and frequently stay until November when sea ice is far from shore (Nikulin, 1946; Tomilin, 1957; Sleptsov, 1961; Pike, 1962). Using projections from these known dates and Pike's (1962) graphic depiction of the migratory timing, it was decided a census at Unimak Pass should be conducted in November and December.

Accordingly, in 1977 a study was undertaken at Cape Sarichef (Rugh and Braham, 1979). Final results from the 1977 study and the two following seasons, 1978 and 1979, are reported here. The objectives were to establish the location of the southward gray whale migration, describe its timing, and provide a population estimate with related parameters.

SITE DESCRIPTION

The east shore of Unimak Pass is bordered by Unimak Island, a large island with 429 km of shoreline (Sekora, 1973; see also Fig. 2). Unimak is nearly contiguous with the Alaska Peninsula, separated by only a narrow strip of water known as False Pass, which is as little as 1 km wide and extremely shallow (<4 m) on its north end. Thus Unimak Island and the Alaska Peninsula form a major projection of the Alaska land mass separating the Bering Sea from the Pacific Ocean. West of Unimak Pass, the Aleutian Islands continue this demarkation in a broken line to the Kamchatka Peninsula, U.S.S.R. (Fig. 1). The ecological significance of Unimak Pass may be eminent considering how this area received one of the highest impact ratings for modeled oil spills in a study conducted on 17 sites along Alaska's coast (Isakson *et al.*, 1975).

Unimak Pass is 18.5 km wide at its narrowest and 48.2 km wide on the north end between Cape Sarichef and Akun Island. It is as much as 180 m deep, but within 4–25 km of Unimak Island, depths are consistently 75–85 m (U.S. Coast and Geodetic Survey, 1939). Tide range from mean lower low to mean higher high water is 1.5 m (National Ocean Survey, 1979). Surface currents during winter are predominately north-

[1]In a letter dated June 14, 1962, to the late Gordon C. Pike of the Fisheries Research Board of Canada.

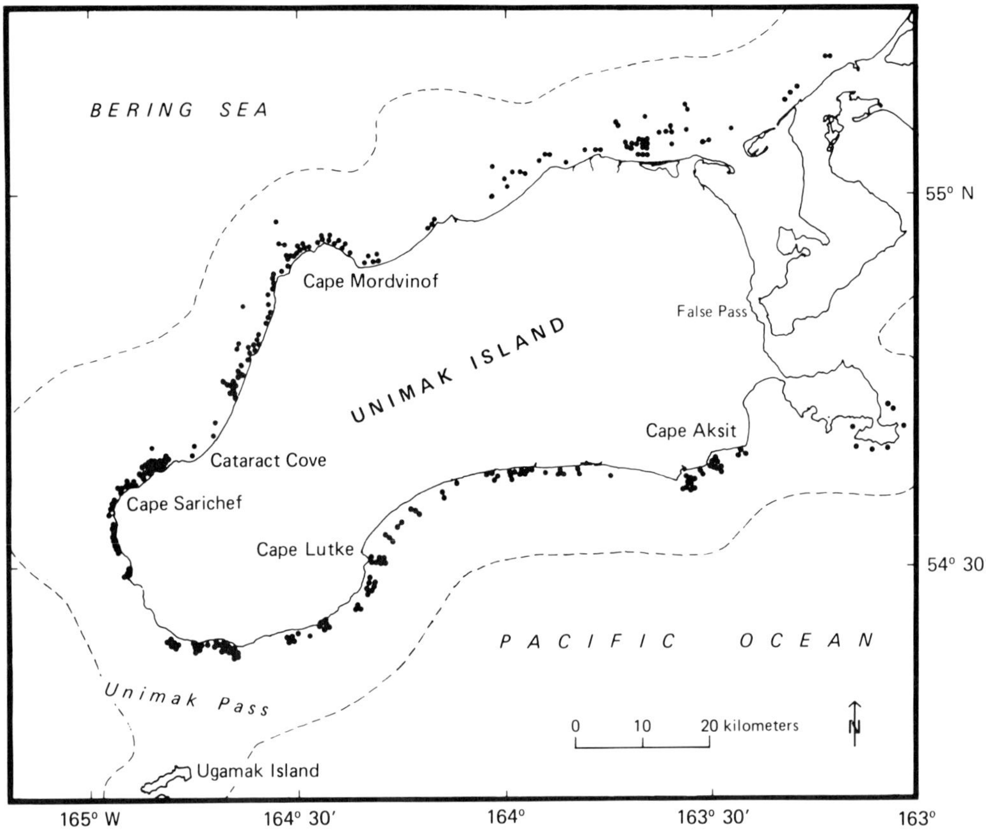

Fig. 2. *Distribution of gray whales around Unimak Island as observed during coastal aerial surveys in November of 1978 and 1979. Outer limit of aerial coverage is depicted as a dashed line. Each dot represents the location of one or more whale sightings.*

erly (U.S. Weather Bureau and U.S. Navy Hydrographic Office, 1961), with flood tides at 300°T averaging 3.7 km/h and ebb tides at 105°T averaging 1.3 km/hr. Currents reach velocities up to 12.2 km/hr (National Ocean Survey, 1975). The continental shelf around Cape Sarichef has a low-to-medium gradient with boulder and black sand seabeds interspersed with areas of gravel and broken shells (Army Map Service, 1958; Sears and Zimmerman, 1977).

Cape Sarichef (54°36′N, 164°56′W) is a promontory on the northwesternmost corner of Unimak Island. Facilities at Sarichef (belonging to the U.S. Coast Guard until July 1, 1979 and then to the U.S. Fish and Wildlife Service) provided shelter, communication, protection from bears, and access via the nearby airstrip. Although extremely isolated, the station's facilities and superb view made it an ideal position for the intended census.

During 1977 and 1978, observations were made from Site A, 31 m above the sea on a west-facing bluff. In 1978 a wood structure, built for our use by the Coast Guard, provided protection from the elements. Site B was established in 1979 at the crest of the bluff 50 m above the sea because the structure at Site A had blown away and Site B had a wider view. During inclement weather, observations were made from a truck parked so that its side windows, when rolled down, could be used for viewing. The sites had viewing arcs of 150° (A) and 165° (B). If the outer viewing perimeter is considered 3.7 km (the outermost conveniently measurable angle, 0.5° below the horizon), the two sites provided coverage of 18 km^2 and 20 km^2, respectively. During adverse weather, observers retreated into the protection of the wood shed (1978) or truck (1979). Doing so reduced the viewing arc to as little as 80° (10 km^2).

In winter, weather in the Aleutian Islands is typified by persistent and heavy winds, clouds, and frequent precipitation. Sekora (1973) suggested that "no other area in the world has weather worse than the Aleutian Islands." During our watches in November and December of 1977–1979, the wind at Cape Sarichef averaged 40 km/hr and frequently reached 75 km/hr. Precipitation, of which 55% was snow, occurred 24% of the time. Temperatures averaged 0.3°C, usually ranging no more than 3°C each day. Sea surface temperatures are usually 4–6°C during early winter (Ingraham, 1981), and sea ice rarely reaches as far south as Unimak Pass. Because of the latitude, daylight during the study period ranged from 9.5 hr/day on October 31 to 7.4 hr/day on December 22 (National Ocean Survey, 1979).

Methods

Each of the 1977–1979 whale counting efforts was timed to coincide with the approximate period in which most of the gray whale population could be observed passing Cape Sarichef. In 1977, I was the sole observer conducting counts from November 20 to December 9, primarily to establish the location and timing of the southward migration. In 1978, four of us maintained a near-continuous watch during all daylight hours from November 7 to December 20. The following season, four to six observers continued the census from October 31, 1979, to January 2, 1980, and conducted experiments to test the reliability of counts.

Distances of whales from shore were recorded with a clinometer which measured degrees below the horizon. Because it is used with only one eye, a clinometer allows for quick readings without interfering with the search. Clinometer angles divided the viewing range into 12 sectors at Site A and 17 sectors at Site B. These sector angles were later converted to distances perpendicular to the shore. Whales passing beyond 3.7 km, the outer limit of the furthest conveniently delineated sector, were thought to be out of sight of observers except during exceptional viewing conditions.

To more precisely calculate positions of passing whales, in 1979 a theodolite was installed in the glass shell used formerly to house the Coast Guard light beacon. This surveying instrument provided azimuth and zenith readings with resolution to 10-sec increments (4.0×10^{-4} deg). Exact elevation and geographical position of the theodolite were determined from benchmark data provided by the U.S. Coast and Geodetic

Survey and by National Ocean Survey, National Oceanic and Atmospheric Administration (NOAA). Bearings were later converted to a metric grid system on which sightings were plotted relative to coastal features. These data were collected separately from the primary census.

To further study gray whale distribution, six aerial surveys (a total of 13.4 hr and 2612 km of trackline) were flown along the shores of Unimak Island and Alaska Peninsula during November and December of 1978 and 1979. Of this total, 8.6 hr and 1376 km were in the vicinity of Unimak Island. From one to four observers searched for whales from a Cessna 180, a Cherokee Six-300, a Cessna Super Cub, and an Aztec. Flights were generally at 220–300 m altitude and 140–190 km/hr. Most surveys were coastal (0.5–1.0 km offshore) although some included coverage to 20 km offshore. Figure 2 depicts the general area covered by these surveys near Unimak Island. Sightings were possible up to 10 km from the aircraft during periods of excellent visibility. More extensive surveys further offshore were not flown for reasons of safety and because of the obvious coastal concentration of whales. For each sighting, the time, the whale's distance from shore (determined by clinometer), and its direction of travel were noted.

All 1978–1979 aerial sightings made between Cataract Cove and Cape Lutke (a distance of 78 km) along the relatively homogenous west shore of Unimak Island were pooled. These were compared to all 1977–1979 land-based data collected with clinometers during standard census procedures when visibility was Good to Excellent and surf was less than 2 m, conditions comparable to those experienced during aerial surveys.

Shore-based watches were usually maintained by two observers, one constantly searching for whales while the other recorded data and searched opportunistically. However, during periods of very low whale counts (<5/hr), only one observer was on watch. Watch periods lasted 2 hr each, the maximum period most observers could effectively continue counts in severe weather. Watches continued in spite of weather but were usually terminated at night. To minimize observer biases, we each overlapped 1 hr with preceding and following observers.

Systematic census records included observer effort (time on and off, visibility, and environmental conditions), whale counts, their distance offshore, and time of crossing an imaginary azimuth due west from the observation site. This azimuth was the area in which whale searches were concentrated. To reduce overcounting, and assuming a southbound migration, whales south of this azimuth were ignored at the start of a day; whales to the north were ignored at the end of a day. When whales deviated from a southbound course, their direction of travel was noted. When opportunity allowed, tallies of observed number of surfacings per whale were recorded. Discrimination of individuals or groups was important in this test, so tallies were not conducted when whale densities were high. These tallies provided a record of the number of independent cues to the presence of a whale that an observer had; accordingly, observers were not to concentrate on visually following groups but instead were requested to maintain systematic search patterns.

Six visibility categories were devised: excellent, very good, good, fair, poor, and unacceptable. In assigning these categories, observers subjectively evaluated and recorded weather, sea state, and light level. Pairing observers and rotating them on overlapping schedules kept categories relatively consistent within each season; however, visibility results were not considered equal between seasons, so those data were not pooled. Data from unacceptable periods were excluded in all analyses.

To establish repeatability of counts from a single site, experiments were conducted in 1978 and 1979 in which simultaneous census records were collected from a second site with virtually equal viewing area. To minimize biases, each team of observers was kept visibly and audibly isolated from each other, and observers rotated between teams. These tests were run only when visibility was good to excellent. Counts were matched between the two data logs, allowing up to 4 min and 2 sectors of discrepancy based on conservative interteam comparison of sighting occurrences after results were reviewed.

In 1978 and 1979, when seas were Beaufort 4 or less (winds under 28 km/hr) and twilight or moonlight was present, night watches were conducted by using U.S. Army night-vision goggles. These goggles amplify reflected visible light in a manner similar to a television screen but without magnification. During night watches, one person observed through the goggles while a second recorded data. Analysis was restricted to periods when vision through goggles was considered fair or better, and it was too dark to see whales without the goggles.

To establish an index of sightings lost due to the narrow field of view (40°) and resolution problems of night-vision goggles, two experiments were run. In the first, observers sat side by side at dusk, one with and one without goggles. When observers considered visibility equal, separate whale counts were maintained. In the second experiment, while one team continued the standard census a second team, nearby but out of audible range, watched for whales with a goggle simulator (night-vision goggles cannot be used under bright conditions). The simulator consisted of a pair of tubes limiting field of view to 40°; plastic covering the tube ends reduced vision by an amount equivalent to the goggles. In this manner, counts from the simulated goggles could be compared to the standard census.

To calculate the extent to which the population estimate might have been affected if there were differences between diurnal and nocturnal travel rates, the proportion of night during the study period, based on sunrise–sunset timings (National Ocean Survey, 1979), was weighted to reflect changes in concentrations in whale counts. Treated in 5-day blocks, daily whale rates were multipled by their mean respective day lengths. These products summed and divided by the total population estimate resulted in an average 7.8 hr/day of daylight. Further, within each 5-day block, twilight duration was established by subtracting the number of hours between sunrise and sunset from the maximum length of watch in fair or better visibility. The mean difference, 0.8 hr/day, was added to the day length calculated above. Therefore, there was an average 8.5 hr/day of daylight, which means 64.0% of the census period was night.

A population estimate was calculated from recorded whale counts. Where count ranges were given by observers, averages were used. Northbound whales (i.e., those deviating from the expected southbound direction) were excluded from the count, and an equal number of southbound whales was subtracted from daily totals on the premise that northbound sightings were of whales temporarily circling north. Daily counts were divided by the number of observation hours in excellent to fair daylight conditions. These rates per hour were multiplied by 24 hr to provide daily estimates which, when summed, resulted in interpolated population estimates for each season. Unsampled days within a season were estimated by averaging 2 preceding and 2 following days. Evaluations of factors affecting the calculated estimate are as described using nonparametric statistical methods (Hollander and Wolfe, 1973) unless otherwise noted.

Results

DISTRIBUTION

As gray whales migrated through Unimak Pass, they passed close to the Sarichef observation station. The median distance from shore for 10,223 sightings with clinometer bearings was 0.50 km. These sightings ranged from 0.03 to nearly 3.70 km offshore.

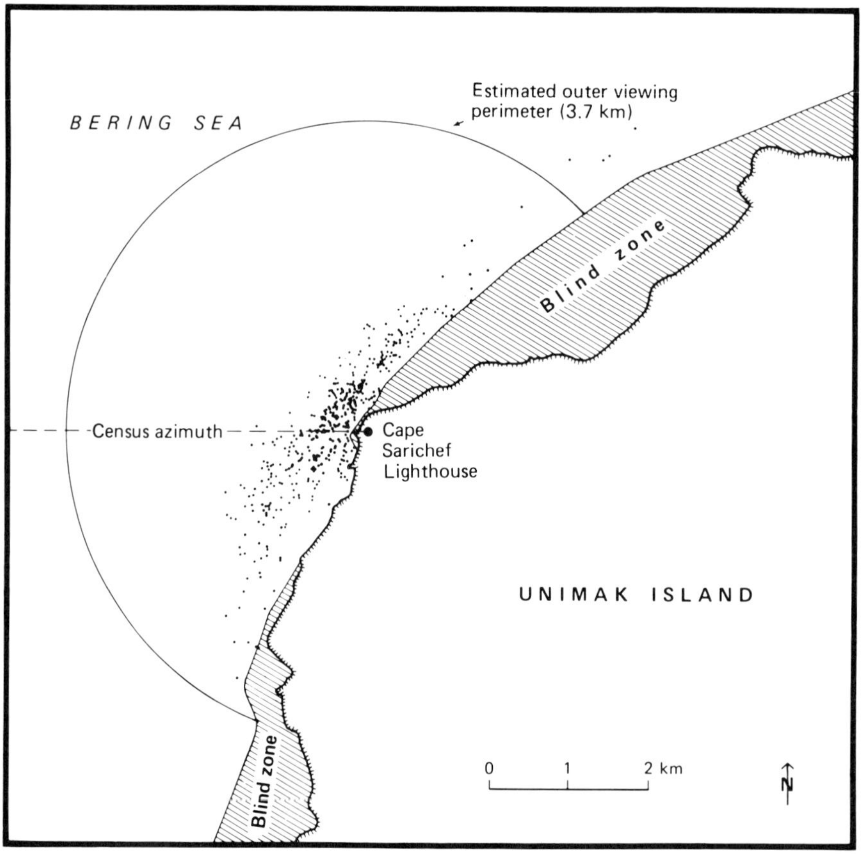

Fig. 3. *Distribution of gray whales migrating past Cape Sarichef, Alaska, during November and December of 1979. Each dot represents one or more whale sightings (n = 413), calculated from theodolite bearings. From the Sarichef lighthouse, certain inshore areas (blind zones) were not visible. The outermost measurable clinometer sector (0.5° = 3.7 km) was considered the standard outer viewing perimeter, but bearings show that under good conditions sightings could be made well beyond 3.7 km.*

Of 413 theodolite bearings, the median offshore distance was 0.48 km, ranging from 0.03 to 1.53 km. Figure 3 demonstrates this nearshore concentration of whales. Although theodolite bearings showed whales were visible from as far away as 6.3 km up the coast to the northeast during exceptional conditions, none were found more than 3.7 km offshore. The furthest offshore sighting made from the air along the west side of Unimak Island was 3.3 km.

The offshore distribution of whales along the west side of Unimak Island was compared between 104 aerial sightings and 3578 land-based sightings (Table I). Using Kendall's nonparametric test, the null hypothesis for independence was rejected ($p = .025$), that is, a close correlation was found between aerial and land-based distributions.

On an aerial survey from King Salmon at the northeasternmost corner of Bristol Bay to Cape Sarichef along the north coast of the Alaska Peninsula on November 7, 1978, no gray whales were found northeast of Port Moller (half-way down the Peninsula, Fig. 1), but 14 were seen to the southwest of Port Moller. Because whales were migrating southwestward, we had expected more whales northeast of Port Moller. Whales were closer to shore near Unimak Pass than they were in the shallower waters to the northeast, as evidenced during an aerial survey on November 29, 1978, when 82% of the 106 whales found between Cape Sarichef and Cape Mordvinof were within 1.4 km of the shore ($\bar{x} = 0.9$ km; $S_{\bar{x}} = 1.1$), while only 5% of 234 seen northeast of Cape Mordvinof were within 1.4 km ($\bar{x} = 3.8$ km; $S_{\bar{x}} = 2.1$).

To further test whether the decline in sightings away from shore was a function of visibility or whale distribution, sightings by shore observers made during fair or better visibility were divided into low (<2 m), medium (2–6 m), and high (>6 m) surf conditions (Fig. 4). When the surf was <2 m, the median offshore distance was 0.29 km ($n = 3578$) where depths were 15 m; with a surf of 2–6 m, the median distance was 0.51 km ($n = 3939$) where depths were 18 m; and in rough seas of >6m surf height, the median offshore distance was 0.84 km ($n = 848$) in water 22 m deep. These distributions were significantly different (χ^2; $p < .001$). Most whales (79% of 10,223 sightings) were seen within 1.0 km of the shore which meant they passed in water less than 20 m deep.

Table I
Comparison between Aerial and Land-Based Observations of Gray Whales' Offshore Distribution along the West Coast of Unimak Island, November–December, 1978–1979[a]

Distance offshore (km)	Aerial observations		Land-based observations	
	Whale count	Percentage	Whale count	Percentage
0.0–0.3	35	33.65	1265	35.35
0.3–1.4	47	45.19	2017	56.37
1.4–2.0	20	19.23	156	4.36
2.0–3.0	1	0.96	134	3.75
3.0–5.3	1	0.96	6	0.17
5.3–9.0	0	0	—	—

[a]A close correlation was evident ($p = .025$).

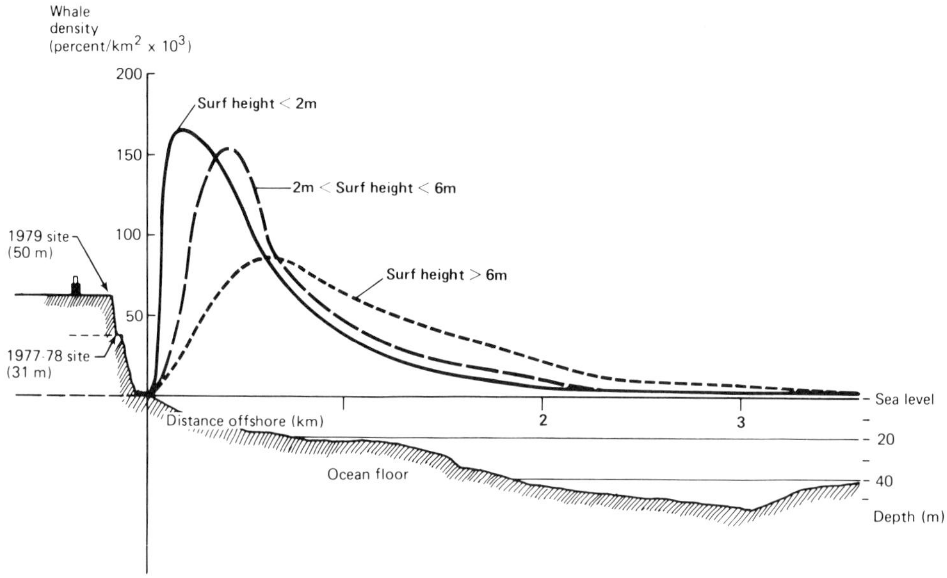

Fig. 4. *Frequency distributions of gray whale sightings offshore of Cape Sarichef during November and December, 1977–1979, as determined by clinometer bearings, using average to smooth curves. Vertical and horizontal scales are not the same. Sea depths were calculated from hydrographic charts provided by National Ocean Survey, NOAA.*

TIMING

In 1977, observations did not begin until November 20, by which time whales were already passing at a rate of >21/hr. Very high rates occurred on November 22–23, when up to 52/hr were observed, after which counts began to decline steadily. Counts were terminated on December 9 when approximately 11/hr were still passing. Estimates of the number of whales passing per day are represented in Fig. 5 as whales per hour (left ordinate scale) and whales per day (right ordinate).

The 1978 survey began earlier and ended later to better depict the onset and termination of the migration; however, many whales were passing throughout this period, as many as 5/hr on the first and last days of the census, increasing to over 15/hr between November 24 and December 18, with a peak of 43/hr near December 1–2. A peak in counts was not as clearly defined as in 1977. After December 2, rates of sightings declined irregularly until the survey was terminated on December 20.

In 1979, the census began well in advance of the principal migration. From October 31 to November 11, sighting rates averaged less than 1/hr. Between November 30 and

Fig. 5. *Chronological distribution of counts of gray whales during their fall migration south through Unimak Pass, November 20–December 9, 1977 (A); November 7–December 20, 1978 (B); and October 31, 1979–January 2, 1980 (C).*

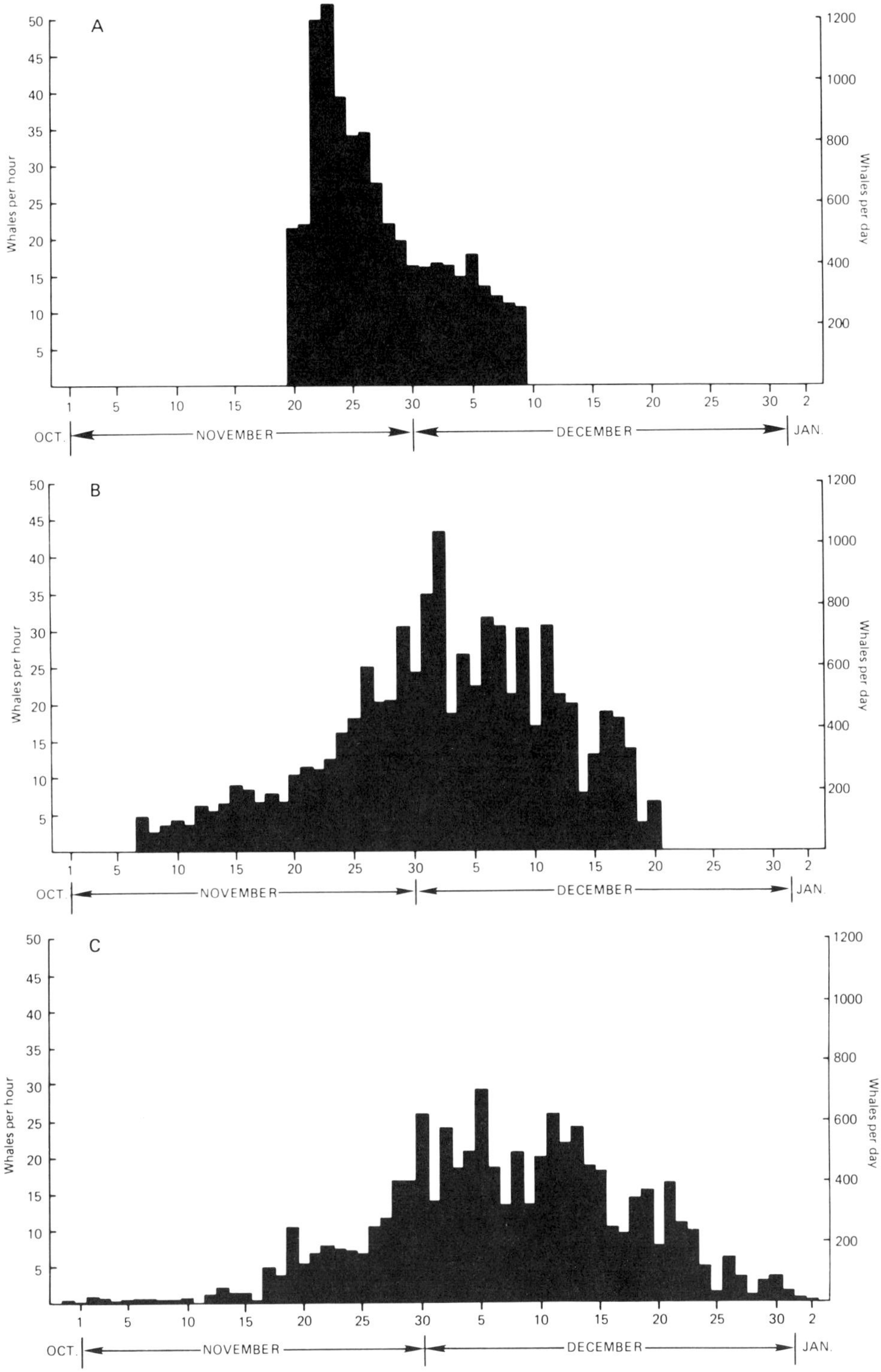
A
B
C
Whales per hour
Whales per day
50
45
40
35
30
25
20
15
10
5
1200
1000
800
600
400
200
1
5
10
15
20
25
30
2
OCT.
NOVEMBER
DECEMBER
JAN.

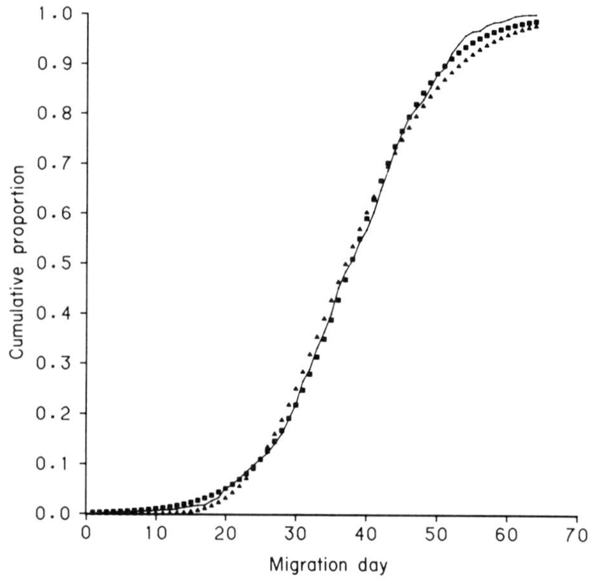

Fig. 6. *Cumulative daily estimates of number of whales passing Cape Sarichef in 1979. Symbols: —, observed; ■, logistic; ▲, gamma. (See text for further explanation.)*

December 21, they were generally more than 15/hr but declined more irregularly than in either 1977 or 1978. By January 2, 1980, rates of sightings had slowed to less than 1/hr.

Data presented in Fig. 5 suggest that in each consecutive year the migration arrived later. Using mutually sampled dates between paired years, median values showed whales in 1978 passed 10–11 days later than in 1977, and in 1979 they passed 2 days later than in 1978. Similar results were found when comparing differences in dates certain rates of whale sightings were reached each year. Averages of these differences showed a 12-day lag between 1977 and 1978 and a 4-day lag between 1978 and 1979.

The cumulative distribution of daily estimates for 1979 whale sightings (using proportions of the total) was fit to both logistic and gamma functions to provide a generalized description of the seasonal trend. The logistic curve provided a better fit (SSE = 0.00905) than the gamma distribution (SSE = 0.03296), as is evident in Fig. 6. The logistic took the form:

$$y = 1/(1 + e^{-a-bx})$$

where $\hat{a} = -6.1988$ and $\hat{b} = 0.1642$. Parameters were estimated by the method of nonlinear least squares. The gamma equation (detailed in Reilly, 1981) is

$$y = \frac{1}{a\Gamma(b)} (j/a)^{b-1} \exp(-j/a)$$

where a and b are scale and shape parameters respectively calculated to be $\hat{a} = 3.3811$ and $\hat{b} = 11.2759$, and where j represents the jth migration day. Parameters were estimated by the maximum likelihood method (Bury, 1975).

Table II
Census Summary of Gray Whales Seen from Cape Sarichef, 1977–1979

Survey dates	Whale counts	Observer hours	Interpolated estimate[a]	Sightings per whale
Nov. 20–Dec. 9, 1977	2,055	82.58	11,207	3.9 (971/247)
Nov. 7–Dec. 20, 1978	5,050	328.45	16,928	5.8 (15,225/2,630)
Oct. 31, 1979–Jan. 2, 1980	4,061	453.43	13,676	4.4 (12,314/2,782)

[a]Interpolated estimates are daily whale counts divided by observation hours and multiplied by 24 hr. Daily estimates are summed for season totals. Days without observation effort were estimated by averaging 2 preceding and 2 following days.

POPULATION SIZE

Whale counts, numbers of observation hours, rates of sightings per whale, and interpolated population estimates for each season 1977–1979 are presented in Table II. Whale counts and sightings per whale were highest in 1978 despite the longer field season in 1979. Further population analysis will focus on 1978 data.

Losses of sightings occurred with deteriorated visibility. For instance, in 1978 rates of sightings per whale dropped successively from 7.6 for excellent, 6.8 for very good, 5.9 for good, 3.7 for fair, to 2.8 for poor visibility. Whale counts per hours of observation effort are shown by year in Fig. 7. Details for 1978 are presented in Table III. All of these results were from daylight periods only and should not reflect differences which might exist between day and night whale movement rates. Between the first and last dates on which high rates of whale sightings occurred (greater than 15/hr), 8.9% of the recorded visibility was poor in 1978 and 12.4% in 1979. Because sighting rates in poor visibility

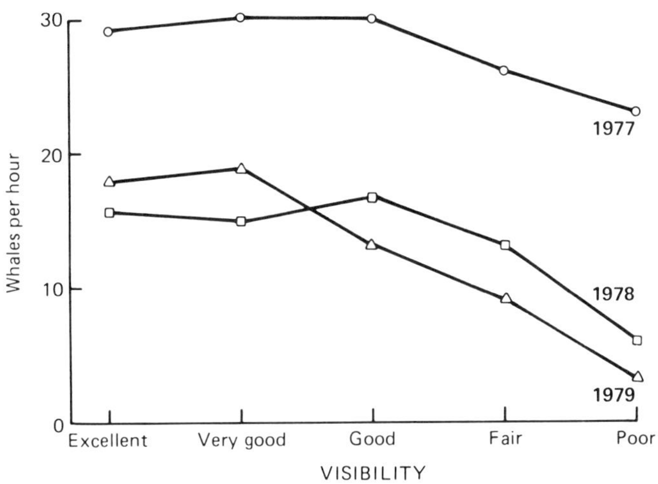

Fig. 7. *Mean number of whales sighted per hour in each visibility category, November–December, 1977–1979.*

Table III
Counts of Gray Whales as a Function of Visibility at Cape Sarichef, November–December, 1978[a,b]

Visibility	Whale counts	Percent counts	Observer hours	Counts per hour	$S_{\bar{x}}$
Excellent	1276	22.5	81.36	15.7	18.3
Very Good	1343	23.7	89.53	15.0	10.6
Good	1808	31.9	107.70	16.8	11.3
Fair	1043	18.4	79.31	13.2	12.0
Poor	200	3.5	34.09	5.9	14.2

[a]Where ranges were given, maximum figures are used.
[b]Results from poor conditions were excluded from the analysis; equivocal differences ($p = .06$; Jonckheere test) for fair conditions precluded applying a correction factor; the other conditions appear to have similar results.

were low, and because this did not represent a large portion of the data, all counts and effort in poor conditions were dropped from further analysis. Although rates of sightings in fair conditions (13.2 whales/hr) appear lower than rates in excellent, very good, and good visibility (15.7, 15.0, and 16.8 whales/hr respectively), the high standard deviations and equivocal statistical difference ($p = .06$; Jonckheere test) preclude applying a correction factor to adequately account for whales missed in fair visibility.

Losses of sightings due to depressed visibility may be biased if whale rates of movement change as storm conditions worsen. To quantify potential changes, data were evaluated for trends in deviations from the southbound migratory course relative to Beaufort condition. Allowing for the assumption observers could detect a whale's direction of movement in spite of sea state, northbound and milling sightings were compared to all sightings (Table IV). Of 13,777 total sightings for the years 1977–1979, there were 1,437 deviations ($\bar{x} = 10.4\%$). No significant trend in the data was evident ($p = .09$; Theil's test for slope coefficient for Beaufort 1 to 8).

Table IV
Course Deviations of Gray Whales Correlated to Beaufort Scale as Observed from Cape Sarichef, November–December, 1977–1979[a]

Beaufort	Deviations	Total sightings	Percentage
1	480	4681	10.3
2	215	1694	12.7
3	253	2492	10.2
4	185	1753	10.6
5	104	1301	8.0
6	144	900	16.0
7	47	740	6.4
8	9	212	4.2
9	0	4	0

[a]A downward trend in percentages is suggested ($p = .09$).

Table V
Frequency of Sightings of Whales as a Function of Group Size as Observed at Cape Sarichef, November–December, 1978–1979[a]

	Number of whales/group					
1978	1	2	3	4–5	6 or more	Sum
Number of sightings	1,777	2,685	2,805	3,696	4,262	15,225
Number of whales	496	625	534	544	431	2,630
Sightings/whale	3.6	4.3	5.3	6.8	9.9	5.8
All sightings (%)[b]	23.7	33.6	22.0	17.8	2.9	100.0
1979						
Number of sightings	2,681	3,049	2,414	2,612	1,558	12,314
Number of whales	736	790	534	488	234	2,782
Sightings/whale	3.6	3.9	4.5	5.4	6.7	4.4
All sightings (%)[b]	29.9	32.8	19.4	12.8	5.1	100.0

[a]The rate of sightings per whale increased with increase in group size ($p = .008$).
[b]Percentage of all whales recorded in the census log, including those without sightings/whale information.

Table V demonstrates frequency of sightings of whales as a function of group size. Because sighting rates were lower in 1979 than 1978, the 2 years were treated separately. In both years, sightings per whale increased with increase in group size ($p = .008$; Theil's test for slope coefficient).

The number of sightings per whale was not significantly affected by the number of observers on watch (Table VI) [Page's statistic for ranked sums (L) = 7.25; $l(0.05,4,3) = 84$].

No upward correction for land-based counts was indicated by aerial survey data. During 8.2 hr of aerial surveys along the coast of Unimak Island, 280 whales were sighted. By calculating their expected time of arrival opposite the Cape Sarichef count-

Table VI
Recorded Sightings per Whale as a Function of Number of Observers at Cape Sarichef, November–December, 1977–1979[a]

	Number of obsrvers			
Visibility	1	2	3	4
Excellent–Very Good	6.9	6.5	8.0	6.5
Good–Fair	4.2	4.7	5.4	3.9
Poor	2.5	3.1	2.7	2.7
Mean	5.2	5.2	6.1	4.6
Sightings/whale (sample size)	8,654/1,656	19,829/3,821	1,607/264	187/41

[a]There was no evident change ($p > .05$) in rates of sightings.

ing station based on 9.0 km/hr [the upper speed calculated by Rice and Wolman (1971), is used here for a conservative comparison], far fewer whales (280) were seen from the air than were seen from the land (1182), a ratio of 1:4.2.

Night-vision goggles were used for 66.53 hr of which only 46.64 hr were compared to day rates. A total of 334 whales (7.16/hr) recorded at night was compared to 4491 whales (14.31/hr) seen during 313.85 matching daylight hr. The mean night rate was 50.0% below daytime rates of sightings (Table VII).

Table VII
Comparison of Night and Day Rates of Whale Sightings at Cape Sarichef, November–December, 1978–1979[a]

		Night			Day	
	Date	Time	Effective watch (hr)	Whale count	Effective watch (hr)	Whale count
1978	Nov. 11	17:45–18:30	0.75	0	9.33	33–35
	12	17:30–19:00	1.00	2–3	10.25	62–64
	12	21:00–23:45	2.75	9	20.33	119–122
	13	6:00–8:00	1.17	0	10.08	57–58
	13	18:30–23:00	4.50	20–22	19.58	116–123
	14	5:15–8:00	2.75	0	9.50	59–65
	15	17:30–18:45	0.75	5	9.83	84–90
	17	5:15–8:00	2.58	28–32	9.42	61–64
	19	17:45–19:00	0.50	11	9.67	63–67
	20	4:00–8:15	1.17	3	8.58	86–91
	21	17:00–18:15	0.75	3–4	9.25	102–108
	22	17:00–18:15	0.50	6	9.58	105–109
	23	6:00–8:20	1.33	12–14	7.17	88–92
	25	7:45–9:00	0.58	5	8.67	152–160
	Dec. 10–11	18:30–2:00	5.87	117–137	17.20	397–427
	11	6:00–8:30	0.50	1–2	8.75	259–276
	15	17:10–17:45	0.25	0	7.45	93–104
	16	0:45–2:30	1.75	6–9	11.95	174–193
1979	Nov. 20	17:05–18:30	1.08	5–7	8.85	45–47
	25	17:37–23:19	4.73	19–21	8.60	54–60
	27	7:35–8:25	0.58	0	5.97	67–71
	Dec. 2	17:00–18:07	0.35	0	8.27	190–209
	4	0:40–2:15	1.58	3–4	15.83	295–326
	4	23:17–24:00	0.72	1–2	17.08	420–444
	5	6:02–18:08	2.47	21–25	8.92	259–265
	8	4:43–18:00	3.78	11–17	9.37	191–198
	10	17:18–18:08	0.83	7–9	8.78	170–181
	11	17:12–18:05	0.42	9–11	9.05	225–244
	12	17:25–17:48	0.22	1–2	8.67	184–198
	14	17:42–18:08	0.43	1–2	7.87	131–150
			46.64	306–362	313.85	4341–4641
				$\bar{x} = 334$		$\bar{x} = 4491$
			(7.16 whales/hr)		(14.31 whales/hr)	

[a]The mean night rate of sightings is 50.0% of the day rate.

In one experiment to examine loss of sightings due to goggles, 4.2 hr of tests were run at dusk in which 20 whales were counted by observers without goggles compared to 10 seen by observers with goggles. In a second experiment during 14.0 hr of tests, teams without goggles counted 414 whales while those with goggle simulators counted 284, a 68.6% reduction. During the same period, 179 whales were seen an average 4.8 times each by those without goggles while 201 were seen an average 3.6 times each by those with goggle simulators, a drop to 75.0%. The number of flukes recorded by the standard team (67) compared to those seen by the team with goggle simulators (40) suggested a 59.7% depression in visibility due to goggles.

To evaluate repeatability of counts, seven tests were run totaling 23.25 hr, during which 485 whales were counted at one site compared to 453 at a second, simultaneously run site. There were 201 of 297 groups of whales recorded by both teams. One team missed 88 whales while the other missed 120. If we assume the combined result (485 + 88 = 573) provides the true number of whales passing during these test periods, then 15–21% of the whales were missed by each team. The average, 18%, provides an estimate of whales unrecorded during standard census periods when viewing conditions were good or better. A frequency distribution of differences in counts between these simultaneously run teams is displayed in Fig. 8. Differences are depicted as percentages using a point-to-point line without averaging ($n = 294$). Note that if data from two teams were identical, all values would lie on a vertical line above zero on the abscissa; if differences were random, they would appear as a horizontal line.

The recorded time a whale spends at the surface averaged 4.9 sec ($S_{\bar{x}} = 1.3$; $n =$

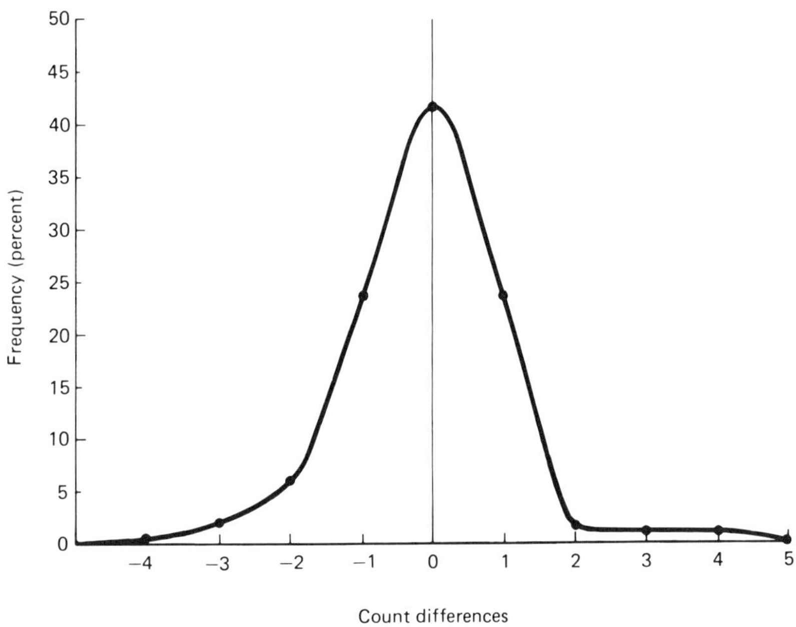

Fig. 8. *Frequency distribution of differences in numbers of gray whales counted at two sites at Cape Sarichef with equal views of Unimak Pass, November–December, 1978–1979. (See text for explanation.)*

335) ranging from 1.4 to 9.2 sec. Of 1000 timed dives, the average was 88 sec ($S_{\bar{x}} = 109$), ranging from 3 to 700 sec. The longer a whale was down during a dive, the less likely it would be recognized on resurfacing, thereby terminating recordings on long dives and biasing average values toward short dives. Whales were observed at the surface only 5.5% of the time (or less allowing that long dives were probably underrated).

Discussion

DISTRIBUTION

As gray whales swam southwest along the Alaska Peninsula and Unimak Island in the fall, they concentrated progressively closer to shore until they went through Unimak Pass (based on aerial surveys in this study and Chapter 11, this volume). The whale's neritic tendency is evident in Figs. 2 and 3 showing plotted sightings relative to the Cape Sarichef observation site. Most (79%) of the 10,223 sightings made with clinometer bearings were within 1.0 km of the shore where water was less than 20 m deep. The median of these sightings was 0.5 km offshore, and all sightings fell within 3.7 km of shore. The close proximity of whales to shore evident in these data was corroborated by (1) aerial surveys which included coverage to 20 km offshore in Unimak Pass but located no whales beyond 3.3 km (Fig. 2 and Table I); (2) theodolite bearings which independently showed a median offshore distance of 0.5 km; (3) an absence of sightings beyond 3.7 km from shore although theodolite bearings showed whales could be sighted 6.3 km away (Fig. 3); and (4) the fact that in worsening surf conditions whales were observed further offshore in deeper water than in calm conditions, showing that the drop in sightings away from shore was more than a function of reduced visibility with distance (Fig. 4). Data from aerial and land-based observations showing whale distribution relative to the shore were not significantly different (Table I).

The gray whale's neritic predilection has been discussed in most publications dealing with this species (cf. Chapter 11, this volume). Rice and Wolman (1971) stated that 73% passed within 1.5 km of the California coast, and Sund and O'Connor (1974) reported 94% were within 1.6 km of Yankee Point, California. All gray whales Hessing (1981) observed in the spring migration were within 3.0 km of the west Unimak shore, even closer than was observed in the fall migration.

TIMING

Gray whales commonly are found in the Bering and Chukchi Seas between June and October (Rice and Wolman, 1971). Natives of the Chukotsk Peninsula described these whales as numerous in August and September but absent in October (Tomilin, 1957). In late October they may start to appear in Unimak Pass. A biologist stationed at Cape Sarichef saw no gray whales throughout the summer and early fall of 1976 even as late as October (J. Nelson, personal communication), and observers found no gray whales on surveys made along the coast of Unimak Island by helicopter and ship during

October 21–25, 1976 (Braham *et al.*, 1977). On October 31, 1979, the first day of my observations for that season at Cape Sarichef, an estimated 5 whales were passing per day (Fig. 5). This rate steadily increased until by November 24, 1978, and November 30, 1979, whales were passing at a rate of more than 360 per day. High rates continued until nearly the end of the census in 1978 (December 18) and until December 21 in 1979. Throughout most of the 1977 survey (November 20 to December 6), rates were greater than 360 per day. In 1979, the year with the most thorough coverage, 90% of the sightings occurred between November 19 and December 23. Judging from the cumulative distribution in 1978 sightings (Fig. 6), few whales traveled through Unimak Pass before November or after December. During 1979–1980, a biologist stationed year round at Cape Sarichef saw no gray whales between January 3 and March 22, after which the spring migration began (R. Bartels, personal communication). Observations made on Kodiak Island confirm the general timings recorded at Cape Sarichef (R. MacIntosh, personal communication).

Although the distribution of sightings in 1977 (Fig. 5) had a distinctive peak and decline, sightings in 1978 and 1979 had much less regular patterns. These irregularities made it difficult to compare migration dates; however, by using median values from concurrently sampled periods among years, it appears the 1977 migration began 10–11 days before the 1978 migration, which was 2 days in advance of that in 1979. Reilly (1981) found these migrations to be 2 and 3 days delayed, respectively, with the 1978 migration mean date equal to the grand mean for 13 seasons of counts from 1967 to 1980. The 1977 migration may have been unusually early due to heavy ice conditions (Kukla and Gavin, 1978). Sketches of ice charts taken from satellite imagery (provided by Jan Zender of the Geophysical Institute, University of Alaska) show the 1977 ice front advanced well south of perimeters evidenced in 1978 and 1979. Where available images start in 1979 (on November 18), the ice distribution looks comparable to that on November 1, 1977, and November 2, 1978. By the first week in December, 1979, ice had caught up with 1978's freezing rate, but not until December 12 did the 1978 and 1979 ice reach south of St. Lawrence Island, an advance which was made as early as November 27 in 1977. These ice advances correlate to differences in median whale migration dates in that each year, 1977 to 1979, arrival dates were progressively later with the biggest difference in both cases being between 1977 and 1978; however, it is evident that variations in ice conditions from year to year are far greater than are whale migration dates.

POPULATION SIZE

Observers at Cape Sarichef recorded 2,055 southbound gray whales in November–December of 1977; 5,050 in 1978; and 4,061 in 1979. If we assume observations made during standard watch periods were representative of each day, then within respective census seasons there were 11,207 whales in 1977, 16,928 in 1978, and 13,676 in 1979. These fundamental estimates lack several important considerations: (1) unrecorded whales passing prior to and after the census; (2) loss of sightings due to visibility problems; (3) diel variations in rates of travel; and (4) whales missed by observers even under ideal conditions.

Of the three study seasons, 1978 had the largest whale count and largest interpolated estimate. High densities of sightings were observed in 1977, but the season was too short to permit adequate estimation of the entire population. Although 1979 was the most thorough season of the three, counts were lower than in 1978 (Table II and Fig. 7). Site B, used in 1979, was 19 m higher than Site A, a difference thought to be beneficial through an increased viewing area. However, rates of sightings per whale dropped from 5.8 in 1978 to 4.4 in 1979. This discrepancy in rates, 84%, is similar to the difference in population estimates, 81%. It is possible visibility was worse in 1979 during critical periods: 8.9% was poor in 1978 compared to 12.4% in 1979. The 1978 mean wind speed was 32 km/hr; the 1979 mean was 42 km/hr.

Differences in visibility may not be the only explanation for the 1978–1979 discrepancy, however. Near Monterey, California, observers found higher numbers in the 1979 migration (17,577) than in 1978 (13,676) (Reilly, 1981). Monterey estimates increased 22% while Sarichef estimates decreased 24%, suggesting a portion of the population may not have completed the migration north in 1979. What we cannot estimate from these data is the portion of the population that did not pass either census site. It is assumed here the number of gray whales wintering in the Bering Sea is negligible. The number staying south of the Bering Sea in summer on the other hand, may be significant (Hubbs, 1959; Carl, 1967; Hatler and Darling, 1974; Darling, 1977).

Differences among years might be considered a function of change in observers; however, I was present and supervised counts each year. No differences were found with changes in number of observers (Table VI). Discrepancies among individual observers, biases, and problems with observer fatigue were dealt with directly in the field and are considered minimal.

Using 1978 as the prime census year, parameters were evaluated to establish the reliability of the population estimate. It is evident whales were missed prior to and after the 1978 census (Fig. 5); however, no attempt is made here to estimate this uncounted portion. Instead, the final estimate will be treated as conservative.

As visibility worsened, fewer sightings/whale occurred; however, whale counts/hr did not drop definitively until visibility was considered poor or unacceptable (Fig. 7 and Table III). Data from these conditions were not used in population analysis. There was a suggestion of loss of sightings in fair visibility as well, but no estimate was made as to how many whales this might have included.

There was no evidence whales traveled slower as storms worsened. If they did, interpolations for periods when strong winds made visibility poor would overestimate the population. Many complications arise when trying to sort visibility from whale behavior in storms; however, minimal bias was assumed in recordings of a whale's direction of travel. Table IV showed no strong upward trend in deviations from the migratory course with rise in Beaufort scale; in fact, except for one high value at Beaufort 6, a strong downward trend was suggested. If whales rest or mill more in calm than in rough weather, interpolations for poor visibility during storms may lead to an underestimate of the true population. The infamous weather around Unimak Pass may indeed have had undetected but significant affects on these data; in 1978 and 1979, 15.5% of the watches recorded poor or unacceptable conditions.

Although several authors (Cummings *et al.*, 1968; Rice and Wolman, 1971; Swartz and Cummings, 1978) found no evidence of nocturnal slowdowns in gray whale movements, Hubbs and Hubbs (1967) and Ramsey (1968) believe they do have such evidence. Reilly (1981) concluded he had no statistical differences in diel rates of sightings, but he did show a slight decrease in whale counts during the first hour of morning which suggested either reduced visibility or a change in rates of movement. Gilmore (1960) and Adams (1968) used arbitrary correction factors of 50% and 70%, respectively. Observations made with night-vision goggles at Cape Sarichef suggested that only 50% as many sightings occurred at night as during corresponding day periods (Table VII); however, the restricted view and resolution of goggles were such that, when corrected, day and night rates appear equal (according to one test using dusk sightings) or (in a second test using goggle simulators) that a 73% nighttime slowdown occurred (50.0/ 68.6). If we accept the test using dusk sightings, no corrections are necessary; however, if whales slow down at night to as much as 73.0%, as the second test suggested, and because this affected 64.0% of the data (the average amount of night during the census), there may be a 17.3% overestimate in fundamental population interpolations.

Rates of sightings at Cape Sarichef were decidedly higher than those made during 13 seasons of study in California, where the average high concentration was 25/hr, and at no time were there more than 30/hr (Reilly, 1981). During 3 seasons at Sarichef, rates were over 30/hr on 12 days, with 52/hr the highest daily rate. These density differences were most probably a function of latitudinal dispersion as the whales traveled south. If there are diel rate differences, however, the shorter average day (8.6 hr) at Unimak Pass might make whale movements more concentrated during daylight hours than off California, where days are longer (10.1 hr).

Within systematic watch periods, even when visibility was not a problem, there were probably some inaccuracies in whale counts. Observers at Sarichef recorded a 4.3% range in counts (total minimum = 9,237; total maximum = 10,075), an indication of ambiguity of the number of whales in a pod. When two counting teams operated simultaneously, there was an 18% discrepancy (Fig. 8). In fact, whales may have been missed by both teams. With whales at the surface only 5.5% of the time, a significant percentage might have been missed.

Losses in sightings are also suggested by low mean sightings per whale (5.8 in 1978). The more times whales are seen on the average, the less likely it is whales will pass without being seen at all. Recorded sightings/whale suffer observer biases in that individual whales or groups need to be consistently identified. These data do serve, however, as an index of observers' perceptions of counts. Assuming no observable change in travel rates or surface timings with differences in group size, frequency of sightings/whale might be expected to remain consistent regardless of the number of whales traveling together. Instead there was a noticeable rise in sightings/whale with increase in group size (Table V). If this is indicative of losses in whale counts, both single whales, with their low sighting rates, and large groups, which were easy to follow but confusing to count, could have been sources of undercounting.

Reilly (1981) performed verification experiments by counting gray whales from a circling plane while others counted from shore. He found that pods of two and three whales each were counted accurately, that 35% of whales recorded as singles were actually pairs, and that groups larger than three were undercounted by 33%. Comparable calculated losses occur in bowhead censuses at Point Barrow, Alaska: 29.6%, 22.2%, 36.4%, and 37.7% of the whales were missed in 1978, 1979, 1980, and 1981, respectively (Krogman *et al.*, 1982).

Aerial surveys did not provide corrections for land-based results at Sarichef, but instead showed the superiority of land counts. They did, however, corroborate the offshore distribution record (Table I), so no correction appears necessary for whales passing out of viewing range of observers. A small component may swim through Unimak Pass directly from the Pribilof Islands, where sightings have occurred (Gilmore, 1960; Rice and Wolman, 1971; Braham *et al.*, 1977), and some may travel through False Pass, but lacking any such evidence, it is assumed here only a small portion of the population migrating out of the Bering Sea does not use the east side of Unimak Pass.

In review of the components of the gray whale population estimate, I have selected 1978 as the primary census year. The 5,050 whales that were counted resulted in a fundamental estimate of 16,928 within the census period. If whales slow down at night as much as one test suggested, there may be a 17.3% overestimate; however, evidence of changes in travel rates are equivocal, and therefore constant day and night rates are assumed here. Factors that indicate the above population estimate is conservative include (1) the lack of estimates for whales missed prior to and after the season; (2) the unknown portion of the population not migrating past Cape Sarichef, that is, staying south of Unimak Island in the summer or traveling other than along the east shore of Unimak Pass; (3) a suggestion of an uncorrected loss in counts during periods of fair visibility; (4) potential increases in travel rates with increasing storm conditions; and (5) unquantifiable losses in sightings even under ideal situations, as was reflected in sightings/whale records.

Summary

Shore-based studies supported by aerial surveys were made at Cape Sarichef, Alaska, to census gray whales migrating through Unimak Pass, the easternmost prominent corridor between the Bering Sea and North Pacific Ocean. Results from November–December 1977–1979 showed that (1) southward migrating gray whales crossing the Bering Sea converged toward Unimak Island where the median of 10,223 shore-based sightings occurred 0.5 km off the west Unimak shore; all of these sightings occurred within 3.7 km; (2) whales migrated south through the Pass from late October to early January but primarily during the last 2 weeks of November and the first 3 weeks of December; and (3) there were approximately 17,000 whales, an estimate based on 5,050 sightings during 328 hr of systematic watch in 1978, the prime census year. Unwatched periods were interpolated within the census season assuming no diel variations in travel rates. No extrapolation for whales missed prior to and after the census was included. Poor visibility effort was excluded; however, no other corrections for loss in sightings due to visibility were applied. Data parameters suggest this population estimate is conservative.

Acknowledgments

Howard Braham helped with the original design and implementation of this study, including appropriation of necessary funds. Field assistants included Liam Antrim, James Cubbage, Thomas Fleischner, Elizabeth Hacker, Pauline Hessing, Ruthe Rugh, Ronald Sonntag, and Andrew Taber. Their ability to perform tedious data collecting in a rigorous winter climate is especially noteworthy. Cubbage, Antrim, and Taber also assisted in data compilation, as did Kristin Fogdall. Computer assistance was given by R. Sonntag, G. Priebe, and C. D'Vincent. S. Reilly, J. Breiwick, and R. Kappenman provided statistical consultation. Research facilities at Cape Sarichef were made available through the hospitality of Lt. H. Lee and Ltjg. T. Vorholt of the U.S. Coast Guard and of the Bartels and J. Sarvis of the U.S. Fish and Wildlife Service. J. Sarvis and Orin Seybert, of Peninsula Airways, King Salmon, Alaska, provided aircraft for aerial surveys and freight service. Night-vision goggles were borrowed from the Night Vision Laboratory of Fort Belvoir, Virginia, Department of the Army. Leola Hietala, Joyce Waychoff, Joanne Wejak, and Muriel Wood deserve special note for their supportive services. Editorial review was provided by H. Braham, M. Tillman, S. Reilly, and D. Rice. This study was partially funded in 1977 by the Bureau of Land Management, U.S. Department of Interior, under an interagency agreement with the National Oceanic and Atmospheric Administration, U.S. Department of Commerce, as part of a multiyear program managed by the Alaska Outer Continental Shelf Environmental Assessment Program offices in Juneau, Alaska.

References

Adams, L. (1968). Census of the gray whale, 1966–67. *Nor. Hvalfangst-Tid.* **57**(2), 41–43.

Army Map Service (1958). "Unimak C-3 SW Alaska (Topographical Chart)," No. 137142, Ser. NN3-5, Q501, Sheet 443011 SW, Edition 1-AMS. Corps of Engineers.

Braham, H., Fiscus, C., and Rugh, D. (1977). Marine mammals of the Bering and southern Chukchi Seas. *Environ. Assess. Alaskan Cont. Shelf Annu. Rep.* **1,** 1–99.

Bury, K. (1975). "Statistical Models in Applied Science." Wiley, New York.

Carl, G. (1967). A gray whale in inside waters. *Murrelet* **48**(3), 56.

Cummings, W., Thompson, P., and Cook, R. (1968). Underwater sounds of migrating gray whales, *Eschrichtius glaucus* (Cope). *J. Acoust. Soc. Am.* **44**(5), 1278–1281.

Darling, J. (1977). The Vancouver Island gray whales. *Waters* **2**(1), 5–19.

Gilmore, R. (1955). The return of the gray whale. *Sci. Am.* **192**(1), 62–67.

Gilmore, R. (1960). A census of the California gray whale. *U.S. Fish Wildl. Serv., Spec. Sci. Rep.-Fish.* **342,** 1–30.

Gilmore, R. (1968). The gray whale. *Oceans Mag.* **1**(1), 9–20.

Hatler, D., and Darling, J. (1974). Recent observations of the gray whale in British Columbia. *Can. Field Nat.* **88,** 449–459.

Hessing, P. (1981). "Gray Whale (*Eschrichtius robustus*) Migration into the Bering Sea, Spring 1981," Final Rep. Contract NA 81SRGA 00080. Outer Continental Shelf Environ. Assess. Program, NOAA, Juneau Project Office.

Hollander, M., and Wolfe, D. (1973). "Nonparametric Statistical Methods." Wiley, New York.

Hubbs, C. (1959). Natural history of the gray whale. *Proc. Int. Congr. Zool., 15th, 1958* pp. 313–316.

Hubbs, C., and Hubbs, L. (1967). Gray whale censuses by airplane in Mexico. *Calif. Fish Game* **53**(1), 23–27.

Ichihara, T. (1958). Gray whale observed in the Bering Sea. *Sci. Rep. Whales Res. Inst.* **13,** 201–205.

Ingraham, W. J. (1981). "Temperature and Salinity Observations at Surface and Near Bottom over the Eastern

Bering Sea Shelf, Averaged by 1° × ½° Squares," NWAFC Process. Rep. 81-09. U.S. Dept. Comm., NOAA, Natl. Mar. Fish. Serv., Northwest and Alaska Fish. Cent.

Isakson, J., Storie, J., Vagners, J., Erickson, G., Kruger, J., and Corlett, R. (1975). "Comparison of Ecological Impacts of Postulated Oil Spills at Selected Alaskan Locations," Vol. 2. Mathematical Sciences Northwest, Inc., Bellevue, Washington.

Krogman, B., Sonntag, R., Rugh, D., Zeh, J., and Grotefendt, R. (1982). Ice-based census results from 1978–81 on the western arctic stock of the bowhead whale. *Rep. Int. Whal. Commn.*, **34,** Doc. SC/34/PS 6, 1–54.

Kukla, G., and Gavin, J. (1978). Snow and sea ice cover fluctuations in 1977–78. *3rd Annu. Climate Diagn. Workshop, 1978* pp. 1–15.

Maher, W. (1960). Recent records of the California grey whale (*Eschrichtius glaucus*) along the north coast of Alaska. *Arctic* **13**(4), 257–265.

National Ocean Survey (1975). "Tidal Current Tables 1976; Pacific Coast of North America and Asia." U.S. Dept. Comm., NOAA, Natl. Ocean Surv., Washington, D.C.

National Ocean Survey (1979). "Tide Tables. 1980 High and Low Water Predictions; West Coast of North and South America Including the Hawaiian Islands." U.S. Dept. Comm., NOAA, Natl. Ocean Surv., Washington, D.C.

Nikulin, P. (1946). O raspredelenii kitoobraznykh v moryakh omyvayushchikh Chukotzkii poluostrov. (Concerning the distribution of whales in the seas adjacent to the Chukotskiy peninsula.) *Izv. Tikhookean. Nauchno-Issled. Inst. Rybn. Khoz. Okeanogr.* **22,** 255–257 (transl. by U.S. Dept Navy, Nav. Oceanogr. Off., Washington, D.C., 1969, Transl. 428).

Pike, G. C. (1962). Migration and feeding of the gray whale (*Eschrichtius gibbosus*). *J. Fish. Res. Board Can.* **19**(5), 815–838.

Ramsey, D. H. (1968). Diurnal fluctuations in censuses of migrating California gray whales. *Nor. Hvalfangst.-Tid.* **57**(5), 101–105.

Reilly, S. B. (1981). Population assessment and population dynamics of the California gray whale (*Eschrichtius robustus*). Ph.D. Thesis, University of Washington, Seattle.

Rice, D., and Wolman, A. (1971). The life history and ecology of the gray whale (*Eschrichtius robustus*). *Spec. Publ.—Am. Soc. Mammal.* **3,** 1–142.

Rugh, D., and Braham, H. (1979). California gray whale (*Eschrichtius robustus*) fall migration through Unimak Pass, Alaska, 1977: A preliminary report. *Rep. Int. Whal. Commn.* **29,** 315–320.

Scammon, C. (1874). "The Marine Mammals of the Northwestern Coast of North America." John H. Carmany & Co., San Francisco, California.

Sears, H. S., and Zimmerman, S. T. (1977). "Alaska Intertidal Survey Atlas." U.S. Dept. Comm., NOAA, Natl. Mar. Fish. Serv., Northwest and Alaska Fish. Cent., Auke Bay Lab., Auke Bay, Alaska.

Sekora, P. (1973). "Aleutian Islands National Wildlife Refuge," Wilderness Study Rep., Preliminary Draft. U.S. Dept. Int., Fish Wildl. Serv., Anchorage, Alaska.

Sleptsov, M. (1961). O kolebanii chislennosti kitov v Chukotskom more v raznyye gody. (Fluctuations in the number of whales of the Chukchi Sea in various years.) *Tr. Inst. Morfol. Zhivotnykh* **34,** 54–64 (transl. by U.S. Dept. Navy, Nav. Oceanogr. Off., Washington, D.C., 1970, Transl. 478).

Sund, P. N., and O'Connor, J. L. (1974). Aerial observations of gray whales during 1973. *Mar. Fish. Rev.* **36**(4), 51–52.

Swartz, S., and Cummings, W. (1978). "Gray Whales, *Eschrichtius robustus*, in Laguna San Ignacio, Baja California, Mexico," Final Rep. Marine Mammal Commission, Washington, D.C.

Tomilin, A. (1957). Kitoobraznye. (Cetacea.) "Zveri SSSR i prilezhashchikh stran. (Mammals of the U.S.S.R. and Adjacent Countries.)," Vol. 9. Publ. House U.S.S.R. Acad. Sci., Moscow (in Russian) (transl. by O. Ronen, Israel Program Sci. Transl., Jerusalem, 1967, available by Natl. Tech. Inf. Serv., Springfield, Virginia)

Turner, L. (1886). Contributions to the Natural History of Alaska." U.S. Gov. Printing Office, Washington, D.C.

U.S. Coast and Geodetic Survey (1939). "United States, Alaska-Aleutian Islands, Unimak and Akutan Passes and Approaches," Nautical Chart Catalog No. 3. U.S. Dept. Comm., NOAA, Natl. Ocean Surv., Washington, D.C.

U.S. Weather Bureau and U.S. Navy Hydrographic Office (1961). "Climatological and Oceanographic Atlas for Mariners," Vol. 2. U.S. Govt. Printing Office, Washington, D.C.

11

Distribution and Migration of Gray Whales in Alaska

Howard W. Braham

Introduction

Pike (1962) discussed the distribution, migration, and feeding of gray whales based on an analysis of the literature and of unpublished sightings along the coast of British Columbia (B.C.), Canada. He speculated that between the Pacific Northwest Coast and Bering Sea, the spring and perhaps autumn migrations were close to shore instead of at sea as had been previously thought. He also surmised, from a few sightings in Unimak Pass and near the Pribilof Islands and from the lack of sightings in the Aleutian Islands, that gray whales enter and leave the Bering Sea through Unimak Pass. The data Pike used to formulate his coastal route hypothesis were from Townsend (1935), Nikulin (1946), Ichihara (1958), Gilmore (1960), Nasu (1960), and Wilke and Fiscus (1961) and from his own work for British Columbia. However, in 1962 Pike had very little empirical data relative to his hypothesis for much of the migration route in Alaska.

In 1975, scientists at the National Marine Mammal Laboratory began a study of the distribution and migration of gray whales in Alaskan waters (Fig. 1) with emphasis on the Bering Sea (Braham, 1977; Braham *et al.*, 1977; Rugh and Braham, 1979; Consiglieri and Braham, 1982). From September 1975 to October 1981 gray whales were observed from air, land, and at sea during their annual migrations and on their northern feeding grounds.

The objective of this research was to determine the occurrence, timing, and movements of gray whales in and adjacent to proposed oil and gas lease areas in Alaska. The results confirmed with only a few albeit important modifications much of the speculation on gray whale migration routes and timing in Alaska first reported by Scammon (1874)

ISBN 0-12-389180-9

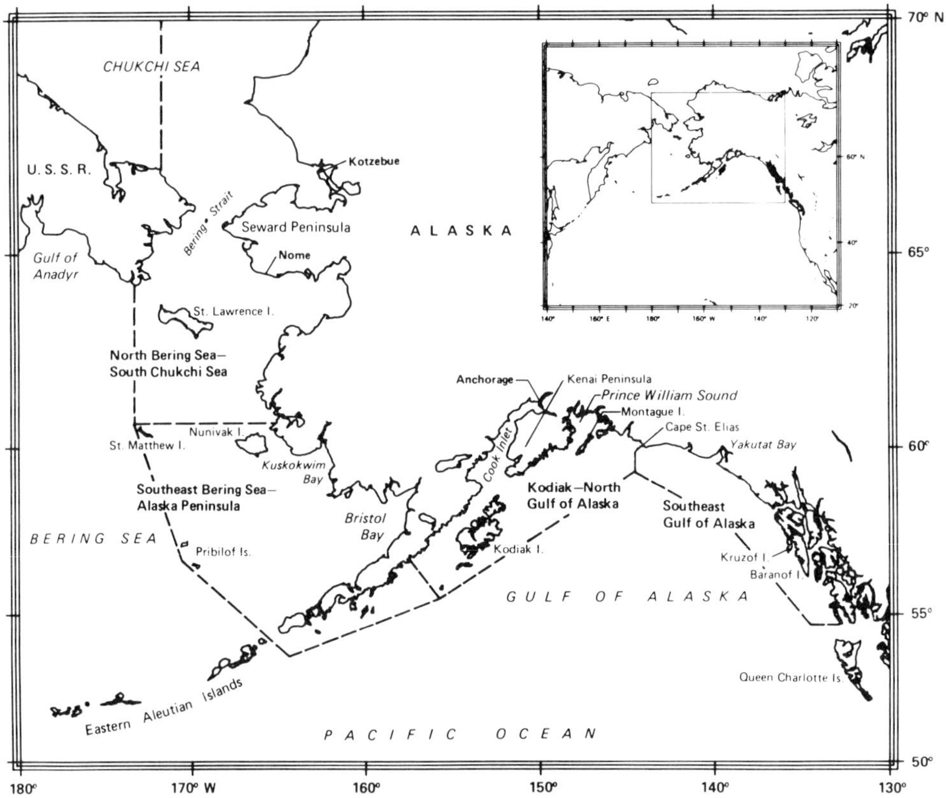

Fig. 1. *Study area indicating regions into which observations are divided for discussion.*

and later refined by Pike (1962) and Rice and Wolman (1971). This chapter presents data on the seasonal distribution of gray whales and evidence of a coastal migration throughout much of Alaska. I also discuss why gray whales migrate in such close proximity to shore.

Methods

The most important data were sightings made in August and October, 1975; April, June, and August, 1976; April through August, 1977; May and June, 1978; and May 1979 during aerial surveys[1] we conducted in the Chukchi Sea, eastern Bering Sea, and along the coast of the Alaska Peninsula, Gulf of Alaska and southeast Alaska. Sightings were made along several thousand kilometers of ship tracklines as well, in the Gulf of Alaska,

[1]Fixed-wing aircraft used were Grumman Widgeon (1975–1978), P2V Neptune (1975–1976), Grumman Goose (1976–1977), and Aerocommander (1978–1979); UH1H (1976–1977) and Bell-206 (1976–1978) helicopters. (Reference to trade names does not imply endorsement by the National Marine Fisheries Service.)

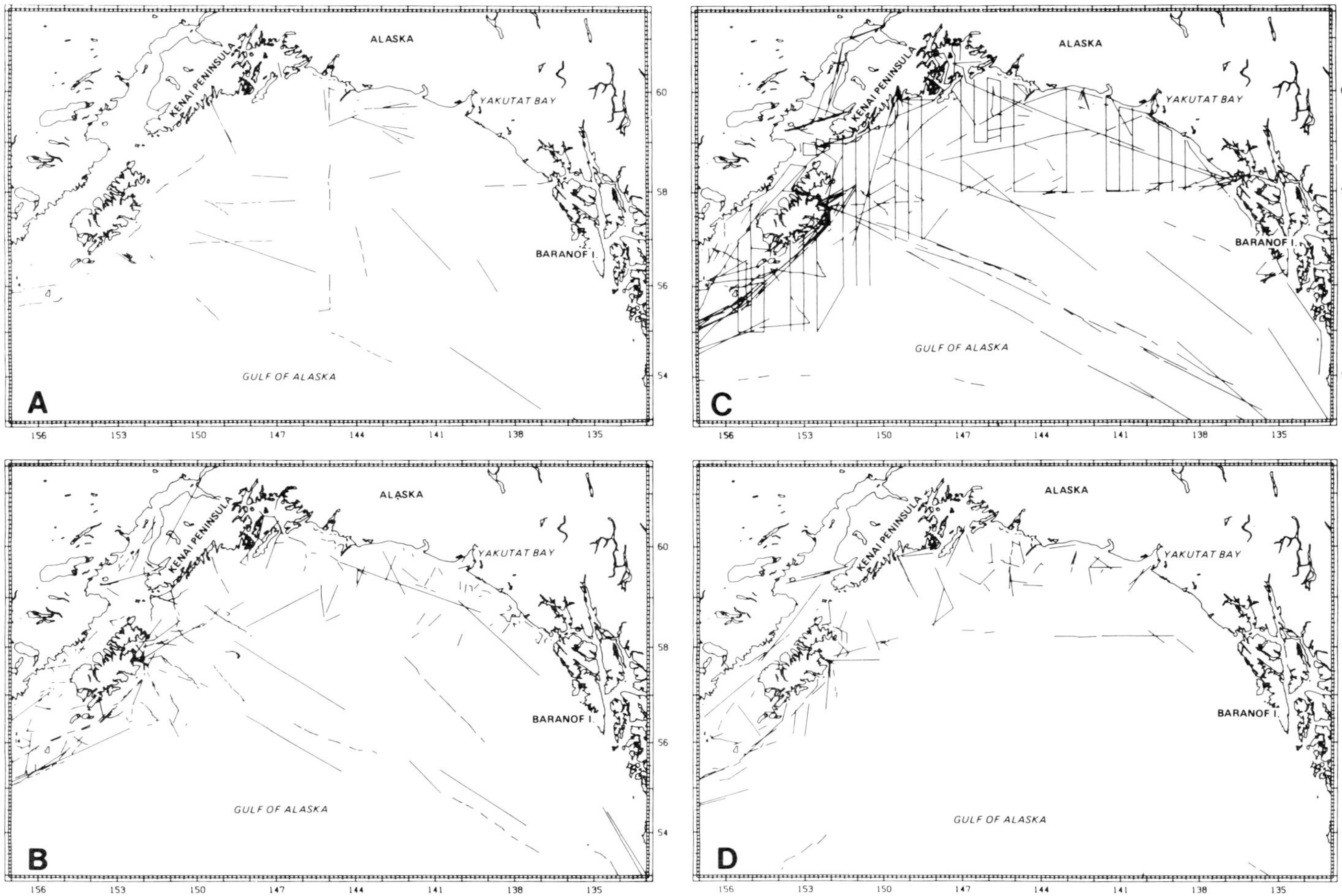

Fig. 2. *Ship track lines in the vicinity of the Gulf of Alaska on which searches were made for whales, 1958–1980: (A) December–February; (B) March–May; (C) June–August; and (D) September–November.*

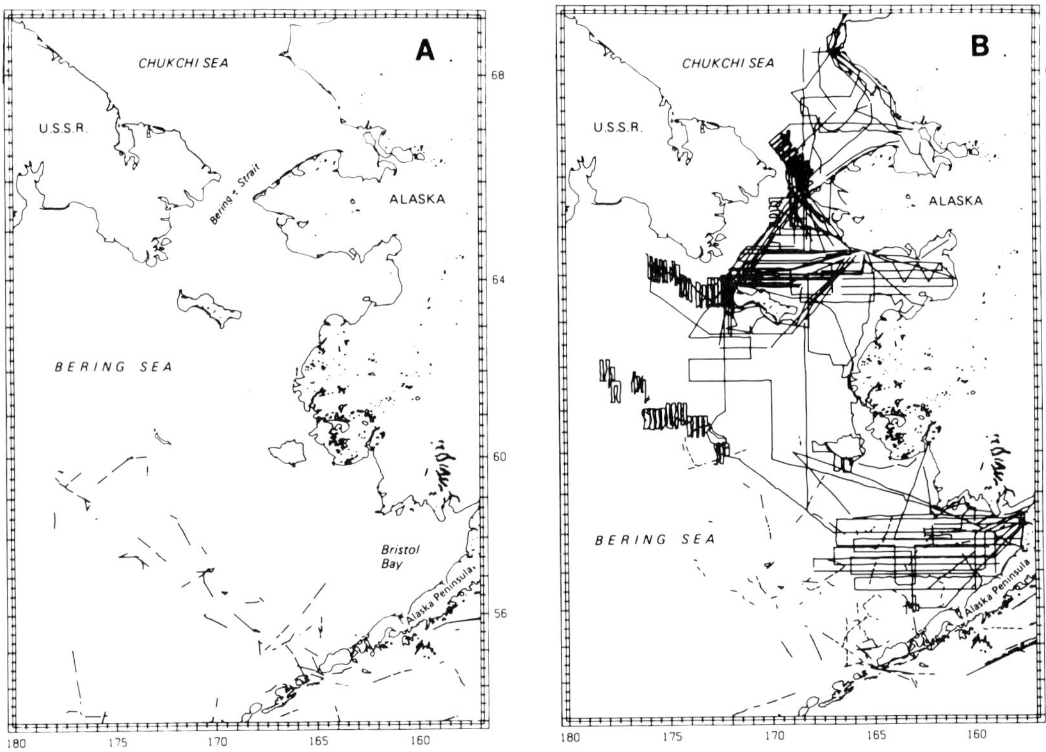

Fig. 3. *Ship and aerial survey track lines along the Alaska Peninsula in the eastern Bering Sea and southern Chukchi Sea on which searches were made for whales, 1975–1980: (A) December–February; (B) March–May; (C) June–August; and (D) September–November.*

Bering Sea, and Chukchi Sea, in 1976–1981. Ship and aerial survey effort are plotted in Figs. 2 and 3. Observations were also made from land at Kodiak Island (April, 1976 and May, 1977), Chignik Bay (June, 1978), Unimak Island (June, 1976–1978, and November and December, 1977–1979) (Chapter 10, this volume), Nelson Lagoon and Port Heiden (June, 1976; May, June, and August, 1977), and St. Lawrence Island (June and July, 1977 and 1978). Previously unpublished sightings from 1958 to 1973 collected during the northern fur seal (*Callorhinus ursinus*) investigations in the North Pacific were also compiled and are reported here. University, State of Alaska, and federal scientists, plus fishermen, pleasure craft operators, and airplane pilots generously provided firsthand, unpublished observations. Virtually all these sightings and general vessel and aerial survey effort are in the National Marine Mammal Laboratory (NMML) Platforms of Opportunity Program (POP) data files used to produce the plotted sightings.[2] Additional

[2]The POP is a NOAA-supported data base management project conducted at the NMML since 1975. Its purpose is to compile a central computerized data base of incidental and systematically collected sightings of marine mammals in the North Pacific Ocean. The total number of whales observed far exceeded several thousand since 1958, and thus for purposes of space it was not practical to list in this chapter (except on plots) individual sightings by time, location, observers, mode of investigation, et cetera.

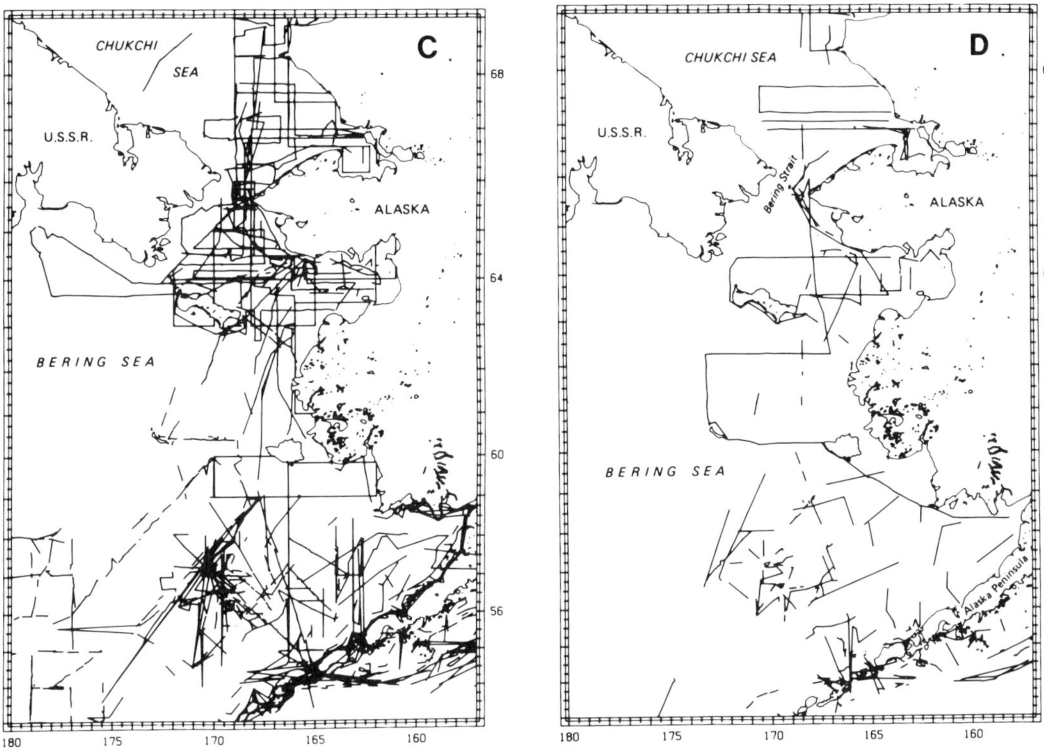

Fig. 3. *Continued*

recently published winter sighting data for Kodiak Island (Forsell and Gould, 1981) are also reported here and included in the plotted results.

Survey effort was not uniform among or within years, seasons or areas, although some effort was expended in all months in most study area regions reported in Fig. 1.

Results and Discussion

DISTRIBUTION

Southeast Alaska and Gulf of Alaska

From March to May, gray whales were observed only along the coast (Fig. 4A) in spite of some 120 vessel surveys totaling more than 3000 days conducted offshore since 1958 (Fig. 2A–D). The highest number of whales was recorded in March and April, with peak abundance during the third week of April at Cape St. Elias (D. Calkins, personal communication; Cunningham and Stanford, 1979). Sightings were numerous along the shore of Kruzof Island, Yakutat Bay to west of Cape St. Elias, and along the south coast

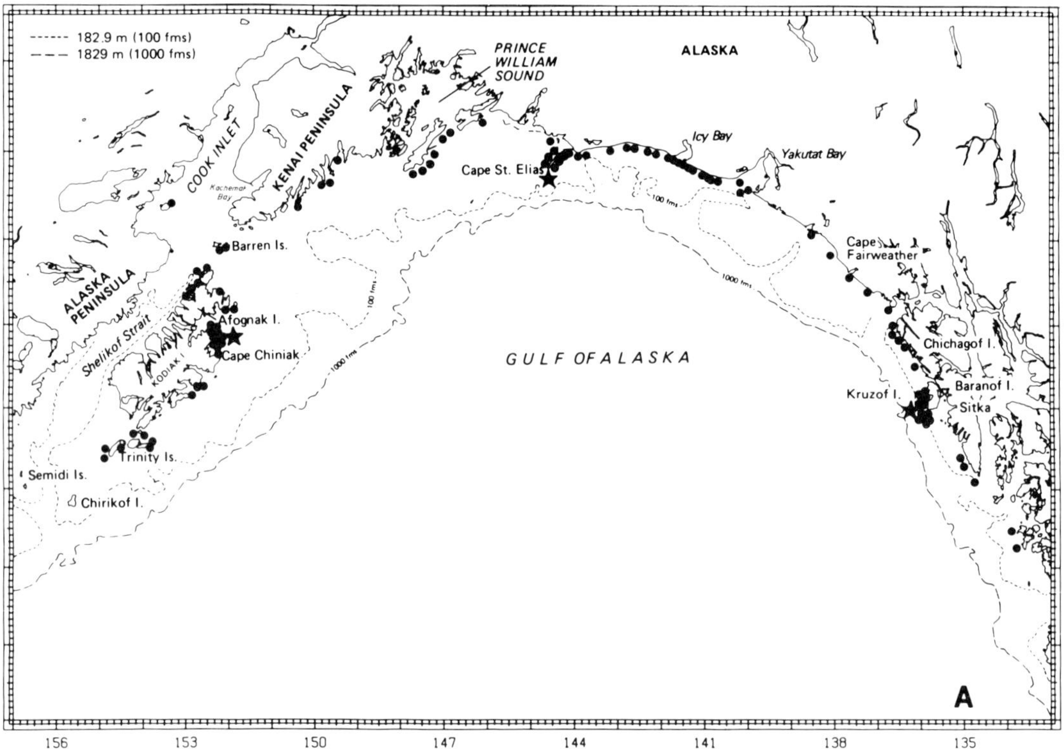

182.9 m (100 fms)
1829 m (1000 fms)
PRINCE WILLIAM SOUND
ALASKA
COOK INLET
KENAI PENINSULA
Kachemak Bay
Icy Bay
Yakutat Bay
Cape St. Elias
Barren Is.
ALASKA PENINSULA
Shelikof Strait
Afognak I.
KODIAK I.
Cape Chiniak
Cape Fairweather
Chichagof I.
Baranof I.
Kruzof I.
Sitka
GULF OF ALASKA
Trinity Is.
Semidi Is.
Chirikof I.
A
156
153
150
147
144
141
138
135
60
58
56
54

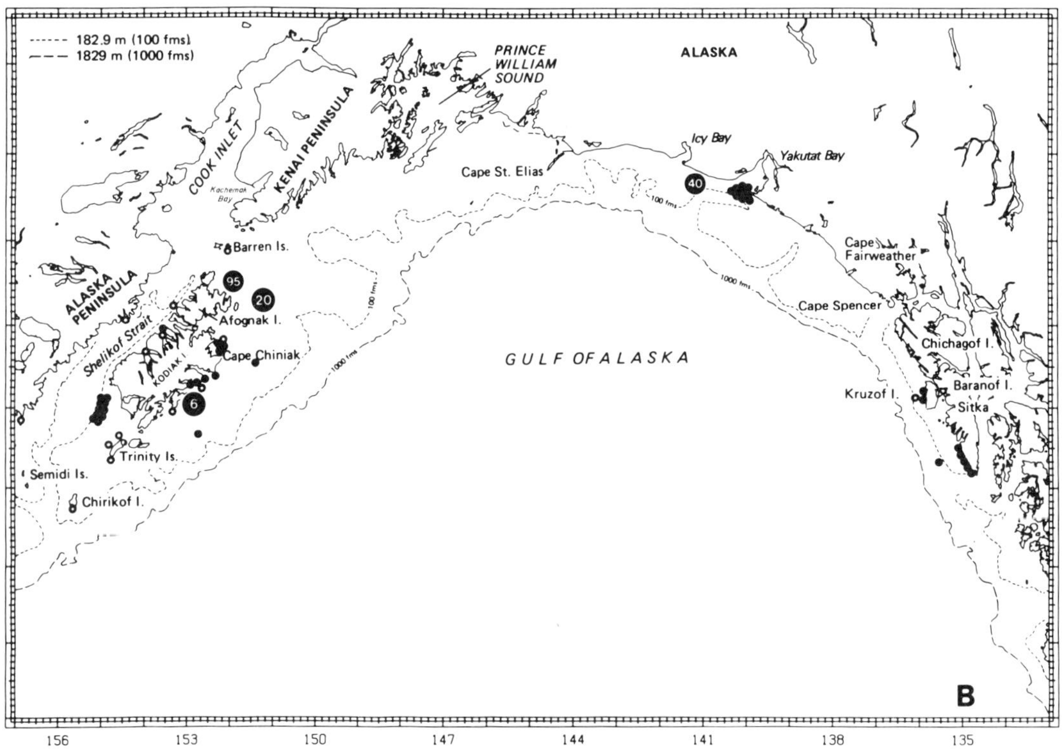

182.9 m (100 fms)
1829 m (1000 fms)
PRINCE WILLIAM SOUND
ALASKA
COOK INLET
KENAI PENINSULA
Kachemak Bay
Icy Bay
Yakutat Bay
Cape St. Elias
40
Barren Is.
95
20
ALASKA PENINSULA
Shelikof Strait
Afognak I.
KODIAK I.
Cape Chiniak
6
Cape Fairweather
Cape Spencer
Chichagof I.
Baranof I.
Kruzof I.
Sitka
GULF OF ALASKA
Trinity Is.
Semidi Is.
Chirikof I.
B
156
153
150
147
144
141
138
135
60
58
56
54

of Kodiak Island. Scattered sightings from late March to mid-May from Chichagof Island to Yakutat Bay, along Kenai Peninsula, at the Barren Islands, along the north coast of Afognak Island in Shelikof Strait, and at the Trinity Islands comprise the remainder of sightings reported (Figure 4A). Disproportionate effort among areas and years precludes any discussion of variability in annual temporal and spatial distributions.

The distribution of gray whales in summer, autumn, and winter (June–February) is not as well documented as that in spring. Although a considerable number of vessel surveys occurred offshore in summer and autumn since 1958, very few offshore sightings of gray whales were made; most animals were seen nearshore. October and November sightings of gray whales are plotted in Fig. 4B as reported by local residents or scientists working on land or along the coast. Rice and Wolman (1982) surveyed the Gulf of Alaska by ship in June–August, 1980, covering the entire coastline from Cape Spencer to the west end of Shelikof Strait. They saw no gray whales. The coast of the Gulf of Alaska is extensive, however, and is marked by numerous small bays and inlets in which gray whales, if present in summer, might have been undetected during vessel surveys of the Gulf.

No whales were reported for winter months (December–February), except those sighted by Forsell and Gould (1981) along the south side of Kodiak Island, which perhaps reflects the low level of effort. Of the several sightings made since 1958, a few animals were seen farther offshore than during spring (Fig. 4B). One group of 20 animals in November, 1979 was 20 km northeast of Kodiak Island (58°20′N, 151°10′W), well out to sea, swimming due east. In another sighting two whales were 25 km off southwest Baranof Island in November (56°20′N, 135°30′W). The sightings in outer Yakutat Bay were also made in November (D. Calkins, personal communication).

Alaska Peninsula and Eastern Bering Sea

Few gray whales were seen along the south side of the Alaska Peninsula because there has been less effort between Shelikof Strait and Unimak Island than in the Gulf of Alaska and southeastern Bering Sea. Fishermen at Chignik Bay report that gray whales cross the mouth of the Bay in April (most), May, and June (few). Vessels westbound from the Gulf of Alaska to Unimak Pass usually pass along the south side of Kodiak and the Trinity Islands. Observers onboard seldom report gray whales; however, most ships transiting that area sail directly from Kodiak to Unimak Pass, thereby avoiding the many islands and rocks which extend to 90 km offshore. If gray whales remain close to shore along the south side of the Alaska Peninsula, as elsewhere, they were probably missed during such ship surveys.

The earliest sightings of gray whales in the southern Bering Sea were a single animal on March 22, 1980 in Unimak Pass (R. Bartels, personal communication); 1 animal on April 10, 1976 at 57°35′N, 165°44′W (F. Fay, personal communication); 5 animals

---

Fig. 4. *Sightings of gray whales in the Gulf of Alaska, 1958–1980: (A) March–May, each dot (•) is one or more whales, stars (★) are areas where whales were observed apparently feeding; (B) June–September (○) and October–January (•).*

at 56°20′N, 166°15′W on April 18, 1976 (Braham *et al.*, 1977); 3 seen in southern Bristol Bay on April 24, 1976 (1 at 56°03′N, 163°12′W and 2 at 56°03.9′N, 163°23.3′W) (POP, NOAA ship *Surveyor*); and an estimated 75 whales seen on April 18, 1977 at 56°52′N, 159°23′W near Port Heiden (POP, NOAA ship *Surveyor*) along the ice front in Bristol Bay. Sightings made in May were generally in northern Bristol Bay; whales seen in April were generally offshore, and most were seen within a few kilometers of the spring pack ice (Fig. 5A). The only time gray whales were seen away from shore in Bristol Bay was in April. In those years when that area was surveyed, as in 1976–1977, shorefast ice was extensive, and the pack ice extended well south into the southeastern Bering Sea. All along the Alaska Peninsula and in Bristol Bay, the whales were extruding mud from their mouths and diving in a manner similar to feeding behavior seen farther north in the Bering and Chukchi Seas (e.g., see Chapter 17, this volume).

Most of our sightings along the north side of the Alaska Peninsula and Bristol Bay were in June (Fig. 5B). From 1975 to 1979, approximately 150 hr of aerial surveys were flown along the north side of the Alaska Peninsula and the eastern Aleutian Islands (Braham *et al.*, 1977, 1980; Everitt and Braham, 1980) to determine spatial and temporal distribution of gray whales along the migration route in the Bering Sea.

Between June 8 and 20, 1976–1978 (6 flights), 511 whales (including 50 calves) were seen between Unimak Pass and Naknek River near the north end of the Alaska Peninsula (Fig. 5B). Whales were seen at a mean rate of 0.14/km (SD = 0.03, $n = 6$) or about 1 whale/7.5 km surveyed. Only 6 animals were farther than 2 km offshore along the Alaskan Peninsula and in Bristol Bay; the remaining 505 were within 1 km of shore. No gray whales were seen among the eastern Aleutian Islands west of Unimak Pass.

The few sightings in July and August (Fig. 5C) and in September and October (Fig. 5C) reflect the fact that essentially the entire population is farther north. No November sightings were made except along the north shore of the southwest end of the Alaska Peninsula and Unimak Island (see also Chapter 10, this volume).

Northern Bering and Southern Chukchi Sea

The earliest gray whales of the year generally reach St. Lawrence Island by May as ice recedes north or when leads or polynyas are extensive (Fay, 1974; McNutt 1981). Whales were seen first in May near Southeast Cape and along the west end of the island, actively feeding presumably on benthic organisms, but were most notably abundant in June. Each year during 1976–1978, in June, we observed over 500 whales actively feeding near St. Lawrence Island (Fig. 5B). I believe this is where they first concentrate to feed each year on the northern feeding grounds, especially at the southeast and west ends of the island, and thereafter disperse north, west and southwest.

Sightings that we made during June 8–14, 1976, May 11–13, 1977, and June 2–10, 1977, for example, were perhaps of the earliest whales to arrive in the northern Bering Sea during those years. On occasion gray whales are seen in May in loose pack ice south of Bering Strait (Fig. 5A) and in early June in leads near Point Hope and Barrow (the late F. Durham and A. Brower, personal communications). The earliest sighting of which I am aware (Fig. 5A) is that of a single whale seen approximately 100 km southwest of Southwest Cape, St. Lawrence Island, at 62°50′N, 172°54′W, on April 25, 1979 (POP, NOAA Ship *Surveyor*).

By June, gray whales are common in the northern Bering Sea in ice-free years, tending to concentrate north and northwest (Nikulin, 1946; Tomilin, 1957) of St. Lawrence Island in very low numbers east into Norton Sound, and through the Bering Strait into the southern Chukchi Sea during summer and autumn (Figs. 5B,C). The distribution of gray whale sightings in the northeastern Chukchi and Beaufort Seas was reported by Marquette and Braham (1982). Reported distributions for these areas in summer and autumn illustrate that gray whales are found in greatest abundance in areas which are apparently preferred feeding habitat, such as the Chirikov Basin north of St. Lawrence Island and at Southeast Cape (see also Chapter 18, this volume).

Sightings in the Soviet Chukchi Sea of 127 whales in mid-October, 1979 (Johnson *et al.*, 1981) and 125 in late September, 1980 (Marquette *et al.*, 1982), and one sighting of an estimated 100 whales heading south 50 km southeast of St. Matthew Island on October 15, 1976 (L. Calkette, personal communication; Fig. 5C), demonstrated that the population is dispersed through much of the northern range during autumn months. Gray whales are present in northern waters until at least November (Chapter 23, this volume), by which time the southbound migration is underway. The latest record of the year of which I am aware is a sighting of 2 gray whales on December 14, 1982 (Conrad Oozeva, personal communication) nearshore to Gambell, St. Lawrence Island.

MIGRATION

Migration Routes

It is clear from data presented herein that the migration route is coastal in Alaska, just as it is from Baja California to British Columbia. Pike's hypothesis (1962) that gray whales circumnavigate the Gulf of Alaska to Unimak Pass is correct. Gray whales migrate very close to shore; often, they can even be seen in the surf zone. The few exceptions occur when they swim across open water, such as from Kenai Peninsula to Afognak Island. Only a few whales were found in Shelikof Strait, so I believe that most follow the south side of Kodiak Island. Their route(s) from the southwestern tip of Kodiak Island to the south side of the Alaska Peninsula remains unknown, but it is probably offshore also, perhaps to Chignik Bay.

Kellogg (1929), Ichihara (1958), and Gilmore (1955, 1960) hypothesized that from the Washington coast northbound gray whales migrated directly across the North Pacific and entered the Bering Sea through the many passes between the Aleutian Islands (Fig. 6). However, no gray whales have been seen in American waters west of Unimak Pass (Ichihara, 1958; Pike, 1962; Chapter 10, this volume; O. Seybert, personal communication[3]; Fig. 5), and few have been sighted more than 3 km offshore Unimak Island in

[3]Orin Seybert of Peninsula Airways, King Salmon, Alaska, has flown the Alaska Peninsula and eastern Aleutian Islands almost weekly for over 20 years; he reported seeing gray whales only along the north coast of the Alaska Peninsula from Unimak Island to the Naknek River.

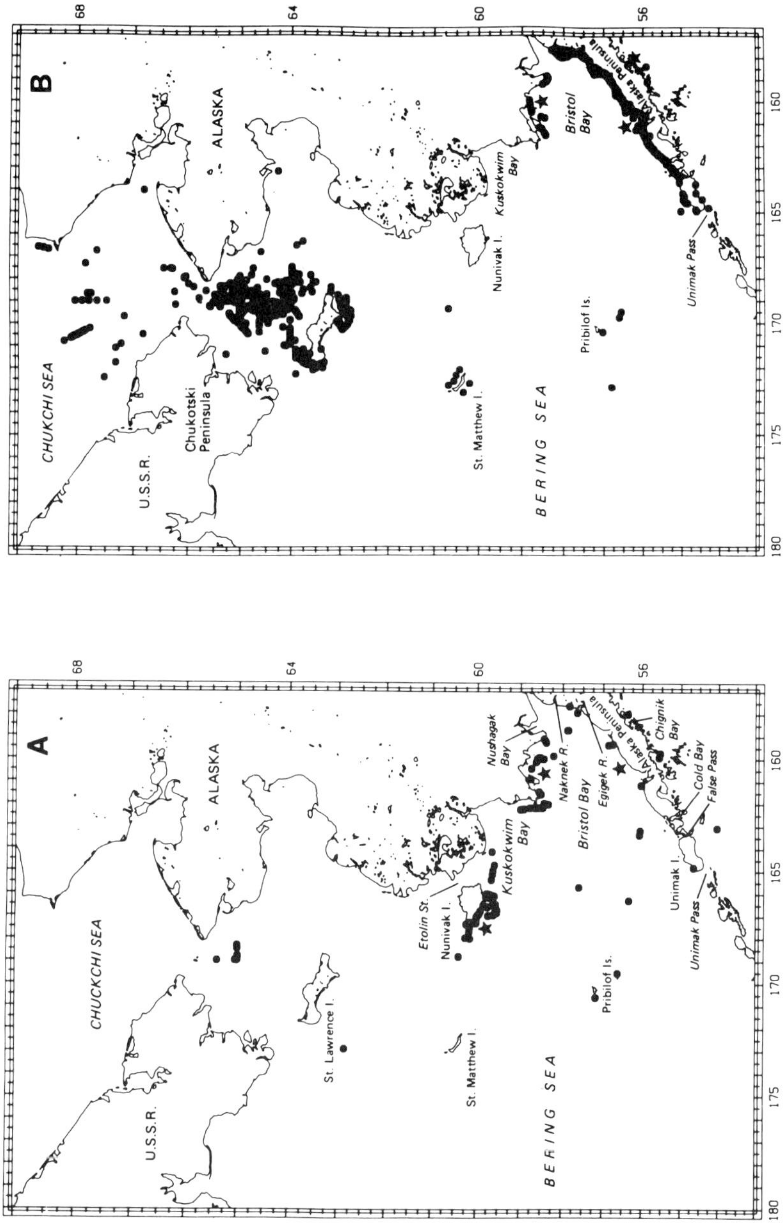
B
CHUKCHI SEA
U.S.S.R.
Chukotski Peninsula
ALASKA
St. Matthew I.
BERING SEA
Nunivak I.
Kuskokwim Bay
Bristol Bay
Alaska Peninsula
Pribilof Is.
Unimak Pass
68
64
60
56
160
165
170
175
180
A
CHUCKCHI SEA
ALASKA
U.S.S.R.
St. Lawrence I.
St. Matthew I.
BERING SEA
Etolin St.
Nunivak I.
Kuskokwim Bay
Nushagak Bay
Naknek R.
Bristol Bay
Egigek R.
Alaska Peninsula
Chignik Bay
Cold Bay
False Pass
Unimak I.
Unimak Pass
Pribilof Is.
68
64
60
56
160
165
170
175
180

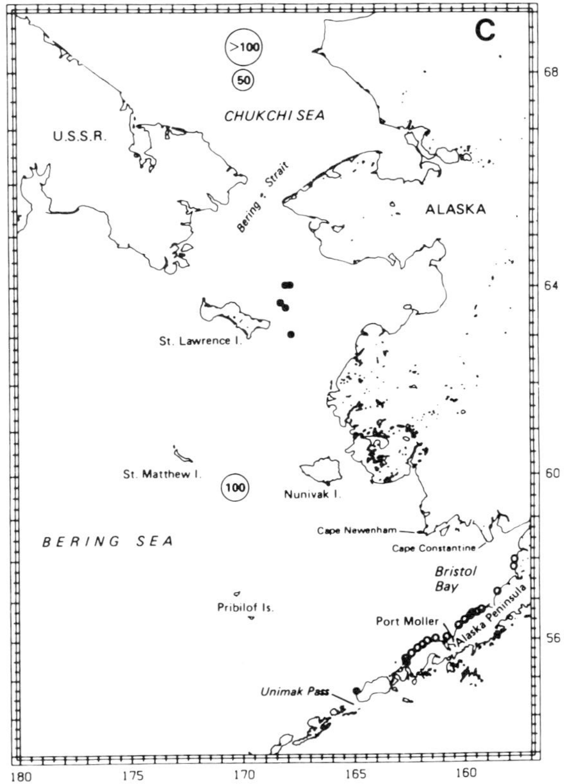

Fig. 5. *Sightings of gray whales along the Alaska Peninsula and in the eastern Bering and southern Chukchi Seas: (A) April–May, 1958–1980 (no March sightings); (B) June–August, 1958–1980; (C) July–August, 1975–1980 (○) and September–October, 1958–1981 (no November sightings) (● plus circled numbers). The Chukchi Sea sightings (C) are from Johnson* et al. *(1981) and Marquette* et al. *(1982).*

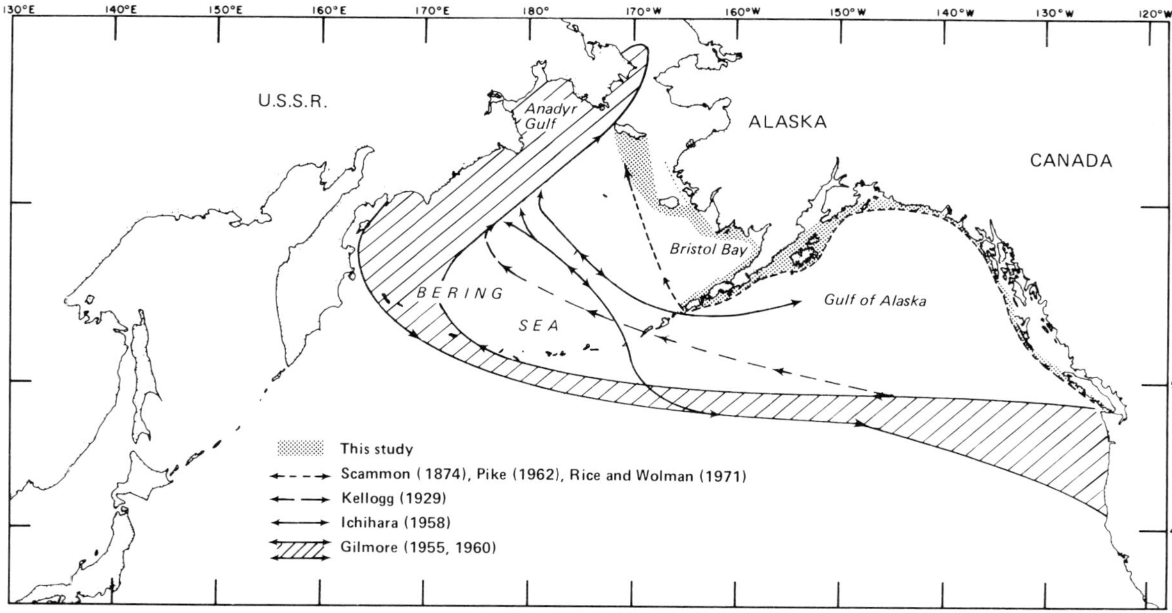

Fig. 6. *Migration route hypotheses for gray whales in the eastern North Pacific Ocean and Alaskan waters.*

Unimak Pass (Turner, 1886; Ichihara, 1958; Hessing, 1981; and Fig. 5). Gray whales reported along the east coast of the Soviet Union from southern Kamchatka Peninsula to Japan were reported by Andrews (1914), Zenkovich (1937), Sleptsov (1955), and Tomilin (1957) to be from the Korean stock.

It is now clear that once gray whales have entered the Bering Sea through Unimak Pass most continue to follow the coast along the north side of the Alaska Peninsula and elsewhere in Bristol Bay (Fig. 6). In April, and sometimes in May, when sea ice persists in Bristol Bay, some whales move offshore along or near the ice front. They generally do not move to the coast in northern Bristol Bay in early spring because shorefast and pack ice can be extensive. When the Bristol Bay coast is free of ice, as was seen during our May 1977 and June 1975–1978 surveys, the whales approach Egegik River and then turn west toward Cape Constantine.

From Cape Constantine they continue nearshore westward to Cape Newenham. From Cape Newenham whales apparently move directly across outer Kuskokwim Bay (a distance of approximately 150 km) to the southeast tip of Nunivak Island where the migration continues along the south side and perhaps to a lesser extent the north side as well. Beyond Nunivak Island the whales probably fan out towards and to St. Lawrence Island, avoiding coastal waters immediately adjacent to Nunivak Island such as Etolin Strait, and to the Yukon Delta. It is in this segment that the migration apparently shifts from a strictly nearshore to an offshore pattern (Fig. 6).

Shallow water (<4 m) and extreme tidal fluctuations near the Naknek, Kvichak, and Nushagak Rivers may explain why gray whales cross Bristol Bay from near Egegik River to Cape Constantine, a distance of about 110 km, rather than remain along the coast. East of Nunivak Island, Etolin Strait is shallow and is generally clogged with ice in spring. This may explain why gray whales avoid this area in spring, but it does not explain why they apparently avoid the area in June during open-water periods, suggesting that there is some other factor restricting their movements.

Gray whales seen near the Pribilof and St. Matthew Islands may demonstrate that not all whales strictly follow the coast once they pass through Unimak Pass. Also, sightings in the northwest Bering Sea, near Cape Navarin, U.S.S.R., and in the outer Gulf of Anadyr from May into autumn (Votrogov and Bogoslovskaya, 1980; Bogoslovskaya *et al.*, 1981; A. A. Berzin and V. Ivashin, personal communications) may be of gray whales moving west and south from St. Lawrence Island, or west and north from Nunivak, St. Mattnew, or the Pribilof Islands.

Although the northbound migration along California and Oregon has been reported to be in at least two phases or "waves," separated as age–class groups, with adults and immatures coming first, followed in 1 or more months by cows with calves,[4] this pattern did not emerge from the data collected for this study or from Cunningham and Stanford (1979). Hessing (1981), working at Unimak Pass from March 23 to June 17, 1981, observed two phases, and although all sizes of whales were observed throughout the study, cows with calves were not seen until May, continuing to increase to mid-June when the study ended.

[4]Reported in two abstracts (see Poole, 1981; Herzing and Mate, 1981).

It is difficult to determine precisely when the southbound migration begins. Maher (1960) and D. J. Rugh (personal communication) noted that gray whales in mid-August near Barrow were predominantly moving south. In the Bering Sea, the migration probably begins in October in most years, when the whales move south across the eastern Bering Sea and leave through Unimak Pass, primarily in November and December (Chapter 10, this volume). Occasional sightings east of St. Matthew Island to central Bristol Bay in October and November (O. Seybert, personal communication, and Fig. 5C), and from a few aerial surveys in November (Chapter 10, this volume), suggest that the southbound migration in southeastern Bering Sea may be farther offshore than the northbound migration. Frost *et al.* (1982) observed no gray whales in Bristol Bay in September–November, or near Port Moller in December, in 1980.

The few autumn–winter sightings of gray whales in the Gulf of Alaska support the hypothesis that the southbound migration is also coastal, but perhaps not as close to shore as in spring.

Feeding during Spring Migration

Gray whales appeared to be feeding[5] at seven locations along the northbound migration route in Alaska (Figs. 4 and 5).[6] Feeding activity was observed at Cape St. Elias in April and at Kodiak Island in April–May (R. MacIntosh, personal communication), and I witnessed this activity along the northside of the Alaska Peninsula and north coast of Bristol Bay in June and August and on the south side of Nunivak Island in May. Other feeding events were reported off Baranof Island and in outer Chignik Bay between April and June. Although behavior interpreted as feeding is only circumstantial evidence of food intake, this behavior suggests gray whales do feed during their spring migration, at least in Alaskan waters. The suggestion (see Chapter 18, this volume, for review) that little important feeding occurs from November to March perhaps means that any foraging during the northbound migration might replace some energy lost as a result of migrating. This may partly explain why the gray whales, which are principally bottom feeders, migrate nearshore along the coast rather than offshore.

The reasons why gray whales migrate along the coast and the environmental factors that aid them are of particular interest. Norris and Prescott (1961) hypothesized that gray whales use ocean floor contours for orientation. Rice (1965) reported that they are seldom seen in water deeper than 180 m, and in Alaska they are usually seen in waters shallower than this (Figs. 4 and 5). Reilly (1981) found that 95% of the population passing at Yankee Point, California, was within 5 km of shore. In the Gulf of Alaska, where they are also seen regularly within 5 km of shore, this would mean that virtually the entire migration is in water less than 30 m deep.

Gilmore (1960) and Nasu (1974) suggested that the gray whale migration follows

[5]As demonstrated by surface skimming with their mouths partially open or when expelling mud from their mouths after a dive.

[6]See Table II in Chapter 17, this volume, for locations and contributors listing where whales apparently were feeding.

and is aided by oceanic (offshore) currents. Pike (1962) found no such correlation and noted that current systems in autumn were opposed to the southbound migration off Washington, Oregon, and California, and that at least along the British Columbia coast, there is no appreciable surface current in spring to aid the northbound migration. Pike (1962) hypothesized that instead the gray whales' affinity for the shore was related to migrational cues such as coastal mountains and promontories, which the whales use to follow the coast. However, gray whales certainly can and do traverse expanses of water. There are places along the northbound migratory route where no landmarks are apparent, such as Queen Charlotte Sound, a distance of 190 km. Presumably they are capable of swimming directly across the Gulf of Alaska to Unimak Pass, but do not. Their choice does not appear to be based on distance alone: The distance from Cape Flattery, Washington, to Unimak Pass via the coastal route is approximately 400 km longer than a direct oceanic route, requiring only about 4 more days of travel assuming equivalent swimming speed.

One alternative to Pike's hypothesis (1962) is that the coastal northbound migratory route is influenced by the availability of sublittoral food resources. For a benthic predator such as the gray whale, such resources can be obtained with less expenditure of energy in shallow than in deeper (offshore) waters. If significant fasting takes place during the southbound migration as well as throughout winter, then gray whales may need to feed after they leave Baja California. Consequently, remaining in shallow coastal waters would allow them to take advantage of readily available and presumably productive food resources. All these factors would favor the evolution of coastal migratory behavior.

Summary

Observations on the distribution of gray whales (*Eschrichtius robustus*) in Alaskan waters, from research conducted during 1975 to 1979 and incidental sightings reported between 1958 and 1981, are presented. The northbound migration occurs from March through June and is roughly divided into two phases or "waves." The first phase apparently includes nonparturient adults and immatures, followed in about 1 month mainly by cows and young-of-the-year calves. Most whales follow the coast, staying within 2 km of shore except as they cross major bays, straits, and inlets from southeast Alaska to Nunivak Island in the eastern Bering Sea. They migrate offshore from Nunivak Island to St. Lawrence Island and major feeding grounds in the northern Bering Sea and southern Chukchi Sea. The peak of migration for the first phase of the migration at Cape St. Elias in the Gulf of Alaska is mid-April, and at Unimak Pass it is in late April and early May.

Summer sightings of whales in the Gulf of Alaska and southeastern Bering Sea are also reported for areas well south of the whales' traditional feeding grounds. These sightings may be of animals that make up the second phase of northbound migrating whales, and (or more likely) of summer residents. The southbound migration probably begins by early October when the whales travel from the northern to the southeastern Bering Sea and intersect the southwest end of the Alaska Peninsula before leaving the

Bering Sea through Unimak Pass, from late October to January. The migration apparently continues along the coast around the Gulf of Alaska. Because feeding behavior was frequently observed during migration, I suggest the availability of sublittoral food resources may have been important in the evolution of a coastal migratory route.

Acknowledgments

I thank C. Fiscus, R. Everitt, D. Rugh, M. Nerini, D. Withrow, B. Krogman, N. Severinhaus, T. Bray, L. Consiglieri, M. Dahlheim, R. Grotefendt, J. Johnson, P. McGuire, R. Mercer, R. Miller, C. Peterson, R. Punsley, D. Rice, R. Sonntag, M. Tillman, J. Wejak, A. Wolman, and M. Wood, current or former fellow employees of the National Marine Mammal Laboratory (NMML), for help in data collection and preparation, and especially D. Rugh, M. Nerini, D. Rice, and the editors for valuable advice and exchange of ideas during manuscript preparation. O. Seybert, Peninsula Airways, King Salmon, Alaska; P. Arneson, J. Burns, D. Calkins, W. Cunningham, J. Faro, L. Johnson, K. Pitcher, K. Schneider, S. Stanford, and N. Steen, Alaska Department of Fish and Game; R. MacIntosh, National Marine Fisheries Service, Kodiak; T. Balleau, R. Bartells, L. Calkette, J. Frank, J. Hall, and A. Sowl, U.S. Fish and Wildlife; F. Fay, University of Alaska, Fairbanks; F. Durham (deceased), University of Southern California, Los Angeles; A. Brower, Sr., Barrow, Alaska; C. Oozeva, Gambell, Alaska; G. Fedoseev, TINRO, Magaden, A. Berzin, TINRO, Vladivostok, and M. Ivashin, VNIRO, Moskow, USSR; and D. Day, G. Lapine, and R. Swope, NOAA OCSEAP, Juneau, Alaska, gave me unpublished sightings or valuable assistance over the years, for which I am grateful. L. Hietala, NMML, C. Hastings, J. Peacock, and E. Zweifel, Northwest and Alaska Fisheries Center, provided invaluable typing, graphics, and processing assistance. This research was partially supported in 1976–1977 and in 1980 by the Minerals Management Service (formerly Bureau of Land Management), U.S. Department of the Interior, under an interagency agreement with the Outer Continental Shelf Environmental Assessment Program, NOAA, Juneau Project Office.

References

Andrews, R. C. (1914). Monographs of the Pacific Cetacea. I. The California gray whale (*Rhachianectes glaucus* Cope). *Mem. Am. Mus. Nat. Hist.* [N. S.] **1,** 227–287.

Bogoslovskaya, L., Votrogov, L., and Semenova, T. (1981). Feeding habits of the gray whale off Chukotka. *Rep. Int. Whal. Commn.* **31,** 507–510.

Braham, H. W. (1977). California gray whale (*Eschrichtius robustus*) spring migration in Alaska. *Proc. Conf. Biol. Mar. Mamm., 1977* Abstracts, p. 59.

Braham, H. W., Fiscus, C. H., and Rugh, D. J. (1977). Marine mammals of the Bering and southern Chukchi Seas. *Environ. Assess. Alaskan Cont. Shelf, Annu. Rep.* **1,** 1–99.

Braham, H. W., Everitt, R. D., and Rugh, D. J. (1980). Northern sea lion population decline on eastern Aleutian Islands, Alaska. *J. Wildl. Manage.* **44,** 25–33.

Consiglieri, L., and Braham, H. W. (1982). "Seasonal Distribution and Relative Abundance of Marine Mammals in the Gulf of Alaska," Final Rep. Res. Unit 68. Alaska Outer Continental Environmental Assessment Program, NOAA, Juneau Project Office.

Cunningham, W., and Stanford, S. (1979). "Observations of Migrating Gray Whales (*Eschrichtius robustus*) at Cape St. Elias, Alaska." Alaska Department of Fish and Game, Anchorage (unpublished).

Everitt, R. D., and Braham, H. W. (1980). Aerial survey of Pacific harbor seals in the southeastern Bering Sea. *Northwest Sci.* **54**(4), 281–283.

Fay, F. H. (1974). The role of ice in the ecology of marine mammals of the Bering Sea. *In* "Oceanography of the Bering Sea" (D. W. Hood and E. J. Kelley, eds.), pp. 383–399. Int. Mar. Sci., University of Alaska, Fairbanks.

Forsell, D. J., and Gould, P. J. (1981). Distribution and abundance of marine birds and mammals wintering in the Kodiak area of Alaska. *U.S., Fish Wildl. Serv., Off. Biol. Serv. [Tech. Rep.] FWS/OBS* **FWS/OBS-81/13.**

Frost, K. J., Lowry, L. F., and Burns, J. J. (1982). "Distribution of Marine Mammals in the Coastal Zone of the Bering Sea During Summer and Autumn," Final Rep. Res. Unit 613. Outer Continental Shelf Environmental Assessment Program, Juneau, Alaska.

Gilmore, R. (1955). The return of the gray whale. *Sci. Am.* **192**(1), 62–67.

Gilmore, R. (1960). Census and migration of the California gray whale. *Nor. Hvalfangst-Tid.* **49,** 409–431.

Herzing, D., and Mate, B. R. (1981). California gray whale migration along the Oregon coast. *Proc. Conf. Biol. Mar. Mamm., 4th, 1981* Abstracts, p. 54.

Hessing, P. (1981). "Gray Whale (*Eschrichtius robustus*) Migration into the Bering Sea, Spring, 1981," Final Rep. Contract NA 81SRGA 00080. Outer Continental Shelf Environ. Assess. Program, NOAA, Juneau Project Office.

Ichihara, T. (1958). Gray whales observed in the Bering Sea. *Sci. Rep. Whales Res. Inst.* **13,** 201–206.

Johnson, J., Braham, H., Krogman, B., Marquette, W., Sonntag, R., and Rugh, D. (1981). Research conducted on bowhead whales, June 1979 to June 1980. *Rep. Int. Whal. Commn.* **31,** 461–475.

Kellogg, R. (1929). What is known of the migration of some of the whalebone whales. *Smithson. Inst., Annu. Rep.* pp. 467–494.

McNutt, S. L. (1981). Remote sensing analysis of ice growth and distribution in the eastern Bering Sea. *In* "The Eastern Bering Sea Shelf: Oceanography and Resources" (D. W. Hood and J. A. Calder, eds.), Vol. 1, pp. 141–165. Univ. of Washington Press, Seattle.

Maher, W. J. (1960). Recent records of the California gray whale (*Eschrichtius glaucus*) along the north coast of Alaska. *Arctic* **13,** 257–265.

Marquette, W. M., and Braham, H. W. (1982). Gray whale distribution and catch by Alaskan Eskimos: A replacement for the bowhead whale? *Arctic* **35,** 386–394.

Marquette, W. M., Braham, H. W., Nerini, M. K., and Miller, R. V. (1982). Bowhead whale studies, Autumn 1980–Spring 1981: Harvest, biology and distribution. *Rep. Int. Whal. Commn.* **32,** 357–370.

Nasu, K. (1960). Oceanographic investigation in the Chukchi Sea during the summer of 1958. *Sci. Rep. Whales Res. Inst.* **15,** 143–157.

Nasu, K. (1974). Movements of baleen whales in relation to hydrographic conditions in the northern part of the North Pacific Ocean, Bering Sea. *In* "Oceanography of the Bering Sea" (D. W. Hood and E. J. Kelley, eds.), pp. 345–361. Inst. Mar. Sci., University of Alaska, Fairbanks.

Nikulin, P. (1946). O raspredelenii kitoobraznzykh v moryakh omyvayushchikh Chukotzkii poluostrov. (Concerning the distribution of whales in the seas adjacent to the Chukotskiy peninsula.) *Izv. Tikhookean. Nauchno-Issled. Inst. Rybn. Khoz. Okeanogr.* **22,** 255–257 (transl. by U.S. Dept. Navy, Nav. Oceanogr. Off., Washington, D.C., 1969, Transl. 428).

Norris, K. S., and Prescott, J. H. (1961). Observations of Pacific cetaceans of California and Mexican waters. *Univ. Calif., Berkeley, Publ. Zool.* **63,** 291–402.

Pike, G. (1962). Migration and feeding of gray whales (*Eschrichtius robustus*). *J. Fish. Res. Board Can.* **19**(5), 815–838.

Poole, M. (1981). The northbound migration of the California gray whale, *Eschrichtius robustus*, off the Central California coast. *Proc. Conf. Biol. Mar. Mamm., 4th, 1981* Abstr., p. 26.

Reilly, S. B. (1981). Population assessment and dynamics of the California gray whale (*Eschrichtius robustus*). Doctoral Dissertation, University of Washington, Seattle.

Rice, D. W. (1965). Offshore southward migration of gray whales off southern California. *J. Mammal.* **46,** 504–505.

Rice, D. W., and Wolman, A. A. (1971). *The life history and ecology of the gray whale (Eschrichtius robustus).* *Spec. Publ.—Am. Soc. Mammal.* **3,** 1–142.

Rice, D. W., and Wolman, A. A. (1982). Whale census in the Gulf of Alaska, June to August 1980. *Rep. Int. Whal. Commn.* **32,** 491–497.

Rugh, D. J., and Braham, H. W. (1979). California gray whale (*Eschrichtius robustus*) fall migration through Unimak Pass, Alaska, 1977: A preliminary report. *Rep. Int. Whal. Commn.* **29,** 315–320.

Scammon, C. (1874). "The Marine Mammals of the Northwestern Coast of North America." John H. Carmany & Co., San Francisco, California.

Sleptsov, M. M. (1955). "Biologiya i promysel kitov dalnevostochnykh morei (Biology of Whales and the Whaling Fishery in Far Eastern Seas)." "Pishch. Prom.," Moscow (in Russian) (transl. by Fish. Res. Board Can., Transl. Ser. 118. Contents and conclusions only).

Tomilin, A. (1957). Kitoobraznye. (Cetacea.) "Zveri SSSR i prilezhashchikh stran. (Mammals of the U.S.S.R. and Adjacent Countries.)," Vol. 9. Publ. House USSR Acad. Sci., Moscow (in Russian) (transl. by O. Ronen, Israel Program Sci. Transl., Jerusalem, 1967, available by Natl. Tech. Inf. Serv., Springfield, Virginia.

Townsend, C. H. (1935). The distribution of certain whales as shown by logbook records of American whaleships. *Zoologica* **19**(1), 3–50.

Turner, L. M. (1886). "Contributions to the Natural History of Alaska." U.S. Govt. Printing Office, Washington, D.C.

Votrogov, L. M., and Bogoslovskaya, L. S. (1980). Gray whales off the Chukotka Peninsula. *Rep. Int. Whal. Commn.* **30,** 435–437.

Wilke, F., and Fiscus, C. H. (1961). Gray whale observations. *J. Mammal.* **42,** 108–109.

Zenkovich, B. A. (1937). Esche o serom Kaliforniiskom kitae (Rhachianectes glaucus Cope, 1864). [More on the California gray whale (Rhachianectes glaucus Cope, 1864).] *Vestn. Akad. Nauk SSSR, Dal'ne-vostochnyi Fil.* **23,** 91–103.

12

Gray Whales off Vancouver Island, British Columbia

James D. Darling

Introduction

Each spring, gray whales (*Eschrichtius robustus*) migrate from calving lagoons in Baja, California, along the west coast of North America to feeding grounds in the Bering and Chukchi Seas (Scammon, 1869; Pike, 1962; Rice and Wolman, 1971). Although censuses conducted at Unimak Pass, Alaska (Hall *et al*., 1977; Rugh and Braham, 1979; Chapter 10, this volume) indicated that the majority reached northern seas, sightings in summer off California, Oregon, Washington, and British Columbia suggested some whales terminate their migration far to the south (Gilmore, 1960; Pike, 1962; Pike and MacAskie, 1969; Rice and Wolman, 1971; Hatler and Darling, 1974; Darling, 1978; J. Sumich, cited in Chapter 13, this volume).

For waters off Vancouver Island, British Columbia (B.C.), three earlier studies demonstrated that (1) north- and southbound migrants pass along the west coast (Pike, 1962; Pike and MacAskie, 1969; Hatler and Darling, 1974; (2) gray whales regularly occur in summer along the outer coast (Hatler and Darling, 1974); and (3) at least one gray

ISBN 0-12-389180-9

whale, recognized in photographs by its unique natural markings, remained off the island throughout one summer and returned for several years (Hatler and Darling, 1974).

Techniques for identifying individual whales from photographs were developed in the early 1970s for southern right whales, *Eubalaena australis* (Payne *et al.*, 1981); for killer whales, *Orcinus orca* (Bigg *et al.*, 1976; Bigg, 1982); for humpback whales, *Megaptera novaeangliae* (Katona *et al.*, 1979); and, as described here, for gray whales. Such techniques have supported long-term studies of abundance and behavior by allowing repeated observations of the same individuals without the need for tagging.

This chapter describes the abundance, stability, distribution, and general behavior of gray whales that spend the summer off Vancouver Island.[1] It is based on direct observations and photoidentifications over the 10 years since the first individuals were identified in 1970–1973 (Hatler and Darling, 1974).

Methods

STUDY AREA

Vancouver Island (Fig. 1) is located approximately halfway between the Mexican calving lagoons and the northernmost Chukchi Sea feeding grounds. Its west coast is characterized by sandy bays and beaches (e.g., Long Beach, Wickaninnish Bay, is 11.3 km in length) separated by rocky headlands or long stretches of rocky coast and cliffs. Along the central west coast, a system of protected waterways exists between coastal islands.

Small craft surveys (1972–1976) initially centered on a 40-km stretch of the west coast (circled in Fig. 1). After 1976, these surveys were gradually extended to cover most areas accessible to small craft between the entrance to the Straits of Juan de Fuca and Cape Scott (Fig. 1). With cooperation from personnel at 10 Vancouver Island lightstations (Fig. 2), much of the west coast of the island was monitored during the migrations. All observation was coastal, within 4 km of shore; no offshore surveys were attempted.

MIGRATIONS

The present study was primarily concerned with gray whales which summer along the Vancouver Island coast. Therefore, migrations past the island were observed mainly to determine the interval between the spring (northbound) and winter (southbound) migrants. Lightstation personnel recorded the following data on gray whales that they saw between November and May: date, number of whales seen, direction of movement, and appropriate comments. Observations were incidental to their other duties, but reports did include records of any whales noticed and the number of days of poor visibility. The reporting lightstations from south to north, as indicated by triangles in Fig. 2, were Sheringham Point, Carmanah Point, Pachena Point, Cape Beale, Amphitrite

[1]This chapter summarizes an unpublished M.Sc. thesis by Darling (1978) which will be referred to throughout.

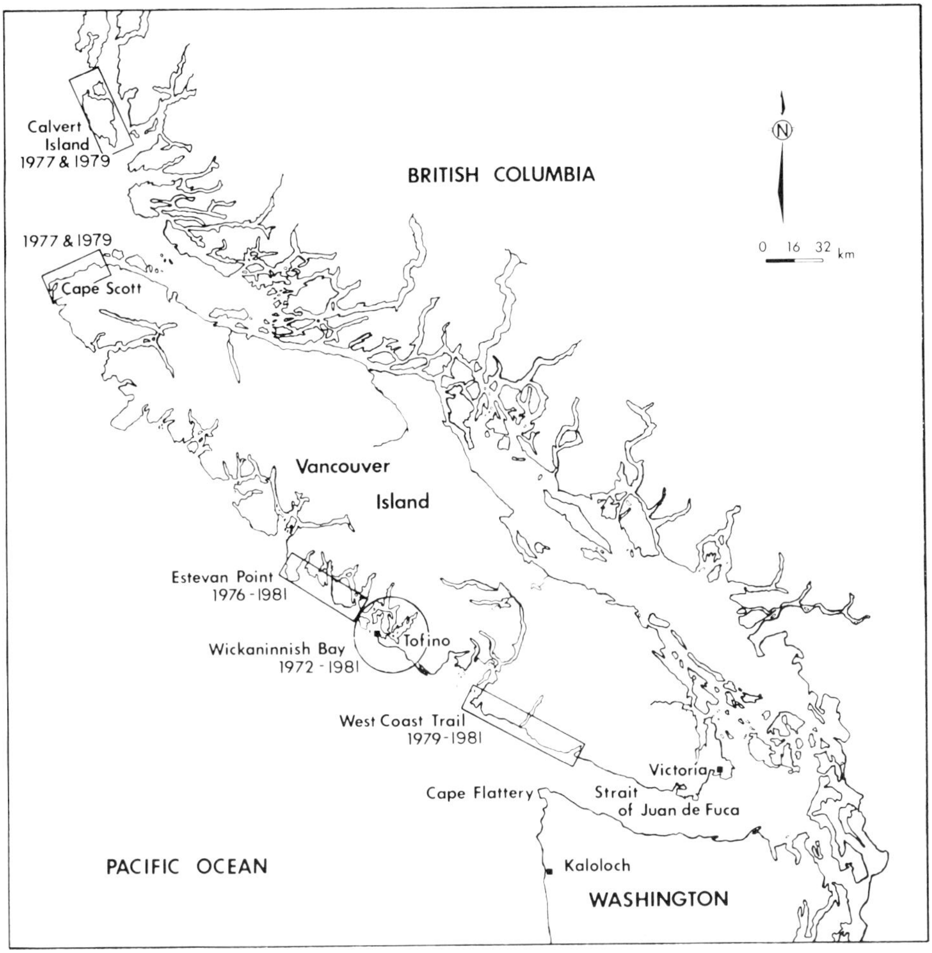

Fig. 1. *Vancouver Island, British Columbia, showing the central study area; areas surveyed frequently (circled) and areas reached less regularly (rectangles) in the years indicated.*

Point, Lennard Island, Estevan Point, Nootka, Kains Island, and Cape Scott. Lightstation records were used to monitor migrations in four seasons; 1972–1973 (reported in Hatler and Darling, 1974), 1973–1974, 1975–1976, and 1976–1977. In 1975 and 1976, the author also observed the spring migrations (February–June) from shore stations in the central study area and from small craft; the winter (November–January) migration was observed sporadically from shore; and incidental sightings reported by some residents of the area, particularly Pacific Rim National Park naturalists, were logged.

To establish timing of the migrations and indicate their peaks, all reporting lightstations were treated as one locus, that is, "the west coast of Vancouver Island." All sightings for each 2-week period, November 15 to May 1, from the four winters 1972–1973, 1973–1974, 1975–1976, and 1976–1977 were averaged. Intensive observations in the study area by the author, beginning in March of the years 1975 and 1976, allowed

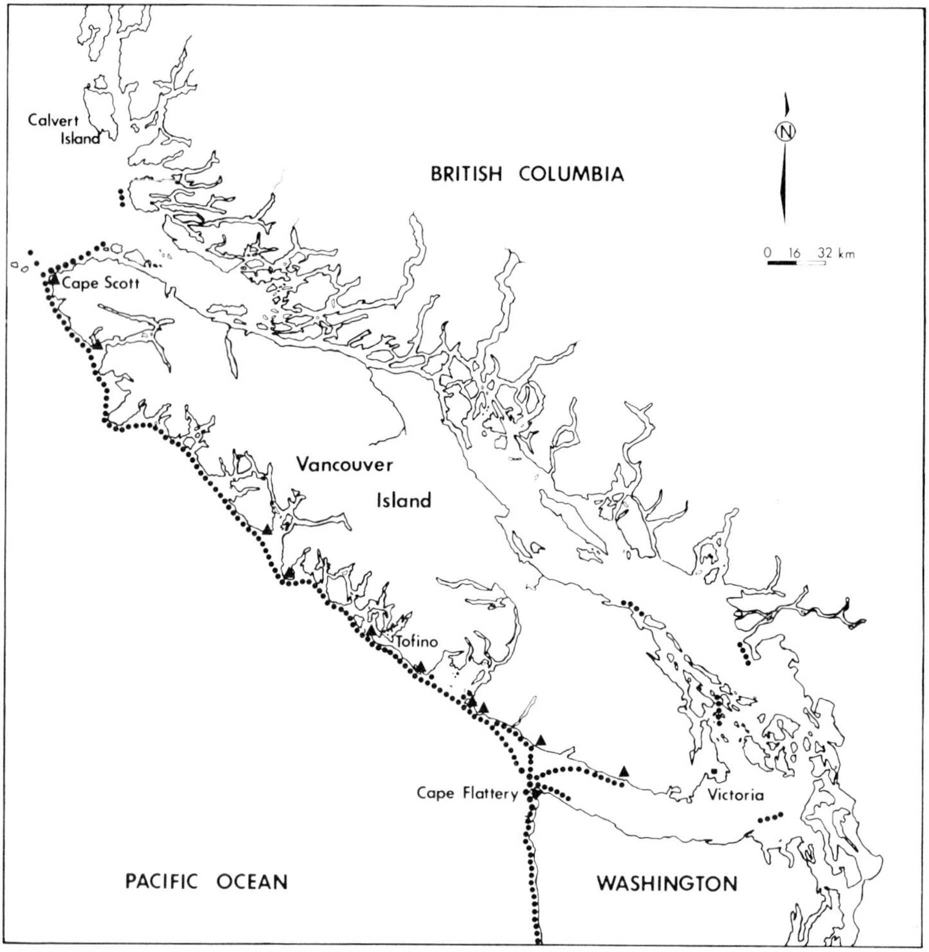

Fig. 2. *The spring (northward) migratory route of gray whales along the west coast of Vancouver Island, based on all data available, is shown by large dots. Triangles mark the position of the following lightstations, south to north: Sheringham Point, Carmanah Point, Pachena Point, Cape Beale, Amphitrite Point, Lennard Island, Estevan Point, Nootka, Kains Island, and Cape Scott.*

further delineation of the northward migratory period; there were no comparable observations of the southward migration.

INDIVIDUAL IDENTIFICATION

Photographs were taken of pigmentation patterns on visible portions of whales' backs[2] with a 35-mm camera with a 200- or 300-mm lens, Kodak Tri-X Pan Film pushed

[2]Examinations of a cross-section of the skin of a dead whale indicated that these patterns were indeed pigmentation patterns and not superficial.

to 1600 ASA, and a shutter speed of 1000th/sec (cf. Bigg *et al.*, 1976; Bigg, 1982). Any scars or other prominent natural markings were also photographed.

Once photographed, the patterns were sketched on a data chart and corresponding film frames noted. These patterns were used for identification because for most animals they were the only part regularly seen. Most of the whales encountered were feeding in shallow water and rarely raised their flukes above the water surface. Therefore, identification by fluke patterns, as has been done in humpback whales (Katona *et al.*, 1979) was impractical.

The need to recognize individuals from side patterns complicated identification work because it required photographs of both sides to avoid duplication in counts. However, once photographs of both sides of an individual were on record, a picture of either side was sufficient. To avoid double counting in abundance analysis, when there were uncertainties about assignment of a photograph to an individual whale only photographs of the left side were used. Identification photographs were cataloged by year and records of repeat sightings established.

SUMMER POPULATION

During the summer, whales were located, observed, and photographed for individual identification from a small outboard-powered boat (5 m in length). Each whale's position, companions, and behavior were noted. Numerous reefs, islands, and other points in the study area allowed accurate positioning by landmark. Observations were made from within 100 m by the author and an assistant, usually during the 8 AM–1 PM period. Westerly winds beginning after noon are usual in the study area and make work after that time more difficult. Flights to locate whales were carried out weekly in 1976 and sporadically in other years. Behavioral observations were incidental to photo-identification and aerial survey work. Summer effort, summarized in Table I, varied from year to year.

Results

MIGRATIONS

Timetable

The first northbound whales passed Vancouver Island in February; numbers peaked the last 2 weeks of March then declined (Fig. 3 and Table II). Lightstation sightings were not reported after April 30. However, there were still some whales moving north in May and early June, as will be shown.

Southbound migrants pass Vancouver Island from November to late January with peak numbers during the last 2 weeks in December (Fig. 3 and Table II). From mid-January to mid-February, sightings were rare. The whales occasionally seen during this interval are discussed below.

Table I
Study Effort off the West Coast of Vancouver Island, B.C., 1972–1981

Year	Months of Observations	Methods				Number of Days Identification Photos Taken	Comments
		Boat	Aircraft	Shore	Lightstation		
1972	July–Dec.	Daily in July and Aug.	—	Sporadic	Nov. 1–Apr. 30, 1972–1973	<5	Observations incidental to operation of whale-watch boat; also Dec. 14 boat survey and identification photographs
1973	Jan.–Dec.	Occasional May–Sept.	—	Sporadic	Nov. 1–Apr. 30, 1973–1974	<10	Observations incidental to other work in the area
1974	May–Sept.	Occasional, May–Sept.	—	Sporadic	—	<10	Observations incidental to other work in the area
1975	Mar.–Dec.	Daily Mar.–May & Aug.–Oct. Frequent, June–Jul.	Occasional (opportunistic)	Regular Mar.–May	Nov. 1–Apr. 30, 1975–1976	53	In June and July, observations were incidental to other work in the area
1976	Mar.–Oct.	Daily	Weekly	Regular Mar.–May	Nov. 1–Apr. 30, 1976–1977	54	By boat *as weather allowed*
1977	June–Sept.	Daily	One survey flight of study region	Sporadic	—	39	Reached Cape Scott and Calvert Island Area (see Fig. 1.)
1978	July–Aug.	Occasional		Sporadic	—	9	Observations incidental to other work in the area
1979	June–Sept.	Daily	One survey flight of study region	Sporadic	—	18	Reached Cape Scott and Calvert Island Area (see Fig. 1.)
1980	June–Sept.	Occasional	—	Sporadic	—	9	Observations incidental to other work, plus one complete boat survey of south coast
1981	June–Sept.	Occasional	—	Sporadic	—	8	Observations incidental to other work, plus one complete boat survey of south coast

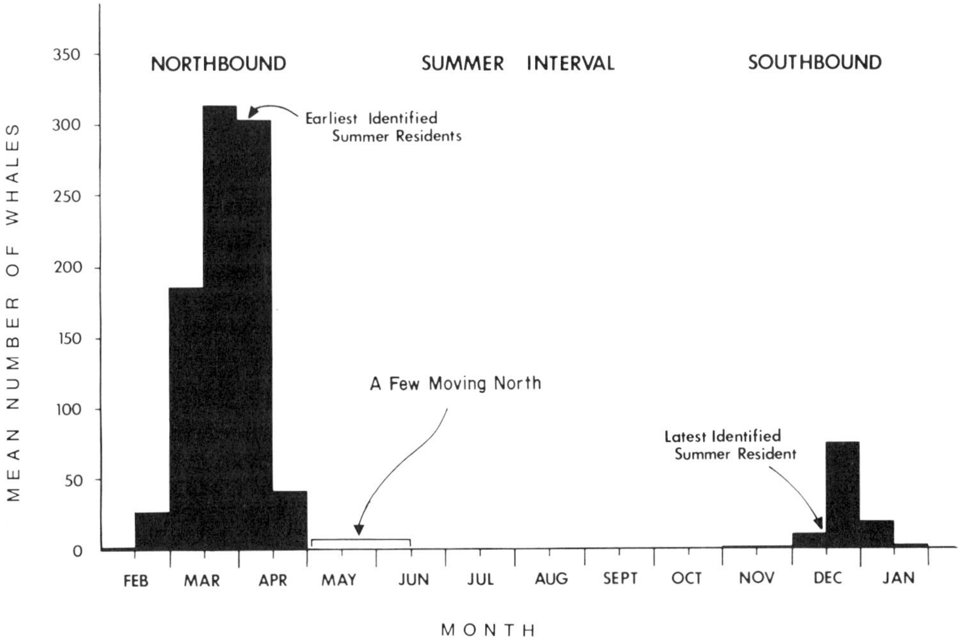

Fig. 3. *Timing of gray whale migrations past Vancouver Island based on lightstation sightings, 1972–1977 (data are unavailable for the period November 1974 through November 1975). See also Table I.*

Fewer whales were sighted during the southward migration than during the northward migration, probably reflecting poor winter weather rather than differences in numbers of whales passing. Two bits of evidence support this conclusion. A note received from the Kains Island lightkeeper in December, 1974 with his sighting forms for that month stated: "On about three days the visibility was clear enough to encourage us to glance out to sea and to see the odd whale . . . at least that is how it would appear from our fog alarm report." On those 3 days, they saw 26 whales. Also, in December, 1977, when weather was unseasonably good, significantly more whales were reported (Table II).

Intensive observations by the author allowed some further delineation of the spring migratory period (Darling, 1978). During this period whales were either (1) moving steadily north; (2) slowly and sporadically, but still obviously, moving north, or (3) lingering in the area. As was expected from lightstation reports, obvious northbound migrants were common through April, although they were fewer after mid-month. In May most whales observed were stationary and feeding; however, a few were still obviously migrating. A photographically identified cow and calf appeared in Wickaninnish Bay on June 2, 1976 and remained until June 5, 1976. They were observed to be feeding and were not traveling. They were gone on June 6 and had probably moved northwards as they were not seen again. This is the latest reasonably clear record of a migrant. Other whales identified in May and June were also not seen a second time, suggesting that they too had destinations farther north. Southbound migrants were sighted in Wickanin-

Table II
Lightstation Reports of Gray Whales Migrating Past the West Coast of Vancouver Island

Dates	Number of whales sighted				Total	Average
	1972–1973	1973–1974	1975–1976	1976–1977		
Nov. 1–15	0	—[a]	0	5	5	1
Nov. 16–30	1	3	0	1	5	1
Dec. 1–15	30	3	4	7	44	11
Dec. 16–31	8	22	36	233	299	75
Jan. 1–15	18	30	30	79	157	39
Jan. 16–31	0	2	—	4	6	2
Feb. 1–15	0	—	5	16	21	5
Feb. 16–28	13	1	22	70	106	27
Mar. 1–15	86	28	230	411	755	189
Mar. 16–31	183	162	271	631	1247	312
April 1–15	364	100	311	445	1220	305
April 16–31	55	35	11	66	167	42

[a]A dash (—) indicates no reports. Averages are rounded to the nearest whole number.

nish Bay in December and January, consistent with lightstation reports (Darling, 1978).

The winter interval between the last southbound and first northbound migrants was not clear cut. Lightstation reports indicate it occurs from approximately mid-January to mid-February, although in some years whales were sighted during that period.

Route

The migratory route observed during this investigation is shown in Figure 3. It conforms with the accounts of earlier researchers (Pike, 1962; Pike and MacAskie, 1969; Hatler and Darling, 1974). Northbound whales leave the Washington coast at Cape Flattery, head generally north until they reach Vancouver Island then change course to follow its west coast. All sightings were within 4 km of shore; most were within 1 km and many within 100 m. Occasional sightings of gray whales in the Straits of Juan de Fuca and Georgia Strait (Darling, field notes, 1975–1981) indicate that a few animals turn southeast at Cape Flattery.

INDIVIDUAL IDENTIFICATION

Approximately 5000 photographs were taken. Over 100 different gray whales have been identified off the west coast of Vancouver Island since 1970. Most markings were light, white or gray on a darker background. Extent, intensity, and location of identifying elements in the pattern varied dramatically (Figs. 4–6). They ranged from large portions of white, to varied combinations of white patches, slashes, dots, and speckles, to almost total black. Most individuals were immediately recognizable; all were distinguishable on close examination.

The longevity of observed patterns is suggested in Fig. 5, animal No. 741, pho-

Fig. 4. *Examples of pigmentation patterns used to identify individual gray whales.*

tographed over 5 years—in which there was no significant change in markings—and in Fig. 6, both sides of animal No. 756 over 2 years, 1975 and 1976—in which the smallest details of the patterns are constant (arrows). Some half-moon-shaped marks (circled) present on the left side of No. 756 in 1976 were not there the previous year. Similar superficial, nonpermanent marks noticed on some whales were not sufficiently large or common to interfere with identification. From a photograph of a near-term fetus with white skin patterns (see Rice and Wolman, 1971, p. 78), it appears whales are born with

Fig. 5. *Photographs of whale No. 741 in 1974 and 1979. The photographs were taken at slightly different angles. Arrows indicate a common reference point.*

these patterns and maintain them throughout their lives although scratches or superficial growths may partially obscure them at times.

When a scar was the dominant marking, it was used as the basis for identification along with skin patterns. Such was the case with the scarred whale first described by Hatler and Darling (1974). The scar on this whale remained obvious from at least 1970 to 1975, the years it was recorded off Vancouver Island. No substantial change in the scar was noted.

Gray whales have patches of barnacles (*Cryptolepas rhachianceti*), especially on the head and the most anterior portion of the back, and cyamids (*Cyamus scammoni, C. ceti,*

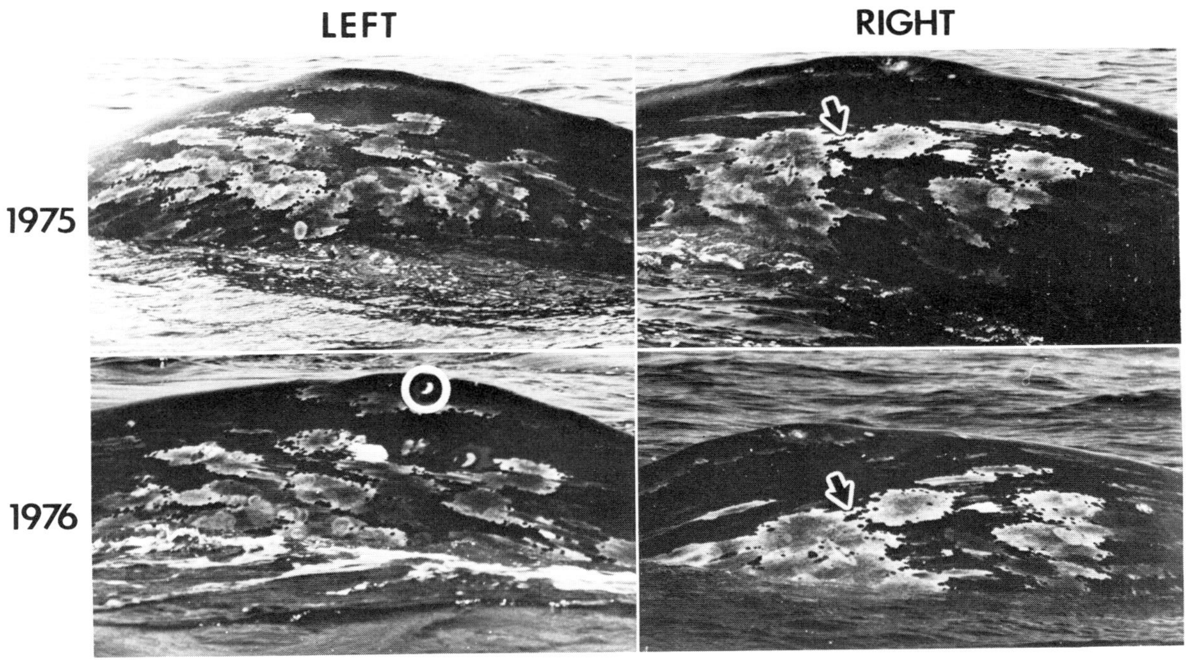

Fig. 6. *Pigmentation patterns of whale No. 756 in successive years. Note the half moon-shaped markings (circled) on the left side in 1976. which are not present in 1975. Arrows indicate a common reference point.*

and *C. kessleri*), particularly around the barnacles and on wounds (Rice and Wolman, 1971). Barnacles and cyamids were not used for individual recognition.

SUMMER POPULATION

Arrival and Departure of Summer Residents

As has been shown, the majority of northward migrants reach Vancouver Island between February and June. Summer residents, animals which remained in the area during the summer interval, were identified as early as April 8 (Whales #701, 721, and 752), April 11 (whale #7513), April 16 (whale #741, and April 17 (whale #754) in 1975, and April 12 (whale 754), and May 13 (whale #759 in 1976. Therefore, summer residents can be present in the area by the time of the local peak of the northward migration in late March and early April (Fig. 3).

Data indicating when summer residents leave for the winter are sketchy because day-to-day observation ended in September when many, if not all, identified whales were still present. The latest sighting of a positively identified summer resident was December 14, 1972 when whale #721 was seen feeding in Wickaninnish Bay. Other summer residents were identified as late as October 12, 1975 (whale #701) and October 23, 1976 (whales #759, 756, and 7511). Small craft searches were not made after these dates.

Apparently the duration of a whale's stay can be from peak to peak of the migrations or about 8–9 months (Fig. 3). The longest recorded stay of a single individual in 1 year was 7 months; whale #701 was first identified on April 8, 1974 and last identified on October 12, 1974.

Abundance

The largest number of resident whales identified in one summer was 34 in 1976 and 1977, the years of maximum effort (Tables I and III). Numbers identified in other years reflect effort level (Tables I and III). In both 1976 and 1977 previously unidentified whales were still being "discovered," albeit rarely, at the end of the season, indicating that not

Table III
Identification of Individual Whales, 1975–1981

Category	1975	1976	1977	1978	1979	1980	1981
A. Number of whales identified in the summer	27	34	34	13	25	10	20
B. Number of A seen in other years	20	25	22	8	14	7	10
C. Percentage of A seen in other years	74	65	65	61.5	56	70	50
D. Number of A seen only that year	7	9	11	5	11	3	10

all residents had been identified. It is not likely, however, that the maximum number of whales which summer in the area is much higher than 34. Aerial surveys carried out in the middle of summer did not indicate that large numbers of animals were "missed" in the identification effort. One of the best such surveys, considering weather and the amount of time spent searching for whales, was that on August 29 and 30, 1977, in which

Table IV
Individual Whales Identified off Vancouver Island in More than 1 Year

Whale ID number	1970	1971	1972	1973	1974	1975	1976	1977	1978	1979	1980	1981
701	x		x	x	x	x						
711		x				x	x	x				
721			x	x	x	x	x	x	x			
741					x	x	x	x		x	x	x
742					x	x	x	x		x		
751						x	x	x		x		x
752						x	x	x		x	x	
753						x	x	x	x	x		
754						x	x	x		x		
755						x	x	x				
756						x	x	x				
757						x	x	x				
758						x	x	x				
759						x	x	x				
7510						x		x	x			
7511						x	x					
7512						x		x				
7513						x	x					
7514						x	x					
7515						x	x					
761							x	x	x			x
762							x	x		x		
763							x	x	x			
764							x	x	x			
765							x			x		
766							x		x			
767							x		x			
768							x	x				
771								x		x		
772								x		x		
791										x	x	x
792										x	x	x
793										x		x
794										x		x
802											x	x
803											x	x
804											x	x

the entire west coast of the island from Port Renfrew at the southern end of the West Coast Trail to Cape Scott was searched. Twenty-nine whales were counted, compared to 34 identified in that year. In 1979, in a survey from the Kaloloch on the Washington Coast to Calvert Island, north of Vancouver Island, we saw an estimated 30–40 whales; 22 individuals were identified that year. From the above, it seems unlikely that more than 50 whales range onto the Vancouver Island coast in any one summer, and a reasonable estimate might be 35–50.

In the six summers from 1975 to 1981, 93 individual whales were identified on this coast. This is the total of the 56 identified whales seen in only one summer [Table III, (D)] plus those 37 seen in more than 1 year (Table IV). It is thought that not all of these return to Vancouver Island each year.

Annual Returns

Since 1970, 37 whales have been seen in more than 1 year (Table IV). Whale #741 was seen for 8 years (1974–1981); whale #711, 7 years (1971–1977); whale #721, 7 years (1972–1978); and whales #701, #742, and #751, 5 years (1970–1975, 1974–1979, and 1975–1981, respectively) (Table IV). Among the animals identified each year there were those which had been identified in other years (average, 63%) and those which were seen only that year (average, 37%). In 15 cases (in Table IV), a whale was "missed" in one or more summers but seen in subsequent years.

Although some identified whales were resighted in the exact location in which they were identified the previous years (e.g., whale #701 in Wickaninnish Bay in 1970, 1973, 1974, and 1975), others were resighted in different years as much as 150 km from previous sighting location(s) (Table VA and Fig. 1).

Distribution and Movements

The combined locations where gray whales were found during the summers of 1975–1981 cover most of the Vancouver Island coast from Victoria to Cape Scott and

Table VA
Changes in Location of Photoidentified Whales from One Summer to the Next

Whale ID Number	First Location and Date	Second Location and Date	Direction	Distance (km)
762	Wickaninnish Bay Aug. 4, 1976	Estevan Point Aug. 11, 1977	NW	80
803	Wickaninnish Bay July 9, 1980	West Coast Trail Aug. 6, 1981	SE	77
804	West Coast Trail Aug. 23, 1979	Wickaninnish Bay July 9, 1980	NW	77
771	Estevan Point Aug. 17, 1977	West Coast Trail Aug. 5, 1979	SE	150

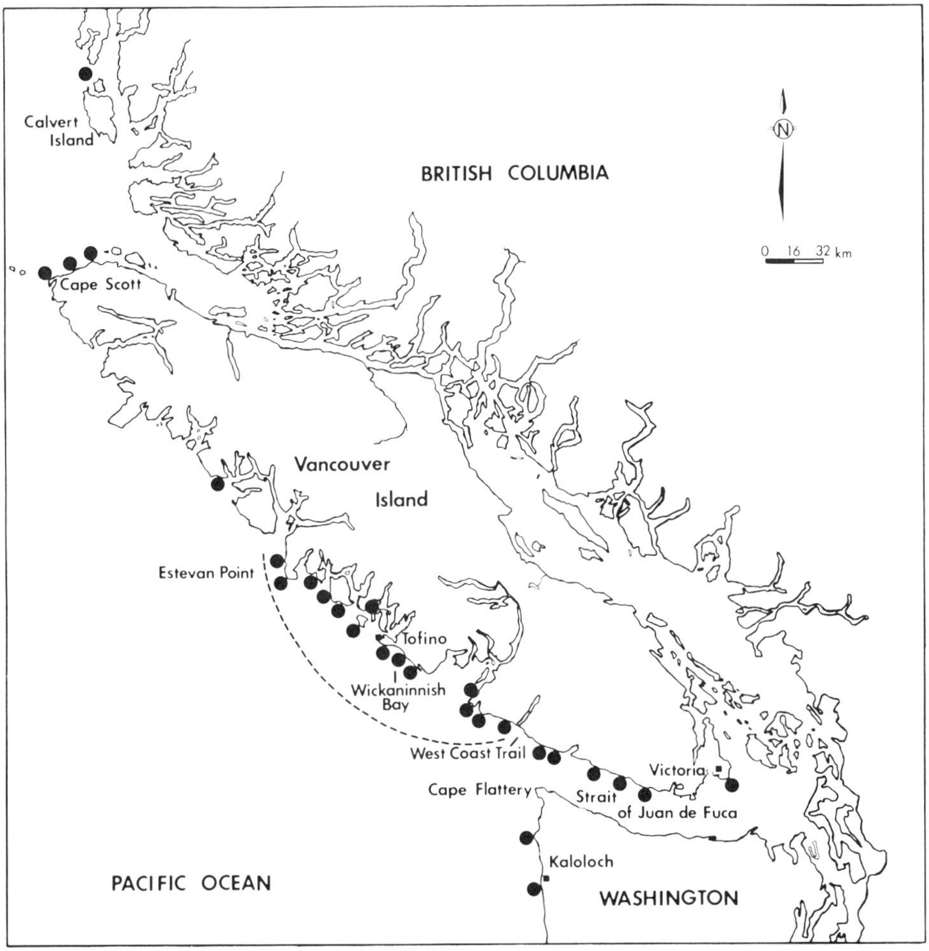

Fig. 7. *Locations of sightings of gray whales in summer (June–October), 1975 to 1981, indicated by small solid circles.*

farther north off Calvert Island along the mainland B. C. coast (Fig. 7). The relatively larger gaps between sightings towards the north may well reflect the spacing of observers. The whales were feeding in all locations. All sightings were within 1 km of the shore; most were within 100 m. Apparently suitable habitat on exposed coast includes shallow sandy bays, areas along rocky coastlines and cliffs, and areas off sandstone shelves. Whales were repeatedly but less commonly seen feeding in protected waterways near Tofino.

Whales traveled among sites at least 77 km apart within a summer [Table VB; Fig. 7 (semicircle)]. The maximum distance traveled is not known but the six whales identified off Cape Scott, 220 km north of Estevan Point on July 20, 1977 were not among whales identified off southern Vancouver Island. Apparently, movements are among feeding

Table VB
Travels of Individual Whales along the West Coast of Vancouver Island, within One Summer

Whale ID Number	First Location and Date	Second Location and Date	Direction	Distance (km)
794	Wickaninnish Bay July 14, 1979 (traveling SE)	West Coast Trail Aug. 23, 1979 (feeding)	SE	77
7913	West Coast Trail Aug. 23, 1979 (feeding)	Wickaninnish Bay Aug. 28, 1979 (traveling NW)	NW	77
761	Wickaninnish Bay June 2, 1977 (feeding)	Estevan Point Aug. 17, 1977 (feeding)	NW	80
791	Wickaninnish Bay July 9, 1980 (traveling NW)	West Coast Trail Sept. 10, 1980 (feeding)	SE	77

areas, although traveling whales may not stop at the next nearest area and may bypass feeding whales en route. Whales were seen traveling alone, in pairs, in trios, and in patterns from slow and wandering to quick and direct. On occasion traveling whales moved to a riptide sand bar area where they remained for at least as long as 30 min to several hours.

In some areas whales remained only momentarily, and some animals had as many as four destinations in 1 week, covering a distance of approximately 40 km (whales #741 and 752 in August, 1975). At the other extreme, whales #701 and #754 each remained in Wickaninnish Bay for 80+ days in 1975 and 1976, respectively (Darling, 1978).

Discussion

MIGRATION TIMETABLE

Migratory times and peaks determined during this study generally agree with the observations of Pike (1962), Pike and MacAskie (1969), and Hatler and Darling (1974). The northern migration past the island begins in February, peaks in late March and early April, and continues through May or early June. There may be some slight variance in the timing among years. Based on sightings off Vancouver and Queen Charlotte Islands, Pike (1962) and Pike and MacAskie (1969) suggested the southbound migration occurs from late September or early October until early January, with a peak in the last 2 weeks of December. The sightings in September and October seem early for Vancouver Island. Sightings reported by Hatler and Darling (1974) occurred from the end of November through mid-January. Observations reported here confirmed that this migration occurs from November to mid-January with a peak the last 2 weeks in December. These findings are consistent with predictions from observations in other areas along the migratory route (Leatherwood, 1974; Reilly *et al.*, 1980; Chapters 10, 13, and 16, this volume).

The interval between the last northbound and first southbound migrants extends from mid-June to November. Summer residents were present off Vancouver Island throughout this period, confirming the speculations of Pike (1962) and Pike and MacAskie (1969) and reports by Hatler and Darling (1974). Many of the same identified individuals were seen repeatedly during the summer, as was one whale observed in 1971 (Hatler and Darling, 1974). Summer residents were present by early April, the peak of the northern migration, and stayed 8–9 months until December and the peak of the southern migration. Whales making the full migration to the Bering and Chukchi Seas must budget 2 months more traveling time than those resident off Vancouver Island, arriving in northern areas 1 month later and leaving 1 month earlier (Pike, 1962). They are left with only 6–7 months to feed. Many migrants do, however, feed sporadically while enroute north (Darling, 1978; Wilke and Fiscus, 1961).

The winter interval between the last southbound and first northbound whales is short and not always clear cut. To date there is no direct evidence that Vancouver Island summer residents migrate and in some years unidentified whales have been sighted in the area throughout the winter (Hatler and Darling, 1974). These authors note that it is possible the whales sighted during this period were successive groups of migrating animals, southbound and northbound. The degree of turnover of identified animals each summer suggests that some, if not all, join the rest of the herd.

INDIVIDUAL IDENTIFICATION

Several researchers have used natural markings to identify individual gray whales over the short term; Sauer (1963), describing apparent sexual activity off St. Lawrence Island, Alaska, stated that individual whales could be recognized by skin patterns. Leatherwood (1974) used natural markings to identify three gray whales on successive days off the California coast. Photographic identification work described in Hatler and Darling (1974) and in this chapter show the technique can be used as a basis for long-term studies in which every animal can be permanently identifiable. Swartz and Jones (1978, 1981) and Bryant and Lafferty (1980) have identified gray whales in Mexican calving lagoons and are constructing identification catalogs. Comparison of identified whales from different parts of the gray whales range may elucidate associations of the Vancouver Island whales with the rest of the eastern Pacific herd.

THE SUMMER POPULATION

Gilmore (1960) suggested that the whales lingering all summer on the northwest coast did so because they found themselves too far behind to reach the Bering Sea on time to store enough fat so stayed on more southerly feeding grounds. This explanation can be ruled out based on the facts that individuals return to Vancouver Island year to year, and that they are present at the peak of the northern migration.

Hatler and Darling (1974) speculated that "pockets" of habitat suitable for regular occupancy by gray whales occur between southern Vancouver Island and Alaska. Such "pockets" apparently do exist throughout the species range, off California (Dohl *et al.*, 1981), Oregon (J. Sumich in Chapter 13, this volume), Washington (Darling, 1979), and British Columbia (this chapter).

The Vancouver Island "pocket" is occupied by 35–50 whales each summer. These animals may well be part of a larger "northwest coast" population. Approximately 75 whales summer off Oregon each year (B. R. Mate, 1979, personal communication), so it is likely there are at least 100 in the British Columbia–Washington–Oregon area. This is a small part of the approximately 17,000 animals estimated in the eastern Pacific herd (Reilly *et al.*, 1980, Chapters 9 and 13, this volume).

Votrogov and Bogoslovskaya (1980) and Bogoslavskaya *et al.*, (1981) found gray whales unevenly distributed in "divisions" along the Asian coast. Between these were empty areas through which whales passed rapidly. Presumably these were not good feeding areas. These reports identify two apparent types of whale assemblage, small groups of 20–50 containing a high proportion of young, found mostly in shallow coastal waters, and larger concentrations of as many as 400 adults, feeding in apparently rich areas farther offshore. From observations of whaling operations along that coast in 1936, Zenkovich (1937) reported areas that he suggested may be permanent feeding grounds for younger, apparently recently weaned animals. Almost all whales captured in these areas were less than 2 years old, from which Zenkovich (1937) concluded that young gray whales form separate schools at weaning. Bogoslovskaya *et al.* (1981) suggested that such separation does exist but may not always be so well defined.

The "divisions" described above may equate with the "pockets" along the North American coast. The Vancouver Island summer population does include young animals, apparently recently weaned (Darling 1978). From individual sighting records reported in this chapter, others proved to be at least 8 years old. A more complete investigation of the age structure of this population is necessary; one tempting hypothesis is that the Vancouver Island summer assemblage is similar in age structure to the inshore "divisions" of predominantly young animals off the Asian coast.

The present observations suggest that a gray whale has a preferred location, extending over at least 150 km of coastline, to which it returns for at least part of its life. There are at least two plausible explanations for the variable turnover patterns of identified whales observed in the present study. The first is that they may result from incomplete and varying identification effort among years and mixing within a larger "northwest coast" population. The second assumes that the Vancouver Island whales are a distinct group and that the observed patterns are significant. Some animals return annually for a series of years (2, to at least 8) then go elsewhere, probably on full migration; others spend only one summer in the area. To speculate, the former group might well be immature whales and the latter related to the younger animals or attracted to the area by chance. This explanation suggests a dynamic relationship with the rest of the herd and is tempting because it is compatible with the observations by Votrogov and Bogoslovskaya (1980) and Bogoslovskaya *et al.* (1981) of separate assemblages of younger animals; however, it is far from proven.

The predominant behavior of the Vancouver Island whales is feeding (Darling, 1977,

1978). Feeding behavior and food of gray whales is discussed for the entire range by Nerini (Chapter 18, this volume) and for Vancouver Island by Murison *et al.* (Chapter 19, this volume). Three nonfeeding behavior patterns of note were observed off Vancouver Island in the summer: traveling, apparent resting, and rubbing at a sand bar–riptide area (Darling, 1978). Traveling, discussed above, essentially involves movement among feeding and riptide areas. Whales apparently resting were observed lying low and motionless in the water with just the top of the back above the surface and with wavelets breaking over them; they blew at regular intervals.

At the seaward entrance of several passages to extensive inlet systems in the study area, a "riptide" occurs with each tide change over a sand bar. Whales in these areas engage in stereotyped behavior; rising head-first out of the water and "falling" over backwards, clapping the jaws, spitting water, and apparently rubbing on the sand bar (Darling, 1978). This is probably "grooming" or "recreational," perhaps similar to the rubbing behavior described for killer whales (*Orcinus orca*) off the east coast of Vancouver Island (Ford and Fisher, 1983).

Summary

This study examines the abundance, distribution, and behavior of gray whales, *Eschrichtius robustus*, which are summer residents along the west coast of Vancouver Island, British Columbia. The migratory timetable was determined by observations during four winters from 10 west coast lightstations. The study of the summer population was based on individual identification of animals from photographs of natural markings, and on repeat observations of "known" whales for 10 years, 1972–1981. Direct observations were made from small boats and aircraft.

The northern migration begins in mid-February, peaks in late March, and continues through early June. It is followed by a summer interval lasting until November when the first southbound migrants pass; the southern migration peaks in late December and continues to approximately mid-January. Each year, 35–50 whales are resident in this area off British Columbia during the summer interval. An average of 63% of these returned for more than one summer; 37% were seen in 1 year only. Some individuals returned for at least 8 years. Whales were found feeding in numerous locations along the coast from Victoria to Cape Scott. Individual whales were identified in locations up to 150 km apart in different years, and within one summer they traveled between feeding areas at least 77 km apart. Feeding is the predominant behavior, although whales were also observed traveling, apparently resting, and rubbing at a sand bar–riptide area.

Acknowledgments

This work was funded by the Vancouver Public Aquarium, Parks Canada, The National Boat Show, Chevron Canada Ltd., the National Marine Fisheries Service, Marine Mammal Laboratory, Seattle, and research grants to Dr. J. McInerney, University of Victoria.

The support and assistance of many people are gratefully acknowledged. Special thanks are due to J. E. McInerney, R. S. Payne, M. Bigg, M. Newman, D. Hatler, S. Proctor, M. Tillman, K. Norris, F. Deloume, D. Arnet, D. Banks, B. Campbell, W. Campbell, A. Dorst, C. Dundon, C. Dunn, K. and M. Fiddy, J. and D. Ford, D. Gamble, T. Goodspeed, the late K. Hobson, M. Horn, B. McIntyre, W. McPetrie, B. Milne, A. Oliver, D. Palfrey, J. and L. Reeves, W. Stetski, G. Trenholm, The Vancouver Aquarium Volunteers; and lightstation keepers along the west coast of Vancouver Island including Captain Exley, J. W. Bruton, N. Evans, I. McNeil, K. Nuttal, A. D. and Pat Thomson, N. Brand, D. Weeden, A. Bablitz, R. Noble, B. Shulldes, S. G. Westhaven, K. Dryburgh, A. W. Holland, S. Anderson, M. D'Een, D. Kidder, S. Schivers, T. Carr, K. D. Fraser, R. H. W. Collins, R. Mae, F. Bergthorson, F. Britton, O. Edwards, I. G. Swaney, and M. E. Slater.

Thanks also go to Dr. M. Bigg, I. B. MacAskie, and Dr. K. Norris for reading and making suggestions on rough drafts of the manuscript. The initial part of this study was conducted as part of a M.Sc. program at the University of Victoria in 1975–1976.

References

Bigg, M. A. (1982). An assessment of killer whales (*Orcinus Orca*) stocks of Vancouver Island, B. C. *Rep. Int. Whal. Commn.* **32,** 655–663.

Bigg, M. A., MacAskie, I. B., and Ellis, G. (1976). "Abundance and Movement of Killer Whales off Eastern and Southern Vancouver Island with Comments on Management," Prelim. Rep. Available from Pacific Biological Station, Nanaimo, B.C. (unpublished).

Bogoslovskaya, L. S., Votgogov, L. M., and Semenova, T. N. (1981). Feeding habits of the gray whale off Chutotka. *Rep. Int. Whal. Commn.* **31,** 507–510.

Bryant, P. J., and Lafferty, C. M. (1980). The gray whales of Laguna Guerrero Negro. *Whalewatcher* **14**(4), 3–5.

Darling, J. (1977). The Vancouver Island gray whales. *Waters. J. Vancouver Public Aquarium* **2,** 5–19.

Darling, J. (1978). Aspects of the behavior and ecology of the Vancouver Island gray whales, *Erchrichtius glaucus*, Cope. M.Sc. Thesis, Department of Biology, University of Victoria, Victoria, British Columbia.

Darling, J. (1979). "Summer Abundance and Distribution of Gray Whales along the Washington and B.C. Coast," Prog. Rep. Available from National Marine Mammal Laboratory, Seattle, Washington.

Dohl, T. P., Duman, M. L., and Helm, R. C. (1981). Distribution and movement of migratory and resident gray whales along the California coastline. *4th Bienn. Conf. Biol. Mar. Mamm., 1981* Abstract, p. 25.

Ford, J. K. B., and Fisher, D. (1983). Group specific dialects in the killer whale (*Orcinus orca*). *In* "Communication and Behavior of Whales" (R. Payne, ed.), AAAS Symp. Ser. Westview Press, Boulder, Colorado (in press).

Gilmore, R. M. (1960). Census and migration of the California gray whale. *Nor. Hvalfangst-Tid.* **49,** 409–431.

Hall, J. D., Harrison, C. S., Nelson, J., and Taber, A. (1977). The migration of California Gray Whales into Bering Sea. *Proc. Conf. Biol. Conserv. Mar. Mamm., 2nd, 1977* Abstract, p. 8.

Hatler, D. F., and Darling, J. D. (1974). Recent observations of the gray whale in British Columbia. *Can. Field Nat.* **88,** 449–459.

Katona, S., Baxter, B., Brazier, O., Kraus, S., Perkins, J., and Whitehead, H. (1979). Identification of humpback whales by fluke photographs. *In* "Natural History of Cetaceans" (H. E. Winn and B. L. Olla, eds.), Vol. 3, p. 334. Plenum, New York.

Leatherwood, J. S. (1974). Aerial observations of migrating gray whales, *Eschrichtius robustus*, off southern California, 1969–72. MFR Paper 1055. *Mar. Fish. Rev.* **36**(4), 45–49.

Payne, R., Brazier, O., Dorsey, E., Perkins, J., Rowntree, V., and Titus, A. (1981). External features in southern right whales (*Eubalaena australis*) and their use in identifying individuals. *U.S. Dep. Comm. NTIS Publ.* **PB 81-161093,** 1–77.

Pike, G. C. (1962). Migration and feeding of the gray whale (*Eschrichtius gibbosus*). *J. Fish. Res. Board Can.* **19,** 815–838.

Pike, G. C., and MacAskie, I. B. (1969). Marine mammals of British Columbia. *Fish. Res. Board Can. Bull.* **171,** 1–54.

Reilly, S. B., Rice, D. W., and Wolman, A. A. (1980). Preliminary population estimate for the California Gray Whale based on Monterey Shore 1967–1968 to 1978–79. *Rep. Int. Whal. Commn.* **30,** SC/31/DOC 37.

Rice, D. W., and Wolman, A. A. (1971). The life history and ecology of the gray whale (*Eschrichtius robustus*). *Spec. Publ.—Am. Soc. Mammal.* **3,** 1–142.

Rugh, D. J., and Braham, H. M. (1979). California gray whale (*Eschrichtius robustus*) Fall migration through Unimak Pass, Alaska, 1977. A preliminary report. *Rep. Int. Whal. Commn.* **29,** SC/30 Doc 46.

Sauer, E. G. F. (1963). Courtship and copulation of the gray whale in the Bering Sea at St. Lawrence Island, Alaska. *Psychol. Forsch.* **37,** 157–174.

Scammon, C. M. (1869). On the cetaceans of the western coast of North America *Proc. Acad. Nat. Sci. Philadelphia* **21,** 40–49.

Swartz, S. L., and Jones, M. L. (1978). The evaluation of human activities on gray whales, *Eschrichtius robustus*, in Laguna San Ignacio, Baja California, Mexico. *U.S., NTIS, PB Rep.* **PB 289737,** 1–34.

Swartz, S. L., and Jones, M. L. (1981). Demographic studies and habitat assessment of gray whales (*Eschrichtius robustus*) in Laguna San Ignacio Baja California, Mexico. *U.S., NTIS PB Rep.* **PB 82-123373,** 1–56.

Votrogov, L. M., and Bogoslovskaya, L. S. (1980). Gray whales off the Chukotka Peninsula. *Rep. Int. Whal. Commn.* **3O,** 435–437.

Wilke, F., and Fiscus, C. H. (1961). Gray whale observations. *J. Mammal.* **42**(1), 108–109.

Zenkovich, B. A. (1937). More on the gray California whale (*Rhachianectes glaucus*, Cope 1864). *Bull. Far East. Branch Acad. Sci. USSR* No. 23, pp. 1–19.

13

Gray Whale Migrations along the Oregon Coast, 1978–1981

Denise L. Herzing and Bruce R. Mate

Introduction

Studies of gray whale (*Eschrichtius robustus*) movements along the eastern Pacific coast of North America previously have emphasized the fall (southward) migration, primarily to develop abundance estimates. No previous studies have monitored fall and spring migrations continuously from the same location. This chapter reports a 3-year study (1978–1981) designed to document migratory timing, pod-size frequency, and distribution relative to shore of gray whales during their fall and spring migration along the Oregon coast. Estimates of abundance were made for comparison with those from other studies.

ISBN 0-12-389180-9

Methods

STUDY AREA

Gray whales were observed from November through May, 1978–1981, from the Yaquina Head Lighthouse, located on a rocky headland 4.8 km north of Newport, Oregon (44°41′N, 124°05′W) (Fig. 1). From the observation platform 50 m above sea level, visibility extends 35 km to the horizon. Prevailing winds are from the southwest in the fall and winter and from the northwest in spring and summer. The 10- and 20-fathom contours are parallel to shore at distances of 0.8 and 2.4 km, respectively, from Yaquina Head.

FIELD STUDY

During the 1978–1979 pilot study, five observers counted whales on 3 predetermined days each week from late November to the end of May. These counts were used

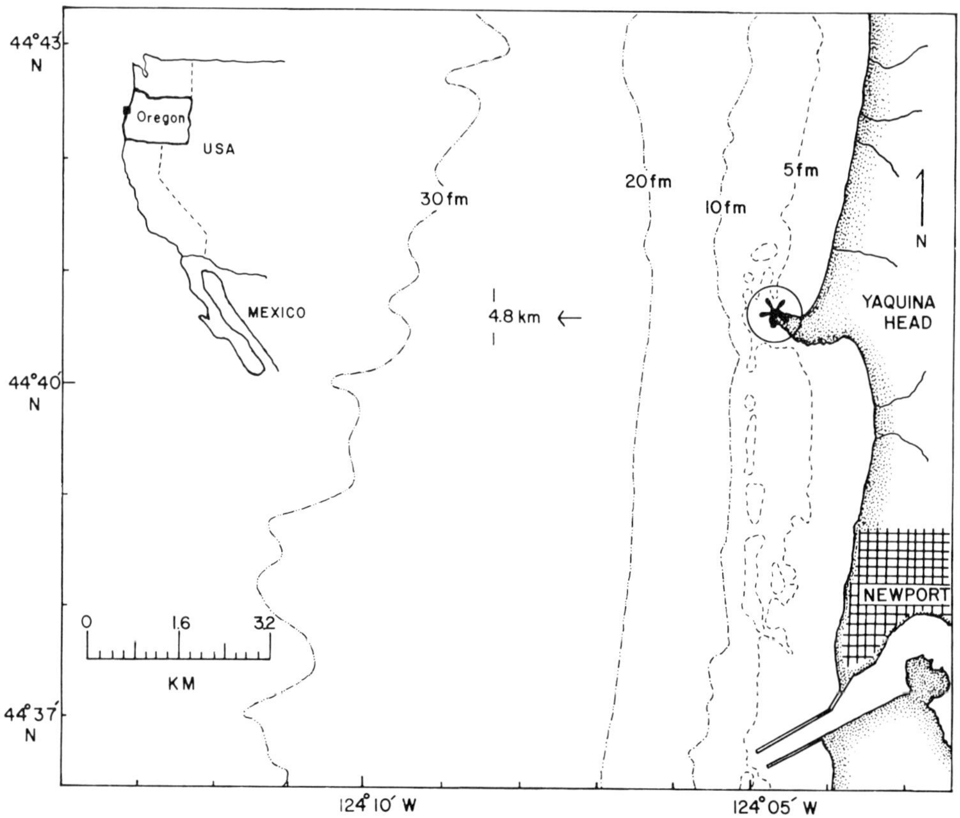

Fig. 1. *Location of study site at Yaquina Head, Oregon.*

to determine migratory timing and to develop methods suitable for the more complete studies which followed in 1979–1980 and 1980–1981. To eliminate interobserver bias, D. Herzing was the sole observer in 1979–1980 and 1980–1981.

Observations were made during five 0.5-hr periods alternated with 0.5-hr rest periods throughout morning hours, generally 0800–1300. Observers recorded the following: (1) viewing conditions, which were monitored by subjectively estimating Beaufort sea state; (2) pod size, which was estimated by continuously observing whales as they entered and passed through the study area, thus allowing multiple surfacing sequences to be observed; (3) distances of whales from shore for each sighting (in 1979–1980 and 1980–1981 only), which were calculated from vertical angles to whales as they passed directly off the lighthouse perpendicular to the coastline (measured with a clinometer) and from the known elevation of the observation platform. Distance-to-whale sightings were grouped as occurring 0–1.6 km, 1.6–3.2 km, and 3.2–4.8 km from shore. We arbitrarily limited our observations to within 4.8 km.

To determine the relative distribution of gray whales migrating within and farther than 4.8 km from shore, in January 1981 16 aerial surveys were flown perpendicular to the coast, extending to 32 km offshore. Flying at 600-m elevation in a Cessna 182, two observers, one on each side of the aircraft counted gray whales within 1.6 km of the aircraft trackline. The strip was defined by strut-mounted streamers. Distance from shore was calculated from flying time and airspeed. Flights were made on calm weather days when winds were negligible. Counts were made on both inbound and outbound transects. Inbound transects were moved a sufficient distance northward of outbound transects to avoid duplicate sightings. As we were interested only in the relative distribution of whales from shore, uniform sightability throughout the transect width was not necessary. We recognized that large groups may be easier to see than are individuals and thus the technique may be biased if group size is correlated with distance from shore.

ANALYSIS

Daily estimates of the number of whales passing through the study area were made from shore observations when the sea state was recorded as Beaufort 3 or less (winds $<$10 knots, occasional white caps) and visibility extended to a minimum of 4.8 km offshore (as determined by clinometer). Such weather conditions are well within Reilly's (1981) categories of "excellent to fair" visibility, under which no significant interference with shore censusing of gray whales from the weather was demonstrated; therefore data in the Oregon study were uncorrected for weather-related visibility categories.

Daily estimates of the numbers of animals passing the study site for 1979–1980 and 1980–1981 were calculated by multiplying the average number of whales observed per hour by 24 hr. Because Reilly (1981) found no significant difference between mean hourly migration rates or between morning versus afternoon migration rates and since no conclusive data exist that demonstrate a significant difference between day and night rates, all hourly rates in the Oregon study were assumed to be equivalent throughout a 24-hr period. Estimates were interpolated for unobserved days by following the linear trend of the migration immediately before and after the unobserved period. All daily estimates were summed to produce an estimate of the total number of whales passing through the study area. Ninety-five percent confidence limits were calculated by

$$\bar{P} \pm t \frac{S\bar{p}}{\sqrt{n}}$$

where $\bar{P}$ = mean estimate, $S\bar{p}$ = SD of the mean, n = sample size, t = Student's T value for $\alpha = .05$ [following Sokal and Rohlf (1969)].

During the pilot study in 1978–1979, estimates of whales passing per hour were averaged weekly and interpolated as above. However, because of observer variation and less accurate methods of calculating distance, data from the pilot study were not used to produce population estimates but only to compare the timing of migratory phases in subsequent years.

For testing purposes we divided the migration into southward and northward, and subdivided northward into that portion without calves (Phase A) and that with calves (Phase B). A G test of independence was used to compare pod size and distribution relative to shore between (1) the southward and Phase A migration, (2) the early and late Phase A migration.

Results

MIGRATORY TIMING

Gray whales migrated along the Oregon coast in three discernible phases (Fig. 2a,b,c): the fall/southward migration which occurred from early December to mid-February; the first portion of the spring/northward migration when calves were not present, which occurred from mid-February through April and is referred to as northward/Phase A; and the second portion of the spring/northward migration, composed primarily of cows with calves, which occurred from late April through May and is referred to as northward/Phase B or the cow/calf migration. The migration timing and peak migration rates for three annual migration cycles are shown in Table I.

Southward

The first southbound migrants were observed in early December. Maximum numbers of whales passed during the first week in January (maximum daily averaged rate was 29 whales/hr) with 90% of the animals passing between December 19 and January 23. Rates gradually declined to zero by mid-February, after which no further whales were observed until the onset of the northward migration.

Northward/Phase A

The northward migration began during the last week in February, and the rate of northbound migrants during Phase A peaked during mid-March (e.g., to a maximum daily average of 14 whales/hr in 1980–1981). After that, rates gradually declined to zero

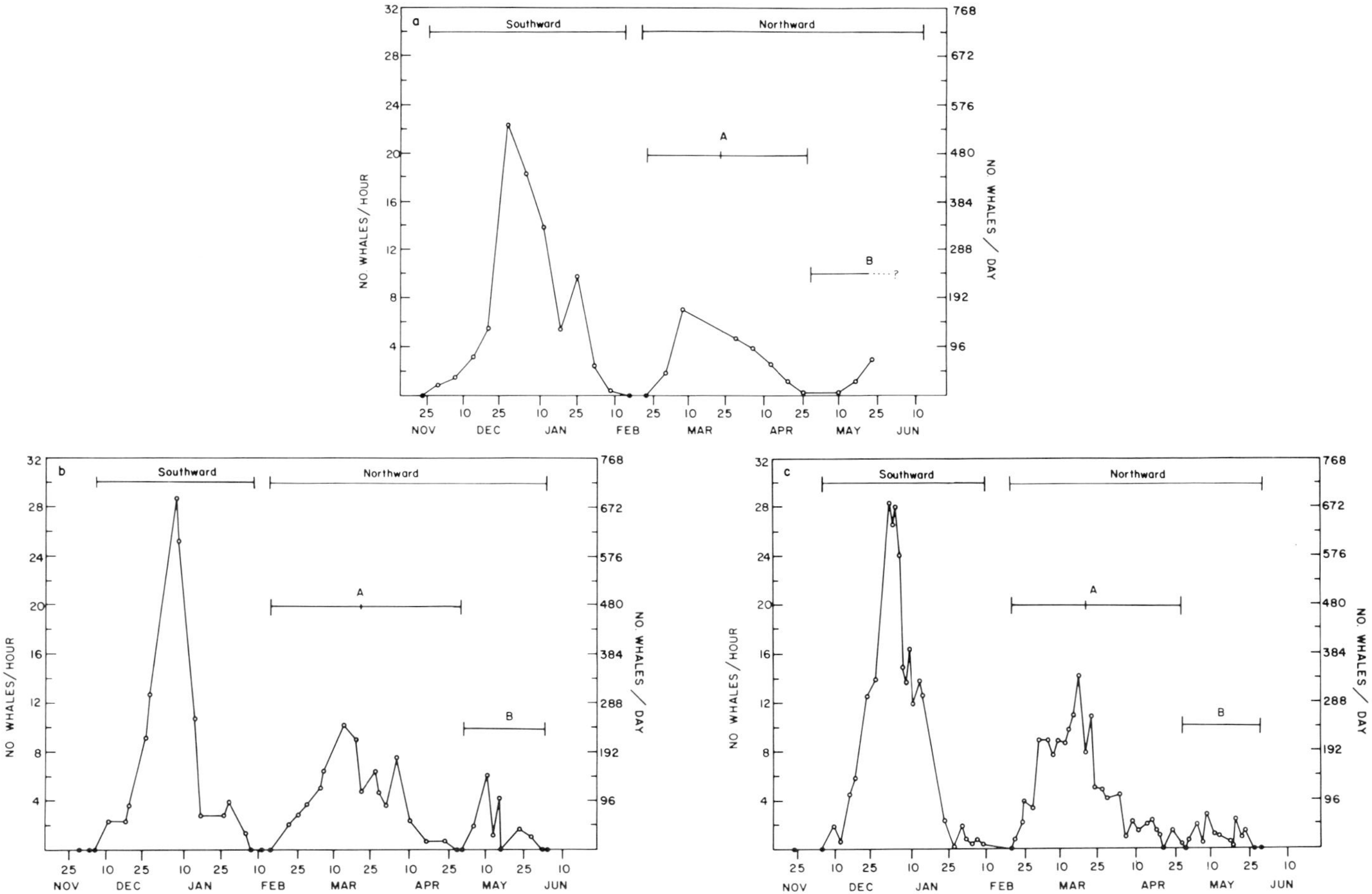

Fig. 2. *Temporal distribution of migrating gray whales during southward and northward/Phase A and B migrations, 1978–1981. Points on the graph denote weekly averages during 1978–1979 (a) and actual observation days during 1979–1980 (b), and 1980–1981 (c).*

Table I
Migration Timing and Peak Migration Rates of Whales Migrating past Yaquina Head during 1978–1981

	Southward			Northward/Phase A			Northward/Phase B		
	1978–1979	1979–1980	1980–1981	1978–1979	1979–1980	1980–1981	1978–1979	1979–1980	1980–1981
First migrant	Dec. 2	Dec. 11	Dec. 10	Feb. 27	Feb. 21	Feb. 22	May 10	May 6	Apr. 24
Last migrant	Feb. 1	Feb. 4	Feb. 9	Apr. 26	Apr. 24	Apr. 20	—	—	—
Peak	Dec. 28	Jan. 6	Jan. 1	Mar. 8	Mar. 15	Mar. 17	May 24	May 11	May 8
Whale/hr rate	22	29	29	8	10	14	3	6	3
First cow/calf pair	—	—	—	Mar. 29	Mar. 28	Apr. 4	May 10	May 6	Apr. 24

by the end of April. These sightings included a single cow/calf pair in late March or early April of each year.

Northward/Phase B

Cows with calves began migrating past the study area in late April. During Phase B the rate of whale passage increased in early May (e.g., to a maximum daily averaged rate of 6 whales/hr in 1979–1980) and this portion of the migration was composed principally of cows with calves (73% in 1979–1980 and 79% in 1980–1981). Single animals either traveling alone or in the presence of a cow/calf pair were observed through mid-May, after which 100% (of the whales observed) were cow/calf pairs.

UNCORRECTED ESTIMATES

Table II summarizes the estimates of the number of whales in each migratory phase for 1979–1980 and 1980–1981. We estimated that in the 1979–1980 annual migration, 11,962 whales passed southward (based on 253 sightings), 7,340 passed northward during Phase A (159 sightings), and 1,464 passed northward during Phase B (57 sightings). In the 1980–1981 annual migration, an estimated 13,627 whales passed southward (653 sightings), 8,020 passed northward during Phase A (333 sightings), and 945 passed northward during Phase B (76 sightings). If the observed percentages of cow/calf pairs (73 and 79%) are applied to the total estimated number of whales passing during northward/Phase B (1,464 and 945), an estimated 534 calves passed Yaquina Head from late April through May during 1979–80, and 373 during the same period in 1980–1981.

Table II
Number of Whales Observed, Estimated Number of Whales Passing Yaquina Head, and Cow/Calf Composition, 1979–1981[a]

	Southward		Northward/Phase A		Northward/Phase B	
	1979–1980	1980–1981	1979–1980	1980–1981	1979–1980	1980–1981
Number of whales observed	253	653	159	333	57	76
Estimated number of whales within 4.8 km	11,962	13,627	7,340	8,020	1,464	945
95% upper confidence limit	18,274	18,571	12,657	12,264	3,427	2,179
95% lower confidence limit	5,550	8,583	2,023	3,776	—	—
Percentage of cow/calf pairs	—	—	—	—	73	79
Estimated number of calves	—	—	—	—	534	373

[a]All dashes mean not applicable.

The estimates given above were not considered total population estimates because they did not include: (1) estimates of the number of whales unobserved because they passed prior to or after the study; (2) estimates of the percentage of whales that traveled outside the 4.8-km observation area; and (3) corrections for probable errors in the estimated pod size.

POD SIZE

As the direction of migration changed from south to north, a general trend of decreasing pod size was observed (Fig. 3a,b). During this transition in the 1979–1980

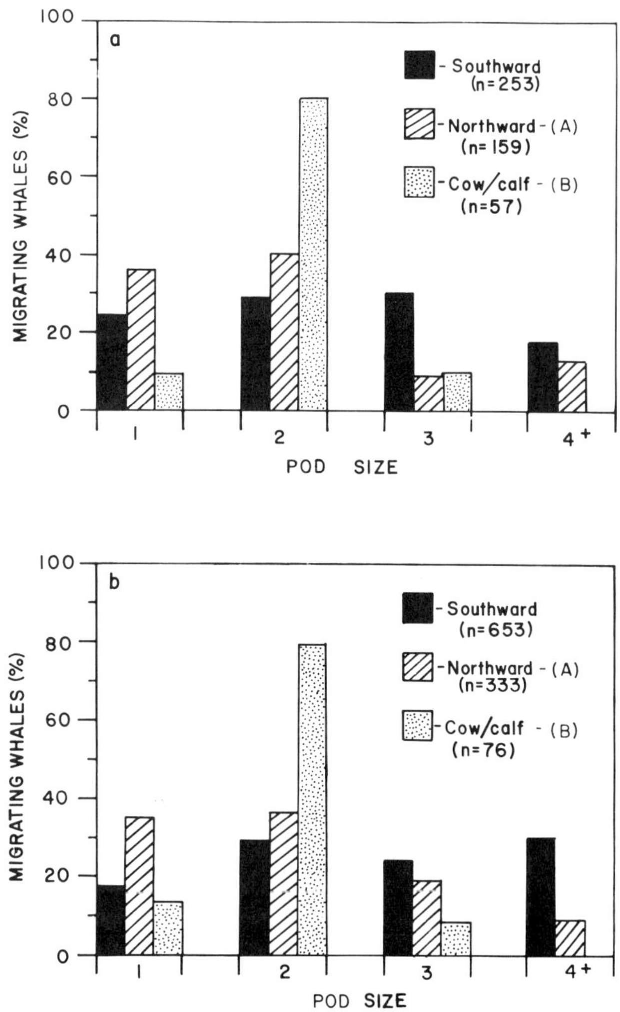

Fig. 3. *Distribution of pod sizes passing Yaquina Head in the three migration phases, (a) 1979–1980, (b) 1980–1981.*

Table III
Percentages of Whales in Small (Singles and Pairs) and Large (Three or More) Groups During Each Migratory Phase, 1979–1980/1980–1981

Pod size	Southward	Northward/Phase A	Northward/Phase B
Small (1–2)	52/47	77/71	90/92
Large (3+)	48/53	23/29	10/8

season there was a significant (α = .01, G = 32.070) increase in the percentage of whales recorded as singles or pairs, from 52% (southward) to 77% (northward/Phase A), and a decrease in the percentage of whales observed in pods of three or larger, from 48% (southward) to 23% (northward/Phase A) (Table III). Further analysis between early and late Phase A revealed that small pods (1–2) were found in a significantly (α = .01, G = 31.956) greater percentage during late Phase A. Thus the change in the late portion of the Phase A migration largely accounts for the observed changes in pod size between the southward and northward/Phase A migrations. Similar trends were observed during the 1980–1981 season. By the time the cow/calf migration (Phase B) had begun in 1980 and 1981, more than 90% of all observed whales were recorded as single animals or pairs.

During the southward migration, 80% of pods composed of four or more whales migrated past Yaquina Head in mid-season (late December to early January), while singles, pairs, and trios were observed regularly from the beginning of December through the end of January. During the northward/Phase A migration, 90% of pods of four or more whales migrated in the beginning of the season (late February to mid-March), and smaller pod sizes were observed throughout the northward movement. At the onset of the cow/calf migration, singles, pairs, cow/calf pairs, and trios (always including a calf) were observed, but by mid-May only cow/calf pairs were observed.

DISTRIBUTION RELATIVE TO SHORE

Over 50% of all whales observed during both the southward and northward/Phase A migrations traveled between 1.6 and 3.2 km offshore (Fig. 4a,b). However, a general trend of decreasing distance of whales from shore was observed as the migration direction changed from south to north. For instance, during this transition in the 1979–1980 season there was a significant (α = .01, G 22.826) increase in the percentage of whales observed less than 1.6 km from shore, from 20% (southward) to 40% (northward/Phase A), and a decrease in the percentage of whales observed further than 3.2 km from shore, from 20% (southward) to 9% (northward/Phase A). Furthermore, analysis between early and late Phase A revealed that a significantly (α = .01, G = 39.652) greater percentage of whales traveled closer to shore in the late than in the early Phase A, which probably accounts for most of the observed difference between the southward and northward/Phase A migrations. Similar trends were observed during the 1980–1981 season. During the cow/calf (Phase B) migration, over 97% of all whales observed traveled within 0.8 km of shore during both years, well within the 1.6-km category.

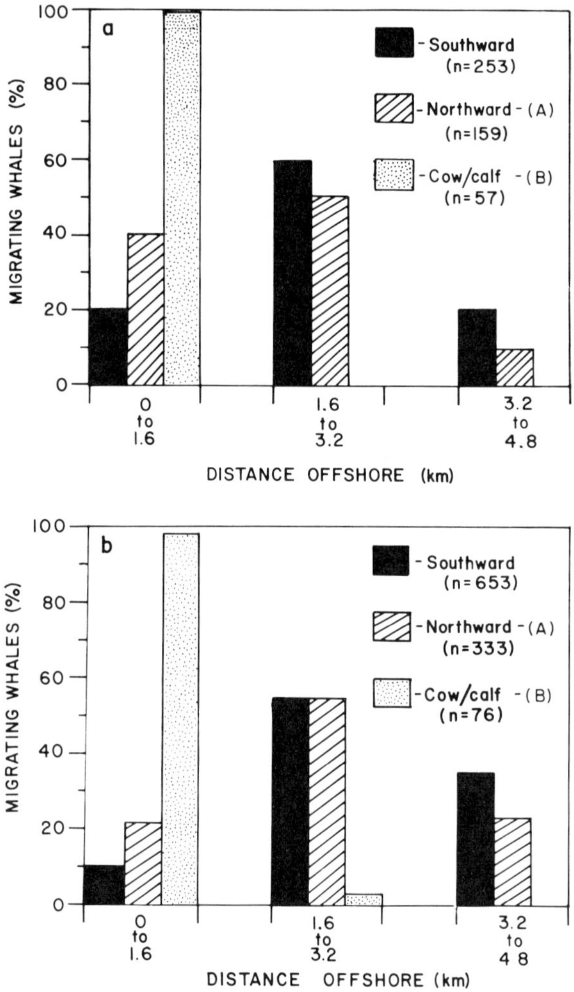

Fig. 4. *Distribution relative to shore of whales in the three migration phases. (a) 1979–1980. (b) 1980–1981.*

POD SIZE VERSUS DISTRIBUTION RELATIVE TO SHORE

Except for the Phase B migration, the majority of all sightings for all pod sizes were made in the 1.6- to 3.2-km region (Fig. 5a,b). During the combined southward and northward/Phase A migrations there was a significantly (α = .01, G = 33.702) smaller percentage of pods of four or more whales observed within 1.6 km than observed between 3.2 and 4.8 km (e.g., 8 versus 30% in 1979–1980). Further analysis between early and late Phase A revealed that a significantly (α = .01, G = 27.146) greater percentage of (1) small pods were observed closer to shore during late Phase A than during early Phase A, and (2) large pods were observed father from shore during early Phase A than during late Phase A. These changes within Phase A reemphasize the

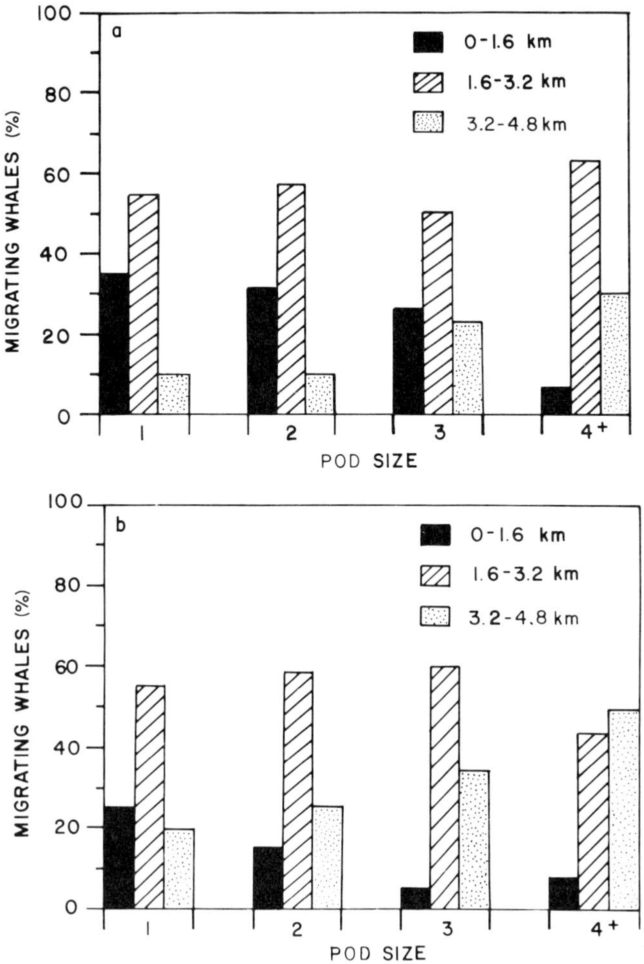

Fig. 5. *Distribution relative to shore of pod sizes for the combined southward and northward/Phase A migrations, (a) 1979–1980, (b) 1980–1981.*

apparent interrelationship of pod size, distribution relative to shore, and migration timing. Similar trends were observed in the 1980–1981 season.

AERIAL SURVEYS

The pod size and distribution relative to shore of migrating gray whales as determined during 16 aerial transects and shore surveys on January 1, 2, and 4, 1981 are summarized in Table IV. Of the 87 whales observed during the aerial transects, 34 (39%) were beyond the 4.8-km boundary, 23 of which were singles or pairs. During aerial surveys to 32 km offshore, whales were seen as far as 20 km from shore. Shore-based observations revealed that 80% of all pods of 4 or more whales that migrated past the

Table IV
Numbers of Whales Observed in Different Pod Sizes and Distance Categories for the Combined Dates of January 1, 2, and 4, 1981, During Aerial/Shore Surveys[a]

Distance (km)	Singles	Pairs	Trios	4+	Percentage within 4.8 km
0–1.6	2/3	0/4	0/3	0/10	3/9
1.6–3.2	8/14	4/20	3/33	0/39	29/45
3.2–4.8	9/3	2/26	15/18	10/64	68/46
4.8–32.0	11/0	12/0	3/0	8/0	—/—[b]

[a]Aerial, $n = 87$; shore, $n = 237$.
[b]Dash (—) means not applicable.

study site did so between December 28, 1980 and January 5, 1981; as viewed from shore, large pods tended to travel farther offshore (Fig. 6). Throughout the entire southward migration an average of 36% of all whales were observed from shore in the 3.2- to 4.8-km region. However, during the period of the aerial surveys, shore-based observations indicated 46% of the whales observed were in the 3.2- to 4.8-km region, and aerial survey data restricted to 4.8 km revealed 68% of the whales in the same region. Problems of insufficient sample size and sampling techniques precluded statistical comparison of these differences by specific pod size and distance categories. However, a comparison of aerial and shore-based counts for all whales observed in each distance region of the 4.8-km study area is made in the Discussion and Conclusions section.

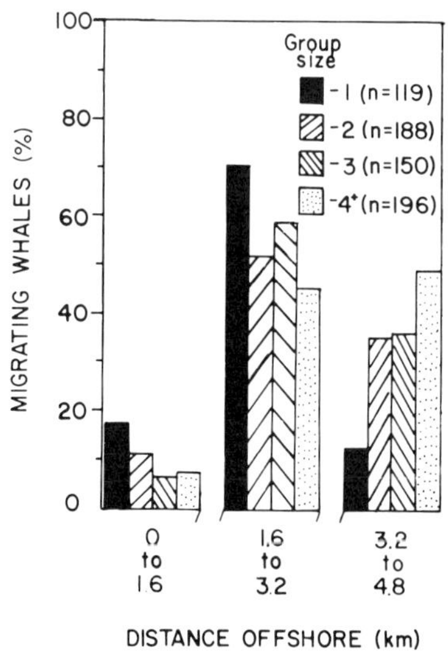

Fig. 6. *Distribution of pod sizes in each distance category during the southward migration in 1980–1981.*

Discussion and Conclusions

MIGRATORY TIMING

Southward

Southward migration along the Oregon coast was similar in duration but approximately 1 month later (early January) than the southward migration at Unimak Pass, Alaska (late November/early December). The southward migration lasted 2 months in both Oregon and Alaska. Ninety per cent of the observed gray whales migrated past Oregon between December 19 and January 23. Rugh (Chapter 10, this volume) reported 90% of the observed whales in Alaska passed between November 19 and December 23. Peak migration dates along Oregon occurred approximately 1 week prior to the average peak passage date (January 9) off Monterey, California (Reilly, 1981).

Observed peak passage dates varied slightly for the 3 years studied along the Oregon coast. For each consecutive year (1977, 1978, 1979) the peak passage date was later, both at Unimak Pass, Alaska and at Monterey, California (Chapter 10, this volume). Rugh suggests such variation may correspond to changes in environmental features such as ice development in northern feeding grounds.

The maximum passage rate of 52 whales/hr observed at Unimak Pass, Alaska (Chapter 10, this volume) was higher than the 29 whales/hr observed along the Oregon coast and the 30 whales/hr reported by Reilly (1981) for Monterey. As whales travel south, they may change rates of travel relative to sex, age, and reproductive status, or they may travel farther offshore (out of view) as suggested by our aerial surveys along the Oregon coast, thus accounting for lower passage rates from shore observations at lower latitudes.

Northward/Phase A

In 1981, peak passage of the Phase A migration occurred along the Oregon coast in mid-March, 2 weeks later than Poole's (Chapter 16, this volume) observations of peak passage in late February off Point Piedras Blancas, California. Rugh (Chapter 10, this volume) reported no whales observed northbound through Unimak Pass, Alaska between January 2 and March 22, 1980. Hall *et al.* (1977) surveyed northbound gray whales between April 7 and May 26, with only a few cow/calf pairs sighted. Recent studies at Unimak Pass in 1981 indicated peak passage of the Phase A migration to be April 27 (Hessing, 1981), which would indicate an approximate lag of 4–5 weeks between the beginning of the northward/Phase A in Oregon and in Unimak Pass, Alaska.

Unlike the rapid increase in whale-passage rate observed during the southward migration, the northward/Phase A migration was characterized by a gradual increase to 14 whales/hr in mid-March and a gradual decrease to 0 whales/hr by late April. Gradual changes in Phase A rates were also observed along central California with a higher observed daily passage rate of 18.5 whales/hr (Chapter 16, this volume). Poole also reported his first sighting of a cow/calf pair at the end of March in 1980 and in early April in 1981, the same as observed in Oregon.

Northward/Phase B

The peak passage rate during the cow/calf migration was lower and therefore less conspicuous along Oregon in 1981 than in 1980. In 1980, this peak rate occurred 2 weeks later (mid-May) than the late April peak at Point Piedras Blancas, California (Chapter 16, this volume). Hessing (1981) reported a peak in cow/calf migration at Unimak Pass, Alaska on June 4, approximately 4 weeks following the cow/calf peak along the Oregon coast. Swartz and Jones (1980a) reported two "departure waves" of gray whales from San Ignacio Lagoon in Baja, California: (1) Whales without calves decreased after mid-February and declined to zero by mid-March, and (2) cow/calf pairs declined after an early April peak. If these two waves of departure from the lagoon are representative of departure of whales from waters outside the lagoon as well, they may correspond to Phases A and B. However, as these dates are only 1 month before those phases peak along the Oregon coast, it seems unlikely.

The most recent attempts to thoroughly census the major calving and breeding lagoons in Baja (Rice *et al.*, 1981) produced estimates well below half the estimated southbound population. Thus it appears most whales are outside the lagoon at any particular time. Some may enter a lagoon and soon leave, as indicated by some the short residency times found by Harvey and Mate (Chapter 24, this volume). Thus departure dates from the lagoons may not be representative of the beginning of northward movement for the population in general.

Some whales may never get far enough south to enter a lagoon before they turn around again to head north. Mate and Harvey (Chapter 25, this volume) also reported a single whale, radio-tagged in San Ignacio Lagoon in late February, which passed Oregon in early May towards the end of Phase A (2 months enroute). Late Phase A along the Oregon coast may include whales from the initial departure wave reported from San Ignacio Lagoon by Swartz and Jones (1980a), indicating a 5- to 6-week migration period between Mexican lagoons and the central Oregon coast.

POD SIZE AND DISTRIBUTION RELATIVE TO SHORE

Southward

The changes in pod size observed throughout the southward migration along the Oregon coast may be explained by the temporal segregation of different age/sex and reproductive classes as reported by Rice and Wolman (1971): pregnant females migrate first, followed by nonpregnant females, adult males, and immatures. They also reported that single whales passed early, large groups passed mid-season, and pod size decreased through the end of the migration. Rugh (Chapter 10, this volume) reported temporal–size segregation at Unimak Pass with large whales (presumably pregnant females with greater motivation for leaving northern feeding areas than other age/sex classes) migrating earlier, and small whales migrating later.

Reilly (1981) reported that pod size was related to distance from shore with larger pods occurring farther from shore. He further identified the middle of the southward migration (December 30 to January 18) as having more animals farther from shore, corresponding to the middle part of the Oregon southward migration when the majority of the large pods traveled farther offshore. Shore-based observations along Oregon concluded that large pods travel more frequently in the 3.2- to 4.8-km region than do small pods. However, aerial surveys revealed numerous single and paired whales beyond 4.8 km and a higher percentage of whales within the 3.2- to 4.8-km region than did shore-based observations. This established that smaller pod sizes pass the study site unobserved from shore in the 3.2- to 4.8-km region as well as outside the 4.8-km viewing area. With a reduced probability of sighting small pods with increasing distance, then the conclusion from shore-based observations that large pods are the predominant pod size in that region may be incorrect.

Migration through water less than 90 m in depth was characteristic along Oregon as well as at other study sites. The maximum distance gray whales traveled from shore along the Oregon coast was 20 km (9O-m water depth); but the majority of whales were observed between 1.6 and 3.2 km (40–60 m). Although Unimak Pass is from 18.5 to 48.2 km wide, Rugh (Chapter 10, this volume) reported that all gray whales observed from shore and by plane were within 3.7 km of Unimak Island in water less than 75 m deep; 79% of these traveled within 1 km (20-m depth). Ninety per cent of the whales off Yankee Point, California were observed within 1.6 km of shore (Sund and O'Connor, 1974), well within the 90-m depth contour.

Northward/Phase A

Changes in pod size observed within Phase A also correspond to trends reported by Rice and Wolman (1971) during the northbound migration along the California coast, where large groups of mature whales migrated early and smaller pods of immature whales migrated later. Poole (Chapter 16, this volume) also reported smaller whales nearer to shore throughout this period. During Phase A, whales migrating along the central California coast moved closer to shore in the later half of the migration (Chapter 16, this volume), corresponding to late Phase A along Oregon during which a significant decrease in pod size and the distance whales migrated from shore was observed. This period corresponds with immature whale migration and suggests that smaller animals show a strong preference for migrating close to shore.

Northward/Phase B

The cow/calf migration was characterized by migration extremely close to shore. Poole (Chapter 16, this volume) observed that over 90% of Phase B whales along central California traveled within 200 m of shore, similar to the observations along the Oregon coast where 90% were found within 800 m. Hessing (1981) reported that after May 21, no whales were seen farther offshore than 800 m and within the last 3 weeks of migration 90% of all whales traveled within 100 m of shore. Rice and Wolman (197i) did not see cow/calf pairs during their shipboard observations off central California through April and postulated that these animals migrated farther offshore than the rest of the migration. This study and others cited in this chapter reveal that cow/calf pairs migrate later than the majority of the population and display a strong tendency to travel extremely close to shore. Rice and Wolman's surveys may have been too early in the season or too far offshore to discern this migratory phase.

POPULATION SIZE

An estimated 11,962 and 13,627 whales passed through the Yaquina Head study area during the southward migration in 1979–1980 and 1980–1981, respectively. These shore estimates were (1) not corrected for weather-related sightability, which was assumed equal during all observation days, based on Reilly's (1981) visibility correction data, (2) not corrected for portions of the migration before or after the observation study period (rather, it was assumed that the census began and terminated well before and after the majority of migrants had passed), and (3) not corrected for animals moving outside the 4.8-km outer boundary of the study area. We attempted a correction for the latter during the short period of December 28, 1980 to January 5, 1981, when aerial surveys revealed 39% of the whales were passing beyond 4.8 km. This resulted in an additional 1,835 whales being added to the estimated 4,705 whales passing through the study area during that time, increasing out initial estimate of 13,627 for 1980–1981 to 15,462. Assuming that our sightability bias for underestimating pod size (especially of single whales) at increasing distance is real, we could apply the corrections developed by Reilly (1981) for northern California to our shore-based data. Reilly found that 35% of the single animals were missed by shore counts, pairs and trios were accurately identified, and pods of four or larger were underestimated by 33%. Thus, if we applied these correction factors to the proportion of whales in each pod size throughout the entire southward migration, the result would be an increase in the total estimated number of whales passing Oregon from 15,462 to 17,671 whales. These corrected estimates are close to those for Alaska of 16,928 whales in 1978 (Chapter 10, this volume), to the central California estimates of 16,304 whales (Reilly *et al.*, 1980) and 15,640 (Poole, 1984) in 1981.

ESTIMATES OF CALF PRODUCTION

We estimated that 534 and 373 calves passed Oregon during Phase B in 1980 and 1981, respectively. Swartz and Jones (1983) have estimated the annual calf production at 6.8% of the population. When this figure is applied to our own best estimate of 15,462 (excluding Reilly's correction for sightability), calf production is estimated at 1,051 individuals. We therefore estimated that when we stopped surveying in late May that approximately 500–700 calves either had died, had moved past offshore beyond our view, or were still south of the study site. However, using the uncorrected estimate of 11,962 whales passing Yaquina Head, Oregon and 534 estimated calves from that population, a calf production rate of 4.7% is estimated in agreement with Poole's (1984) estimate of calf production of 4.7% off Pt. Piedras Blancas, California and Hessing's

(1981) estimate of a rate of 4.6% of calf production from Unimak Pass, Alaska. These calf production rates are much lower than the 13% that Rice and Wolman (1971) have estimated. Lower calf production rates may be due to a variety of factors.

Swartz and Jones (1983) estimated a 5.3% mortality rate for calves from the time of birth and initial lagoon occupation through departure from the lagoon. Following the departure of calves from the lagoons to a central point in California a second critical period is suggested. Poole (1984) estimated a total mortality rate of 25% for calves from their birth in the lagoon to their migration off central California. Such a high mortality seems unlikely in view of the extremely small number of dead calves found along California and Oregon beaches (Scientific Event Alert Network).[1] Poole (1984) discusses observations of a high number of dead calves found in a single spot along the northern Baja coast. Such high numbers of unreported calves may indicate that calves dying after leaving the lagoons in Mexico and washing up on shore in Baja would normally remain undetected and therefore not be considered in estimations of calf mortality rates. Twenty-one percent of Phase B migrants off the Oregon coast in 1981 were single whales either traveling alone or in the presence of a cow/calf pair. An unknown number of these may have been postpartum females that had lost their calves. Even if all Phase B single whales were postpartum females, it would account for only a maximum of 198 additional calves, a total of 571 calves, still only slightly more than half of the estimated production of 1,051 calves.

As a result of their extremely close proximity to shore and distinctive behavior, it is unlikely that large numbers of calves passed Yaquina Head undetected during the study period. No correction factors exist for whales outside the 4.8-km viewing area during Phase B, but the 3% found in the 1.6- to 3.2-km region in 1981 and the 0% found in the 3.2- to 4.8-km region in 1980 and 1981 suggests the number traveling outside 4.8 km would be very small. Poole (Chapter 16, this volume) observed all of the cow/calf pairs at Pt. Piedras Blancas, CA within 400 m of shore during aerial surveys, indicating that most, if not all of the cow/calf pairs travel very close to shore.

Another explanation for low calf numbers during this study may be that a significant proportion of cow/calf pairs had not yet reached the study area before termination of the study. Swartz and Jones (1980b) reported that females with calves may remain at the lagoon entrances through late May and early June. These late migrant cow/calf pairs and others still migrating along southern California would be undetected along central California and Oregon before late May. Many whales including cow/calf pairs may spend the entire summer along the Baja California and southern Oregon coasts and never migrate past the Oregon study site. Summering whales including cow/calf pairs have been noted along the Baja coast, (A. Aguayo, personal communication), the Oregon coast (J. Sumich, personal communication), and along Vancouver Island, British Columbia, Canada (Darling, 1977). Numbers of whales summering along the California coast are unknown. Using our uncorrected Oregon estimates of 8,804 northbound migrants in 1979–1980 and 8,965 in 1980–1981 as a comparison with 11,962 southbound

[1]Scientific Events Alert Network, National Museum of Natural History MRC 129, Smithsonian Institution, Washington, D.C.

migrants in 1979–1980 and 13,627 in 1980–1981, we estimate that 3,000 to 4,500 southbound migrants are unaccounted for on the return migration northward. These discrepancies and the low number of calves may be partially explained by the summering behavior of gray whales as well as higher mortality rates of calves during their northward migration.

Summary

To determine the temporal and spatial distribution and pod size of the California gray whale *Eschrichtius robustus* along the Oregon coast, whale movement was monitored from November through May during 1978–1981. Primary observations were made from shore, and supplemental observations were from aerial surveys. The migration occurred in three major phases: (1) southward migration (early December to mid-February) which peaked in early January at a maximum rate of 29 whales/hr; (2) northward/Phase A migration (mid-February through April) which did not include calves and peaked in mid-March at a maximum rate of 14 whales/hr; (3) northward/Phase B (cow/calf) migration (late April through May) which peaked in mid-May at a maximum rate of 6 whales/hr. Over 50% of all whales in the combined southward and northward/Phase A migrations passed by Yaquina Head between 1.6 and 3.2 km from shore. Pod size decreased and whales moved progressively closer to shore throughout the northward migration. The cow/calf migration was characterized by a high percentage of singles and pairs (90%) and travel within 1.6 km of shore (97%). In the 1980–1981 season, an estimated 15,462 whales migrated past Yaquina Head during the southward migration. Discrepancies in the estimated number of calves and total estimated number of whales passing south and north by Oregon may be attributable to late migrants, animals summering south of the study area, and higher mortality rates of calves during their northward migration.

Acknowledgments

Special thanks to Robin Brown, Jim Harvey, and Jim Sumich for their suggestions and field assistance, to Dr. Debbie Duffield and Dr. Bernd Würsig for reviewing this manuscript, to Nancy Brown for typing, to the U.S. Coast Guard for the use of Yaquina Head Lighthouse, and to the many friends and volunteers who contributed time and energy. We are grateful for the support of the Office of Naval Research (Grant no. 700014-79-C-004) and the Oregon State University Foundation.

References

Darling, J. D. (1977). Aspects of the behavior and ecology of Vancouver Island gray whales, *Eschrichtius glaucus* (Cope). M.A. Thesis, University of British Columbia, Vancouver, Canada.

Hall, J. D., Harrison, C. S., Nelson, J., and Taber, A. (1977). The migration of California gray whales into the Bering Sea. *Proc. Conf. Biol. Mar. Mammal., 2nd,* (Abstr.), p. 8.

Hessing, P. (1981). Gray whale, (*Eschrichtius robustus*), migration into the Bering Sea, Spring 1981. Final Report. NOAA/OMPA Contract: NA 8IRGA 00080.

Poole, M. M. (1984). Preliminary assessment of annual calf production of the California gray whale, *Eschrichtius robustus*, from Pt. Piedras Blancas, California. *In* "Cetacean Reproduction: Estimating Parameters for Stock Assessment and Management" (W. Perrin, R. D. Brownell, Jr., and D. DeMaster, eds.). La Jolla, California. (in press)

Reilly, S. B. (1981). Population assessment and population dynamics of the California gray whale (*Eschrichtius robustus*). Ph.D. Thesis, University of Washington, Seattle.

Reilly, S. B., Rice, D. W., and Wolman, A. A. (1980). Preliminary population estimates for the California gray whale based on Monterey shore censuses, 1967–68 to 1978–79. *Rep. Int. Whal. Commn.* **30,** 359–368.

Rice, D. W. and Wolman, A. A. (1971). The life history and ecology of the gray whale, *Eschrichtius robustus*. *Spec. Publ.—Am Soc. Mammal.* **3,** 1–142.

Rice, D. W., Wolman, A. A., Withrow, D. E., and Fleisher, L. A. (1981). Gray whales on the winter grounds in Baja California. *Rep. Int. Whal. Commn.* **31,** 477–489.

Sokal, R. R., and Rohlf, E. J. (1969). "Biometry: The Principles and Practices of Statistics in Biological Research." Freeman, San Francisco, California.

Sund, P. N., and O'Connor, J. L. (1974). Aerial observations of gray whales during 1973. *Mar. Fish. Rev.* **36**(4), 51–52.

Swartz, S. L., and Jones, M. L. (1980a). "Gray Whales, *Eschrichtius robustus*, during the 1977–1978 and 1978–1979 Winter Seasons in Laguna San Ignacio, Baja California Sur, Mexico." U.S. Dept of Commerce Report to the Marine Mammal Commission. NTIS PB80-202989.

Swartz, S. L., and Jones, M. L. (1980b). "Gray Whales, *Eschrichtius robustus*, during the 1980 Winter Season in Laguna San Ignacio, Baja California Sur, Mexico." Submitted to the July 1980 meeting of the International. Whaling Commission.

Swartz, S. L., and Jones, M. L. (1983). Gray whale calf mortality and production in the winter range. *Rep. Int. Whal. Commn.* **33,** 503–507.

14

Demography and Phenology of Gray Whales and Evaluation of Whale-Watching Activities in Laguna San Ignacio, Baja California Sur, Mexico

Mary Lou Jones and Steven L. Swartz

Introduction

As whale-watching tourism, local fishing, and industrial activities increased throughout the southern range of the gray whale in the 1970s, the government of Mexico expressed a desire for detailed information on gray whales in their winter range with which to formulate management policy. The U.S. government raised the concern that

ISBN 0-12-389180-9

visits to the gray whale breeding lagoons by U.S.-based excursion companies may create environmental disturbance detrimental to the whales and thus constitute "harassment" and a violation of the U.S. Marine Mammal Protection Act[1] and the Endangered Species Act (Reeves, 1977; Bean, 1980; U.S. Marine Mammal Commission, Washington, D.C., personal communication. Literature on gray whale breeding biology and on the possible effects of whale-watching activity (or human activities) on gray whales was extremely limited, mostly anecdotal, and not quantified. One line of reasoning suggested that whale-watching activities or the mere presence of humans within the breeding lagoons would displace the whales and cause them to seek alternate, perhaps suboptimal, areas in which to mate and calve (Rice and Wolman, 1971; Reeves, 1977). Gard (1974) stated that the gray whales may abandon important habitats and shift their distribution from lagoon to lagoon in response to human activities, but the results of his surveys were not sufficiently conclusive to demonstrate any shift. During the 1960s, gray whales apparently abandoned the small lagoon of Guerrero Negro while it was used as a port for salt barges, and then reoccupied the lagoon following the departure of the salt company and its associated vessel traffic (Reeves, 1977; Chapter 15, this volume). Several investigators suggested "that disturbance from whale-watching may draw needed energy from the finely tuned energy budget of these animals" (Kaza, 1980); however, there were no data to substantiate this point. Still others proposed that if whale-watching operations (and other human activities) resulted in general disturbance that interrupted mating behavior frequently enough to lower reproductive success and increase calf mortality and was not benign over the long term, a decrease in the whale population and its productivity would be expected (Swartz and Cummings, 1978). None of the proposed consequences of disturbance due to whale watching within the lagoons could be evaluated as there was a scarcity of information available on the sensitivity of the whales to human activites.

In full recognition of the paramount importance of the breeding lagoons to the reproductive success and continued recovery of gray whales and of the economic, educational, and aesthetic value of controlled visitation to wildlife areas, the government of Mexico enacted as protective measures a series of legal and administrative provisions that established Laguna Ojo de Liebre (Scammon's Lagoon) and Laguna Guerrero Negro (Black Warrior Lagoon) as national refuges for gray whales, the first of their kind for any cetacean (Vargas, 1981). Beginning in 1972, all commercial vessels, domestic and foreign, were required to obtain permits before entering Laguna Ojo de Liebre, and in 1974 they were restricted to one lower channel near the lagoon inlet (Kaza, 1981).[2] With the restrictions on entry to Laguna Ojo de Liebre, the focus of whale-watching tourism shifted to the largely undisturbed Laguna San Ignacio approximately 150 km further south.

Lagura San Ignacio, the third largest breeding area of gray whales, is the least developed of all the major lagoons. Local Mexican fishing from outboard-powered, 5- or

[1]United States Marine Mammal Protection Act 1972, 16 U.S.C. 1361 et seq. and the Endangered Species Act 1973, 16 U.S.C. 1531 et seq.

[2]Diario Oficial de la Federation de los Estados Unidos Mexicanos, 14 January 1972.

6-m "pongas" was the only commercial activity prior to 1975, when large-vessel traffic associated with whale-watching tourism began to visit the lagoon. Nearly 100% of the visitors to Laguna San Ignacio utilized vessels to enter the lagoon or to explore the interior, after arriving by aircraft or overland. Considering the dearth of information on gray whale abundance and distribution in this lagoon and because of the recent influx of tourism, we conducted a reconnaissance study in 1977 to assess the need for and feasibility of an intensive long-term research program in Laguna San Ignacio (U.S. Marine Mammal Commission Contract No. MM7AC008). This initial investigation indicated that the majority of females and their calves occupied the middle and upper lagoon "nursery" areas early in the winter when the calves were young. These areas were separated from the lower lagoon and inlet area by a constriction of the main entry channel at Punta Piedra. The width of this constriction (1.5 km) relative to the middle and upper lagoon suggested that human activities there, if adverse, could impede the movement of pregnant near-term females and females with newborn calves into and out of the nursery areas. Also, disturbance in the lower lagoon could disrupt courtship and mating activity of whales without calves and possibly reduce reproductive success. It was clear that knowledge of "normal" gray whale abundance, distribution, and behavior was a prerequisite for understanding the biological significance of responses to any human disturbance (Swartz and Cummings, 1978).

In 1978 we initiated the first long-term systematic study of gray whales in Laguna San Ignacio with the two-fold aim of providing baseline information on the demography and phenology of breeding gray whales and evaluating the impact of whale-watching activities within the lagoon on the whales. It was hoped that the information generated by this research, in conjunction with on-going monitoring programs, would be used to detect and evaluate changes in the wintertime abundance, distribution, and movements of gray whales that may be correlated with future human development.

Following the second year of this study, Mexico declared Laguna San Ignacio a refuge for gray whales and enacted regulations to control the number of whale-watching vessels that visited the lagoon and restrict their activities to specific areas within it. The intent of this action was to protect areas essential to the females and their calves (nursery areas), while at the same time permitting controlled commercial activity in the lower lagoon nearest the sea.

The Laguna San Ignacio research program continued for five consecutive winters, beginning in 1978 and concluding in 1982. We felt that 5 years was the minimum period required to develop a meaningful data base on gray whale trends and to determine whether frequent or continuous vessel disturbance caused changes in gray whale utilization of the lagoon.

The specific objectives of the Laguna San Ignacio research program were to

1. Document the seasonal timetable of lagoon occupation by whales
2. Determine the number and distribution of gray whales inhabiting the lagoon interior and inlet
3. Estimate the proportion of females with calves to whales without calves that utilize the lagoon throughout the season

4. Identify specific nursery areas and the abundance of calves in these areas
5. Ascertain the seasonal mortality of calves, immatures, and adult whales
6. Monitor the level of whale-watching tourism and other human activities and evaluate its impact on the abundance and distribution of whales wintering in the lagoon.

In this chapter we report new findings on gray whale demography and phenology in Laguna San Ignacio; we provide detailed data on patterns of movement into and out of the lagoon, compare changes in the timing of the whales' occupation of the lagoon among all years, evaluate trends in relative abundance across the 5-year study, and present information on the impact of whale-watching activity on the whales. These findings are also compared with studies of gray whales in other breeding lagoons within their winter assembly area. This research establishes a data base on which to evaluate changes in abundance and habitat utilization of gray whales wintering in Laguna San Ignacio which may occur as a result of increasing vessel traffic, industrial development, and petroleum and natural gas production throughout virtually the entire range of this cetacean.

This research was conducted under Permit No. 163 issued by the U.S. Department of Commerce, National Oceanic and Atmospheric Administration, National Marine Fisheries Service. Permission to conduct scientific research in Mexico was authorized under permits No. 14203 (1977), No. 12-459-498 (1978), No. 920 (1979), No. 10919 (1980), No. 10455 (1981), and No. 4048 (1982) issued by the Secretaria de Pesca, Direccion General De Regulacion Pesquera de Mexico.

Methods

STUDY SITE

Description

Laguna San Ignacio is located on the Pacific coast of Baja California adjacent to Bahia Ballenas, approximately 680 km south of the International Border between 26°42′ and 27°00′N and 113°7′ and 113°18′W (Fig. 1). The lagoon is part of the Vizcaino Desert and borders a gently sloping, dry, coastal flood plain the sediments of which are presumed to be Cretaceous and Tertiary, capped by Pleistocene alluvium (Mina, 1957; Phelger and Ewing, 1962). The climate is arid with an annual average rainfall of 56 mm (2 in.) which falls mainly during winter months; there is no runoff and no freshwater streams or rivers flow into the lagoon (Bostic, 1975).

Prior to our study, the most recent hydrographic chart dated from 1896 (U.S. Hydrographic Chart No. 1494) and did not accurately represent the lagoon in its present state. Therefore, it was necessary for us to conduct a habitat study to establish the physical characteristics of the lagoon and to construct an accurate chart. The present-day shoreline, the location of major shoals, and an accurate estimate of the total lagoon acreage were determined from satellite imagery (LANSAT). Lagoon bathymetry was

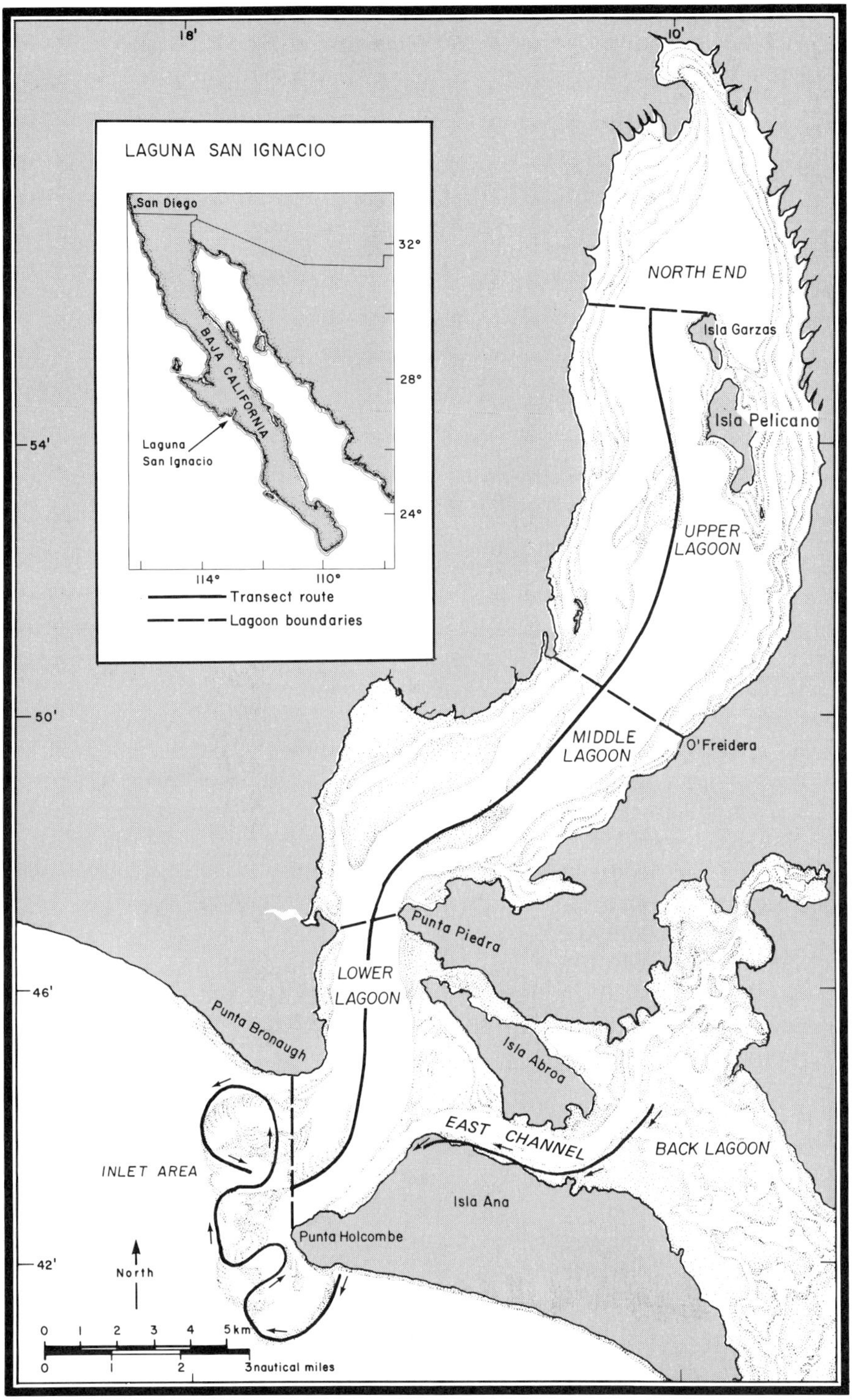

Fig. 1. *Laguna San Ignacio showing the boat transect trackline in the lower, middle, and upper lagoon areas and the aerial transect trackline in the east channel and inlet area. Arrows indicate the direction of flight of the aircraft.*

discerned from "echograms" made by a continuous-recording fathometer. Scuba divers made visual observations of the lagoon floor and sampled bottom sediments (Swartz and Jones, 1981).

The lagoon is a system of narrow, relatively deep channels surrounded by large intertidal flats. It lies on a north-to-south axis, extends inland approximately 32 km, ranges from 1.8 to 6.5 km wide, and has an area of 152 km^2, of which only approximately 57% or 87 km^2 is of sufficient depth (>2 m) to be navigable by whales. The area of the inlet is approximately 20.5 km^2, and the east channel adds approximately 3.6 km^2. The semidiurnal tide ranges from 0.9 to 2.4 m, and the tidal currents in the inlet and channels are very turbulent.

Laguna San Ignacio has a distinctive topography characteristic of lagoons with an appreciable tidal range. The bathymetry and sedimentology may be divided into five areas: the inlet, east channel, and the lower, middle, and upper lagoons (Swartz and Jones, 1981).

The 3.8-km-wide inlet is defined by a breaker line and includes a 1-km-wide entry channel with a maximum depth of 16.8 m, bordered by the inner inlet delta that averages 3.6 m in depth (Fig. 2, Profile 1). This steep-walled channel runs northward and parallels the west shore of the large barrier island, Isla Ana. The delta ends 1.5 km inside the inlet and the channel becomes deep over its entire width (Fig. 2, Profile 2). Sediments range from well-sorted, hard-packed, medium-grained sand with large shell fragments on the inner delta to irregular rock rubble covering the floor of the inlet channel. The delta is devoid of macroscophic plants, but the channel supports sparse algal vegetation in the rocky areas.

The lower lagoon consists of a steep-walled channel with a maximum depth of 25.9 m. It narrows from 3-km wide near the inlet to a constriction 1.8-km wide at Punta Piedra, where it terminates as the lagoon trends northeast. This segment is the widest relatively deep channel in the lagoon (Fig. 2, Profiles 3 and 4). West of Isla Ana, the 0.5-km-wide and 9.1-m-deep east channel tributary leaves the main lagoon, turns southeast behind Isla Ana, and runs 5.1 km to a shallow lagoon 1.5- to 2.4-m deep (Fig. 2, Profiles 13–15). Here the bottom is composed of poorly sorted coarse- and medium-grained sand with crushed shells. Extending north from Isla Ana, the main channel floor varies from a series of irregular sand ridges perpendicular to the lagoon axis to rocky areas (*coquina*) overlain with sand. The deepest regions of the channel are covered with poorly sorted, hard-packed, fine- to coarse-grained sand. West of Punta Piedra, medium-grained sand ridges are interspersed with occasional rock outcroppings 4- to 6-m tall.

The middle lagoon is characterized by a system of three channels ranging from 7.6 to 21.3 m deep, which become shallower and irregular in profile west of O'Freidera (Fig. 2, Profiles 5 and 6). Sediments are poorly sorted, fine- to coarse-grained sand with crushed shell, and some rocky areas. Extensive sand bars support stands of eelgrass (*Zostera marina*), but the channels contain little plant life.

The upper lagoon is a gently sloping basin with an average depth of 4.6 m. Two islands, Isla Garzas and Isla Pelicano, are separated by a shallow isthmus and located approximately mid-lagoon (Fig. 2, Profiles 8–12). The areas at the head of the lagoon

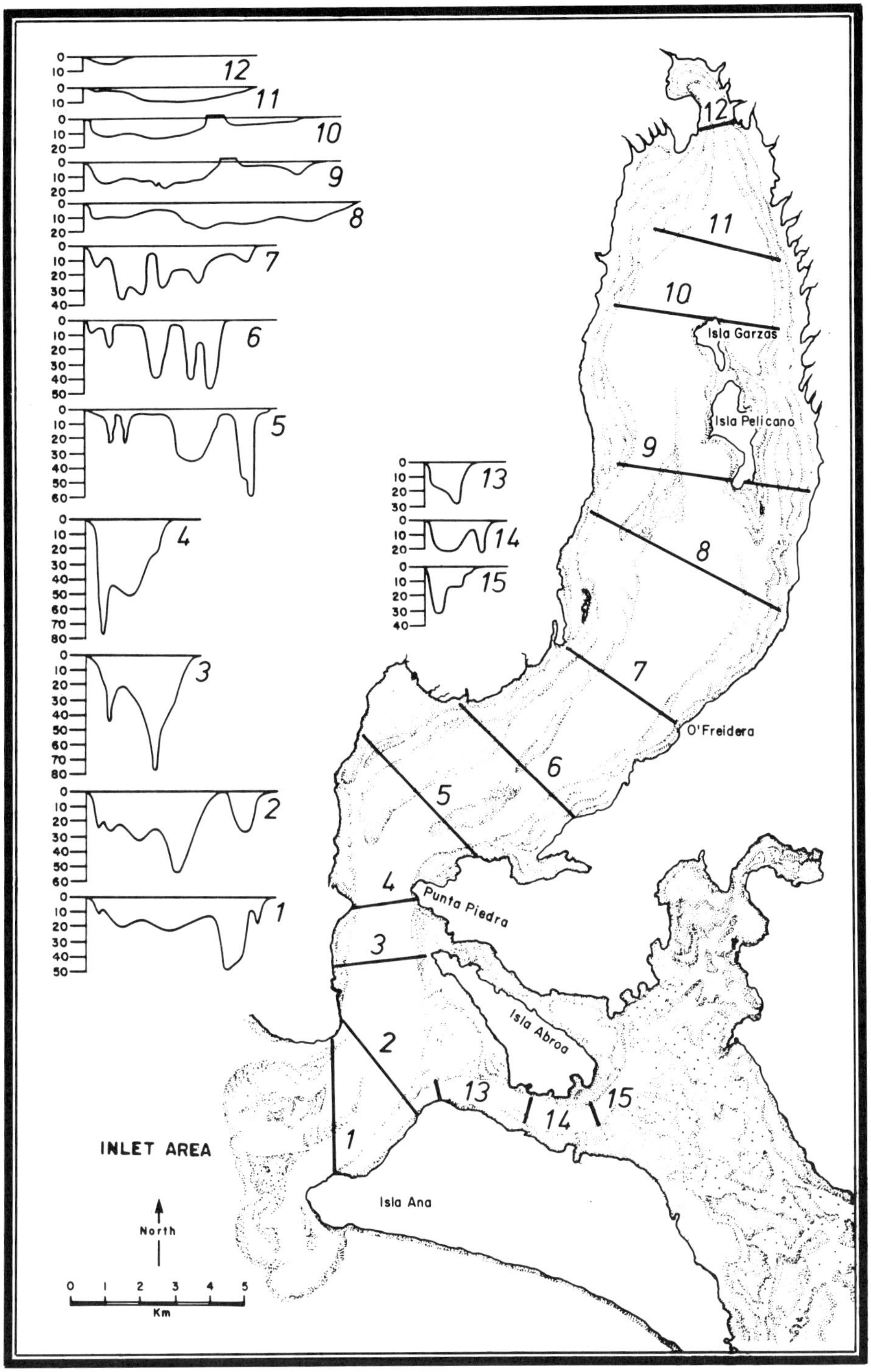

Fig. 2. *Depth profiles of Laguna San Ignacio based on echograms. Horizontal scale approximate. Soundings are in feet, and vertical exaggeration is 100×.*

north and east of the islands contain fine sand, silt, and clay. The bottom in the north is dominated by dense stands of eelgrass containing a variety of crustaceans, gastropods, pelecypods, encrusting hydroids, sponges, and tunicates. Red and brown algal epiphytes and encrusting invertebrates are attached to rocks, shells, and eelgrass blades.

Refuge Regulations

By 1975 almost all whale-watching tourism had switched from Laguna Ojo de Liebre to Laguna San Ignacio, and the need for regulation of commercial activities in this lagoon became apparent (Reeves, 1977). Our first two winters of research, 1977 and 1978, indicated the majority of females and their calves utilized the middle and upper lagoon areas early in the winter when the calves were young. In 1979 Mexico declared Laguna San Ignacio as a gray whale refuge, and our initial findings were used to designate the boundaries of the "closed" nursery area (middle and upper lagoon areas) and the "open area" in the lower lagoon where controlled tourism and fishing were permitted and to formulate regulations for the kinds of activities that would be permitted within the lagoon.[3] The regulations specify that between December 15 and March 15 each year all vessel traffic is restricted to the lower lagoon south of Punta Piedra, thus providing the upper two-thirds of the lagoon as a sanctuary free of human activities. In addition, no more than two commercial vessels are allowed within the lagoon at any one time. Individual permits issued to each vessel specify the number of days a vessel may remain within the lagoon, the number of passengers it may carry, the number of skiffs it may launch, and the kinds of activities permitted, such as whale watching, shore explorations, etc.

BOAT SURVEYS

Each of the field seasons was timed to coincide with the period during which most of the gray whales that utilized the lagoon could be observed. We began the first field season (1977–1978) on December 23, 1977, in an attempt to document the arrival of the earliest whales to the lagoon. In subsequent winters (1979–1982), field work began the second week in January, and in all years it continued until mid-April.

Each week, line transects (Eberhardt, 1978) of the entire lagoon interior were conducted to estimate whale abundance, to elucidate whale distribution and movement, and to document the seasonal timetable of lagoon occupation. Transects were executed in a 4.7-m inflatable boat powered by a 20-hp outboard engine and traveling at approximately 11 km/hr. Each transect required approximately 7 hr to complete. The crew included a left and a right observer and a vessel operator. Transects were run along an imaginary line drawn midlagoon from Punta Piedra south to the breaker line near the lagoon inlet, north to Isla Garzas at the northern end of the lagoon, and then south again to the starting point at Punta Piedra. In this way, the entire transect line was traversed twice. The maximum distance from the boat to the 2-m depth contour along shore was 2.6 km and the minimum 0.8 km; thus, whale-inhabitable water and both

[3]Diario Oficial de la Federation de los Estados Unidos Mexicanos, 16 July 1979.

shorelines were clearly visible at all times within the lagoon. Whales in the "North End" of the upper lagoon north of the transect termination were observed through binoculars from a 5-m-high bluff on the northern tip of Isla Garza (Fig. 1).

All whales sighted from the boat were counted when they were approximately abeam of the boat, and all observations were recorded on a cassette tape recorder. The higher number of whales counted from the two replicate transects was used as a minimum abundance count, an index of the whale population for that week. Because only two replicates could be obtained per census day, we did not estimate errors for these abundance counts. Whale distribution was determined by recording transect starting time, time each whale or group of whales was encountered, passage of landmarks, progression of tidal stages, and time of completion. To prevent observer anticipation of whales previously counted, observers viewed a different side of a transect leg on each replicate. Transects were discontinued when wind and wave conditions exceeded Beaufort Sea State No. 3 (winds 12–18 km/hr), which obscured whale blows, diminished the sightability of inconspicuous behaviors, and lowered the overall probability of sighting whales (particularly calves). Corrections for whales missed are discussed in the results.

Unless otherwise noted, "female–calf pairs" were considered a single unit, and counts of these pairs are equivalent to calf counts; "single whales" refer to nonparturient females, adult males, and immatures. Single whales often occurred in pairs or groups.

AERIAL SURVEYS

Aerial surveys augmented the boat surveys in 1980, allowing simultaneous estimation of whale abundance and distribution in areas not accessible by boat, namely the east channel, inlet, and nearshore waters adjacent to the lagoon. These surveys were flown in a Piper Super Cub (a two-passenger, single-engined, high-wing aircraft) at an altitude of 330 m and an average ground speed of 123 km/hr. Total coverage (100%) was attempted (Eberhardt, 1978) by flying a serpentine course over the inlet with observers making counts from each side of the aircraft. The east channel was surveyed by flying along its southeastern shore (Fig. 1). As with boat surveys, all observations were recorded on a cassette tape recorder. For larger whale groups, the aircraft diverted from course and circled until the observer's count stabilized, then returned immediately to the original course. Practice flights were flown to familiarize the pilot and observers with the transect route and survey procedure. Surveys were conducted during Beaufort Sea State No. 3 conditions or better, and surveys were flown between 0900 and 1500 hr to reduce the effects of surface glare on the water. Sighting cues indicating the presence of whales, such as bubble bursts, mud upwellings, or a series of whale "foot-prints," were counted as single animals.

MORTALITY

Beaches were searched by boat or aircraft for stranded dead whales at least twice per week, and additional information was requested of local fishermen who participated by reporting sightings of whale carcasses. The sex of dead whales was determined, and standard measurements were recorded. Length was measured as the distance from the tip of the snout (rostrum) to the notch of the flukes. Whale carcasses were marked and photographed, and their locations were recorded to prevent duplicate sightings of individuals in the event that the tide and winds moved bloated carcasses from their original site of discovery. The calf mortality rate for Laguna San Ignacio was calculated from the number of dead calves discovered and the estimated total calf production. Finally, we researched the records on natural mortality in other breeding areas for comparison with the Laguna San Ignacio data.

WHALE-WATCHING ACTIVITIES

When we began the gray whale study in Laguna San Ignacio in 1977, there was little information available from which to judge the possible reactions of the whales to the growing tourist industry within the lagoon. Gray whales seemed particularly vulnerable to human activities within the confinement of the lagoons because they cannot readily flee from an adverse disturbance as they can in the open ocean. Also, given that the lagoon is a critical habitat necessary for successful reproduction, exclusion from one or more lagoon areas could cause overcrowding in remaining areas or force whales to seek alternate and perhaps suboptimal areas in which to breed that may not provide the protective features offered by the lagoons.

To document the relative amount of disturbance introduced into the lagoon each winter by human activities, our first step was to qualitatively and quantitatively characterize human disturbance. A log was kept of the number, location, and hours of activity of all vessels entering or transiting through the lower lagoon, the area designated for commercial activities. For the whale-watching excursions, we determined the number of hours that vessels were present in the lagoon, the number of hours they were underway and at anchor, and the number of hours whale-watching skiffs were in operation. The sum of the hours of each disturbance category was then expressed as a percentage of each entire season, and these percentages were compared among years. It was, however, beyond the scope of the program to identify the specific attributes of vessels that might cause significant changes in the utilization of the lagoon by gray whales.

In our study, we employed natural experimental and observational approaches to evaluate the demographic reactions of gray whales to potential sources of disturbance from whale-watching activities over the 5 years. To determine whether whales showed a response to human disturbance in the lagoon among years, seasonal trends in whale abundance and distribution were statistically compared among years as levels of tourism increased and decreased. Specifically, we sought to identify the following possible responses to disturbance:

1. A decrease in seasonal abundance of gray whales utilizing the entire lagoon
2. A decrease in the number of females and calves utilizing the nursery area
3. A decrease in the seasonal abundance of gray whales in the lower lagoon, the area utilized by tour vessels, and
4. An increase in the seasonal mortality of whales, especially calves.

Although interactions between whales and people have a variety of forms, in designing our study we believed that the consequences of changes in these parameters would be immediately "meaningful" from a management point of view. Findings on gray whale behavioral reactions to vessel disturbance such as changes in the relative frequency, duration, and context of social behavior, aerial behavior, sexual behavior, swimming direction, and breathing rates are reported elsewhere (Swartz and Jones, 1978, 1981; Jones and Swartz, 1984).

In addition to the analysis of seasonal demographic trends among years, we compared gray whale abundance and distribution within years during "control" periods without human activity and "experimental" periods with human activity (specifically tourism) in the lagoon. Here, the potential adverse effects of whale-watching tested for were responses 1, 2, and 3 mentioned earlier.

To evaluate these potential effects, boat transect counts from consecutive pairs of days were compared; these pairs included a day preceding and a day following the arrival of a vessel, two days with vessel activity, and a day before and a day following a vessel departure. We tested the null hypothesis that whale-watching activities did *not* cause significant changes in gray whale abundance, density, and distribution in the lagoon as a whole, in the nursery area, and in areas where whale and human activities occurred simultaneously.

The proposed consequences of exposure to human disturbance (i.e., whale-watching activities) constituted our operational definition of "harassment" or disturbance to whales.

Results

ABUNDANCE, DISTRIBUTION, AND OCCUPATION TIMETABLE

Correction for Unseen Whales

Raw boat transect counts may be biased toward underestimating the actual number of whales in an area for two primary reasons, (1) some whales that are present but are below the water's surface while the transect vessel passes through the survey area are not detected, and (2) some whales that are at the surface may be missed by the observer (Caughley, 1977; Eberhardt, 1978; Eberhardt *et al.*, 1979; Davis *et al.*, 1982). The first source of bias is a function of the length of time the observer has to view the survey area and the whale's surface interval and duration of dives. A correction factor to determine the probability that a whale will surface while the observer is viewing the survey area is given by Eberhardt (1978) as

$$P = \frac{s}{s+u} + \frac{t}{s+u} = \frac{s+t}{s+u} \tag{1}$$

where s is the mean duration of surfacings (i.e., the mean surface interval), u is the mean duration of dives, and t is the length of time a point along the transect line is visible to the observer. To allow for submerged animals, raw whale counts should be divided by P. Two of the assumptions of this formula are that t is less than or equal to u and that s and u are constant; however, because our gray whale values for u were very much less than t and because the duration of surface intervals and dive durations varied with the behavior of the whales in Laguna San Ignacio, these assumptions were not valid for our boat surveys. Under these conditions, Eberhardt (1978), and Eberhardt *et al.* (1979) suggest that $P = (s + t)/(s + u)$ is inappropriate to compute the probability that a whale will be detected, but, to our knowledge, no alternate method has been published.

We employed a different approach to determine whether to adjust our raw whale counts for whales missed while they were below the surface. First, to calculate t, the duration of potential detectability during our boat survey, we divided the observer's field of view by the boat survey speed. We estimated that the length of the observer's field of view was 1.5 km, beginning approximately 1.0 km ahead of the boat and ending 0.5 km behind. The survey speed of 11 km per hour was selected for three reasons: (1) it was faster than the observed range of normal gray whale swimming speeds of up to 6 km per hour (Gilmore, 1960; Pike, 1962; Sumich, 1983) and thereby reduced the possibility of counting the same whale twice; (2) it allowed adequate time to view most whales more than once (see below); and (3) it allowed us to complete two transect replicates per census. Thus, our value of t was

$$t = \frac{1.5 \text{ km}}{11 \text{ km/hr}} = 8.18 \text{ min}, \tag{2}$$

the period during which a point along the transect line was within the observer's view.

Our next step was to determine how many times a whale was likely to surface within the period of potential detectability. To do this we used the mean surfacing rate and mean surfacing interval calculated by Mate and Harvey (1981) and Harvey and Mate (Chapter 24, this volume) for 10 radio-tagged gray whales in Laguna San Ignacio. Because their radio tag only transmitted signals when the antenna broke the surface, Harvey and Mate obtained the most precise estimates available for these parameters during 307 hr of monitoring.[4] Their mean surfacing rate for adult whales was 35.6 $\pm$ SE 0.8 surfacings/hr ($n = 202$), or 0.59 surfacings/min. In addition to the data for radio-tagged whales, from their shore observations they reported a rate of surfacing for gray whale calves of 100.2 surfacings/hr, or 1.67 surfacings/min (sample size not reported). Adult whales, then, would be expected to surface once every 1.69 min (1 min and 41 sec) and calves once every 0.59 min (35 sec). From these surfacing rates, the mean number of times an adult whale would be expected to surface within the observer's duration of detectability was calculated as 8.18 min $\times$ 0.59 surfacings/min = 4.8 surfacings. Simi-

[4]Surface intervals may have been slightly longer than those reported by Mate and Harvey because their radio tags were placed behind the whale's blowholes; thus, as the animals arched and rolled while beginning a dive the antenna would submerge a moment before the entire animal.

larly, the mean number of surfacings expected for a calf would be 8.18 min × 1.67 surfacings/min = 13.7 surfacings.

To determine how many seconds a whale would be visible during the period of potential detectability, we used Harvey and Mate's mean surface interval for adult whales of 0.07 ± SE 0.01 min (n = 11,229 surfacings) or 4.2 sec. They did not report a mean surface interval for calves. The mean proportion of the duration of potential detectability that an adult whale would be visible was calculated as the number of surfacings multiplied by the surface interval, divided by the duration of potential detectability or

$$\frac{4.8 \text{ surfacings} \times 0.07 \text{ min}}{8.18 \text{ min}} = 0.041 \quad (3)$$

or approximately 20 sec during every 8.18 min.

Our estimate of a mean of 4.8 surfacings and a mean total surface duration of 20 sec during the period of potential detectability suggests that we had an opportunity to see most if not all whales (including calves) one or more times during a transect. For this reason, we did not adjust our raw whale counts for whales missed because they were below the surface.

The second source of bias, whales at the surface but missed by the observer, is a function of environmental conditions affecting visibility during the survey and the observer's sighting ability and experience (Leatherwood *et al.*, 1978; Davis *et al.*, 1982). Again, we did not correct for this bias, but, to minimize this source of error, transects were conducted only during ideal conditions, Beaufort Sea State No. 3 or less, and excellent through-air visibility (no rain, mist, or fog). In addition, all observers were experienced at surveying gray whales in Laguna San Ignacio; new observers participated in training transects and accompanied experienced observers on actual transects before they conducted counts on their own. Despite these precautions, some whales were undoubtedly missed. For this reason, we chose the highest count from the two transect replicates as the minimum whale abundance estimate for that day.

Inlet and Lagoon Interior

Counts of gray whales in the inlet and the east channel of the lagoon (Fig. 1) were obtained from 16 systematic aerial surveys conducted between January 16 and March 28, 1980. Seasonal counts of single whales and female–calf pairs in the inlet are shown in Fig. 3. An overview of the total seasonal whale abundance for the entire Laguna San Ignacio system was obtained for 1980 by combining aerial counts with corresponding boat counts of the lagoon interior (Fig. 4A).

The maximum combined count (single whales + female–calf pairs) for the entire system occurred on February 16 and was 458 whales (333 single whales and 125 female–calf pairs). These whales were distributed as follows: 27% (119 single whales and 5 female–calf pairs) were in the inlet, 72% (214 single whales and 118 female–calf pairs) were in the lagoon interior, and 1% (2 female–calf pairs) was in the east channel (Fig. 4A).

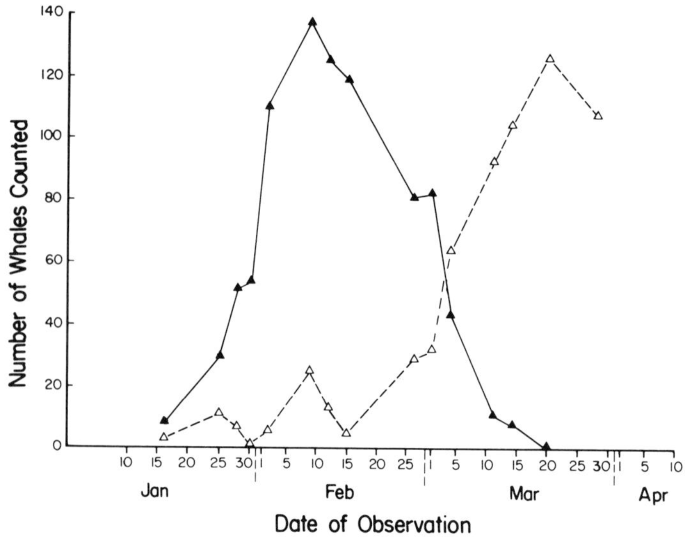

Fig. 3. *The number of single whales (solid triangles) and female-calf pairs (open triangles) counted in 16 aerial transects of the inlet of Laguna San Ignacio during the 1980 winter.*

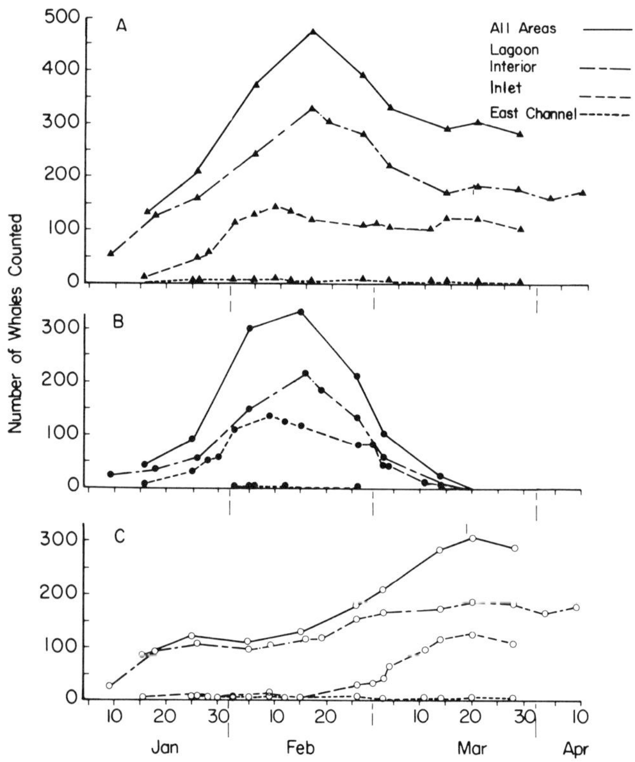

Fig. 4. *Total number of whales counted in the entire Laguna San Ignacio system in 1980. (A) Triangles are total adult whales. (B) Closed circles are single whales. (C) Open circles are female–calf pairs.*

The maximum single whale count of 351 animals also occurred on February 16, 1980. Thirty-nine percent of these animals were in the inlet, and 61% were in the lagoon interior. No single whales were in the east channel (Fig. 4B). By March 3, single whales in the inlet had declined to 26% of their maximum count for the entire lagoon system, and by March 14, only 3% remained.

In contrast to the single whales, at the maximum combined count only 8% of the 125 female–calf pairs counted were in the inlet, while 90% were in the lagoon interior, concentrated in the upper lagoon nursery area (Fig. 4C). The east channel, the smallest inner-lagoon area, never contained more than 2% of the lagoon whale population.

Following the departure of single whales from the lagoon, counts of female–calf pairs continued to increase and reached a maximum of 315 pairs on March 19. At this time, 40% were in the inlet, 59% were in the lagoon interior, and 1% was in the east channel. During the final aerial survey on March 28, 108 female–calf pairs were counted in the inlet. Local fishermen informed us that some females and their calves may be

Table I
Abundance of Gray Whales in the Inlet and Interior of Laguna San Ignacio at Two Key Time Periods in 1980

	Lagoon area					
	Inlet[a]	Lower[b]	Middle[c]	Upper[d]	East channel[e]	Total
Maximum combined counts (February 16)						
Total whales						
Number	124	153	69	110	2	458
Density[f]	7.75	9.28	3.40	4.50	1.11	—
Percentage of total	27	33	15	24	1	100
Single whales						
Number	137	144	60	10	0	351
Density	6.68	8.25	2.62	0.21	0	—
Percentage of total	39	41	17	3	0	100
Female–calf pairs						
Number	11	9	9	100	2	131
Density	1.07	1.03	0.79	4.29	1.11	—
Percentage of total	8	7	7	76	2	100
Maximum female–calf counts[g] (March 20)						
Number	126	163	23	0	2	314
Density	12.29	18.68	2.01	0	1.11	—
Percentage of total	40	52	7	0	1	100

[a]Inlet area = 20.50 km^2.
[b]Lower lagoon area = 17.45 km^2.
[c]Middle lagoon area = 22.91 km^2.
[d]Upper lagoon area = 46.62 km^2.
[e]East channel area = 3.60 km^2.
[f]Whales/km^2. Female–calf pairs were considered as two animals in density calculations.
[g]No single whales present.

seen in the surf areas of the inlet through April and May of some years, but they are rarely seen in June.

To illustrate the utilization of the inlet relative to the lagoon interior at the maximum combined count and at the maximum female–calf pair count, a more detailed breakdown of the number of whales utilizing the complete lagoon system is presented in Table I. The inlet was utilized extensively throughout the season and was clearly a major area. On the date of the maximum combined count, 39% of the single whales occupied the inlet at a density of 6.68 whales/km^2, while only 8% of the female–calf pairs were present in this area. Coinciding with the departure of single whales, however, the number of female–calf pairs gradually increased in the inlet and at their maximum count, the inlet contained 40% of these animals at a density of 12.29 whales/km^2. or double the density of single whales (female–calf pairs were counted as two animals for densities).

Lagoon Interior

To identify trends in gray whale abundance, distribution, and duration of season, the counts of whales were analyzed at two key time periods. The first period was the annual date of maximum combined counts, which coincided with the period of single whale maximum counts. For this time category, an overview of the total whale abundance and distribution in the lagoon is presented first. Then, the total whale population is subdivided into single whales and female–calf pairs and these two components are analyzed as separate but interacting groups. The second time period was the date of maximum female–calf pair counts which occurred after the departure of single whales from the lagoon each year. To determine whether whale abundance at the two time periods changed significantly among years, a least squares exponential regression line was fitted to the yearly maximum counts for each whale category (i.e., total whales, single whales, and female–calf pairs); the regression (significance of trend) and significance of slope (significance of increase or decrease) were evaluated with analysis of variance (ANOVA) (Sokal and Rohlf, 1981).

During the 1977 reconnaissance study, we learned that gray whales were distributed differentially within the lagoon. Whales occupied some areas throughout the season, while other areas were occupied only part of the winter. Therefore, to more precisely elucidate whale distribution, the lagoon interior was divided into three sections, the lower, middle and upper lagoon, based on our preliminary observations of whale distribution, and on the bathymetric features of the lagoon. The seasonal distribution of whales within each section was determined by subdividing the weekly censuses of the entire lagoon into the number of whales counted per area.

The dynamics of gray whale distribution in the lagoon interior was assessed by investigating three parameters: abundance, density, and the percentage of the total lagoon population that was within each area. We first tested for differences among lagoon areas, and then tested for changes within each area during the 5 years. A Kruskal–Wallis nonparametric test was used to evaluate differences in the three variables among the lower, middle, and upper lagoon areas. In the situation where the null hypothesis of the Kruskal–Wallis was rejected, nonparametric multiple comparisons

(Zar, 1974) were done to determine between which of the locations significant differences occurred.

Differences in the abundance of whales within each lagoon area over the 5 years were evaluated by using G-tests to compare each yearly count to the mean count for all years. Trends in abundance within each area were evaluated by fitting least squares, exponential regression lines to the counts for all years, and the significance of regression was evaluated with ANOVA.

To examine whether whale density (whales/km^2) varied significantly within each of the three areas, counts were analyzed with three separate Kolmogorov–Smirnov tests, where the expected frequency was the mean density in each location for the 5 years. In addition, least squares regression lines were fitted to the densities for each year and the significance of each regression was evaluated with ANOVA.

To determine whether there was a pattern to the whale distribution in the lagoon over the 5 years, the number of whales within each of the three areas was expressed as a percentage of the total number of whales in the lagoon, and changes in the proportion of the whales in each area were evaluated by calculating G-statistics for the log-likelihood ratios of observed percentages versus the mean percentage for each area.

Maximum Combined Counts. Gray whale counts from 60 boat transects, 10–14 each winter between 1978 and 1982, were used as an index of the weekly whale abundance within the lagoon interior. In all years, whales were present in low numbers when field work began. During the 1977–1978 season when transects were begun in December, the first count (5 single whales and 1 female–calf pair) on December 26, was only 2% of the eventual maximum count, and the second count (29 single whales and 2 female–calf pairs) on January 4, was 9% of the eventual maximum count for that year. In subsequent years, the first transects were conducted during the second week in January, and these counts averaged 20% (mean = 67 whales) of the annual maximum count.

The seasonal distribution of the overall whale population in the lagoon was bimodal: the major mode represented the maximum combined counts for each year and was composed primarily of single whales and some female–calf pairs (Fig. 5). Maximum combined counts occurred as early as February 2 in 1978 and as late as March 1 in 1979 with a mean day for the five winters of February 15 ± SE 3.9 days, and a 95% confidence interval of 21 days from February 5 to February 26. In 1978 and 1979, the dates of maximum combined counts were the most widely separated (27 days) and were significantly different from the mean day; the dates of maximum combined counts for the 1980 to 1982 seasons, however, were clustered about the mean day; these dates were February 16 in 1980, February 12 in 1981, and February 14 in 1982

Maximum combined counts increased significantly, 7.3% per year, from 300 adult whales in 1978 to 407 whales in 1982 [$F_{(1,3)} = 32.88$, $p < .025$, $r^2 = 0.916$] (Fig. 6). This increase was primarily attributable to greater counts of female–calf pairs each year. Counts of these whales increased significantly, 18.3% per year [$F_{(1,3)} = 16.13$, $p < .05$, $r^2 = 0.843$], from 65 pairs in 1978 to 137 pairs in 1982. In contrast, the 2.7% per year increase in the single whale counts was not significant [$F_{(1,3)} = 0.52$, $p > .50$, $r^2 = 0.147$].

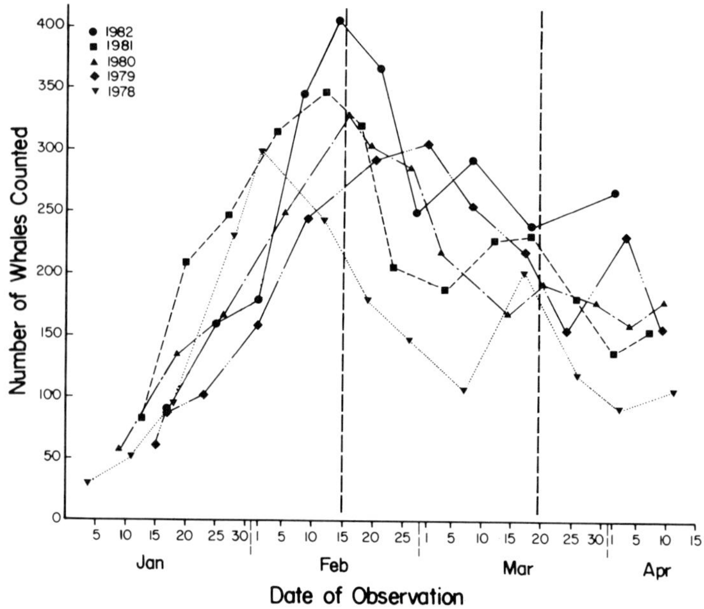

Fig. 5. *The number of whales (except calves) counted in 60 vessel transects of Laguna San Ignacio between 1978 and 1982. Broken vertical lines represent the mean date of combined maximum counts (February 15) and the mean date of maximum female–calf pair counts (March 19).*

The seasonal abundance and distribution of gray whales in the lower, middle, and upper lagoon areas between 1978 and 1982 are shown in Table II and in Figs. 7–9. A Kruskal–Wallis test of equality among lagoon areas revealed that whale abundance was not the same in all three areas [$X^2_{(2)} = 12.02$, $p < .01$]. The nonparametric multiple comparison test disclosed that abundance was significantly different in each lagoon area.

Whale abundance was greatest in the lower lagoon during all years. A G-test

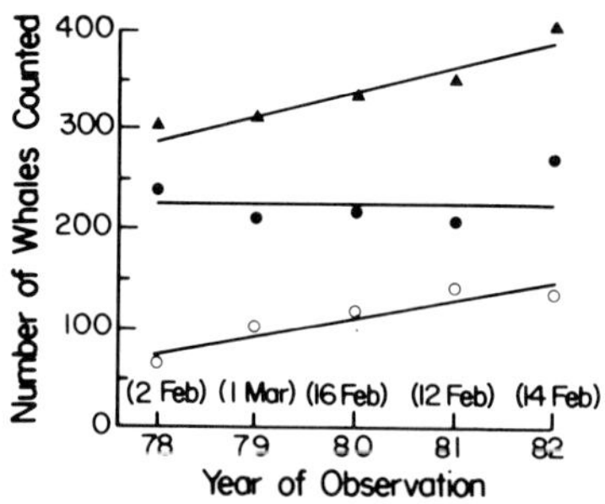

Fig. 6. *Maximum combined counts of gray whales (triangles) in Laguna San Ignacio between 1978 and 1982. Closed circles are the single whale component and open circles are the female–calf pair component of the maximum combined count. Dates of maximum counts are in parentheses. Black lines are least squares regression lines.*

Table II
Abundance of Gray Whales in the Three Lagoon Interior Areas and Inlet Between 1978 and 1982

Lagoon area	Year	Maximum combined counts: Total adults	Single whales	Female–calf pairs	Maximum[a] female–calf pair counts
Inlet	1980	124	119	5	126
Lower lagoon	1978	137	129	8	179
	1979	168	144	24	198
	1980	153	144	9	163
	1981	158	134	24	190
	1982	202	187	15	200
	G	13.97**[b]	13.61**	15.54**	5.11
	$P_{(G)}$	<.01	<.01	<.005	>.10
	F	4.97	3.43	1.57	0.38
	$P_{(F)}$	>.10	>.10	>.10	>.50
	r^2	.62	.53	.34	.11
Middle lagoon	1978	73	69	4	16
	1979	50	33	17	29
	1980	69	60	9	23
	1981	72	56	16	32
	1982	67	52	15	49
	G	5.63	14.11**	11.76*	19.90***
	$P_{(G)}$	>.25	<.01	<.025	<.001
	F	0.03	3.90	2.43	11.80*
	$P_{(F)}$	>.90	>.10	>.05	<.05
	r^2	.01	.57	.45	.80
Upper lagoon	1978	92	37	55	7
	1979	90	31	59	0
	1980	110	10	100	0
	1981	118	17	101	7
	1982	138	31	107	33
	G	14.15**	22.24***	31.61***	74.59***
	$P_{(G)}$	<.01	<.001	<.001	<.001
	F	42.45**	0.25	14.96*	1.44
	$P_{(F)}$	<.01	>.50	<.05	>.50
	r^2	.90	.08	.85	.32

[a]No single whales present.
[b]* = significant at $\alpha \leq .05$, ** = significant at $\alpha \leq .01$, and *** = significant at $\alpha \leq .001$.

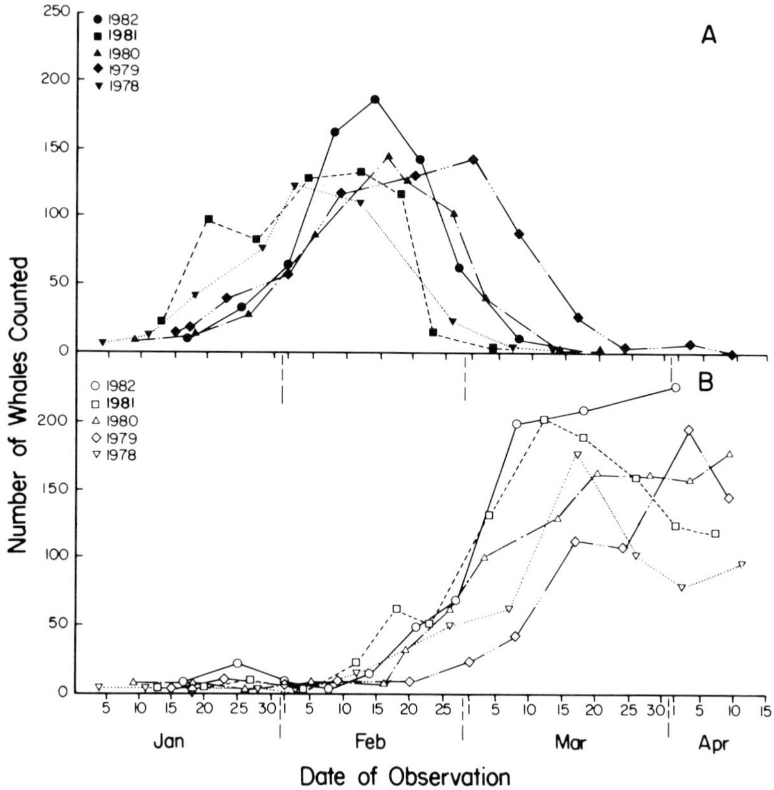

Fig. 7. *The number of single whales (A) and female–calf pairs (B) counted in 60 vessel transects of the lower lagoon between 1978 and 1982.*

indicated that the number of whales in this area at the date of maximum combined counts was significantly different each year [$G = 13.97 > X^2_{.05(4)} = 9.49$, $p < .01$] (Table II). The slope of the least squares, exponential regression line fitted to these counts indicated that the population in the lower lagoon increased 7.1% per year from 137 whales in 1978 to 202 in 1982, but, this increase was not statistically significant [$F_{(1,3)} = 4.97$, $p > .10$, $r^2 = 0.624$] (Fig. 10A, top). Thus, the heterogeneity in the counts among years was not explained by the regression.

The number of whales in the middle lagoon did not change significantly among years [$G = 5.63 < X^2_{.05(4)} = 9.49$, $p > .10$], and the 3.8% per year decrease in these counts from 73 in 1978 to 67 in 1982 was also not significant [$F_{(1,3)} = 0.03$, $p > .90$, $r^2 = 0.010$].

Finally, whale counts in the upper lagoon nursery did change significantly [$G = 14.15 > X^2_{.05(4)} = 9.47$, $p < .01$] showing a significant 8.9% per year increase from 92 in 1978 to 138 whales in 1982 [$F_{(1,3)} = 42.45$, $p < .01$, $r^2 = 0.934$]. Notably, the 7.3% per year overall increase in the whale population in the lagoon between 1978 and 1982 was

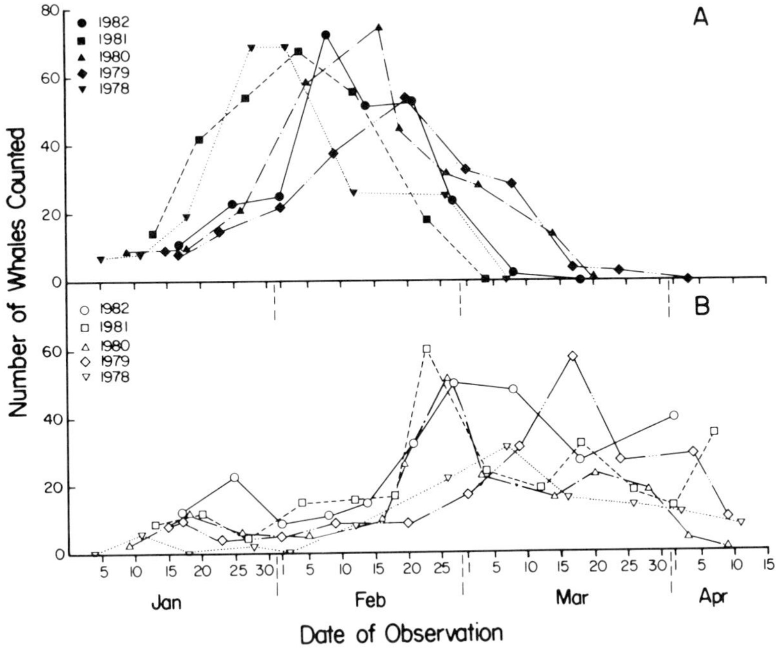

Fig. 8. *The number of single whales (A) and female–calf pairs (B) counted in 60 vessel transects of the middle lagoon between 1978 and 1982.*

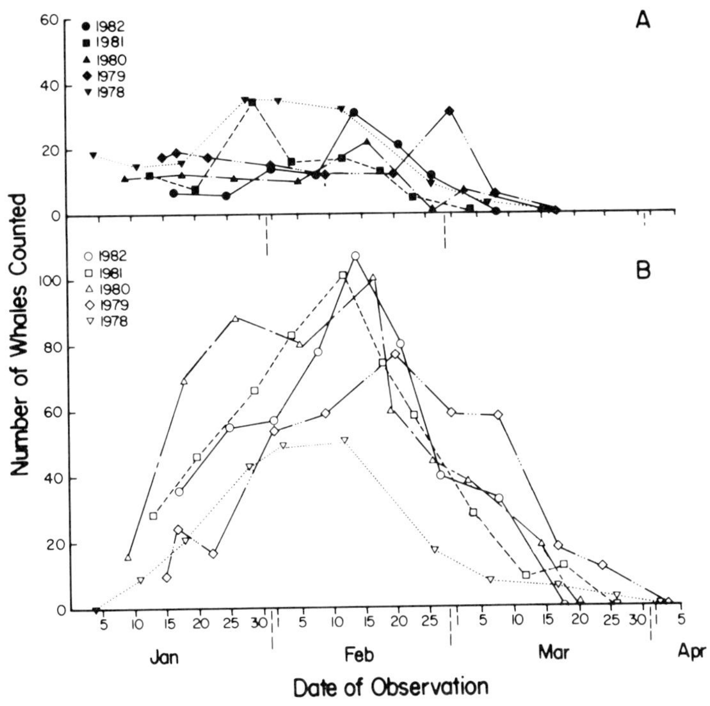

Fig. 9. *The number of single whales (A) and female–calf pairs (B) counted in 60 vessel transects of the upper lagoon between 1978 and 1982.*

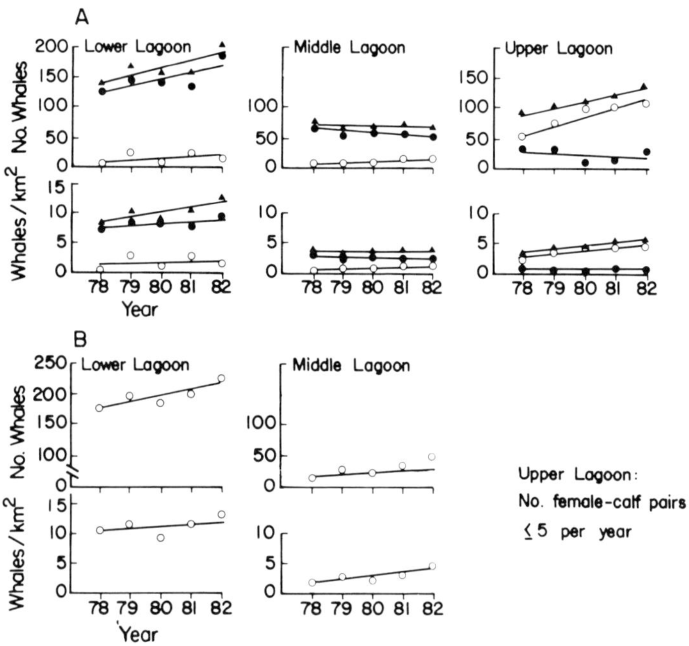

Fig. 10. *The abundance and density of whales in the lower, middle, and upper areas of Laguna San Ignacio between 1978 and 1982. Section A represents whale counts (top) and densities (bottom) on the date of the maximum combined count each year. Section B represents the female–calf count (top) and density (bottom) on the date of their maximum count each year. Triangles are combined counts (single whales + female–calf pairs), solid circles are single whales, and open circles are female–calf pairs. Black lines are least squares regression lines. Female–calf pairs were counted as two animals in density calculations.*

seen primarily as an increase in the use of the upper lagoon area, while whale abundance in the lower and middle areas remained relatively unchanged.

As with whale abundance, densities of whales in the lower and middle lagoon areas did not change significantly between 1978 and 1982, and averaged 10.29 and 3.42 whales/km^2 respectively. The density of whales in the upper lagoon, however, did increase significantly from 3.16 whales/km^2 in 1978 to 5.25 whales/km^2 in 1982 [$F_{(1,3)}$ = 27.14, $p < .02$, $r^2 = 0.900$].

The densities of whales within each area by year and the results of the Kolmogorov–Smirnov comparisons and ANOVAs for significance of regression are presented in Table III. The abundance and densities of whales per lagoon area and the least squares regression lines are shown in Figure 10A.

At the maximum combined count, the percentage of the total whale population in the lagoon that was present in each area did not change significantly between 1978 and 1982. These percentages and the results of the G-tests for significant differences within each area among years are summarized in Table IV. The lower lagoon contained an average of 48.3 ± SE 5% of the population, the middle lagoon contained an average of 20 ± SE 3.7%, and the upper lagoon contained an average of 32 ± SE 2.5% of the

Table III
Densities of Gray Whales in the Three Lagoon Interior Areas and Inlet, 1978–1982

Lagoon area	Year	Maximum combined counts: Total whales	Single whales	Female–calf[a] pairs	Maximum[b] female–calf pair counts
Inlet (20.5 km^2)	1980	7.75	6.68	1.07	12.29
Lower lagoon (17.5 km^2)	1978	8.31	7.39	0.92	20.52
	1979	11.00	8.25	2.75	22.69
	1980	9.28	8.25	1.03	18.68
	1981	10.43	7.68	2.75	21.78
	1982	12.44	10.72	1.72	22.92
	$D_{(n)}$	0.044	0.053	0.099	0.019
	$p_{(D)}$	>.50	>.50	>.50	>.50
	F	4.28	3.43	0.51	0.39
	$p_{(F)}$	>.10	>.10	>.50	>.10
	r^2	.59	.53	.14	.11
Middle lagoon (22.9 km^2)	1978	3.36	3.01	0.35	1.40
	1979	2.92	1.44	1.48	2.53
	1980	3.40	2.62	0.79	2.01
	1981	3.84	2.44	1.40	2.79
	1982	3.58	2.27	1.31	4.28
	$D_{(n)}$	0.033	0.055	0.134	0.143
	$p_{(D)}$	>.50	>.50	>.50	>.50
	F	1.95	0.00	2.44	11.88*
	$p_{(F)}$	>.25	>.99	>.25	<.05
	r^2	.39	.00	.45	.80
Upper lagoon (46.6 km^2)	1978	3.15	0.79	2.36	0.30
	1979	3.20	0.66	2.53	0.00
	1980	4.50	0.21	4.29	0.00
	1981	4.70	0.36	4.33	0.30
	1982	5.25	0.66	4.59	1.42
	$D_{(n)}$	0.095	0.141	0.130	0.501
	$p_{(D)}$	>.50	>.50	>.50	>.50
	F	27.14*	0.25	14.59*	0.58
	$p_{(F)}$	<.02	>.50	<.05	>.50
	r^2	.90	.08	.83	.16

[a]Female–calf pairs were counted as two whales in densities.
[b]No single whales present.
[c]* = significant at $\alpha \leq 0.05$.

Table IV
Percentages of the Total Gray Whales in the Lagoon Areas Between 1978 and 1982

Lagoon area	Year	Maximum combined counts: Total adults	Single whales	Female–calf pairs	Maximum female–calf pair counts
Lower lagoon	1978	46	55	10	89
	1979	54	69	10	87
	1980	46	67	8	88
	1981	46	65	17	88
	1982	50	69	11	85
	$\bar{X} \pm$ SE	48 ± 5	65 ± 2	11 ± 1	87 ± 6
	G	1.04	2.17	3.87	0.10
	$P_{(G)}$	>.50	>.50	>.10	>.95
Middle lagoon	1978	24	29	6	9
	1979	16	16	9	13
	1980	21	28	7	12
	1981	21	27	11	9
	1982	17	20	11	15
	$\bar{X} \pm$ SE	20 ± 4	24 ± 2	9 ± 1	12 ± 1
	G	2.17	5.70	2.42	2.35
	$P_{(G)}$	>.50	>.10	>.50	>.50
Upper lagoon	1978	30	16	84	2
	1979	29	15	81	0
	1980	33	5	85	0
	1981	34	8	72	3
	1982	34	11	78	0
	$\bar{X} \pm$ SE	32 ± 2	11 ± 2	80 ± 2	1 ± 0.6
	G	.69	8.31	1.39	9.32
	$P_{(G)}$	>.90	>.05	>.50	>.05

population. Thus, despite the significant annual increase in the abundance of whales in the upper lagoon, the proportion of the total yearly population that utilized each section of the lagoon remained nearly constant.

Following the maximum combined count each year, single whale abundance declined in all areas (see Single Whales), while female–calf pair abundance continued to increase, resulting in a second smaller peak in abundance around March 19, the minor mode of the population curves. At this time, the lagoon was occupied entirely by female–calf pairs (see Female and Calf Pairs).

Single Whales. Each year single whale counts increased rapidly during the 6-week period approximately from the last week in December through the second week in

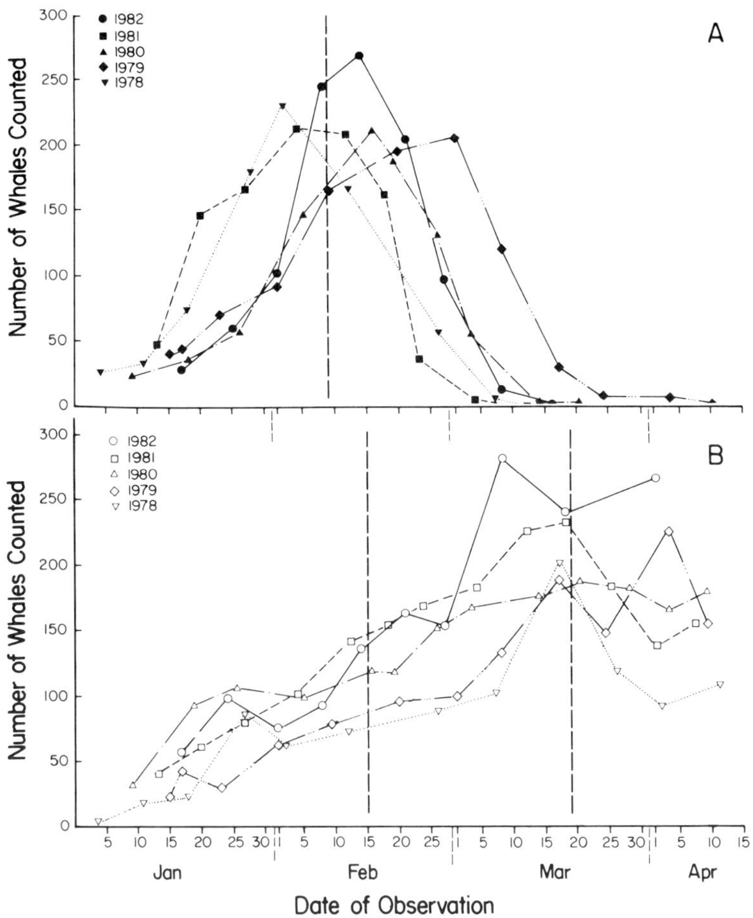

Fig. 11. *The number of single whales (A) and female–calf pairs (B) counted in 60 vessel transects of Laguna San Ignacio between 1978 and 1982. Broken vertical lines indicate the mean date of the maximum single whale counts on February 9 (A), the female–calf counts at the end of the birth period on February 15, and the mean date of maximum female–calf pair counts on March 19 (B).*

February (Fig. 11A). During our earliest census on December 26, 1977, five single whales or 2% of the eventual maximum single whale count for that year were already present. Single whale counts reached 50% of their eventual maximums as early as January 17 in 1981 and as late as February 2 in 1979 and 1982, with a mean date of January 27 ± SE 3 days.

The maximum single whale counts each year occurred on the same day as the combined maximum count; the average date of maximum single whale counts for all years was February 15 ± SE 3.9 days, with a range from February 2 in 1978 to March 1 in 1979 (Fig. 4). Both of these dates fell outside the 21-day 95% confidence interval for the mean date. Maximum single whale counts ranged from a low of 207 animals on February 12, 1981 to a high of 270 whales on February 14, 1982, with a mean maximum count of 227 whales. The slope of the exponential regression line fitted to these counts for all years indicated a 2.7% per year increase in single whale maximum counts, but this slope was not significantly different from 0, indicating there was no significant change in the abundance of single whales over the 5 years [$F_{(1,3)} = 0.516$, $p > .50$, $r^2 = .147$] (Fig. 6). Following peak abundance, single whale counts rapidly declined to 50% of their maximum counts on dates which ranged from February 19 in 1978 to March 11 in 1979, with a mean date of February 27 ± SE 3 days. Single whale counts had dropped further to 2–4% of their maximums as early as March 4 in 1981 and as late as March 24 in 1979, with a mean date of March 11 ± SE 3 days.

The duration of the single whale season was defined as the time elapsed between the first and last sighting of a single whale during the boat transects each year. We assumed that the single whale season began on December 26 each year (see p. 325). The weekly counts were scaled so that December 26 was equal to day one each year, rather than the calendar date.

Single whale seasons ranged from 76 days in 1981 to 104 days in 1979 with a mean duration of season of 84 days, and a 95% confidence interval of 72 to 97 days (Table V). Only the 1979 season of 104 days was significantly different from the mean length of season. That year, approximately seven whales, 3% of the maximum single whale count, remained in the lagoon 20 days longer than the overall mean season length. The mean length of the season for the four similar years was 79 ± SE 2.9 days, with a range of 6 days among years. The last day that single whales were counted in the boat transects ranged from March 12 in 1981 to April 9 in 1979 (Fig. 11). In summary, the length of the single whale season in Laguna San Ignacio averaged 11 weeks for the four similar years, but was 14.8 weeks in 1979.

To evaluate shifts in the timing of the single whale seasons, we calculated the mean day and its accompanying variance for each year and compared these among years (Reilly, 1981) (Table V). The mean day $\bar{j}$ is expressed as

$$\mu = \bar{j} = \frac{\Sigma j f_j}{n} \tag{4}$$

Table V
Summary of the Single Whale Seasonal Occupation of Laguna San Ignacio Between 1978 and 1982

Year	Number of transects	Maximum count day[a]	Duration of season	Mean day[a]	Variance of mean
1978	8	Feb 2	81	Feb 1	154.01
1979	9	Mar 1	104	Feb 16	292.74
1980	9	Feb 16	78	Feb 13	155.42
1981	12	Feb 12	76	Feb 4	130.68
1982	8	Feb 14	82	Feb 12	101.02
Average		Feb 15	84	Feb 9	

[a]Maximum count and mean days are scaled so that December 26 = day 1.

where $\bar{j}$ = the day of the season expressed with December 26 = day 1, f_j = the single whale count (as a proportion of the total count for the year), and n = the total number of days. The variance, (σ^2), is

$$\sigma^2 = \frac{\Sigma j f_j^2 - \frac{(\Sigma j f_j)^2}{n}}{n - 1} \quad (5)$$

The mean days ranged from February 1 in 1978 to February 16 in 1979. A Bartlett's test indicated that the variances for each year were homoscedastic [$B_c = 2.71$, $p = .61$], and the ANOVA indicated the means were not significantly different [$F_{(4,39)} = 2.07$, $p < .05$], permitting the calculation of a pooled mean for all years of February 9. The 95% confidence interval for this pooled mean was 14 days, from February 2 to 16. The February 1, 1978 mean day was the only date to fall outside this confidence interval. The greatest shift in timing of the single whale season was 15 days between February 1, 1978 and February 16, 1979.

The seasonal distribution of single whales in the lower, middle, and upper lagoon areas between 1978 and 1982 are shown in Figs. 7A, 8A, and 9A. At their maximum counts, single whale abundance was not the same in the three lagoon locations [Kruskal–Wallis test: $X^2_{(4)} = 12.00$, $p < .01$]. Multiple comparison tests at $\alpha = .05$ confirmed that the abundance was unique in each area; moreover, counts of single whales were heterogeneous within each area over the 5 years, but did not show a significant increase or decrease. The G-tests of equality of abundance and ANOVAs for significance of regression in each area among years are summarized in Table II.

Single whales were most abundant in the lower lagoon, where their counts were different each year [$G = 13.61 > X^2_{.05(4)} = 9.49$, $p < .01$]. Maximum counts in the lower lagoon increased 6.7% per year from 129 in 1978 to 187 in 1982 (Fig. 7A); however, this increase was not statistically significant [$F_{(1,3)} = 3.43$, $p > .10$, $r^2 = .533$] (Fig. 10A, top).

Maximum counts of single whales in the middle lagoon were also significantly different each year [$G = 14.11 > X^2_{.05(4)} = 9.49$, $p < .01$] (Fig. 8A). The slope of the regression line fitted to these counts indicated a nonsignificant 5.3% per year decrease [$F_{(1,3)} = 3.90$, $p > .10$, $r^2 = .565$] (Fig. 10A, top).

Finally, single whales were least abundant in the upper lagoon nursery area where their counts were significantly different each year [$G = 22.24 > X^2_{.05(4)} = 9.49$, $p < .05$] (Fig. 9A).

In general, single whale maximum counts were heterogeneous in all areas in all years, and their slight increase in the lower lagoon and decrease in the middle and upper lagoon areas were not significant.

The density of single whales was homogeneous within each section of the lagoon; thus, the five yearly densities were pooled and a mean density calculated for each area (Table III). In the lower lagoon, single whale density averaged 8.46 ± SE 0.53 whales/km^2, in the middle lagoon it averaged 2.36 ± SE 0.23 whales/km^2, and in the upper lagoon single whales were least dense averaging 0.54 ± SE 0.09 whales/km^2 (Fig. 10A, bottom).

The Kruskal–Wallis test revealed that the proportion of single whales among areas was significantly different [$X^2_{(2)} = 12.29$, $p < .01$]. Single whales were distributed in a gradient each year being most abundant in the lower lagoon nearest the inlet and becoming less abundant toward the upper lagoon. As would be expected, since there were no significant changes in single whale abundance or density within each area during the 5 years, a G-test of the proportions of the total single whales that were within each area did not indicate any significant changes. An average of 65 ± SE 2% of the single whale population was in the lower lagoon, 24 ± SE 2% was in the middle lagoon, and 11 ± SE 2% was in the upper lagoon (Table IV).

In summary, the distribution of single whales at their maximum counts analyzed as abundance, density, and proportion of whales in each area did not change significantly between 1978 and 1982.

Female and Calf Pairs. The pattern of abundance and distribution of female–calf pairs in Laguna San Ignacio was more complex than that of single whales. Counts of female–calf pairs increased continuously between early January and late March or early April in all years (Fig. 11B). However, following approximately mid-February, the increase was not caused by continued births; rather, late season calves were judged by their size, color, and degree of barnacle infestation to be at least 1.5 to 3.0 months old. For this reason, the analysis of female–calf pairs was conducted in two stages.

The first stage was the time during which most births occurred and females with newborn calves occupied the lagoon. It is believed that most gray whale births occur during a period of at least 66 days from December 26 to March 1 (Swartz and Jones, 1983) with a mean birth date of about January 27, and that the majority of births have taken place by February 15 (Rice *et al.*, 1981). During 6 years we observed only two newborn calves after February 15; therefore, we assumed that late births were relatively rare events and that February 15 was best representative of the date by which a high percentage of the births had taken place. Counts of female–calf pairs on February 15 were used to estimate the number of females using Laguna San Ignacio as a nursery for newborn calves.

The second part of the analysis was the time of maximum female–calf pair counts. The increase in the number of female–calf pairs that utilized the lagoon after the birth period was representative of the number of whales that used the lagoon as a staging area (see below).

The Birth Period. The seasonal distributions of females with calves from January 4 to February 15 between 1978 and 1982 are shown in Fig. 11B. The number of female–calf pairs that occupied Laguna San Ignacio on February 15, the end of the birth period. increased significantly each winter. Female and calf counts rose from 81 pairs in 1978 to 137 pairs in 1982, a significant increase of 14.3% per year [$F_{(1,3)} = 33.56$, $p < .025$, $r^2 = .914$] (Fig. 12).

The significant increase in the number of female–calf pairs was accompanied by a significant increase in their abundance in one of the three lagoon areas. The seasonal distribution of females and calves in the lower, middle, and upper lagoon are shown in

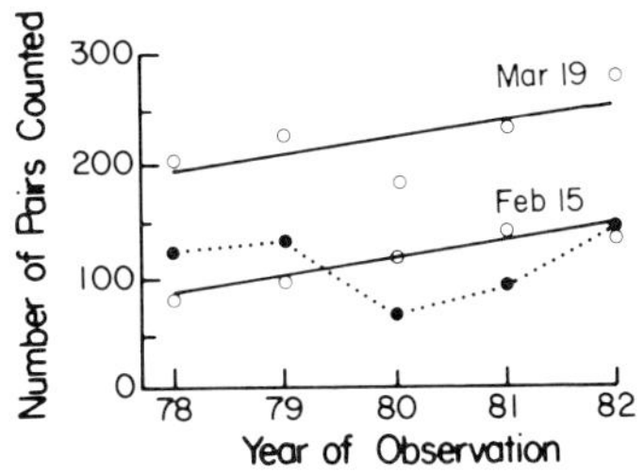

Fig. 12. *A comparison of the maximum female–calf pair counts (average date of March 19) and February 15 female–calf pair counts between 1978 and 1982 (open circles). Closed circles are the differences between these counts and represent minimum estimates of whales utilizing the lagoon as a staging area. Fitted lines are least square regression lines.*

Figures 7B, 8B, and 9B. A Kruskal–Wallis test revealed that the female–calf pair abundance was not the same in all three areas [$X^2_{(2)} = 9.555$, $p < .01$]; abundance in the upper and lower, and upper and middle lagoon areas was significantly different, but in the lower and middle lagoon it was not significantly different.

In all years, females with newborn calves were least numerous in the lower lagoon nearest the inlet. Counts were generally low, ranging from 8 pairs in 1978 to 24 pairs in 1979 and 1981, and were significantly different [$G = 15.54 > X^2_{.05(4)} = 9.49$, $p < .001$] (Table II). The slope of the least squares regression line fitted to these counts indicated that female–calf pairs increased 32.2% per year, but this trend was not statistically significant [$F_{(1,3)} = 1.57$, $p > .10$, $r^2 = 0.345$] (Fig. 10A, top).

Counts of female–calf pairs in the middle lagoon were also significantly different among years [$G = 11.76 > X^2_{.05(4)} = 9.49$, $p < .025$] (Table II), but their 25.8% per year increase was not significant [$F_{(1,3)} = 2.43$, $p > .05$, $r^2 = .448$] (Fig. 10A, top).

The increase in the annual female–calf pair counts was seen primarily as a significant positive rate of increase in their numbers in the upper lagoon nursery (Table II). Counts in this area were different each year [$G = 31.61 > X^2_{.05(4)} = 9.49$, $p < .001$] increasing significantly 16.2% per year from 55 pairs in 1978 to 107 pairs in 1982 [$F_{(1,3)} = 14.96$, $p < .05$, $r^2 = .833$] (Fig. 10A, top).

The densities of female and calf pairs within each area and the values for the Kolmogorov–Smirnov tests and ANOVAs are presented in Table II. Densities of female–calf pairs in the lower and middle lagoon did not change significantly among years and averaged 9.83 ± SE 0.72 and 1.06 ± SE 0.17 whales/km^2 respectively (Fig. 10A, bottom). For the upper lagoon, the Kolmogorov–Smirnov test suggested that the differences in whale densities among years were nonsignificant; nevertheless, the ANOVA for the least squares regression analysis of density showed that the steady increase in density from 2.36 whales/km^2 in 1978 to 4.59 whales/km^2 in 1982 was statistically significant [$F_{(1,3)} = 14.59$, $p < .05$, $r^2 = .829$] (Fig. 10A, bottom). The conflicting results of the Kolmogrov–Smirnov test and the ANOVA can best be explained by the fact that the regression analysis is sensitive to the small but steady increase in density from year to year.

On February 15 each year, female–calf pairs were concentrated in the upper lagoon nursery area and became less abundant toward the inlet. In addition, the percentage of the total female–calf pairs in the lagoon that occupied each area remained essentially the same each year (Table IV). The percentage of female–calf pairs was smallest in the lower lagoon where they averaged 11 ± SE 2% of the total, the middle lagoon contained an average of 9 ± SE 9%, and the upper lagoon had the highest percentage of female–calf pairs, averaging 80 ± SE 2%. Although the number of female–calf pairs utilizing the upper lagoon increased significantly, the proportion of the population utilizing this area remained the same each year.

Maximum Female–Calf Pair Counts. The maximum count of female–calf pairs occurred late each season following the departure of single whales from the lagoon. After February 15, females and calves continued to increase steadily to their maximum counts that occurred as early as March 2 in 1982 and as late as April 3 in 1979, with a mean date of March 19 (Fig. 11B). The maximum counts increased approximately 6.9% per year from 202 pairs in 1978 to 281 pairs in 1982; however, this increase was not statistically significant [$F_{(1,3)} = 2.69$, $p > .25$, $r^2 = .473$] (Fig. 12).

To determine the degree to which Laguna San Ignacio was utilized as a staging area, we fitted a least squares regression line to the annual February 15 counts and the annual maximum female–calf counts and compared their slopes and elevations. There was no significant difference between the slopes of the lines [$t_{(6)} = 0.058$, $p > .80$], but the elevations were significantly different [$t_{(6)} = 13.17$, $p << .001$]. The difference between the elevations of the lines averaged 110 ± SE 0.7 female–calf pairs and represented a minimum estimate of the number of female–calf pairs that were gathering at Laguna San Ignacio late in the season and using it as a staging area. The maximum counts, the February 15 counts, and the difference between them are plotted in Fig. 12. The average difference was 111 ± SE 12 female–calf pairs and ranged from 67 in 1980 to 144 female–calf pairs in 1982. As mentioned previously, this late season increase included older calves and not newborn animals.

Following the combined maximum counts, single whales began to decrease and averaged only 3% of their maximum count by March 14. As the single whales departed, the female–calf pairs continued to increase and gradually shifted their distribution to the lower and middle lagoon (Figs. 7B and 8B); they were seen only rarely in the upper lagoon nursery after the third week in March (Fig. 9B). At their maximum counts, the abundance of females and calves was not the same in all three areas [Kruskal–Wallis test: $X^2_{(2)} = 10.86$, $p < .005$]. The multiple comparison tests of equality of abundance among years indicated that the lower and upper, and lower and middle lagoon areas were significantly different from each other, while the middle and upper areas were not significantly different.

In all years, females and calves were most abundant in the lower lagoon where their counts averaged 186 ± SE 6 pairs and did not change significantly (Table II; Fig. 10B, top).

In the middle lagoon, the number of female–calf pairs was significantly different each year [$G = 19.90 > X^2_{.05(4)} = 9.49$, $p < .001$], and their counts increased significantly 22.9% per year from 16 pairs in 1978 to 49 pairs in 1982 [$F_{(1,3)} = 11.80$, $p < .05$, $r^2 = .797$] (Table II; Fig. 10B, top).

Female–calf pairs were least abundant in the upper lagoon nursery area in late March each year (Fig. 9B). Here, counts ranged from no whales in 2 years to 33 pairs in 1982, and were significantly different each year [$G = 74.59 > X^2_{.05(4)} = 9.49$, $p < .001$] (Table II; Fig. 10B, top). The slope of the regression line fitted to the upper lagoon counts was not significantly different from zero [$F_{(1,3)} = 1.44$, $p > .50$, $r^2 = .324$].

Densities of female–calf pairs in the lower and upper lagoon areas did not change significantly between 1978 and 1982, averaging 21.3 ± SE 0.7 whales/km^2 and 0.4 ± SE 0.2 whales/km^2 respectively (Table II; Fig. 10B, bottom). As would be expected from the significant increase in abundance of female–calf pairs in the middle lagoon, the density of these whales also increased significantly in this area from 1.4 whales/km^2 in 1978 to 4.28 whales/km^2 in 1982 [$F_{(1,3)} = 11.88$, $p < .05$, $r^2 = .80$].

The percentage of the female and calf population in the lagoon that was in each area remained essentially the same (Table IV). The lower lagoon averaged 87.4 ± SE 6.0%, the middle lagoon contained 11.6 ± SE 1%, and the upper lagoon averaged 1.0 ± SE 0.6% of the population each year.

To summarize, at their maximum counts, female calf pairs were distributed in a gradient being most dense in the lower lagoon near the inlet and becoming less dense with increasing distance from the inlet. This gradient was consistent each year, as the abundance of female–calf pairs in each area and the percentage of the lagoon female–calf pair population that utilized each area remained the same.

Female and calf counts generally declined after mid-March, although a substantial number of female–calf pairs remained in the lagoon at the termination of our field work each April. The percentage of the maximum female–calf pair count remaining in the lagoon at the last transect each winter was 53% in 1978, 68% in 1979, 96% in 1980, 66% in 1981, and 96% in 1982.

Time limitations of field studies prevented us from documenting the entire female–calf pair occupation of the lagoon. Therefore, to estimate their minimum abundance and duration of season, we utilized late April and early May sightings of whales reported by other observers. J. Harvey and B. R. Mate (personal communication) reported that 25–35 female–calf pairs remained in Laguna San Ignacio and that an unknown number were outside the lagoon inlet in the breakers on April 26, 1980. In addition, Storro-Patterson (1982) reported that a few (no number specified) females and calves remained in the inlet in early May; and Mexican fishermen (Abreojos Fishing Cooperative, personal communication) indicated that gray whales could be found outside the lagoon in the inlet as late as early June of some years. The earliest arrival of female–calf pairs in Laguna San Ignacio during our study was December 26. The combination of our early season observations with these late season sightings yielded an estimate of a 4- to 4.5-month (16- to 18-week) duration of season for female–calf pairs in Laguna San Ignacio. This was 1–1.5 months longer than single whales that arrived at approximately the same time, but departed from the lagoon by mid-March.

Table VI
Summary of Length Measurements and Sex Composition of Dead Gray Whale Calves in Four Breeding Areas, 1954–1983

Location	Year	Number of calves: Total	Number of calves: Number measured	Body length (m): $\bar{X} \pm$ SE	Body length (m): Range	Sex: M	Sex: F	Sex: ?	Reference
Laguna San Ignacio	1977	4	4	4.56 ± 0.13	4.40–5.00	—	—	4	Swartz and Cummings (1978)
	1978	5	4	4.45 ± 0.12	4.10–4.80	1	3	1	Swartz and Jones (1979)
	1979	4	3	4.63 ± 0.14	4.42–4.98	2	1	1	Swartz and Jones (1980)
	1980	4	4	4.87 ± 0.14	4.40–5.10	4	0	0	This chapter
	1981	4	4	4.43 ± 0.22	4.04–5.16	3	1	0	Swartz and Jones (1981)
	1982	4	4	4.20 ± 0.22	3.84–4.95	2	2	0	This chapter
	1983	9	6	4.42 ± 0.29	3.76–5.94	5	2	2	Dahlheim (1983)
	Total	34	29	$\bar{X}_p$ = 4.47 ± 0.09					
Laguna Ojo de Liebre	1954–57	12	7	4.05 ± 0.14	3.54–4.51	2	—	10	Gilmore (1960)
	1962	6	6	4.68 ± 0.19	3.95–5.40	4	2	0	Eberhardt and Norris (1964)
	1975	20	13	4.54 ± 0.10[a]	3.96–5.05	5	8	7	White and Griese (1978)
	1979	4	4	4.47 ± 0.16	4.00–4.88	1	2	1	D. W. Rice (personal communication)
	1980	12	8	4.30 ± 0.09	3.84–4.60	3	3	2	Rice *et al.* (1981)
	1982	10	10	4.31 ± 0.15	3.45–4.94	4	3	3	Fleischer *et al.* (1983)
	1983	12	7	4.48 ± 0.18	4.00–5.47	7	3	2	Fleischer *et al.* (1983)
	Total	76	55	$\bar{X}_p$ = 4.40 ± 0.05					
Isla Arena[b]	1980	20	0	— —	—	—	—	20	Chapter 15, this volume
	1981	10	10	4.46 ± 0.13	3.76–5.05	5	1	4	Bryant *et al.* (1981), Chapter 15, this volume
	1982	20	18	4.42 ± 0.15	3.5 –6.4	3	1	16	Bryant and Lafferty (1982), Chapter 15, this volume
	1983	2	1	4.37 —	—	0	2	0	Fleischer *et al.* (1983)
	Total	52	29	$\bar{X}_p$ = 4.43 ± 0.10					
Laguna Guerrero Negro	1980	7	2	3.77 ± 0.23	—	—	2	5	Bryant (1980), Chapter 15, this volume
	1981	4	4	4.49 ± 0.13	4.27–4.95	2	2	0	Bryant *et al.* (1981), Chapter 15, this volume
	1982	2	2	3.85 ± 0.11	3.7 –4.0	1	—	1	Fleischer *et al.* (1983)
	1983	1	1	3.90 —	—	0	1	0	Fleischer *et al.* (1983)
	Total	14	9	$\bar{X}_p$ = 4.13 ± 0.13					
Boca de Soledad	1982	6	4	4.50 ± 0.16	4.13–4.92	1	3	2	Lawson (1983)
				Grand $\bar{X}$ = 4.43 ± 0.06	Σ =	55	42	81	

[a]Individual lengths were not reported; mean length was used in the calculation of the pooled mean.
[b]Isla Arena separates the inlets of Guerrero Negro and Ojo de Liebre lagoons.

MORTALITY

The basic unit of the mortality section is the Laguna San Ignacio data from 1977 to 1982. In brief, we give a general account of mortality and then examine four aspects of mortality, (1) the number of dead whales, (2) whale lengths, (3) age class composition, and (4) the calf mortality rate. In addition, data on natural mortality from 1954 to 1983 which were available for Laguna Guerrero Negro, Laguna Ojo de Liebre, and the Boca de Soledad area of the Bahia Magdalena complex are compared and integrated with the Laguna San Ignacio information to determine (1) the age class composition of dead whales, (2) the proportion of carcasses in each age class among years, (3) the mean length of dead calves for comparison with the length of near-term fetuses, and (4) the sex ratio of calves and all whales combined. A summary of calf mortality is presented in Table VI, and immature and adult mortality is shown in Table VII.

Age frequencies were determined by grouping the carcasses into three broadly defined categories—calves, immatures, and adults. Animals between 3 and 8.99 m were considered to be calves of the year, those measuring 9–10.99 m were regarded as immatures, and whales 11 m and larger were classified as adults (based on Rice and Wolman, 1971; Zimushko and Ivashin, 1980; and Chapter 21 , this volume).

Table VII
Summary of Length Measurements and Sex Composition of Dead Immature and Adult Gray Whales in Three Breeding Areas[a]

Location	Dates of observations	Dates of sighting	Age[b] class	Length (m)	Sex[c]	Source
Laguna San Ignacio	1977–1983	1978	I	8.41	M	Swartz and Jones (1979)
		1978	A	13.80	U	Swartz and Jones (1979)
		1979	I	10.40	U	Swartz and Jones (1980)
		1980	I	8.00	U	This chapter
		1980	I	7.40	U	This chapter
		1981	I	7.51	U	Swartz and Jones (1981)
		1981	I	7.50	U	Swartz and Jones (1981)
		1983	I	10.41	F	Dahlheim (1983)
Laguna Ojo de Liebre	1980, 1982, 1983	1980	A	—	U	Bryant (1980)
		1982	I	7.82	U	Fleischer *et al.* (1983)
		1982	A	11.00	U	Fleischer *et al.* (1983)
		1983	I	—	U	Fleischer *et al.* (1983)
Isla Arena[d]	1980–1983	1980	A	—	U	Bryant (1980)
		1982	A	—	U	Bryant and Lafferty (1982)
		1983	A	11.00	F	Fleischer *et al.* (1983)
Laguna Guerrero Negro	1980–1983	1980	I	8.4	M	Bryant (1980)

[a]No dead immature or adult gray whales were sighted in Boca de Soledad in 1982.
[b]A = adult, I = immature.
[c]M = male, F = female, and U = sex unknown.
[d]Isla Arena separates the inlets of Guerrero Negro and Laguna Ojo de Liebre lagoons.

Thirty-two recently dead gray whales including 25 calves, 6 immatures, and 1 adult were examined in Laguna San Ignacio. The cause of death was not determined for any of these whales; however, it appeared that one immature died as the result of a killer whale (*Orcinus orca*) attack and that many calves were stillborn or had died shortly after birth as evidenced by intact umbilical cords or bleeding umbilical wounds that had not healed prior to death. In addition, the 23 calves that were discovered in January and February lacked epizoic barnacles compared to 2 stranded calves discovered in March which possessed these parasites.

The minimum calf mortality rate for each 97-day season in Laguna San Ignacio from 1978 to 1982 was estimated by dividing the number of dead calves discovered per season by the estimated gross calf production per season (i.e., the number of living calves at the maximum combined count + the number of dead calves per season) (Caughley, 1966, 1977). The calf mortality rate was 5.81% in 1978, 4.00% in 1979, 3.25% in 1980, 2.78% in 1981, and 2.84% in 1982.

The average, minimum, calf mortality rate for the 5-year period was calculated as the sum of the dead calves discovered in all years divided by the sum of the estimated gross production for all 5 years, or

$$\frac{21}{573 + 21} = 0.0354 \qquad (6)$$

A G-test indicated that the seasonal calf mortality rates were not significantly different from the mean rate for all five seasons [$G_{(4)} = 3.67 < X^2_{.05(4)} = 9.49$, $p > .25$].

To determine whether the number of dead calves, immatures, and adults discovered per year in Laguna San Ignacio was the same over the 5 years, each age category was tested for heterogeneity by computing a G-statistic of the difference between the observed mortality per year and the mean number of carcasses per class. All categories were homogenous among years; the number of dead calves ranged from 4 to 5 per year and averaged 4.17 whales per year [$G = 0.15 < X^2_{.05(4)} = 9.49$, $p > .99$] (Table VI). The number of immatures ranged from 0 to 2 per year with a mean of 0.67 whales per year [$G = 1.64 < X^2_{.05} = 9.49$, $p > .75$]. And finally, the number of dead adults ranged from 0 to 1 with a mean of 0.17 whales per year [$G = 3.49 < X^2_{.05(4)} = 9.49$, $p > .10$] (Table VII). Because the frequency of carcasses within each category was not significantly different during the 5 years, we pooled the frequencies per class, and calculated the percentage of the total sample that was in each age class. Seventy-eight percent were calves, 19.5% were immatures, and 2.5% were adults.

In order to compare age class frequencies of dead whales from different breeding areas in which there was unequal search effort within years, we converted the total number of whales in each category to a percent of the total number of whales counted in each area. For each age class, we pooled the percentages from all areas, including Laguna San Ignacio, and calculated a mean percentage. These means were used as the expected frequencies in three separate G-tests of the null hypothesis of equality between the mean proportion and the observed proportion per class for each area. The percentages of calves among areas were homogeneous, ranging from 78 to 100% with a mean of 91.4% [$G = 3.13 < X^2_{.05(4)} = 9.49$, $p > .50$]. The percentages of immatures, however, were significantly different in each area, and ranged from 0 to 19.5% [$G = 42.07 > X^2_{.05(4)} = 9.49$, $p < .001$]. The percentages of dead adults were also significantly different in each area, ranging from 0 to 5% [$G = 12.73 > X^2_{.05(4)} = 9.49$, $p < .025$]. Due to the small sample sizes for dead immatures and adults, these estimates may not be representative.

Sex was determined for 101 of the 194 dead gray whales reported from all the lagoons. These included 57 males and 44 females (Tables VI and VII). The sex ratio of calves, immatures, and adult whales combined was 56.4% males to 43.6% females. A G-test with a William's correction (Sokal and Rohlf, 1981) indicated that the observed ratio of dead male to female whales was not significantly different from 1:1. Here, the expected sex ratio was 1:1, as reported by Rice and Wolman (1971, p. 118).

To determine whether male and female calves died in equal numbers, we tested for a difference in the sex ratio of these animals. Of the 125 calves from all areas combined, sex was determined for 97. The ratio of 55 males (56.7%) to 42 females (43.3%) was not significantly different from unity [$G_{adj}(1) = 1.747 < X^2_{.05(1)} = 3.841$, $.10 < p < .20$]. This suggests that the slight preponderance of male calves was not indicative of differential mortality between the sexes.

Sample sizes of the immature and adult age classes were too small to evaluate the significance of their observed sex ratios.

A Bartlett's test for homoscedasticity of calf carcass lengths for samples among years from Laguna San Ignacio between 1977 and 1983 indicated that the variances were not significantly different [$B_c = 6.99 < X^2_{.05(6)} = 12.59$, $p = .32$]. The ANOVA disclosed that the mean lengths for each year were also not significantly different [$F_{(6,22)} = 0.866 < F_{.05(6,22)} = 2.55$, $p = .53$]. This permitted the calculation of a pooled mean for calf carcass lengths in Laguna San Ignacio which was 4.47 ± SE 0.086 m. The mean calf lengths per year are shown in Table VI.

The lengths of dead whales in different breeding areas were tested for significant differences. A Bartlett's test revealed that the samples from each breeding area were highly heteroscedastic [$B_{(c)} = 200 >> X^2_{.05(4)} = 12.59$, $p < .001$]; but, the Kruskal–Wallis test of the null hypothesis that the "location" of the sample populations were the same was not rejected. Therefore, we concluded that the lengths of the dead calves from each area were not significantly different [$H_c = 4.79 < X^2_{.05(4)} = 9.49$, $p > .25$]. The grand mean of the lengths of dead calves for all areas was 4.43 ± SE 0.06 m.

The lengths of dead calves have been used as indicators of gray whale calf length at birth (Rice and Wolman, 1971; Rice *et al.*, 1981; Rice, 1983). To determine the validity of extrapolating the mean length of calves at birth from dead calf data, we compared the lengths of dead calves reported from the lagoons to lengths of near-term fetuses.

The series of length measurements for near-term fetuses reported by Rice and Wolman (1971) was the only sample presently available that was large enough to provide statistically reliable data for comparison with the sample of length measurements of recently born, dead calves from the breeding lagoons. Other data on lengths of near-term fetuses taken from females in the Korean or California populations were reported by Scammon (1874), Townsend (1887), Andrews (1914), Risting (1928) and Mizue (1951). These accounts, however, were of limited use because many of the calf lengths were estimated and not actually measured, ranges of lengths were often given rather than individual lengths, and the collection dates of individual whales were frequently not reported.

Rice and Wolman (1971) measured 55 near-term fetuses taken from female gray whales collected off San Francisco during a 38-day period from December 15 to January 20 from 1959 to 1969. These ranged in length from 3.60 to 5.31 m with a mean of 4.62 ± SE 0.007 m. To ascertain whether the "location" of the calf lengths from the lagoons was significantly different from the length of near-term fetuses, we tested the null hypothesis that the lengths of dead gray whale calves in the southern range were greater than or equal to the length of near-term fetuses collected during migration. A one-tailed Mann–Whitney U-test failed to accept the null hypothesis [$Z = 3.819 > t_{.05(1),\infty} = 1.96$, $p < .005$]. Therefore, we concluded that the mean length of 4.43 m for dead neonates from the southern range in January and February was highly significantly smaller than the mean length of 4.62 m for near-term fetuses collected during migration.

Seven (78%) of the nine immature whales measured in three breeding areas fell into the yearling (8.2 to 9.6 m) age category, suggesting that next to calves, yearlings had the second highest mortality rate. The lengths of immatures ranged from 7.4 to 10.41 m with a mean length of 8.20 ± SE 0.43 m (Table VII).

The analysis of the lengths of some immatures was problematic. According to Rice and Wolman (1971), Zimushko and Ivashin (1980), and Blokhin (Chapter 21, this volume), the length of a gray whale at 1 year of age should range from 8.2 to 9.6 m; but, four of the seven dead immatures in Laguna San Ignacio were too small to be classified as yearlings and two were too large to be classified as calves of the year (>7.3m). All of these carcasses were infested with adult barnacles, further suggesting they were yearlings, as newborn calves do not develop adult parasites while in the breeding lagoons.

Only one dead adult whale was discovered in the lagoon. In addition to the whale mortality in Laguna San Ignacio, five adult and three immature carcasses were reported from other lagoon areas, and all are summarized in Table VII. Sample sizes of these whales were too small for statistical comparisons.

WHALE-WATCHING ACTIVITIES

Nature of Human Activities

Human activities within Laguna San Ignacio during the winter included local fishing, U.S. based overland and seagoing commercial whale-watching excursions, and visitation in private yachts.

Local Mexican fishermen operated 4- to 6-m fiberglass and wood pongas powered by 40- to 65-hp outboard engines throughout each winter, but their numbers and working days fluctuated with the condition of the fisheries and the weather. During the winters of 1978 and 1979, six to eight pongas operated approximately 5 days each week in the lower lagoon. Following the designation of the lagoon as a refuge in 1979, only three pongas operated in the lower lagoon between 1980 and 1981. In 1982, again only three pongas regularly operated in the lower lagoon; however, between 15 and 18 pongas with hooka-assisted divers operated in the middle and upper lagoon areas in response to an emergent bay-scallop (*Pectin* sp.) fishery in the shallow eelgrass beds north of Punta Piedra.

Individuals aboard private yachts continued to visit the lagoon following its designation as a refuge, averaging six visits each winter lasting from 1 to 10 days each.

Between 1977 and 1979, three or four land-based whale-watching excursions visited Laguna San Ignacio, but these companies ceased operation when the refuge was established in 1979 because the regulations prevented their access to the lower lagoon whale-watching area which can only be reached by boat.

Seagoing whale-watching excursions based in San Diego, California accounted for 90% or more of the human activities within the lagoon each winter. These excursions were conducted aboard 100-ton vessels that carried an average of 25 passengers each and typically spent 2½ days (3 days and 2 nights) or approximately 60 hr in the lagoon each visit. These vessels are powered by two or three diesel engines (V-12 to V-16, 550 to 965 hp range) with two 60- to 100-kW diesel generators (Fig. 13). During a typical visit to the lagoon, these vessels were underway with their main engines running approximately 1.5 hr (2% of 60 hr) and moved around the lagoon at speeds between 5 and 8

Fig. 13. *Typical whale-watching excursion vessel in Laguna San Ignacio. Note gray whale spyhopping in foreground.*

knots; total running time within the lagoon included entry (30 min), at least one change of anchorage (30 min), and their departure (30 min). The remainder of the time (97.5% of 60 hr) these vessels were at anchor with only their power generators operating. While underway they produced broadband waterborne noise ranging from 125 Hz to 20 kHz at source levels ranging from 124 to 170 dB *re.* μPa at 1 m depending upon hull construction and engine type (Dahlheim, 1984; Chapter 22, this volume).

Each excursion vessel launched three or four inflatable, wooden, or aluminum 4- to 5-m skiffs powered by 20-hp outboard engines to take their passengers whale watching and to visit various places of interest within the lagoon. These skiffs normally operated 8 hr per day between 07:00 and 16:00 hr for approximately 24 running hr per visit (40% of 60 hr); during the remaining 36 hr the skiffs were not in operation. Besides the presence of vessels at anchor (generators on), whales did not have to contend with skiff activity at night. Therefore, whales were followed less than 50% of the day. While whale watching at idle speeds, skiff waterborne noise source levels ranged from 75 to 120 dB *re.* μPa at 1 m, and while motoring at half throttle, source levels ranged from 105 to 120 dB *re.* μPa at 1 m (Dahlheim, 1984; Chapter 22, this volume).

Whale-Watching Excursion Visitation

The number of whale-watching excursions that called at Laguna San Ignacio during the 97-day period from January 5 to April 15 between 1978 and 1982 is summarized in Fig. 14.

In 1978, 28 commercial whale-watching excursions visited Laguna San Ignacio during 50 days (51%) of the 97-day winter season. The number of these excursions increased 28% in 1979 to 36 excursions visiting during 56 days (58%) of the winter, and in 1980 they increased another 5.5% to 38 excursions during 72 days (73%) of the winter.

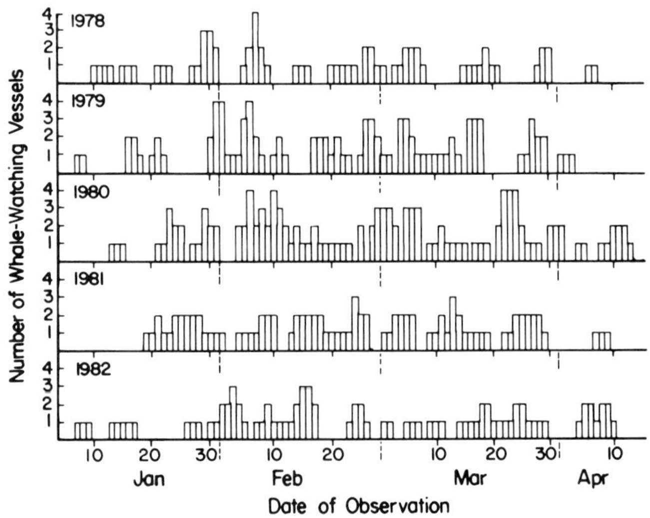

Fig. 14. *The number of whale-watching excursion vessels that visited Laguna San Ignacio from January 5 to April 12 between 1978 and 1982.*

Table VIII
Whale-Watching Excursion Vessel Visitations to Laguna San Ignacio during each 97-day[a] Winter Season between January 5 and April 12 for the Years 1978 to 1982

Year	Number of trips/passengers	Vessel days		Number of vessels per vessel day				Vessels with engines engaged[b]		Vessels at anchor[c]		Skiffs operating[d]	
		Number	% season	1	2	3	4	hr/season	% season	hr/season	% season	hr/season	% season
1978	28/700	50	51	36	11	2	1	42	1.8	800	34	400	17
1979	36/900	56	58	29	15	9	3	54	2.3	896	38	448	19
1980	38/950	72	73	34	23	10	5	57	2.4	1152	49	576	25
1981	28/700	60	62	31	27	2	0	42	1.8	960	41	480	21
1982	30/750	63	65	46	14	3	0	45	1.9	1008	43	664	29
Average	32/800	60	62	35	18	5	2	45	2.0	963	41	514	22

[a]Ninety-seven days = 2328 hr.
[b]Engines engaged based on 1.5 hr per trip.
[c]At anchor with generators running based on 16 hr per vessel day.
[d]Skiffs in operation based on 8 hr per vessel day.

In 1981, the number of whale-watching excursions declined by 28% (primarily in response to a 125% fare increase from 1978) to 28 excursions visiting during 60 days (62%) of the winter and then increased slightly in 1982 to 30 excursions visiting during 63 days (65%). The number of days without vessel activity within the lagoon declined from 49% in 1978 to 27% in 1980 and increased to 35% in 1982. Between 1978 and 1982, the number of excursion vessels within the lagoon at any one time averaged one vessel during 35 days (40%), two vessels during 18 days (20%), three vessels during 5 days (5%), four vessels during 2 days (2%), and no vessels in the lagoon the remaining 33% of the time (Table VIII).

Although the Mexican refuge regulations and commercial vessel permits allow only two commercial vessels in the lagoon at any time, three to four vessels occupied the lagoon together on 32 days over the 5 years. Because of the extensive shoals and strong currents at the lagoon inlet, multiple entries and exits occurred when one vessel followed another into or out of the lagoon, particularly if one operator was more familiar with the inlet channels. At these times excursion trips overlapped and when a third or fourth vessel arrived on the day that another vessel was scheduled to depart. This overlap resulted in three to four vessels occupying the lagoon simultaneously for approximately ½ day. On other occasions more than two vessels occupied the lagoon for one to three days. In both situations, there was a concomitant cumulative increase in the noise disturbance.

Ambient noise in the lagoon comes from both biological and nonbiological sources. Dahlheim *et al.* (Chapter 22, this volume) reported that levels of biological noise were lowest below 2 kHz, increased to high levels between 2 and 5 kHz, and declined gradually through 20 kHz. The average ambient noise levels attributable to biological sources (excluding cetaceans) ranged from 94 to 110 dB *re* μPa. The amount of nonbiological waterborne noise of human origin introduced into the lagoon habitat each season was dependent upon the number of hours that vessels were in the lagoon, the class of the vessel, and the amount of time each vessel or skiff was engaged in various activities.

One-hundred-ton class excursion vessels while underway produced the greatest levels of noise; however, on the average, these vessels ran their main engines only 48 hr of each 2328 hr- (97-day) season or approximately 2.0% of the winter. While at anchor, with only their power generators running, excursion vessels produced noise levels of 70 to 110 dB re. μPa at 1 m (Chapter 22, this volume) for approximately 963 hr or 41% of each winter. Outboard-powered skiffs were active an average of 514 hr or 22% of each winter season (Table VIII) and produced noise levels between 75 and 120 dB re. μPa at 1 m (Chapter 22, this volume).

In 1978, excursion vessels introduced noise distrubance during 51.5% (50 days) of the 97-day winter season. By 1980 these trips were introducing noise during 73% (72 days) of the winter, but the noise disturbance declined to 65% (63 days) by 1982. The 22.5% increase in the amount of time vessel noise was present in the lagoon between 1978 and 1980 represented a substantial change in the acoustical environment of the lagoon, particularly in the low-frequency range characteristically used by gray whales to communicate (Dahlheim, 1984). The 1980 winter saw the greatest amount of visitation and the greatest amount of nonbiological noise introduced into the lagoon.

Vessel Versus Nonvessel Day Comparison

Our goal of comparing gray whale abundance and distribution during control and experimental conditions was not fully realized. Almost continuous vessel visitation in some years combined with unstable winter weather severely hindered the opportunity to obtain data for pairwise comparisons between consecutive days with and without whale-watching activity. Only seven pairs of days were satisfactory for analysis.

To evaluate the short-term impact of whale-watching disturbance on gray whale demography in Laguna San Ignacio, we tested the hypotheses that gray whale abundance and distribution on days with and without tour vessel activity remained the same in the lagoon as a whole, in the lower lagoon where whales and human activities occurred simultaneously, and in the upper lagoon nursery area.

There was no evidence that gray whales moved out of the lagoon on days during the presence of whale-watching activities. In our three samples of paired-days prior to and following the arrival of a vessel, lagoon whale abundance increased in two instances and declined in one. In four samples of paired-days prior to and following the departure of a vessel, whale abundance decreased three times and increased only once. A Wilcoxon's paired-sample test (Sokal and Rohlf, 1981) of the nonvessel control days

Table IX
Comparison of Whale Abundance on Paired Days with and without Vessel Activity in Laguna San Ignacio, 1978–1982

Test comparison days	Date	Number of vessels	Number of whales: Entire lagoon	Lower lagoon	Upper lagoon
Preceeding vessel arrival versus vessel present	Jan 15	0	64	18	28
	Jan 16	2	86	24	43
	Mar 24	0	155	113	15
	Mar 25	1	139	92	15
	Apr 1	0	219	174	—
	Apr 2	2	234	205	—
Vessel present versus following vessel departure	Feb 1	2	129	64	45
	Feb 2	0	138	54	41
	Feb 9	1	245	126	61
	Feb 10	0	235	92	53
	Feb 26	2	137	74	23
	Feb 27	0	108	50	19
	Mar 7	1	107	66	10
	Mar 8	0	82	56	11
Wilcoxon's paired-sample test			$T_s = 5$ $p > .20$ $n = 7$	$T_s = 4$ $.10 < p < .20$ $n = 7$	$T_s = 1$ $.05 < p < .10$ $n = 6$

versus the vessel disturbance, experimental days revealed that the observed differences were not statistically significant [$T_s = 5 > t_{.05(2)7} = 2$, $.10 < p < .20$] (Table IX).

We suspected that any change in gray whale abundance in the area designated for vessel activity might be local, possibly masked, and not detected when the overall lagoon abundance was used as an estimator. We therefore repeated the above analysis using whale abundance data for only the lower lagoon. Whales were slightly more abundant in the lower lagoon while vessels were present. With the arrival of a vessel in the lower lagoon, whale abundance increased twice and decreased once. Following the departure of a vessel from the lower lagoon, whale abundance decreased in four instances; however, these differences in abundance were not significant [$T_s = 4 > t_{.05(2)7} = 2$, $.10 < p < .20$].

Finally, there was no conclusive evidence of an alteration in abundance or distribution of whales in the nursery area relative to the presence or absence of vessel activity in the lower lagoon. In two samples of paired days prior to and following the arrival of a vessel, whale abundance increased once and was unchanged once. In four samples of days prior to and following the departure of a vessel, female-calf pair abundance decreased in three instances and increased only once. The Wilcoxon's test revealed these changes were not significant [$T_s = 1 > t_{.05(2)7} = 0$, $.05 < p < .10$].

Friendly Whales

One of the unexpected effects of commercial whale watching in Laguna San Ignacio was the appearance of curious or "friendly whales." Friendly whales, as nicknamed by Gilmore, were first described for Laguna San Ignacio in 1975–1976 by Gilmore (1976), Wolfson (1976), and B. Cauble (personal communication), who reported that a few gray whales were soliciting human attention by deliberately approaching tour boats and whale-watching skiffs and allowing the passengers to pet them (Fig. 15).

During the course of this study, the friendly whale phenomenon rapidly expanded. In 1977–1978 friendly or curious whale behavior was a rare event involving a small number of individual whales (Swartz and Cummings, 1978; Larsen, 1978). But, by 1982 the incidence of this behavior had increased dramatically such that we personally had encounters with at least 200 friendly whales during 70 days of our 97-day field season. Moreover, virtually all whale-watching excursions that visited the lagoon between 1980 and 1982 have had encounters with friendly whales. For instance, in 1981, 26 of the 28 excursion vessels had contacts with curious whales, and nearly all 700 passengers were able to "pet a whale." Similarly, in 1982, each of the 30 excursions had repeated encounters with friendly whales (Swartz and Jones, 1981; Jones and Swartz, 1984).

Reports by investigators working in other areas indicate that friendly whale behavior has spread and is no longer unique to Laguna San Ignacio. At La Entrada (Bahia Magdalena) in 1979, Norris *et al.* (1983) reported two instances of friendly whale behavior when whales approached their research vessel, the *Regina Maris*, while she lay at anchor. In 1982, a chronic friendly whale that had been photographed for four consecutive winters in Laguna San Ignacio was encountered by researchers working in Laguna Guerrero Negro (Chapter 15, this volume). During the same year in Laguna Ojo

Fig. 15. *Friendly gray whale calf allowing whale watchers to pet it on the head. Note the tip of the mother whale's rostrum below and between calf and skiff.*

de Liebre, a team of Mexican scientists reported an encounter with a "curious" whale (Anonymous, 1983[5]). Similarly, in 1983, Lawson (1983) encountered a friendly whale in the Boca de Soledad area of Bahia Magdalena that nudged and rubbed against the skiffs tied astern a large excursion vessel. Most recently, a gray whale that was feeding along the west coast of Vancouver Island, British Columbia approached a skiff and exhibited friendly behavior (Hudnall, 1983). This whale remained in the area for a week or more where it was "friendly" to numerous vessels (J. D. Darling personal communication). Lastly, while recording gray whale vocalizations on the Bering Sea feeding grounds in July of 1982, M. Dahlheim (personal communication), who has experienced friendly whales in Laguna San Ignacio (Chapter 21, this volume), encountered a friendly whale that followed her boat for approximately 20 min.

Salient features of friendly whale behavior in Laguna San Ignacio have included: (1) participation by whales of all sex–age classes, (i.e., females and calves, males, females, groups of adults and immatures, and single individuals, (2) a seeming attraction toward a skiff with its engine idling in neutral gear, (3) frequent approaches to skiffs from astern, as though curious about the source of the engine sound, (4) frequent departures when the engines were shut off, (5) blowing large bubble bursts under and around the skiffs,

[5]Cetacean workshop discussion, VIII Reunion Internacional Sobre Mamiferos Marinos de Baja California, La Paz, Baja California Sur, Mexico.

(6) producing airborne "Bronx cheer" sounds from slightly pursed blowholes, (7) allowing physical contact (patting, rubbing, stroking, etc.) by passengers, (8) probing and frequently lifting the skiff, (9) occasionally jostling skiffs hard enough to knock a passenger overboard, (10) approaching and remaining near or directly alongside excursion vessels at anchor, (11) an accumulative effect where the presence of a single curious whale or female–calf pair appeared to attract additional passing whales creating groups of friendly whales of up to 20 individuals, (12) whales leaving the company of others to follow a skiff, (13) chronic curious behavior from some individual whales which continued throughout an entire day, and from week to week, month to month, and year to year for up to 6 years, (14) following slow moving whale-watching skiffs for up to 5 km and resuming friendly behavior when the skiffs stopped, (15) following skiffs attempting to leave an area, even at speeds in excess of 11 km/hr (6 knots).

The large majority of curious whale encounters have occurred without incident. Most friendly whales were cautious to the point of being timid. Whales when first making contact, generally probed the skiffs with the tip of their rostrum. We discovered that the whales were very sensitive to the slightest touch and often flinched if touched unexpectedly before they made the initial contact with the boat. On occasion, however, energetic "friendlies" have prodded skiffs with sufficient force to cause concern for the safety of the passengers. During six encounters, whales bumped whale-watching skiffs hard enough to knock six people overboard and cause slight damage to one of the skiffs and outboard engines. Skiff operators reported that the initial behavior of these whales toward the boats was gentle, but over time became increasingly rambunctious. In no instance, however, did this behavior appear as either aggressive or defensive behavior as described by Scammon (1874), Norris and Gentry (1974), or Norris *et al.*, (1977). On subsequent encounters with the same whales, the more forceful behavior was pronounced throughout and the whales did not revert to the formerly gentle contact. In addition, twice during curious encounters adult whales were observed breaching and lob-tailing within a distance of less than one body length from the skiffs, sending a cascade of water over the occupants. Furthermore, instances of near collisions with whales surfacing directly in the path of skiffs underway were observed.

Discussion

ABUNDANCE, DISTRIBUTION, AND OCCUPATION TIMETABLE

The demographic study of gray whales in Laguna San Ignacio provides a detailed picture of whale abundance and use of this breeding area between 1978 and 1982. An important finding is that the Laguna San Ignacio whale population at maximum combined counts has increased moderately at a statistically significant rate of 7.3% per year over the 5 years. This increase was chiefly due to significantly greater numbers of females and calves that primarily occupied the upper lagoon nursery during the birth period. The 18.3% per year increase in their counts indicated that greater numbers of females were either utilizing Laguna San Ignacio for parturition or were entering the lagoon with newborn calves. This increase, however, was not accompanied by a change in the timetable of female–calf pair occupation or their duration of stay, which remained the same each year. In contrast to the female–calf pairs, our results provide good evidence that the demography and phenology of the single whale population has remained essentially stable since 1978. There were no significant changes in the maximum abundance, distribution, or density of single whales, and, with the exception of one season, their timetable of occupation remained the same.

Prior to our research, mid-February aerial surveys of gray whale abundance in Laguna San Ignacio were conducted between 1947 and 1965 by Hubbs and Hubbs (1967) and later by Gilmore (1960). A second series of aerial counts was conducted by Gard (1978) between 1974 and 1976. Although all of these investigators attempted 100% coverage, each utilized unstandardized procedures and transect track lines; consequently, their results preclude quantitative comparisons. These early surveys are valuable, however, because they provide an index of gray whale abundance in Laguna San Ignacio over the past 35 years. An exponential curve fit to the counts since 1947 along with our boat counts between 1978 and 1982 suggests that the number of gray whales utilizing the lagoon has increased at least 7.9% per year [$F_{(1,18)} = 97.01$, $p << .001$, $r^2 = .851$] (Fig. 16). This rate of increase agrees closely with the 7.3% per year rate of increase of our maximum combined counts between 1978 and 1982; however, it also suggests an annual rate of increase for the lagoon in excess of the population growth. The best estimate for the annual rate of increase of the California–Chukotka gray whale population is 2.5% per year, based on Reilly's (1981) analysis of 13 years of shore censuses of the fall migration past Monterey, California between 1968 and 1980; this was concurrent with an annual exploitation of 1.2% per year, yielding a 3.7% per year net realized rate of increase. During our 5-year study, then, gray whale counts in the lagoon increased approximately 4.8% per year above the growth rate of the population itself. At the beginning of our study in 1977–1978, the size of the gray whale herd was estimated to be 14,811 ± SD 2,272 whales and would have increased to approximately 16,000 by 1982 (Reilly *et al.*, 1983).

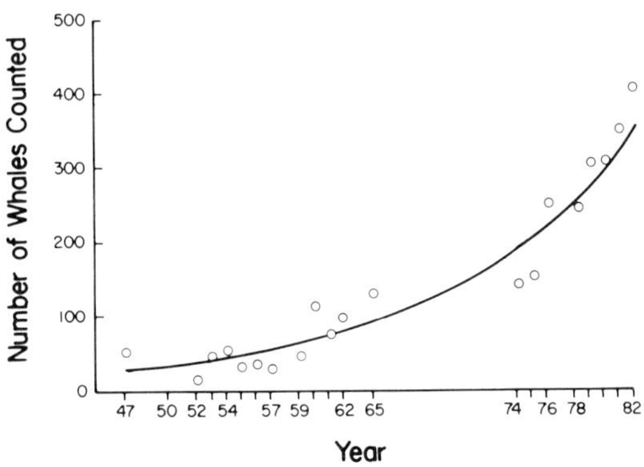

Fig. 16. *Gray whale abundance counts in Laguna San Ignacio from surveys conducted between 1947 and 1982. Aerial surveys were flown between 1947 and 1976 by Gilmore (1960), Hubbs and Hubbs (1967), and Gard (1978). Boat counts between 1978 and 1982 were conducted by the authors. Black line is a least squares exponential regression line fitted to the counts.*

There are at least two possible explanations for the overall increase in whale abundance in Laguna San Ignacio between 1978 and 1982. The first is that the gray whale herd is continuing to grow following its recovery from exploitation, and, therefore, counts in the lagoons continue to increase. An alternate explanation is that the higher overall whale counts could be due to a larger fraction of the total herd utilizing this lagoon. With the exception of Laguna Guerrero Negro, where whale counts did not increase significantly between 1980 and 1982 (Chapter 15, this volume), we do not know if increases in abundance occurred in other breeding areas, or whether Laguna San Ignacio is unique in attracting larger numbers of whales per year.

Notably, Laguna San Ignacio supports the second largest number of whales of all the major winter breeding areas. This is demonstrated by the results of boat surveys of gray whales (not including calves) reported from other breeding areas during the same time period (Fig. 17). Total adult whales in Laguna Ojo de Liebre reached a maximum of 829 animals on February 1, 1980 (Rice *et al.*, 1981); in Laguna San Ignacio the maximum count was 407 whales on February 14, 1982; in Boca de Soledad the maximum count was 225 adults on February 8, 1982 (Lawson, 1983); and maximum counts ranged from 120 to 207 whales between 1980 and 1982 in Laguna Guerrero Negro (Chapter 15, this volume). In addition, Rice *et al.* (1981), who conducted extensive aerial surveys of the west coast of Baja California, showed that two other areas were utilized by substantial numbers of adult grays; their unadjusted counts revealed that 170 whales were in San Juanico Bight on February 4, 1980, and 139 whales were in Bahia Almejas on February 5, 1980.

A main difference in the composition of the whale population in Laguna San Ignacio compared with other lagoons is that Laguna San Ignacio, at the time of maximum combined counts, is a single-whale-dominated system, whereas female–calf pairs predominate in the other lagoons during the same time period. For example, single whales averaged 70% of the population in Laguna San Ignacio at maximum combined counts, compared to 44% in Laguna Ojo de Liebre, 32% in Boca de Soledad, and an average of 19% in Laguna Guerrero Negro. Thus, Laguna San Ignacio supports a relatively small proportion of female–calf pairs while these animals outnumber single whales in the other areas at maximum combined counts.

To evaluate the relative importance of Laguna San Ignacio as a calving area for females and as a breeding area for single whales, we compared maximum counts of both groups in Laguna San Ignacio with maximum counts of these whales in other breeding areas. In terms of overall female–calf pair usage, Laguna San Ignacio appears to host the second highest number of females and calves with a maximum of 281 counted on March 8, 1982 (Fig. 17). The area with the greatest abundance of female–calf pairs was clearly Laguna Ojo de Liebre with 571 pairs counted there on February 4, 1980 (Rice *et al.*, 1981). One hundred and fifty-four mothers and calves were counted in Boca de Soledad on February 4, 1982, making it the third largest female–calf area (Lawson, 1983). Finally, Laguna Guerrero Negro contained the least number of these whales, averaging 127 pairs between 1980 and 1982 (Chapter 15, this volume). Laguna San Ignacio also appears to host the second greatest number of single whales, with an average maximum count of 227 (Fig. 17). Again, this is second only to Laguna Ojo de

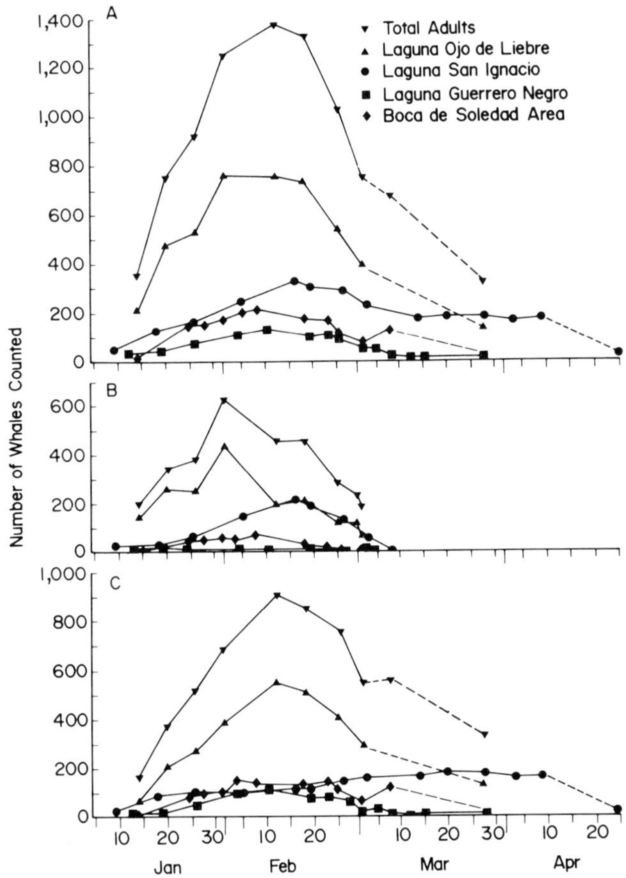

Fig. 17. *The number of gray whales (excluding calves) counted in four major breeding areas:* ▼, *total adults;* ▲, *Laguna Ojo de Liebre (Rice* et al., *1981);* ●, *Laguna San Ignacio (Swartz and Jones, 1981);* ■, *Laguna Guerrero Negro (Chapter 15, this volume);* ◆, *Boca de Soledad area (Lawson, 1983). All counts are for the 1980 winter except the data for Boca de Soledad which are from the 1983 season. (A) Total whales for all areas, (B) single whales, (C) female–calf pairs. Counts connected by solid lines are from boat surveys; those connected by broken lines are from 100% aerial surveys (see text for explanation).*

Liebre which contained 438 single whales at their maximum count on February 1, 1980 (Rice *et al.*, 1981). Boca de Soledad contained the third largest number of single whales, with 75 on February 8, 1982 (Lawson, 1983), and Laguna Guerrero Negro supported the smallest number of single whales of all these areas; single whale counts in this lagoon ranged from 19 to 43 in mid-February between 1980 and 1982 (Chapter 15, this volume).

We cannot explain why the single whale maximum abundance remained the same each year rather than increasing significantly; however, we believe that the total number of single whales using Laguna San Ignacio during a winter season is far greater than the number counted in the lagoon at maximum combined counts. The boat surveys provide an index of the minimum number of whales in the lagoon at specific points in time throughout a season; but, we do not mean to imply that the population is a static group or that the same individuals are being counted repeatedly each week as the number of whales builds to its maximum each year. In other words, we suspect that single whales have a high turnover rate in the lagoon. Evidence supporting the notion of a relatively continuous exchange of whales through the lagoon during each season comes from our photographic identification study between 1977 and 1982, and our shore-based observations of the main entry channel of the lagoon inlet in 1982. Photographic identification results revealed that 81% of the single whales photographed in Laguna San Ignacio had a duration of stay of 1 week or less. In addition, the fact that two single whales photographed by D. Withrow (personal communication) in Laguna Oje de Liebre were resighted approximately 1 week later in Laguna San Ignacio during the same season suggests the possibility of a relatively rapid circulation of whales between breeding areas as well (Jones, 1984). Finally, our census of whale movement through the inlet demonstrated that substantial numbers of single whales enter and exit the lagoon every day. For example, at the time of the maximum combined count in 1982, when 270 whales were counted within the lagoon, a minimum of 341 others entered and 185 exited the lagoon during a 7-hr. period. Because the majority of the singles present in the inlet throughout the season were actively transiting and their movement was independent of the tide, counts of these whales through the inlet were indicative of a real exchange of animals rather than being repeat counts of the same individuals passively moving into and out of the lagoon with the tide (Jones and Swartz, 1984).

The duration of the single whale season in Laguna San Ignacio was statistically consistent during 4 of the 5 years when it averaged 79 days (11 weeks). In 1979, however, the season was significantly longer, as seven whales (3% of the maximum count) remained in the lagoon approximately 2 weeks longer that usual. Since the number of whales present during the tail end of the deviant year was small, we think that the 4 year average is the best estimate for the single whale duration of stay in Laguna San Ignacio between 1978 and 1982. Nevertheless, there are two obvious reasons why this estimate may be somewhat biased toward overestimating the actual length of the single whale season. First, late-pregnant females are believed to be the first animals to arrive at the lagoons (Rice and Wolman, 1971), and because these females are indistinguishable from single whales until they give birth to their calves, census counts early in the winter probably extend the length of the single whale season beyond its actual length. Similarly, whale counts late in the season may include some females that lost their calves and would be counted as single whales.

The durations of stay of single whales were nearly the same in Laguna San Ignacio, Laguna Ojo de Liebre, and Boca de Soledad, and each of these areas hosted single whales longer than Laguna Guerrero Negro. In Laguna Ojo de Liebre, the minimum length of the single whale season in 1980 was 75 days (10.7 weeks). This period is based on an estimated beginning date for the season of December 20 which Rice *et al.* (1981) extrapolated from the timing of the southbound migration past Point Loma, California, and from the date of their last lagoon boat census on March 3. Because their final count included 67 single whales, and our March 28 aerial survey of the same area revealed no single whales, we suspect that the season probably extended to at least March 10, 1980 and, therefore, was at least 80 days (11.4 weeks) in duration. Single whales were present in Boca de Soledad for a minimum of 78 days (11.1 weeks) from January 9 to March 28, 1982 (Lawson, 1983). Lawson stated that approximately 15 whales were already present at the beginning of her study, thus, the duration of the season was probably a few days longer. Lastly, Bryant *et al.* (Chapter 15, this volume) reported that the single whale season in Laguna Guerrero Negro began on approximately January 3 and usually ended by March 10 for the 3 years from 1980 to 1982. This yields a 66-day (9.4-week) duration of single whale season, which is approximately 2 weeks shorter than that reported for any of the other breeding areas. This shorter single whale season may be related to Bryant *et al.*'s (Chapter 15, this volume) finding that Laguna Guerrero Negro consistently hosted only a small number of single whales that averaged 16.2% of the total adult whales. In summary, the period of lagoon occupation for single whales in these areas ranged from 9.4 to 11.4 weeks, beginning in late December or early January and ending by approximately mid-March. Their departure from these lagoons occurred approximately 1 month before the departure of female–calf pairs.

Knowledge of the duration of a behavior contributes to the understanding of an animal's time budget. In the case of the gray whale, the duration of stay in the lagoons has important implications in terms of the whales' energy requirements. Because single whales leave the breeding areas approximately 1 month earlier than many of the female–calf pairs, they presumably arrive on the summer feeding grounds earlier and spend more time there than females that give birth to calves during the same year (Rice and Wolman, 1971). Given that females normally reproduce biennially, this suggests that adult females would alternate between a longer period on the summer range when newly pregnant and a shorter period on the summer range when lactating. Males, however, would be able to spend the same amount of time feeding on the summer range each year. For newly pregnant females, this extra summer feeding time is of obvious energetic significance related to their need to acquire more fat reserves for the gestation of a calf and a 7-month lactation period.

The variation in the timing of the single whale seasonal occupation of Laguna San Ignacio is best expressed as differences between the mean day per season, rather than as changes in the dates of the maximum counts. The timing of maximum counts may be more a function of the weekly sampling schedule than of actual variations in the timing of single whale abundance. For example, the dates of single whale maximum counts in Laguna San Ignacio differed as much as 27 days between winters and were significantly different, whereas the mean days differed by 15 days and were not significantly different from one another or the pooled mean day. For predictive purposes, the pooled mean day of February 9 is the best estimate of the midpoint of the single whale season between 1978 and 1982. The variation in the seasonal timing, expressed as differences between mean days, may occur as a result of environmental factors such as changes in the timing of the advance of the Arctic ice sheet (Chapter 10, this volume) or extremely favorable or unfavorable weather in the summer range and along the migratory route.

Estimates of the mean day of the single whale season are not available for other breeding areas. At present, the timing of peak abundance of single whales among breeding areas can only be evaluated by comparing the dates of maximum counts. A comparison of the 5-year average day of maximum counts in Laguna San Ignacio with the date of maximum counts in other areas indicated that the timing of single whale maximum counts is variable among breeding areas. Peak counts coincided in the northernmost lagoon of Guerrero Negro and the southernmost lagoon of Bahia Magdalena, but occurred a few days earlier than in Laguna San Ignacio; for example, in Laguna Guerrero Negro maximum counts were obtained between February 8 and 10 during the period 1980 to 1982 (Chapter 15, this volume), and in Boca de Soledad they occurred on February 8, 1982 (Lawson, 1983). In Laguna Ojo de Liebre maximum counts were obtained on February 1, 1980 (Rice *et al.*, 1981), 2 weeks earlier than in Laguna San Ignacio. The variation among the different areas may be the result of real changes in the single whale season or of the survey schedules employed in the various breeding areas. Clearly though, the timing of single whale counts from these lagoons is not suggestive of a sequential occupation of areas from north to south (Norris *et al.*, 1983); rather, the timing of the occupation appears to be independent of geographical location.

The pattern of temporal distribution of female–calf pairs in Laguna San Ignacio was virtually the same each year, but was strikingly different from that in other lagoons. In Laguna San Ignacio, the number of females with calves increased steadily throughout the birth period from late December through mid-February, and reached peak numbers in late March or early April. At maximum counts, abundance was nearly double that in mid-February and averaged 226 pairs compared to 116 pairs. In Laguna Guerrero Negro, Laguna Ojo de Liebre, and Boca de Soledad, however, maximum counts of females and calves occurred much earlier. In these areas, counts increased in early January to maximums in mid-February and then decreased. For instance, maximum counts of female–calf pairs were obtained between February 8 and 10 between 1980 and 1982 in Laguna Guerrero Negro (Chapter 15, this volume), on February 12, 1980 in Laguna Ojo de Liebre (Rice *et al.*, 1981), and on February 4, 1982 in Boca de Soledad (Lawson, 1983).

One explanation for the difference in timing of female and calf peak counts between Laguna San Ignacio and the other areas is that this lagoon serves as a staging area for female–calf pairs prior to their departure from the southern range. There are two sources of evidence supporting this idea. First is the observation that the late season influx is not due to continued births; rather, these late season females have calves that are approximately 2 to 3 months old (Swartz and Jones, 1981). Second, the results of our 6-year photographic identification study, in combination with a 3-year program in Laguna Guerrero Negro by Bryant and co-workers and a 1-year project in Boca de Soledad by Lawson, revealed that some females with older calves were immigrating to Laguna San Ignacio after leaving these breeding areas. Prior to the extension of photographic identification studies to other breeding areas, we and other investigators had speculated that the late season influx of whales consisted of females with calves that were moving northward from Bahia Magdalena (Swartz and Jones, 1978; Rice *et al.*, 1981). But this is only partly true. In addition to northbound immigrants, some females with calves photographed early in the season in more northerly lagoons, vacated these areas and moved south to Laguna San Ignacio late in the season (Jones, 1984).

Recently, a late season influx of female–calf pairs was reported for Boca de

Soledad; however, this influx occurred on March 8 and was preceded by a decrease in abundance from their maximum count on February 4, rather than showing the continual increase seen in Laguna San Ignacio (Lawson, 1983). Lawson speculates that this late season influx in Boca de Soledad consists of whales moving north from the main bay of Bahia Magdalena or from the Gulf of California. Additional observations will be required to determine if this influx is an annual event in Boca de Soledad.

The reason why females and their calves are attracted to Laguna San Ignacio late in the season is unknown, but may be related to a combination of factors. For example, females may utilize potential food resources in Laguna San Ignacio, as there is a growing body of direct and indirect evidence indicating that a subgroup of gray whales, particularly females with calves, may opportunistically consume appreciable amounts of prey within the lagoon and its nearshore waters (Swartz and Jones, 1981; Jones and Swartz, 1984). Another possibility is that this aggregation of whales is related to the social structure of the herd or to the "socialization" of older calves. Gray whales have been characterized as being unsocial animals, partly because of the solitary nature of mothers and calves during the early part of the winter (Gilmore, 1960). However, the contention that these gatherings may have a social function is supported by the observation that in the latter part of the season in Laguna San Ignacio when calves are older, mothers and young frequently join to form large groups characterized by intense physical contact (Swartz and Jones, 1981). Group members cavort, rolling and rubbing against each other, wheeling and diving at the surface and below. Multiple bubble bursts frequently mark their location while submerged. The size and duration of these groups vary, but can involve up to 20 mother–calf pairs at a time, and can last from a few minutes to over 3 hours. The groups are fluid, with pairs coalescing and exchanging continually (Jones and Swartz, 1984). Another hypothesis is related to the physical development of the calves; that is, Laguna San Ignacio may serve as a resting area where females stop over to rest and nurse their growing young after periods of oceanic swimming. Additional study will be required to evaluate the validity of these tentative hypotheses.

A major objective of the study was to learn how gray whales use Laguna San Ignacio. This lagoon is an important calving and breeding area for single whales and female–calf pairs, and during our study, the whales exhibited a regular distribution that did not vary significantly over the 5 years. During the period from late December to late February, when single whales and females with calves occupied the lagoon jointly, 50% of the total whale population was distributed in the lower lagoon, 20% in the middle, and 30% in the upper lagoon. However, there was a consistent separation between the calving and breeding portions of the lagoon population. Females with calves and single whales utilized the lagoon quite differently. As whales entered the lagoon, the population became segregated spatially and temporally such that the distribution, gross movement, and timetable of occupation differed for each group. Single whales were dispersed in a gradient being most dense near the inlet and becoming less dense with increasing distance from the inlet, while most female–calf pairs were concentrated in the upper lagoon nursery farthest from the inlet. To illustrate, at the date of combined maximum counts each year, the average distribution of single whales was 65% in the lower lagoon, 24% in the middle, and 11% in the upper lagoon. In contrast, the average female–calf pair distribution at the same time was 11% in the lower lagoon, 9% in the middle, and 80% in the upper lagoon nursery.

The single whales' definite preference for the lower lagoon may be related to access to the relatively deep and wide channel, which is unrestricted even at low tide, or to close proximity to the inlet and open ocean areas, or both. While in this area, single whales are predominantly engaged in social activities associated with courtship and mating. The lower lagoon is 2–4 times deeper than the middle and upper lagoon respectively, and may be more conducive to sexual behavior by allowing the whales greater mobility.

The localization of single whales in the inlet and lower lagoon region of Laguna San Ignacio is a feature of lagoon use by single whales that also occurs in other breeding areas. Gilmore (1960) reported that lone adults, courting adults, and immatures also predominate at the entrances to Laguna Ojo de Liebre and Bahia Magdalena, and suggested that courting and mating are the primary activities of the whales in these areas. In addition, Samaras (1974) stated that adult male and female grays use the inlet of Laguna Ojo de Liebre as a staging area for precopulatory behavior, while pregnant females occupy the inner lagoon calving area. Norris *et al.* (1983), who studied lagoon entrance aggregations of gray whales at La Entrada, the main entrance to Bahia Magdalena, and Canal Rehusa, the entrance of the southernmost calving lagoons at Bahias Almejas and Santa Maria, reported that the entrance aggregations were composed entirely of males, nonparturient females, and juveniles whose primary behaviors appeared to be courtship, mating, and possibly feeding. On the other hand, Bryant *et al.* (Chapter 15, this volume) described Laguna Guerrero Negro as hosting very few single whales and they did not appear to aggregate in the area nearest the inlet. They also noted, however, that this lagoon does not have any deep-water areas comparable to those in the larger breeding lagoons, but further remarked that a large number of singles have been observed in the deep-water area just offshore. This finding, then, is reminiscent of the pattern of single whale distribution in other breeding areas.

The predominance of female–calf pairs in the upper lagoon nursery area of Laguna San Ignacio is undoubtedly related to a number of features that are conducive to the rearing of very young calves. First, the upper lagoon area is relatively shallow and least affected by tidal currents, but the area available to the whales is greatly reduced at low tide. Evidently, these inner lagoon features are attractive to females with newborn calves since they consistently utilize the innermost areas when single whales are present in the lagoon. Mothers and calves in the upper lagoon are primarily engaged in the more tranquil activities of resting, nursing, and moving about with the changing tides. Moreover, it is generally accepted that the nursery area furthest from the inlet provides an area of relative solitude for the females following parturition and during early calf development. In addition, we feel that harassment by courting whales is probably a major factor in the segregation of females and calves away from single whales inhabiting the outer lagoon areas. Females with calves avoid courting groups; despite their avoidance of single whales, they are occasionally pursued by groups of single whales that appear to harass and attempt to mate with these females (Swartz and Jones, 1981; Jones and Swartz, 1984). Mating bouts appear to be very disruptive events and may involve as many as 15 individual whales thrashing at the surface and throwing their flukes and flippers about as they jocky for mating position. It is certainly reasonable to think that a calf might be separated from its mother or could sustain injuries or even be killed in these circumstances. We agree with Norris *et al.* (1977) that "courting–mating aggregations are almost certainly disruptive places for mother–calf pairs."

The pattern of female–calf pair distribution in Laguna San Ignacio during the first half of each winter is consistent with the findings of investigators working in other lagoons who report that the uppermost lagoon areas farthest from the open sea are utilized extensively by female gray whales with newborn calves (Scammon, 1874; Gilmore, 1960; Rice and Wolman, 1971; Samaras, 1974; Gard, 1974; White, 1975; Norris *et al.*, 1977). The dynamics of female–calf pair distribution in Laguna Guerrero Negro appears to correspond to that observed in the upper portions of the larger breeding areas of Laguna Ojo de Liebre, Laguna San Ignacio, and Bahia Magdalena; that is, the duration of female–calf pair occupation in Laguna Guerrero Negro and in the upper regions of these areas begins and ends at approximately the same time, their periods of peak abundance coincide, and very few single whales are found in these areas.

The middle lagoon consistently hosted the lowest density of whales early in the season. One explanation for the decrease in single whales in the middle lagoon and nursery is that the significantly larger numbers of female–calf pairs using these areas, in effect, may have displaced single whales. In addition, this region may serve as a "buffer" between courting single whales and females with their newborn calves. Norris *et al.* (1983) commented that an almost whale-free "buffer-zone" existed at Bahia Magdalena between single whales in the area nearest the inlet and the female–calf pairs deepest within the lagoon interior.

Following the period of combined maximum counts, and with the gradual departure of single whales from Laguna San Ignacio, female–calf counts continued to increase to their maximum counts and these whales gradually shifted their distribution to the lower lagoon and inlet, essentially abandoning the upper lagoon nursery. Thus, throughout each season, gray whales are most numerous in the lower lagoon and inlet. The average distribution of female–calf pairs at this time is: 87% in the lower lagoon, 12% in the middle lagoon, and 1% in the upper lagoon.

The trend for female–calf pairs to occupy the area nearest the sea following the end of the birth period and the departure of single whales from the lagoon was prevalent during all winters in Laguna San Ignacio, and suggests that females prefer the inlet and lower lagoon over other areas when their calves are 2–3 months old. We conducted aerial surveys (100% counts) of Lagunas Guerrero Negro and Ojo de Liebre on March 28, 1980 and found that female–calf pairs had abandoned the innermost regions of these lagoons as well and were concentrated in the lower lagoon and inlet areas; in Laguna Guerrero Negro, we counted 17 mother–calf pairs in the lower lagoon entry channel and 73 pairs in the inlet, and in Laguna Ojo de Liebre we counted 22 pairs in the upper portion of the lagoon, 83 pairs in the lower portion, and 121 pairs in the inlet, indicating that a late-season shift in the distribution of female–calf pairs toward the inlet also occurs in these lagoons. The tendency for whales to shift their distribution from the inner regions of the breeding areas to the lower and inlet areas is also seen in Boca de

Soledad (Lawson, 1983) and Bahia Magdalena (Norris *et al.*, 1983). Scammon (1874) and Norris *et al.* (1977) interpreted this shift as being the first stage of the migration preparatory to leaving the lagoons.

MORTALITY

Age-specific mortality is a vital parameter in any demographic study. The pattern of gray whale mortality in the winter range that emerged from the evaluation of mortality data in Laguna San Ignacio, Laguna Guerrero Negro, Laguna Ojo de Liebre, and Boca de Soledad between 1954 and 1983 can be characterized by three age-class components: (1) a very high proportion of calves, consistent in all areas, that averaged 91.4% of the total dead whales, (2) a relatively high, but unstable proportion of immatures that ranged from 0–19.5% among areas, and (3) a consistently low, but unstable proportion of adults ranging from 0 to 5% between areas.

This mortality pattern supports both the contention that calves are more susceptible than adults to fatal strandings in the lagoons and the idea that calf and yearling mortality is higher than that of adults (Eberhardt and Norris, 1964; Gard and Gard, 1972; Gard, 1976) and further implies that gray whale age specific mortality is typically mammalian. That is, in most mammal species, young and immatures have higher mortality rates in relationship to older individuals, whose mortality rates remain relatively constant following maturity (Allen, 1981). Caughley (1977) states that mammal mortality rates typically follow a U-shaped trend with age, where prepubertal mortality varies greatly compared to adult mortality rates which remain less variable with age. Gray whale mortality appears to fit such a trend, with calf and immature mortality rates significantly greater than that of adults. Since the gray whale population is changing demographically as it continues to grow, records of strandings, particularly calf mortality, are valuable for monitoring changes in age-specific mortality and will provide a more accurate profile of the population age structure.

The preponderance of dead gray whale calves discovered in the breeding areas in all years suggests that the period including birth and the time immediately following birth within and near the breeding lagoons is critical for the survival of the young. Calf carcasses were commonly discovered in the lagoons throughout the birth period from January 5 to February 15, but were infrequently encountered in March and April, suggesting that the probability of dying is greatest for a calf just after birth and decreases with time.

The mean calf mortality rate of 3.54% calculated for whales in Laguna San Ignacio (1978–1982) is comparable to the 2.5% mean rate for calves in Laguna Guerrero Negro (1980–1982) and the 2.9% rate in Laguna Ojo de Liebre in 1980 (Swartz and Jones, 1983). Unfortunately, data on calf mortality prior to the exploitation and subsequent depletion of gray whales are nonexistent, making it impossible to determine whether the mortality rates for Laguna San Ignacio and the other areas are high, low, or typical for the species. Counts of dead calves undoubtedly underestimate actual calf mortality because not all carcasses are discovered: some dead whales are washed out to sea with the tides, some are consumed by scavengers, others strand on beaches not visited by investigators, and still others are missed by researchers altogether. Nonetheless, mortality rates based on dead whales are useful as minimum estimates.

From our analysis of the body size of dead calves stranded in Boca de Soledad, and Guerrero Negro, Ojo de Liebre, and San Ignacio lagoons, we concluded that the mean lengths of the calves in all these areas were consistent and together averaged 4.43 m. It has been generally assumed that the lengths of dead calves in the lagoons are representative of the lengths of gray whales at birth (Rice and Wolman, 1971; Rice *et al.*, 1971; Rice, 1983). However, our comparison of the lengths of dead calves from the breeding areas with the lengths of near-term gray whale fetuses collected by Rice and Wolman (1971) clearly indicated that the dead neonates were significantly shorter than the fetuses (4.43 to 4.62 m respectively) and suggests that the typical length at birth of most viable gray whale calves would be longer than the lengths of the dead lagoon calves. Presumably, had the fetuses developed to full term, only a small percentage of them, equal to the calf mortality rate in the lagoons, would be expected to die, and we would expect the lengths of *these* calves to be similar to the lengths of dead neonates from the lagoons.

Rice *et al.* (1981) also noted the smaller size of dead neonates in the breeding lagoons compared to near-term fetuses, and proposed that ". . . the body length shrinks after birth, possibly due to compression of the intervertebral disks." An alternate explanation for the dead neonates being shorter is that they might be premature, and as premature animals their lengths would not be representative of full term neonates. The fact that they are dead, also suggests that these whales are abnormal and not representative of normally developed young. Another possibility is that neonatal mortality is higher for smaller calves. It is generally known that among mammal species, the bigger the offspring, the better are its chances of survival (Ralls, 1976), and that for humans and some species of nonhuman primates and some ungulates, small offspring have higher mortality than larger individuals. Although we have no hard evidence to support any of these ideas for gray whale neonate mortality, we feel that the deaths of these smaller whales are probably related in some way to reduced fitness at birth.

Theoretically, gray whale yearlings should attain a length of about 8.2–9.6 m by the time they make their first return trip to the southern range. The dead immatures from Laguna San Ignacio that were either too small to be yearlings or too large to be calves of the year, perhaps, were animals that became ill before or following separation from their mothers, or failed to learn how to forage for themselves, or did not spend their first summer in a suitable feeding area with the result that their growth was stunted. These "runt" animals could then have followed the fall migration of the population to the southern range where they died as undersized yearlings.

Most population biologists agree that accurate aging of animals is important because knowledge of the dynamics of a population depends heavily on the ability to age individuals (Caughley, 1977). Many also agree that errors are inevitable especially when the criterion of age is a morphological feature, like body size, that changes continuously with age. In the case of gray whales where body length is used to age individuals, the problem of correctly categorizing small gray whales as yearlings rather than calves points out an obvious bias with this or any fixed length–age interval classification scheme. Alternative methods of aging small whales are limited primarily because small whale carcasses decompose rapidly (Eberhardt and Norris, 1964), and short-term clues as to their age, such as the presence of adult barnacles, are lost relatively soon after death. Thus, small yearlings could be misidentified by age as calves of the year. One important effect of this error is that calf mortality rates might be biased upward. In this instance, it would be important to know what proportion of dead calves is actually undersized yearlings. A second effect is that yearling lengths misclassified with calf lengths would contribute error when these data are used to assess calf growth rates (i.e., it would bias calf growth rates upward). In the future, it will be valuable to know how often the gray whale age classification index leads to error, so that researchers can reduce bias in their results and better evaluate the significance of age-specific mortality in this species.

Sex Ratio

The 1:1 sex ratio of dead gray whale calves, immatures, and adults combined reported from the four breeding lagoons is in agreement with the findings of Rice and Wolman (1971), who report that gray whales have an equal sex ratio at all age classes. Based on the sex ratio of gray whales collected during migration, Rice and Wolman (1971) state that ". . . there is probably little sex difference in mortality rate, a conclusion further supported by the essentially equal sex ratio at all ages." While this may be true for gray whale calves, the sample of dead immatures and adult whales from the lagoon was too small to permit the evaluation of differential mortality between males and females of these age classes. If gray whale mortality is typically mammalian, we would expect proportionately greater male mortality in older age classes (Caughley, 1966; Ralls *et al.*, 1980). However, additional data are needed to clarify this issue.

WHALE-WATCHING ACTIVITIES

Concern about the potential detrimental consequences of the intrusion of human activities into the breeding lagoons of the gray whale has been widespread since whale-watching excursions began to visit these areas in the early 1970s. Gard (1974) and Villa-Ramirez (1975) maintained that the presence of people seeking to view the whales close-up displaces gray whales from their preferred breeding areas, but they provided no conclusive evidence. Rice (1975) stated that "considerable harassment is caused by commercial cruise boats which take people into the calving lagoons to see the whales." Additional speculation came from Gilmore (1978) who wrote that the "entry of man with his industrial or recreational activities into the calving and courting lagoons . . . , could only have an adverse effect on the reproduction and survival of the newborn [whales]."

In seeking a biological basis for the argument that whale-watching excursion vessels pose a significant threat to breeding gray whales in Laguna San Ignacio, we attempted to assess the long and short-term demographic effects on gray whales including changes in abundance, distribution, and mortality that may occur as the result of repeated encounters with excursion vessels and their related activities. To our surprise, none of the adverse consequences that we proposed might result from exposure to human activities were substantiated. Rather, our findings to date suggest that the gray whales possess sufficient resiliency to tolerate the physical presence and activities of whale-watching vessels and skiffs and the noise produced by this level of activity without major disruption.

Gray whales did not shift their distribution from Laguna San Ignacio in response to human activities. They continued to return to the lagoon, and there was no decrease in whale abundance concurrent with the growth of the thriving whale-watching industry within the lagoon. In fact, the whale population in Laguna San Ignacio has steadily increased since whale-watching activities began in 1974, as demonstrated by the comparison of previous aerial surveys over the past 35 years with our 5 years of abundance counts. Moreover, the increase in gray whale abundance in the lagoon each year exceeded the estimated rate of growth of the population, suggesting that gray whales were preferentially utilizing this lagoon even with the presence of whale-watching activity. Further evidence of the grays' apparent tolerance of whale-watching disturbance, which is of major importance to their reproductive success, is the fact that the number of female–calf pairs inhabiting the lagoon during the birth period also increased significantly at a rate approximately 10% per year above the population growth between 1978 and 1982. We feel that a key factor responsible for maintaining the stability of the whale population in Laguna San Ignacio was the establishment of the gray whale refuge in the lagoon, which serves to regulate the number of vessels operating in the lagoon and provide an area free of all vessel activity as a sanctuary for the use of the whales (particularly females with calves) (Vargas, 1981).

Our index of whale abundance in all regions of the lagoon throughout the season revealed that patterns of use of the three lagoon interior areas have been essentially the same in all years despite increased levels of tourism. The upper lagoon and the lower lagoon merit special consideration. For instance, the number of females and calves that utilized the upper lagoon nursery area showed a significant increase, indicating that human activities in the constriction of the main entry channel at Punta Piedra did not impede their access to and from the upper area or prevent them from entering the lagoon. If tranquility within the lagoon is important to females with calves, then perhaps containment of whale-watching activities in the lower lagoon has had a beneficial influence on the whales in that the closed middle and upper lagoon nursery areas provide a refuge where females with calves remain relatively undisturbed.

In the lower lagoon, gray whales have continued to co-exist next to whale-watching activities. The whale population has made no detectable shift away from the lower lagoon toward areas of less human contact. Although the lower lagoon is the smallest of the three lagoon sections, it was the region most heavily utilized by the whales. This area consistently contained 65% of the single whales during the first half of each winter, and later was occupied by 87% of the female–calf pairs. In fact, each year, whale density was greatest in this area when female–calf pairs occupied it exclusively during the last half of each season. Thus, it is reasonable to conclude that the single whales and female–calf pairs have not been displaced from their preferred habitat as a result of whale-watching activities in the lower lagoon from 1978 to 1982.

The whales' continued utilization of the lower lagoon during the presence of whale-watching activities may be related to the fact that whale-watching skiffs were operated only during 8 daylight hours, or approximately 33% of each day that they were present in the lagoon, while the remaining 66% of the day, the whales were free to move about the lower lagoon unencumbered by interruptions caused by whale-watching skiff activity. On a seasonal basis, in fact, skiffs were only in use during approximately 22% of the time. Similarly, excursion vessels were underway only during daylight hours, and operated their main engines (which produced the greatest amount of waterborne noise) for only 2% of each season. If night movement is an important element in the gray whale's winter life cycle, then the lessened vessel traffic during this time is likely to contribute to their apparent ability to tolerate vessel activity within the lagoon. Indeed, there is some evidence from behavioral observations in Laguna San Ignacio which demonstrates that the whales are active at night. For instance, from nighttime observations with a starlight scope, Swartz and Jones (1981) determined that there were no outstanding differences in the types of behaviors exhibited by females and calves, lone single whales, and groups of single whales during the night and the day in Laguna San Ignacio. At night, whales jumped out of the water (breached), held their heads above the surface (spy-hopped), moved around the lagoon as singles or in groups, rolled at the surface, or remained motionless, drifting with the current (Swartz and Cummings, 1978). Furthermore, Mate and Harvey (1981) and Harvey and Mate (Chapter 24, this volume) reported that "radio-tagged whales generally left the lagoon during the night and returned in the morning hours."

Another factor that may contribute to the continued use of the lower lagoon by gray whales is the method of whale-watching employed by the skiff operators. The primary objective of the whale-watching excursions is to provide an opportunity for the passengers to view free-ranging gray whales in their natural habitat. Paying passengers and crew generally show a genuine respect for the whales and a desire to minimize their disturbance (Gilmore, 1976). Most skiff operators have had years of experience at operating skiffs around whales and have learned that gray whales are not seen by forceful pursuit (Storro-Patterson, 1981). Thus, practices that overtly disrupt whale behavior or cause the whales to flee from whale-watching skiffs are not generally employed. On the contrary, when following whales, skiff operators typically move at slow speeds, usually following a whale for only a short distance or remaining stationary with the engine running in neutral waiting for whales to pass by. It is to their advantage not to "run" a female with a newborn calf or to "spook" the whales and cause them to leave the whale-watching area. Additionally, because gray whales are so numerous in Laguna San Ignacio, if one individual shows avoidance behavior, many other whales are available for skiff operators to follow instead.

It is also possible that because of their continued exposure to boat noise and whale watching at the controlled level observed during our study, gray whales may have, to some degree, become habituated to the vessel activity in the lower lagoon. In other words, gray whales may be growing accustomed to the presence of tourist vessels in this lagoon. A finding that lends support to this idea is the widespread increase in the incidence of friendly whales in this lagoon. Previously, Gilmore (1976) remarked that friendly whale behavior "suggests that sport-fishing [i.e., whale-watching] boats and their skiffs have not been harassing the whales, disturbing their calving, nursing, and courting, as have been contended." Likewise, Reeves (1977, p. 20) quotes Hubbs (in litt.) as stating that the San Ignacio whales "have shown very definite indications of becoming accustomed to the party boats [i.e., whale-watching vessels] and the small craft, and have actually, in some instances, given every indication of inviting attention and even physical, friendly contact." At the time these encounters were first reported, one could argue that the friendly whale phenomenon was aberrant behavior by a few animals. During our project, however, we witnessed the incidence of friendly whale behavior grow from a few isolated encounters each season to a commonplace event that involved numerous whales and was experienced by nearly every whale watcher aboard the whale-watching excursions to the lagoon.

Finally, we obtained no conclusive evidence that whales move out of the general area of whale-watching activities during or on the day following human disturbance, which further suggests that whale-watching did not significantly disturb gray whales in the lagoon. Support for this comes from the pairwise comparison of the whale abundance on days with and without the presence of excursion vessels. These activities did not seem to alter the overall distribution of whales in the lagoon or change the distribution of whales in the area of chronic disturbance. Gray whale demography in the upper and lower lagoon, and in the lagoon as a whole did not change significantly with the arrival or departure of excursion vessels in the lower lagoon. Unfortunately our sample size of paired days with and without disturbance was relatively small and, therefore, may not be sufficiently representative to allow the formulation of a conclusion regarding the short-term cause–effect relationship between the whales and whale-watching activities. Given the small sample size, caution should be used when interpreting these results as additional data will be required to fully evaluate the short-term significance of vessel disturbance on whales in the lagoon.

A very important consequence of the disruption of gray whales by whale-watching activities in the lagoon as a whole would be the potential for increased mortality of calves, either as the direct result of collisions with vessels or separation from their mothers, or as the indirect result of severe vessel-induced stress. Reeves (1977) speculated that whale-watching activity may cause calves to be separated from their mothers and possibly abandoned, which would result in the death of the abandoned calf. From our findings in Laguna San Ignacio, it does not appear that whale-watching activity has caused an increase in calf mortality. Calf mortality was highly consistent at four dead calves per year during five winters and five calves in one winter between 1977 and 1982. There was no indication that vessel activity contributed to any of these deaths. Additional evidence indicating that the effects of whale watching on calf mortality in Laguna San Ignacio must not be too severe is the finding that calf mortality in this lagoon was proportionately the same as calf mortality in three breeding areas not visited by significant numbers of whale-watching excursions.

There have been several studies in recent years which suggest that grays and other species of whales are able to accommodate to various kinds of human disturbance. Reeves (1977), who investigated the problem of whale harrassment at the breeding lagoons, stated that the continued high level use of Laguna Ojo de Liebre by gray whales following relocation of commercial salt shipping operations from Laguna Guerrero Negro to Laguna Ojo de Liebre after 1967, suggests that the movement of salt barges "may not be such a nuisance to them." Laguna Ojo de Liebre is, by far, the largest and perhaps the most important single breeding area of the gray whale, and Reeves contends that because it is a large area, "the whales there simply may have been able to 'move-over' and co-exist next to the barges." Based on his aerial surveys of this lagoon in the 1970s, Storro-Patterson (1981) agrees that the whale distribution in Laguna Ojo de Liebre seems to be only minimally effected by salt barge traffic. A similar apparent accomodation to vessel disturbance was reported by Fraker *et al.* (1982) who concluded that bowhead whales (*Balaena mysticetus*) did not permanently vacate areas where they had been exposed to short-term noise from seismic surveys, dredges, and vessel disturbance and that "flight" response seemed brief in duration and was usually followed by a return to normal activities. Watkins and Schevill (1979) reported that feeding humpback whales (*Megaptera novaeangliae*) were not disrupted by the passage of an oil tanker to within 800 m of the animals. Humpback whales wintering off Maui, Hawaii also appear to habituate to routine vessel traffic activity associated with the harbor (Darling, 1983).

On the other hand, although the current levels of whale-watching activity in Laguna San Ignacio do not appear to be having a detrimental effect on the demography of gray whales in this lagoon, there is some evidence from other geographical areas that gray whales and other mysticetes are sensitive to and may be displaced by human disturbance under different circumstances. For example, during the gray whales' recent recovery from depletion, their utilization of Laguna Guerrero Negro has varied widely (Storro-Patterson, 1981). By 1967, the evaporative salt works located in the town of Guerrero Negro was shipping approximately 3 million tons of salt per year out of the lagoon. Prior to the growth of the salt industry, Gard (1974) stated that the number of gray whales using Laguna Guerrero Negro had increased steadily during the 1950s. As salt-barge traffic and dredging activities increased with the growth of the salt industry, Gard's aerial counts of gray whales decreased to zero by 1964. Following the relocation of the salt works to Laguna Ojo de Liebre in 1967, salt shipping ceased in Laguna Guerrero Negro and counts of whales increased to 104 animals by 1973. Gard speculates that the "shipping of salt from this lagoon between 1957 and 1967 apparently drove out the whales, but they reinvaded the lagoon in substantial numbers after shipping was eliminated." In assessing the effects of human activities on the whales, however, it has been very difficult to determine whether salt shipping was entirely responsible for the whales abandonment of the lagoon. Concurrent dredging operations to maintain the channel of the lagoon, natural variations in water temperature or chemistry in the habitat, or biological changes within the population itself may have also contributed to the changes in lagoon use (Storro-Patterson, 1981). In the case of humpback whales, it has been proposed that the apparent sudden abandonment of Glacier Bay, Alaska, by humpback whales in 1978 may be due to a dramatic increase in whale watching and other fishing and recreational vessel traffic within the confines of the bay, suggesting that the whales' tolerance for vessel activity had been exceeded (U.S. Marine Mammal Commission, 1980; Baker *et al.*, 1982).

During the course of this study from 1978 to 1982 we have accrued no statistically significant evidence to substantiate the contention that whale watching has had a detrimental effect on the demography of gray whales in Laguna San Ignacio. However, it may well be that human activities in the lagoon, if they negatively impact the whales, do so in an indirect manner. For instance, the cumulative effects of continual exposure to increased water-borne noise, visual stimuli, restriction of space available to whales, or stress associated with vessels being in and passing through whale areas could have serious biological implications if, for example, gray whale reproductive fitness is reduced over the long term (U.S. Marine Mammal Commission, 1980). Studies of other species of large mammals suggest that stress may have distinct physiological and population effects (Geraci and St. Aubin, 1980; Eisenberg, 1981), and that significantly loud noise could mask important environmental noise and/or interfere with communication between whales (Davis *et al.*, 1983). Clearly, the degree to which the gray whale is affected over the long term by habitual exposure to human activities will require continuous evaluation of their abundance, distribution, and productivity. The breeding lagoons provide an excellent opportunity to continue to obtain this information.

In conclusion, to date, whale-watching activity in Laguna San Ignacio does not seem to pose a serious threat to the gray whales, but its potential for becoming one should be acknowledged so that future developments can be monitored carefully for the benefit of whales and for humans.

Friendly Whales

The apparent interest in human activities exhibited by friendly whales in Laguna San Ignacio confounds the argument that whale-watching activity harasses gray whales. The friendly whale phenomenon has also given rise to many important questions. For instance, if whale-watching activity is not benign and seriously disturbs gray whales, why has curious behavior developed? Why has this behavior persisted year after year? And, are recent reports of friendly whale behavior outside Laguna San Ignacio indicative of the spread of this behavior throughout the gray whale's range? The interaction involving calves is perhaps the single most astounding case, as calf participation contradicts most previous beliefs about maternal protection. Friendly mother gray whales seemingly do not perceive the interaction between the boats and their calves as a threatening situation.

Speculation as to why whales appear to demonstrate curiosity is widespread among the scientific and lay communities alike. In a letter to the editor of *Science Magazine*, Wolfson (1976) suggested that the present gray whale population, recovering from near depletion, and composed of primarily young animals, has never been severely harassed by the whaling industry, and therefore has never become conditioned to avoid boats.

Gilmore (1976) proposed that the whales are curious about the small boats that hover over them in the lagoons. He felt that continued exposure to boat noise has probably conditioned most gray whales to tolerate the vessels and has encouraged some whales to approach and even explore whale-watching skiffs.

Storro-Patterson (1981) commented that this behavior may be the result of the method of whale-watching employed by the skiff operators in the lagoon. In his opinion, the skiff operators do not forcefully pursue the whales, but instead "almost seem to 'tease' the whales by refusing to follow them for more than 20–50 yards. They move at very slow speeds or stay in one place waiting, but always with the motor running, often in neutral."

Finally, it is widely accepted that the acoustical sense of cetaceans is their most important sensory process, and we believe that gray whales are initially attracted to some sounds of the outboard engines. They appear to show an interest in the submerged portion of the outboard engine running in neutral, such as the nonrevolving propeller and exhaust ports from which sound emanates. Some whales repeatedly bumped the still propeller of the engine with their rostrum and even took it into their open mouth. The whales avoided Mexican fishing boats with 40-hp engines running at high speed. However, they approached and even followed 20-hp engines running at moderate or low speed. Curious whales usually left the vicinity of idling outboard engines when they were shut down, and they also avoided nonmotorized vessels including kayaks, canoes, and small sail boats. Many of the sounds from gray whales fall within the frequency range of slow-running (or idling) outboard engines (Swartz and Cummings, 1978; Chapter 22, this volume). We agree with Dahlheim *et al.* (Chapter 22, this volume) that in addition to engine sounds, other factors may be involved in maintaining the friendly behavior once the whales physically contact the boat. We suspect that tactile stimulation from rubbing against the boat and from the "strokes" of the passengers act in concert with engine sounds and other stimuli to reinforce curious behavior over the long term.

To determine whether gray whales had displayed this behavior before the 1975–1976 reports from Laguna San Ignacio, we and our assistant Mike Bursk searched the literature and spoke to individuals that had worked in the breeding lagoons prior to 1975. There were no reports of friendly whale behavior in the literature prior to 1975, and investigators that had studied gray whales in the lagoons had not experienced these animals (R. M. Gilmore, C. L. Hubbs, K. S. Norris, T. J. Walker, and others, personal communications). At the San Diego Natural History Museum, we found a chart of Laguna Ojo de Liebre made by Jim Mills, a Santa Barbara Pharmacist. The chart, dated in the early 1960s, noted "curious and friendly whales" in the lagoon. Captain Bill Lang, an excursion vessel operator, spoke of a few experiences with curious gray whales while running whale-watching trips to Laguna Ojo de Liebre in the late 1960s. Captain Lang recalled that small whales lingered behind the excursion vessel, apparently inspecting the empty skiffs tied astern. And, occasionally whales approached the skiffs filled with whale watchers, as they cruised about in the lagoon (Swartz and Bursk, 1979). But, at this time, gray whales were not to be trusted. From the days of commercial whaling, the whales, particularly females with calves, held a notorious reputation for inflicting violent attacks on pursuers. The name "devilfish" was well earned; and, consequently, close contact with whales was avoided. Perhaps the forerunners of friendly whales were in

Laguna Ojo de Liebre, but cautious skiff operators were hesitant to let any whale get too close to his skiff full of passengers.

With the closing of Laguna Ojo de Liebre, attention focused on Laguna San Ignacio as the prime whale-watching location. Because Laguna Ojo de Liebre was closer and easily accessible for U.S.-based excursions operators, Laguna San Ignacio had received few visitors; thus, we found no records of past expeditions. We spoke to the residents of La Laguna, a San Ignacio fishing village, and none were familiar with friendly whales. However, for these people any whale encounter was unlikely because the local fishermen avoid all contact with them. When traveling across the lagoon, the fishermen usually motor their pongas over sand bars to minimize the risk of colliding with a whale. While crossing deep channels, they pound on the hulls of their boats and sing loudly as if to alert the whales (Swartz and Bursk, 1979). The first bonafide reports of friendly whales in Laguna San Ignacio, then, were the encounters during the 1975–1976 winter.

In summary, the phenomenon of friendly gray whales in Laguna San Ignacio between 1975 and 1982 suggests that not all interactions with boats are necessarily disturbing to the whales. The bottom line that virtually all whale watchers have agreeded upon is that whether or not a whale exhibits curious behavior is entirely up to the animal; in other words, friendly whales find you, you don't find them. The educational, recreational, and economic importance of carefully controlled whale-watching activities is evident; however, the observation that grays occasionally jostle boats or appear to act rambunctiously (Swartz and Jones, 1981) signifies that there can be some degree of potential threat to humans as well as harassment of whales. As noted by Baker *et al.* (1982), the increased incidence of "jostling" of boats by gray whales "may indicate aggressive behavior or may only reflect the actions of the laws of probability: that accidental contact is increasingly more likely with increased human use of the whale habitat." In any event, it is imperative that the whale-watching community be reminded that gray whales are wild and immensely powerful animals and need to be treated with great caution.

Acknowledgments

Field research on free-ranging cetaceans and the analysis of the results requires the help of many hands and many minds for long periods. We were exceptionally fortunate to have had the assistance of many talented people and agencies throughout this project. In particular we wish to thank our full-time field assistants (in order of appearance) Mike Bursk, Gayle Dana, Debbie Miller, Dennis Miller, Bruce Reitherman, Jim Sumich, and Mike Symons. Special thanks are due the Mexican biologists Geriminal Marcet, Carlos Maya, and Ricardo Sanchez who participated in the study. We wish to offer our sincere appreciation to Raymond M. Gilmore of the San Diego Society of Natural History and to Laura and Carl Hubbs of Scripps Institution of Oceanography for sharing their extensive knowledge of gray whales in the breeding lagoons and for their continuous and enthusiastic support of this study. Scientific research permits were issued by the Secretaria de Pesca de Mexico. The work was funded by the American Cetacean Society, Hubbs–Sea World Research Institute, the International Union for the Conservation of Nature, the International Whaling Commission, the Committee for Research and Exploration of the National Geographic Society, the National Marine Mammal Laboratory of the National Marine Fisheries Service, the U.S. Marine Mammal Commission, and the World Wildlife Fund—U.S. Cetacean Research Associates of San Diego, California administrated this program. Our heartfelt thanks to all.

References

Allen, K. R. (1981). Application of population models to large whales. *In* "Dynamics of Large Mammal Populations" (C. W. Fowler and T. D. Smith, eds.), pp. 263–276. Wiley, New York.

Andrews, R. C. (1914). Monographs of the Pacific Cetacea. I. The California gray whale (*Rhachianectes glaucus* Cope). *Mem. Am. Mus. Nat. Hist.* [N. S.] **1,** 227–287.

Baker. C. S., Herman, L. M., Bays, B. G., and Stifel, W. F. (1982). The impact of vessel traffic on the behavior of humpback whales in Southeast Alaska. Natl. Mar. Fish. Ser. Contract No. 81-ABC-00114. University of Hawaii, Kewalo Basin Marine Mammal Laboratory, Honolulu, 39 p.

Bean, M. A. (1981). Legal aspects of gray whale conservation. *In* "Background Papers for Workshop on International Cooperation for Conservation of Gray Whales" (S. Kaza, ed.), pp. 276–298. World Wildlife Fund—U.S., I.U.C.N. Whale Center, Oakland, California.

Bostic, D. L. (1975). "A Natural History Guide to the Pacific Coast of North Central Baja California and Adjacent Islands." Biological Educational Expeditions, Vista, California.

Bryant, P. J. (1980). Gray whales of Guerrero Negro lagoon, Baja California. Department of Developmental and Cell Biology, University of California, Irvine, California (unpublished report).

Bryant, P. J., Lafferty, C. M., and Lafferty, S. K. (1981). Gray whales (*Eschrichtius robustus*) in Laguna Guerrero Negro, Baja California: 1981 season. Department of Developmental and Cell Biology, University of California, Irvine, California (unpublished report).

Bryant, P. J., and Lafferty, S. K. (1982). Gray whales, *Eschrichtius robustus*, in Laguna Guerrero Negro, Baja California: 1982 season. School of Biological Sciences, University of California, Irvine, California (unpublished report).

Caughley, G. (1966). Mortality patterns in mammals. *Ecol.* **47,** 906–918.

Caughley, G. (1977). "Analysis of Vertebrate Populations." Wiley, New York.

Dahlheim, M. E. (1983). Acoustrical studies on the gray whale (*Eschrichtius robustus*) in Laguna San Ignacio, Baja California Sur, Mexico. Final report to La Secretaria de Pesca, Mexico (unpublished report).

Dahlheim, M. E. (1984). Sound production by gray whales (*Eschrichtius robustus*) and response to playback of biological and non-biological noise. Ph.D. Dissertation, University of British Columbia, Vancouver (in prep.).

Darling, J. (1983). Migrations, abundance, and behavior of Hawaiian humpback whales. Ph.D. Dissertation, University of California, Santa Cruz, California.

Davis, R. A., Koski, W. R., Richardson, W. J., Evans, C. R., and Allison, W. G. (1982). Distribution, numbers and productivity of the western Arctic stock of bowhead whales in the eastern Beaufort Sea and Amondson Gulf, summer 1981. Report by L.G.L. Limited, Toronto, Ontario, to Dome Peteroleum Limited, Calgary, Alberta, and to SOHIO Alaska Petroleum Company, Anchorage, Alaska.

Eberhardt, R. L. (1978). Transect methods for population studies. *J. Wildl. Manage.* **42,** 1–31.

Eberhardt, R. L., and Norris, K. S. (1964). Observations of newborn Pacific gray whales on Mexican calving grounds. *J. Mammal.* **45**(1), 88–95.

Eberhardt, R. I., Chapman, D. G., and Gilbert, J. R. (1979). A review of marine mammal census methods. *Wild. Monogr.* **63,** 1–46.

Eisenberg, J. F. (1981). "The Mammalian Radiations, an Analysis of Trends in Evolution, Adaptation, and Behavior." Univ. of Chicago Press, Chicago, Illinois.

Fleischer, L. A., Manzanilla, S., and Fuentes, R. (1983). Estudios de mortalidad incidental de ballena gris en Baja California Sur, Mexico. Secretaria de PESCA, La Paz, Baja California Sur, Mexico (unpublished report).

Fraker, M. A., Richardson, W. J., and Wursig, B. (1982). Disturbance responses of bowheads. *In* "Behavior, Disturbance Responses and Feeding of Bowhead Whales (*Balaena mysticetus*) in the Beaufort Sea, 1980–1981" (W. J. Richardson, ed.). Report by LGL Ecol. Res. Asoc., Inc., Bryan, Texas, for BLM, Washington, D.C. (unpublished).

Gard, R. (1974). Aerial census of gray whales in Baja California lagoons, 1970 and 1973, with notes on behavior, mortality, and conservation. *Calif. Fish Game* **60**(3), 132–144

Gard, R. (1976). Aerial census, behavior and population dynamics study of gray whales in Mexico during the 1974–75 calving and mating season. Report U.S. Marine Mammal Commission No. MM5AC006.

Gard, R. (1978). Aerial census and population dynamics study of gray whales in Baja California during the 1976 calving and mating season. *U.S., NTIS, PB Rep.* **PB-275297,** 1–24.

Gard, R., and Gard, S. (1972). El desierto de Vizcaino: un proyecto de reserva nacional de fauna silvestre (The Vizcaino desert: a national wildlife refuge project). Subsecretaria Forestal y de la Fauna Direccion General de la Fauna Silvestre. Mexico.

Geraci, J. R., and St. Aubin, D. J. (1980). Offshore petroleum resource development and marine mammals: A review and research recommendations. *Mar. Fish. Rev.* **42**(11), 1–12.

Gilmore, R. M. (1960). A census of the California gray whale. *U.S. Fish Wildl. Serv., Spec. Sci. Rep.–Fish.* **342,** 1–30.

Gilmore, R. M. (1976). The friendly whales of Laguna San Ignacio. *Terra* **15,** 24–28.

Hubbs, C. L., and Hubbs, L. C. (1967). Gray whale census by airplane in Mexico. *Calif. Fish Game* **53,** 23–27.

Hudnall, J. (1983). Gray whale feeding behavior along the Vancouver Island coast. *Whalewatcher* **17**(3), 3–5.

Jones, M. L. (1984). Gray whale reproduction, distribution, and duration of stay in Laguna San Ignacio, and inter-lagoon exchange of whales in the winter range. Master of Science Thesis, Moss Landing Marine Laboratory, Moss Landing, California.

Jones, M. L., and Swartz, S. L. (1984). Mothers and calves. *Oceans* **17**(2), 11–19.

Jones, M. L., and Swartz, S. L. (1984). Demography, phenology, and evaluation of the impact of human activities on gray whales (*Eschrichtius robustus*) in Laguna San Ignacio, Baja California Sur, Mexico between 1977 and 1982. *U.S. NTIS, PB. Rep.* (in press).

Kaza, S. (1981). Socioeconomic aspects of gray whales. *In* "Background Papers for Workshop on International Cooperation for Conservation of Gray Whales" (S. Kaza, ed.), pp. 166–256. World Wildlife Fund—U.S., I.U.C.N. Whale Center, Oakland, California.

Larsen, K. (1978). Close encounters (of the whale kind). *Sea Front.* **23**(4), 195–202.

Lawson, S. (1983). A population study of the gray whale (*Eschrichtius robustus*). *J. Undergrad. Res. Biol. Sci.* **13,** 75–86.

Leatherwood, J. S., Gilbert, J. R., and Chapman, D. G. (1978). An evaluation of some techniques for aerial censuses of bottlenosed dolphins. *J. Wildl. Manage.* **42,** 239–250.

Mate, B. R., and Harvey, J. T. (1981). An evaluation of three radio tags applied to gray whales and their utility in assessing whale movements and dive characteristics. *U.S. Dep. Int. BLM Contract Rep.* **AA 730 79 4120 0109,** 1–68.

Mina, U. F. (1957). Bosquejo geologico del Territorio Sur de la Baja California. *Bol. Asoc. Mexicana de Geologos Petroleros.* **9**(3–4), 139–270.

Mizue, K. (1951). Gray whales on the east sea area of Korea. *Sci. Rep. Whales. Res. Inst. (Tokyo)* **5,** 71–79.

Norris, K. S., and Gentry, R. L. (1974). Capture and harnessing of young California gray whales (*Eschrichtius robustus*). *Mar. Fish Rev.* **36**(4), 58–64.

Norris, K. S., Goodman, R. M., Villa-Ramirez, B., and Hobbs, L. J. (1977). Behavior of the California gray whale (*Eschrichtius robustus*) in southern Baja California, Mexico. *Fish. Bull.* 75(1), 159–172.

Norris, K. S., Villa-Ramirez, B., Nichols, G., Wursig, B., and Miller, K. (1983). Lagoon entrance and other aggregations of gray whales (*Eschrichtius robustus*). *In* "Behavior and Communication of Whales" (R. Payne, ed.), pp. 259–294. AAAS Select Symp. Ser. 76, Westview Press, Boulder, Colorado.

Phleger, F. B., and Ewing, G. C. (1962). Sedimentology and oceanography of coastal lagoons in Baja California, Mexico. *Geologic. Soc. Amer. Bull.* **73,** 145–182.

Pike, G. C. (1962). Migration and feeding of the gray whale (*Eschrichtius gibbosus*). *J. Fish. Res. Board Can.* **19,** 815–838.

Ralls, K. (1976). Mammals in which females are larger than males. *Quart. Rev. Biol.* **51,** 245–276.

Ralls, K., Brownell, R. L., Jr., and Ballou, J. (1980). "Differential mortality by sex and age in mammals, with specific reference to the sperm whale. *Rep. Int. Whale. Comm., Spec. Issue* **2,** SC/SP78/18.

Reeves, R. R. (1977). The problem of gray whale (*Eschrichtius robustus*) harassment: At the breeding lagoons and during migration. *U.S. NTIS, PB Rep.* **PB-272 506,** 1–60.

Reilly, S. B. (1981). Population assessment and population dynamics of the California gray whale (*Eschrichtius robustus*). Ph.D. Dissertation, College of Fisheries, University of Washington.

Reilly, S. B., Rice, D. W., and Wolman, A. A. (1983). Population assessment of the gray whale, *Eschrichtius robustus*, from California shore censuses, 1967–80. *Fish. Bull.* **81**(2), 267–281.

Rice, D. W. (1983). Gestation period and fetal growth of the gray whale. *Rep. Intl. Whal. Commn.* **33,** 539–544.

Rice, D. W., and Wolman, A. A. (1971). Life history and ecology of the gray whale (*Eschrichtius robustus*). *Spec. Publ.—Am. Soc. Mammal.* **3,** 1–42.

Rice, D. W., Wolman, A. A., Withrow, D. E., and Fleischer, L. A. (1981). Gray whales on the winter grounds in Baja California. *Rep. Int. Whal. Commn.* **31,** 477–493.

Risting, S. (1928). Whales and whale fetuses. *Cons. Perm. Internat. Explor. Mer. Rapp. Proc.—Verb. Reun.* **50,** 1–122.

Samaras, W. F. (1974). Reproductive behavior of the gray whale (*Eschrichtius robustus*), in Baja California. *Bull. South. Calif. Acad. Sci.* **73,** 57–64.

Scammon, C. M. (1874). "The Marine Mammals of the North western Coast of North America." John H. Carmany & Co., San Francisco, California (reprinted by Dover, New York, 1968).

Sokal, R. R., and Rohlf, F. J. (1981). "Biometry, 2nd Edition." Freeman, San Francisco, California.

Storro-Patterson, R. (1981). Biological aspects of the eastern Pacific stock of the gray whale, (*Eschrichtius robustus*). *In* "Background Papers for Workshop on International Cooperation for Conservation of Gray Whales" (S. Kaza, ed.), pp. 276–298. World Wildlife Fund—I.U.C.N. Whale Center, Oakland, California.

Sumich, J. L. (1983). Swimming velocities, breathing patterns, and estimated costs of locomotion in migrating gray whales. *Eschrichtius robustus. Can. J. Zool.* **61**(3), 647–652.

Swartz, S. L., and Cummings, W. C. (1978). Gray whales, (*Eschrichtius robustus*) in Laguna San Ignacio, Baja California, Mexico. *U.S. NTIS PB Rep.* **PB-276319,** 1–38.

Swartz, S. L., and Jones, M. L. (1979). The evaluation of human activities on gray whales (*Eschrichtius robustus*) in Laguna San Ignacio, Baja California, Mexico. *U.S. NTIS PB Rep.* **PB-289737,** 1–42.

Swartz, S. L., and Bursk, M. K. (1979). Gray whales of Laguna San Ignacio; after two years. *Whalewatcher* **13**(1), 7–9.

Swartz, S. L., and Jones, M. L. (1980). Gray whales (*Eschrichtius robustus*) during the 1977–1978 and 1978–1979 winter seasons in Laguna San Ignacio, Baja California Sur, Mexico. *U.S. NTIS PB Rep.* **PB 80-202989,** 1–34.

Swartz, S. L., and Jones, M. L. (1981). Demographic studies and habitat assessment of gray whales (*Eschrichtius robustus*), in Laguna San Ignacio, Baja California Sur, Mexico. *U.S., NTIS PB Rep.* **PB82-1233773,** 1–56.

Swartz, S. L., and Jones, M. L. (1983). Gray whale (*Eschrichtius robustus*) calf production and mortality in the winter range *Rep. Int. Whal. Comm.* **33,** 503–508.

Townsend, C. H. (1887). Present condition of the California gray whale fishery. *Bull. U.S. Fish. Comm.* **6,** 340–350.

U.S. Marine Mammal Commission (1980). Humpback whales in Glacier Bay National Monument, Alaska. *U.S. NTIS, PB Rep.* **PB-80-141559,** 1–44.

Vargas, J. (1981). History of the gray whale in Mexico. *In* "Background Papers for Workshop on International Cooperation for Conservation of Gray Whales" (S. Kaza, ed.), pp. 257–275. World Wildlife Fund—U.S., I.U.C.N. Whale Center, Oakland, California.

Villa-Ramirez, B. (1975). Las ballenas grises de Baja California. *Bioconservation y Supervivencia* **1**(1), 8–12.

Watkins, W. A., and Schevill, W. E. (1979). Aerial observations of feeding behavior in four baleen whales: *Eubalaena glacialis, Balaenoptera borealis, Megaptera novaeangliae*, and *Balaenoptera physalus. J. Mammal.* **60,** 155–163.

White, S. B. (1975). Population and behavioral patterns of gray whales wintering in Scammon Lagoon. Master of Science Thesis, Colorado State University, Fort Collins, Colorado.

White, S. B., and Griese, H. J. (1978). Notes on lengths, weights, and mortality of gray whale calves. *J. Mammal.* **59**(2), 440–441.

Wolfson, F. (1976). Letters to the editor. *Science* **195**.

Zar, J. H. (1974). "Biostatistical Analysis." Prentice-Hall, Englewood Cliffs, New Jersey.

Zimushko, V. V., and Ivashin, M. V. (1980). Some results of Soviet investigations and whaling of gray whales (*Eschrichtius robustus*, Lilljeborg, 1861). *Rep. Int. Whal. Comm.* **30,** 237–246.

15

Reoccupation of Laguna Guerrero Negro, Baja California, Mexico, by Gray Whales

Peter J. Bryant, Christopher M. Lafferty, and Susan K. Lafferty

Introduction

Laguna Guerrero Negro (LGN) is the most northern and most easily accessible of all the winter assembly areas of the California gray whale (*Eschrichtius robustus*) in Baja California. The lagoon has never been an important site of whaling activity; in fact, the few whales taken there in the mid-nineteenth century were taken by boats from whale ships lying outside (Chapter 7, this volume). However, the lagoon was subject to disturbance by commercial shipping and associated dredging activities between 1957 and

ISBN 0-12-389180-9

1967. The whales apparently deserted the lagoon during the latter part of this decade but began to reestablish themselves there during the 1970s (Gard, 1974, 1976).

We have monitored the gray whales of LGN by carrying out studies during the 1980, 1981, and 1982 seasons. The objectives were to determine the extent to which LGN supports breeding and calving activities, to determine whether the occupancy timetable for this area differs significantly from that described for the more southern areas, and to record whale mortality during the calving seasons.

Methods

STUDY AREA

Laguna Guerrero Negro is located on the west coast of Baja California, straddling the 28°N parallel immediately north of Laguna Ojo de Liebre (LOL). The lagoon (Fig. 1) is composed of an entrance channel (about 6.5 km long), an east arm (about 6 km long), and a west arm (about 12 km long). At high tide, water covers approximately 70 km^2, of which only about 12 km^2 is deep enough to be used by the whales. The narrow channels are flanked by salt marshes, numerous sand bars, and mud flats which are partly covered with eel grass (*Zostera marina*). The mouth of the lagoon is partly obstructed by a large sand bar, leaving a narrow deep-water entrance. The southern end of the west arm is connected to LOL by a network of shallow, winding channels that are not navigable by the whales.

We determined depth profiles at 14 locations divided among the three areas of the lagoon. The depth profiles were determined during the 1981 season using an EPSCO ProSearch 60/60 portable depth finder[1] with the transducer held about 30 cm below the water surface at the side of an inflatable boat. For each profile, the boat was piloted at constant speed and compass course from a known location on one side of the channel to the other side. Depth readings were taken every 10 sec, and times were converted to estimated positions on the basis of the total length of the profile which was measured from the chart (Fig. 1). Our results (Fig. 2) in general confirmed the description by Phleger (1965). The 11-m deep, narrow channel which connects the lagoon to the ocean (Fig. 2, Profile 1) branches into two 7- to 10-m deep channels separated by a central shoal area (Fig. 2, Profiles 2 and 3). The easternmost of the two channels (Fig. 2, Profile 4) then branches into the main channels of the east and west arms. An 8- to 10-m deep channel follows the outer curve of the east arm to a shallow basin at its upper end (Fig. 2, Profiles 5–9). A channel that is uniformly about 9 m deep (presumably dredged) connects the intersection of the three arms of the lagoon with the west-arm channel just off the pier at Puerto Venustiano Carranza (Fig. 2, Profile 5). The west arm of the lagoon is deepest (10 m) just off the pier (Fig. 2, Profile 10). South of the pier the deepest part of

[1]The use of trade names does not imply endorsement by the authors, editors, or publishers of this book.

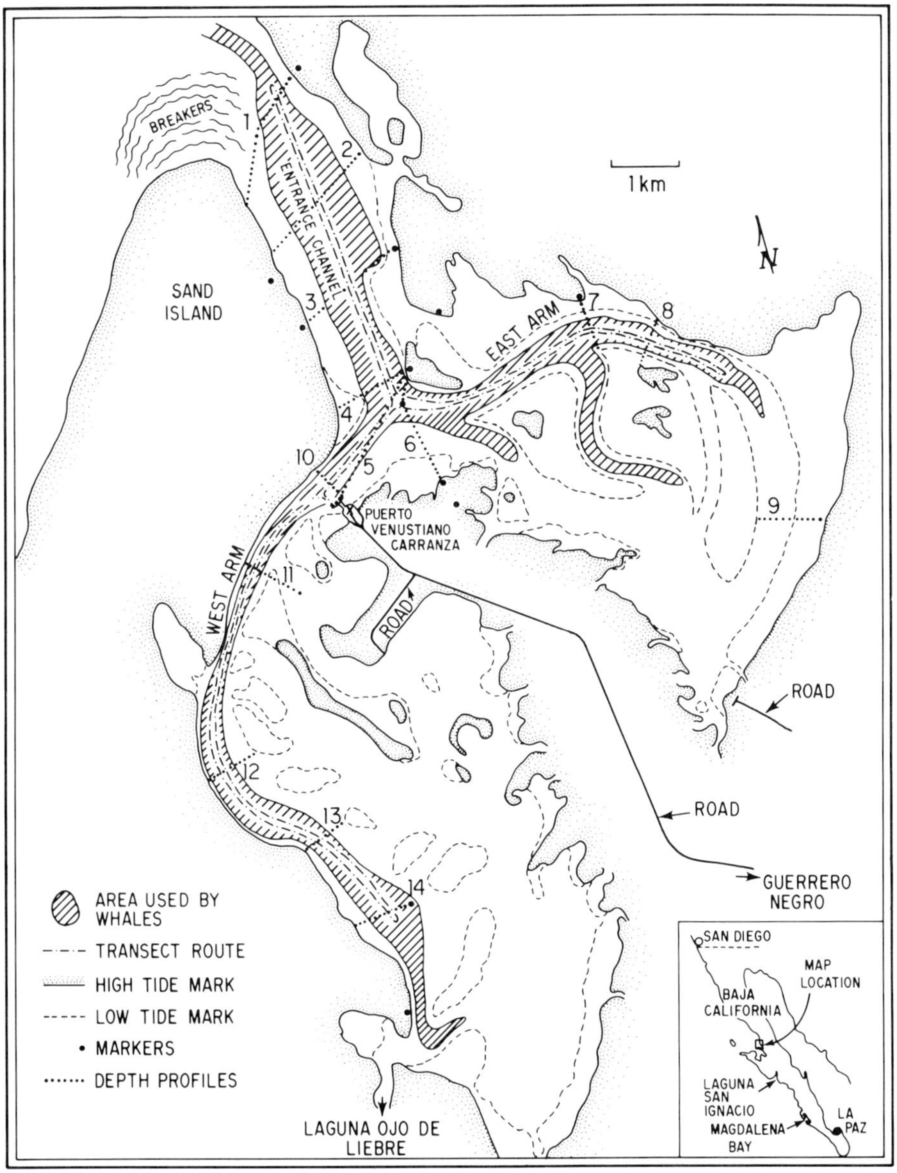

Fig. 1. *The study area (Laguna Guerrero Negro, Baja California, Mexico), showing the approximate areas used by gray whales, the transect routes used in counting the whale population, and the locations of the depth profiles shown in Fig. 2.*

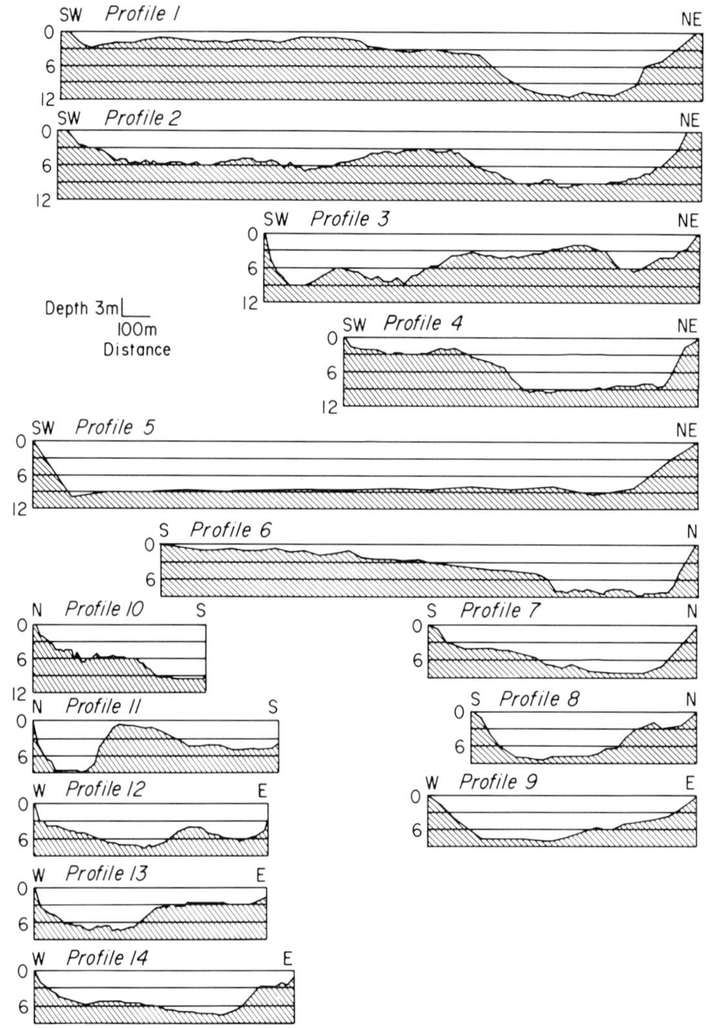

Fig. 2. *Depth profiles of LGN at the locations shown in Fig. 1. Depth is in m below mean low water; vertical exaggeration × 19.*

the channel (about 8 m deep) traverses to the outside of the curve of the west arm and continues all the way to a large, fairly shallow basin marked by a buoy near the connection to LOL (Fig. 2, Profiles 11–14).

FIELD SEASONS

Field studies were conducted January 9–March 15, 1980, January 1–March 25, 1981, and January 30–March 10, 1982.

POPULATION COUNTS

The number of whales in the lagoon was estimated from boat transects which were run every third day, weather permitting. Whenever the estimated sea state exceeded Beaufort 3 (winds greater than 18 km/hr and occasional white caps), the transect was postponed until the following day. The boat was an Avon S400 Sportboat[1] equipped with a Mercury 20 hp outboard motor.[1] An observer on each side of the boat counted the whales between the boat and the shore in all three arms of the lagoon; a recorder, on the opposite side of the boat, logged the whales as singles or cow/calf pairs, assuming that all whales accompanied by calves were cows. The transect lines were traversed at an estimated speed of 16–20 km/hr, one transect outbound from camp and one inbound (Fig. 1), providing replicate estimates of the population for each survey day.

MORTALITY

The shoreline was inspected during each transect for the presence of dead gray whale carcasses. Carcasses were examined and measured as soon as possible after discovery.

Results

THE GRAY WHALE POPULATION AT LAGUNA GUERRERO NEGRO

Counts of gray whales for all three seasons are shown in Fig. 3 as replicates and means. The earliest gray whales seen in LGN were three single animals and two cow/calf pairs on January 3, 1981. In the 1980 season, 13 singles and 18 pairs were present at the first count on January 13. We were unable to carry out field work during early January 1982. The number of cow/calf pairs increased rapidly to a peak of 115 on February 10 in 1980, to 164 on February 7 in 1981, and to 101 on February 10 in 1982. In all three seasons, the first peak of cow/calf pairs in early February was followed by a dip and then a second, smaller peak in late February. Thereafter the numbers decreased until mid-March, by which time almost all of the pairs had left the lagoon. Single animals were present in relatively small numbers throughout all three seasons, their populations peaking at 22 and 19 animals on January 19 and February 19, 1980, respectively, at 43 animals on February 7, 1981, and at 23 animals on February 10, 1982. All of the single animals in 1980, and all but one in 1981 and 1982, had left the lagoon by March 10. Single animals constituted 16.2% (1980), 15.2% (1981), 18.6% (1982), and 16.2% (overall), of all adult whales counted.

"FRIENDLY" WHALES

Whales exhibiting behavior similar to that of the "friendly whales" of Laguna San Ignacio (LSI) (Swartz and Cummings, 1978; Swartz and Jones, 1978, 1980, 1981) were first encountered in LGN on February 16, 17, 18, 19, and 20, 1981. These whales approached the boat and remained close enough to be easily touched by its occupants. On three occasions the friendly animal was single; on one occasion it was the mother, and on another the calf, of a cow/calf pair. Two adult whales were observed showing friendly behavior in LGN during the 1982 season. The first animal, the mother in a cow/calf pair, was encountered only once, on February 2. The second animal, also the

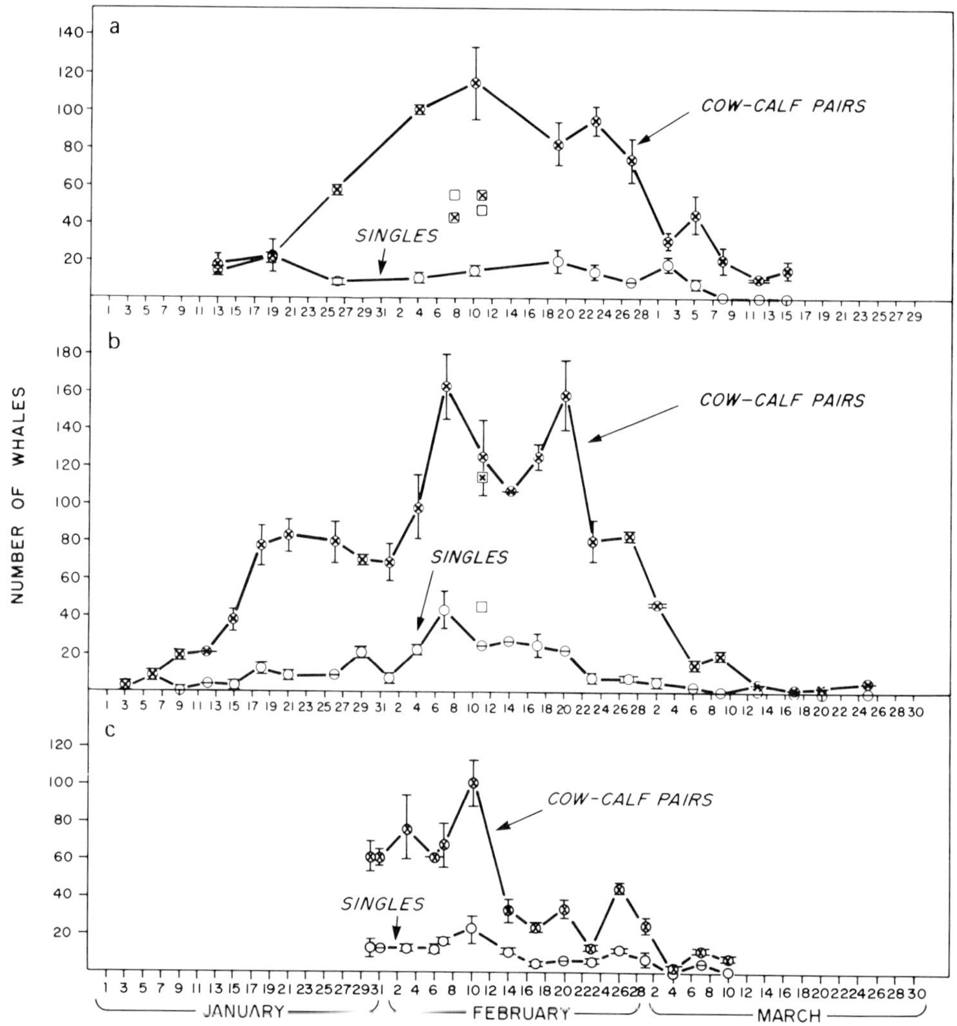

Fig. 3. *Number of cow/calf pairs (⊗) and single whales (○) counted during each transect from (a) January 13 to March 16, 1980, (b) January 1 to March 25, 1981, and (c) January 30 to March 10, 1982. Symbols are plotted as averages of high and low counts, which are indicated by ends of the vertical bars. Also presented are aerial counts of cow/calf pairs (⊠) and single whales (□) reported by Rice et al. (1981) for 1980, and by Rice et al. (1982) for 1981.*

Table I
Number, Location, and Length (Snout to Notch of Flukes) of Dead Gray Whales[a]

Season	Lagoon shoreline	Ocean coast of Sand Island
1980	7 calves (2 measured = 4.0 m and 3.53 m; $\bar{x}$ = 3.77 m, SD = 0.33 m) 1 yearling (8.3 m)	20 calves[b] 1 adult[b]
1981	4 calves (4.34 m; 4.27 m; 4.42 m; 4.95 m; $\bar{x}$ = 4.50 m, SD = 0.31 m)	10 calves (3.76 m; 4.24 m; 4.95 m; 3.95 m; 5.03 m; 4.55 m; 5.05 m; 4.39 m; 4.50 m; 4.14 m; $\bar{x}$ = 4.46 m, SD = 0.45 m)
1982	2 calves (4.0 m; 3.7 m; x = 3.85 m, SD = 0.21 m)	20 calves (18 measured = 4.1 m; 5.1 m; 4.1 m; 4.2 m; 4.1 m; 4.5 m; 5.0 m; 4.2 m; 4.71 m; 4.13 m; 4.43 m; 6.4 m; 4.3 m; 3.5 m; 4.4 m; 4.4 m; 3.5 m; 4.43 m; $\bar{x}$ = 4.42 m, SD = 0.64 m) 1 adult[b]

[a]Found on the shoreline of LGN and on the ocean coast of Sand Island during the 1980, 1981, and 1982 field seasons.
[b]Not measured.

mother in a cow/calf pair, was encountered on February 8, 9, 10, 14, 15, 16, 26, and 27. In all these contacts, the whale stayed with the boat for at least 1 hr and as long as 2.5 hr. The calf of the second pair became increasingly friendly with successive encounters, and during the last two encounters it approached the boat as eagerly as did the mother.

MORTALITY

During the study we found a total of 13 dead calves on the lagoon shoreline. Also during the time of the study, L. Muñoz (personal communication) found a total of 50 dead calves on the ocean coast of Sand Island, which separates LGN and LOL from Bahia Sebastian Vizcaino (Table I). The 8 dead calves which were measured in the lagoon were an average of 4.15 ± 0.45 (S.D.) m long (range, 3.53–4.95 m), whereas the 28 calves measured on the ocean coast averaged 4.43 ± 0.58 m (S.D.) (range, 3.50–6.40 m). The overall average length was 4.37 ± 0.55 m (S.D.). Shark bites were sometimes evident on the carcasses, but usually they were not severe enough to have been the obvious cause of death. A dead yearling was found in the lagoon in 1980, and single dead adults were found on Sand Island shoreline in 1980 and 1982 (Table I). No dead adults were found in the lagoon during the study.

Discussion and Conclusions

THE GRAY WHALE POPULATION OF LAGUNA GUERRERO NEGRO

Our survey method provided two replicate counts of the population for each time point. The two counts often differed from each other, presumably because the three arms of the lagoon were counted sequentially and whales that moved between arms during the count could have been counted more than twice, only once, or not at all. This source of error, of course, is inevitable with single-boat transects.

Aerial counts of whales in LGN in the 1980 season reported by Rice *et al.* (1981) agreed with our boat-based counts of the adult population within 10% but showed less than half the number of calves that were counted from the boat (Fig. 3; Bryant and Lafferty, 1980). However, because aerial and boat-based counts were not performed on exactly the same days, a strict comparison cannot be made.

Two aerial counts (Rice *et al.*, 1982) on February 11, 1981 (Fig. 3) allowed direct comparison with two boat-based counts performed on the same day. The aerial censuses (mean of two estimates) showed 159.5 adults and 114 calves, whereas the boat count (also mean of two estimates) indicated 151 adults and 125.5 calves. Thus, in both 1980 and 1981 aerial and boat counts agreed fairly closely on counts of adults, but aerial surveys underestimated the number of calves. Presumably, more nursing (and therefore submerged) calves are missed in a high-speed aerial survey than in a boat survey during which each whale is often observed several times. Swartz and Jones (1981) have found that under good conditions aerial and boat counts of both adults and calves can agree within 2%.

At the present time, LGN appears to be an important winter assembly area for gray whales, supporting a population that peaks in early February at 101–164 cow–calf pairs (202–328 individuals) and 19–43 single adults. However, according to the literature this has not always been the case.

Whaling records provide an account of the LOL, LSI, and Bahia Magdalena populations, but data pertaining to LGN are quite fragmentary. Scammon (1859) reported that a total of three whaling vessels visited LGN, all in the winter of 1858–1859, but that only one gray whale was taken. This led Henderson (1972) to suggest that the lagoon was not a major calving area for this species. The bark *Black Warrior*, which arrived in the lagoon on November 28, 1858, was wrecked when the whalers attempted to tow her out on December 10. The *Metropolis* and *Lark* were in the lagoon on December 23 but had already moved to LOL and were successfully whaling there by February 13, 1859 (Scammon, 1859). Because these vessels may have been present at LGN too early in the season to find whales there, the possibility that the lagoon supported a substantial gray whale population in the nineteenth century cannot be excluded.

Results from aerial censuses conducted at LGN between 1952 and 1976 were reviewed by Gard (1976) and are shown in Fig. 4. The population increased in the early 1950s but then declined dramatically during the late 1950s. On February 20–24, 1964, Gard was unable to find any whales in the lagoon and reported that the whales were

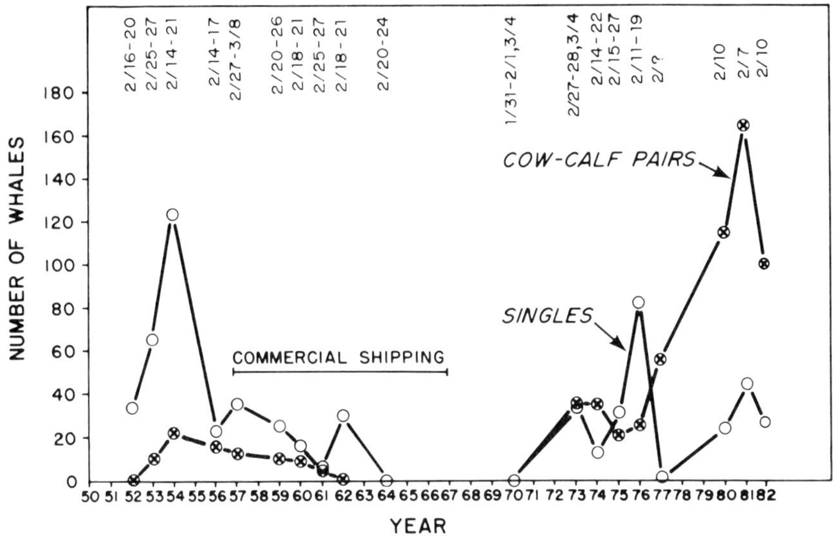

Fig. 4. *Censuses of gray whales in LGN, 1952–1982. Data for 1972 to 1976 are from single aerial surveys conducted during February and early March of each year by C. L. and L. C. Hubbs, R. M. Gilmore, G. C. Ewing and co-workers (1952–1964), and R. Gard and co-workers (1970–1976). See Gard (1976) for detailed sources of data. Data of 1977 are from a boat-based count conducted in February by W. F. Samaras (personal communication). Data of 1980, 1981, and 1982 are peak counts from the present study. Dates of each survey are shown along the top of the figure. "Commerical shipping" refers to a period of export of salt from LGN which necessitated almost continuous dredging of the entrance channel (Eberhardt, 1966).*

probably absent "at least through 1970." This conclusion is apparently based on the fact that no whales were seen in 1970, and that local residents interviewed in 1970 said that only an occasional whale had been seen in the lagoon in recent years. Gard (1976) felt that the decline was probably caused by the commercial operation of shipping salt from LGN, which was carried out from 1957 to 1967. The constant dredging operation necessary to keep the channel open (Eberhardt, 1966) may have been the main source of disturbance. Whales (69 adults and 35 calves) were again seen in the lagoon by Gard and co-workers in 1973. In the ensuing three seasons (1974–1976) the numbers fluctuated from 46 to 105 adults and from 20 to 34 calves. In February, 1977, W. F. Samaras (personal communication) counted 57 cow/calf pairs and 1 unaccompanied calf in LGN from a boat.

In the present study, peak population counts of both cow/calf pairs and single whales in 1980 and 1981 seemed to extend the trend towards higher numbers which started after 1970 following transfer of the salt-moving operations to LOL in 1967 (Fig. 4). However, the peak number of cow/calf pairs in LGN was lower in 1982 than in either 1980 or 1981. There may, of course, be considerable turnover of the lagoon population during the season, and differences in the times that individual whales enter and leave could easily affect the peak counts. However, the low peak count in 1982 does not seem to be simply a result of greater variation in the times that individual whales occupied the lagoon in that season, because the population peak in that year was actually sharper than in 1980. Thus, there seem to have been fewer animals utilizing LGN in 1982, and the steady population increase seen between 1975 and 1981 has not continued (Fig. 4). Year-to-year fluctuations in population levels have been reported for several of the lagoons (Gard, 1976) and would be expected in view of the finding from LSI that some whales enter the lagoons in successive years whereas others return after longer intervals (Swartz and Jones, 1981). Population studies over the next few years should show whether or not the number of whales using LGN is still increasing over the long term and whether the decrease seen in 1982 was a short-term fluctuation.

The proportion of adult whales that is recorded as accompanied by calves is much higher in the recent data from boat surveys in LGN (the 1977 count as well as our own data from 1980, 1981, and 1982) than for most of the earlier aerial surveys (Fig. 4). Because direct comparisons of the two methods suggest that calves are often missed by aerial observers, the significance of this difference is difficult to evaluate. Our data provide no indication of a change in calving rate over the 3 years studied, but such a change might not be detectable in LGN as the whales that enter it are in any case mainly calving females.

OCCUPATION TIMETABLES

The population curves for gray whales in LGN were quite similar in the three seasons studied. In late January, 50.4 (1980), 42.7 (1981), and 60.9% (1982) of the seasonal peak number of cow/calf pairs were already in the lagoon. Peaks of 115 (1980), 164 (1981), and 101 (1982) cow/calf pairs occurred in early February. A second peak of 82.6 (1980), 96.3 (1981), and 44.6% (1982) of the number of cow/calf pairs seen in the first peak occurred in late February, and only 13.9 (1980), 2.7 (1981), and 6.9% (1982) of the peak number of pairs remained in mid-March. The curve for single whales also showed an early February peak of 43 and 23 animals in 1981 and 1982, respectively, and two peaks of 22 and 19 animals in mid-January and mid-February, 1980. In all three seasons, the percentage of adults that were without calves was quite small (16.2, 15.2, and 18.6%, in 1980, 1981, and 1982, respectively).

Our results, in comparison to those of other workers, indicate that the different breeding lagoons are utilized quite differently by the whales. The two northern lagoons, LGN and LOL, are occupied by cow/calf pairs that show a peak population during February (Bryant and Lafferty, 1980; Rice *et al.*, 1981), whereas in LSI to the south, the peak does not occur until mid- to late March or early April (Swartz and Jones, 1978, 1980, 1981; Chapter 14, this volume), at which time LGN is practically devoid of whales.

One possible explanation for the different timings of the cow/calf population peaks is that the breeding females show lagoon fidelity, and that for some unknown reason the populations associated with the northern lagoons are out of phase with the LSI population. Lagoon fidelity was suggested by the experience of the lagoon whalers, who after severely depleting LOL by several seasons of whaling were still able to find plenty of whales in LSI (Henderson, 1972; also this volume). Swartz and Jones (1981) used photoidentification to show that some whales returned to LSI in consecutive seasons and that others returned after longer intervals. The exact degree of lagoon fidelity, however, will be established only after several seasons of photoidentification work are completed in several of the lagoons.

An alternative explanation for the late cow/calf pair population peak at LSI is that the late-appearing pairs are actually northbound migrants who spent the earlier part of the season in Bahia Magdalena or further south (Chapter 14, this volume). This idea is supported by preliminary observations suggesting that the population peak in Bahia Magdalena is earlier than that in LSI (S. Lawson, personal communication). It is, of course, equally possible that some of the whales seen early in the season at LGN may continue their southward migration and contribute to the later peak at LSI.

Another major difference between the lagoon populations is that few single whales are found in LGN in comparison to LOL and LSI. LOL shows a large population of single whales in late January and early February (Rice *et al.*, 1981) and LSI shows a large peak of single whales in February (Swartz & Jones, 1978, 1980, 1981). However, LGN supports only a relatively small number of single whales (an average of about 16.2% of the total number of adults), and they never outnumber cow/calf pairs.

These differences suggest that LGN can be considered as primarily a nursery lagoon, similar to the upper part of LSI (Swartz and Jones, 1981). LGN apparently has no area analogous to the lower part of LSI, which is favored by the single whales. The lower part of LSI contains channels more than twice as deep (20–27 m) as the deepest parts of LGN (Swartz and Jones, 1981), which might suggest that single whales prefer deeper water, presumably for some reason connected with their different behavior patterns compared to nursing mothers. Consistent with this interpretation is the finding (Rice *et al.*, 1981) that many single whales inhabit the deep-water area immediately offshore from LGN during the calving season.

MORTALITY

Until the total number of calves produced during a season is determined, counts of dead calves cannot provide an estimate of neonatal mortality. However, our results show about four times as many dead calves on the ocean coast of the island separating LGN from the ocean as are found in the lagoon itself. This suggests that many births take place in the open ocean rather than in the lagoon. The carcasses measured averaged 4.37 m in length, and Rice *et al.* (1981) reported that the average length of dead calves found in the lagoons was 4.41 m. This is substantially shorter than the mean length of near-term fetuses, which has been reported to be 4.62 m (Rice and Wolman, 1971), suggesting either than many of the dead calves were born prematurely, or that they shrank either before or after death (Rice, 1982), or that the neonatal mortality rate is higher among the smaller calves.

Summary

Gray whales (*Eschrichtius robustus*) were studied in Laguna Guerrero Negro, Baja California, during the 1980, 1981, and 1982 calving seasons. The lagoon population included peaks of 115 cow/calf pairs and 22 single adults in 1980, 164 pairs and 43 singles in 1981, and 101 pairs and 23 singles in 1982. Cow/calf pairs began to appear in the lagoon in early January, reached peak numbers in early February, and almost all had departed by mid-March of each season. Whales unaccompanied by calves were always in the minority, reaching peak numbers in late January and late February in 1980 and in early February in 1981 and 1982. "Friendly" whales were observed in mid-February in both 1981 and 1982. During the study periods, 13 dead calves were found on the lagoon shoreline, and 50 more were found on the ocean coast of the barrier island separating the lagoon from the ocean.

Acknowledgments

This study was supported by grants from the Los Angeles and Orange County Chapters of the American Cetacean Society, the National Marine Fisheries Service, and the University of California Research Expeditions Program. It was carried out under the auspices of Permits Nos. 12151 and 139 from the Secretaria de Pesca in Mexico City, to whom we are grateful. Data for the 1981 season were collected under a permit issued to Sr. Luis Fleischer. We are, of course, deeply indebted to the many dedicated volunteers who spent long hours on the project without tangible compensation, and most of all we would like to express our heartfelt thanks to our good friends Leysl and Francisco Muñoz, who made us welcome and helped us in so many ways during the course of this study.

References

Bryant, P. J., and Lafferty, C. M. (1980). The gray whales of Laguna Guerrero Negro. *Whalewatcher* **14**(4), 3–5.

Eberhardt, R. L. (1966). Coastal geographical features of Laguna Guerrero Negro. *Calif. Geogr.* **7,** 29–35.

Gard, R. (1974). Aerial census of gray whales in Baja California lagoons, 1970 and 1973, with notes on behavior, mortality, and conservation. *Calif. Fish Game* **60,** 132–143.

Gard, R. (1976). "Aerial Census and a Population Dynamics Study of Gray Whales in Baja California during the 1976 Calving and Mating Season," Contract No. MM6AC014, pp. 1–23. Report to the U.S. Marine Mammal Commission, Washington, D.C.

Henderson, D. A. (1972). "Men and Whales at Scammon's Lagoon," Baja Calif. Travel Ser. 29. Dawson's Book Shop, Los Angeles, California.

Phleger, R. B. (1965). Sedimentology of Guerrero Negro Lagoon, Baja California, Mexico. *Colston Pap.* **17,** 205–237.

Rice, D. W. (1982). Gestation and fetal growth of the gray whale. *Int. Whal. Commn. Sci. Comm. Doc.* **SC/34/PS11,** 1–14.

Rice, D. W., and Wolman, A. A. (1971). The life history and ecology of the gray whale (*Eschrichtius robustus*). *Spec. Publ. Am. Soc. Mammal.* **3,** 1–142.

Rice, D. W., Wolman, A. A., Withrow, D. E., and Fleischer, L. A. (1981). Gray whales on the winter grounds in Baja California. *Rep. Int. Whal. Commn.* **31,** 477–493.

Rice, D. W., Wolman, A. A., and Withrow, D. E. (1982). Distribution and numbers of gray whales on their Baja California winter grounds. *Int. Whal. Commn. Sci. Comm. Doc.* **SC/34/PS12,** 1–22.

Scammon, C. M. (1859). "Journal Aboard the Bark *Ocean Bird* on a Whaling Voyage to Scammon's Lagoon, Winter of 1858–1859" [reprinted as Baja Calif. Travel Ser. 21, by Dawson's Book Shop, Los Angeles, 1970 (D. A. Henderson, ed.)].

Swartz, S. L., and Cummings, W. C. (1978). Gray whales, (*Eschrichtius robustus*), in Laguna San Ignacio, Baja California, Mexico. *U.S., NTIS PB Rep.* **PB-276319,** 1–38.
Swartz, S. L., and Jones, M. L. (1978). The evaluation of human activities on gray whales (*Eschrichtius robustus*), in Laguna San Ignacio, Baja California, Mexico. *U.S., NTIS PB Rep.* **PB-289737,** 1–34.
Swartz, S. L., and Jones, M. L. (1980). Gray whales, *Eschrichtius robustus*, during the 1977–1978 and 1978–1979 winter seasons in Laguna San Ignacio, Baja California Sur, Mexico. *U.S. NTIS, PB Rep.* **PB80-202989,** 1–35.
Swartz, S. L., and Jones, M. L. (1981). Demographic studies and habitat assessment of gray whales (*Eschrichtius robustus*), in Laguna San Ignacio, Baja California Sur, Mexico. *U.S., NTIS, PB Rep.* **PB82-123373,** 1–56.

16

Migration Corridors of Gray Whales along the Central California Coast, 1980–1982

M. Michael Poole

Introduction

The spring northward migration of gray whales (*Eschrichtius robustus*), from their winter calving lagoons in Baja California and mainland Mexico to their summer feeding grounds in the Bering, Chukchi, and Beaufort Seas, is not as well studied or understood as their fall southward migration. It is known that they follow the coast in spring (Kellogg, 1929) and that historically native American Indians, such as the Makahs, hunted them chiefly during that season (Scheffer and Slipp, 1948; Chapter 4, this volume). The northward migration is of great significance, as it includes females that have recently conceived as well as new calves of the year. Since the late 1800s, the migration route used by females with calves (cow/calf pairs) has been reported to be farther offshore than the route used by the rest of the population (Jordan, 1887; Town-

ISBN 0-12-389180-9

send, 1887; Rice and Wolman, 1971), although it is reported that originally it was very close to shore, along the kelp beds (Scammon, 1874).

Hayes' "Emigration Notes" (unpublished manuscript), cited by Starks (1922), stated that during the 1800s the San Diego whaling station took gray whales in the early spring (see Chapter 6, this volume) as they traveled quite close to the shoreline along the kelp beds. Scammon (1874) reported that before the advent of coastal shore whaling ". . . both male and female, with their young, will be seen . . . following the shore so near that they often pass through the kelp near the beach. The mother, with her young . . . makes the best of her way . . . avoiding the rough sea by passing between or near the rocks and islets that stud the points and capes." He then remarked, however, that due to increased pressures from coastal whaling stations operating near the kelp beds the whales altered their migration patterns, avoiding ". . . these fatal regions, making wide deviations in their course." Since that report more than 100 years ago, most published summaries have similiarly stated that females with calves utilize a more distant offshore migrational corridor.

Captain Clark, of the San Simeon whaling station, was reported by Townsend (1887) as having seen only one cow/calf pair during the 10-year period of 1877–1887. The spring whaling season was profitable at San Simeon, but the catch consisted mostly of males, this, it was reported, was because females with calves were "keeping farther offshore." Another published report (Jordan, 1887) indicated that males returned northward nearshore, and cow/calf pairs offshore. During a 10-year study (1959–1969) in which 316 gray whales were collected by whaling vessels under scientific permit, only three cow/calf pairs were observed by boat surveys and none by concurrent shore and aerial surveys (Rice and Wolman, 1971). The authors concluded "The route taken by females with calves during the spring migration is unknown. . . . Unlike the other members of the population, they must travel farther offshore."

Beginning in the spring of 1980 and continuing in the springs of 1981 and 1982, the

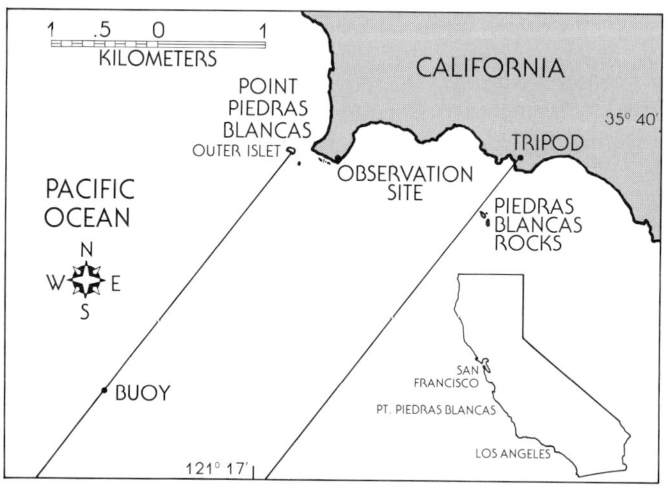

Fig. 1. *Shore survey site at Pt. Piedras Blancas, California showing observation zone.*

author studied the northward migration of gray whales along the central California coast between Point Buchon and Cape San Martin near the old whaling station of San Simeon (Fig. 1). The primary objective was to test the long-held hypothesis that cow/calf pairs use a distant, offshore migrational route.

Methods

STUDY SITE

It should be noted that in Figs. 1, 2, 7, and 8 compass headings are with respect to true north. All angle measurements within this paper are with respect to magnetic north. Magnetic variation is approximately $16\frac{1}{2}°$ easterly over the entire area.

Point Piedras Blancas (35°40′N; 121°17′W; Fig. 1) was chosen as the site for shore-based surveys as well as for a base camp. Lying approximately 8 km northwest of San Simeon, it is the only point to intersect a straight-line course from Point Buchon to Point Sur along 160 km of the central California coast. On its east (leeward) side a cove is protected from prevailing northwestern winds. This cove is the habitat of extensive beds of *Macrocystis* and *Nereocystis* kelp during the spring.

A 1.6 km-long study zone was demarcated using natural landmarks, a large tripod, and a buoy. At 90° east from Point Piedras Blancas, two large islets known as Piedras Blancas Rocks are 400 m offshore. The rocks are hosts to numerous birds, and one serves as a haul-out for as many as 300 California sea lions (*Zalophus californianus*). A tripod of unknown origin and ~25 m in height stands on the shore across from Piedras Blancas Rocks. A line extending from the tripod through the rocks demarcated the southeastern boundary of the study zone. At 270° from the observation site and 100 m offshore is another natural islet, Outer Islet. Approximately 35 m high, Outer Islet is also a California sea lion haul-out. Beyond the islet, 210° from the observation site and 3.2 km offshore, is a Coast Guard buoy. The northwestern boundary of the study zone was demarcated by a line extending from Outer Islet through the buoy. Dimensions of the study zone were verified with a surveyor's transit. Observers were situated on a rocky outcropping 11 m above sea level and 7 m from the cliff edge.

SHORE SURVEYS

Data were collected from the shore site 10 hr per day, 6 days per week, for 12 weeks in 1980 (March 10–May 31) and for 16 weeks in 1981 (February 9–May 30) using methods similar to those described by Rice and Wolman (1971) and Reilly *et al.* (1980). Two 5-hr shifts (0700–1200 and 1200–1700) were conducted. Two observers per shift monitored movements of whales passing through the study zone. Whales could be detected with 7×50 binoculars and a $20\times$–$60\times$ spotting scope prior to their entrance into the study zone. The standard procedure was to record the time a whale (or "pod" of whales) entered the zone, to measure the angle of approach with an engineer's compass, and to keep the whales under observation by one or both observers as they passed through the zone. Data recorded were number of whales in each pod and method for determining the number (by observing blows, backs, flukes, or combinations); positions of individuals; and conspicuous behavior. When the whales were nearest the shoreline, their distance offshore was estimated subjectively using offshore landmarks where distances had been previously determined with a surveyor's transit as references. In addition, a rangefinder was used periodically to cross-check offshore estimates. As the whales passed out of the study zone, observers noted the time and determined their final compass angle of departure. Environmental conditions affecting visibility, including sea state and wind velocity, were noted hourly or as they changed. A visibility code with relative values assigned to various weather conditions was adapted from Reilly *et al.* (1980) (Table I).

Hours of observation and sightings were grouped according to visibility conditions. Sightings in Codes 4–6 were deleted from all calculations; those in Codes 1–3 were grouped by week, estimated offshore distance, and composition (cow/calf pairs or other whales). Sightings with distance estimates were pooled into four distance intervals: ≤200 m from the shoreline; >200 and ≤400 m; >400 and <800 m; and ≥800 m extending to over 3.2 km. For each week, the relative percentages of each interval were calculated in addition to weekly mean distance ($\bar{d}$), with standard error (SE), 95% confidence limits, and weekly median distance. Data were compared among weeks and between two sets of data pooled for 8-week periods (February 9–April 4 versus April 6–May 31). Finally, distribution of distances offshore of cow/calf pairs was compared with that of other whales, which were referred to as "singles."

Table I
Visibility Codes[a]

Code	Effective sighting distance	Condition	Description
1	4.8+ km	Excellent	Clear day or high clouds No glare Horizon visible
2	3.2 km–4.8 km	Very good	Clear or some cloud cover Some glare Surface ripple
3	1.6 km–3.2 km	Good	Some fog, haze, low clouds Some interference from chop, surf, or glare
4	800 m–1.6 km	Fair	Fog, full overcast, light rain, haze with glare Frequent whitecaps
5	400 m–800 m	Poor	Moderate rain or fog, large surf, bad glare
6	0–400 m	Terrible	Combination of conditions makes it very difficult or impossible to see even the closest whales

[a]Modified from Reilly *et al.*, 1980a.

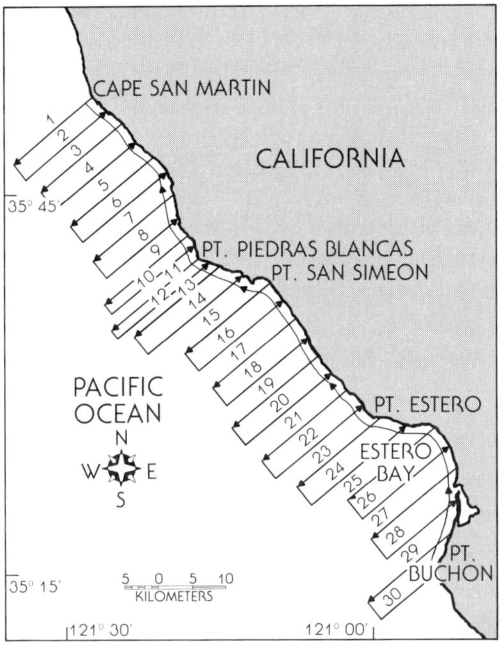

Fig. 2. *Aerial survey transects from Cape San Martin to Pt. Buchon.*

AERIAL SURVEYS

1981

Two weeks of aerial surveys were conducted using methods modified from Reilly *et al.* (1980, 1983) and Swartz and Jones (1980). During March 3–11, a Cessna 172[1] high-wing aircraft flew at an altitude of 330 m and an average ground speed of 128km/hr over a series of 30 transect lines perpendicular to the coastline between Cape San Martin and Point Buchon, which are separated by 96 km of coastline (Fig. 2). Each transect extended 16 km from shore and was 3.2 km from the next nearest transect, except at Point Piedras Blancas where the interval between transect lines was reduced to 1.6 km for 5 transects (Nos. 9–13). An assistant acting as data recorder sat in the aircraft's right front seat, and observers were in the left and right rear seats. The 90°-angle distance of whales from the plane's track line was determined from markings on the plane's wing struts and windows corresponding to angles measured by clinometer. Distance offshore was calculated from the plane's ground speed and the shore intersect time. During April 21–29, transects were flown parallel to one another following the contour of the coast at 400-, 800-, and 1200-m intervals offshore, except at Point Piedras Blancas where 5 perpendicular transects (9–13) 1.6 km apart were conducted as in March. These 5

[1]The use of trade names does not imply endorsement of products.

transects extended 16 km offshore during the first run of the day and 8 km during subsequent flights.

1982

In 1982, aerial surveys were flown only over Estero Bay between Points Estero and Buchon from April 30–May 3 and on May 12. These consisted of eight perpendicular transects (Nos. 23–30) to 16 km offshore. Additional flights were made parallel to the coastline between Point Estero and Point Buchon at 400-, 800-, and 1200-m intervals, thereby providing replicate coverage of the nearshore areas.

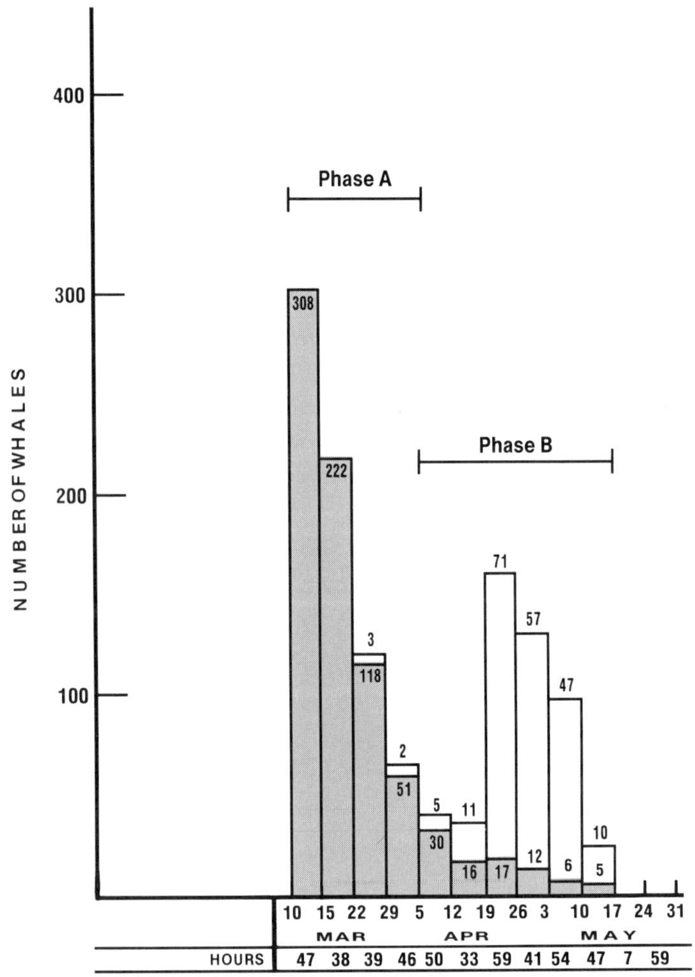

Fig. 3. *Weekly counts, 1980. Singles, solid bar; cow/calf pairs, open bar (illustrated as individuals, enumerated as single units).*

Results

SHORE COUNTS

1980

During 659 hr of observation between March 10 and May 31, 1980, 1496 whales were observed migrating northward; 228 were calves (Fig. 3). Of the 659 hr, 520 were conducted in visibility conditions of Codes 1–3 and 139 in Codes 4–6. The effort in Codes 1–3 produced 1197 whales for which there were distance estimates; 785 were

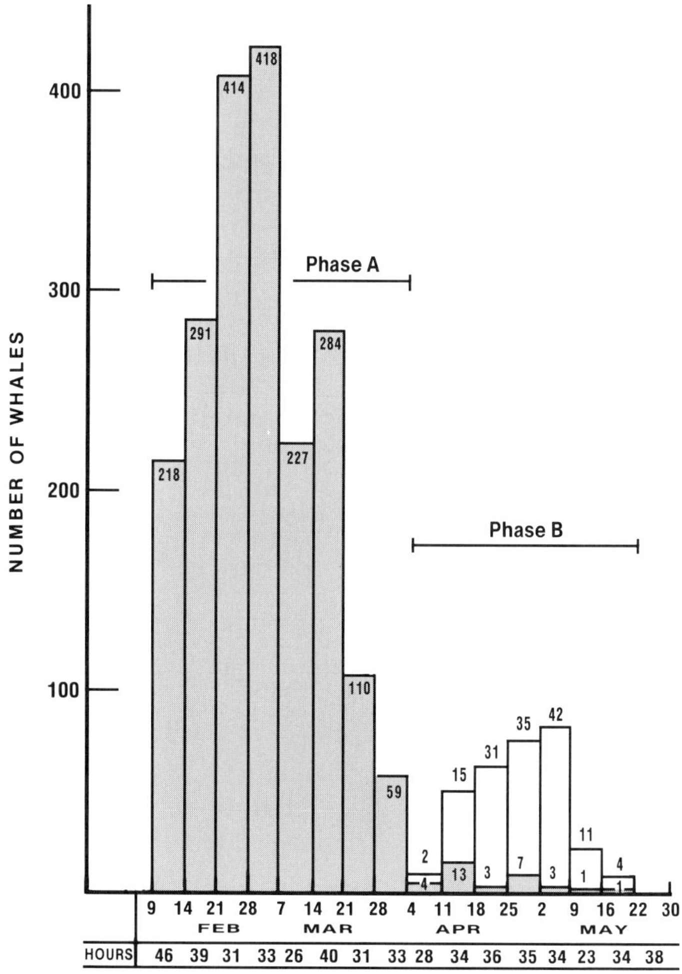

Fig. 4. *Weekly counts, 1981. Singles, solid bar; cow/calf pairs, open bar (illustrated as individuals, enumerated as single units).*

singles and 206 were cow/calf pairs. The migration was bimodal and the two modes were dubbed Phase A and Phase B (Poole, 1980; 1984). The first cow/calf pair was observed the week of March 24–29. From that week until the end of the migration, the percentage of cow/calf pairs increased and that of singles decreased. Cow/calf pairs were 82% of Phase B whales.

1981

During 755 hr of observation between February 9 and May 30, 1981, we observed 3296 northbound whales; 209 were calves (Fig. 4). Of the 755 hr of observation, 540 were conducted in visibility Codes 1–3 and 215 in Codes 4–6. The effort in Codes 1–3 produced 2333 whales which were given distance estimates, 2053 singles and 140 cow/calf pairs. The migration was clearly bimodal (Poole, 1981a). Between February 9–April 4, Phase A, the migration of singles (whales other than cow/calf pairs) took place. The first cow/calf pair was observed during the week March 30–April 4. The second pulse of the migration (April 6–May 30), Phase B, contained 90% cow/calf pairs.

ESTIMATED DISTANCES OFFSHORE

1980

Sightings occurred at distances of 10 m to 3.2 km from shore. Table II gives the relative percentages of whales in each of the four distance intervals. As the migration season progressed, the distance of whales from shore decreased. Seventy-one percent (220) of those passing March 10–15 were in Distance Interval 4 (800 m to 3.2 km), while only 18% (56) were in Interval 1 (≤200 m). But from April 14–May 31, only three whales

Table II
Weekly Distribution of Whales Offshore in Four Distance Intervals During Codes 1–3 Visibility Conditions[a]

Week	Number of whales	Interval (m) 0–200	>200–400	>400–<800	≥800
3/10–3/15	308	56 (18%)	32 (10%)	0	220 (71%)
3/17–3/22	222	93 (42%)	40 (18%)	0	89 (40%)
3/24–3/29	124	73 (59%)	21 (17%)	0	30 (24%)
3/31–4/05	55	30 (54%)	13 (24%)	0	12 (22%)
4/07–4/12	40	23 (57%)	8 (20%)	0	9 (22%)
4/14–4/19	38	33 (87%)	5 (13%)	0	0
4/21–4/26	159	158 (99%)	1 (1%)	0	0
4/28–5/03	126	126 (100%)	0	0	0
5/05–5/10	100	100 (100%)	0	0	0
5/12–5/17	25	22 (88%)	3 (12%)	0	0
5/19–5/24	0	0	0	0	0
5/26–5/31	0	0	0	0	0

[a]Data from 1980.

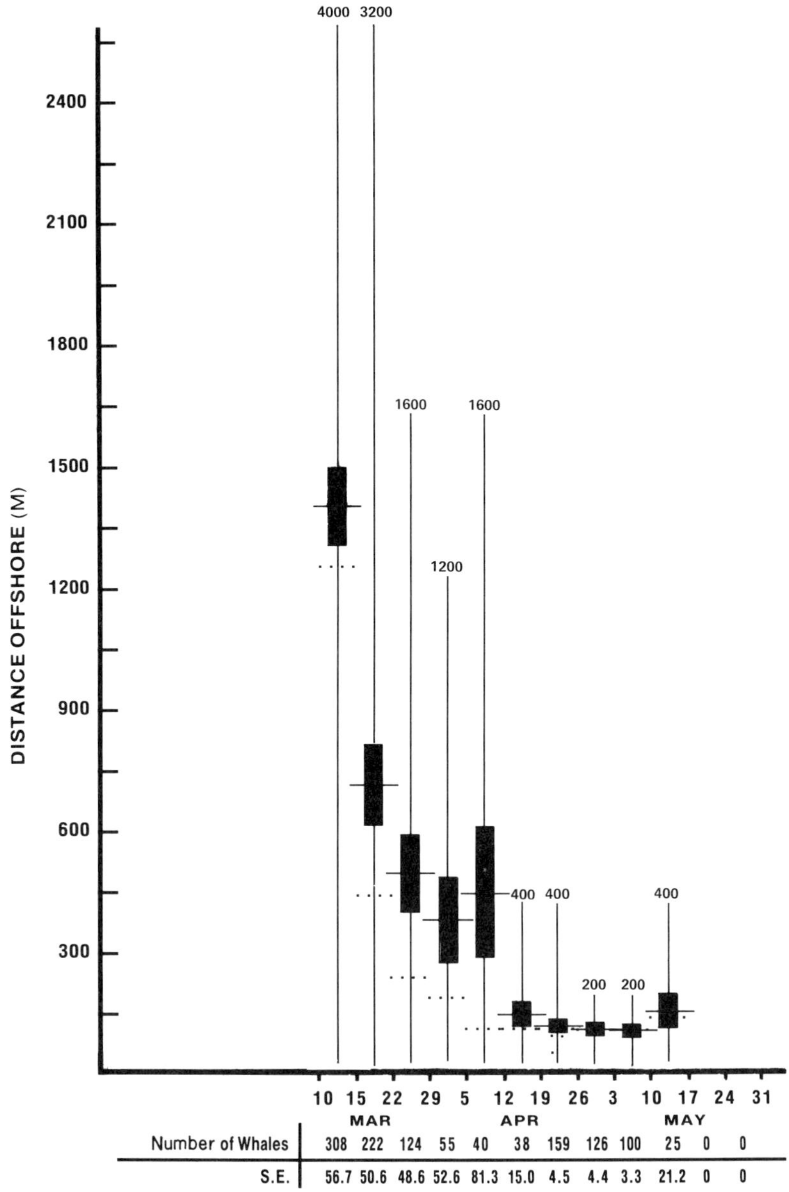

Fig. 5. *Distance offshore as estimated from shore site, 1980. Symbols: horizontal solid line, mean; vertical bars, 95% confidence limits on either side of mean; horizontal dots, median; vertical lines, range.*

were observed farther offshore than Interval 1. Figure 5 gives the weekly mean distances, with standard error of the mean and 95% confidence limits, the range, and the weekly median. Weekly means declined steadily from 1350 m during the first week (March 10–15) to 79 m (May 5–10), and medians fell from 1200 to 75m.

Phase A. From March 10–April 5, 50% (351) of all whales observed (709) were in Interval 4, 800 m to 3.2 km offshore; 15% (106) were in Intervals 2 and 3 between 200 and 800 m; and 35% (252) were in Interval 1 within 200 m of the shore.

Phase B. In contrast, 95% (462) of the whales observed passing from April 7–May 31 (488) were within Interval 1, 204 cow/calf pairs and 54 singles. Only 3% (17 whales, 2 cow/calf pairs + 13 singles) observed were passing in Intervals 2 and 3 between 200 and 800m, and 2% (9 whales; all singles) passed in Interval 4 at 800 m. Of the 206 cow/calf pairs that passed in 1980, 204 (99%) were between 10 and 200 m.

1981

Again in 1981, whales passed Point Piedras Blancas between 10 m and 3.2 km offshore. Table III gives the relative percentages of whales in each of the four distance intervals. The pattern remained the same as in 1980. Weekly percentages by distance intervals shifted from 97% in the furthest offshore interval (February 9–14) to at least 75% in the most coastal interval (April 6–May 30). Only one whale was sighted in Interval 4

Table III
Weekly Distribution of Whales Offshore in Four Distance Intervals During Codes 1–3 Visibility Conditions[a]

Week	Number of whales	Interval (m)			
		0–200	>200–400	>400–<800	≥800
2/09–2/13	218	0	6 (3%)	0	212 (97%)
2/15–2/21	291	29 (10%)	21 (7%)	0	241 (83%)
2/23–2/28	414	25 (6%)	23 (6%)	13 (3%)	353 (85%)
3/02–3/07	418	37 (9%)	69 (16%)	12 (3%)	300 (72%)
3/09–3/14	227	35 (15%)	81 (36%)	22 (10%)	89 (39%)
3/16–3/21	284	59 (21%)	70 (25%)	18 (6%)	137 (48%)
3/23–3/28	110	36 (33%)	44 (40%)	9 (8%)	21 (19%)
3/30–4/04	59	33 (56%)	10 (17%)	4 (7%)	12 (20%)
4/06–4/11	8	6 (75%)	2 (25%)	0	0
4/13–4/18	43	40 (93%)	2 (5%)	0	1 (1%)
4/20–4/25	65	62 (95%)	3 (5%)	0	0
4/27–5/02	77	74 (96%)	3 (4%)	0	0
5/04–5/09	87	77 (88%)	5 (6%)	5 (6%)	0
5/11–5/16	23	23 (100%)	0	0	0
5/18–5/23	9	4 (44%)	5 (56%)	0	0
5/25–5/30	0	0	0	0	0

[a]Data from 1981.

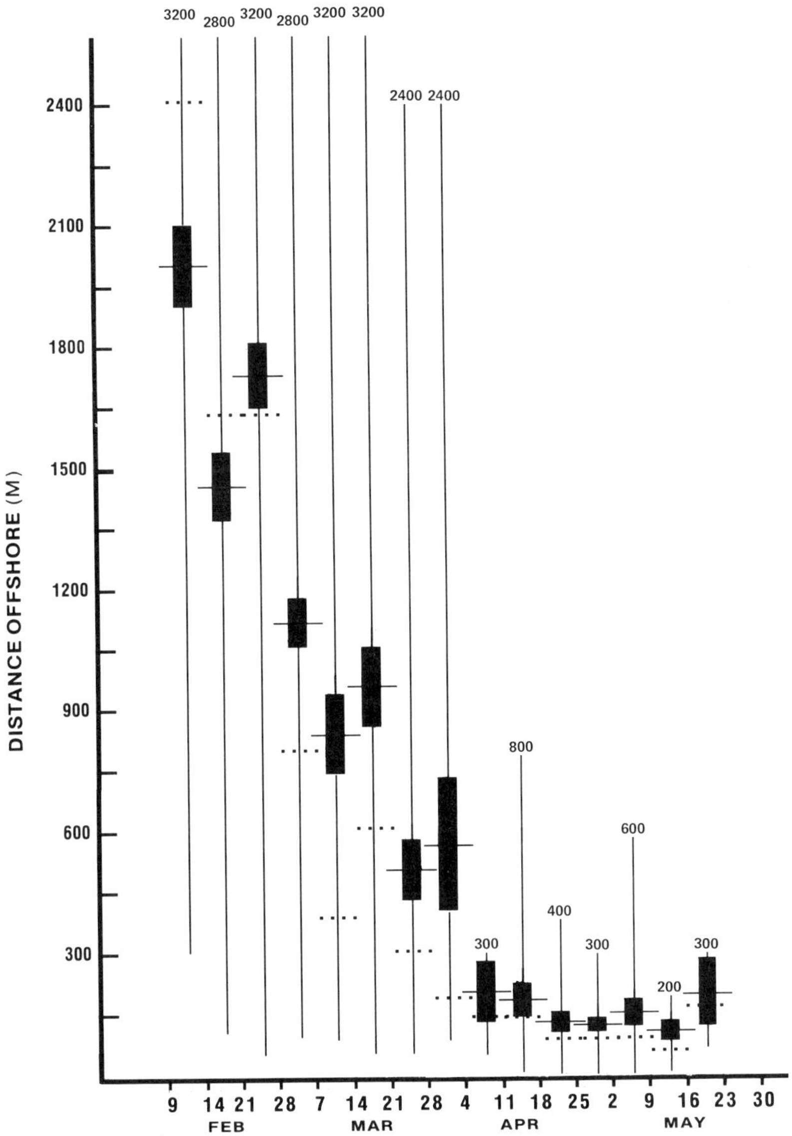

Number of Whales	218	291	414	418	227	284	110	59	8	43	65	77	87	23	9	0
S.E.	52.2	42.9	40.9	31.5	49.3	49.3	36.4	82.1	32.6	20.4	9.8	6.7	14.2	12.4	36.6	0

Fig. 6. *Distance offshore as estimated from shore site, 1981. Symbols: Horizontal solid line, mean; vertical bar, 95% confidence limits on either side of mean; horizontal dots, median; vertical line, range.*

(800 m–3.2 km) after April 4. Mean distance dropped from 1972 m (February 2–14) to 91 m (May 11–16). Weekly medians decreased from 2400 to 75 m (Fig. 6).

Phase A. From February 2–April 4, 67% [1365 of all (2021) passing whales] observed were in Interval 4 (800 m to 3.2 km). Twenty percent (402) were between 200 and 800m in Intervals 2 and 3, and only 13% (254) passed in Interval 1.

Phase B. During the last half of the migration (April 6–May 30), 312 whales passed Point Piedras Blancas. Ninety-two percent [287 whales (135 cow/calf pairs + 17 singles)] passed in Interval 1 (10–200 m), and only 8% (24 whales, 5 cow/calf pairs + 14 singles) were in Intervals 2 and 3 between 200 and 800 m; 1 whale was observed in Interval 4 at 800 m. Of 140 cow/calf pairs in 1981, 135 (96%) passed between 10 and 200 m.

AERIAL SURVEYS

1981

During the March flights, 225 northbound whales were observed. These whales took rather direct routes between points of land, avoiding bights and indentations in the coastline and thus confirming conclusions reached from previous shore observations (Poole, 1980). Whales approached Point Piedras Blancas from Points San Simeon,

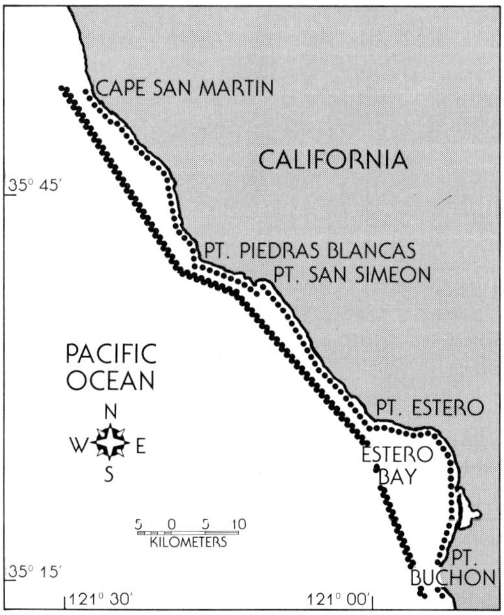

Fig. 7. *Migrational corridors used by northward-migrating gray whales between Point Buchon and Cape San Martin. Phase A, wide, offshore, straight-line corridor; Phase B, narrow, nearshore, following coastline contour.*

Estero, or Buchon, avoided Estero Bay, and utilized a wide, straight-line corridor from Point Buchon to Point Estero (Fig. 7). This corridor placed them almost 12 km offshore at Estero Bay (Poole, 1981a). Flights along transects 9–13 at Point Piedras Blancas yielded observations of 119 whales at a mean distance of 2.4 km offshore.

During April, 59 whales (27 cow/calf pairs, one duo, 3 singles) were counted in the flights between Cape San Martin and Point Buchon. All were within 400 m of the shoreline, following the major contours of the coastline (Fig. 7). They were frequently observed on the edge of kelp beds and within them, not more than 25–200 m offshore (Poole, 1981b). They approached Point Piedras Blancas by passing between Piedras Blancas Rocks and the shore along the kelp lines and were observed diving repeatedly near and in the kelp beds (Fig. 8).

1982

Because the 1981 surveys yielded few whales in the Estero Bay region, aerial surveys in 1982 were conducted only over Estero Bay. Of 50 whales observed (24 cow/calf pairs, 2 singles), 38 (76%, 18 cow/calf pairs, 2 singles) were in Intervals 1 and 2 (25–400 m offshore); one pod of 8 whales (16%, 4 cow/calf pairs) was 600 m offshore and heading toward shore; one cow/calf pair (4%) was 1200 m offshore inside a kelp bed; and one cow/calf pair (4%) was approximately 10.4 km offshore. This last sighting was quite unexpected, as was the one at 1200 m.

Most cow/calf pairs were repeatedly observed within or just outside breaking waves in Estero Bay, only 50–100 m off the shallow sandy beaches that line the interior of the Bay. Several pods were followed throughout the entire Bay as they progressed from Point Buchon to Point Estero following the sandy beaches all the way; they were never more than 400 and at times not more than 50 m offshore (Fig. 7).

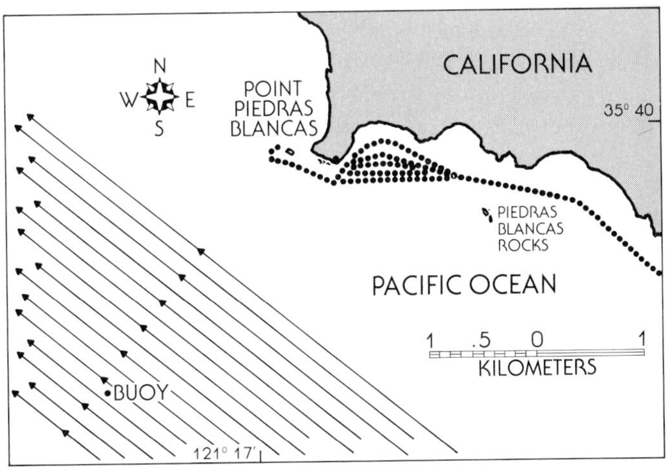

Fig. 8. *Migrational corridors at Point Piedras Blancas. Phase A, wide, offshore, straight-line corridor; Phase B, narrow, nearshore, following coastline contour.*

Discussion

HISTORICAL REVIEW

Was the hypothesis that gray whale females with calves migrated north far offshore correct at one time? It is most certainly not the case at present. But have the behavior and migratory patterns of gray whales altered over the last 100 years? Over the last 20 years?

Scammon (1874) stated that cow/calf pairs did indeed migrate north very close to shore in the past, as they do now, but he concluded that they altered their migratory course to an offshore corridor because of the pressures of coastal whaling stations. Unfortunately, he offered no actual sightings of cow/calf pairs in support of this claim. He did however add that by the first of April all whales had passed the coastal whaling stations of California. Townsend (1887) reported the whaling season lasted until April at San Simeon, and that the head of the station had observed only one cow–calf pair in the 10 years from 1877 to 1887 as a result of females with calves "keeping well out to sea." Like Scammon, Townsend presented no data to substantiate his statement.

The present study at Point Piedras Blancas showed that not only do cow/calf pairs migrate in a nearshore corridor, but as importantly they do not arrive at San Simeon until April; they do not pass in appreciable numbers until the first of May, and not all have passed until the end of May (Poole, 1981b). Therefore, if Scammon and Townsend were correct in their summary of the timing and routes of the northward migration of cow/calf pairs, then not only has their corridor been altered, but their timing has been shifted at least 1–2 months later. When and why might such alterations have occurred?

Based on southern California sightings, Hubbs (1959) reported "On the northward journey the grays move less consistently close to shore and the cows with calves seem to take a more offshore path." For the coasts of Washington and British Columbia, Pike (1962) stated the northward migration began in February, ended in May, and peaked during the ". . . first two weeks of April." In 1971 Rice and Wolman hypothesized the migration route for females with calves to be farther offshore than that used by the rest of the population because only three cow/calf pairs were observed in 10 years of observations. One cow/calf pair was observed by a whaling catcher-boat crew near Point Reyes; the other two pairs were observed during transect cruises near San Clemente Island. However, the whaling vessels used in the project ceased collection in April, and they operated between Half Moon Bay and San Francisco, or approximately 480 km north of Point Piedras Blancas. The transect cruises were conducted in January and February off southern and central California. As demonstrated here, cow/calf pairs do not arrive at Point Piedras Blancas until the end of March or beginning of April, and they do not pass in appreciable numbers until the first of May. Therefore, if previous investigators had continued their observations beyond April and May, they might well have observed more cow/calf pairs.

Recent evidence off the northern California coast (Dohl *et al.*, 1981) and the coast of Oregon (Chapter 13, this volume) indicated patterns similiar to those demonstrated in the present study. As T. Dohl (personal communication) remarked, regarding our two observations of gray whale cow/calf pairs 1200 m and 10.4 km offshore, the number of cow/calf pairs traveling offshore off the central California coast is in the "tenths of one percent."

Cow-calf pairs have been observed previously near the shore (Leatherwood, 1974; K. Norris, personal communication), and the pattern of a bimodal migration with a nearshore corridor used by cow/calf pairs was observed at Point Piedras Blancas in 1979 by U.S. Fish and Wildlife Service biologists (R. Jameson and G. VanBlaricom, personal communication). Therefore, if cow/calf pairs were utilizing an offshore corridor from the late 1800s through 1969, then between 1969 and 1979 their migratory corridor was moved much closer to the shoreline and the timing of Phase B was delayed by 1 to 2 months.

It is more likely cow/calf pairs were not seen in previous studies because observations were terminated prior to the arrival of these whales in the study areas, not because they migrated through a distant offshore corridor. Indeed, for more than 100 years females with calves may have been missed because of the timing of Phase B, and in times past because there was a very low number of cow/calf pairs in a seriously depleted stock.

DECREASING OFFSHORE DISTANCES IN PHASE A

As Phase A progressed through February and March, whales passed progressively closer to shore. Early migrants used outside portions of an offshore corridor; later migrants traveled in middle and nearshore portions of that corridor. Although we have no data, our subjective impressions are that the mean size of passing whales decreased through Phase A. Storro-Patterson (1981) reported a similar pattern of decreasing size off the northern California coast, also based on subjective impressions. Rice and Wolman (1971) reported that the northward migration was segregated by age and sex: first to pass were impregnated females, followed in sequence by adult males, adult females, immature females, and immature males. They also reported the size differences between each class. It appears that these classes of northbound whales not only migrate in a temporal segregation, but as our data indicate, also utilize different parts of the offshore corridor.

Examination of the weekly mean and median offshore distances (Figs. 5 and 6) reveals that between April 14–May 31, 1980 and April 6–May 30 1981 the mean and median distances are nearly identical and remain rather constant for the remainder of the migration. This suggests that a uniform, homogeneous group of individuals passes during that 8-week period, and in fact this is Phase B, in which cow/calf pairs utilize a narrow nearshore corridor.

Between March 10–April 12, 1980 and February 9–April 4, 1981, however, this similiarity of means and medians does not exist. During the first 2 weeks of 1981 (February 9–21) the medians are greater than the means, indicating that a larger percentage of whales travel on the outside than inside portions of the offshore corridor. From February 23 to April 4, the medians are all less than the means, suggesting a greater percentage of whales utilizing the nearshore portions of that offshore corridor. These 8 weeks then suggest a nonuniform, heterogeneous mixture of individuals of different classes. This conforms to Rice and Wolman's data (1971), and is Phase A, the migration of whales other than females with calves.

ADVANTAGES OF A NEARSHORE CORRIDOR

What advantages might there be to a cow/calf pair using a nearshore migrational route? First, food for the cow is more readily available in shallow nearshore waters. There is evidence that some whales feed, at least sporadically, during the migration (Hubbs, 1959; Herzing and Mate, 1981; Poole, 1980, 1981b; Chapter 12, this volume). One likely food resource available along shore is mysid shrimp, which exist in dense swarms near or within kelp beds (Wellington and Anderson, 1978; Hudnall, 1981; Chapter 19, this volume). Recent studies in bioenergetics suggest the cow needs to feed to support the calf and herself during the migration (Sumich, 1982). Several reports have suggested gray whales may ingest eel grass (*Zostera marmia*) in Laguna San Ignacio (Swartz and Jones, 1981) and kelp off Point Piedras Blancas (Poole, 1981b, 1982).

A nearshore route may also afford protection against predators such as killer whales (*Orcinus orca*). Andrews (1914) reported gray whales retreated into shallow water and attempted to "hide" behind rocks when approached by killer whales. On May 2, 1967, a trio of gray whales (one cow/calf pair, one single) was approached by killer whales; the cow and calf immediately headed for shore and into breaking waves. After the killer whales left the area, the cow/calf pair resumed northward travel along the breakers (Morejohn, 1968). On May 12, 1972, killer whales attacked a female with calf, which retreated in the kelp off Carmel, California. The killer whales successfully attacked and killed the calf, while the female exhaled underwater and passed through the kelp beds in shallow water to safety (Baldridge, 1972). Exhaling underwater in response to killer whales was also reported by Hubbs (1950). On April 21, 1980, two gray whales exhibited avoidance behavior when five killer whales approached; these grays also apparently exhaled underwater while remaining submerged for 17 min. When the killer whales left the area, the gray whales surfaced in the same position but without visible exhalations (Poole, 1980, 1981a). It is interesting to note that during the same aerial survey flight in 1982 in which a cow/calf pair was observed 10.4 km offshore off Estero Bay, one large adult male killer whale (determined by dorsal fin shape and size) accompanied by a second (sex undetermined) was also traveling northward 10.8 km offshore and 3.2 km due south of the grays.

Kelp beds may offer a physical and an acoustic "screen" for protection of gray whales against predators. Killer whales utilizing echolocation may not detect whales within kelp beds, because the gas-filled pneumatocysts, coupled with canopy noise and surf noise, may present an acoustic barrier (S. L. Swartz, personal communication). Predators may also be reluctant to enter a kelp bed after locating grays nearby. In sandy-bottom areas such as Estero Bay, where the absence of rocks and reefs offers no suitable habitat for the formation of kelp beds, the presence of surf noise may offer protection for traveling grays.

Conclusions

1. In the northward migration along the central California coast, gray whales utilize two separate migration corridors.

2. Whales other than females with calves (Phase A) utilize a wide, offshore, nearly straight-line corridor, swimming from one major point of land to the next and avoiding bights and indentations in the coastline. This corridor extends 800 m to 3.2 km offshore at Point Piedras Blancas to 12 km in the Estero Bay region. As Phase A progresses through February and March, the offshore distance of whales decreases as subpopulation composition shifts from adults to juveniles. Adults utilize the outside and juveniles the nearshore portions of the offshore corridor.

3. Cow/calf pairs (Phase B) do not presently utilize an offshore migration route, as has been reported in the literature for more than 100 years, but take a more nearshore route following the contour of the coastline. They travel primarily within 200 m of the shoreline at Point Piedras Blancas, and within 400 m along the rest of the 96 km of coastline surveyed. Cow/calf pairs may often pass between islets and shore and travel along the edge of or through kelp beds and near or through breaking surf.

Summary

The northward migration of the gray whale (*Eschrichtius robustus*) was monitored along the central California coast by shore-based surveys at Point Piedras Blancas (1980 and 1981) and by aerial surveys between Cape San Martin and Point Buchon (1981 and 1982). Whales were observed as follows: from shore, 1496 (including 228 calves) during 12 weeks in 1980 (March 10–May 31); and 3296 (including 209 calves) during 16 weeks in 1981 (February 9–May 30); from aircraft, 284 (including 27 calves) in 1981, and 50 (including 24 calves) in 1982.

Northward-migrating whales utilized two distinct corridors; whales other than females with calves (Phase A, February–March) moved offshore within a nearly straight-line corridor from one major point of land to the next, avoiding bights in the coastline. In 1981, Phase A whales seen in good to excellent visibility conditions (the only data analyzed) were distributed as follows relative to Point Piedras Blancas: 67% between 800 m and 3.2 km; 20% between 200 m and 800 m; and 13% within 200 m of shore. Average distance offshore decreased continuously through Phase A, early migrants using outside portions of the offshore corridor and later migrants the nearshore portions.

Phase B (April–May), composed of 90% females with calves, passed Point Piedras Blancas 2 months later than Phase A. The cow/calf pairs traveled in a narrow corridor near and closely following the coastal contour. Ninety-nine per cent (1980) and 96% (1981) passed within 200 m of Point Piedras Blancas. Aerial surveys confirmed that the above patterns persisted for the 96 km examined.

Acknowledgments

I wish to offer my sincere appreciation to the following individuals and organizations for their assistance and support in this project: field assistants of 1980 and 1981—Tracy Lee Finkbine, Chris Henry, Helen Jones, Mei Ting (Tina) Mark, Janet Lee Patterson, Laura Patty, and David Peterson; aerial survey assistants of 1982—

Alan Baker, Kathi Barry, Michael Garguilo, Craig Harris, and Brian Thurmond; pilots—William Cook and James Caine; and Lori Pascual and Barbara Philpott (Northwestern Graphics) for their graphics. James Bodkin, Howard Braham, Robert Brownell, Thomas Dohl, Denise Herzing, Donald Isaac, Ronald Jameson, Mary Lou Jones, Stephen Leatherwood, Bruce Mate, Kenneth Norris, Stephen Reilly, Aryan Roest, Robert Rubin, Jacqueline Schonewald, James Sumich, Steven Swartz, and Glenn VanBlaricom all provided encouragement, support, and constructive criticism throughout the project. Sonoma State University, University of California at Santa Cruz, and California State Polytechnic University provided research assistants and equipment; Unruh's Photography provided film and equipment; U.S. Fish and Wildlife Service provided housing and the research site at Point Piedras Blancas. Partial funding was provided by National Marine Fisheries Service/National Marine Mammal Laboratory Grant #81-ABD-51, the American Cetacean Society, and Free to Live Productions, Ltd. I would also like to thank my family and friends: IXOYE, Tom and Jean Gray, Everett and Laurena Hornocker, Maurice Hornocker, Wayne and Phyllis Hornocker, Andrew Lamb, Roger and Camilla McGuinn, Jeanette and William Poole, Mark Poole, Nancy and Phillip Pope, the Sackrisons, and William and Naomi Thurmond.

References

Andrews, R. C. (1914). Monographs of the Pacific cetacea. I. The California gray whale (*Rhachianectes glaucus* Cope). *Mem. Am. Mus. Nat. Hist.* [N. S.] **1,** 227–287.

Baldridge, A. (1972). Killer whales attack and eat a gray whale. *J. Mammal.* **53**(4), 898–900.

Dohl, T. P., Duman, M. L., and Helm, R. C. (1981). Distribution and movements of migratory and resident gray whales along the California coastline. *Proc. Bienn. Conf. Biol. Mar. Mamm., 4th, 1981* Abstracts, p. 25.

Hayes, B. I. "Scrapbooks, Southern California. Los Angeles County, Monterey, Santa Barbara and San Luis Obispo Counties, 1851–61." Bancroft Library, Berkeley, California.

Herzing, D. L., and Mate, B. R. (1981). Gray whale migration and summer activities along the Oregon coast. *Mar. Mamm. Info.* Dec., p. 16.

Hubbs, C. L. (1950). The gray whale. *West. Fish.* **39**(5), 15.

Hubbs, C. L. (1959). Natural history of the gray whale. *Proc. Int. Congr. Zool., 15th, 1958* pp. 313–316.

Hudnall, J. (1981). Population estimate, feeding behavior and food source of gray whales (*Eschrichtius robustus*), occupying the Straits of Juan de Fuca, British Columbia. *Proc. Bienn. Conf. Biol. Mar. Mamm., 4th, 1981* Abstracts, p. 58.

Jordan, D. S. (1887). Coast of California. *In* "The Fisheries and Fishery Industries of the United States, 1884–1887" (G. B. Goode, ed.), Sect. 5, Vol. 2, Part 15, pp. 52–61.

Kellogg, R. (1929). What is known of the migration of some of the whalebone whales. *Smithson. Inst., Annu. Rep.* pp. 467–494.

Leatherwood, J. S. (1974). Aerial observations of migrating gray whales, *Eschrichtius robustus*, off Southern California (1969–1972). *Mar. Fish. Rev.* **36**(4), 45–49.

Morejohn, G. V. (1968). A killer whale–gray whale encounter. *J. Mammal.* **49,** 327–328.

Pike, G. C. (1962). Migration and feeding of the gray whale (*Eschrichtius gibbosus*). *J. Fish. Res. Board Can.* **19,** 815–838.

Poole, M. M. (1980). "Research on the Spring Migration of the California Gray Whale, *Eschrichtius robustus*, from Pt. Piedras Blancas, California" (unpublished circulated manuscript).

Poole, M. M. (1981a). The northward migration of the California gray whale, *Eschrichtius robustus*. *Proc. Conf. Assoc. North Bay Sci., 1981* 17pp.

Poole, M. M. (1981b). The northward migration of the California gray whale, *Eschrichtius robustus*, off the central California coast. *Proc. Bienn. Conf. Biol. Mar. Mamm., 4th, 1981* 25pp. Abstracts, p. 96.

Poole, M. M. (1982). Differential migrational corridors during the northward migration of the California gray whale. *Proc. South. Calif. Acad. Sci., 1982* Abstracts, p. 31.

Poole, M. M. (1984). Preliminary assessment of annual calf production of the California gray whale, *Eschrichtius robustus*, from Pt. Piedras Blancas, California. In "Reproduction in Whales, Dolphins, and Porpoises." *Rept. Int. Whal. Commn.* (Spec. Issue 6): 223–231.

Reilly, S. B., Rice, D. W., and Wolman, A. A. (1980). Preliminary population estimate for the California gray whale based upon Monterey shore census, 1967/68 to 1978/79. *Rep. Int. Whal. Commn.* **30,** 359–368.

Reilly, S. B., Rice, D. W., and Wolman, A. A. (1983). Population assessment of the gray whale, *Eschrichtius robustus*, from California shore censuses, 1967–80. *Fish. Bull.*, **83**(2): 267–281.

Rice, D. W., and Wolman, A. A. (1971). The life history and ecology of the gray whale (*Eschrichtius robustus*). *Spec. Publ.—Am. Soc. Mammal.* **3,** 1–142.

Scammon, C. M. (1874). "The Marine Mammals of the Northwestern Coast of North America." John H. Carmany & Co., San Francisco, California (reprinted by Dover, New York, 1968).

Scheffer, V. B., and Slipp, J. W. (1948). The whales and dolphins of Washington state with a key to the cetaceans of the west coast of North America. *Am. Midl. Nat.* **39,** 257–337.

Starks, E. C. (1922). A history of California shore whaling. *Fish. Bull.* **6,** 38pp.

Storro-Patterson, R. (1981). "The California Gray Whale." *Proc. Symposium, Whales: Past, Present and Future*, Davis, California. 10pp.

Sumich, J. L. (1983). Swimming velocities, breathing patterns, and estimated costs of locomotion in migrating gray whales, *Eschrichtius robustus*. Can. J. Zool. **61,** 647–652.

Swartz, S. L., and Jones, M. L. (1980). Gray whales, (*Eschrichtius robustus*) during the 1977–78 and 1978–79 winter seasons in Laguna San Ignacio, Baja California Sur, Mexico, *U.S., NTIS, PB Rep.* **PB80-202989,** 1–34.

Swartz, S. L., and Jones, M. L. (1981). Demographic studies and habitat assessment of gray whales (*Eschrichtius robustus*), in Laguna San Ignacio, Baja California Sur, Mexico. *U.S., NTIS, PB Rep.* **PB82-123373,** 1–56.

Townsend, C. H. (1887). Present condition of the California gray whale fishery. *Bull. U.S. Fish Comm.* **6,** 340–350.

Wellington, G. M., and Anderson S. (1978). Surface feeding by a juvenile gray whale, *Eschrichtius robustus*. *Fish. Bull.* **76**(1), 290–293.

17

Soviet Studies on the Distribution and Numbers of the Gray Whale in the Bering and Chukchi Seas from 1968 to 1982

Alfred Antonovich Berzin

Introduction

The main task of Soviet research of the Chukotka–California population of gray whales was to estimate their present abundance. A review and analysis of the literature and material from Soviet scientific research cruises to study the distribution and numbers of gray whales during the period of 1958 to 1964 is presented in the work of Berzin and Rovnin (1966). The present chapter does not intend to present and analyze population estimates for the Chukotka–California gray whale population produced by U.S. researchers during the past few years as this information is presented in numerous publications. At the same time, it is perfectly obvious that an exact determination of gray whale abundance must include a census of all regions the whales inhabit, particularly the summer range. The study of the whale's abundance, especially on the summer grounds, is inseparably linked to the study of their distribution. The scientific colleagues of TINRO (Pacific Research Institute of Fisheries and Oceanography) conducted systematic surveys of gray whale abundance and distribution from aircraft and vessels in the Bering and Chukchi Seas from 1968 to 1982.

ISBN 0-12-389180-9

Table I
Summary of Soviet Research Effort for Gray Whales: 1968 to 1982

	Aerial surveys			Marine surveys	
Year	Date	Extent of route ($km^2 \times 10^3$)	Area ($km^2 \times 10^3$)	Date	Area ($km^2 \times 10^3$)
1968	July 10–30	8	108		
1973	Aug. 3–23 Oct. 17–23	40	216		
1974				Sept. 26 Oct. 23	21.5
1975	Sept. 24–25 Oct. 1–9	22	162		
1979				Sept. 28 Oct. 25	37.8
1980				Sept. 17 Nov. 15	33.0
1982				Aug. 23 Oct. 13	90.0

[a]No data available.

Methods

The first aerial surveys of the gray whale's abundance and distribution in the coastal regions of the Bering and Chukchi Seas were conducted in 1968 and 1973 as part of the research program of TINRO. Subsequent surveys were conducted in 1974 and 1975 as part of the Soviet national research program within the framework of the "U.S.S.R.–U.S.A. Agreement for Collaboration in the Field of Environmental Conservation."

In accordance with the joint program, a research plan was formulated in 1975 for a simultaneous aerial census of Soviet and American portions of the Bering and Chukchi Seas for the purpose of surveying the entire summer range of the gray whale population (Table I). Unfortunately, the Americans were unable to conduct their surveys, which significantly decreased the effectiveness of the surveys conducted by Soviet scientists in the Soviet sector. Vessel surveys to estimate gray whale abundance and distribution were conducted jointly by Soviet and American scientists in 1979 to 1982 as part of this same program (Project 02.05-61, "Marine Mammals"). Participants from TINRO who participated in this research included V. V. Zimushko, A. P. Shustov, N. V. Doroshenko, V. N. Kolesnikov, and the author.

Aerial surveys were conducted in a twin-engine aircraft, the IL-14. Survey altitude was dependent upon weather conditions, but the optimum altitude selected was 300 m. Surveys followed parallel tracks separated by 15 km and extended a maximum distance of 250 km from shore. Survey strip width was determined by measuring the time from when a whale was spotted until the aircraft was over the animal and correlating this with the aircraft's ground speed. Maximum strip width was 5 km, but it was frequently less during surveys because of the weather conditions. In 1975 surveys were conducted in a flying "photosurveyor" which had "bubble" windows on both sides. This allowed simultaneous observation from both sides of the aircraft and doubled the effective survey zone. During each meeting with whales, V. V. Zimushko noted visibility, altitude, aircraft speed, etc. From these factors the survey strip width was determined; however, strip width changed relatively rapidly (within 15–20 min) from 270 m to 2 km.

The number of whales observed at any one moment does not reflect their true numbers as individuals may be below the surface; therefore, counts must be corrected for animals missed below the water's surface. In 1973 several systematic surveys were undertaken to determine a correction factor. Control counts were conducted from a vessel while an aircraft circled above the same group of whales. In our experience, this was the first comparison of its kind done for these whales. The counts from these surveys are a bit contradictory and insufficient, but they have the advantage of being actual counts and not estimations. The comparison of counts of the gray whales conducted simultaneously from Soviet vessels and aircraft in the coastal waters of Chukotka suggested that aerial observers detect approximately two-thirds of the whales. A correction factor is dependent upon the behavior of the whales: the group structures, their dive times, and their surface intervals. Weather conditions are not predictable and may affect the visibility of whales profoundly; in our opinion they cannot be introduced into a correction factor as this would produce uncertainties. The following relationship, recommended by Arsen'yev *et al.* (1973), was used to determine the abundance of whales in the survey area:

$$N = kn(1 + t_1/t_2)$$

where N is the general quantity of whales, n is the number of detected whales, t_1 is the time the whale is submerged underwater, t_2 is the time the whale remains at the surface, and k is a function of the aircraft's speed, visibility, and the distance of detection. V. V. Zimushko utilized this relationship in his analysis of the 1973 aerial census and estimated the population at three times the base (raw) count.

The summer distribution of gray whales is very irregular; thus, the survey area was divided into several regions each with different concentrations of whales to increase accuracy. The flight tracks were plotted on charts and the area that had been surveyed was determined, as was the number of whales encountered along a track line. From these data, whale densities in each region were calculated, and the total population was extrapolated region by region. I participated in the aerial observations of gray whales in the Bering and Chukchi Seas in 1973; based on my experience, I consider an aircraft flying at 250–300 km/hr to be a poor platform for observations of this kind. The aircraft passes over areas too rapidly to detect many great whales whose dive times are long, and because of the aircraft's large turning radius, it is difficult to return to whales once they have been detected. One must, however, utilize fixed-wing aircraft in certain conditions, like the Chukchi Sea, when a helicopter cannot be utilized.

When whales were spotted from survey vessels, the vessel followed the whales to obtain a precise count. Sighting information from years of vessel surveys in various regions of the northwest Pacific indicated that while vessels are underway the effective observation zone is up to 10 miles wide (5 miles on each side).

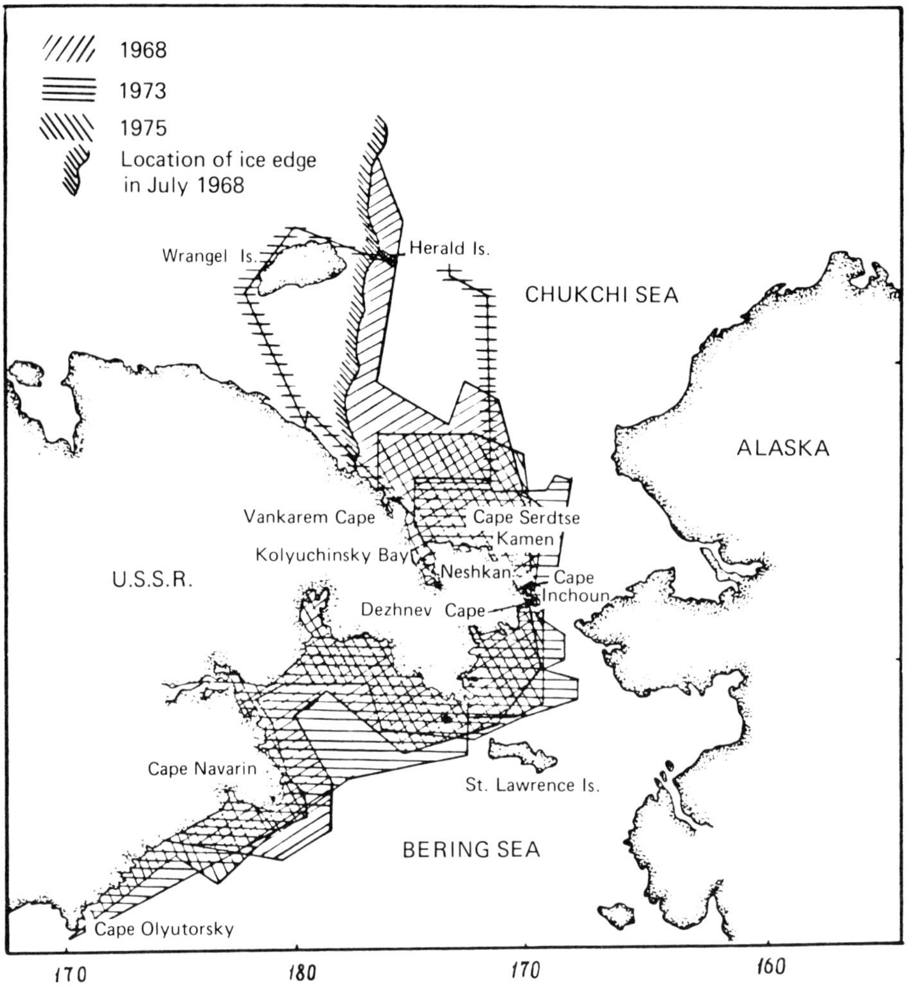

Fig. 1. *Regions of aerial observations during 1968, 1973, and 1975.*

Results

Aerial surveys of the Bering and Chukchi Seas conducted in 1968 included the coast and up to 50 km from shore and the ice-edge zone by Herald Island (Fig. 1). In all, 124 whales were counted. The largest group seen at one time included 17 animals observed in the region of Cape Inchoun (in the Chukchi Sea). The majority of the whales were seen along the coast no more than 35 km offshore. During the 1968 survey, the distribution of gray whales in the open sea was documented: two groups of 3 animals each were counted on a transect off Cape Serdtse Kamen at a distance of 90 km from shore.

Gray whales were observed in the Chukchi Sea only in the area of Cape Dezhnev (the Bering Strait) to the settlement of Neshkan. No additional whales were counted in the remaining area north from the settlement of Neshkan along the ice edge to the pack ice near Herald Island. Only a few lone whales were observed in the area of Cape Navarin and in the area to the south at Cape Khatyrka (Zimushko, 1970).

In 1973, aerial surveys were conducted of the northwest Bering Sea and the portion of the Chukchi Sea from Cape Dezhnev to Wrangel and Herald Islands (Fig. 1). A total of 290 whales was counted. The greatest density of whales was noted in the Chukchi Sea from Cape Dezhnev to Kolyuchinsky: the density was 0.11 whales/km^2. As in 1968, no whales were observed in the northern area from Cape Kolyuchin along the ice edge to Wrangel and Herald Islands. Five individual whales were counted by Cape Olyutorsky, although no whales were seen in Olyutorsky Bay itself. Whales were seen at Cape Pouten in the Bering Sea. In the Chukchi Sea (68°N, 169°05′W) 44 whales were counted during a 12-km long, 3-min flight. Later, during a 30-min flight along the same track, 132 whales were counted. During that same year whales gathered in significant numbers further offshore, which was explained by a corresponding change in their foraging areas. Sixty-eight percent of the sightings were of whales in groups of 2 to 3 animals. The largest group included 24 animals.

In 1974, the main region of Soviet studies was the Okhotsk Sea; therefore, studies in the Bering and Chukchi Seas were limited to short cruises in a limited area by the R/V *Tamango*. The count of whales along the Koryaksk coast and the Chukotka peninsula was 198 animals. This was extrapolated to a population size of 1800 to 2000 whales as a guide to the whale density in this area (Kuz'min and Berzin, 1975). The general movements of the gray whales confirmed that a massive departure of animals from the Chukchi Sea occurs in mid-October.

In 1975 aerial surveys included the coastal regions of the northwest Bering Sea; in the Chukchi Sea an investigation was conducted from Cape Dezhnev to Cape Vankarem and along the ice edge to 68°40′N (Fig. 1). Some 873 gray whales were counted. As in previous years in the Chukchi Sea, the following traditional regions contained aggregations of gray whales:

1. From 67°30′ to 68°15′N and 171°30′ to 172°00′W.
2. From 68°10′ to 68°30′N and 171°30′ to 172°00′W.
3. From 66°30′N and 169°55′ to 170°10′W.

A dense aggregation of whales usually occurred in the Bering Strait near Cape Pouten, where 50 whales were counted in 1975 during a 3-min flight along a 12-km transect line. One of the main groups of gray whales was located, as in previous years, in the Bering Strait at 64°50′ to 65°00′N and 170°00′ to 170°20′W. Anadyrsky Bay continues to be a region with the lowest density of whales. Solitary whales were observed near Cape Olyutorsky. This same year, no whales were observed beyond the 25-km wide coastal zone.

According to the aerial observations of V. V. Zimushko, gray whales on their feeding grounds are usually observed as single animals or small groups. Out of our 232 sightings of whales, single animals were observed 102 times (44%), groups of 2 to 5 animals were seen 113 times, and groups greater than 5 animals were observed on 17 occasions (7.8%). A group size of 2 to 3 animals was most frequently seen (67.6%). As previously stated, the largest group included 24 whales.

During a Soviet–American cooperative expedition on board the R/V *Avangard* to study the abundance and distribution of gray whales in 1979, large quantities of grays were confirmed up to 100 miles from shore over depths of 50 m (Fig. 2). On October 19 at 67°47′N and 172°72′W, 260 whales were counted. All of these animals were large, and they swam as singles or in groups of 2 to 3 animals as they do in other portions of their summer range. Occasionally (eight observations), compact groups of 5 to 6 animals were observed. The largest group contained 13 whales. Large concentrations of whales (70 animals) were encountered in the region of King Island, and on October 25, 50 whales were counted north of Saint Lawrence Island. A total of 411 whales was counted during this cruise.

The results of the second Soviet–American expedition on the R/V *Razyashchy* contributed substantially to our present-day understanding of the distribution of whales in these regions (Fig. 2). Whales were first encountered in the coastal regions of Long Strait and in the eastern part of the East Siberian Sea. Large groups were encountered in the region of Cape Dvukh Pilotov (Cape of Two Pilots), and a smaller quantity was near Cape Shmidta. At the end of September, gray whales were encountered near the coastal shelf of the East Siberian Sea from Cape Yakan to Cape Billings and further west. The westernmost sighting of gray whales was at 178°30′E. The results of this cruise defined the boundaries of the gray whale distribution and identified locations of aggregations of whales in the Chukchi Sea 100 miles from shore at approximately 67°40′ to 68°15′N and 169°40′ to 172°00′W. The approximate number of whales in this region was ~2000 animals, although undoubtedly these data are only approximate and demand more precise determination. Because 588 gray whales were counted over a 4-day period, it is apparent that gray whales are abundant in this area. During the 1980 cruise 1021 gray whales were counted in this same area.

Information gathered during the 1979 and 1980 cruises confirmed that the gray whale fall migration out of the summer grounds begins in the middle of October and ends with the formation of the ice sheet. Newly pregnant females are the first to enter the summer grounds and the last whales to leave the Chukchi Sea, departing with the polar whales or bowheads (*Balaena mysticetus*).

The program of cruises on board the Soviet vessel R/V *Enthusiast* in 1982, which included American scientists, differed from the 1979 and 1980 cruises in that it started earlier (July 15) and covered a wider area in the Chukchi Sea (to 158°W) (Fig. 3).

The 1982 research may be divided into three stages. During the first stage of the work (from July 23 to August 8), gray whales were encountered in groups of 4 animals on July 25 north of Saint Lawrence Island. Upon entering the Chukchi Sea on July 27 between the settlements of Uelen and Inchoun, groups of 2 to 20 gray whales were encountered. That same day 300 gray whales were counted in an area of approximately 120 square miles north of Cape Serdtse Kamen. The center of this aggregation contained a dense group of 70 to 80 whales, all of which were feeding intensively. The following day, three groups of 40, 20, and 60 animals, respectively, were discovered

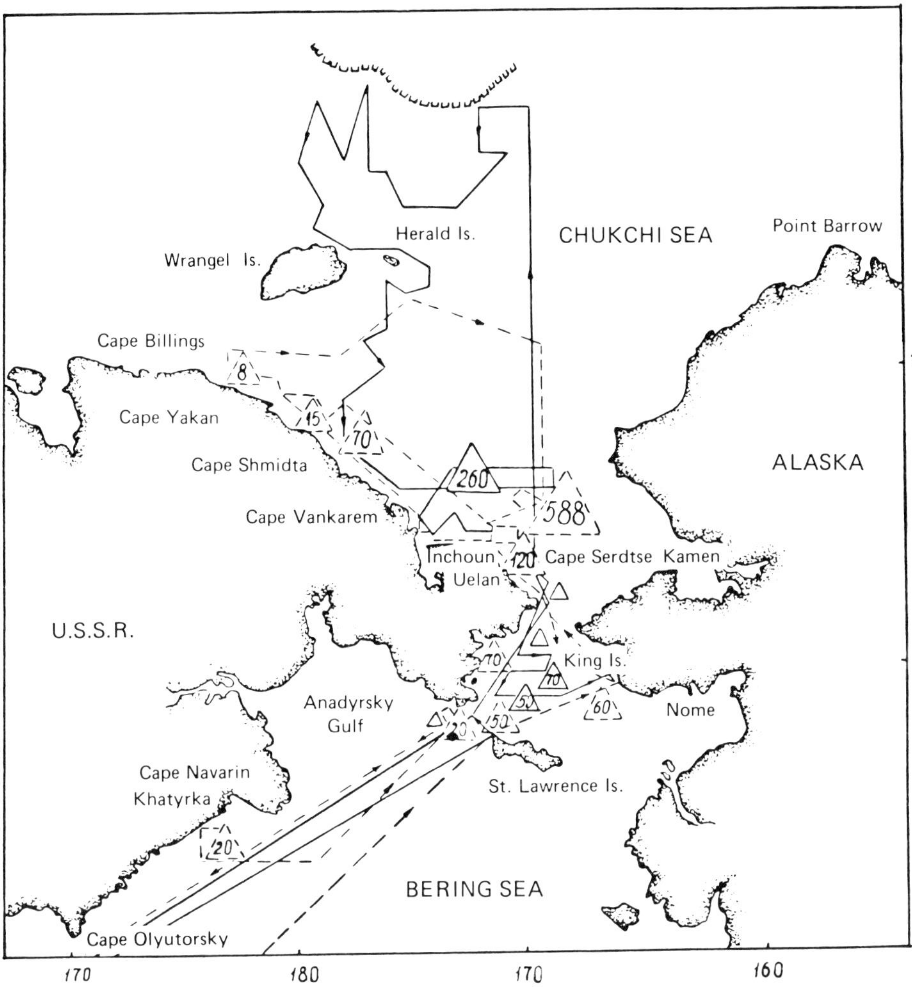

Fig. 2. *The distribution of gray whales and the routes of the R/V* Avangard *and R/V* Razyashchy *in 1979 and 1980. Symbols: solid arrow, route of R/V* Avangard, *1979; broken arrow, route of R/V* Razyashchy, *1980; zigzag area at top, ice edge in October, 1979. Numbers in triangles are counts of whales.*

between Cape Serdtse Kamen and Cape Vankarem. No gray whales were encountered at the ice edge to the west–northwest at Herald Island in the Chukchi Sea. On August 4 near Point Barrow (71°08′N 158°19′W), a group of 40 gray whales was encountered, our easternmost sighting of gray whales. Dense ice forced the ship to turn to the west–southwest where a group of 200 gray whales was encountered.

During the second stage (August 6–25), the ship traveled to the southwest and encountered a group of 60 gray whales in the same region. No large groups of whales were seen while traveling in a southerly direction along the coast of Alaska. Significant numbers of gray whales were again encountered in the same regions of the Chukchi coast visited during the first stage of this cruise: two groups of whales numbering

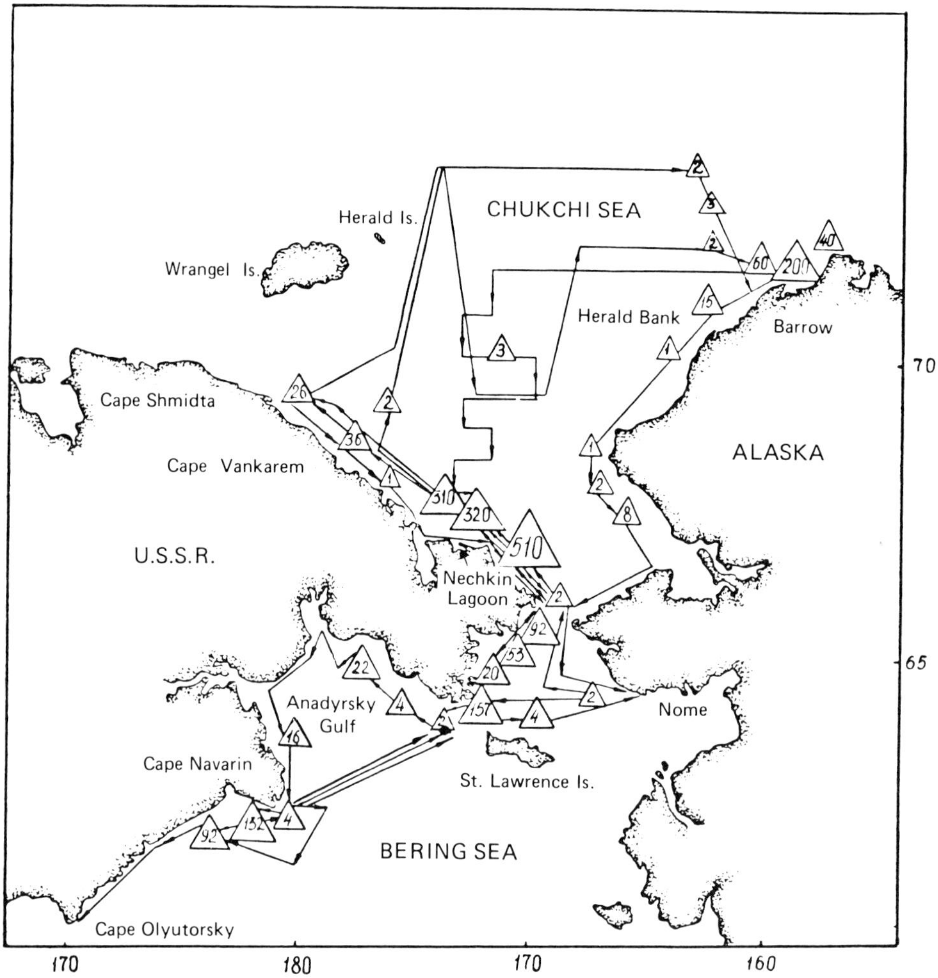

Fig. 3. *Distribution of gray whales and the route of R/V* Enthusiast *in 1982. Numbers in triangles are counts of whales.*

approximately 90 animals each were encountered in the region of Cape Serdtse Kamen. No whales were encountered during a return visit to the ice edge north of the Chukchi Sea. Five animals, a pair and a group of 3, were encountered near the ice edge at 71°56′N and 161°47′W (Fig. 3). Only 3 whales were seen during the vessel's westward track in the central Chukchi Sea; these animals were encountered southeast of Herald Banks on August 20. On August 22–23, 190 whales were counted in an area of 1000–1100 square miles located at 68°24′ to 68°00′N and 172°174′W.

In the third stage (September 2 to October 13), a 12- to 15-mile coastal zone of the Chukchi Sea from the Bering Strait to Long Strait was investigated. Near Cape Shmidta, two groups of 13 and 10 gray whales were sighted. Single whales and groups of 8 to 30 animals were observed in all the traditional areas near Cape Shmidta and Cape Dvukh Pilotov. From Nechkin Lagoon to Cape Shmidta, 259 whales were encountered during a 2-day period. Three whales were encountered at the westernmost point of the survey at the edge of the ice near Long Strait (69°27′N and 179°19′E).

Groups of 2 to 20 gray whales were encountered throughout the northern part of the Bering Sea from Cape Dezhnev to Arakemchechen Island. Near Cape Chaplino, 127 whales were counted in a 120- to 140-square mile area, and approximately 60 whales were encountered in the coastal waters of Anadyrsky Bay. At the end of September near Cape Navarin (in the region of Maino-Pilgyn), a gathering of 130 whales was observed, and in the region of Khatyrka ~90 whales were seen. This cruise confirmed the continuous distribution of gray whales in these areas. In all, 2142 whales were counted during the 1982 cruise.

Discussion and Conclusions

Studies of the Chukotka–California population of gray whales in their summer grounds have been limited basically to the collection and analysis of material generated by the whaling industry and to nearshore observations where these whales were collected. As the result of the systematic study conducted according to the Soviet national plan, as well as the joint Soviet–American program, the morphobiological aspects of the gray whale were significantly investigated and new information was obtained on the summer habits and distribution of these whales.

Analysis of the information collected over the last few years on the abundance and distribution of the Chukotka–California population of gray whales indicates that their summer grounds contain regular aggregations of animals in specific areas determined by their forage (prey) distribution. It has been established that in the Chukchi Sea, a significant part of which is relatively shallow, gray whales form aggregations a great distance from shore (in part up to 180 km) to the north of Cape Serdtse Kamen. Thus, gray whales are not a coastal species of whales in the full sense of the word, but only inasmuch as shallow depths as a rule occur nearshore.

According to Tomilin (1957), gray whales do not go farther west than Cape Serdtse Kamen. According to the latest data, gray whales do form some of their largest aggregations in this area. Our observations indicated that the gray whales westernmost distribution is off the lagoons of Tenkergyn–Pilgyn (177°30′W) and that this defines the western boundary of their range (Berzin and Rovnin, 1966). Recent studies suggest that gray whales begin to be encountered in the Long Strait and the Eastern Siberian Sea along the edge of the pack ice. Our westernmost sighting of a gray whale was 178°30′E. This suggests that the western distribution of the gray whales is limited by the pack ice, and because of the increase in the population the whales have recently expanded westward (Doroshenko, 1981). Studies of the eastern distribution of gray whales in the Chukchi Sea clearly suggest that their eastern distribution is also limited by pack ice. Our easternmost sighting of gray whales was at 158°19′W.

The indications in the literature of periodic penetrations of gray whales into the high latitudes near Herald and Wrangell Islands is not supported by contemporary information. Additionally, replicate surveys of the central and northern regions of the Chukchi Sea have demonstrated an absence of whales in a large area of this basin; the only exception was the sighting of a small group of whales in the shoal water to the southeast of Herald Bank in the central Chukchi Sea (Fig. 3). The extreme northernmost sighting of gray whales was in the western area at 69°27′N and on the east at 72°N.

Information on the abundance of gray whales (in groups of 100 or more) in the coastal regions to the south of Cape Navarin (Tomilin, 1957; Berzin and Rovnin, 1966) relate to the period of commercial whaling. Up until the 1970s only a few gray whales were ever encountered in these areas during the summer. Only during the migration were sightings of gray whales indicative of an increase. Gray whales have not been seen south of Glubokaya Bay for the past 20 years, and this latitude (61°12′N) is considered the southern limit of gray whale distribution along the northwest coast of the Bering Sea. Frequent recent observations of large aggregations of gray whales south of Cape Navarin, near the Maino–Pilgyn lagoon, the settlement of Khatyrka, and others, along with the discovery of gray whales near Cape Olyutorsky, is indicative of the increase of the population of whales and the southward expansion of their range to 60°N. The distribution and southern limit of the gray whale's summer grounds on the eastern shelf of the Bering Sea are not known.

The extrapolation of the census counts of gray whales in the areas investigated in 1982 allowed the calculation of the first numbers of whales arriving on the feeding grounds, although these first estimates are undoubtedly imprecise and should be improved. Approximately 10,000 gray whales inhabit an area of 45,000 square miles of the Chukchi Sea during the summer, according to V. N. Kolesnikov, and within a 14,000 square mile area of the northwest and northern part of the Bering Sea there are approximately 3,300 to 3,500 whales. As previously mentioned, we have no data on the numbers of gray whales in the eastern part of the Bering Sea. Previous information suggests the benthic communities in these areas are relatively poor, and it is unlikely that any large quantities of whales live in these areas. If one postulates a minimum number of 1,000 whales, then we may assume that approximately 14,500 to 15,000 gray whales inhabit the northern waters of the Bering and Chukchi Seas to feed during the summer. If this is compared with current American estimates of 17,000 whales in the population (Chapter 10, this volume), then we may assume that approximately 2,000 whales linger in the winter range and along the migratory route.

The Chukotka–California population of gray whales inhabits a limited and relatively easily studied area and has reestablished its abundance, following its catastrophic decline, to a current population size nearing the present carrying capacity of the ecosystem of the North Pacific. Because most of the biological parameters of the gray whale have been studied more intensively than for any other species of great whale, the gray whale may be considered a model population. Research on the Chukotka–California population of gray whales that inhabit the territorial waters of several nations need to be coordinated by all nations in all portions of its range, from the winter calving grounds, along the migratory route, and in the summer grounds. This is within our present technical capabilities and is in the best interest of the species.

Summary

This chapter presents the results of Soviet studies on the distribution and abundance of the Chukotka–California population of gray whales on their summer grounds between 1968 and 1982. Aerial and vessel observations established that gray whales regularly concentrate in the same general areas of the Bering and Chukchi Seas. The formation of permanent groups was observed at distances up to 100 nautical miles or more from the shallow coastal regions of the Chukchi Sea. Large aggregations of whales were encountered on the area's eastern boundary near the ice at Point Barrow. The area occupied by gray whales has expanded in a westerly direction and they have appeared in the Eastern Siberian Sea. An increase in abundance of a permanent population of gray whales was noted in the coastal zone to the south of Cape Navarin. Gray whales have appeared for the first time in several years near Cape Olyutorsky. These studies have defined more exactly the boundaries of the gray whale's summer habitat. Population estimates of the number of gray whales summering in the Bering and Chukchi Seas are between 14,000 and 15,000 animals, of which 10,000 inhabit the Chukchi Sea. The detailed information available for the Chukotka–California population of gray whales is the result of many years of multidisciplined investigations by Soviet and American specialists and allows us to consider a model for this population of great whales.

References

Arsen'yev, V. A., Zemsky, V. A., and Studenetskaya, I. S. (1973). Gray Whale, *Eschrichtius robustus* Gray, 1864. *In* "Marine Mammals," pp. 30–35. Pishulevaya Promyshienost. Moscow.

Berzin, A. A., and Rovnin, A. A. (1966). The distribution and present condition of the number of migrating whales in the northwest part of the Pacific Ocean, in the Bering and Chukchi Seas. *Izv. TINRO* **58,** 179–207.

Doroshenko, N. V. (1981). Concise results of cetacean research during the cruise of R/V Razyashchy in the Bering, Chukchi, and East Siberian seas. *In* "A Collection of Research Works on Marine Mammals of the Northern Pacific Ocean in 1980–1981," Proj. 02.06-61, pp. 13–16. VNIRO, Moscow.

Kuz'min, A. A., and Berzin, A. A. (1975). Distribution and present condition of the number of sleek and gray whales in the far eastern seas. *In* "Biological Resources of the Seas of the Far East," pp. 121–122. Vladivostok.

Tomilin, A. G. (1957). *Mammals of the USSR and Adjacent Countries.* Vol. IX Cetacea, pp. 328–358. Izd. Akad. Nauk. SSSR, Moskva.

Zimushko, V. V. (1970). An aerial-visual count of the numbers and observations of the distribution of gray whales in the coastal waters of the Chukchi. *Izv. TINRO* **71,** 289–294.

IV

Biology and Behavior

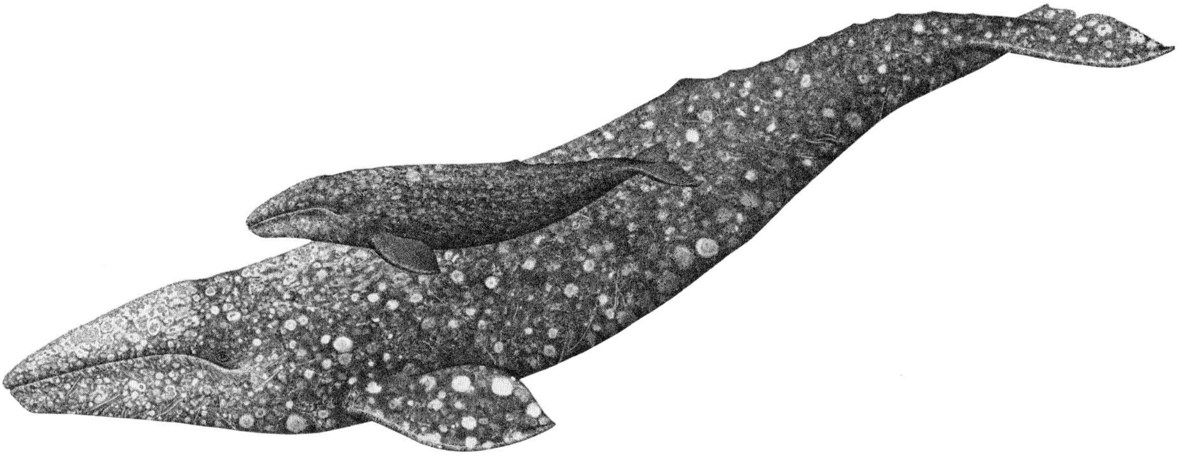

18

A Review of Gray Whale Feeding Ecology

Mary Nerini

Introduction and Methods

The gray whale (*Eschrichtius robustus*) is unique among large cetaceans in that it feeds primarily upon benthic organisms. In so doing, it leaves a record in the sediments, however obscure, of where it has been and what it has eaten. Because it is difficult to observe feeding whales directly, and because of the paucity of stomach content data, such records offer valuable clues to the interactions between the gray whale and the benthic community.

This chapter reviews information available on such questions as How do gray whales feed, where do they feed, what do they eat, how much do they eat, and how do they influence their prey community? As a review, it is based largely on published information. It also, however, incorporates unpublished observations of feeding gray whales recorded by many researchers along the west coast of North America and results of my own research in the Bering Sea in June–July and September 1980.

My research was conducted aboard the NOAA ship *Surveyor* operating primarily in the Chirikov Basin of the northern Bering Sea. It was designed to study both feeding of gray whales and dynamics of the benthic amphipod communities on which they were known to feed. Methods and findings were presented in detail in Nerini *et al.* (1980) and are only summarized here.

ISBN 0-12-389180-9

A cruise June 23 through July 17 operated off Southeast Cape, St. Lawrence Island, in the band between about 168 and 169°W from the northeast tip of the island to about 65°N. A cruise September 10–30 operated exclusively off Southeast Cape. During the June–July cruise, we towed a Klein and Associates Side–Scan Sonar unit with a 500-kHz transducer.[1] The sonar, recording a strip of ocean floor 50 m wide, proved capable of detecting bottom features such as depressions or mounds in the surface of the sediments. Over the approximately 787 km it was towed, the sonar's traces provided insight into sizes and shapes of feeding depressions made by whales.

During both cruises we used a 0.1-m^2 Smith-MacIntyre Grab, a 0.025-m^2 box corer, and scuba divers to collect bottom samples. Each diver-collected sample contained 0.0188 m^2 of material. The grab samples were sieved through a 1-mm mesh; the diver collected cores through a 0.5-mm mesh. Materials remaining on the screens were fixed in 5% formalin and transferred to 70% ethanol.

Feeding Mechanism

Although other whales may occasionally sample the bottom fauna, the gray whale is the only baleen whale known to regularly consume benthic resources (Nemoto, 1970). Numerous authors, beginning with Scammon (1874), have noted gray whales surfacing with bottom sediments clinging to the rostrum and spewing from the mouth. The baleen is sturdy and adapted to contact with bottom sediments; the plates are thicker and the hairs are coarser and less numerous than those of any other mysticete (Nemoto, 1959).

The mechanism by which the whales collect the benthic organisms is still unclear although several hypotheses have been advanced. Walker (1971) made the unlikely suggestion that the whales acted as bulldozers, engulfing "power-shovel helpings" of benthic fauna and the associated sediments. As the prey organisms of gray whales live in the upper 2 cm of the sediments, a deep scoop would be unnecessary. Furthermore, cetacean skin is easily abraded and would not withstand extensive plowing through sandy sediments. From the available data it seems more probable that gray whales somehow suck in their prey and at least partially separate them from the sediments. Kasuya and Rice (1970) document greater wear of the baleen on the right side of most gray whales coinciding with fewer barnacles and more skin abrasions of the head region. From this evidence they inferred that most whales feed while on their right sides taking in some sediment with their prey items and occasionally coming in contact with the bottom.

Gigi II, a captive juvenile gray whale, provided an opportunity to make detailed observation of feeding. Although one could argue that Gigi's behavior in a small cement enclosure may not have been identical to the natural behavior of gray whales, it is probably instructive; I recount the salient details presented in Ray and Schevill (1974). Approaching squid strewn on the floor of her pool, Gigi first rolled 120° onto her left side

[1]The use of trade names in this report is intended as documentation and does not imply endorsement of products by the author or the National Marine Fisheries Service.

so that her mouth was parallel to the bottom and 10–20 cm above it. As she traveled over the squid, she apparently created a pulsating suction with her mouth (probably by depressing her muscular tongue) leaving a clear swath of 30–50 cm in the food. Her direction of travel while on her side was dorsad of straight ahead. In other words, she swept across the squid at an angle 30° to her mouth. A similar behavior was documented in the waters off southern California by two scuba divers (Grigg and Dana, 1969), although they attributed the roll to a fright response. More recently, Hudnall (1981) filmed a feeding whale in the Straits of Juan de Fuca. This whale is reported to have rolled onto its side, swept very near the bottom, righted itself, and disappeared behind a cloud of sediment. In its wake the whale left depressions in the bottom sediments that although unmeasured were estimated to be approximately the size of its head.

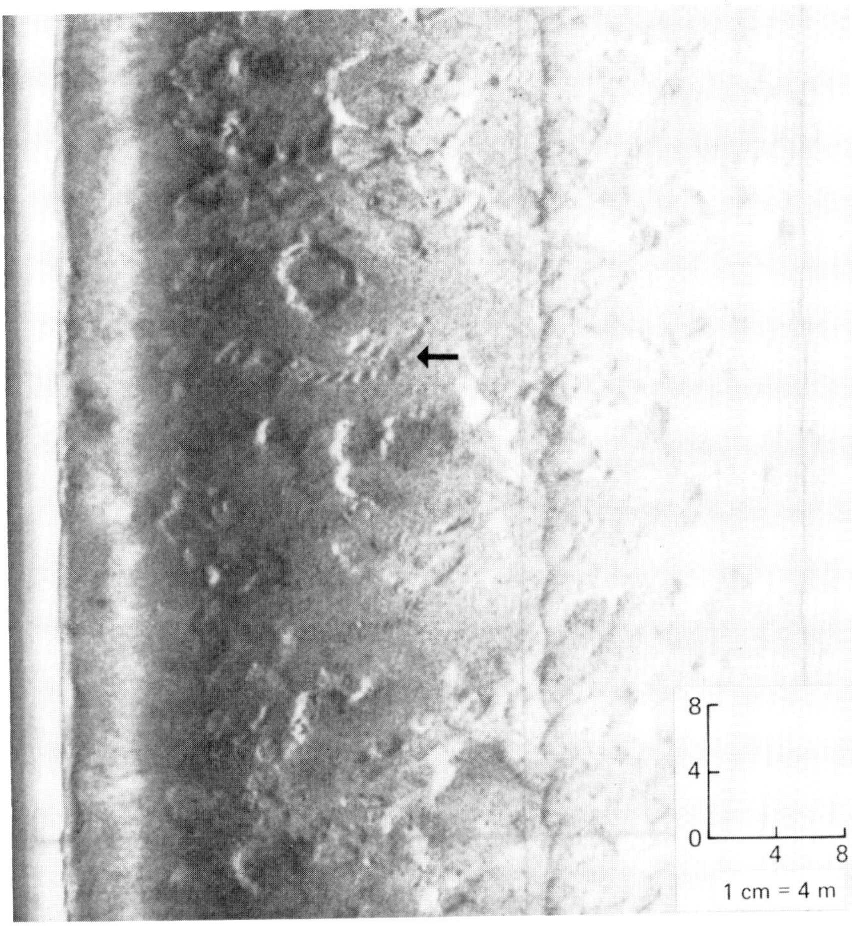

Fig. 1. *Side-scan sonar record of the Bering Sea floor made during July 1980. The patterns at the arrow are probably depressions produced by a feeding gray whale.*

We had hypothesized that whales feeding near the bottom would leave a record of their activities, albeit a complex record, in the sediments, from which we might infer aspects of their feeding ecology. This has proved to be the case: The side-scan sonar was able to distinguish bottom features that may have been produced by foraging gray whales in the Bering Sea (Fig. 1).

These bottom features were unique to the Chirikov Basin and the nearshore regions

Fig. 2. *Schematic figure of feeding gray whale and resultant bottom depressions. Although depicted as open in this figure, the left side of the whale's mouth is probably closed while feeding.*

of St. Lawrence Island as seen on our cruise. Their occurrence and orientation were irregular, suggesting the features were biogenic in origin. Furthermore, these features were only seen when feeding gray whales were in the immediate area and their shape and size were consistent with the previously described feeding mechanism. That is, a whale moving while on its side and sucking up infauna in pulses should theoretically leave a series of oblong, mouth-sized depressions from which only the top layers of sediment were removed (Fig. 2). The only other large benthic predator (excluding the bowhead whale) capable of creating such a massive disturbance is the walrus (*Odobenus rosmarus*). Walrus however feed differently than gray whales, leaving long, narrow sinuous tracks or small pits (Oliver *et al.*, 1983) which we were able to differentiate on the side-scan sonar (Fig. 3).

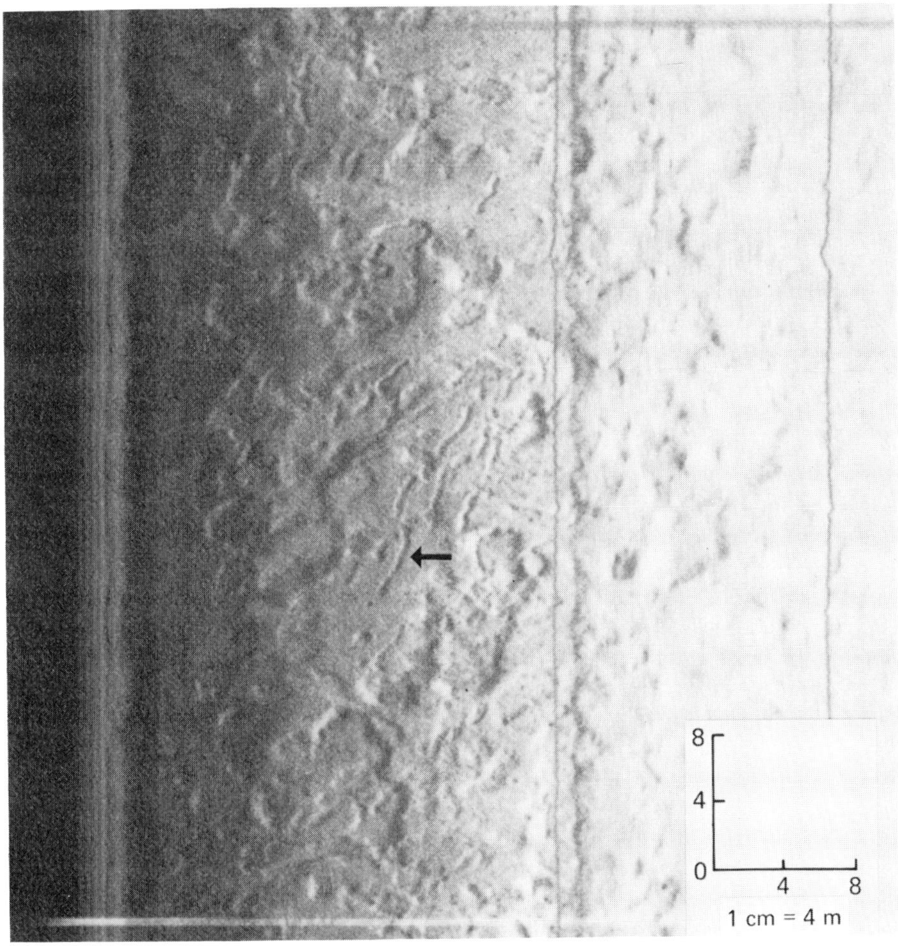

Fig. 3. *Side-scan sonar record of the Bering Sea floor collected during July 1980 with a 500-kHz transducer. The arrow indicates furrows probably produced by a feeding walrus.*

Each bottom trace is composed of a series of small depressions often arrayed in a slight curve (Fig. 4), where the entire series of depressions probably represents the feeding by one whale in one dive. The mean size of the component depressions measured from the side-scan records was 1.6 × 0.6 m (n = 13). The entire feature was composed of (a mean of) 6.4 such depressions (n = 15) and ranged in total length from 2.1 to 6.9 m (n = 13). The maximum abundance of the features was 9/km. Because these features were relatively small and because the edges of the depression were recorded indistinctly these measurements have an accuracy of approximately 0.3 m.

Recently, more extensive side-scan information has been analyzed from the northern Bering Sea. From this data it appears that the bottom features produced by recently foraging whales can be distinguished from older, current-scoured feeding depressions on the basis of size and shape. In this fashion fresh feeding depressions were defined as those ranging from 1 to 3 m long and from 0.5 to 1.5 m wide (Nelson *et al.*, 1983). Depressions measured *in situ* by scuba divers were elliptical with mean dimensions of 1.1 m (n = 14, SE + 0.50) × 1.65 m (n = 14, SE = 0.77).

In addition to its bottom-feeding abilities, the gray whale is also capable of feeding on pelagic prey by surface skimming and engulfing. Hubbs (as reported in Pike, 1962) reported gray whales circling tightly, apparently feeding on spawning squid; J. Sumich (personal communication) and S. Leatherwood (personal communication, 1982) have seen similar circling of bait fish. Gilmore (1961) observed gray whales "criss-crossing" through a dense school of small fish, and Sund (1975) reported two occasions of a group of whales forming a tight circle around a school of small fish; each whale in turn would dive and surface through the fish school with its mouth open in a fashion similar to the balaenopterid whales.

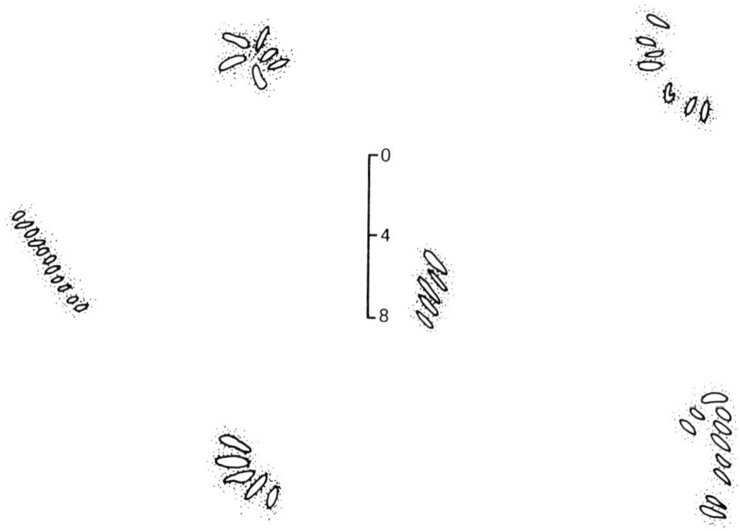

Fig. 4. *Shapes of six bottom features produced by feeding gray whales and detected by the side-scan sonar.*

Norris *et al.* (1983) offered evidence that gray whales feed in the Baja lagoons by orienting themselves against a tidal current or rip and allowing the water to funnel planktonic food items to them. R. Gill (personal communication, 1982) reports similar behavior from Nelson Lagoon, Alaska. Murison *et al.* (Chapter 19, this volume) documented whales feeding in an inverted position off Vancouver Island on mysiid swarms located approximately 1 m above a rocky bottom. Hudnall (1981), also off Vancouver Island, filmed a foraging whale engulfing a swarm of mysids hemmed in by a rock face. Wellington and Anderson (1978) reported on a use of surface skimming in the center of a dense kelp bed, and Swartz and Jones (1981) reported similar skimming of windrows of eel grass (*Zostera marina*) in one of the breeding lagoons.

Thus, with three modes of feeding, benthic suction, engulfing, and skimming, the gray whale has perhaps a greater range of foraging techniques than any of the other great whales (cf. Nemoto, 1970). This diversity may lend the gray whale greater dietary flexibility, lessening its reliance on any single prey item, and consequently providing it with greater resilience with regard to changes in its food resources.

Evidence of Feeding

The best and the only irrefutable proof that a whale is feeding is derived from its stomach contents, but there are other less drastic ways of identifying feeding whales. As mentioned earlier, the whale takes in some sediment as it feeds on the bottom; this sediment, finer than the associated prey items, is expelled through the baleen as the whale surfaces, producing a sediment trail. These mud plumes are clearly visible from an aircraft (Fig. 5), the elevated shoreline, or a nearby vessel. Most reports of feeding whales are so categorized because of the presence of mud plumes; however, the presence of mud plumes is not unequivocal evidence of feeding activity. Plumes of sediment could be created by whales investigating a prey community or contacting the bottom for some other reason, and thus activities can be erroneously reported as feeding. Sightings of defecating whales, fecal slicks, reports of whales with foul-smelling breath, or of whales trailed by seabirds are other indicators of feeding that are less frequently noted. Some investigators have identified feeding whales by extended observations of an animal "working" an area (i.e., systematically and repeatedly diving in a small locality). Other cues include the presence of bottom disturbances produced by foraging whales, either observed directly using scuba or indirectly using a closed-circuit video system or a side-scan sonar.

Feeding Areas

Gray whale feeding behavior can be categorized by three areas: in and near the Baja lagoons, along the migratory corridor, and in the nothern Bering, Chukchi, and (rarely) Beaufort Seas. Available accounts, summarized in Table I are discussed next.

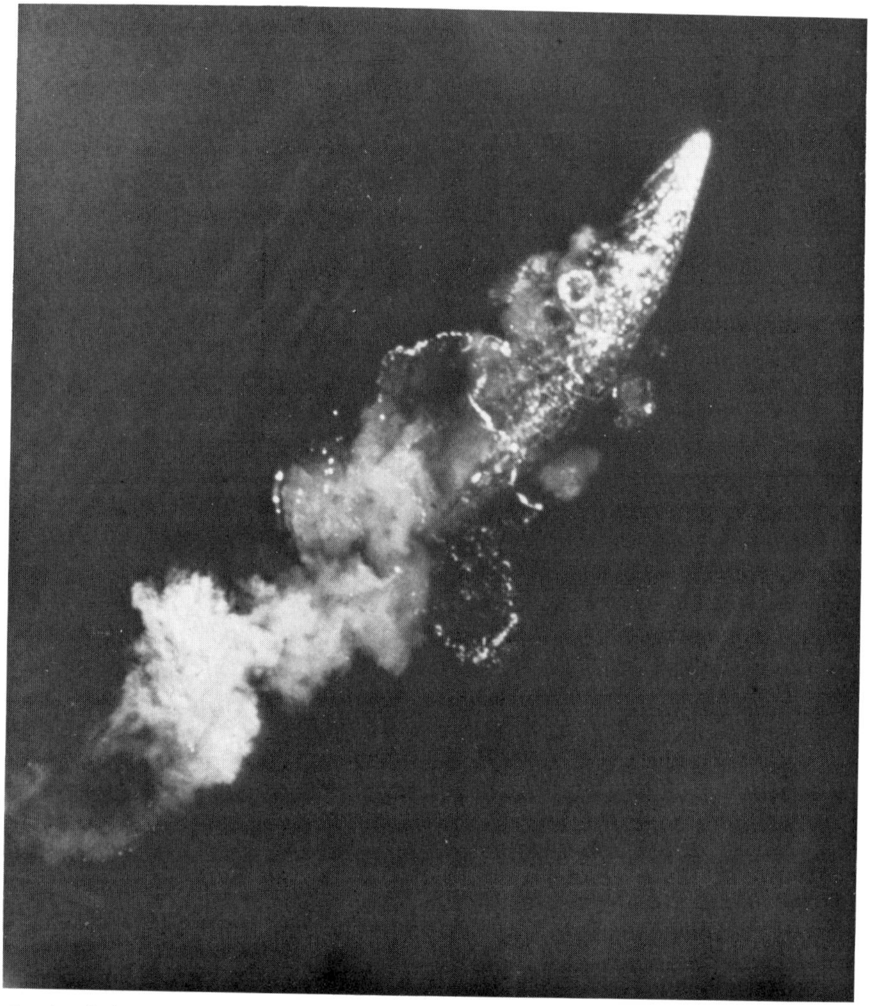

Fig. 5. *Aerial photograph of a feeding gray whale illustrating the "mud plume." Photograph taken on June, 1976, in the Bering Sea, by H. Braham.*

THE BAJA LAGOONS

There are very few reports of feeding whales from the breeding lagoons. Walker (1949) reported finding sardines in a stranded gray whale found in Laguna Ojo de Liebre (LOL). Norris *et al.* (in press) reported feeding on *Pleuroncodes* spp. and *Nyctiphanes* spp. at the mouth of Bahia Magdalena, and J. Sumich (personal communication) twice saw gray whales in Laguna San Ignacio (LSI) criss-crossing through schools of bait fish. Hubbs (as reported in Pike, 1962) learned from fisherman in the area that the stomach of a gray whale stranded in LSI was filled with "sardines," a collective term that he thought

Table I
Locations Known to be Frequented by Foraging Gray Whales Exclusive of the Northern Bering and Chukchi Seas

Location	Probable prey item	Feeding and other behavior	Bottom substrate	Reference
San Ignacio Lagoon, Baja California, Mexico	Eelgrass mats, Associated small crustaceans	Surface skimming	Sand	Swartz and Jones (1980)
San Ignacio Lagoon, Baja California, Mexico	Unidentified bait fish	Tight circling and engulfing	Sand	J. Sumich[a] (personal communication)
Magdalena Bay, Baja California, Mexico	*Pleuroncodes* (pelagic red crab), *Nyctiphanes*	Regular dives, attending seabirds, tidal feeding	—	Matthews (1932), Norris *et al.* (1983)
Punta San Juanico, Baja California	Unknown	Sediment trails	Unknown	D. Rice[b] (personal communication)
Laguna de San Quentin, Baja California, Mexico	*Ampelisca*	Unknown	Sand	Sprague *et al.* (1978); Mate and Harvey (Chapter 25, this volume)
La Jolla, California	Spawning squid	Unknown	Unknown	Pike (1962)
Point Loma, California	Unknown	Mouthing kelp	Kelp bed	S. Leatherwood[c] (personal communication)
Point Mugu, California	Unidentified bait fish	Circling and swimming erratically through schools	Sand	S. Leatherwood[c] (personal communication)
Santa Barbara, California	*Acanthomysis*	Surface feeding	Kelp bed	Wellington and Anderson (1978)
San Miguel Island, California	Unknown, possibly small pelagic red crustaceans	Mouthing kelp at surface	Kelp bed	G. Antonelis[d] (personal communication)
Port San Louis Harbor, California	Unknown	Churning bottom	Sand, kelp bed	A. Roust[e] (personal communication)
Piedras Blancas, California	Unknown	Sediment trails, mouthing kelp	Sand, kelp bed	M. Poole[f] (unpublished manuscript)
Monterey, California	Unidentified bait fish	Circling and engulfing	Unknown	Sund (1975)

(*continued*)

Table I (*Continued*)

Location	Probable prey item	Feeding and other behavior	Bottom substrate	Reference
Moss Landing, California	*Diapatra*	SCUBA observations	Sandy mud	C. Goebel[g] (personal communication)
Farrallon Island, California	Unknown	Sediment trails	Unknown	D. Ainley[h] (personal communication)
Point Reyes, California	Unidentified amphipods	Sediment trails	Unknown	B. Jones[i]
Russian River, California	Unknown	Side-scan sonar depressions	Sand	G. Tate[j] (personal communication)
Eureka, California	Unknown	Sediment trails	Unknown	J. Heyning (personal communication)
Crescent City, California	*Euphausia*	Unknown	Rocky shore	Howell and Huey (1930)
Klamath River, California	Unknown	Unknown	Unknown	T. Dohl[k] (abstract)
Port Orford, Oregon	Unknown	Moving back and forth near breakwater	Unknown	T. Loughlin[l] (personal communication)
Cape Falcon, Oregon	Unknown	Sediment trails	Unknown	S. Jeffries[m]
Sea Lion Rock, Oregon	Unknown	Moving back and forth	Unknown	T. Loughlin[l] (personal communication)
Grays Harbor, Washington	Unknown	Shallow dives, flukes out	Unknown	S. Jeffries[m] (personal communication)
La Push, Washington	Unknown	Diving in same spot, no sediment trails	Sand	M. Dahlheim[n] (personal communication)
Neah Bay, Washington	Unknown	Shallow dives on side, milling in kelp bed; no sediment trails	Rocky, kelp bed	G. Joyce[o]
Tapaltos Bay, Vancouver Island, British Columbia	*Holmesimysis*	Inverted position, fecal material, no sediment trails	Kelp bed, rocky shore	Murison *et al.* (Chapter 19, this volume)
Wickaninish Bay, and along the west coast of Vancouver Island, British Columbia	*Onuphis*	Diving in the same spot, sediment trails	Sand	Darling (1977)
Pachaena Bay, British Columbia	*Ampelisca* spp.	Feeding excavations	Sand	Oliver *et al.* (in press)

Table I (*Continued*)

Location	Probable prey item	Feeding and other behavior	Bottom substrate	Reference
Pachina Point, Vancouver Island, British Columbia	Unknown	Diving in same spot, no sediment trails	Unknown	Hart (1977)
Rose Spit, Queen Charlotte Island	Unknown	Sediment trails	Unknown	Hatler and Darling (1974)
Chichagof Island, Alaska	Shrimp-like organisms	Near surface, passes, attending seabirds	Rocky, kelp bed	C. Johnstone[f] (personal communication)
Cape St. Elias, Alaska	Unknown	Surface swimming and gulping	Rocky point	Cunningham and Stanford[g] (unpublished manuscript)
Nelson Lagoon, Alaska	*Crangon*	Tidal feeding, no sediment trails	Sand, rock	R. Gill[r] (personal communication)
Port Heiden to Port Moller, Alaska	Unknown	Mud trails very near surf zone	Unknown	Braham *et al.* (1977)
Ugashik, Alaska	*Anonyx*	Rolling about near surface	Sand	K. Haflinger[s] (personal communication)
Pack ice edge, N. Central Bristol Bay	Unknown	Mud trails	Unknown	S. Leatherwood[c] (personal communication)
Goodnews Bay, Alaska	Herring	Unknown	Unknown	Frost *et al.* (1982)
St. George Island, Alaska	Unidentified amphipods	Mud trails, foul breath	Sand	R. Gentry[t]
St. Matthews Island, Alaska	Unknown	Shallow dives, nearshore	Sand	A. Sowls[u] (personal communication); G. Joyce[n] (personal communication)
South of St. Lawrence Island, Alaska	Unknown	Sediment trails	Unknown	E. Biggs (personal communication)

[a]J. Sumich, Grossmont College, El Cajon, California.
[b]D. Rice, National Marine Mammal Laboratory, Seattle, Washington.
[c]S. Leatherwood, Hubbs Sea World Research Institute, San Diego, California.
[d]G. Antonelis, National Marine Mammal Laboratory, Seattle, Washington.
[e]A. Roust, California Polytechnical State Univ., San Luis Obispo, California.
[f]M. Poole, Biology Dept., Sonoma State Univ., Rohnert Park, California.
[g]C. Goebel, Univ. of Washington, College of Fisheries, Seattle, Washington.
[h]D. Ainley, Pt. Reyes Bird Observatory, Stinson Beach, California.
[i]B. Jones, Museum of Vertebrate Zoology, Berkeley, California.
[j]G. Tate, U.S. Geological Survey, Menlo Park, California.
[k]T. Dohl, p. 25 in *Abstr. of Fourth Conf. on Biology of Marine Mammals, 1981.*

(*continued*)

Table I (*Continued*)

[l]T. Loughlin, National Marine Mammal Laboratory, Seattle, Washington.
[m]S. Jeffries, Washington State Dept. of Game, Astoria, Oregon.
[n]M. Dahlheim, Univ. of British Columbia, Vancouver, B.C.
[o]G. Joyce, National Marine Mammal Laboratory, Seattle, Washington.
[p]C. Johnstone, Sitka, Alaska.
[q]W. Cunningham and S. Stanford, Alaska Dept. Fish and Game, Anchorage, Alaska.
[r]R. Gill, Fish and Wildlife Service, Anchorage, Alaska.
[s]K. Haflinger, Univ. of Alaska, Institute of Marine Science, Fairbanks, Alaska.
[t]R. Gentry, National Marine Mammal Laboratory, Seattle, Washington.
[u]A. Sowls, Fish and Wildlife Service, Anchorage, Alaska.

the fisherman used for any small bait fish. Swartz and Jones (1980, 1981) sighted whales with mud plumes in LSI; however, they did not feel this was indicative of feeding because when they sampled the bottom fauna they found a depauparate benthic fauna.

The pelagic realm appears to be richer. From various plankton tows taken in LSI, copepods and mysids appeared abundant in the vicinity of eel grass mats (Swartz and Jones, 1980, 1981). Crab larvae (150/m^3) and euphausiid juveniles (182/m^3) were also reported in dense concentrations at the mouth of Bahia Magdalena (Norris *et al.*, in press). Thus, the benthic resources available to the gray whale appear to be minimal in the lagoons, a hypothesis that is consistent with data on stomach contents. The feeding that does occur is probably on pelagic food items.

THE MIGRATORY CORRIDOR

The coastal migratory corridor from Baja California to the Bering Sea is approximately 6000 km long. Most whales appear to terminate the northward migration in the Bering Sea, presumably drawn by the rich benthic communities of the continental shelf. In evidence of this, Rugh (Chapter 10, this volume) estimated that 17,648 whales moved south through Unimak Pass and out of the Bering Sea in the fall of 1979, while Reilly (1981) projected a total population size of 17,557 animals from his 1979 count of whales passing Yankee Point, California.

Some whales do appear to linger along the Pacific coast to feed, rather than participate in the complete migration. In recent years there has been an increase in the number of records of gray whales summering south of the Bering Sea; off California (Dohl *et al.*, 1981); off Oregon (Chapters 12 and 13, this volume; Sumich, 1982); and off Washington and British Columbia (Chapter 12, this volume). An estimated 100 whales summer off these three last-named areas. Those summering off British Columbia, at least are principally engaged in feeding (Chapters 12 and 19, this volume; Oliver *et al.*, in press).

Whales destined to summer in northern waters also feed on both benthic and pelagic prey at select locations along the migratory route. Migrating whales have been seen feeding on bottom fauna mostly at the mouths of rivers or estuaries. For example, J. I. Sumich (personal communication) estimated that over 50% of the sightings of feeding whales (identified by mud plumes), made during aerial surveys of the Oregon coast were at river mouths. Similarly, S. Jeffries (personal communication, 1982) noted most feeding whales in the vicinity of rivers along the Washington coast. (However, most sightings of whales are near river mouths so the sighting effort confounds the data.) G. Tate (personal communication, 1982) also noted bottom features on the side-scan sonar near river mouths on the California coast that appear similar to those made by foraging whales in the Bering Sea. Sediments near river mouths may be organically enriched by comparison with the adjacent substrate, and consequently the benthic community may be richer.

Along the north side of the Alaska Peninsula, northward-migrating whales are commonly seen trailing mud plumes. Between Port Moller and Cape Greig, many seem to travel and feed in the area just seaward of the surf zone (personal observation). The sediments in this region are composed of coarse sand and gravel (Sharma, 1979) and the nearshore and littoral zones are often scoured by shorefast ice. Because of this ice scour and the course sediments, the prey items are possibly highly motile organisms, such as the scavenging amphipods and isopods. Whales found feeding in the vicinity of Ugashik, Alaska during early June, 1982, were believed to be preying on motile amphipods such as *Anonyx*. Large numbers of *Anonyx* sp., clinging to hydrozoans, were dragged up by fishing nets in the shallow areas where the whales were foraging and fishermen were working (K. Haflinger, personal communication).

The areas where whales are not seen feeding along their route are also of interest. For example, there was only one sighting of a feeding whale during 11 years of census operations conducted at Point Loma and Yankee Point, California (D. Rice, personal communication, 1982), and no sightings at Yaquina Head, Oregon during 2 years (Chapter 13, this volume). Similarly, very little feeding activity has been noted in the Unimak Pass area (D. Rugh, personal communication, 1982; Hessing, 1981) despite extensive observational effort during the migration. All of these observations were made from high bluffs with sharp escarpments and rocky bases. Possibly such headlands, ordinarily high-wave-energy regions, do not support appropriate benthic resources.

Sightings of whales feeding in the water column have been made along the entire range but are concentrated in the southern regions (Table I). They can be divided into those whales (juveniles?) feeding in kelp beds and those exploiting pelagic resources such as small schooling fish or crab larvae. The latter prey items are unpredictable in space, and consequently their exploitation by whales is probably opportunistic and certainly rare. S. Leatherwood (personal communication), who estimates having observed several hundreds of gray whales off California and northern Baja California by air and from vessels in 14 years, has recorded only one reliable feeding incident. In short, whales feeding while migrating or summering along the northern half of the migration route are nearly always consuming benthic resources.

THE NORTHERN SEAS

The northern feeding grounds of the Bering and Chukchi Seas are found on the expansive continental shelf. The region is shallow, generally less than 50 m deep in the northern Bering Sea and less than 68 m in the southern Chukchi Sea. The distribution of feeding whales in these waters has been reported by Braham (Chapter 11, this volume) and Moore and Ljungblad (Chapter 23, this volume). The gray whale distribution plot from Braham (Chapter 11, this volume) is reproduced in Fig. 6. In the Chirikov basin of the northern Bering Sea, this distribution coincides with that of an extensive infaunal

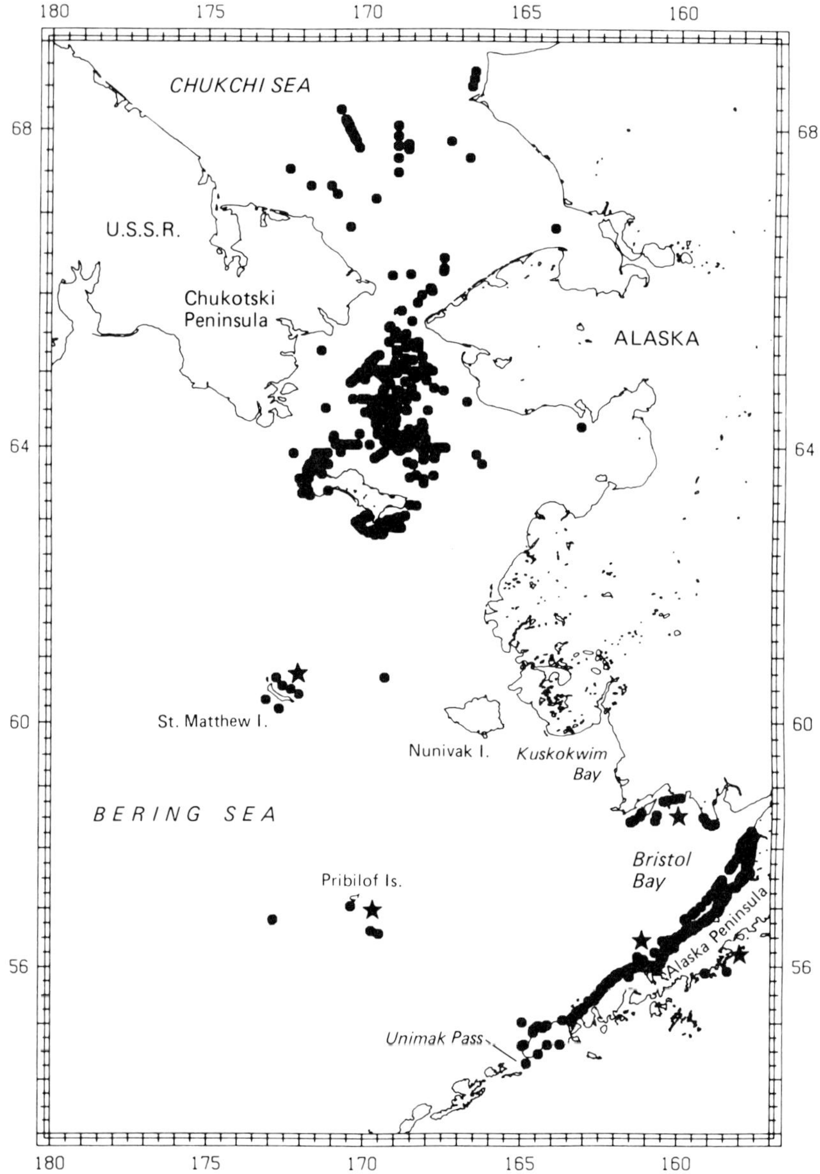

Fig. 6. *Distribution of feeding gray whales in the northern Bering Sea (Fig. 5 in Chapter 11, this volume).*

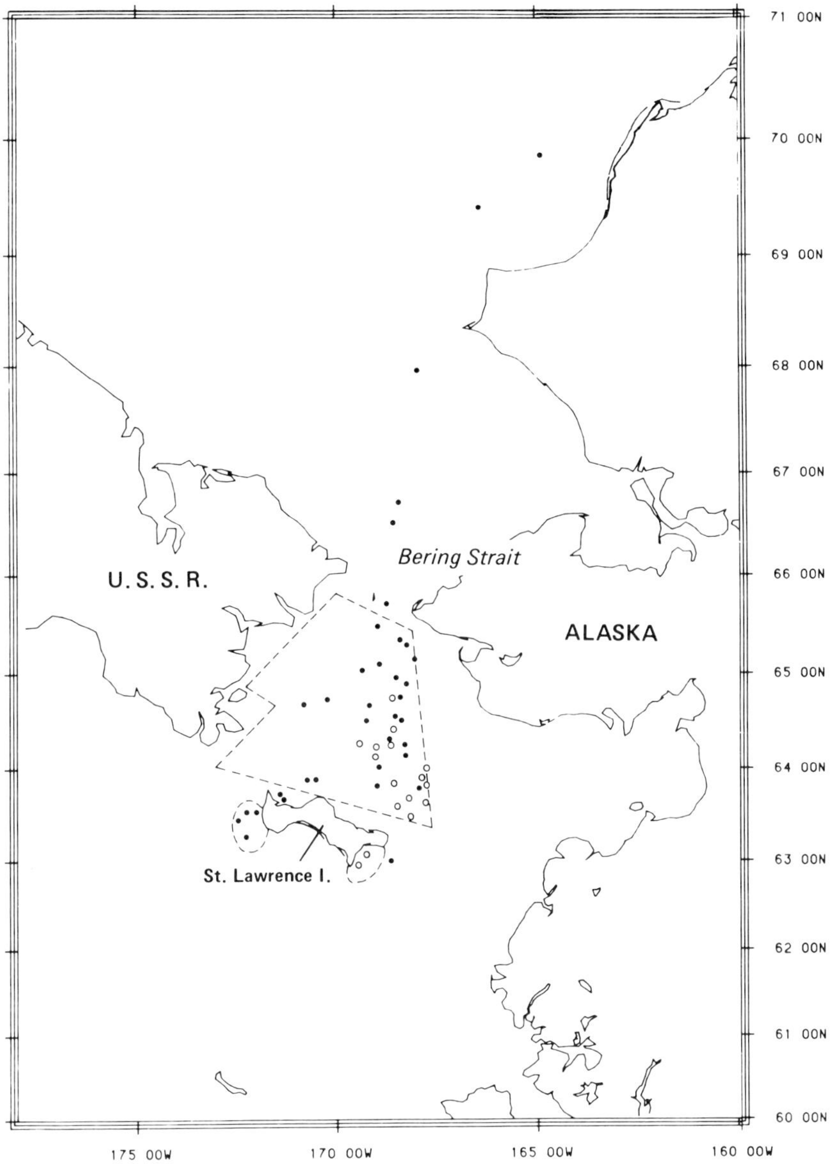

Fig. 7. *Distribution of the amphipod community in the Northern Bering Sea. Dotted line encloses the* 3.7×10^4 km^2 *area estimated to be of greatest importance. Black circles (●), dominated by amphipods [in benthic samples collected by Stoker (1978)]; open circles (○), dominated by amphipods (in benthic samples collected by Nerini in 1980).*

amphipod community (Fig. 7), which is dominated by the ampeliscid amphipods, most notably *Ampelisca macrocephala*. Densities of ampeliscids alone in grab samples taken in this area during the 1980 research cruises ranged from 400 to 23,780 individuals/m^2. The mean combined amphipod biomass for this community was 161.5/m^2 ($n = 133$, SE $= 142.4$), and the maximum combined amphipod biomass in any single grab was 941/m^2. Stoker (1978) gave a mean total biomass for this region of 482 $\pm$ 286/m^2 wet weight in which *A. macrocephala* alone supplies 31% of the organic carbon biomass (calculated by author as 7.2 g/m^2). Earlier biomass estimates from this region ranged from 200 to 1000 g/m^2 (Kuznetsov, 1964).

The variance in amphipod densities and concomitant biomass is high throughout the region (C. V. $\approx$ 30%) and is partly related to the effect of the bottom depressions. That is, samples collected by scuba divers from within bottom depressions had lower densities of *Ampelisca* spp. than those taken outside the depressions ($p < .01$, Wilcoxon–Mann–Whitney rank–sum test, $n = 18$). The mobile scavengers were, however, in greater abundance inside of recently formed depressions. Furthermore, experimentally defaunated patches of the benthos created to simulate whale feeding pits showed a rapid recruitment to the denuded areas by the mobile scavengers (Nerini and Oliver, 1983). This has led to the belief that the gray whale foraging disturbance is instrumental in the structuring of the benthic community.

Other northern areas where whales have been noted feeding on the benthos include the nearshore waters of the western Bering Sea (Votrogov and Bogoslovskaya, 1980; Blokhin and Pavlyuchkov, in press), the southern capes of St. Lawrence Island (F. H. Fay, personal communication, 1977), the southern Chukchi Sea (Wilke and Fiscus, 1961), and the north side of the Chukchi Peninsula (Fedoseev, 1966; Miller *et al.*, 1984; Nasu, 1960). Benthic communities in these areas, clearly of importance to the whales, have not been well studied. The benthic community in the nearshore waters of the western Bering Sea is an amphipod assemblage (Kuznetsov, 1964), but the dominant amphipods are unknown. The southeastern cape of St. Lawrence Island has an amphipod assemblage dominated by *Ampelisca* in some localities and by *Photis* in others (personal observation). The area off the southwest side of St. Lawrence Island is apparently similar (Thompson and Martin, 1983). The subtidal area of St. Lawrence Island where we noted whales feeding in the surf zone supports a dense, probably seasonal community of isopods (*Gnorimosphaeroma* and *Synidotea* spp.) and scavenging amphipods (*Anonyx* and *Atylus* spp.) (Fig. 8). The southern Chukchi Sea had the highest biomass (1195 g/m^2) and levels of organic carbon (56.5 g/m^2) found on the Bering/Chukchi shelf (Stoker, 1978). Along the north side of the Chukchi Peninsula, the bottom community is variable; in areas where whales were found feeding, however, the dominant organisms were amphipod species of *Photis*, *Ampelisca*, *Pontoporeia*, and *Ischyrocerus*, and a cumacean, *Diastylis* sp. (Coyle, 1981).

The benthic foraging areas utilized by the whales seem to have one characteristic in common; they are underlain by dense crustacean infaunal communities. On the northern grounds, the predominant community form is an amphipod assemblage. Previous authors have expounded upon the importance of a particular amphipod, *Ampelisca*

Fig. 8. *Epifaunal benthic community in the nearshore area (<10 m deep) of St. Lawrence Island. Isopods are genus* Synidotea. *Photo taken by L. Consiglieri, July 1980.*

macrocephala, to the diet of the gray whale. Because the whales feed in several different amphipod communities, I expect that the importance of this single species may be exaggerated; however, the complex of amphipods inhabiting the Chirikov Basin is certainly critical to the well-being of the population.

Stomach Contents and Prey Items

Stomach content data, although sparse, does exist from various regions of the gray whale's range. Although available data do suggest a general pattern of feeding in northern waters and fasting in southern waters (Rice and Wolman, 1971), the overall picture is more complex and has long been a subject of dispute. Whales taken in the Bering and Chukchi Seas usually have full stomachs. Zimushko and Lenskaya (1970) reported food in the stomachs of 85% of the whales taken in the Soviet whale fishery in the Bering and Chukchi Seas, whereas whales landed during migration on either side of the Pacific have rarely displayed evidence of recent of extensive feeding (Andrews, 1914; Mizue, 1951; Rice and Wolman, 1971); stomachs of whales taken from the winter grounds in the lagoons of Baja California have similarly been devoid of substantial food (Scammon, 1874). Although data on stomach contents are available from migrating whales which stranded along the west coast of the United States, one cannot assume these stomach contents are representative of the feeding by normal healthy animals. Further, stomach contents of stranded animals are usually severely decomposed rendering some prey items difficult to identify.

The list of prey items is extensive for gray whales (Table II), reflecting both their opportunistic approach to feeding and the nonselective nature of their feeding mechanism. Both pelagic and benthic fauna are consumed, although these two feeding modes are probably important in different areas of the gray whale's distribution. Prey inferred from the mode and area where whales were feeding are listed in Table I.

On the northern feeding grounds, benthic amphipods are clearly of paramount importance, with one or two species often comprising 90% of the food remains and many other species occurring in small quantities. Stomachs from 324 gray whales taken by Soviet whalers in the northern Bering Sea contained six dominant amphipod genera representing four families: the Ampeliscidae (*Ampelisca macrocephala, A. eschricti, Byblis gaimardi, Haploops* sp.); Atylidae (*Atylus*); Lysianassidae (*Anonyx*), and Haustoriidae (*Pontoporeia*) (Zimushko and Lenskaya, 1970; Bogoslovskya *et al.*, 1981; Blokhin and Pavlyuchkov, 1983, in press). A closer look at the ecology of these amphipods can provide some information on gray whale feeding.

All these species are relatively large bodied, ranging from 13 to 27 mm in length. The Ampeliscidae are tube builders and were found in dense concentrations in our samples, reaching a maximum of 937 g/m^2 and 23,780 individuals per m^2 in the Chirikov Basin. They thrive in physically disturbed, sandy sediments; on the east coast of North America ampeliscids colonize sandy bars and stabilize the bottom sediments by their network of tube dwellings (Mills, 1967). They have a short life span of 1 to 2 years (Kanneworff, 1964) and grow to maturity rapidly (Thorsen, 1957; Mills, 1967), thereby engendering a high productivity. Juvenile animals settle in uncolonized areas (Mills, 1967). Samples collected by scuba divers revealed greater numbers of newly hatched juvenile ampeliscids (3 mm in length) and larger mature adults within bottom depressions (Fig. 9). Thus, by its feeding in the Chirikov basin the gray whale may play a part in the formation of cleared areas needed by colonizing juveniles.

By contrast, species such as *Atylus* and *Anonyx*, the mobile scavenging amphipods, ordinarily do not occur in dense concentrations. In our grab samples from the Chirikov Basin, the lysianassids (of which *Anonyx* is a member) were only numerous in one sample, in which they comprised nearly 15% of the amphipods (290 individuals). In all other samples they represented less than 7% of the amphipods. *Atylus* appeared in still lower numbers and was ordinarily absent. We did note elevated concentrations of *Anonyx* in the shallow water near St. Lawrence Island swarming on small masses on detritus. As the areas shallower than 18 m are disturbed frequently by ice and are readily affected by tidal currents and wind-generated waves, it may be the nearshore areas support a mobile community of epifauna. It is therefore interesting to note that the whales which had been feeding primarily on *Anonyx* were taken only in nearshore areas along the Soviet coast (Bogoslovskya *et al.*, 1981). Similarly, *Atylus, Pontoporeia*, and *Synidothea* (an isopod) were prominent only in the stomachs of whales taken nearshore (Zimushko and Lenskaya, 1971; Bogoslovskaya *et al.*, 1981; Blokhin and Pavlyuchkov, in press). *Atylus*, like many other amphipods, annually enters the water column to breed

Table II
Genera[a] of Food Items Reported from Gray Whale Stomachs

Food item	Reference[b]	Dominance in stomach	Location of sample taken if other than Bering Sea[c]
Porifera			
Spongia	1, 20		
Hydrozoa			
Hydropolipae	1, 4, 16		
Leptolida	20		
Anthozoa			
Actinia	1, 16, 20		
Polychaeta			
Ampharetidae	20		
Brada	1		
Eunicidae	1		
Flabelligera	1		
Lumbriconereis	20		
Maldanidae	16, 20		
Nephthys	1, 20		
Onuphis	20		
Oweniidae	20		
Pectinaria	1, 2, 20		
Potamilla	20		
Sabellidae	20		
Stylarioides	1		
Terebellidae	20		
Travisia	1, 4, 20	XX[d]	
Priapulida			
Priapulus	1		
Echiura			
Echiurus	20		
Sipuncula			
Golfingia	20		
Phascolosoma	1		
Sumacea	1		
Isopoda			
Idothea	20		
Synidothea	1, 20		
Amphipoda			
Acanthostepheia	1, 3, 20		
Ampelisca	1–5, 16, 17, 20	XX	B, C
Ampithoe	20		
Anisogammarus (*Echinogammarus*)	1, 16, 20		
Anonyx	1–5, 16, 17, 20	XX	B, C
Arrhis	16		
Atylus (*Nototropis*)	1, 3, 4, 6, 16, 20	XX	
Bathymedon	20		

(*continued*)

Table II (*Continued*)

Food item	Reference[b]	Dominance in stomach	Location of sample taken if other than Bering Sea[c]
Byblis	1, 16, 20		
Caprella	20		
Corophiidae	16		
Dulichia	3, 16		
Erichthonius	20		
Eusirus	4, 16, 20	X[e]	
Haploops	1, 16, 20		
Hippomedon	1, 3, 20		
Ischyrocerus	1, 3, 16, 20		
Lembos	1, 4, 5, 16, 17, 20	X	B, C
Lepidepecreum	20		
Maera	20		
Melita	1, 17, 20		B, C
Monoculodes	16, 20		
Odius	20		
Orchomene	1, 3		
Paroediceros	1		
Paraphoxus (*Pontharpinia*)	1, 3		
Paradulichia	1		
Photis	1, 20		
Pleustes	3		
Pontoporeia	1, 4, 5, 16, 17, 20	XX	B, C
Pseudolibrotus	1		
Rhachotropis	16, 20		
Socarnes	1, 20		
Stegocephalus	1, 16, 20		
Stenopleustes	20		
Mysidacea	4, 16		
Cirripedia			
Balanus	20		
Cumacea	2, 3, 20		
Euphausiacea	15		Ca
Decapoda			
Chionocetes	1, 20		
Fabia (planktonic)	3	XX	Ca
Hyas	1, 20		
Nectocrangon	1, 20		
Nephrops (planktonic)	8	XX	K
Pachycheles (planktonic)	3	XX	Ca
Pleuroncodes (planktonic)	7		Ba
Sabinea	1		
Spirontocaris	1, 20		
Gastropoda			
Buccinidae	1, 5, 16		
Buccinidae (egg masses)	1, 4, 20		

Table II (*Continued*)

Food item	Reference[b]	Dominance in stomach	Location of sample taken if other than Bering Sea[c]
Margarites	1, 20		
Neptunea	16, 20		
Nucella	1		
Polinices	1, 2, 16		
Trichotropis	1, 20		
Velutina	1		
Bivalvia			
Hiatella	1, 20		
Macoma	1, 20		
Montacuta	1		
Musculus	1, 20		
Mya	1, 20		
Mytilus	1		
Serippes	1, 20		
Holothuroidea			
Cucumaria	1, 20		
Unidentified	2, 3, 16		
Ascidiacea			
Dendrodoa	1		
Pelonaia	1, 20		
Sijnascidia	1		
Unidentified	2, 3, 16, 20		
Pisces			
Ammodytidae	1		
Clupeidae	9, 10, 11	XX	Ba, W
Plant material	1, 5, 4, 125, 13, 14, 16, 185, 195	XX	B, W, Ba, K, Ca
Mud, sand, silt, gravel	1, 4, 2, 3, 125, 6		

[a]Note that only italicized names are genera. Other categories are family, order, class, or phylum.
[b]Sources: (S denotes information from a stranded whale)

(1) Zimushko and Lenskaya, 1970 (n = 70).
(2) Pike, 1962 (n = 3).
(3) Rice and Wolman, 1971 (n = 317).
(4) Tomilin, 1957 (n = 57).
(5) Zenkovich, 1934 (n = 2).
(6) Zenkovich, 1937 (n = 192).
(7) Matthews, 1932 (n = 1).
(8) Mizue, 1951 (n = 545).
(9) Klumov, 1963 (n = 1).
(10) Walker, 1949 (n = 1).
(11) Balcomb, personal communication (n = 1).
(12) Nerini, unpublished data (n = 3).
(13) Scammon, 1874 (n = several).
(14) Andrews, 1914 (n = unknown).
(15) Howell and Huey, 1930 (n = 1).
(16) Bogoslovskaya *et al.*, 1981 (n = 113).
(17) Coyle, 1981 (n = 1).
(18) Jones, personal communication (n = 1).
(19) Harvey, personal communication (n = 1).
(20) Blokhin and Pavlyuchkov, in press (n = 120).

[c]B, Bering Sea; C, Chukchi Sea; Ba, Baja; Ca, California; W, Washington; K, Korea.
[d]XX, comprised >40% by volume in at least one stomach.
[e]X, frequent occurrence, not dominating.

(C. Staude, personal communication) and perhaps it is at this time that the whales consume them. In 1980 we noted a cloud of juvenile *Atylus* sp. distributed from the surface to the bottom in water 9 m deep. In our samples, *Pontoporeia femorata* was found only in water shallower than 34 m, where in one sample it reached a maximum of 49% (3618 individuals) of the total amphipods.

The composition of the prey items taken by gray whales, as evidenced by stomach contents, is similar to the benthic infaunal composition observed in the area of catch. For example, stomach contents of whales taken in Soviet whaling area 12 (the Chirikov basin) consisted of ampelisciid amphipods (95%), *Lembos* spp. (3%), and *Anonyx* spp. (2%) (Bogoslavskaya *et al.*, 1981). Our westernmost grab sample in the Chirikov basin contained ampelisciid amphipods (95%), *Lembos* spp. (1.8%), and *Anonyx* spp. (0.3%). Assuming the gray whale does suck up sediments and associated fauna, one might reasonably expect heavier organisms such as the large or deep-living bivalves which are present in the community (e.g., *Serripes* and *Mya* spp.) to be underrepresented in the stomach contents, as they are. Thus, the extraordinary diversity of prey items displayed in Table II is merely a by-product of the whales' inability to sort out perhaps less desirable or rare items occurring with their preferred prey. Some selection based on size and shape of the prey may occur due to the coarseness or "sieve size" of the baleen. Rice and Wolman (1971) measured prey items varying from 6 to 25 mm in length, and Coyle (1981) only found items 8–10 mm. I have measured crab zoae collected from the stomach of a migrating whale which were 3 mm in length. The shape of the organism

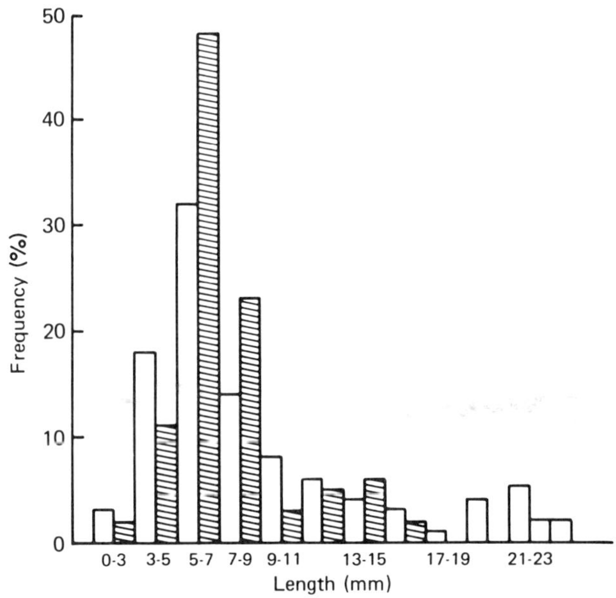

Fig. 9. *Length frequency distribution of amphipod,* Ampelisca macrocephala, *taken inside (white bars) and outside (shaded bars) of a feeding depression in July 1980, off Southeast Cape, St. Lawrence Island.*

may be as important as its size; Those animals with many appendages are more likely to be snared by the baleen than relatively smooth-bodied items such as polychaetes.

The planktonic items known to have been eaten by gray whales (Table II) occur in swarms or schools. Their presence in a stomach is usually exclusive of any other species, and they only occur in stomachs of whales taken or stranded south of the Bering Sea.

Plant material has also been reported from the stomachs of gray whales. Most authors, excepting Tomilin (1957), have discounted the value of plant material and attributed its presence among the prey items to incidental ingestion. I suggest that ingestion of plant material may at least in some cases be deliberate. It appears frequently in the stomach contents, sometimes in large quantities. For example, 120 liters of kelp (*Laminaria* sp.) and algae were found in the stomach of a gray whale stranded at Wauna, Washington (personal observations) and a lesser amount associated with unidentified crustaceans was found in the stomach of a whale stranded at Neah Bay, Washington. Although stranded animals are not representative of normal animals, the quantity of plant material found in this case suggests it was not an accidental ingestion. Plant material is also found in normal animals. Bogoslavskya *et al.* (1981) reported that as much as 35% of a stomach of a whale harvested in the Bering Sea was filled with algae, and Scammon (1874) reported most stomachs of gray whales in the Baja lagoons contained a "sea moss," which I assume was eelgrass. Andrews (1914) states that every stomach examined from Korea was filled with dark green water containing bits of kelp and sea grass. Fecal material in these whales was dark green and the consistency of thick cream. Recent analyses of three samples from three stranded gray whales for the occurrence of seven volatile fatty acids has revealed significant amounts of acetic, propionic, and butyric acids (Herwig *et al.*, 1984). The levels of these compounds, found in the forestomachs, could be produced only during microbial fermentation. One of the stomachs tested contained only kelp; thus, although microbial fermentation has not been previously described in cetaceans, there is some evidence that this process occurs in the stomach. Such a process would allow the gray whale to obtain an undetermined amount of energy from plant material, rendering the whale a partial facultative herbivore.

Rates of Consumption

No discussion of the feeding ecology of the gray whale would be complete without attention, however sketchy, to the energetics of the whale and the total amount of food consumed by the population. As discussed, the gray whale does not feed solely in arctic waters; however, for simplicity and because of our ignorance about the extent of feeding in southern waters, I assume that most of the annual food ration is taken in the Bering and Chukchi Seas. Migrating whales weigh more when taken on the southward leg of their annual journey than they do on the northward leg (Rice and Wolman, 1971). Much of this weight increase is attributable to fat stored in the tissues—in fact, there is no appreciable increase in blubber thickness (Rice and Wolman, 1971)—and this can be measured very grossly by the total oil content.

During the approximately 5 months between the northward and southward migration (June–October), a gray whale is estimated to gain, on an average, 5,063 kg in rendered oil (Rice and Wolman, 1971), or about 16–30% of its body weight. If we assume that the predominant prey are benthic amphipods, which have a lipid content of 7–22% (Percy and Fife, 1981)[2] and that only the lipid fraction of the prey is diverted to fat storage with an associated transfer efficiency of 75% (cf. Brodie, 1975), one whale will ingest, conservatively, an estimated minimum of 61,370 kg of prey during the five months. If we further assume that only 3 of the 5 summer months are spent feeding in the Bering Sea and the remaining 2 are spent feeding in the Chukchi Sea, then the average gray whale may be predicted to remove 36,821 kg/year (409 kg/day) from the Bering Sea sediments. Since the mean amphipod biomass observed in the whale foraging area was 161 g/m^2, to obtain the predicted ration from the Bering Sea an average whale would need to remove the amphipods from 228,707 m^2 (57 acres) of sediment. The entire population, conservatively estimated at over 15,500 gray whales, would then turn over 3,565 km^2/yr of sea bottom or an estimated 9% of the available amphipod community (Fig. 7). Zimushko and Lenskaya (1971) estimated the average adult gray whale consumed 1,200 kg/day.[3] If we substitute that figure for 409 kg/day in the preceding calculation, the estimated bottom consumption is 27% of the total available. If, however, we substitute into the preceding model Stoker's (1978) figures on the average total biomass, for this region of 498 g/m^2, then the whales consume 9% of the total available benthos.

Frost and Lowry (1981) and Rice and Wolman (1971) made similar calculations and arrived at annual consumptions of 0.3–2 and 0.2–1% respectively, of the total standing stock of the benthos. However, the first estimates were based on a total summer range of 1×10^6 km^2, an area several orders of magnitude greater than the circumscribed amphipod community. Lastly, Nelson *et al.* (1981) calculated from their side-scan data that at a minimum, 3.4% of the Chirikov basin shows evidence of recent feeding disturbance. Without better data on the parameters involved in the calculations, it is difficult to determine which of the estimates is more realistic. In any case, the gray whale must be viewed as having a significant impact on the benthic community.

Conclusions

The population of gray whales relies primarily upon the shallow benthic communities of the northern seas for its annual food supply. But the importance of peripheral feeding areas is less clear, and many questions relating to the distribution of the

[2]This was reported as 7–22% of the dry weight of the amphipods, which converts to 5–16% of the net weight. Thus I have considered 11% as the average lipid content and used it in the ensuing calculations.

[3]Rice and Wolman (1971) and Brodie (1975) separately calculated daily food rations of 1 metric ton/day for large cetaceans.

southern food resources and the metabolic needs of the "resident" whale populations are unanswered. Taking the recent sightings of whales feeding in southern waters into account, it now seems likely that the winter fast is broken by ancillary feeding on schools of bait fish or pelagic resources inside and outside the breeding lagoons.

The prey items themselves are generally organisms found naturally in large aggregations. Like other animals that consume prey much smaller than themselves, the cost to the gray whale of catching and handling their prey is small relative to the cost of finding enough of the prey species. Consequently, whales rely on locating patches of prey items (cf. Brodie *et al.*, 1978). It is not surprising therefore that several of the staple prey items, for example, *Ampelisca* and *Photis* spp., are pioneer species capable of recruiting quickly onto unoccupied substrate and also capable of growing rapidly into nearly monospecific assemblages. It is not known how the whales locate patches of food, or how completely they may utilize a patch before continuing on their way. The gray whale may be unusual among whales in that, in addition to being an important predator, it is also a major source of physical disturbance to the exploited community. In concert with currents, sea ice, and storm waves, the gray whale may be responsible for clearing space which can be later colonized by the prime prey species. In this way, it may help to maintain the very amphipod community it exploits.

Finally, perhaps the most extraordinary aspect of the gray whale's feeding ecology is its apparent dietary flexibility. For an animal which has developed highly specialized feeding structures such as baleen, the gray whale has remained a generalist. If trophic specialization is in fact a selective disadvantage as Fowler and MacMahon (1982) argue, then perhaps it is for this reason that the Pacific gray whale population has persisted through geologic time, has recovered from severe exploitation, and remains today one of the least endangered of the great whales.

Summary

Data on feeding gray whales (*Eschrictius robustus*) collected during a 1980 research cruise in the northern Bering Sea are combined with published and unpublished observations to produce a more comprehensive view of gray whale feeding ecology. This review discusses feeding mechanisms, feeding areas, prey items, and rates of food consumption. Although the gray whale population relies primarily upon the benthic communities of northern seas for its annual food ration, peripheral areas and pelagic resources are also exploited. Calculation of food consumption in the northern Bering Sea is estimated to be 3.4–27% of the available benthic community.

Acknowledgments

I am especially grateful to D. Ainley, G. Antonelis, H. Braham, M. Dahlheim, A. Geiger, R. Gentry, R. Gill, C. Goebel, K. Haflinger, S. Jeffries, C. Johnstone, B. Jones, M. L. Jones, G. Joyce, S. Leatherwood, T. Loughlin, L. Martin, D. Rice, A. Roust, A. Sowls, G. Sumich, S. Swartz, G. Tate, D. Withrow, and A. Wolman for allowing me to use their unpublished observations of feeding gray whales in this chapter. M. Dahlheim, K. Johnson, and D. Rice offered many fruitful discussions, and H. Braham, L. Jones, and M. Tillman provided encouragement and discussion throughout this work. All three editors as well as K. Frost and L. Lowry substantially improved an earlier manuscript by their comments. Lastly, the graphics were prepared by C. Hasting and the manuscript by M. Wood. The funding for the 1980 field work was provided by the Bureau of Land Management under interagency agreement to the National Oceanic and Atmospheric Administration Outer Continental Shelf Environmental Assessment Program.

References

Andrews, R. C. (1914). Monographs of the Pacific cetacea. I. The California gray whale (*Rhachianectes glaucus* Cope). *Mem. Am. Mus. Nat. Hist.* [N. S.] **1,** 227–287.

Blokhin, S. A. and Pavlyuchkov, V. A. (1983). Feeding of gray whales off Chukotka. *Rep. Int. Whal. Commn.* **33,** 549–552.

Blokhin, S. A. and Pavlyuchkov, V. A. (in press). Feeding of gray whales in summer–fall in the inshore Chukotka waters. Paper submitted to U.S.–U.S.S.R. Protection of the Environment Cooperative Research Program. 25pp.

Bogoslovskaya, L. S., Votrogov, L. M., and Semenova, T. N. (1981). Feeding habits of the gray whales off the Chukotka peninsula. *Rep. Int. Whal. Commn.* **31,** 507–510.

Braham, H. W., Fiscus, C. H., and Rugh, D. J. (1977). Marine mammals of the Bering Sea and Southern Chukchi Sea. *Environ. Assess. Alaskan Cont. Shelf Annu. Rep.* **1,** 1–99.

Brodie, P. F. (1975). Cetacean energetics: An overview of intraspecific size variation. *Ecology* **56**(1), 152–161.

Brodie, P. F., Sameoto, D. D., and Sheldon, R. W. (1978). Population densities of euphausiids off Nova Scotia as indicated by net samples, whale stomach contents, and sonar. *Limnol. Oceanogr.* **23**(6), 1264–1267.

Coyle, K. (1981). "The Oceanographic Results of the Cooperative Soviet-American Cruise to the Chukchi and East Siberian Seas Aboard the Soviet Whale-hunting Ship *Razyashchii* September–October 1980." University of Alaska, Fairbanks.

Darling, J. (1977). Aspects of the behaviour and ecology of Vancouver Island gray whales, *Eschrichtius glaucus* Cope. M.Sc. Thesis, University of Victoria, Victoria, British Columbia.

Dohl, T. R., Duman, M. I., and Helm, R. C. (1981). Distribution and movements of migratory and resident gray whales along the California coastline. *Proc. Bienn. Conf. Biol. Mar. Mamm., 4th, 1981* Abstracts, p. 25.

Fedoseev, G. A. (1966). Aerovizual'nye nablyudeniya za morskimi mlekopitayushchimi v Beringovom i Chukotskom moryakh (Aerial observations on marine mammals in the Bering and Chukchi Seas). *Izv. Tikhookean. Nauchno-Issled. Inst. Rybn. Khoz. Okeanogr.* **58,** 173–177.

Fowler, C. W., and MacMahon, J. A. (1982). Selective extinction and speciation: their influence on the structure and functioning of communities and ecosystems. *Am. Nat.* **119,** 480–498.

Frost, K. J., and Lowry, L. F. (1981). Foods and trophic relationships of cetaceans in the Bering Sea. *In* "The Eastern Bering Sea Shelf: Oceanography and Resources" (D. W. Hood and J. A. Calder, eds.), pp. 825–836. Off. Mar. Pollut. Assess. Natl. Ocean. Atmos. Admin., Washington, D.C.

Frost, K. J., Lowry, L. F., and Burns, J. J. (1982). Distribution of marine mammals in the coastal zone of the Bering Sea during summer and autumn, Alaska. *Final Rep. to Outer Continental Shelf Environ. Assessment Program.* 166p.

Gilmore, R. M. (1961). "The Story of the Gray Whale," 2nd ed. Pioneer Printers, San Diego, California.

Grigg, R. W., and Dana, T. (1969). Accidental encounter with a whale. *J. Mammal.* **50,** 818–819.

Hart, F. G. (1977). Observations on the spring migration and behavior of gray whales near Pachena Point, British Columbia. *Murrelet* **58,** 40–43.

Hatler, D. F., and Darling, J. D. (1974). Recent observations of the gray whale in British Columbia. *Can. Field Nat.* **88,** 449–459.

Herwig, R., Staley, J., Nerini, M., and Braham, H. (1984). "Baleen whales: Preliminary evidence for forestomach microbial fermentation. *Appl. Environ. Microbiol.* **47,** 421–423.

Hessing, P. (1981). "Spring Census of Gray Whales through Unimak Pass." Fairbanks, Alaska.

Howell, A. B., and Huey, L. M. (1930). Food of the gray and other whales. *J. Mammal.* **11,** 321–322.

Hudnall, J. (1981). Population estimate, feeding behavior and food sources of gray whales (*Eschrictius robustus*), occupying the Straits of Juan de Fuca, British Columbia. *Proc. Bienn. Conf. Biol. Mar. Mamm., 4th, 1981* Abstracts, p. 58.

Kanneworff, E. (1965). Life cycle, food, and growth of the amphipod *Ampelisca macrocephala* liljebörg from the Øresund. *Ophelia* **2,** 305–318.

Kasuya, T., and Rice, D. W. (1970). Notes on baleen plates and on arrangement of parasitic barnacles of gray whale. *Sci. Rep. Whales Res. Inst.* **22,** 39–43.

Klumov, S. K. (1963). Pitanie i gelmin to fauna usatykh kitov (Mystacoceti) v osnovnykh promyslovykh raionakh mirovogo okeana. (Food and helminth fauna of whalebone whales (Mysticoti) in the main whaling regions of the world ocean.) *Tr. Inst. Okeanol. im. P.P. Shirshova, Akad. Nauk SSSR* **71,** 94–194 [transl. by *Fish. Res. Board Can.* (1965)].

Kuznetsov, A. P. (1964). Raspredeleniye donnoy fauny zapadnoy chasti Beringova Morya po troticheskim zonam i Nekotoryye obshchiye voprosy troticheskoy zonal' nosti (Distribution of benthic fauna in the western Bering Sea by trophic zones and some general problems of trophic zonation.) *Tr. Inst. Okeanol. im. P.P. Shirshova, Akad. Nauk SSSR* **69,** 98–177 [transl. by *U.S. Nav. Oceanogr. Off.* (1964)].

Matthews, L. H. (1932). Lobster-krill, anomuran crustacea food of whales. *'Discovery' Rep.* **5,** 467–484.

Miller, R. V., Johnson, J. H., and Doroshenko, N. (in press). Grey whales (*Eschrichtius robustus*) in the western Chukchi and east Siberian Seas. *Arctic.*

Mills, E. L. (1967). The biology of an ampeliscid amphipod crustacean sibling species pair. *J. Fish. Res. Board Can.* **24,** 305–355.

Mizue, K. (1951). Gray whales in the east sea area of Korea. *Sci. Rep. Whales Res. Inst.* **5,** 71–79.

Nasu, K. (1960). Oceanographic investigation in the Chukchi Sea during the summer of 1958. *Sci. Rep. Whales Res. Inst.* **5,** 143–157.

Nelson, C. H., Johnson, K. R., Mitchell, H.-L. (1983). Assessment of gray whale feeding grounds and sea floor interaction in the northeastern Bering Sea. *Draft Rep. to Outer Continental Shelf Environ. Assessment Program.*

Nemoto, T. (1959). Food of baleen whales with reference to whale movements. *Sci. Rep. Whales Res. Inst.* **14,** 149–290.

Nemoto, T. (1970). Feeding pattern of baleen whales in the ocean. *In* "Marine Food Chains" (J. H. Steele, ed.), pp. 241–252. Oliver & Boyd, Edinburgh.

Nerini, M., Jones, L., and Braham, H. (1980). "Gray Whale Feeding Ecology," Final report to NOAA, OCSEAP. (RU 593), Contract No. R7120828. Natl. Mar. Mammal Lab., Seattle, Washington.

Nerini, M. K. and Oliver, J. (1983). Gray whales and the structure of the Bering Sea benthos. *Oecologia* **59,** 224–225.

Norris, K. S., Villa-Ramirez, B., Nichols, G., Würsig, B., and Miller, K. (1983). Lagoon entrance and other aggregations of gray whales, *Eschrictius robustus*. *In* "Behavior and Communication of Whales" (R. Payne, ed.), pp. 259–293. Westview Press, Boulder, Colorado.

Oliver, J. S., Slattery, P. N., O'Connnor, E. F., Lowry, L. F. (1983). Walrus feeding in the Bering Sea; a benthic perspective. *Fish. Bull.* **81,** 501–512.

Oliver, J. S., Slattery, P. N., Silberstein, M. A., O'Connor, E. F. (1984). Gray whale feeding on dense ampeliscid amphipod communities near Bamfield, British Columbia. *Can. J. Zool.* **62,** 41–49.

Percy, J. A., and Fife, F. J. (1981). The biochemical composition and energy content of Arctic marine macrozooplankton. *Arctic* **34,** 307–313.

Pike, G. C. (1962). Migration and feeding of the gray whale (*Eschrichtius gibbosus*). *J. Fish. Res. Board Can.* **19,** 815–838.

Ray, G. C., and Schevill, W. E. (1974). Feeding of a captive gray whale, *Eschrichtius robustus*. *Mar. Fish. Rev.* **36**(4), 31–38.

Reilly, S. B. (1981). Population assessment and population dynamics of the California gray whale (*Eschrictius robustus*). Ph.D. Thesis, University of Washington, Seattle.

Rice, D. W., and Wolman, A. A. (1971). The life history and ecology of the gray whale (*Eschrichtius robustus*). *Spec. Publ.—Am. Soc. Mammal.* **3,** 1–142.

Scammon, C. M. (1874). "The Marine Mammals of the Northwestern Coast of North America." John H. Carmany & Co., San Francisco, California (reprinted by Dover, New York, 1968). 319 p.

Sharma, G. D. (1979). "The Alaskan Shelf: Hydrographic, Sedimentary and Geochemical Environment." Springer-Verlag, Berlin and New York.

Sprague, J. G., Miller, N. B., and Sumich, J. L. (1978). Observations of gray whales in Laguna de San Quintin, Northwestern Baja California. *J. Mammal.* **59,** 425–427.

Stoker, S. W. (1978). Benthic invertebrate macrofauna of the eastern continental Shelf of the Bering and Chukchi Seas. Ph.D. Thesis, University of Alaska, Fairbanks.

Sumich, J. I. (1982). "Recent Summer Occurrences of Gray Whales (*Eschrichtius robustus*) Along the Oregon Coast." Oregon State University, Marine Science Center, Newport, Oregon (unpublished manuscript).

Sund, P. N. (1975). Evidence of feeding during migration and of an early birth of the California gray whale (*Eschrichtius robustus*). *J. Mammal.* **56,** 265–266.

Swartz, S. L., and Jones, M. L. (1980). Gray whales (*Eschrictius robustus*), during the 1977–1978 and 1978–1979 winter sessions in Laguna San Ignacio, Baja California Sur, Mexico. *U.S., NTIS, PB Rep.* **PB80-202989,** 1–35.

Swartz, S., and Jones, M. L. (1981). Demographic studies and habitat assessment of gray whales, (*Eschrichtius robustus*), in Laguna San Ignacio, Baja California Sur, Mexico. *U.S., NTIS, PB Rep.* **PB82-123373,** 1–56.

Thompson, D., and Martin, L. (1983). Feeding ecology of gray whales (*Eschrictius robustus*) in the Chirikov basin in the summer of 1982. *Draft Rep.* to *Outer Continental Shelf Environ. Assessment Program.*

Thorsen, G. (1957). Bottom communities. *In Mem.—Geol. Soc. Am.* **671,** 461–534.

Tomilin, A. G. (1957). "Zveri SSSR i prilezhaschikh stran kitoobraznye" (Mammals of the U.S.S.R. and adjacent countries—cetacea). Izv. Akad. Nauk SSSR, Moskva (transl. by Israel Program Sci. Transl., 1967).

Votrogov, L. M., and Bogoslovskaya, L. S. (1980). Gray whale off the Chukotka peninsula. *Rep. Int. Whal. Commn.* **30,** 435–438.

Walker, L. W. (1949). Nursery of the gray whales. *Nat. Hist. (N.Y.)* **58**(6), 248–256.

Walker, T. J. (1971). The California gray whale comes back. *Natl. Geogr. Mag.* **139**(3), 394–415.

Wellington, G. M., and Anderson, S. (1978). Surface feeding by a juvenile gray whale, *Eschrichtius robustus. Fish. Bull.* **76,** 290–293.

Wilke, F., and Fiscus, C. H. (1961). Gray whale observations. *J. Mammal.* **42,** 108–109.

Zenkovich, B. A. (1934). Materialy k pozhaniyu kotoobraznykh dal'nevostochnykh morei (Seryi Kaliforniiski kit—*Rhachianectes glaucus* Cope) (Research data on Cetaceans of far eastern seas [the California gray whale—*Rhachianectes glaucus* Cope]). *Vestn. Dal'nevost. Fil. Akad. Nauk SSSR* **10,** 9–25.

Zenkovich, B. A. (1937). Esche o serom Kaliforniiskom kite (*Rhachianectes glaucus* Cope, 1864) (More on the California gray whale [*Rhachianectes glaucus* Cope, 1864]). *Vestn. Dal'nevost. Fil. Akad. Nauk SSSR* **23,** 91–103.

Zimushko, V. V., and Lenskaya, S. K. (1970). O pitanii serogo kita (*Eschrichtius gibbosus* Erx.) na mestakh nagula (Feeding of the gray whale [*Eschrichtius gibbosus* Erx.] at foraging grounds). *Ekol. Akad. Nauk SSSR* **1**(3), 26–35.

19

Foraging of the Gray Whale along the West Coast of Vancouver Island, British Columbia

Laurie D. Murison, Debra J. Murie, Karen R. Morin, and Jeannette da Silva Curiel

Introduction

Prey of the gray whale (*Eschrichtius robustus* Lilljeborg) have been determined primarily by examination of stomach contents. Zimushko and Lenskaya (1970) reported the predominant food item of gray whales taken in the coastal waters of the Chukchi Peninsula to be nectobenthic amphipods, with any one of six species accounting for as much as 80–90% of the total food mass. In reviewing gray whale feeding data collected from Arctic summer grounds, Rice and Wolman (1971) similarly concluded that all organisms found in the stomachs of gray whales were infaunal benthic species. They noted that gammaridean amphipods represented 95% of the food species found in stomach contents from gray whales taken in the Bering Sea. These accounts suggest

ISBN 0-12-389180-9

that gray whales forage selectively, and unlike other baleen whales (Nemoto, 1970), are primarily, if not exclusively, bottom feeders.

Scammon (1874), Wilke and Fiscus (1961), Pike (1962), Walker (1975), Darling (1977), Norris *et al.* (1977), Sprague *et al.* (1978), Leatherwood *et al.* (1982), Nerini and Oliver (1983), Braham (Chapter 11, this volume), Moore and Ljungblad (Chapter 23, this volume), and Nerini (Chapter 18, this volume) have provided observational evidence of bottom feeding by gray whales. They have observed feeding whales surfacing with muddy snouts and expelling clouds of muddy water from the baleen. The occurrence of

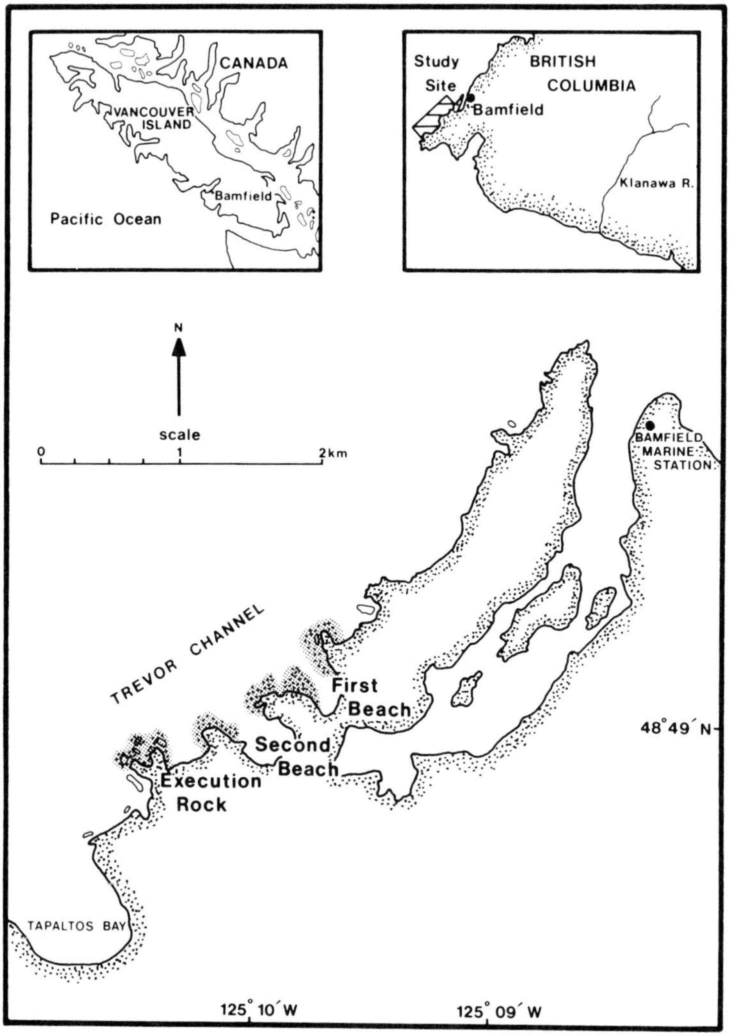

Fig. 1. *Vancouver Island, British Columbia, showing location and detail of the study site and distribution of kelp,* Nereocystis leutkeana *(shaded area).*

sand, silt, and gravel in stomach contents offers additional evidence that gray whales feed on or close to the benthos (Rice and Wolman, 1971; Bogoslovskaya *et al.*, 1981).

The west coast of Vancouver Island, British Columbia, has a summer resident population of gray whales (Darling, 1977, also Chapter 12, this volume; Hudnall, 1981) which in part is represented by whales lingering in nearshore waters between First Beach and Tapaltos Bay (Fig. 1). We hypothesized that the whales were foraging in these waters and thus would present an opportunity to study patterns of diving and feeding behavior and to identify prey species.

Methods

STUDY SITE

The study site included the waters extending from First Beach 3.6 km southwest to Tapaltos Bay (Fig. 1). Limited observations were also carried out near the mouth of the Klanawa River (48°43′N, 124°57′W).

First Beach is a small cove with a gravel beach fringed by a rocky shore and islets. Water depth averages 7 m in the central portion of the cove. Kelp beds of *Nereocystis luetkeana* occupy approximately three-quarters of the cove area, as determined at ebb tide (tidal range, 3.4 m) when the kelp fronds are exposed (Fig. 1). Surf in the cove is moderated by the wave-damping effect of the outcropping rocks and extensive kelp beds.

Execution Rock consists of jagged rock pinnacles and deeply creviced rock faces that drop 18 m below the water surface. The rocky substrate is strewn with boulders and interrupted by a few sand patches. The surf zone is turbulent, and back-eddies are prevalent. Kelp (*Nereocystis luetkeana*) grows subtidally in a narrow band close to the shore (Fig. 1).

DIVING PATTERNS

Whale diving patterns were studied during a 3-week period between July 31 and August 17, 1981. Observations were logged between 0800 and 1400 hr when whales were present. A 4.0- or 5.2-m motorboat (with the motor turned off during observation times) was allowed to drift 3 to 12 m from the whales.

Gray whales have characteristic pigmentation patterns on their backs and flukes which allow repeated observation of recognized individuals (Darling, 1977; Swartz and Cummings, 1978). Diving rhythms for whales were determined by recording dive durations, defined as the time between two consecutive blows accompanied by interim submergence of the whale. During a dive, the location of the whale relative to the shore and to the kelp beds was estimated subjectively and noted. In addition, the proportion (if any) of the flukes that appeared above the water and their aerial movements were recorded.

INVERTEBRATE FAUNA

Subsurface, midwater, and epibenthic horizontal plankton tows were carried out at Execution Rock, and First Beach on August 7 and 14, respectively. Ten tows were taken between 0900 and 1000 h. Tows within kelp areas consisted of two subsurface, two midwater, and four epibenthic; tows outside of kelp areas consisted of two epibenthic. To check for vertical migration of invertebrate fauna, three tows were taken in kelp areas between 2100 and 2200 hr, one each in subsurface, midwater, and epibenthic waters.

Tows were conducted near a diving whale from the bow of a reversing boat. Duration of net tows varied from 1 to 5 min, depending on the substrate irregularities and consequent interruptions, especially of the epibenthic tows. The distances of all tows ranged from 10 to 15 m. Depths of tows were controlled by using an 11-kg weight attached to the mouth of the net while simultaneously regulating the tow rope length. Depths were varied from just below the surface to 18 m. Plankton net dimensions were mouth diameter, 0.47 m; total length, 2.05 m; cod-end diameter, 5.0 cm; cod-end length, 9.5 cm; and net mesh size, 0.8 mm.

Collected specimens were identified to the level of species, and coloration was noted before preservation in 10% neutral formalin. The number of individuals in a 50-ml subsample was counted and used to estimate the total number of individuals in the sample by volumetric extrapolation. Mean lengths and percentage of gravid individuals per sample were similarly determined, using a subsample of 50 individuals.

FECAL MATTER

Fecal samples were obtained by following a whale's "footprint," an upwelling of water visible at the surface caused by motion of the flukes. Feces were collected with a small dip net and a bucket as they floated to the surface. The samples were preserved in 10% neutral formalin and examined microscopically (100× magnification). Photomicrographs were taken of the preserved invertebrates and of the fecal matter.

Results

DIVING PATTERNS

A total of 26.5 hr of observation was made with whales present. Dive durations for 13 of 22 whales present in the study area were recorded over five sessions, for a total recording time of 268 min. Whales (n = 12) sighted for only one session accounted for 50% (134/268) of the combined total, while one whale recorded during two sessions accounted for the remainder. Recording periods for individual whales ranged in duration from 5 to 119 min, depending on sea conditions, observer ability to recognize individuals, and boat traffic. Whales ranged from 8 to 10 m in length and were of undetermined sex.

In nearshore waters off First Beach and Execution Rock, 53% (197/372) of observed

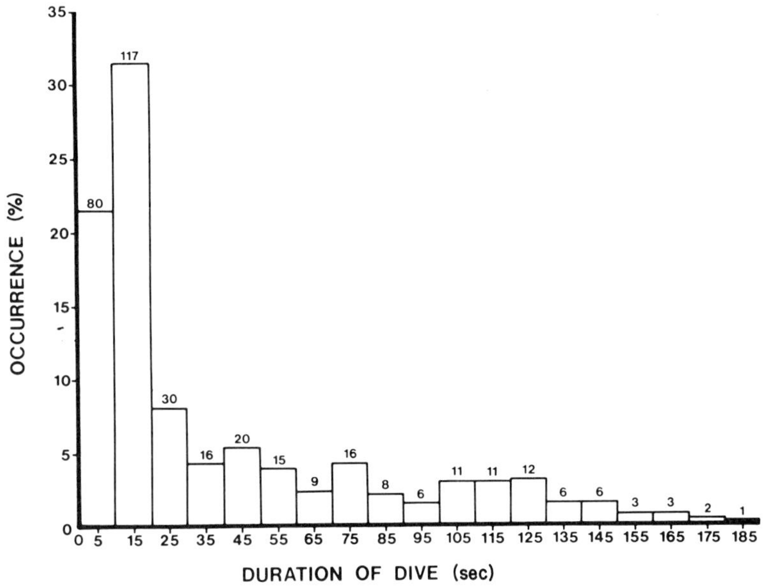

Fig. 2. *Frequency distribution of gray whale dive durations.*

dives by whales were 20 sec or less in duration ($\bar{x}$ = 11.77 sec ± 3.75 (SE), n = 197] (Fig. 2). Dives longer than 20 sec averaged 76.13 sec ± 42.35 (SE) (n = 175) and ranged from 21 to 184 sec. Three dives of 208, 240, and 279 sec, respectively, were recorded from whales moving offshore into the deeper waters of Trevor Channel.

In general, one to five short-duration dives preceded one to four long-duration dives. Cues allowing the prediction of the duration of a forthcoming dive occurred. Surfacing and slow submergence of a whale, when accompanied by presentation of a relatively broadened back, was followed by a short-duration dive (20 sec or less) (72.7%, 16/22). Fifty-six per cent (43/77) of long-duration dives (greater than 21 sec) were accompanied by rapid surfacing and resubmergence of a whale in conjunction with a deeply arched body.

Nearshore diving excursions generally began with a whale surfacing and resubmerging rapidly. Because of its deeply arched body, its flukes were lifted vertically above the surface of the water just before complete submergence. Immediately thereafter, and from 0 to 15 m away from the dive entry, the tips and lateral edges of the flukes repeatedly appeared above the surface for up to 15 sec at a time or were readily visible in near-surface waters, as the flukes and tail stock were rotated and arched back and forth. On one occasion we were able to photograph a diving sequence (Fig. 3,a–h) which demonstrated this rotational and arching movement. The whale, approximately 8–9 m in length, was diving off First Beach in relatively shallow water (7–8 m deep), its flukes breaking the surface regularly. In this sequence the whale was inverted with its flukes arched over backwards. In Fig. 3, frames a–e and g show the ventral surface of the tail stock and part of the dorsal surface of the flukes; frame 3f shows only a portion

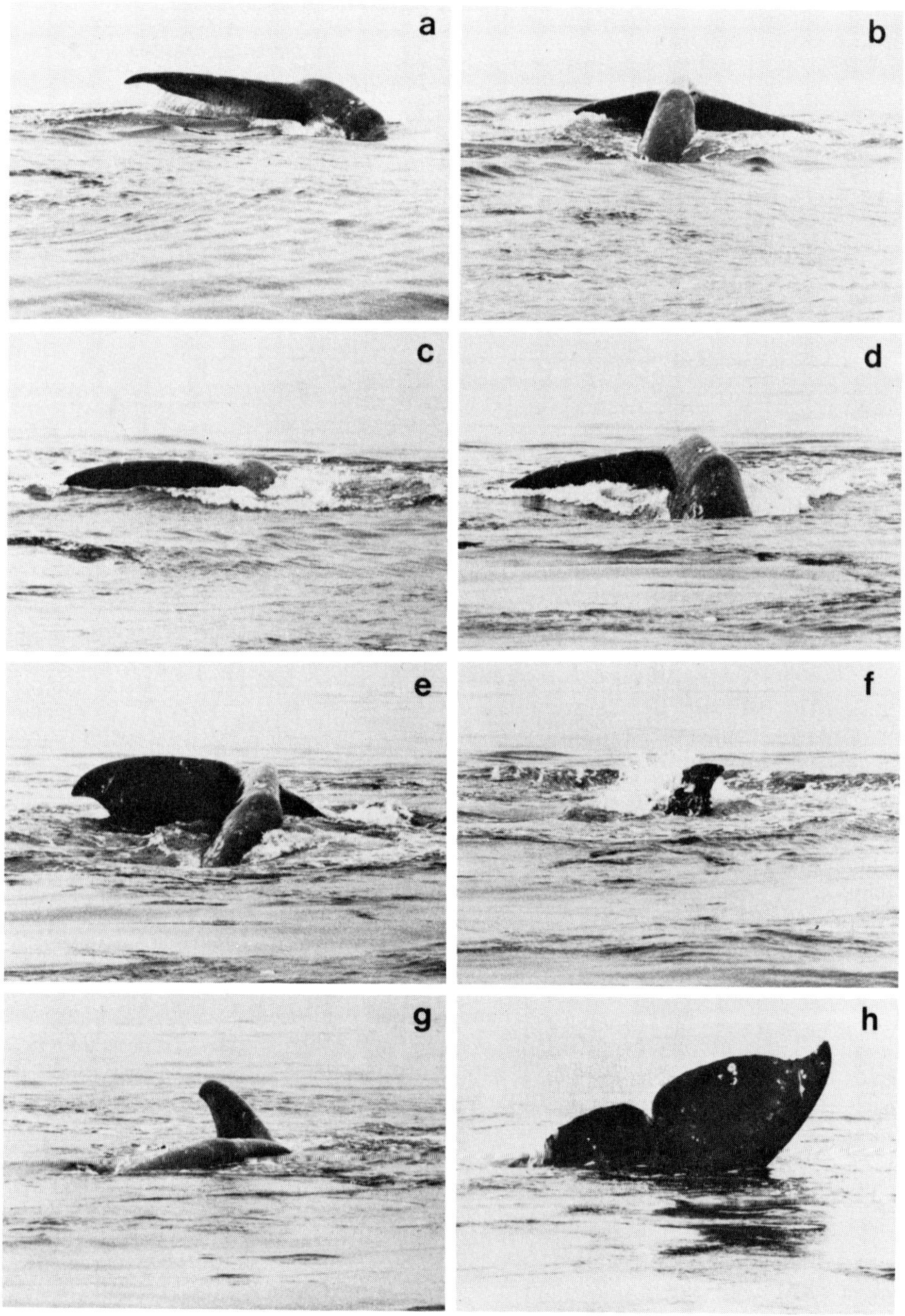

Fig. 3. *Sequential photographic record of gray whale feeding pattern, First Beach, August 14, 1981 (see text for explanation of a–h).*

of the ventral surface of the left fluke. It is not until frame 3h that the whale completes a rotation of the tail stock, bringing the full dorsal surface of the flukes into view.

Whales in the shallow water areas consistently dived, irrespective of the tidal cycle, while in or immediately around kelp beds. Surfacing whales were often observed with kelp fronds draped over the head or back. On three separate occasions, following the surfacing of a whale during a diving sequence, whole mysids or portions thereof, were observed in the surrounding water.

INVERTEBRATE FAUNA

Each daytime subsurface and midwater plankton tow produced small quantities of copepods (<100 individuals) and medusae (<10 individuals). Daytime epibenthic tows made outside the kelp areas resulted in low numbers of copepods, medusae, and small shrimps (less than 10 individuals). Subsurface, midwater, and epibenthic tows conducted in kelp areas at night produced low numbers of mysids (<25 individuals), in addition to low counts of copepods and shrimps. In contrast, daytime epibenthic sampling in areas of kelp yielded large numbers of mysids, with other genera representing insignificant percentages of the total number of individuals collected (Table I).

Two species of mysids were taken in the epibenthic tows at First Beach and Execution Rock. The predominant species was identified as *Holmesimysis sculpta* (E. Brinton, personal communication) (Fig. 4). The second species was tentatively identified as *Acanthomysis pseudomacropsis* (Tattersall, 1951). When first taken from the water the mysids were similar in color to the dark brown kelp, *Nereocystis luetkeana*. In the

Table I
Prominent Invertebrates Collected in Daytime Epibenthic Plankton Tows in Kelp Areas

Location Date Tow number	Fauna	Total number of all individuals	Occurrence (%)
Execution Rock		1,601	
August 7, 1981	*H. sculpta*[a]		100
Tow 1			
Execution Rock		531	
August 7, 1981	*H. sculpta*		95
Tow 2	*A. pseudomacropsis*[b]		2
	Small medusae and worms		3
Execution Rock		318	
August 7, 1981	*H. sculpta*		96
Tow 3	*A. pseudomacropsis*		4
First Beach		95,760	
August 14, 1981	*H. sculpta*		100
Tow 1	Small shrimp and kelp fleas		<0.3

[a]*Holmesimysis sculpta*.
[b]*Acanthomysis pseudomacropsis*.

Fig. 4. *The mysid,* Holmesimysis sculpta. *Gravid female is shown. Length,* $\bar{x}$ = *12 mm (range = 9–15 mm,* n = *50).*

three samples obtained from Execution Rock (Table I) 12, 17, and 23% of the individuals of *H. sculpta*, respectively, were gravid (carrying larvae in marsupia).

In both fresh and preserved plankton samples, the mysids extruded large amounts of an unidentified substance which superficially resembled liquid fat (oil); the substance was water insoluble and formed a slick at the surface of the water.

Dense aggregations of *H. sculpta*, forming a schooling layer approximately 1 m deep over the bottom, were recorded during underwater observations (SCUBA) around Execution Rock. This faunal layer extended horizontally and vertically into the rock crevices and created large clouds when carried by the tidal surge.

FECAL MATTER

While following whales to collect fecal samples, it became apparent that they occasionally travelled a recurring cyclical path, moving from the nearshore waters of First Beach into the deeper, open water of Trevor Channel. Defecation was frequent during these open-water excursions. Whales also frequently defecated before long-duration dives, upon breaching, or when startled. The reddish-brown feces were readily visible as they began to disperse in surface waters.

Analyses of samples of fecal matter, using a compound microscope, revealed a preponderance of recognizable mysid exoskeleton fragments: antennal scales, telsons, endopodites and exopodites of uropods, and thoracic appendages. Telsons and endo-

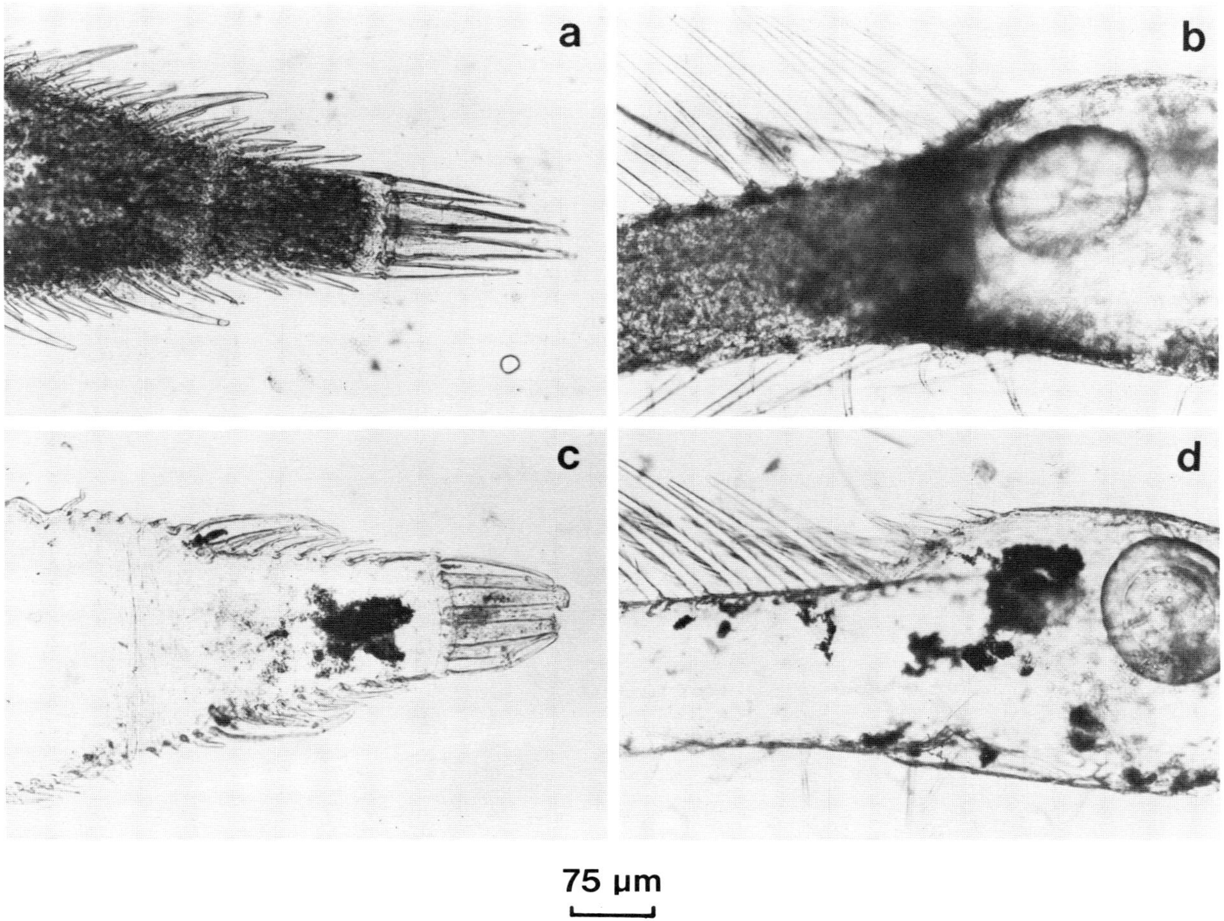

Fig. 5. *Mysid structures: (a) telson and (b) endopodite of uropod with statocyst from* Holmesimysis; *(c) telson and (d) endopodite of uropod with statocyst from fecal material.*

podites of uropods (complete with statocysts) from the fecal matter were found to be identical to comparable structures in *H. sculpta* with respect to size, structure, and spinal arrangement (Fig. 5, a–d).

Discussion and Conclusions

Abundant phytoplankton and zooplankton contribute to the poor underwater visibility of the coastal waters of Vancouver Island during the summer. In addition, legal restrictions on scuba diving prevented direct underwater observation of foraging gray whales. Therefore, to determine whether gray whales were feeding while in the nearshore waters of the study area, it was necessary to examine indirect evidence.

Initially, gray whales were observed diving in or immediately near kelp beds. It is reasonable to assume that if they were feeding, the potential prey would also be in or immediately near these kelp areas. Daytime plankton tows taken to test this assumption determined that a potentially large mysid food source was present in the epibenthic waters in areas of kelp. The finding of these epibenthic plankton tows was further substantiated by direct underwater observations of a large schooling layer of mysids covering the seabed. Furthermore, epibenthic plankton tows taken outside of the kelp areas did not yield any mysids or any other species in large quantities.

In addition, gray whales were not present in the nearshore waters of our study area during the late afternoon and night (S. Insley and R. Rey, personal communication), but were present in the morning. Concomitantly, subsurface, midwater, and epibenthic plankton tows carried out at night in nearshore areas of kelp produced relatively few mysids.

The occurrence of whole mysids and portions thereof in the waters immediately around surfacing whales indicated that mysids were being forcibly expelled or incidentally lost from the baleen of the whales. Fecal samples collected from whales that had been diving in the nearshore kelp areas contained many recognizable fragments of mysid exoskeletons. Collectively, these data are consistent with the hypothesis that gray whales were feeding on mysids while in the nearshore waters between First Beach and Execution Rock.

Having established that the gray whales were feeding in the nearshore kelp areas, we reasoned that the observed diving patterns were coincident with the whales' feeding patterns. The short-duration dives could have little significance in actual foraging because of time constraints. J. Darling (personal communication) has speculated that they may simply be related to the breathing sequence of the whale. It is the long-duration dives that appear to play an active role in feeding. During these dives it was evident from above-surface activity, such as recorded in Fig. 3,a–h, that whales were undoubtedly maintaining a bottom-feeding aspect in their foraging behavior. For the whale represented in Fig. 3, the relationship between the length of the whale (8–9 m), the depth of the water (7–8 m), and the proportion of the whale's flukes and tail stock above the surface suggests the whale was nearly vertical in the water column with its snout in direct contact with the substrate.

Scammon (1874), Wilke and Fiscus (1961), Pike (1962), Walker (1975), Darling (1977), Norris *et al.* (1977) Sprague *et al.* (1978), Leatherwood *et al.* (1982), Nerini and Oliver (1983), Braham (Chapter 11, this volume), Moore and Ljungblad (Chapter 23, this volume), and Nerini (Chapter 18, this volume) have reported gray whales feeding very close to the bottom, as evidenced by leaving mud trails and expelling cloudy water from their baleen upon surfacing. Ray and Schevill (1974) observed that a captive gray whale fed near the bottom of its tank, sweeping parallel to and 10–20 cm above the bottom while sucking up squid in a pulsing motion. Nerini (Chapter 18, this volume) reviewed all data available on gray whale feeding. The major findings of previous research are (1) feeding patterns of gray whales have only been reported for whales feeding over sand/mud bottoms; and (2) gray whales have been recorded to feed by moving over the seabed while sucking or scooping up prey (broad sweeping).

The patterns of diving that typify the foraging activity of gray whales observed in the study differed markedly from those reported in previous studies, with the possible exception of Hudnall (1981). Substrate at Execution Rock, known from underwater observations (SCUBA), is solid rock with a few small sand patches. At First Beach the substrate may be inferred from the surrounding rocky shores and from the knowledge that kelp, which occupies approximately three-quarters of the cove, forms extensive beds by attaching its holdfasts to rock (Abbott and Hollenberg, 1976). Therefore, sweeping of the seabed is probably an inappropriate behavior pattern, given the irregular rocky substrate predominating in the study area. Consistent with this was the absence of mudtrails and clouds as were observed with whales feeding over sand/mud bottoms.

Gray whale feeding patterns in our area suggested whales can forage in areas of rocky shores and kelp by covering only a relatively small forage area per dive. This can be accomplished by remaining in one location per dive, as depicted by the whale in Fig. 3, or by briefly pausing in several locations per dive, as evidenced by occasional observations of intermittent but recurring fluke emergences during a given dive. It should be considered, therefore, that depending on the physical parameters of the seabed and surrounding shore, the gray whale may use different foraging patterns.

The apparent association of gray whales and the kelp may be conditional upon the presence of mysids of the genus *Holmesimysis*. *Holmesimysis* is known to occur in shallow coastal waters inhabited by algae (Holmquist, 1979). During the daytime they form epibenthal schools under the kelp canopy and in lagoon eel grass. Hence if gray whales feed on *Holmesimysis*, they would have to remain near or enter the kelp beds. This three-dimensional association is consistent with the observations of gray whales leaving the kelp beds during the night. Clutter (1967) reported that there is a tendency for the mysids to disperse at night, which is supported by the low number of mysids taken in our nighttime tows.

The mysid, *Holmesimysis*, may supply a valuable energy source for gray whales (see also Chapter 18, this volume). The occurrence of 12 to 23% gravid individuals in our samples indicates a fair proportion of mysids carrying larvae in marsupia, and therefore they may represent an enriched food source. Further research should be directed at determining whether the distribution and seasonal timing of the gray whales is related to the timing and presence of breeding populations of mysids. In addition, the composition and possible role in gray whale nutrition of the unidentified substance superficially resembling an oil (liquid fat), should be explained.

Summary

Gray whales, *Eschrichtius robustus*, were observed in and around beds of kelp, *Nereocystis luetkeana*, nearshore between First Beach and Execution Rock, on the west coast of Vancouver Island, British Columbia, from July 31 to August 17, 1981. Whale diving patterns consisted of a series of brief submergences during breathing sequences ($\bar{x} = 11.77$ sec $\pm$ 3.75 (SE), $n = 197$) followed by dives up to 184 sec in duration in which feeding occurred ($\bar{x} = 76.13$ sec $\pm$ 42.35 (SE), $n = 175$). Foraging whales either remained in one or briefly paused in several locations per dive. Aerial displays of the flukes and tail stock during feeding suggested that the snout of the whale was directly above, or touching, the rocky substrate. Horizontal epibenthic plankton tows in the feeding areas produced almost exclusively the mysid, *Holmesimysis sculpta*, ($\bar{x} = 98\%$ occurrence $\pm$ 2% (SE), $n = 4$ tows) with minor occurrences of the mysid, *Acanthomysis pseudomacropsis*. Underwater observations verified that dense aggregations of the predominant mysid species were schooling close over the bottom. Fecal matter collected directly behind feeding whales contained a preponderance of recognizable mysid exoskeleton fragments. Collectively these data indicate that gray whales summering along the west coast of Vancouver Island feed in close proximity to the substrate on large schools of mysids.

Acknowledgments

We gratefully acknowledge Dr. Kenneth Norris and James Darling for guiding us in this research. The identification of *Holmesimysis sculpta* by Dr. Edward Brinton is greatly appreciated. Our gratitude is also extended to the staff of the Bamfield Marine Station, especially Sabina Leader, and to the SCUBA divers, Flip Nicklin and Jan Versendahl. We thank J. Baker, K. Kovacs, D. Lavigne, N. Munteanu, P. Reynolds, R. Stewart, B. Webb, and G. Worthy for critically reviewing the manuscript. Graphics were prepared by B. Webb and Illustrative Services, University of Guelph, Ontario.

References

Abbott, I. A., and Hollenberg, G. J. (1976). "Marine Algae of California." Stanford Univ. Press, Stanford, California.

Bogoslovskaya, L., Votrogrov, L. M., and Semenova, T. N. (1981). Feeding habits of the gray whale off Chukotka. *Rep. Int. Whal. Commn.* **31,** 507–510.

Clutter, R. I. (1967). Zonation of nearshore mysids. *Ecology* **48**(2), 200–208.

Darling, J. D. (1977). Aspects of the behaviour and ecology of Vancouver Island gray whales (*Eschrichtius glaucus*) Cope. M.Sc. Thesis, University of Victoria, Victoria, British Columbia.

Holmquist, C. (1979). *Mysis costata* Holmes, 1900, and its relations (Crustacea, Mysidacea). *Zool. Jahrb., Abt. Syst. (Oekol.), Geogr. Biol.* **106,** 471–499.
Hudnall, J. (1981). Population estimates, feeding behavior and food source of gray whales, *Eschrichtius robustus*, occupying the Straits of Juan de Fuca, British Columbia. Fourth Biennial Conference on the Biology of Marine Mammals, December 14–18, 1981, Abstracts, p. 58.
Leatherwood, S., Reeves, R. R., Perrin, W. F., and Evans, W. E. (1982). Whales, dolphins and porpoises of eastern north Pacific and adjacent arctic waters. *NOAA Tech. Rep., NMFS Circ.* **444,** 1–296.
Nemoto, T. (1970). Feeding patterns of baleen whales in the ocean. *In* "Marine Food Chains" (J. H. Steele, ed.), pp. 241–252. Univ. of California Press, Berkeley.
Nerini, M. K., and Oliver, J. S. (1983). Gray whales and the structure of the Bering Sea benthos. *Oecologia* **59,** 224–225.
Norris, K. S., Goodman, R. M., Villa-Ramirez, B., and Hobbs, L. (1977). Behaviour of California gray whale, *Eschrichtius robustus*, in southern Baja California, Mexico. *Fish. Bull.* **75**(1), 159–172.
Pike, G. C. (1962). Migration and feeding of the gray whale (*Eschrichtius gibbosus*). *J. Fish. Res. Board Can.* **19**(5), 815–838.
Ray, C., and Schevill, W. E. (1974). Feeding of a captive grey whale, *Eschrichtius robustus*. *Mar. Fish. Rev.* **36**(4), 31–38.
Rice, D. W., and Wolman, A. A. (1971). The life history and ecology of the gray whale (*Eschrichtius robustus*). *Spec. Pub. Am. Soc. Mammal.* **3,** 1–142.
Scammon, C. M. (1874). "The Marine Mammals of the Northwestern Coast of North America." John H. Carmany & Co., San Francisco, California (reprinted by Dover, New York, 1968).
Sprague, J. G., Miller, N. B., and Sumich, J. L. (1978). Observations of gray whales in Laguna de San Quintin, northwestern Baja California, Mexico. *J. Mammal.* **59,** 425–427.
Swartz, S. L., and Cummings, W. C. (1978). Gray whales (*Eschrichtius robustus*) in Laguna San Ignacio, Baja California, Mexico. *U.S., NTIS, PB REP.* **PB-276319,** 1–38.
Tattersall, W. M. (1951). A review of the Mysidacea of the United States National Museum. *Bull. U.S. Natl. Mus.* **201,** 1–292.
Walker, T. J. (1975). "Whale Primer." Cabrillo Historical Society.
Wilke, F. and Fiscus, C. H. (1961). Gray whale observations. *J. Mammal.* **42**(1), 108–109.
Zimusko, V. V., and Lenskaya, S. A. (1970). Feeding of the gray whale (*Eschrichtius robustus* Erx.) at foraging grounds. *Ekologiya (Sverdlovsk)* **1**(3), 205–212.

20

A Review of Russian Research on the Biology and Commercial Whaling of the Gray Whale

Alexey V. Yablokov and Luydmila S. Bogoslovskaya

Introduction

Scientific study of the gray whale (*Eschrichtius robustus*) in the Asiatic waters of the north Pacific was begun in the eighteenth century by S. P. Krasheninnikov (1755) and was renewed episodically in subsequent years by Ditmar (1890–1900), Vasil'yev (1891), Grebnitskii (1902), and Suvorov (1914). Gray whale research became specialized between 1933 and 1936 during the period of active pelagic whaling in the Bering Sea by the Soviet fleet *Aleut* (Zenkovich, 1934a,b,c, 1935a–d, 1937a,b,c, 1938a,b; Tomilin, 1937a, 1946). The results of research of the whaling industry by Soviet scientists were summa-

ISBN 0-12-389180-9

rized in the work of Sleptsov (1952), Zenkovich (1952), Tomilin (1957, 1962), and Klumov (1963).

Following World War II, gray whale research was principally conducted in association with aboriginal whaling in the waters of Chukotka Peninsula, mostly the opportunistic research of Zimushko (1969a,b, 1970a–d, 1972a,b; Zimushko and Lenskaya, 1970; Zimushko and Ivashin, 1980). In recent years, various aspects of gray whale research have been explored by the scientists of TINRO (Pacific Ocean Institute of Fisheries and Oceanography): Berzin (1974, 1978), Berzin and Kuz'min (1975), Berzin and Rovnin (1966), Berzin and Vladimirov (1981), Blokhin (1979, 1981), Blokhin and Pavlyuchkov (1981), and Doroshenko (1981). Additional research has been conducted by members of the Academy of Sciences of the U.S.S.R. and Magadanrybprom: Votrogov and Bogoslovskaya (1980), Bogoslovskaya and Votrogov (1981), Bogoslovskaya *et al.* (1981a,b, 1982), and Bogoslovskaya and Lemberg (1979), Bogoslovskaya et al. (1981b).

The limited volume of this contribution does not allow us to describe in detail the results of Soviet studies on the gray whale, which have been presented in more than 80 scientific publications, or even to mention all of them. Our main goal was to summarize data that is not well known or is mostly lacking in the English scientific literature.

Okhotsk–Korean Population

According to data compiled by earlier researchers, the whales of this population and the humpback whales (*Megaptera novaeangliae*) were very common and even abundant in the coastal waters of the northern part of the Okhotsk Sea and off the western shores of the Kamchatka Peninsula (von Ditmar, 1890–1900; Krasheninnikov, 1755; Vasil'yev (1891) indicated that the main mass of these whales migrated twice each year along the shores of the Far East Seas. Gray whales penetrated into the Okhotsk Sea from the Japanese Sea through the Tartar Strait and probably through Laperuz Strait (Sleptsov, 1961b); relatively recently they have been observed occasionally in Akademiya and Sakhalin Bays (Sleptsov, 1952).

Berzin (1978) summarized all the scattered data on the Okhotsk–Korean population of the gray whale, their wintering grounds with indications of their southern borders, their feeding grounds, and migration. He also commented on the incidental sightings of this animal in the Okhotsk Sea and ascertained the catastrophic condition of the entire population (Berzin, 1974). Solitary gray whales were encountered in 1967 and 1974 in the coastal waters of the Okhotsk Sea (Berzin and Kuz'min, 1975). The gray whales were observed in Peter the Great Bay: in July 1978 near Vladimir's Bay (one whale) and at the end of April, 1979 near Furugelm Island (two whales) (L. M. Votrogov, personal communication). Aerial surveys of the northwest coastal part of the Okhotsk Sea in the summer of 1979 demonstrated an absence of gray whales in the Sakhalin–Shantar region (Berzin and Vladimirov, 1981). Zoologist M. K. Maminov encountered one gray whale near the southeastern tip of the Kamchatka Peninsula at the beginning of October, 1979. Berzin and Vladimirov speculated that this individual was from the Japanese portion of the

Okhotsk–Korean population and possibly part of an independent population that breeds in the inner Sea of Japan. They believed that this population migrates along the eastern shores of the Japanese Islands where sightings of gray whales were previously recorded (see Chapter 3, this volume). The number of whales in the Okhotsk–Korean population prior to the beginning of the whaling industry for this species was estimated to be 1,500–10,000 animals. Contemporary censuses indicate there are no more than 100–200 animals remaining (Berzin and Vladimirov, 1981). According to Berzin (1978), the size of the Okhotsk–Korean population has diminished to its minimum; they are not recovering and their numbers continue to decline. Therefore, any whaling of this population would be catastrophic to their survival and should be banned.

Chukotka–California Population

MIGRATIONS

The spring migration of the gray whale is extended in time (Sleptsov, 1952, 1961a; Berzin and Rovnin, 1966) and is closely related to the ice condition and direction of the ocean currents (Fedoseev, 1981; Votrogov and Bogoslovskaya, 1980; Bogoslovskaya and Votrogov, 1981). These authors believe that there are several independent migratory routes of the gray whales; in the beginning of May the gray whales appear simultaneously and independently in two different regions divided by the vast area covered with ice, one near the coast of Koryak from Cape Khatyrka to Cape Navarin and the other at the southeast tip of Chukotka Peninsula in the waters between the settlement Sireniki, Cape Chaplino, and Saint Lawrence Island. Apparently animals utilize the powerful streams of the Poperechny current for their migratory movements. These streams deviate on the northwest from the Aleutian Islands and flow toward the permanent coastal polynya (an unfrozen patch of water in the midst of ice) which form in these areas (Votrogov and Bogoslovskaya, 1980). Votrogov and Bogoslovskaya (1980) proposed one more, the southernmost migratory route towards the Asian coast, along the Aleutian range and the Commander Islands. This author's opinion on the existence of possible migratory routes is based on the findings by Grebnitskii (1902) and Barabash–Nikiforov (1938) as well as some other results of recent findings (Votrogov and Bogoslovskaya, 1980).

However, the main core of the gray whale population stock moves north along the shores of North America and usually arrives in the Bering Strait by the end of May. Whales are distributed along the cracks of ice throughout the large water areas free of ice packs; one part of the herd begins its movements southward along the Asiatic coast and another part penetrates into the Chukchi Sea where the whales split off into two opposite directions, west and east, towards the Chukotka Peninsula and Alaska (Tomilin, 1937a,b; Berzin and Rovnin, 1966; Fedoseev, 1981; Votrogov and Bogoslovskaya, 1980).

The massive departure of whales from the northern seas during the fall migration begins in the middle of October (Kuz'min and Berzin, 1975; Doroshenko, 1981) to November of some years (Sleptsov, 1961a), when they share these waters with the bowhead whale (*Balaena mysticetus*). In the region of Sireniki polynya, near leads in the ice, and at the entrance to the shore of Providenya Bay, some gray whales hold out until mid-November and sometimes until mid-December. Local natives indicate the gray whale leaves the Koryak Coast by September (Bogoslovskaya *et al.*, 1982). Most of the migratory routes of the gray whales are through coastal waters, and it remains unclear whether the same animals utilize the same routes during the spring and fall migrations.

DISTRIBUTION ON THE FEEDING GROUNDS

Practically all researchers involved in the study of summer distribution of the gray whales were involved in this problem, including Zenkovich (1934a, 1937a) and Tomilin (1937a, 1957). In Soviet waters, the main concentration of the gray whale can be observed in two different regions: the southern area, from Cape Olyutora to Cape Barykov and the Lagoons of Tymne in the Bering Sea, and a northern area, from the Kresta Bay in the Bering Sea to Cape Sertse-Kamen' and the lagoon of Neshkan in the Chukchi Sea (Arsen'yev, 1961; Sleptsov, 1961a; Berzin and Rovnin, 1966; Zimushko, 1970a; Votrogov and Bogoslovskaya, 1980). Based on the data of Berzin and Rovnin, the gray whales in the Chukchi Sea do not go further west than the lagoon of Nutang'ye and Mayino-Pylgyn or further north than 69° latitude, although Sleptsov (1952) indicated that in some years the gray whales may move along the ice leads and drifting ice as far as Wrangel Island (Ostrova Vrangelya).

Recently, Doroshenko (1981) encountered gray whales for the first time in the coastal waters of the Long Strait and the eastern portion of the Eastern Siberian Sea, from Point Yakan to Point Billings. In addition, he discovered in the Chukchi Sea a large concentration (approximately 2000 whales) of animals about 160 km offshore. This concentration was located between 67°40′ and 68°15′N and 169°40′ and 172°W. The author believes that gray whales have expanded their distribution to the west as a result of the continued increase of the Chukotka–California population.

As a rule, gray whales distribute themselves in shallow waters near the shore and rarely go beyond 35 to 50 km offshore (Zimushko, 1970a; Votrogov and Bogoslovskaya, 1980). They choose to frequent areas where the salinity of the water is somewhat lower than 30%, where the wind turbulence reaches to the bottom, and the oxygen content is near 150% (Kuz'min and Berzin, 1975). They congregate and remain in areas where the concentration of benthic organisms is not less 100 g/m^2 (Berzin and Povnin, 1966). In the summer feeding grounds, the whales are unevenly distributed along the Asian coast (Fig. 1). The whales are constantly moving; their distribution in areas 3-21 varies not only from year to year, but even from month to month within the same season, as a result of constant local migrations of the gray whales between the feeding grounds (Votrogov and Bogoslovskaya, 1980; Bogoslovskaya *et al.*, 1981a, 1982).

One can assume that at the present time the populations of California and Korean gray whales do not overlap (Zenkovich, 1937a; Zimushko, 1972b; Zimushko and Ivashin, 1980); however, the question remains open whether or not these two populations were

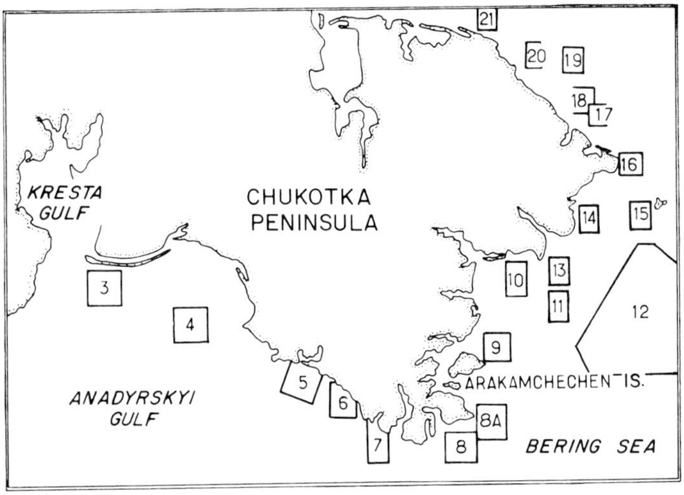

Fig. 1. *Feeding grounds of the Chukotka–California gray whales in Soviet waters, summer to early autumn. From whaling cruises 1964–1980, concentrations of whales have been discovered in areas 3–21 while other regions remain "empty."*

mixed in the past; the data of earlier studies indicated that the gray whales were usually common in the waters around Kamchatka Peninsula and the Commander Islands (Krasheninnikov, 1755; von Ditmar, 1890–1900; Grebnitskii, 1902).

There is a great deal of opinion in the literature that the gray whales of all ages, including both males and females, spend their summer in the Bering Sea. At the same time a great age differentiation of the stocks is also observed: young animals, as a rule, congregate around the Koryak coastline, from the Glubokiy Bay to Cape Navarin, but the adult groupings are in more northern waters, around Chukotka Peninsula (Zenkovich, 1934a, 1937a; Tomilin, 1937a; Sleptsov, 1952). However, working with the results obtained from the whaling industry from 1969 to 1981, as well as from direct observations, predominantly females and young whales distribute themselves mainly in the shallow coastal waters, but the large whales are found further offshore in waters up to 50–60 m deep (Votrogov and Bogoslovskaya, 1980). Moreover, there are four regions where the small animals are predominant in numbers: near Achechen lagoon, and at the mouth of Kurupka River (regions 5 and 6 on Fig. 1), near Cape Kygynin of the Arakamchechen Island (region 9 on Fig. 1), and the west portion of the Mechigmen Bay, including region 10 and the adjacent portion of the Mechigmen Inlet as well as the northern area from Cape Nunyamo, near Cape Chaplino (Bogoslovskaya *et al.*, 1981; Bogoslovskaya *et al.*, 1982). Out of 514 adult gray whales taken from the needs of the aboriginal population of Chukotka Peninsula, 69.8% were females and 30.2% were males (Rezvanov, 1981).

It is important to mention that Zenkovich (1954, 1969) reported two sightings of gray whales around the Chilean–Peruvian coastal waters of South America. This observation may support the hypothesis that some part of the gray whale population may cross the equator in the search for suitable feeding grounds.

ABUNDANCE

In 1942 the Soviet whaling vessels sighted 1197 gray whales within 46 days in the area of Anadyr Gulf and the Bering Strait (Vadivasov, 1947). In August and September of the same year, P. G. Nikulin counted 1154 whales from shore stations in the same regions, although it was known that only in August 1012 animals were observed. According to Vadivasov's data (1947) during the period between 1932 and 1941, whalers would sight every year on the average about 466 gray whales in 46 to 53 days of observations. Apparently, all animals observed belonged to the Chukotka–California population. We want to call attention to observations conducted by Zenkovich (1937a, 1954); on September 3, 1936 he encountered in the Chukchi Sea a herd of 200 gray whales, which formed five or seven different groups. Later, in one of the fijords of the Koryak coastline he counted more than 60 gray whales cleaning themselves of their ectoparasites in the brackish water.

Sleptsov (1961a) believed that by 1946 there were not less than 1500 gray whales in the Chukchi Sea and in the northwestern part of the Bering Sea. In this very same region, the author observed about 1200 gray whales from August 21 until September 11 of 1948 while aboard the whaling vessel *Purga*. He believes that the population of gray whales doubled by the beginning of 1960 and that their numbers reached 2500–3000 individuals. In 1962, 1033 gray whales were sighted in the coastal waters of the Chukotka Peninsula (Berzin and Rovnin, 1966).

V. V. Zimushko conducted three aerial surveys of the Chukotka–California population of gray whales. From July 10 to July 30, 1968, he observed 124 whales in 107,900 km^2 with an average distance from the Koryak shore and Chukotka Peninsula of 35 km; from these counts the population was estimated to be 4800–5000 whales (Zimushko, 1970a). During 159 hours of aerial surveys between August 3 and 23 and October 17 to 27 they surveyed an area of 215,932 km^2 within 240 km offshore and calculated a population estimate of 7700–7800 whales. From September 24 to 25 and October 1–9, 1975, V. V. Zimushko counted only 2000 gray whales during 89 hours of aerial surveys over 162,000 km^2 as much as 250 km from shore. The author correlated these counts with the migration of the whales during this period (Zimushko and Ivashin, 1980).

Zimushko's last population estimates agreed with the recent counts conducted during scientific expeditions in September and October, 1974, when the total number of gray whales estimated in Soviet waters was 1800–2000 (Kuz'min and Berzin, 1975). In the spring of 1978, Fedoseev (1981) conducted aerial surveys from an elevation of 100 m at different locations along the Koryak coast. He counted 30 gray whales, which, when extrapolated from these aerial surveys, suggested that not less than 1200 gray whales occupied the areas in the Bering Sea clear of ice at the end of May.

Surveying from the vessel *Zvezdnyi*, I. M. Votrogov counted 2200 gray whales between the end of June and the beginning of November, 1980 and 1438 whales from the second half of August until November 9, 1981 (Bogoslovskaya *et al.*, 1982).

Comparison of all these data obtained during aerial and vessel surveys allowed us to assume that from the time of the complete abolishing of commercial boat whaling, that is, by 1946, the numbers of the Chukotka–California population were not so low as was suggested by the majority of researchers, and accordingly, the increase of numbers of animals within the period of 1940 to 1960 was not as significant as believed. For example, Blokhin (1979) estimated a rate of increase of 18% per year.

MORPHOLOGY

The average measurements of the gray whale body are summarized in Table I. In spite of some scatter in the data, we may assume that on the average an adult female whale is approximately 0.3 m longer than a male.

Zimushko (1972b) compared pecularities of the external morphology of Chukotka–California and Okhotsk–Korean (data by Andrews, 1914) populations of gray whales. This comparison raises some doubts in the previous conclusion that the whales of the Okhotsk–Korean population are generally larger than individuals of the Chukotka–California population. On the other hand, the distance from the tip of the rostrum to the base of the pectoral fins, from the tip of the rostrum to the eye, the maximum width of the pectoral fin, and the length of the baleen plates were statistically greater in the Okhotsk–Korea whales than similar measurements in Chukotka–California whales. Also, the whales from the waters of Chukotka Peninsula have smaller numbers of baleen plates and on the average fewer throat grooves. The authors at present believe that these differences may indicate the existence of two distinctive groups of gray whales in the northern Pacific, which may allow them to be designated as subspecies.

Zimushko (1972b) revealed that male gray whales differed from females in the relative size of the body and head. In addition, females have significantly longer tail sections in relation to the smaller width of the flukes. The average number of baleen plates in both sexes was practically the same: 323 ± SE 4.9 in females ($n = 111$) and 322 ± SE 3.4 in males ($n = 70$).

The vibrissae of the gray whale embryos are distributed almost equally on upper

Table I
Average Length of Gray Whales Taken in the North Pacific[a]

Males (m)	Females (m)	Comments
11.3 (8.5–14.3) $n = 88$	11.9 (8.6–15.0) $n = 104$	Pelagic whaling in the Bering Sea, 1933–1936
11.5 (8.2–13.0) $n = 39$	12.2 (7.7–14.2) $n = 55$	Coastal (in-shore) whaling off the Chukotka Peninsula, 1965
12.2 (9.3–14.1) $n = 26$	12.2 (8.8–13.8) $n = 15$	Same as above, 1966
12.6 ± SE 0.09 $n = 70$	12.9 ± SE 0.07 $n = 111$	Same as above, 1965–1971

[a]From Zenkovich (1937a) and Zimushko (1969b, 1972a).

Table II
Average Number of Vibrissae on the Head of the Fetus of the Gray Whale[a]

	Upper jaw	Lower jaw
Males	131	141 (117–165)
Females	155	131 (100–172)

[a]From Zimushko (1972b) and Zimushko and Ivashin (1980).

and lower jaws (Table II). It was learned that the total number of vibrissae is slightly less in near-term fetuses, and their number is significantly less on the upper jaw of the adult whales. Thus, Tomilin (1937c) counted just 60 vibrissae on the upper jaw and 120 vibrissae on the lower jaw of an adult female.

Bogoslovskaya and Lemberg (1979) showed that the structure of the gray whale brain is typical of that of other baleen whales; except for the more simplified picture of the arched convolutions, the brain of the gray whale resembles the brain of the sei whale (*Balaenoptera borealis*).

Zimushko (1970d) examined the reproductive system of the gray whale, including 70 ovaries collected during 1965–1968 in the waters of Chukotka Peninsula. He describes in detail the morphology of the ovaries (ovaries of sexually mature females weighed from 318 to 1830 g, but together with the corpora lutea were up to 4820 g) and also the stages of development of follicles and the corpora lutea. Zimushko concluded

Table III
Dimensions of Some Internal Organs of the Gray Whale[a]

Organ	Size	Comments
Heart	131 ± 16 SE (90–166) kg	Sexually mature males and females
	172 kg	Pregnant female 13.4 m
Liver	519 (455–580) kg	Sexually mature males and females
Spleen	3.2 (2.5–4.3) kg	Same
Lungs	333 kg	Same
Testes	23 (14–27) kg	Sexually mature
	50–74 cm length	
	43 kg	Both testes, 12.1 m
	54 kg	Both testes, 13.6 m
Brain	4376 kg	Pregnant female, 13.4 m
	4200 kg	Unknown sex, 11.0 m
	4300 kg	Unknown sex, 12.0 m
Intestine (total length)	177.3 m	Pregnant female, 13.4 m
Urinary bladder (maximum volume)	12.0 liters	Adult male and female

[a]From Bogoslovskaya and Lemberg (1979), Zenkovich (1937b), Zimushko (1972b), and Zimushko and Ivashin (1980).

that the atrophied follicles remain in the ovary for a relatively short time compared to the traces of the corpora lutea and scars from ovulation; between these, there is a distinct histological difference.

There is only one known instance of the complete weighing of a gray whale by Soviet researchers (Zenkovich, 1937b). This pregnant female, carrying a fetus 207 cm long and weighing 126 kg, was caught on August 19, 1936 in the Bering Sea. The female's length was 13.35 m and she weighed 31,460 kg not including the weight of the blood; the blubber weighed 9,100 kg; the muscles, 6,729 kg; the head, 1,692 kg; the lower jaw, 1,265 kg; the tongue, 1,278 kg; the vertebral column plus the ribs and scapulae (with remnants of the muscles), 6,163 kg; the pectoral flippers, 612 kg; and the internal organs, 4,493 kg. Zenkovich concluded that the approximate weight of a gray whale may be determined by $V = (L D^2)/3$ where L = the body length and D = its maximum height. Some measurements of the internal organs of the gray whale are presented in Table III.

In conclusion, on the morphological peculiarities of the gray whale we wish to note that according to Tomilin's data (1937a, 1957) the depigmented white spots, some of which develop from the attachment of parasitic barnacles (*Cryptolepas rhachianecti*) can be repigmented following the removal of the barnacles. As this process continues on different parts of the body at different times, the general color of the whale's skin is mottled with some dark and some light areas. In 1957 Tomilin reported sighting a partially albino gray whale in the Bering Sea. The anterior part of the body of this whale was completely white.

GROWTH AND AGE

Among the gray whales studied by the Soviet researchers in the waters of the North Pacific, they came across whales ranging from 7.7 to 14.3 m long. After comparing whale

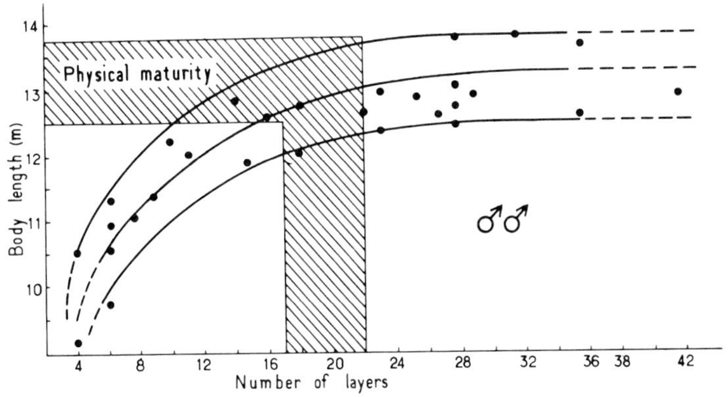

Fig. 2. *Correlation between the body length and the number of ear plug layers (1 layer = 1 year) of male gray whales (n = 29), which allows the determination of the age of physical maturity. [Data by Zimushko (1970b) and Zimushko and Ivashin (1980).] Lines are theoretical curves fit to the mean values (middle line), largest values (upper line), and smallest values (lower line). Hatched area indicates the range of absolute values for body length and number of ear plug layers for the onset of physical maturity.*

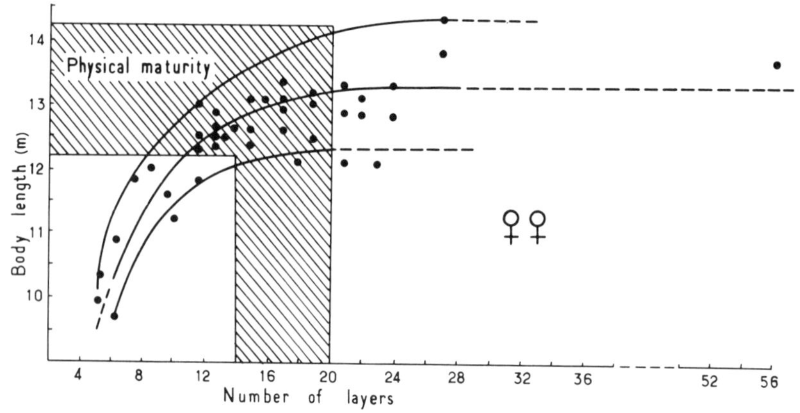

Fig. 3. *Correlation between the body length and the number of ear plug layers (1 layer = 1 year) of female gray whales (n = 41), which allows the determination of the age of physical maturity. [Data by Zimushko (1970b) and Zimushko and Ivashin (1980).] See Fig. 2 for symbol designation.*

body size to the number of layers in the wax ear plugs of whales (one layer comprises one light and one dark lamina), Zimushko (1970b) concluded there is no significant difference in the rate of growth between male and female whales under 12.2 m; the average body length of males and females at the age of 1 year is 8.9 meters (range 8–9.5 m), and at 2 years it is 10.4 m (range 9.5–10.7 m). Further growth of both males and females is slower (Fig. 2), without considering the slightly larger dimensions of the females (sexually mature females on the average are larger than males by 0.3–0.4 m). The 3-year-old whales average about 11 m (range 10.7–11.6 m).

The analysis of relatively large amounts of data, obtained by Zimushko in the Chukotka Peninsula coastal waters in 1965–1976 and summarized in Figs. 2 and 3, allow us to estimate the age of physical maturity of the gray whale. It is determined by the cessation of the growth rate of the body and occurs in males when they reach an average length of 13.2 m; in the case of some smaller animals, it occurs when they reach a length of 12.5 m. This occurs when the animals reach an average age of 19 years (range, 17–22 years), when one considers that each layer of the ear plug is equivalent to 1 year of a whale's life. In females, physical maturity on the average occurs at a body size of 13.2 m, but apparently with a greater range among individuals (the smallest mature female was 12.2 m in length). The data indicated that the age of attaining physical maturity for females encompasses a range of 14 to 20 years, averaging 17 years. We want to add that the maximum number of layers in the ear plug of the gray whale was shown by Zimushko (1970b) to be 56 layers for a female 13.8 m long and 42 layers for a male of 12.9 m.

The data on the length of gray whale embryos and young whales, when presented as a circular graph, demonstrate the range of the size during embryonic growth and during the first year of life (Fig. 4).

Measurements conducted by Zimushko (1972b) on 12 skulls of the gray whale showed that with age the length of the rostrum increased relative to the length of the lower jaw.

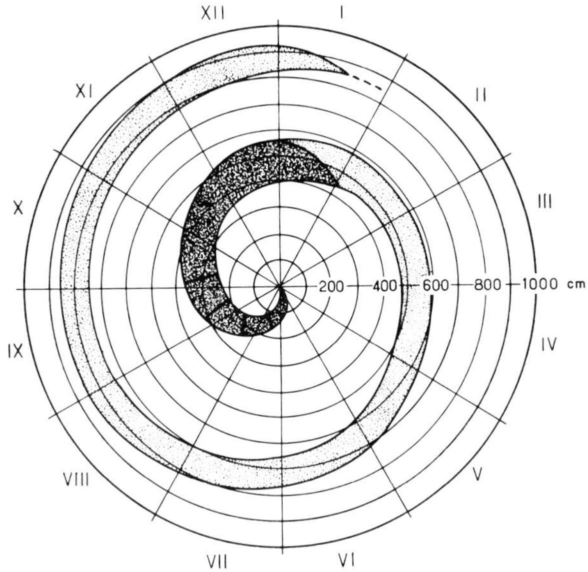

Fig. 4. *Growth of gray whale embryos and neonates according to combined data by Tomilin, 1937c; Zenkovich, 1937a; Rice and Wolman, 1971; and Zimushko and Ivashin, 1980. HEAVY STIPPLING indicates embryo length during pregnancy; embryos 25–140 mm long (not shown) can be found in whales taken in the second half of February (Rice and Wolman, 1971). LIGHT STIPPLING shows observed dimensions of body length during the first year of life. Roman numerals are months from Jan.–Dec.; Arabic numerals are body length in centimeters.*

REPRODUCTION

Attainment of sexual maturity occurs at different ages in female and male gray whales. Sexual maturity in females was determined by the presence of an embryo, corpus luteum, corpus albicans, or evidence of ovulation. Age of sexual maturity of females varied from 8 to 12 years (8–11 layers in the earplugs) when they reached a length of approximately 12 m (Zimushko, 1969b; Zimushko and Ivashin, 1980). The smallest female with a corpus of pregnancy was 11.3 m long although one 12.5 m female had no corpora. The largest number of corpora (11) was found in a 13.2 m female, and the average number of corpora for mature females was 5.1 (Blokhin, 1982). Males aged 6 to 7 years were found to be either immature or mature whereas those males aged 8 years (not less than 8 layers in the earplugs) with lengths of 11.5 m were usually mature (Zimushko, 1969b; Zimushko and Ivashin, 1980). There was a considerable increase in size and weight of the testes when the body length reached almost 11 m (Zimushko, 1969a). The investigation of mature male whales collected from July to October revealed that males had no spermatozoids in their epididymides, indicating a marked seasonality of the mating condition in male gray whales (Zimushko, 1969a).

The sexual cycle in female gray whales is believed to last 2 years and includes copulation, pregnancy, lactation, and a resting period after reproduction. The duration of pregnancy is nearly a year (Tomilin, 1957). Bogoslovskaya *et al.* (1981) found that pregnant females are the first to arrive off Chukotka and they are the first to leave, which corresponds with the findings of Rice and Wolman (1971) that pregnant females were the first to begin the northerly spring migration to the summer range, and they were the first to arrive at the breeding grounds after the fall southerly migration. This migratory timetable is supported by the findings of Blokhin (1982) who reported that the percentage of pregnant females in the 1980 catches was highest in the summer (July) and steadily declined during the 4-month fishing season and was lowest in autumn (October).

Analysis of the correlation between the number of corpora lutea and ovulation in females of reproductive age showed that the average biennial rate of ovulation in physically mature females was 1.4 (Blokhin, 1981). It also was shown that the first ovulation in the young females results in pregnancy, since there were some females whose ovaries showed some signs of pregnancies but no signs of ovulation. On the other hand, there were ovaries that indicated multiple ovulations, but no signs of pregnancy. The overall ratio of pregnant to nonpregnant female whales was 1:1 and supports the 2-year reproductive cycle proposed by Rice and Wolman (1971). On the average, according to Zimushko (1972a) and Zimushko and Ivashin (1980), in females 12.0–12.5 m long the ratio between signs of ovulation and signs of pregnancy was 1:1; in females 13.0–13.5 m long the ratio was 2.5:1. If this finding can be confirmed in the future it may indicate that with maturity (age) the intervals between pregnancies increase.

It is very important to note that some lactating females could also be pregnant at the same time. Zimushko (1969b) encountered seven such females. The occurrence of females simultaneously pregnant and lactating may serve as an indirect confirmation that some females give birth yearly. Determining the exact number of these females is difficult because of the ban on whaling females with calves; collection of this material can only be accidental. However, during observations in the waters of Chukotka Peninsula from 1965 to 1968 about 20% of such females were encountered; this may indicate a relatively widely distributed phenomenon of simultaneous lactation and pregnancy among female gray whales.

In his work, Zenkovich (1938a) described the composition of the milk in lactating gray whale females taken in September (the time when lactation has usually ceased). The milk had a fat content of 53%, a water content of 40.6%, and a dry content of 6.4%. Based on Tomilin's (1946) calculation, during the period of nursing the calf consumes 36.6 kg of milk per day on the average; for the total lactation period this would be approximately 5000 kg of milk. In July to September 1965 to 1970 some nonpregnant females were found to be lactating and others had recently finished lactating (Zimushko and Ivashin, 1980). Blokhin (1982) reports that many of the nonpregnant females taken in October 1980 had large corpora and well developed mammaries, indicating that they had recently weaned their calves. From these data it appears that lactation lasts approximately 7 months and that gray whale calves are weaned when their body length is about 7 m (Zimushko and Ivashin, 1980). Examination of the stomachs of whales from 7.7 to 9.0 m long showed that they were feeding on amphipods. Ascidians, polychaetes, algae, and mud were also found in their stomachs (Zimushko and Ivashin, 1980).

FEEDING

Zenkovich (1934c, 1937a,b) and Tomilin (1937a) were the first researchers who determined that along the Asian coast the main food source of the California gray whales is benthic amphipods; while in the Bering Strait and the Chukchi Sea, they feed predominantly on *Ampelisca macrocephala* but near the Koryak shore they feed on *Pontoporeia* sp. Further study of gray whale feeding by Sleptsov (1952) determined the maximum volume of the stomach of the gray whale was 350 liters; Klumov (1963) compiled the first list of the food species of the gray whale which consisted of 17 different species of invertebrates.

Furthermore, a detailed study of the differences in the aspects of feeding by gray whales off the coast of Chukotka Peninsula was done by Zimushko and Lenskaya (1970) from the analysis of 70 stomach-content samples of gray whales between 1965 and 1969. Blokhin (1981) and Blokhin and Pavlynchkov (1981, 1983) collected stomach samples from 38 whales and Bogoslovskaya and Votrogov (1981, 1982) also sampled 230 stomach contents from whales taken between 1979 and 1980 by the whaling vessel *Zvezdnyi.*

The list of the prey species of the gray whale, based on the data from all authors, consists of 60 species of amphipods and 80–90 other species of invertebrates, as well as algae and significant amounts of sand or pebbles which are present in the stomachs of these whales at all times. According to data by Bogoslovskaya *et al.* (1981a, 1982) the primary prey species of gray whales are 18 species of amphipods from the 12 families of the suborder Gammaridae. No age or sex difference of these prey species has been determined (Zimushko and Lenskaya, 1970; Blokhin and Pavlynchkov, 1981, 1983), but there is a clear difference between the southern (from the Island of Kosa Meechken up to the Providenya Bay in the Bering Sea) and the northern (from the Cape Nunyamo in the Bering Sea up to the Cape Serdtse-Kamen' in the Chukchi Sea) regions of feeding according to the dominant species of amphipods (Fig. 1) (Bogoslovskaya *et al.*, 1981a, 1982). The same authors showed that the aquatic region between Providenya Bay and Cape Nunyamo is characterized mostly by the variety of composition of invertebrates; *Pontoporeia femorata* as a dominating species in the southern regions is gradually replaced by *Ampelisca macrocephala* in the northern regions.

Gray whales congregate in "feeding spots," regions where amphipods accumulate. These areas probably provide a main food resource for other marine mammals and other species as well. For instance, in the aquatic region near the walrus (*Odobenus rosmarus*) hauling-out ground on the Arakamchechen Island gray whales and walruses graze in close association (Bogoslovskaya *et al.*, 1981b, 1982).

Tomilin (1946) calculated that a single fully grown gray whale would consume between 379 and 2496 kg of benthic organisms within a 24-hr period. Recently Zimushko and Lenskaya (1970) determined a 24-hr feeding bout would result in the ingestion of 1200 kg of the food species, 300 kg of which would be freshly ingested. These same authors came to the conclusion that the Chukotka–California population would consume 850,000 tons of food species on their feeding grounds during one season. According to the type of feeding, Tomilin (1954) separated gray whales into absolutely specific groups of benthic foragers and provided a description of morphological characteristics indicative of their benthic feeding habits (head, oral cavity, filtering apparatus, etc.).

GROUPING AND SOCIAL BEHAVIOR

Many aspects of the behavior of whales are touched upon in the works of Zenkovich (1934a, 1937a, 1954), Tomilin (1937a,b,c, 1954, 1957), Sleptsov (1952, 1961a), Zimushko and Ivashin (1980), Bogoslovskaya *et al.* (1981a, 1982), and Bogoslovskaya and Votrogov (1981). These investigators reported that gray whales feed in small groups of 2–4 whales, or more often as single whales (commonly pregnant females). It is not unusual for a herd to consist of 30–40 animals and sometimes 100–400. During the times of such large congregations, lasting from 1–4 days, strange and unexplained behavior is observed which is presumed to be connected with courting and mating behavior, as has been suggested by Tomilin (1937a) and the whalers of the Chukchi Sea.

Zimushko stated (Zimushko and Ivashin, 1980) that in 1973, of 232 whales encountered 44% were single, 24% were in pairs and 13% were in threes; 13 groups of 4 and 12 groups of 5 were also sighted. Groups of 7 and 8 animals were encountered three times each; the largest congregation included 24 whales. In 1980 Bogoslovskaya and Votrogov (1982) encountered 2447 whales of which singles were 47%; pairs, 31%; threes, 13%; and other groupings 9%; correspondingly in 1981 from 1595 whales, singles included 65%, pairs, 25%, threes, 6%; and others, 4%. These observations were conducted from whaling vessels, small motor boats, and Eskimo kayaks.

Among the groupings, the pairs are the most frequent associations. Adult animals are very often strongly attracted to each other during grazing and migrating when they are constantly together and very close; they observe the same breathing regime, surfacing synchronously or with a slight difference in breathing rhythm. The adult animals help each other, often reaching self-sacrificing measures; also there is a mutual attachment between paired females as strong as between male and female (Bogoslovskaya *et al.*, 1982). Groups of larger than three animals are usually unstable. Sex composition of the groups varies: there are different types of pairs, female and male, two females, and two males. Trios include two females and one male (or vice versa), and in the groups of five animals there were two males (Zimushko and Ivashin, 1980; Bogoslovskaya *et al.*, 1982).

It is important to describe the relationship between the mother and the calf. Nursing apparently ceases when the calf reaches a length of 8.5 m, but 9.6-m juveniles were observed swimming with their presumed mothers although they (the juveniles) had been feeding on benthic organisms (Tomilin, 1937a). Bogoslovskaya et al. (1982) assumed that the majority of the young whales separated from their mothers in July to August and congregated in certain areas, usually shallow waters and lagoons. Zenkovich (1937a) pointed out an example when the mother whale attacked the whale boat of the *Uelen* whalers after they wounded her calf. The same authors indicated that when the mother whale is killed, the juvenile whale usually leaves; therefore, in the gray whales at this stage in the calf's life, there is a one-sided attachment of the mother to the calf.

Tomilin (1937a,b) and Zenkovich (1937a,b, 1954) described in detail [after von

Ditmar (1890–1900) and apparently Krasheninnikov (1755)] a specific migration of whales with heavy infestations of skin parasites into the freshwater lagoons and shallow waters off the Koryak coast. The whales entered the lagoons, and dozens of them filled the shallow lakes connected with the sea. Some whales were just lying on the sand bars; with the tide they would move out to sea, and begin to feed as if nothing had happened. Some whales would lie immobile at the surface not paying attention to the approaching boats or even prods of the oars, as if they were sleeping (Zenkovich, 1954).

According to the observations of the Eskimos of Naukan the gray whale is capable of getting off the sand bars if next to a deep channel. He then arches, pushing with the rostrum and tail into the ground, which is possible because of the free neck vertebrae (Sleptsov, 1952).

Many whalers consider the gray whale the most intelligent of all large whales. He knows how to leave when pursued, changing his course drastically or surfacing at the same place where he dove, zig-zagging, entering shallow waters to escape, or surfacing without blowing although revealing only their blowholes (Zenkovich, 1934a, 1954). While being pursued, the gray whales increase their usual speed from 3–4 knots to 7–8 knots (Tomilin, 1937c). They can hold on in the unfrozen patches of water among the pack ice and shallow ice fields and at the very edge of the Arctic ice (Sleptsov, 1961b; L. S. Bogoslovskaya and T. N. Votrogov, unpublished notes). Gray whales coexist peacefully with the other mammals inhabiting the Chukchi Sea. On the feeding grounds they are found with the humpback whales and walruses. For example, in region 9 near the largest hauling ground of the walrus on Arakamchechen Island, whales and walrus graze in very close proximity to each other. Observers never recorded any aggressive behavior between these two species (Bogoslovskaya *et al.*, 1981b). Some authors indicated that the gray whales feeding near the entrance to Providenya Bay do not react to or feel threatened by the passing of nearby boats. They ceased to feed and demonstrated evasive behavior after persistent pursuit by vessels, however.

The behavior of gray whales suggests a plasticity or flexibility which allows them to utilize successfully the entire variety of coastal resources, from subtropical to polar regions.

POSITION OF THE SPECIES IN THE ECOSYSTEM

The position of the gray whale in the North Pacific ecosystem is determined by its interrelationships with food organisms (see above), parasites, enemies, and commensals. Some of these relationships were studied by Soviet researchers for the Chukotka–California population.

Practically all whales observed in the wild or killed by commercial whaling are covered with parasites such as *Cryptolepas rhachianecti* and whale lice (*Cyamus scammoni, C. ceti, and C. kessleri*) (Tomilin, 1937b; Zenkovich, 1935a; Zimushko and Ivashin, 1980). All authors noticed that the majority of the parasites accumulated on the head, around the blowhole, on the jaws, in the regions of the anal and genital orifices, and on the fins. Gray whales enter (brackish) water in coastal lagoons and the mouths of rivers to get rid of these parasites. Apparently this cleaning procedure is very successful. In

1978, a great number of of large black whales (body length between 12.5 and 13.5 m) approached the Chukotka Peninsula coast; all of them were free of parasites (Votrogov and Bogoslovskaya, 1980).

Soviet helminthologists found new species of endoparasites of the gray whale, which included helminths, acanthocephalans, and trematodes (Treshchev *et al.*, 1969; Zimushko and Ivashin, 1980). Most of these helminths were found only in the intestines of the gray whale. Of all the whales studied by Zimushko, 68.3% were infested by the cestode *Priapocephalus eschrichtii*, 58.7% by the trematode *Ogmogaster pentalineatus* (up to 1000 organisms in one whale), and by the acanthocephalans *Corynosoma semerne* and *C. septentrionalis* (29.9%), *C. validum* (24.4%), and *C. strumosum* (9.7%); the trematodes *Orthosplanchus pigmaeus* and *O. plicatus* were found only once. There is a similarity between the helminth fauna of the gray whales and pinnipeds, especially between the walrus and the bearded seal (*Erignathus barbatus*) who also feed on benthic organisms.

Apparently the gray whale has no other enemies than the killer whale (*Orcinus orca*), which pursue the gray whale and attack by grasping them by the pectoral flippers and the flukes and trying to open the whales' mouths to bite into the tongue (Zenkovich, 1934a, 1954). Gray whales escape from killer whales by swimming into shallow water, often inside the surf zone (Zenkovich, 1934a, 1954).

Many researchers commented on the strong correlation between feeding gray whales and large congregations of sea birds, including *Fulmarus glacialis, Larus hyperboreus, Rissa tridactyla, Aethia pusilla, Fratercula corniculata*, and *Puffinus tenuirostris*, which often landed on the whales' backs before they dove (Tomilin, 1937b; Zenkovich, 1954; Bogoslovskaya *et al.*, 1981a). Apparently whale feeding is responsible for bringing to the surface benthic invertebrates which then are easy prey for these birds.

WHALING

From ancient times, the gray whale was the object of aboriginal hunting by the native inhabitants of Kamchatka, the Koryak region, and the Chukotka Peninsula (von Ditmar, 1890–1900; Krasheninnikov, 1755). The most intensive whaling of young gray whales was common on the coast of Mechigmensky Bay, as indicated by many hundreds of the skulls of these animals that were discovered in the ruins of the ancient settlements (Bogoslovskaya *et al.*, 1981b; Bogoslovskaya and Votrogov, 1982a; Chapter 5, this volume). Species identification of the skulls was conducted by L. S. Bogoslovskaya with the assistance of A. G. Tomilin. The history and specific characteristics of the different aboriginal whaling tribes was described by I. I. Krupnik (Chapter 5, this volume).

From 1932 until the complete ban on whaling for gray whales in 1946, the Soviet whalers conducted the hunt on the Chukotka–California population in the Bering and Chukchi Seas (Zenkovich, 1947; Vadivasov, 1947). Whales from the Okhotsk–Korean population were hunted very sporadically during World War II in the region of Peter the Great Bay (Sleptsov, 1961b). In 1969, the aboriginal hunt of the gray whale ceased as a result of a large number of "struck-and-lost" whales. Aboriginal whaling in the Chukotka Peninsula was replaced by the modern whaling vessel *Zvezdnyi*. Statistics of aboriginal and vessel catches of gray whales from 1932 to 1980 are shown in Table IV.

Table IV
Whaling Statistics for the Gray Whale from the Chukotka–California Population[a]

Year	Number taken	Year	Number taken
1932[b]	—	1957	56
1933	2	1958	145
1934	54	1959	187
1935	34	1960	156
1936	102	1961	207
1937	14	1962	147
1938	54	1963	179
1939	29	1964	188
1940	47	1965	175
1941	57	1966	194
1942	101	1967	125
1943	99	1968	135
1944	—	1969[d]	199
1945	30	1970	146
1946	22	1971	150
1947	1	1972	181
1948[c]	19	1973	173
1949	26	1974	181
1950	10	1975	171
1951	12	1976	163
1952	42	1977	186
1953	37	1978	182
1954	36	1979	178
1955	59	1980	179
1956	121		

[a]From Zenkovich (1947), Sleptsov (1952), Ivashin *et al.*, (1972), Ivashin and Mineev (1978), and Ivashin and Mineev (1981).

[b]From 1932 until 1946; data is only from whaling groups.

[c]From 1948 until 1968; data only on aboriginal whaling from small boats.

[d]Catch data from the whaling vessel *Zvezdnyi*.

PROTECTION

All Soviet researchers recognize the critical condition of the Okhotsk–Korean population of the gray whale, which they feel is on the brink of extinction. The status of the Chukotka–California population is satisfactory; however, its existence is threatened by disturbance in the breeding grounds, by pollution along the migration route, and by extensive fisheries along the migratory route and the summer and fall feeding grounds (Bogoslovskaya *et al.*, 1982). Thus, beginning in 1974 the whaling vessel *Zvezdnyi* took one or two whales each year that bore marks of civilization on their bodies; remnants of synthetic ropes, fishing nets, etc. (L. M. Votrogov and L. S. Bogoslovskaya, unpublished notes). Fishing nets that became attached to the peduncles of the whales were probably acquired during the migration along the North American coast, and from their condition these wounds appeared to be 3–5 months old. We wish to emphasize that all the whales seen with fishing gear remnants on their bodies were adult, large animals, which were able to free themselves from the gear. Young animals and yearling calves are probably not as successful.

In conclusion, we wish to call attention to the intensive utilization by man of the natural resources in Alaska and Far Northeastern Asia. The utilization of natural resources along the coastal shelf and coastline jeopardizes the benthic communities and marine fauna in which the gray whale is playing a very visible role. In this regard, it has been proposed that a portion of the Bering Strait, which is important to gray whales as a feeding area, should be set aside as a protected area (Bogoslovskaya *et al.*, 1982).

Conclusions

The present short survey could not include many current observations of the biology of the gray whale contained in numerous Russian research publications. We have not included here any statistical data of the whaling industry, most of which has been made available due to the International Whaling Commission. The authors tried to illuminate those areas of research on gray whales that are known to a lesser degree in the English-speaking scientific community.

Summary

This article provides information on observations of the Okhotsk–Korean population of gray whales. Also, fundamental results of studies of the migration, summer–fall distribution, abundance, morphology, growth, reproduction, feeding, parasites, and behavior of the Chukotka–California population of the gray whale are summarized. Position of the gray whale in the ecosystem of the North Pacific is considered, as are problems associated with commercial whaling and protective measures of these two populations of this species.

Acknowledgments

The authors gratefully acknowledge the help given to us during the compiling of this review. We wish to thank A. A. Berzin for helping to prepare the bibliography of the majority of Russian research on the gray whale, as well as L. M. Votrogov who provided his field notes.

The editors wish to express their thanks to Elena Baskin and Vladimir Gurevich for their assistance in translating this chapter.

References

Andrews, R. C. (1914). Monographs of the Pacific cetacea. I. The California gray whale *Rhachianectes glaucus* Cope), *Mem. Am. Mus. Nat. Hist.* [N. S.] **I,** 227–287.

Arsen'yev, V. A. (1961). The whales distribution in the Bering Sea and the perspective of the whaling development. *Tr. Soveshch. Ikhtiol. Kom.* **12,** 112–124.

Barabash-Nikiforov, T. (1938). Mammals of the Commander Islands and the surrounding sea. *J. Mammal.* **19,** 423–429.

Berzin, A. A. (1974). Problems of studying whales (Cetacea). *In* "Zoology of Vertebrates. Results of Science and Technology," Vol. 6, pp. 158–189. VINITI Acad. Sci. U.S.S.R.

Berzin, A. A. (1978). Distribution and number of the protected whales in the Pacific ocean. *Biol. Sea* **4,** 22–29.

Berzin, A. A., and Kuz'min, A. A. (1975). The gray and the bowhead whales of the Okhotsk sea. *In* "Biological Resources of the Far East Seas, Abstracts of the All-Union Conference on Marine Mammals," pp. 30–32. Vladivostok.

Berzin, A. A., and Rovnin, A. A. (1966). Distribution and migration of whales in the north-western Pacific, the Bering and Chukchi Seas. *Izv. Tikhookean' Nauchno-Issled. Inst. Rybn. Khoz. Okeanogr.* **58,** 179–209.

Berzin, A. A., and Vladimirov, V. L. (1981). Changes in the abundance of whalebone whales in the Pacific and Antarctic since the cessation of their exploitation. *Rep. Int. Whal. Commn.* **31,** 495–498.

Blokhin, S. A. (1979). On the status of the gray whale stocks in the coastal waters of the Chukotka Peninsula. *Rep. Int. Whal. Commn.* **29,** 335–336.

Blokhin, S. A. (1981). Results of the study of the gray whale of the California-Chukotka population in the 1980s. *In* "Scientific Study of Marine Mammals of the Northern Part of the Pacific Ocean in 1980/1981," Proj. 02-05-61, pp. 78–87. VNIRO, Moscow.

Blokhin, S. A., and Pavlyuchkov, V. A. (1981). Feeding of the gray whales of the California-Chukotka population in the waters of the Chukotka Peninsula in 1980. *In* "Scientific Study of Marine Mammals in the Northern Part of the Pacific Ocean in 1980/1981," Proj. 02-05-61. pp. 88–98. VNIRO, Moscow.

Blokhin, S. A., and Pavlyuchov, V. A. (1983). Feeding of gray whales off Chukotka. *Rep. Intl. Whal. Commn.* **33,** 549–552.

Bogoslovskaya, L. S., and Lemberg, A. A. (1979). Morphology of cerebral brain of the gray whale (*Eschrichtius robustus*). *All-Union Conf. Status Prospective Dev. Morphol.* pp. 238–239.

Bogoslovskaya, L. S., and Votrogov, L. M. (1981). Mass wintering of the birds and whales in polynyas of the Bering Sea. *Priroda* **1,** 42–43.

Bogoslovskaya, L. S., Votrogov, L. M., and Semenova, T. N. (1981a). Feeding habits of the gray whale off Chukotka. *Rep. Int. Whal. Commun.* **31,** 507–510.

Bogoslovskaya, L. S., Chlenov, M. A., and Yurtsev, B. A. (1981b). Islands of the Senyavin Strait, unique natural cultural complex. *Priroda* **9,** 91–97.

Bogoslovskaya, L. S., Votrogov, L. M., and Semenova, T. N. (1982). Distribution and feeding of gray whale in the waters of the Chukotka Peninsula in summer and autumn of 1980. *Rep. Int. Whal. Commn.* **32,** SC/33/PS15.

Doroshenko, N. V. (1981). Summary on the study of Cetacea during the expedition on the whaling ship "Razyashchiy" into the Bering, Chukchi, and Eastern-Siberian Seas. *In* "Scientific Study on Marine Mammals in the Northern Part of the Pacific Ocean in 1980/1981," Proj. 02-05-61. pp. 13–16. VNIRO, Moscow.

Fedoseev, G. A. (1981). Aerial survey of the distribution and number of the ice forms of seals, walruses and migration of the whales in the ice packs of the Bering Sea in the spring of 1979. *In* "Scientific Study on Marine Mammals in the Northern Part of the Pacific Ocean in 1980/1981," Proj. 02-05-61. VNIRO, Moscow.

Grebnitskii, V. A. (1902). "Komandorskyi Ostrova (The Commander Islands)." Izd. Dept. Zemled., St. Petersburg.

Ivashin, M. V., and Mineev, V. N. (1978). To the problem of the stock condition of gray whales. *Rybn. Khoz. (Moscow)* **3,** 15–17.

Ivashin, M. V., and Mineev, V. N. (1981). The history of gray whale harvesting off Chukotka. *Rep. Int. Whal. Commn.* **31,** 503–505.

Klumov, S. K. (1963). Feeding and helminthofauna of whales in the main areas of the whaling in the world ocean. *Tr. Inst. Okeanol. im. P. P. Shirshova, Akad. Nauk SSSR* **71,** 94–194.

Krasheninnikov, S. P. (1755). "Description of the Kamchatka Peninsula," Vols. 1 and 2. Imper. Akad. Nauk., St. Petersburg.

Kuz'min, A. A., and Berzin, A. A. (1975). Distribution and contemporary status of the population stock of the bowhead and gray whales in the Far East Seas. *In* "Biological Resources of the Far East Seas, Abstracts of the All-Union Conference on Marine Mammals," pp. 121–122. Vladivostok.

Nikulin, P. G. (1946). On the distribution of Cetacea in the seas surrounding the Chukchi Peninsula. *Izv. Tikhookean. Nauchno-Issled. Inst. Rybn. Khoz. Okeanogr.* 22:255–257.

Rezvanov, G. M. (1981). Data on the spatial structure and sex composition of the groups of the gray whales on the feeding grounds. *In* "Scientific Study on Marine Mammals in the Northern Part of the Pacific Ocean in 1980/1981," Proj. 02-05-61, pp. 99–100. VNIRO, Moscow.

Rice, D. W., and Wolman, A. A. (1971). The life history and ecology of the gray whale (*Eschrichtius robustus*). *Spec. Publ. Am. Soc. Mammal.* **3,** 1–42.

Sleptsov, M. M. (1952). Cetacea of the Far East seas. *Izv. Tikhookean. Nauchno-Issled. Inst. Rybn. Khor. Okeanogr.* **38.**

Sleptsov, M. M. (1961a). Fluctuation in population stocks of the whales in the Chukchi Sea in various years. *Proc. A. N. Severtsoff Inst. Anim. Morphol. Akad. Nauk SSSR* **34,** 54–64.

Sleptsov, M. M. (1961b). Distribution of the feeding grounds and the whales in the Okhotsk Sea. *Proc. A. N. Severtsoff Inst. Anim. Morphol. Akad. Nauk SSSR* **34,** 79–92.

Tomilin, A. G. (1937a). The whales of the Far East. *Uchen. Zap. Mosk. Univ., Ser. Biol. Nauka* **13,** 119–167.

Tomilin, A. G. (1937b). Some peculiarities in the behavior of whales approaching the shores and in "mass strandings." *Bull. Mosk. O-va. Ispyt. Prir., Otd. Biol.* **46**(4), 194–203.

Tomilin, A. G. (1937c). Gray whales in the lagoons of eastern shore in the middle of the Bering Sea. *Priroda* **7,** 132–136.

Tomilin, A. G. (1939). Some remarks to the systematization, anatomy, biology and migrations of Cetacea. *Tr. Rostov.-na-Donu Biol. Obshch.* **3,** 147–182.

Tomilin, A. G. (1946). Problem of lactation and nourishment in Cetacea. *Byull. Mosk. O-va. Ispyt. Prir., Otd. Biol.* **51**(4-5), 44–57.

Tomilin, A. G. (1954). Adaptations of Cetacea. *Zool. Zh.* **33**(3), 677–692.

Tomilin, A. G. (1957). "Animals of the USSR and the Adjacent Countries," Vol. 9. USSR Acad. Sci., Moscow.

Tomilin, A. G. (1962). "Cetacean Fauna of the Soviet Seas," pp. 3–211. Publ. House Akad. Nauk, SSSR.

von Ditmar, K. (1890–1900). "Reisen und Aufenthalt in Kamchatka in den Tahren: 1851–1855," Vols. 1 and 2. Acad. Sci., St. Petersberg.

Treshchev, V. V., Yurakhno, M. V., and Zimushko, V. V. (1969). To the problem on the helminthofauna of the gray whales inhabiting the Bering and Chukchi Seas. *In* "Problems in Marine Biology" (V. A. Vodianitskii, ed.), pp. 130–131. Akad. Nauk. Ukr. SSR, Kiev.

Vadivasov, M. P. (1947). The whaling in the USSR (Far East) in 1941–1944. *Izv. Tikhookean. Nauchno-Issled., Inst. Rybn. Khoz. Okeanogr.* **22,** 239–254.

Vasil'yev, M. (1891). Our East and its industry. *Morsk. Sb.* **5.**

Votrogov, L. M., and Bogoslovskaya, L. S. (1980). Gray whales off the Chukotka Peninsula. *Rep. Int. Whal. Commn.* **30,** 435–437.

Zenkovich, B. A. (1934a). Research data on Cetacea of far eastern seas (the California gray whale, *Rhachianectes glaucus* Cope). *Vestn. D. V. Fil. Akad. Nauk. SSSR* **10,** 9–25.

Zenkovich, B. A. (1934b). Some data on whales of the Far East. *Dolk Akad. Nauk. SSSR* **2(6),** 388–392.

Zenkovich, B. A. (1934c). Whaling in the Kamchatka and Bering Seas (the season of 1933). *Dolk. Akad. Nauk. SSSR* **1–2,** 113–118.

Zenkovich, B. A. (1935a). Ectoparasites of some large whales of the Far East. *Vestn. Dal'nevost. Fil. Akad. Nauk SSSR* **13,** 117–122.

Zenkovich, B. A. (1935b). Whaling in the USSR in 1934. *Rybn. Khoz. Dal'nego Vostoka* **13,** 65–66.

Zenkovich, B. A. (1935c). Respiration and the movements of the large whales. *Rybn. Khoz. Dal'nego Vostoka* **13,** 67–82.

Zenkovich, B. A. (1935d). Embryos of the whales. *Rybn. Khoz. Dal'nego Vostoka* **13,** 83–88.

Zenkovich, B. A. (1937a). Once more to the problem of gray California whale. *Vestn. Dal'nevost. Fil. Akad. Nauk SSSR* **23,** 91–103.

Zenkovich, B. A. (1937b). Weighing of the whales. *Dokl. Akad. Nauk SSSR* **16**(3), 185–190.

Zenkovich, B. A. (1937c). The food of the Far-eastern whales. *Dokl. Akad Nauk SSSR* **16**(4), 231–234.

Zenkovich, B. A. (1938a). Milk of large-sized cetaceans. *Dokl. Akad. Nauk SSSR* **20**(2–3), 203–206.

Zenkovich, B. A. (1938b). Development of marine mammal industry on the Chukotka Peninsula. *Priroda* **11–12.**

Zenkovich, B. A. (1947). Whaling in USSR and its perspectives. *Fish. Mag.* **10,** 15–20.

Zenkovich, B. A. (1954). "Around the World after the Whales." Geographgizizdat, Moscow.

Zenkovich, B. A. (1969). Whales and whaling in the Antarctica waters and in the north part of the Pacific ocean; condition of the population stock. *In* "Marine Mammals," (V. A. Arseniev, B. A. Zenkovich, and K. K. Chapskii, eds.), pp. 5–23. Akad. Nauk. Moskva.

Zimushko, V. V. (1969a). Some data on the biology of gray whales. *In* "Marine Mammals," (V. A. Arseniev, B. A. Zenkovich, and K. K. Chapskii, eds.), pp. 93–97. Akad. Nauk. Moskva.

Zimushko, V. V. (1969b). Materials on the gray whale's reproduction. *Tezisy Dokl.—Vses. Soveshch. Izucheniyu Morskikh Mlekopitayushchikh, 4th*, pp. 24–28. Nauka, Moskva.

Zimushko, V. V. (1970a). Aerial visual censusing and observations on the distribution of gray whales in the inshore waters of Chukotka. *Izv.* TINRO **71,** 289–294.

Zimushko, V. V. (1970b). Age determination of the gray whale (*Eschrichtius robustus*). *Izv. Tikhookean. Nauchno-Issled. Inst. Rybn. Khoz. Okeanogr.* **71,** 295–299.

Zimushko, V. V. (1970c). To the problem on the rudiments of the gland in gray whale. *Izv. Tikhookean. Nauchno-Issled. Inst. Rybn. Khoz. Okeanogr.* **74,** 341–343.

Zimushko, V. V. (1970d). The determination of the yellow bodies traces on the ovaries in gray whale *Eschrichtius gibbosus* (Cetacea Eschrichtiidae). *Zool. Zh.* **49,** 1073–1080.

Zimushko, V. V. (1972a). Materials on the gray whale's reproduction. *Tr. AlantNIRO* **39,** 44–53.

Zimushko, V. V. (1972b). On the morphology of the gray California whale. *Tezisy Vses. Soveshch. Izucheniyu Morskikh Mlekopitayushchikh, 5th*, Part 1, pp. 172–177.

Zimushko, V. V., and Ivashin, M. V. (1980). Some results of Soviet investigations and whaling of gray whales (*Eschrichtius robustus* Lilljeborg, 1861). *Rep. Int. Whal. Commn.* **30,** 237–246.

Zimushko, V. V., and Lenskaya, S. A. (1970). On the feeding of gray whales in their feeding areas. *Ekologiya (Sverdlovsk)* **3,** 26–35.

21

Investigations of Gray Whales Taken in the Chukchi Coastal Waters, U.S.S.R.

Sergei Alekseevich Blokhin

Introduction

The investigation of the Chukotka–California population of gray whales (*Eschrichtius robustus*) began with the taking of two animals in 1933 at the time of the first voyage of the whaling fleet *Aleut* in the Bering Sea (Zenkovich, 1934a). In 1934, 54 gray whales were taken in the waters off the Chukotka Peninsula, and the studies conducted on these whales yielded the first data on gray whale morphology, physiology, and feeding by Soviet scientists (Zenkovich, 1934a, 1935a,b).

During the period of pelagic whaling for gray whales (1933–1946), a collection of material was obtained on the sex, size, and physiological composition of the animals taken, as well as information on various aspects of their biology (Zenkovich, 1937a,b, 1938a,b, 1952; Belopolsky and Maximova, 1934; Tomilin, 1937, 1946; Vadivasov, 1946, 1947; Sleptsov, 1955). All of the data obtained during this period was fully summarized by Tomilin (1957).

The cessation of pelagic whaling of the gray whale in 1946 signaled the end of its research, which was resumed by the Magadansky division of TINRO (Pacific Research Institute of Fisheries and Oceanography) in 1965. Knowledge of the biology of the Chukotka–California gray whale population was expanded significantly by the works of this institute's scientific colleague V. V. Zimushko (Zimushko, 1969, 1970a,b,c, 1971, 1972; Zimushko and Lenskaya, 1970).

Since 1977 the TINRO laboratory of cetacean studies has conducted investigations

ISBN 0-12-389180-9

of the gray whale under the direction of laboratory manager A. A. Berzin. During this time, data has been obtained on the sex and size composition of the population, the physiological condition of the gray whales harvested, and individual aspects of their morphology and biology (Blokhin, 1979, 1981; Blokhin and Pavlyuchkov, 1981).

At the present time, research on the Chukotka–California gray whale population is focused on the assessment of the population size and dynamics. In addition, studies on little-known aspects of the whale's biology are continuing.

Materials and Methods

From 1977 to 1981, 299 gray whales were examined (Fig. 1), a large proportion of which were taken in Mechigmensky Bay and its nearby waters (Fig. 2, areas 7–13). Whales were examined at the "flensing" sites of the Chukotka coastal settlements of Lorino, Yanrakynnot, Sireniki, and Uel'kal. In addition to the author, members of the TINRO cetacean study laboratory participated in the collection of material. Animals were measured with a metal tape to the nearest 0.1 m. The physiological condition of the females was determined from the condition of the ovaries: females having one or more yellow bodies (corpora lutea) were regarded as sexually mature. All females that had reached 11.5 m were also considered sexually mature (Zimushko, 1969). The whale's fatness was determined by blubber thickness (as percentage of body length) at a specific location on one side at the midline on the body directly below the anterior edge of the dorsal ridge. The stomach contents and composition of food were examined in 78 animals. Standard measurements were obtained from 90 adults and 25 fetuses. Ear plug measurements were taken along their shortest edge.

Fig. 1. *A gray whale on the beach prior to processing.*

Table I
The Morphophysiological Characteristics of Gray Whales Taken in the Waters of the Chukotka Peninsula in 1977–1981

Time of examination	Number of whales examined				Average length (m)			Physiological condition						Percentage of catch examined
	Female		Male					Females[a]				Males[a]		
								Immature	Mature[b]					
	n	%	*n*	%	Female	Male	Total		I	II	III	Immature	Mature	
1977 September–October	16	84.3	3	15.7	12.6	11.65	12.5	37.5	70.0	30.0	—	33.3	66.7	10.7
1978 May 25–September 16	23	74.0	8	26.0	11.8	11.9	11.8	56.5	30.0	70.0	—	37.5	62.5	17.0
1979 July–September	38	64.9	19	35.1	11.9	11.5	11.8	36.8	79.1	16.8	4.2	42.1	57.9	32.0
1980 July–October	70	71.4	28	28.6	12.0	11.7	11.9	37.1	52.2	47.8	—	32.1	67.9	54.7
1981 September–October	69	73.4	25	26.6	11.9	11.1	11.7	43.5	55.3	44.7	—	64.0	36.0	69.6

[a]All numbers are percentages.
[b]I = pregnant, II = barren, III = lactating.

Results

COMMERCIAL AND BIOLOGICAL CHARACTERISTICS OF WHALES

Sex and Size Composition

Females comprised the majority of the animals taken off the Chukotka Peninsula during the period when gray whales were harvested (Table I). From the results of the 1980 studies we may determine that male and female whales are segregated during certain periods. Industry data also revealed that more males occur in areas far from shore than in coastal waters. For example, in 1980 among the animals taken in areas 9–10 (Fig. 2), males composed 14.3 and 18.3% of the catch, respectively, whereas in area 11 males were more numerous, 43.7%. During the 1981 whaling season, these proportions were 18.0%, 22.7%, and 35.3%, respectively.

The smallest female examined was 8.2 m long and the largest was 14.1 m. The smallest male was 9.6 m and the largest 13.9 m. Individuals measuring 11.6 to 13.0 m comprised the majority of all females taken (51.8%), and the majority of the males (77%) ranged from 10.6 to 12.5 m (Fig. 3). The average size of the whales examined ranged from 11.1 to 12.6 m (Table I).

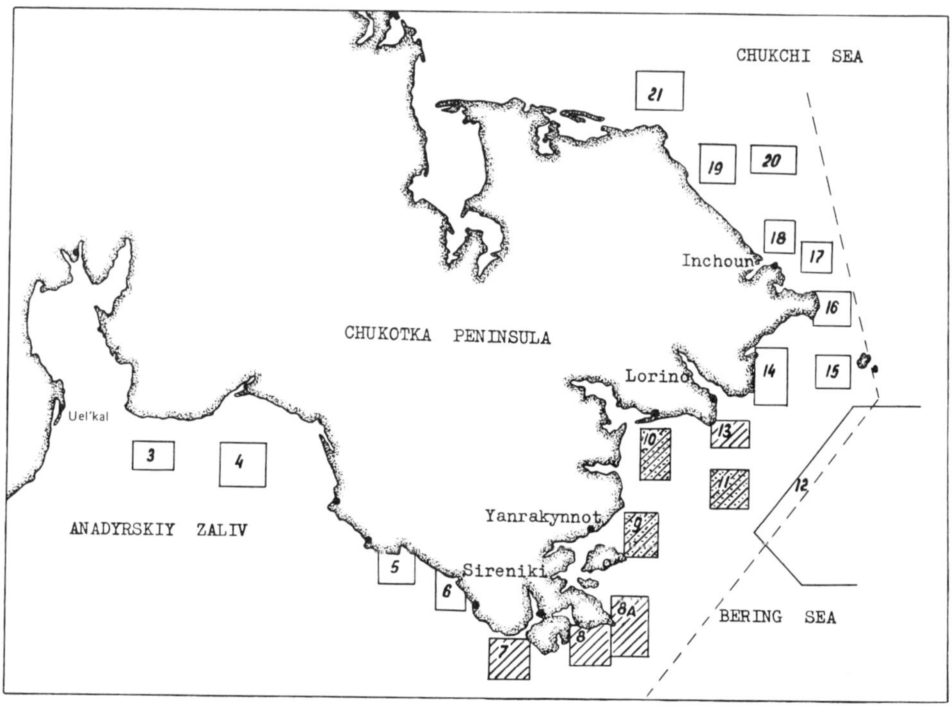

Fig. 2. *Areas in which gray whales were taken off the Chukotka Peninsula. Hatched area, areas of catches of gray whales examined in 1977–1980; stippled areas, areas of catches of gray whales examined in 1981.*

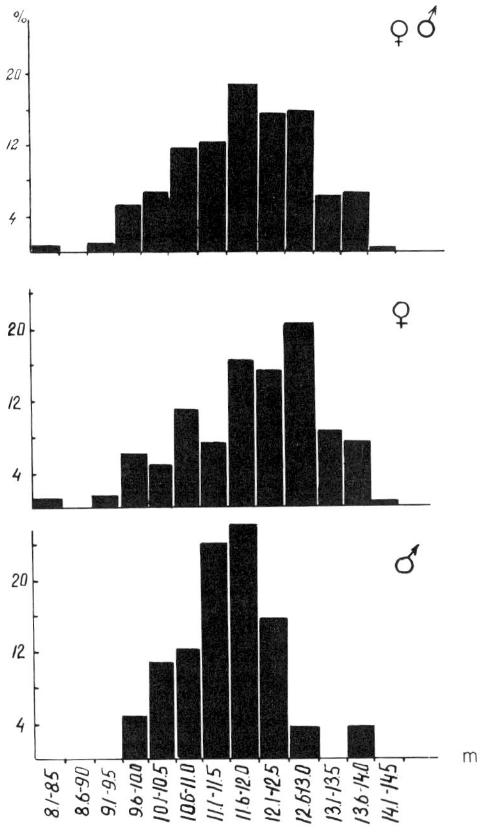

Fig. 3. *Size composition of gray whales taken in 1977–1981.*

Physiological Condition

Data on the physiological condition of the whales taken is presented in Table I. The majority of the sexually mature females was pregnant and barren (not pregnant) individuals. Ninety percent of the barren females were individuals that had recently ceased nursing, as evidenced by the presence of an unresorbed corpus lutea of lactation in the ovaries and by the enlarged mammary tissue.

Pregnant females comprised the majority of the sexually mature animals found in the Mechigmensky Bay waters in July and August. In September and October, barren females approach this region from more southerly portions of their summer range where they had ceased lactation. This suggests that females in similar physiological condition are found together in the waters of Chukotka (Blokhin, 1981). In 1981, however, this pattern was not observed and we feel this resulted from complex ice conditions that occurred in the Chukchi and Bering Seas in October of that year. The proportion of pregnant to barren females in the different size classes was approximately equal (Table

Table II
The Percentage of Females of Different Physiological Conditions in Various Size Classes

	Size class (m)				
	11.6–12.0	12.1–12.5	12.6–13.0	13.1–13.5	13.6–14.0
Immatures[a]	42.3	5.0	3.4	—	—
Pregnant[b]	40.0	47.3	50.0	44.4	57.1
Barren[b]	60.0	52.7	50.0	55.6	42.9

[a]Percentage of all females examined (n = 91).
[b]Percentage of sexually mature females (n = 46).

II), suggesting that female gray whales do not decrease their reproductive rate with increasing age. This data supports the contention of a 2-year reproductive cycle for the majority of sexually mature females (Rice and Wolman, 1971).

Age

The established methods of determining the age of baleen whales by the growth rings of the baleen plates, or by the foliation of laminae in the wall of the tympanic bulla, are not applicable for gray whales. We utilized the number of growth layers in the wax ear plugs to determine the average age of 38 females comprising five size classes (Table III). For comparison, a second method was utilized to characterize the age composition of the whales taken. Rice and Wolman (1971) suggested that the most exact method of age determination for mature females is to count the number of corpora albicantia and corpora lutea in the ovaries. They determined that females become sexually mature at approximately 8 years and become pregnant every 2 years; thus, the age of a mature female whale is judged from the number of corpora on her ovaries using the relationship $2n + 8$. Age determinations from corpora counts in the whales we examined were equivalent to age estimates from ear plug growth layers (Table IV). When these data were regressed against length of various size classes of animals, the age indices derived by both methods were essentially the same in 1980 but differed in 1981.

Fatness

The average blubber thickness of the whales equalled 12.70 cm among females and 12.05 cm among males. Barren females had the thinnest blubber; blubber thickness was greatest in pregnant females (Table V). As described above, most barren females had recently ceased to lactate, and in our opinion this explains their reduced blubber thickness compared to that of males and females of other physiological conditions. Because newly pregnant females are the first to arrive in the waters of Chukotka, they are able to feed for the longest time and thus increase their blubber thickness significantly.

Table III
Average Age of Female Gray Whales of Various Size Classes[a]

Size classes (m)	9.1–9.5	9.6–10.0	10.1–10.5	10.6–11.0	11.1–11.5	11.6–12.0	12.0–12.5	12.6–13.0	13.1–13.5	13.6–14.0
Age by ovaries (n = 91)	1.5[b]	2.0[b]	3.0[b]	4.5[b]	6.5[b]	9.2	13.5	20.5	21.5	27.5
Age by ear plugs (n = 38)	—	—	—	—	—	7.1	14.5	22.4	23.0	29.0

[a]Ages in years.
[b]The age of sexually immature animals (from 9.1 to 11.5 m) are data from Rice and Wolman (1971).

Table IV
The Average Age of Female Gray Whales Taken in 1977–1981[a,b]

Year	Age by ovaries	Age by ear plugs
1977	16.4	15.1
1978	11.6	11.9
1979	13.2	13.6
1980	12.8	13.3
1981	12.7	15.5
1977–1981[c]	12.9	14.0

[a]Ages in years.
[b]Sample size same as Table III.
[c]Weighted mean.

MORPHOLOGY AND BIOLOGY

External Characteristics

Throat Pleats. Gray whale throat pleats are similar to the abdominal pleats found on rorquals (Fig. 4). These pleats allow the increase in the capacity of the mouth cavity at the moment the whale engulfs food organisms, and their number, in all probability, determines the extent of the capacity increase [for a description of the mechanism of rorqual feeding see Lamertsen (1983).] The number of pleats in the whales examined ranged from two to seven (Table VI). Both the number and distribution of the pleats varied (Fig. 5). We determined 10 different combinations of pleats, with three pleats being the most common.

Dorsal Spinal Knobs. Bumps or knobs on the dorsal ridge of the tail ranged from 6 to 12 and varied in form from smooth and hardly noticeable to pointed and clearly distinct. Whales with 8 to 9 knobs were most frequent (Table VI).

The Sebaceous Gland. Andrews (1914) first described this feature among gray whales taken along the Asian coast; Zimushko (1970b) described this structure in detail

Table V
Indices of Fatness of Gray Whales Taken in July–October, 1980–1981[a]

	July	August	September	October	July–October[b]	n
Immature females	0.70	1.00	1.07	1.10	1.03	25
Pregnant females	1.18	1.22	1.16	1.21	1.18	21
Barren females	—	0.97	0.91	0.98	0.96	23
Males	1.10	1.07	1.05	1.05	1.06	28

[a]Blubber thickness expressed as a percentage of body length.
[b]Values are weighted means.

Fig. 4. *Throat pleats of a gray whale.*

and determined it was rudimentary and similar in structure to the sebaceous gland of terrestrial mammals. This sebaceous gland was present in 73.0% of the females and 54.5% of the males examined. Zimushko (1970b) described it in 77.0% of the females and 28.0% of the males he examined in 1965–1966 (Fig. 6). [For a description of the gray whale's "tail stock cyst" see Spencer (1973).]

Table VI
Frequency of Occurrence of Several External Features of Gray Whales Taken in the Waters of Mechigmensky Bay and Adjacent Areas (1980–1981)

Year		Number of submaxillary pleats						Number of spinal–caudal bumps							"Sebaceous" gland			
															Present		Absent	
		2	3	4	5	6	7	6	7	8	9	10	11	12	F	M	F	M
1980	*n*	39	18	2	1	—	—	6	13	29	15	9	6	2				
	%	65.0	30.0	3.3	1.7	—	—	7.5	16.3	36.2	18.8	11.2	7.5	2.5				
1981	*n*	45	41	3	1	—	1	5	11	19	19	9	9	4	27	6	10	5
	%	49.4	45.1	3.3	1.1	—	1.1	6.6	14.5	25.0	25.0	11.8	11.8	5.3	73.0	54.5	27.0	45.5
1980–1981	*n*	84	59	5	2	—	1	11	24	48	34	18	15	6				
	%	55.6	39.1	3.3	1.3	—	0.7	7.1	15.4	30.8	21.8	11.5	9.6	3.8				

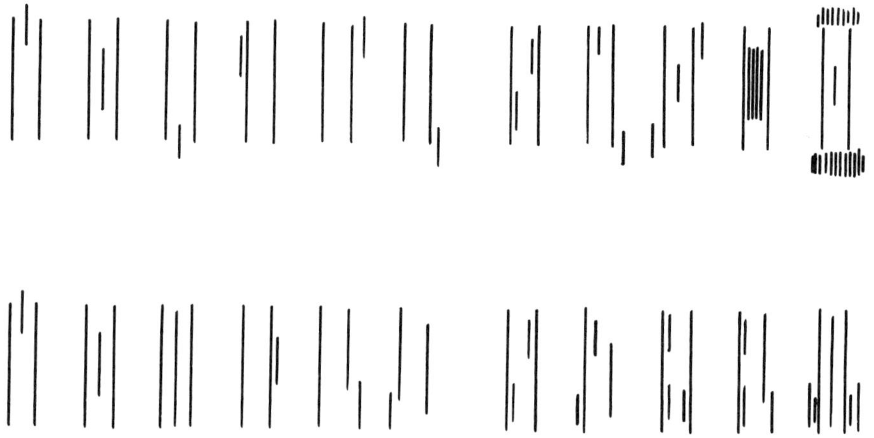

Fig. 5. *Variations in the arrangement of throat pleats in gray whales taken in the waters of the Chukotka Peninsula in 1980 (top) and 1981 (bottom).*

Body Proportions

Body proportions of various size classes and sexes of gray whales do not differ during postnatal growth (Table VII). A comparison of body proportions of fetuses (Table VIII) between 100 and 375 cm long and adult whales revealed significant differences ($p > .999$) in the following measurements: the end of the rostrum to the center of the blowhole, the notch of the tail to the anus, and the center of the eye to the external ear opening.

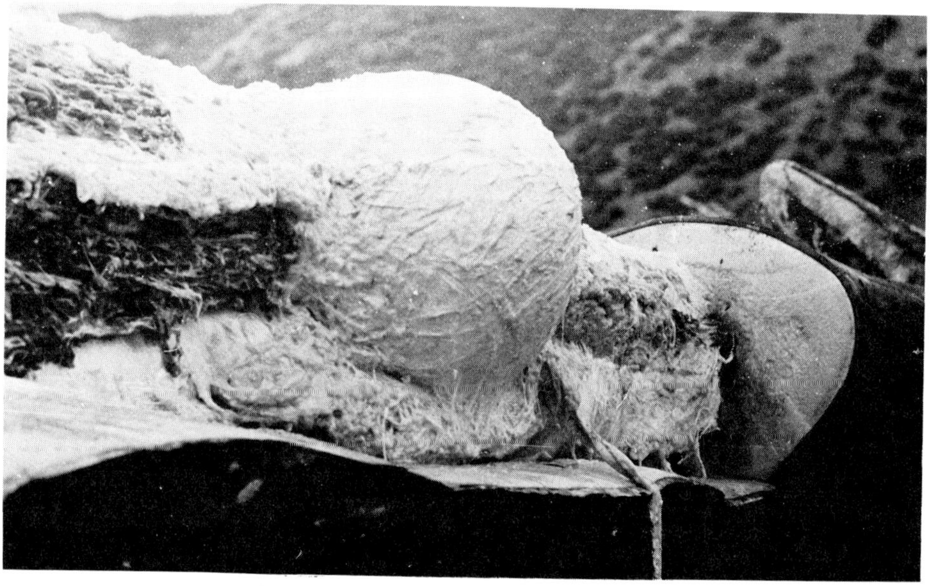

Fig. 6. *"Sebaceous" gland of a gray whale.*

Table VII
Body Measurements as a Percentage of Total Length of Gray Whales Taken in the Waters of Chukotka Peninsula[a]

Females (m)						Females (m) (9.0–14.0)			Males (m) (9.0–13.0)			All whales		
9.0–12.0			12.1–14.0											
M ± SE	CV[a]	*n*	M ± SE	CV	*n*	M ± SE	CV	*n*	M ± SE	CV	*n*	M ± SE	CV	*n*
[b]16.55 ± 0.26	8.32	28	16.89 ± 0.16	5.36	31	16.72 ± 0.19	8.56	59	16.38 ± 0.24	6.87	22	16.48 ± 0.14	8.22	90
[c]27.23 ± 0.04	6.20	32	27.41 ± 0.19	4.26	38	27.33 ± 0.18	5.40	70	27.07 ± 0.05	3.00	24	27.25 ± 0.14	5.15	94
[d]4.20 ± 0.05	4.78	20	4.32 ± 0.06	6.89	22	4.26 ± 0.04	5.97	42	4.55 ± 0.06	4.64	12	4.33 ± 0.04	6.01	54
[e]17.47 ± 0.27	6.98	21	18.18 ± 0.17	4.82	26	17.76 ± 0.16	6.20	47	17.89 ± 0.23	6.14	23	17.78 ± 0.12	5.78	69
[f]6.36 ± 0.07	5.11	21	6.45 ± 0.06	4.78	26	6.41 ± 0.03	4.20	47	6.59 ± 0.06	5.63	23	6.47 ± 0.04	5.53	69
[g]28.4 ± 0.50	8.85	27	29.3 ± 0.22	4.1	29	28.90 ± 0.30	7.46	56	28.30 ± 0.60	9.30	23	28.68 ± 0.33	10.1	79
[h]7.91 ± 0.11	6.18	21	7.72 ± 0.07	4.84	25	7.81 ± 0.06	5.21	46	8.01 ± 0.14	7.16	17	7.86 ± 0.06	6.25	63

[a]Coefficient of variation.
[b]The end of the rostrum to the center of the blowholes.
[c]The end of the rostrum to the base of the pectoral flipper.
[d]The center of the eye to the ear opening.
[e]The length of the pectoral fin along its greatest edge.
[f]The maximum width of the pectoral flipper.
[g]The fork of the tail to the anus.
[h]The maximum width of the flukes.

Table VIII
Body Measurements as a Percentage of Total Length of Gray Whale Fetuses

	100–190 cm			210–290 cm			310–375 cm			All fetuses		
	M ± SE	CV[a]	n	M ± SE	CV	n	M ± SE	CV	n	M ± SE	CV	n
[b]	13.66 ± 0.13	4.96	8	14.0 ± 0.11	8.80	10	14.42 ± 0.42	5.84	4	13.91 ± 0.21	6.94	22
[c]	28.00 ± 0.13	7.12	8	27.54 ± 0.09	5.00	12	28.45 ± 0.87	6.12	4	27.85 ± 0.28	5.00	24
[d]				5.72 ± 0.13	8.35	8						
[e]	17.47 ± 0.12	6.34	9	18.99 ± 0.09	4.11	12	20.2 ± 0.28	2.78	4	18.65 ± 0.27	7.46	25
[f]	5.82 ± 0.12	10.4	9	6.44 ± 0.09	6.69	12	7.05 ± 0.31	8.77	4	6.32 ± 0.16	12.7	25
[g]	29.31 ± 0.12	7.67	9	30.95 ± 0.09	4.24	12	30.15 ± 0.53	3.54	4	30.41 ± 0.35	5.75	25
[h]	7.89 ± 0.15	8.12	7	7.92 ± 0.10	5.69	11	8.39 ± 0.42	9.37	3	7.97 ± 0.11	6.16	21

[a] Coefficient of variation.
[b] The end of the rostrum to the center of the blowholes.
[c] The end of the rostrum to the base of the pectoral flipper.
[d] The center of the eye to the ear opening.
[e] The length of the pectoral fin along its greatest edge.
[f] The maximum width of the pectoral flipper.
[g] The fork of the tail to the anus.
[h] The maximum width of the flukes.

Red Bone Marrow and Blood

The major mass of red bone marrow in gray whales is located in the spinal column; the neck, jugular, pectoral, and lumbar sections include roughly 90–95% of the marrow. The quantity of red bone marrow starts to decline markedly in the last lumbar vertebrae: in lumbar vertebrae 9 and 10, marrow occupies only the central portion, and in the fifth caudal vertebrae marrow is totally absent. Approximately the same marrow distribution is noted in the neural spines and transverse processes of the vertebrae. Small quantities of marrow are located in the ends of the ribs and in the first three chevron bones. Marrow is also absent from the anterior extremities and the bones of the cranium.

The number of red corpuscles in the blood of two fetuses (male, 152 cm; female, 285 cm) equalled $2.00–2.29 \times 10^6/\mu l$. For females 9.8 m long and males 11.0 m long, the red cell counts were 3.05 and $3.35 \times 10^6/\mu l$, respectively. The average diameter of red corpuscles of adult gray whales is 8.17 μm (n = 1200 cells).

Ear Plugs

Ear plugs of gray whales may be divided into three types according to their consistency: hard, soft, and gelatinous. The frequency with which each type was encountered was 63.5, 22.4, and 14.1%, respectively. The color of the plugs varied from dark brown (hard) to light beige (gelatinous).

If one assumes that ear plugs increase in size proportionally to a whale's age, then we may assume that the rate of growth in males slows markedly after they attain a length of 11.1–11.2 m and that for females growth slows after they reach lengths of 11.4–11.5 m (Fig. 7). No layers were found in longitudinal sections of gelatinous ear plugs, and only 61 of 68 hard plugs examined possessed layers (57% of all 107 plugs examined).

Age at Sexual and Physical Maturity

The weights and lengths of single ovaries from female gray whales are presented in Figs. 8 and 9. The smallest female with evidence of at least one pregnancy was 11.3 m

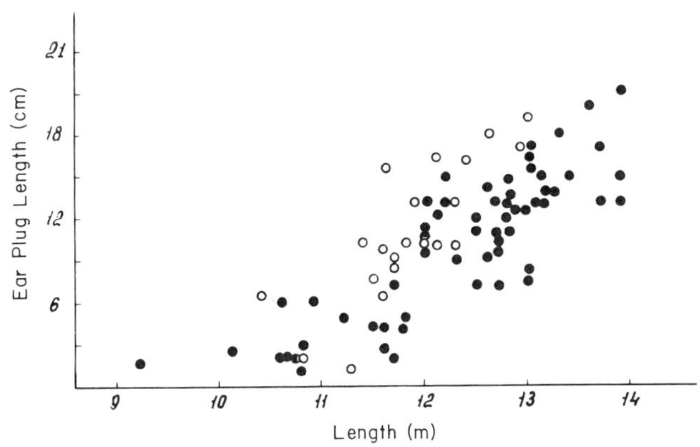

Fig. 7. *Change in the sizes of gray whale ear plugs in relation to body size. Open circles are males; solid circles, females.*

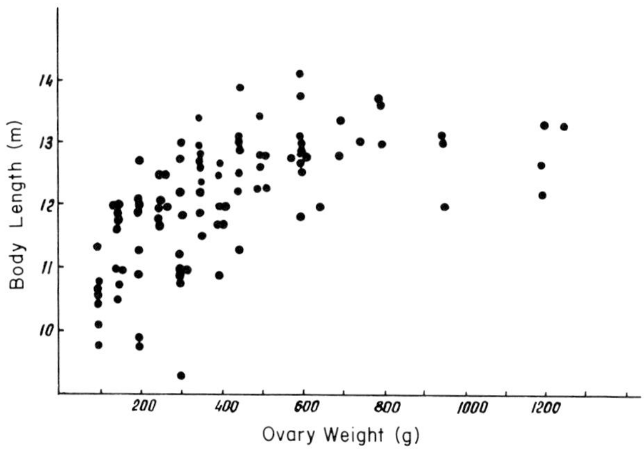

Fig. 8. *Change in the mass of female gray whale ovaries in relation to body length.*

long, and the largest without any trace of corpora lutea was 12.9 m long. The maximum number of corpora (17) was from the ovaries of a 12.5-m female, and the average for sexually mature females was 5.7 corpora. The average size of 13 females with 1 corpus from a past or present pregnancy was 12.0 m, and the average weight of a single ovary was 287 g.

Rice and Wolman (1971) determined that gray whale females became sexually mature at body lengths of 11.7 m, and Zimushko (1969) concluded that whales attained sexual maturity at 12 m. Our analysis of the corpora counts on the ovaries of the females we studied suggested that sexual maturity occurs when these animals attain a length of 11.7 to 11.8 m, which is in agreement with Rice and Wolman's estimate. Our data suggested that after females attain a size of 13.0 to 13.3 m, the correlation between changes in the ovaries' mass and the length of the whales ceases (Fig. 8). In addition, following the accumulation of 9 to 10 corpora on their ovaries, the average size of the females does not increase (Fig. 10). This suggests that with the attainment of an average

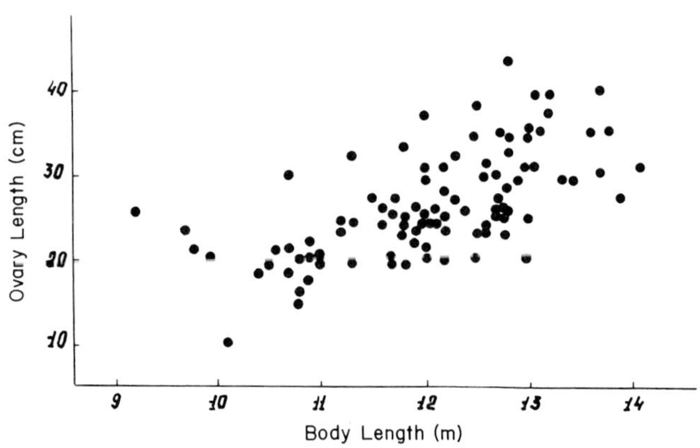

Fig. 9. *Change in the length of female gray whale ovaries in relation to body length.*

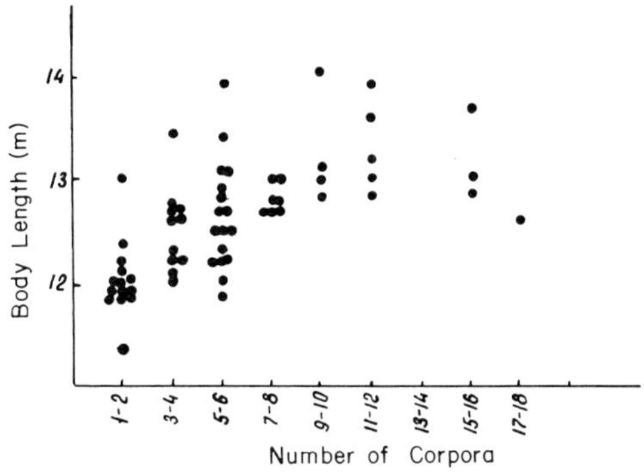

Fig. 10. *Accumulation of traces of corpora lutea on female gray whale ovaries in relation to body size (length).*

length of 13.3 m and an age of 26–28 years (as determined from the ovaries), gray whale females are physically mature.

In male gray whales, a marked increase in the weight and length of the testicles is observed among individuals between 11.2 and 11.4 m in length (Figs. 11 and 12). Histological analysis of testicles weighing 1.5 and 1.6 kg from males 11.3 and 11.6 m, respectively, showed 20% of the seminal canals were open; among males 11.5 m long with testicles weighing 4.1 kg, all seminal canals were open.

Zimushko (1969) examined the testicles from 29 whales taken in the waters of the Chukotka Peninsula and concluded that male gray whales become sexually mature at a body length of approximately 11.5 m; Rice and Wolman (1971) established age at sexual maturity for males at a length of 11.1 m and a testicle weight of 5.0 kg. Our data suggested that males reach sexual maturity when one testicle weighs 4.0 kg and an average body length of 11.2–11.4 m has been reached (at the same time, body length may fluctuate significantly.) Thus, 100% of the seminal canals were open in 2 males examined that were 10.4 and 10.5 m long with testicle weights of 7.2 and 7.5 kg, respectively.

The material collected by Rice and Wolman (1971) was taken from whales collected

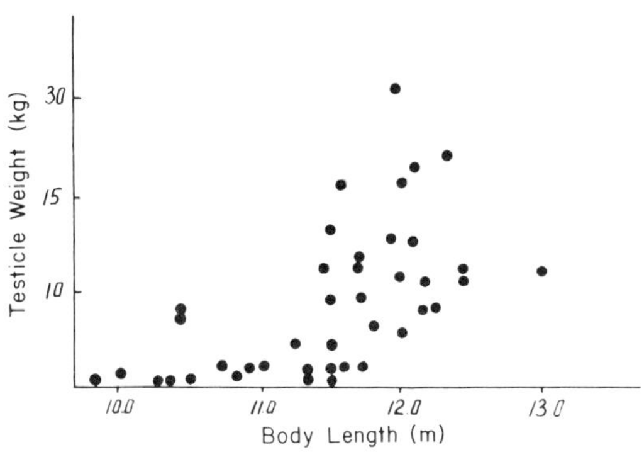

Fig. 11. *Change in the mass of male gray whale testicles in relation to body size (length).*

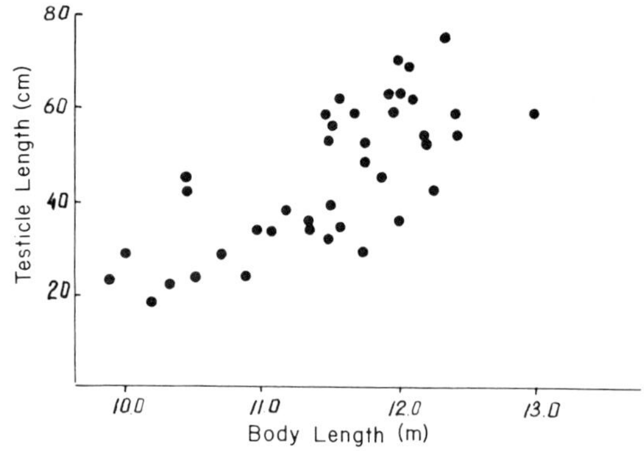

Fig. 12. *Change in the length of male gray whale testicles in relation to body length.*

during the winter breeding season, when the weight of the testicles increases significantly in conjunction with mating. Thus, we suggest that among whales taken during the summer in the waters of Chukotka, males with a single-testicle weight of 4.0 kg or greater should be considered sexually mature. All sexually mature males examined had seminal fluid in their seminal ducts, but spermatozoa was found in only a few individuals. This supported the contention of Zimushko (1971) that sperm production declines or even ceases during the summer months. Our data suggested that gray whales reach sexual maturity when their ear plugs have accumulated 7–8 layers (one layer is composed of one dark and one light stripe) (Fig. 13).

Ectoparasites

All whales examined were infected to some degree with cyamids [Cyamidae, whale lice (Amphipoda)] and barnacles (Cirripedia). The major localization of cirriped barnacles was on the head; they are distributed on the upper and lateral surfaces of the lower jaw. On the head the distribution is uneven: the right side of the head, as a rule, is free of

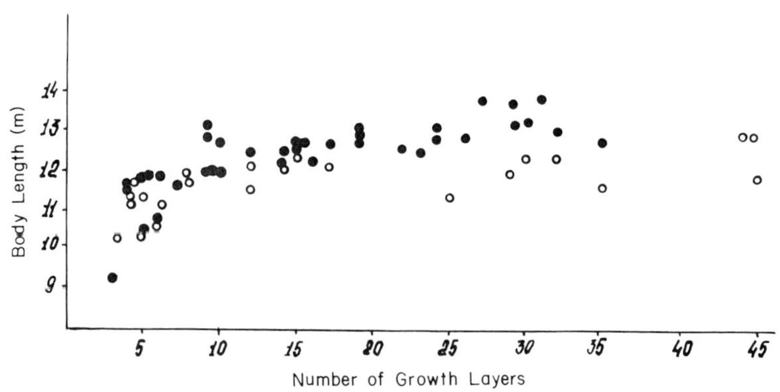

Fig. 13. *Change in the number of layers in the ear plugs of gray whales in relation to body length. Solid circles are females; open circles are males.*

Fig. 14. *Right side of a gray whale's head.*

barnacles (Fig. 14). A similar observation was reported by Kastya and Rice (1970), which they connected with a peculiarity of gray whale feeding behavior. They felt that when the whales fed on benthic organisms they oriented themselves on their right side and subsequently eroded off the attached barnacles. Barnacles (*Cryptolepas rhachianecti*) are found as individuals and small groups on the entire surface of the whale's body. Several whales taken in the autumn had traces of ulcers or wounds from recently dislodged adult barnacles, and juvenile barnacles that ranged in size from 1 to 2 cm (Fig. 15).

The distribution of cyamids on gray whales is more extensive than that of barnacles, and significant numbers of cyamids are found associated with the barnacle clusters. In addition, cyamids are found in the blowhole region, the throat grooves, around the eyes, the ear openings, the pleats of the genital slit, the anus, and the notch of the tail. Cyamid colonies may be found where there are grooves, slits, and skin lesions.

Feeding

Benthic and epibenthic organisms, most of which are amphipods, are the prey of gray whales in the waters of the Chukotka Peninsula (Zenkovich, 1937c; Zimushko and Lenskaya, 1970; Blokhin and Pavlyuchkov, 1981). The stomach contents of whales taken in 1980 and 1981 in Mechigmensky Bay and nearby waters contained as many as 122 species of benthic organisms. Most frequent in occurrence were amphipods: *Pontoporeia femorata* (57.9%), *Ampelisca macrocephala* (53.9%), *Byblus longicornis* (51.3%), and *Lembos arcticus* (47.3%). Additional organisms included annelid worms of the

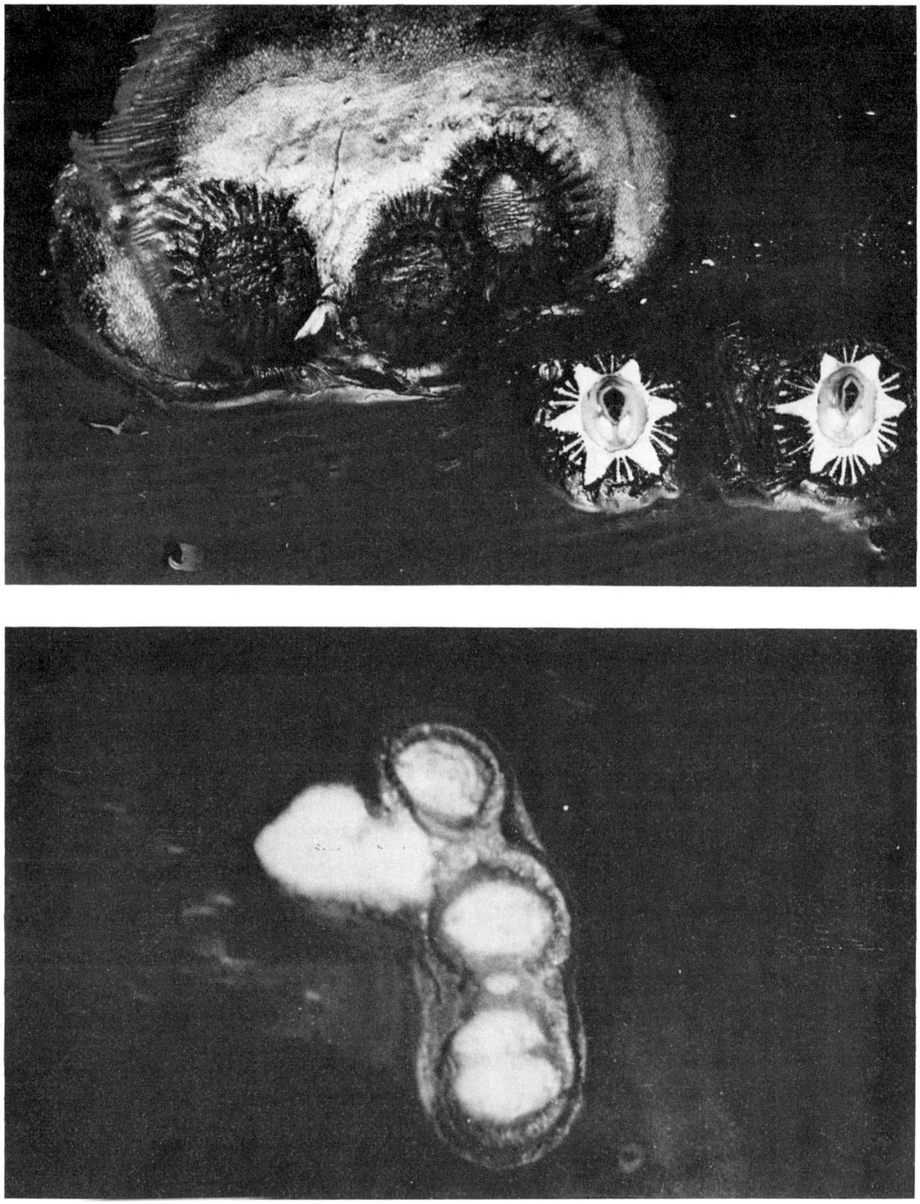

Fig. 15. *Barnacles* (Cryptolepas rhachianecti) *on skin of a gray whale (top). Traces on the skin where adult barnacles have fallen off (bottom).*

Table IX
Sex and Size Compositions of Gray Whale Fetuses Found in Females Taken off Chukotka Peninsula in 1980–1981

	July	August	September	October	November 19	*n*
Females	63, [a]131	170, 190, 238, 210	235, 225, 258, 285	278, 375	—	12
Mean size	97.0	202.0	250.7	326.5		
Males	65, 128, 130	152, 208, 190, 190	220, 275, 189	325, 318	325	13
Mean size	107.7	185.0	228.0	321.5		
Male plus female, Mean size	103.4	193.5	241.0	324.0	—	25

[a]Body lengths in cm.

family Oweniadae (53.9%). The greatest in abundance (by specific weight) were the amphipods *Pontoporeia femorata* (32.8%), *Ampelisca macrocephala* (14.2%), and the annelid worm *Oweniada* sp. (10.8%). Among the whales taken in the region of Providence Bay (Fig. 2, area 7), the composition of the stomach contents was less varied and was dominated by the amphipod *Pontoporeia femorata* (95%). The full stomachs of the whales examined speaks for the abundant forage resource in the waters off the Chukotka Peninsula.

Embryology

During the 4 months from July to October, the size of a gray whale's fetus increases (on the average) by 200 cm (Table IX). The dimensions and weights for the internal organs of two fetuses are presented in Tables X and XI. The intestinal length of females, 210 and 225 cm, and of males, 190 and 240 cm, exceeded the body sizes by 9.0, 9.8, 7.5, and 7.8 times, respectively. The relative blubber thickness of the fetuses is close to that in adult animals (Tables V and X). The number of vibrissae on eight fetuses was greater on the upper rostrum than on the lower jaw; these averaged 108 (range, 72–160) on the rostrum and 94 (range, 64–120) on the lower jaw.

During fetal development of embryos between 130 and 375 cm long, the relative length of the cervical (neck) section decreases from 6.3 to 4.6% of the total body length and the size of the lumbar section of the spinal column increases from 28.3 to 31.9%.

Conclusions

The number of gray whales taken from the Chukotka–California population includes an average of 1.1% of the current estimated population size. This study indicates that females comprise the majority of the animals taken by the whaling industry and

Table X
Size Characteristics of the Internal Organs of Gray Whale Fetuses

Fetus length (cm)		Lungs				Kidneys				Liver				Heart		Blubber thickness
		Right		Left		Right		Left		Right		Left				
		I[a]	II[b]	I	II	I	II	I	II	I	II	I	II	I	II	
275	cm	44.0	20.0	40.0	18.0	28.0	13.5	27.0	13.0	44.0	25.0	45.0	30.0	18.0	26.0	2.5
	%[c]	16.0	7.3	14.5	6.5	10.2	4.9	9.8	4.7	16.0	9.1	16.4	10.9	6.5	9.5	0.9
278	cm	50.0	17.5	47.0	18.0	32.0	14.0	29.0	16.0	—	—	—	—	15.0	24.0	3.0
	%	17.9	6.3	16.9	6.4	11.5	5.0	10.4	5.7	—	—	—	—	5.4	8.6	1.08

[a]I, length (cm).
[b]II, width (cm).
[c]Percentage of body length.

Table XI
Weights (in kg) and Percentages of Total Body Weight of Internal Organs of Gray Whale Fetuses

		Head		Right lung		Left lung		Right kidney		Left kidney		Liver		Intestine		Heart		Tongue	
Fetus length	Fetus weight[a]	Weight	%	Weight	%	Weight	%	Weight	%	Weight	%	Weight	%	Weight	%	Weight	%	Weight	%
275 cm	256 kg	54	21.1	3.6	1.40	2.9	1.13	1.5	0.54	1.5	0.54	12.5	4.54	10.5	3.8	2.7	0.98	8.0	2.90
278 cm	251 kg	49	19.5	3.5	1.40	2.9	1.16	1.8	0.70	1.8	0.70	13.0	5.20	—	—	2.2	0.87	—	—

[a]Fetus weight does not include 5–7 liters of lost blood.

demonstrates that males and females are not found in equal distribution in the areas in which they were collected.

The small average sizes and the percentages of sexually immature individuals taken are the result of the bias of the whaling industry and subsequently are not representative of the age structure of the population. Among sexually mature females, the correlation of pregnant to barren individuals averaged 1:1, suggesting there is a sufficiently high level of reproduction in the Chukotka–California population.

In the summer, the greatest blubber thickness is observed among the pregnant females, and barren females have the least amount of blubber.

The average age of the whales taken is 13–14 years. These most recent studies suggest that the current take of gray whales by the whaling industry presently does not have a negative effect on the population or its reproductive capacity.

Summary

Presented in this work are the results of studies of the Chukotka–California population of gray whales (*Eschrichtius robustus*) taken for the needs of the local population of Chukotka during the years 1977–1981.

Females comprised from 64.9 to 84.3% of the whales, and males comprised from 15.7 to 35.1% of the whales examined. The majority of the females taken (51.8%) ranged from 11.6 to 13.0 m and the majority of the males (77.0%) ranged from 10.6 to 12.5 m in length. The large proportion of immature whales taken reflects the distribution of age–sex classes of whales around Chukotka Peninsula and cannot serve as an index of the age distribution of the entire population. The ratio of pregnant to barren (not pregnant) mature females in 1980–1981 was 1:1. Reproduction in mature females does not decrease with age, and females appear to reproduce every 2 years. The average age of the females was 13–14 years. Pregnant females were well fed in the summer and fall as judged from their blubber thickness; whales with the least thick blubber were barren females. The present data suggests that the existing level of exploitation of the gray whale is not endangering the population. Additional data on external characteristics, reproductive biology, feeding, and development are presented.

References

Andrews, D. W. (1914). Monographs of the Pacific cetacea. I. The California gray whale (*Rhachianectes glaucus* Cope). *Mem. Am. Mus. Nat. Hist.* [N. S.] **1,** 227–287.

Belopolsky, M. B., and Maximova, C. B. (1934). Blubber. *Rybn. Khoz. Dal'nego Vostoka* **1–2,** 118–125.

Blokhin, S. A. (1979). On the status of the gray whale stocks in the coastal waters of the Chukotka Peninsula. *Rep. Int. Whal. Commn.* **29,** 335–336.

Blokhin, S. A. (1981). The results of studies of the California–Chukotka population of gray whales, taken in the Chukotka peninsula's waters in 1980. *In* "Studies of Marine Mammals in the North Pacific in 1980–1981," pp. 78–87. Nauka, Moscow.

Blokhin, S. A., and Pavlyuchkov, V. A. (1981). The diet of the California–Chukotka population of gray whales in the Chukotka Peninsula's waters in 1980. *In* "Studies of Marine Mammals in the North Pacific in 1980–1981." Nauka, Moscow.

Kasuya, T., and Rice, D. W. (1970). Notes on baleen plates and on the arrangement of parasitic barnacles of gray whales. *Sci. Rep. Whales Res. Inst.* **22,** 39–43.

Lambertsen, R. H. (1983). The internal mechanism of rorqual feeding. *J. Mammal.* **64**(1), 76–88.

Rice, D. W., and Wolman, A. A. (1971). The life history and ecology of the gray whale (*Eschrichtius robustus*). *Spec. Publ.—Am. Soc. Mammal.* **3,** 1–142.

Sleptsov, M. M. (1955). "Cetaceans of the Far Eastern Seas." Primorskoe Knizhroe Isdatel'stvo, Vladivostok.

Spencer, M. P. (1973). Scientific studies on the gray whales of Laguna Ojo de Liebre (Scammon's Lagoon), Baja California, Mèxico. *Natl. Geogr. Soc. Res. Rep. 1966 Proj.* pp. 235–253.

Tomilin, A. G. (1937). The whales of the Far East. *Uchen. Zap. Mosk. Univ. Ser. Biol. Nauk.* **12,** 119–167.

Tomilin, A. G. (1946). Toward the question of cetacean lactation and feeding. *Byull. Mosk. O-va. Ispyt. Prir., Otd. Biol.* **51**(4–5), 44–45.

Tomilin, A. G. (1957). Mammals of the USSR and adjacent countries. *In* "Cetacea," Vol. IX, pp. 328–358. Izv. Akad. Nauk. SSR, Moscow.

Vadivasov, M. P. (1946). Soviet whaling in the Far East in 1941–1944. *Izv. TINRO* **22,** 239–254.

Vadivasov, M. P. (1947). Some data on whaling done by the fleet "Aleut" in 1945. *Izv. TINRO* **25,** 229–230.

Zenkovich, B. A. (1934a). Research data on Cetacea of far eastern seas (the California gray whale, *Rhachianectes glaucus* Cope). *Vestn. Dal'nevost. Fil. Akad. Nauk SSSR* **10,** 9–25.

Zenkovich, B. A. (1934b). Some data on whales of the Far East. *Dokl. Akad. Nauk. SSSR* **2** (6), 388–392.

Zenkovich, B. A. (1934c). Whaling in the Kamchatka and Bering Seas (the season of 1933). *Dokl. Akad. Nauk. SSSR* **1** (2), 113–118.

Zenkovich, B. A. (1935a). About whale embryos. *Rybn. Khoz. Dal'nego Vostoka* **13,** 83–88.

Zenkovich, B. A. (1935b). Ectoparasites of a few of the great whales. *Vestn. Dal'nevost. Fil. Akad. Nauk SSSR* **113,** 117–122.

Zenkovich, B. A. (1937a). The weighing of whales. *Dokl. Acad. Sci. USSR* **16**(3), 185–190.

Zenkovich, B. A. (1937b). More about the gray California whale (*Rhachianectes glaucus* Cope 1864). *Vestn. Dal'nevost. Fil. Akad. Nauk SSSR* **23,** 91–103.

Zenkovich, B. A. (1937c). Food of the Far East Whales. *Dokl. Acad. Sci. USSR* **16**(4), 231–234.

Zenkovich, B. A. (1938a). Whale body temperatures. *Dokl. Acad. Sci. USSR* **18**(9), 685–687.

Zenkovich, B. A. (1938b). Milk of the great whales. *Dokl. Acad. Sci. USSR* **20**(2–3), 203–205.

Zenkovich, B. A. (1952). "Whales and Whaling." Pishchpromizdat (Food Industry Publications), Moscow.

Zimushko, V. V. (1969). Some data on the biology of the gray whale. *In* "Marine Mammals" (V. A. Arseniev, B. A. Zenkovich, and K. K. Chapskii, eds.), pp. 93–97. Akad. Nauk., Moscow.

Zimushko, V. V. (1970a). Toward the question of determining gray whale age (*Eschrichtius gibbosus*). *Izv. TINRO* **71,** 295–299.

Zimushko, V. V. (1970b). About the rudimentary sebaceous glands of gray whales. *Izv. TINRO* **74,** 341–343.

Zimushko, V. V. (1970c). Determining traces of yellow bodies (corpora) on female gray whale ovaries. (*Eschrichtius gibbosus* Exr. 1777). *Zool. Zh.* **49**(7), 1073–1080.

Zimushko, V. V. (1971). Material on gray whale reproduction. *Tr. AtlantTINRO* **39,** 44–53.

Zimushko, V. V. (1972). On the morphology of the gray California whale. *Tezisy Vses. Dokl. V Soveshch. Izucheniya Morskikh Mlekopitayushchilah*, Part 1, pp. 172–177.

Zimushko, V. V., and Lenskaya, S. A. (1970). On the diet of the gray whale on its feeding grounds. *Ekhologiya (Sverdlovsk)* **3,** 26–35.

22

Sound Production by the Gray Whale and Ambient Noise Levels in Laguna San Ignacio, Baja California Sur, Mexico

Marilyn E. Dahlheim, H. Dean Fisher, and James D. Schempp

Introduction

The literature on gray whale (*Eschrichtius robustus*) sound production provides inconclusive documentation on the acoustical behavior of these whales. Aldrich (1889) reported that devilfish (gray whales) were known to "sing" on their northern feeding grounds. Tomilin (1957) reported that Soviet whalers operating in the Bering and Chukchi Seas detected "low pitched roars" produced by gray whales. Eberhardt and Evans (1962), recording in Laguna Ojo de Liebre, noted "croaker-like grunts" and "low frequency rumbles" which they ascribed to gray whales. Wenz (1964), in collaboration with Asa-Dorian, reported having detected "clicks" from migrating gray whales. Painter (1963) tentatively identified a low "grunting" sound recorded in the coastal lagoons of

ISBN 0-12-389180-9

Table I
A Summary of Published Data on Sounds Produced by the Gray Whale (1967–1978)

Reference	Study area	Sound type	Frequency range (Hz)	Peak energy (Hz)	Duration (sec)	Pulses/ burst	Pulse rate	Pulse duration (sec)	Comments in text
Asa-Dorian ɛnd Perkins, 1967	In Laguna Ojo de Liebre, Laguna Guerrero Negro, and off San Diego, California	Pulses	70–3000	400–800		5–22	—	—	Pulses may be used in echolocation
		Moans	<2000						
		Whistles	<2000						
Cummings *et al.*, 1968	Off San Diego, California	Moans	20–200	—	1.54	—	—	—	Soniferous day and night
		Blow sounds	15–175		1.00				Assumed sounds were communicative
		Bubble-type	15–305		0.70				
		Knocks	Up to 350						
Poulter, 1968	In Laguna Ojo de Liebre	Grunts		—		—		—	Bong was unusual and may be alert signal
		Rumbles							
		Cries							
		Rasps	Up to 12,000		1.00				
		Pulses					5–18/sec		
		Chirps			2.0–5.0		8/sec		
		Clicks	Up to 12,000						
		Bong	Up to 12,000						
Fish *et al.*, 1974	(a) In captivity and during open-water release	Growl or moan		100–200,1500	1.00			—	
		Pulse train		100–10000,1400	2.00	8–14	14/sec		
		Grunt		200–400,1600	0.20				
		Blowhole rumble							
		Clicks	2000–6000	3400–4000	0.001–0.002	1–833	9–36/sec		
	(ɛ) In Wickaninnish Bay, Vancouver Is and	Click trains	2000–6000	3500–4000	<0.002	1–96	8–40/sec		
Ray and Schevill, 1974	In captivity	Noisy pulsation				—		—	Noisy pulsation heard during feeding
Norris *et al.*, 1977	In Bahia Magdalena	Pulses			<1.00	—		—	
		Clicks	100–20000		0.25		2/sec		
		Grunts							
Swartz and Cummings, 1978	In Laguna San Ignaɔio	Knocks	200–1500	350 and 800	4.00	1–11		0.10	
		Pulses		400	0.20				
		Snorts	66–600		1.00		≈35/sec		
		Slamming	50–2000		0.30				
		Bubble bursts	Up to 1400	400–1200	4.00				
		Squeaking moan	700–1200		0.60				

Mexico as having been produced by gray whales. Recordings made by Rasmussen and Head (1965) in Laguna Ojo de Liebre and off San Diego, California contained no signals attributable to this species. However, recordings they made in the presence of gray whales off Todos Santos Bay, Baja California Norte, Mexico included a "series of intense sounds" they likened to sounds produced by hammering against the hull of a wooden ship. Hubbs (1966) obtained no recordings of gray whale sounds during work around solitary individuals and cow/calf pairs in Laguna San Ignacio. Between 1967 and 1978, various researchers attempted to record sounds produced by the gray whale (Table I).

In reviewing the literature, it became apparent that additional research was needed to accurately understand sound production by the gray whale. Important in such research was the documentation of ambient noise levels, both natural and manmade, necessary for determining the impact such noise had on the gray whale.

Accordingly, an acoustical investigation on the gray whale was initiated in Laguna San Ignacio, Baja California Sur, Mexico. The primary objectives were (1) to determine the levels of naturally occurring underwater biological ambient noise prevalent in the lagoon; (2) to profile the contribution of underwater manmade noise to the ambient; and (3) to determine the types of acoustical signals emitted by the gray whale. This chapter reports these acoustical investigations.

Methods

STUDY AREA

Laguna San Ignacio, Baja California Sur, Mexico, located along the Pacific Coast of Baja California approximately 680 km south of San Diego, California, is one of the major breeding areas for gray whales (Fig. 1). Laguna San Ignacio was a nearly "ideal" location for the proposed work. A dense concentration of whales occurs in a small area, ensuring a large sample size; it is a focal point for tourism, allowing us to monitor phonations and ambient noise on days with and without vessel activity; and areas within the lagoon are relatively undisturbed, facilitating identification of nonbiological sources. In addition, Laguna San Ignacio has been the site of long-term investigations on the gray whale (Swartz and Cummings, 1978; Swartz and Jones, 1980, 1981).

1981 FIELD SEASON (MARCH 8–25)

A calibrated KSP (HS-107/222) hydrophone was connected through a 2-dB step attenuator to a Nakamichi 550 tape recorder.[1] The frequency response of this system was 40 Hz–19 kHz $\pm$ 3 dB. In all sampling procedures, the hydrophone was placed at midwater depth and the same gain and attenuator settings were used. Pertinent environmental and behavioral data were recorded on a narrative channel. Biological ambient

[1]Use of trade names does not imply endorsement by authors, editors, or publishers of this book.

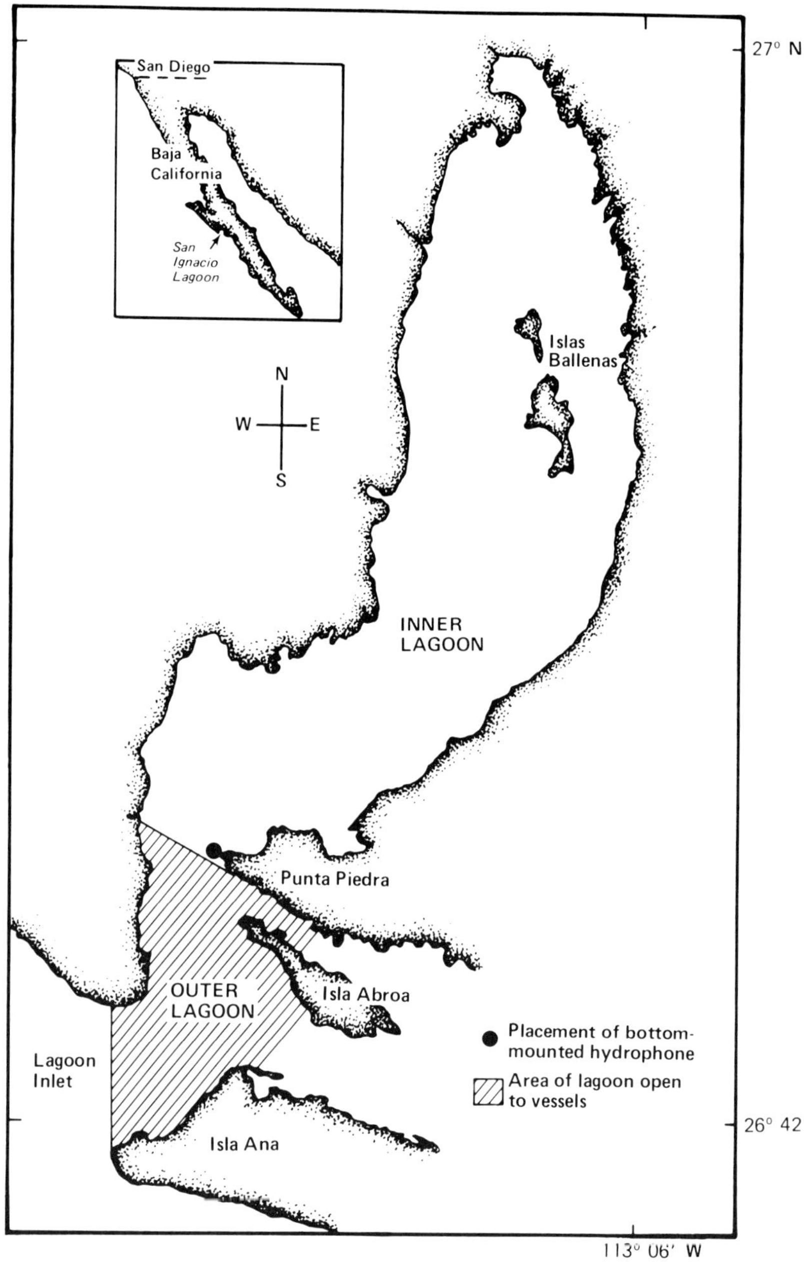

Fig. 1. *Laguna San Ignacio, Baja California Sur, Mexico.*

noise was sampled throughout the lagoon from a 4.2-m inflatable boat. Bearings to three landmarks were taken at the beginning and end of each 15-min recording session to determine locations of sampling stations.

Sound spectra of large vessels (25–35 m in length) were recorded at known distances from the port or starboard beam aspect when these vessels were underway and at anchor in the outer lagoon. Underwater sounds from small skiffs powered by outboard engines were also recorded as they operated at various speeds and at idle in the outer lagoon. These skiffs included inflatable, wooden, and aluminum boats ranging from 3 to 7 m in size. Various types of outboard engines were used: 18-hp Mercurys, 20-hp Mercurys, 35-hp Johnsons, and 40-hp Evinrudes.

To calculate biological ambient noise levels, one-third octave average sound pressure levels were computed using a Spectral Dynamics 345 spectrum analyzer set for 0- to 20-kHz bandpass and either 32 or 64 linear averages. The reported average sound pressure levels were calculated by averaging as many linear averages as possible from a single biological station. Using this method, the reported levels for the stations represent a range of actual sampling times equivalent to a flat weighted average sound pressure meter sampling from 1.28 to 14.0 sec.

One-third octave levels for vessels ranging in length from 25 to 35 m were calculated using a single output of 64 linear averages from a Spectral Dyanmics 345 spectrum analyzer set at a 20-kHz bandpass. These data were not range corrected; thus, they represent levels received at the hydrophone. The outboard engine one-third octave levels were sampled using a General Radio 1995-9008 equipped with a 10-Hz to 80-kHz bandpass. These levels were corrected assuming spherical spreading attenuation (20 log R) to 1 yd. Although Laguna San Ignacio represents a shallow water environment (<30 m), we predicted free-field conditions (spherical spreading) because of the close proximity of the outboard engines to the recording instrumentation.

Gray whale sounds were recorded during ambient noise sampling periods. The narrative accompanying such recordings included estimated distance of whales from the recording platform, behavior, group composition, and the total number of whales within visual range. Signals produced by bottlenosed dolphins (*Tursiops truncatus*) were also recorded incidentally during these sessions.

1982 FIELD SEASON (FEBRUARY 12–MARCH 2)

A calibrated Nagra IV-SJS tape recorder was coupled with the KSP hydrophone and attenuator used in 1981. This system provided an effective frequency response of 25 Hz to 10 kHz ± 2 dB at 9.5 cm/sec and 25 Hz to 20 kHz ± 2 dB at 19 cm/sec. Recordings were made at both speeds to determine whether gray whale signals extended above 10 kHz.

A bottom-mounted hydrophone was used to record whale sounds. The hydrophone was suspended by a float at a depth of 3 m, and its associated cable extended 25 m along the bottom of the lagoon to the shore at Punta Piedra (see Fig. 1). Gray whales were observed to pass within 2 m of the hydrophone. This bottom-mounted system allowed for 24-hr recording and monitoring of gray whale signals and collection of synoptic ambient noise data. As in previous sessions, a simultaneous narrative accompanied recordings.

Sounds produced by gray whales and bottlenose dolphins during 1981 and 1982 were analyzed on a Kay Elemetrics Corporation Sona-Graph (Model 7029A). Narrow filtering was used during all analyses. Using a numbering system, sound categories were established for the gray whale. These categories were visually and aurally compared to published accounts with sufficient quantitative data.

Results

BIOLOGICAL AMBIENT NOISE

Biological ambient noise was recorded at 32 stations, 14 in the outer and 18 in the inner lagoon (Fig. 2). Although noise spectra varied among the stations, typically they followed the pattern depicted in Fig. 3. Levels were lowest below 2 kHz, increased to high levels between 2 and 5 kHz, and then declined gradually through 20 kHz. The average ambient noise levels attributable to biological sources (excluding cetaceans) ranged from 94 to 110 dB re 1 μPa. These levels and associated octave-band frequencies for each of the 32 stations are listed in Table II. A spectrograph of the biological ambient noise is shown in Fig. 4.

NONBIOLOGICAL AMBIENT NOISE

The only sources of nonbiological noise analyzed for this report were vessel and skiff traffic. Data were collected on six large vessels (25 to 35 m in length) and 14 skiffs powered by outboard engines. Because each vessel and skiff had its own sound signature, only representative samples are shown.

Noise from large vessels was broadband, ranging from 125 Hz to 20 kHz, with most energy concentrated below 10 kHz. Tonal peaks were typically observed below 1 kHz. Received levels of large vessels underway averaged 89 to 105 dB re 1 μPa (Fig. 5). Received levels from vessels at anchor were 10 to 20 dB below those from the same vessel underway.

Fourteen skiffs powered by five different models of outboard engines were recorded. Spectrum profiles were obtained for each model of engine. Overall noise levels were similar among models, but variations were prevalent in the sound signature of each engine. Noise levels and emphasized frequency bands also varied with the rpm of the engine. Source levels for a 20 hp Mercury outboard engine operating at idle, quarter throttle, half throttle, and three quarter throttle are shown in Fig. 6A–D. The overall noise contributed by the outboard engine at idle is generally lower than that of the engine in gear. Estimated levels at idle occurred at frequencies below 1 kHz, with a noise level range of 92–117 dB re 1 μPa at 1 yd. Ranges of source levels (re 1 μPa at 1 yd) for speeds other than idle were quarter speed, 104–124 dB; half speed, 108–126 dB; and three-quarter speed, 108–134 dB. The levels reported for the three-quarter speed repre-

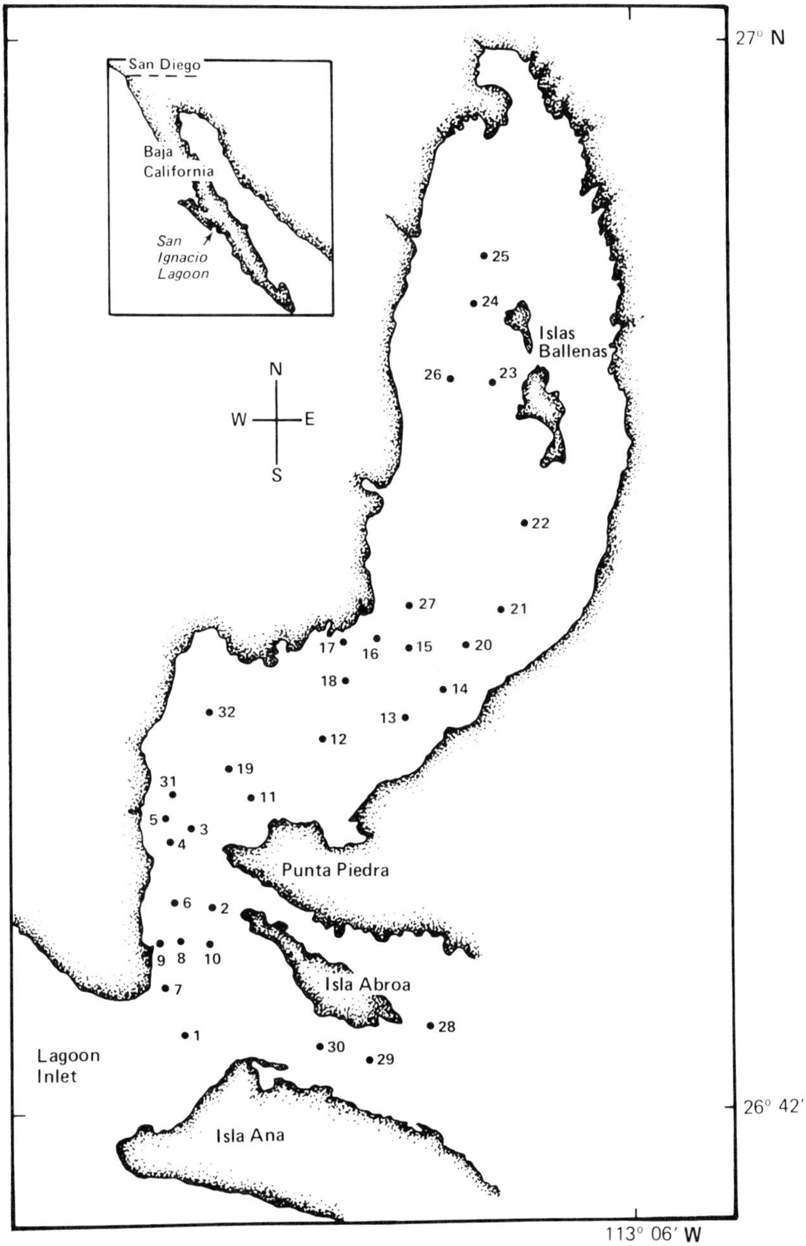

Fig. 2. *Locations of biological ambient noise stations in Laguna San Ignacio.*

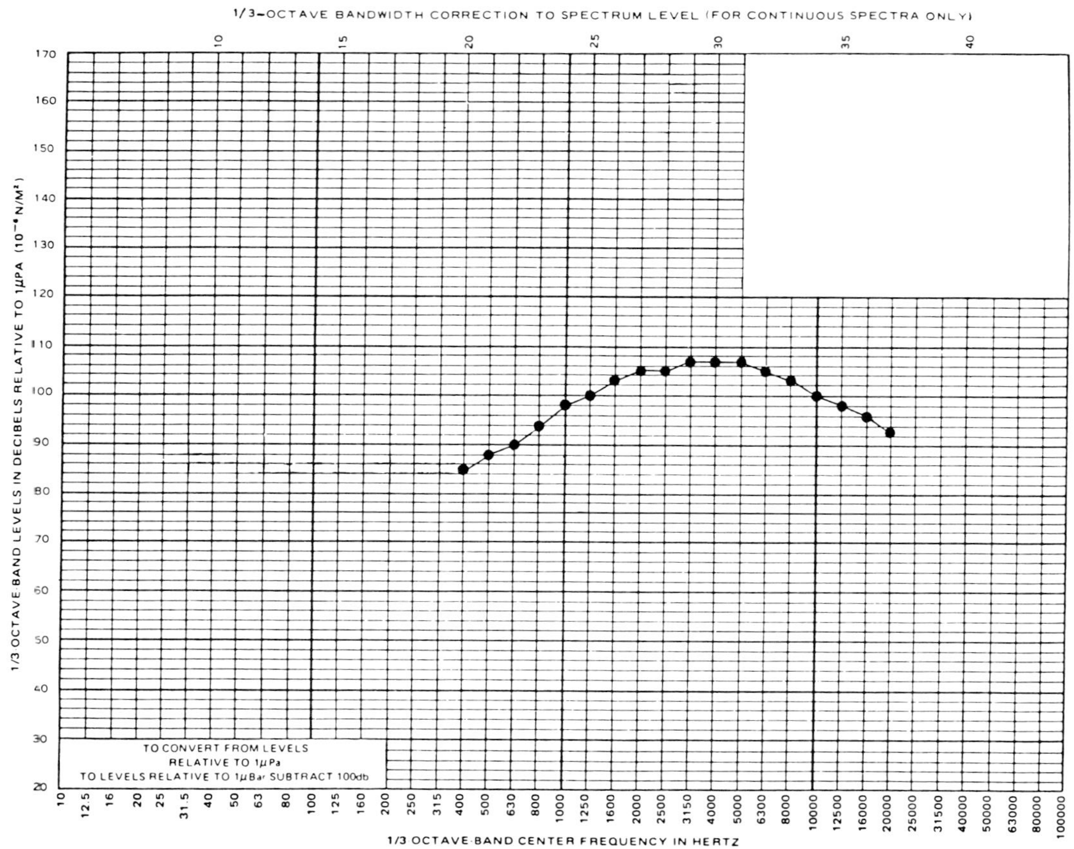

Fig. 3. *Spectra of typical biological ambient noise levels in Laguna San Ignacio.*

Table II
Estimated Levels (dB re 1 μPa) and Their Associated Frequencies (Hz) for Biological Ambient Noise Stations #1–32[a]

Biological noise stations	Estimated level (db re 1 μPa)	Frequency (Hz)						
		1250	1600	2000	2500	3150	4000	5000
1	97						X	X
2	94						X	X
3	103					X	X	
4	107					X	X	X
5	98					X	X	X
6	109						X	X
7	97					X	X	X
8	94					X	X	X
9	99						X	
10	98						X	
11	105					X	X	X
12	102			X	X	X	X	
13	100			X	X	X	X	
14	103						X	
15	102			X				
16	105			X				
17	106			X				
18	104			X				
19	99			X				
20	96			X				
21	103			X	X		X	
22	102					X	X	
23	102				X	X		
24	106			X	X	X	X	
25	103			X				
26	110	X	X	X				
27	106			X	X		X	
28	105						X	
29	109						X	
30	108					X	X	X
31	99					X	X	X
32	95						X	

[a]For example, Station #1 had a maximum level of 97 dB in both the 4- and 5-kHz bands.

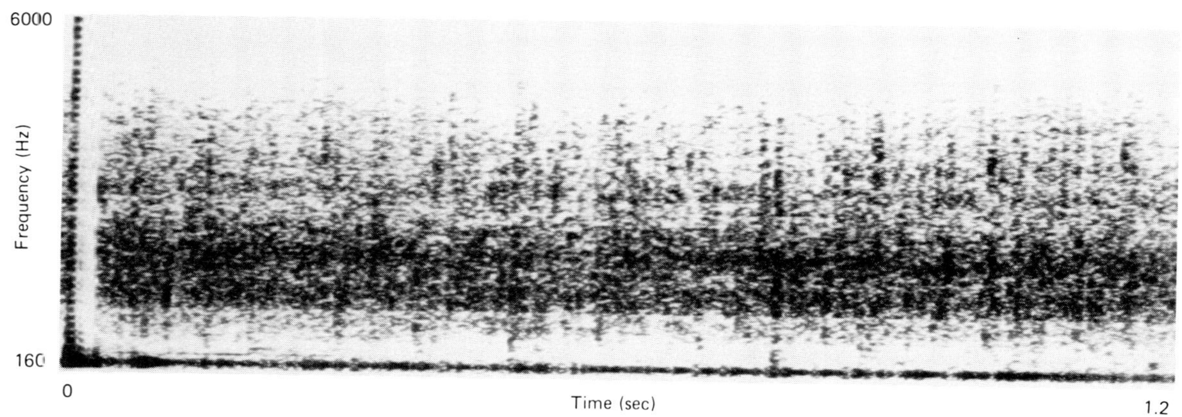

Fig. 4. *Spectrograph of typical biological ambient noise (excluding cetaceans) prevalent in Laguna San Ignacio.*

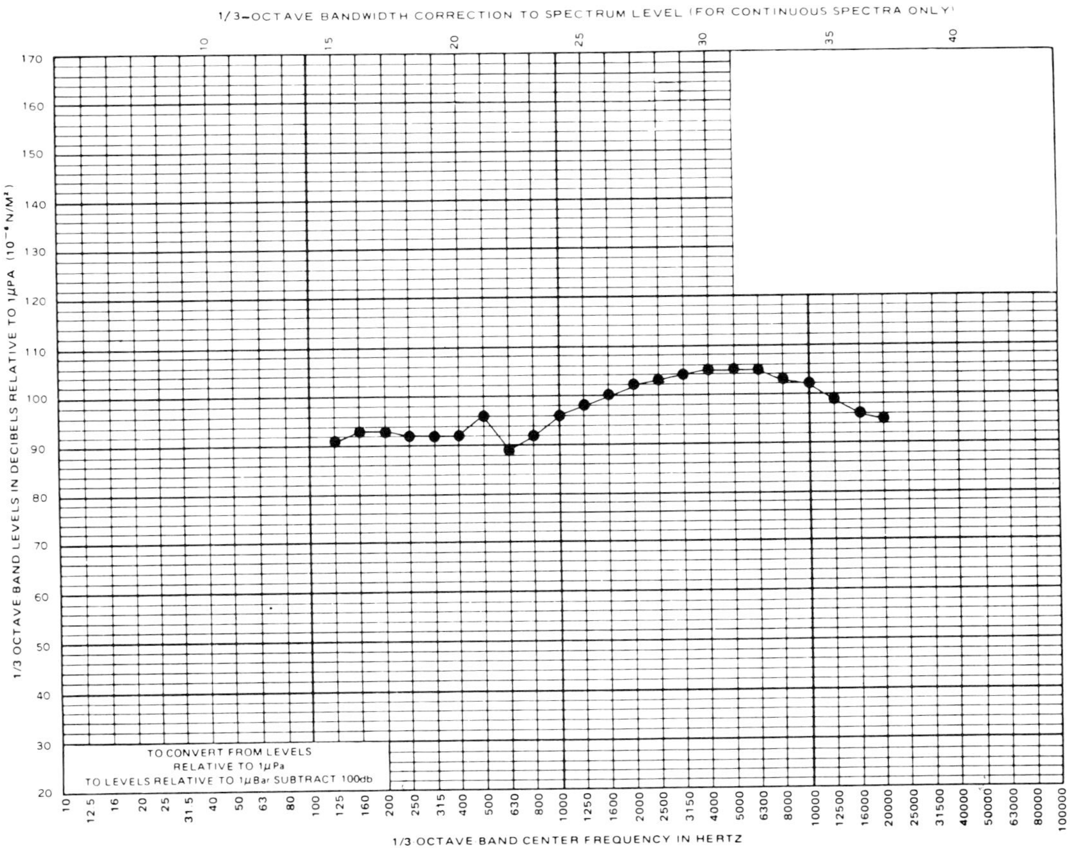

Fig. 5. *Spectra of typical large vessel (25–35 m in length) noise occurring in Laguna San Ignacio.*

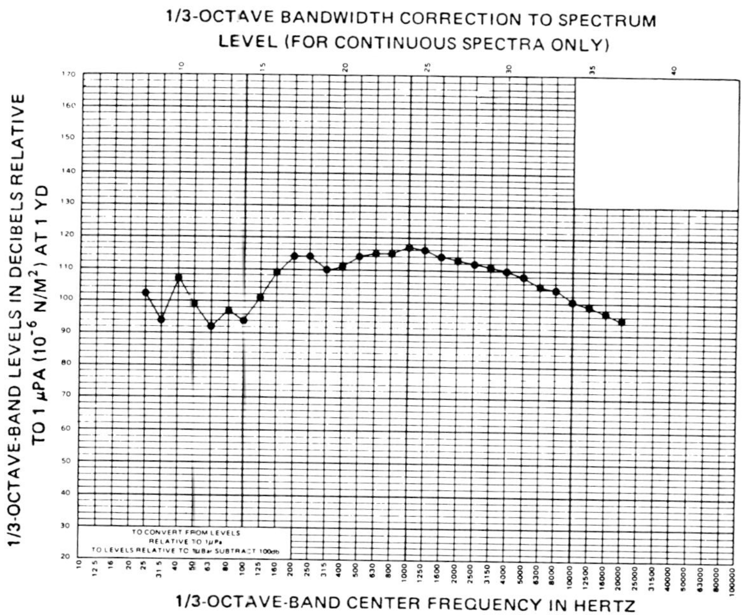
1/3-OCTAVE BANDWIDTH CORRECTION TO SPECTRUM LEVEL (FOR CONTINUOUS SPECTRA ONLY)
1/3-OCTAVE-BAND LEVELS IN DECIBELS RELATIVE TO 1 μPA (10^-6 N/M^2) AT 1 YD
1/3-OCTAVE-BAND CENTER FREQUENCY IN HERTZ
TO CONVERT FROM LEVELS RELATIVE TO 1μPa TO LEVELS RELATIVE TO 1μBar SUBTRACT 100db

A

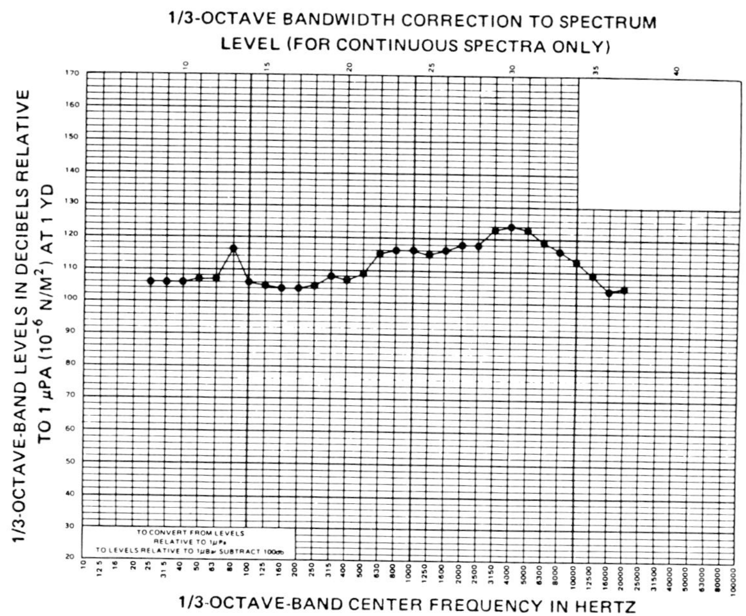
1/3-OCTAVE BANDWIDTH CORRECTION TO SPECTRUM LEVEL (FOR CONTINUOUS SPECTRA ONLY)
1/3-OCTAVE-BAND LEVELS IN DECIBELS RELATIVE TO 1 μPA (10^-6 N/M^2) AT 1 YD
1/3-OCTAVE-BAND CENTER FREQUENCY IN HERTZ
TO CONVERT FROM LEVELS RELATIVE TO 1μPa TO LEVELS RELATIVE TO 1μBar SUBTRACT 100db

B

Table III
A Classification of Sounds Produced by the Gray Whale in Laguna San Ignacio[a]

Sound type	*n*	Frequency range		Concentration of energy		Duration	Number of pulses per series	Pulses per sec	Individual pulse duration
		Low	High	Low	High				
S1	563	90 (20–1250)	1940 (1000–3000)	332 (100–1250)	824 (400–1600)	1.8 (series) (0.3–4.6)	9.4 (2–30)	5.9 (2.2–14.7)	<0.05
S2	30	250 (100–300)	300 (200–350)	250 (200–275)	300 (250–350)	0.3 (0.2–0.4)	—	—	—
S3	40	125 (80–200)	1250 (750–1800)	170 (125–625)	430 (300–750)	2.0 (1.0–4.0)	—	—[b]	—
S4	30	150 (125–200)	1570 (1500–1600)	225 (125–300)	600 (450–750)	0.94 (0.7–1.4)	—	—[b]	—
S5	30	130 (20–250)	840 (600–1500)	200 (20–250)	500 (300–750)	3.2 (1.8–4.5)	—	—	—
S6	15	250 (125–300)	850 (600–1000)	250 (125–300)	700 (300–800)	3.3 (1.2–3.6)	—	—	—

[a]Mean values are given and their associated ranges are shown in parentheses. All frequency values are given in Hz and time values in seconds.
[b]Pulse modulation occurring.

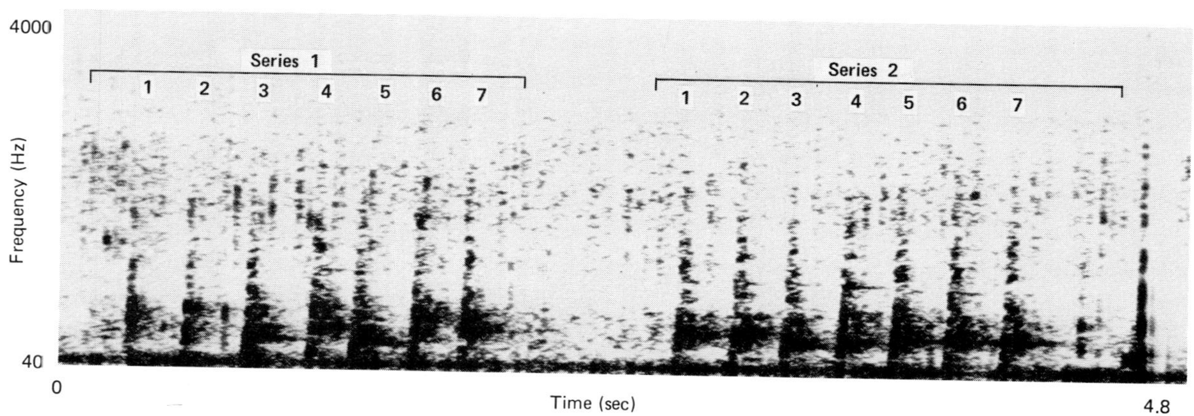

Fig. 7. *Spectrograph of a gray whale S1 series.*

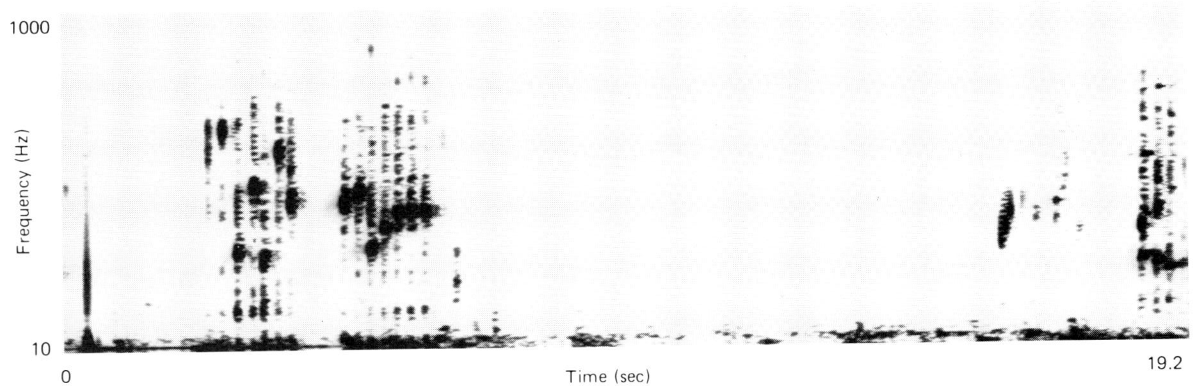

Fig. 8. *Spectrograph of two S1 series, depicting frequency modulation.*

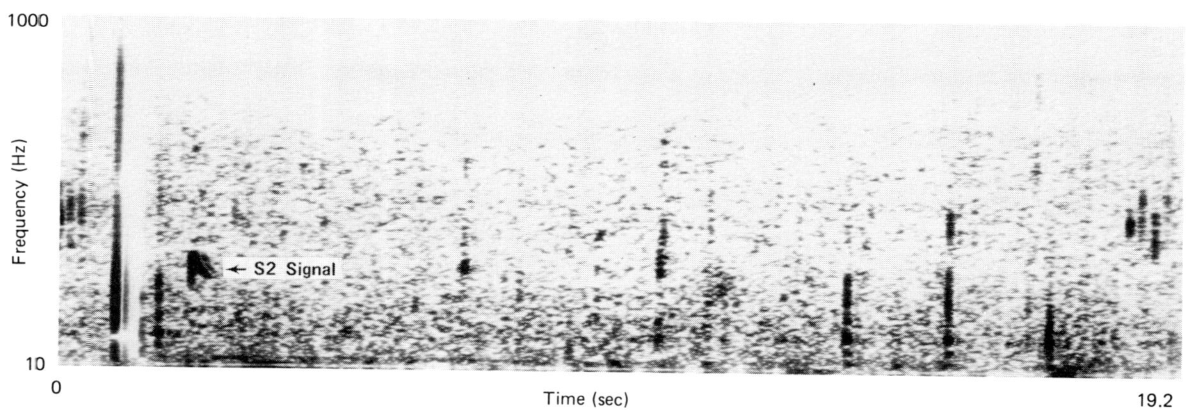

Fig. 9. *Spectrograph of a gray whale S2 signal.*

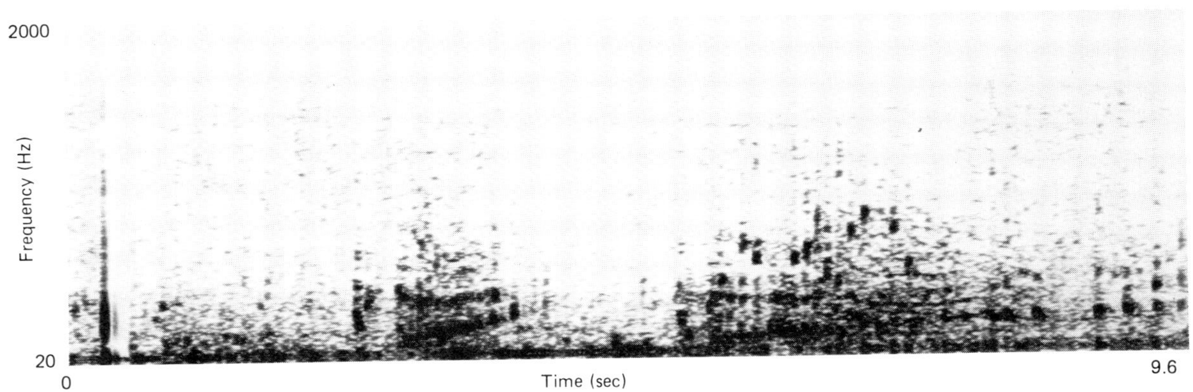

Fig. 10. *Spectrograph of a gray whale S3 signal.*

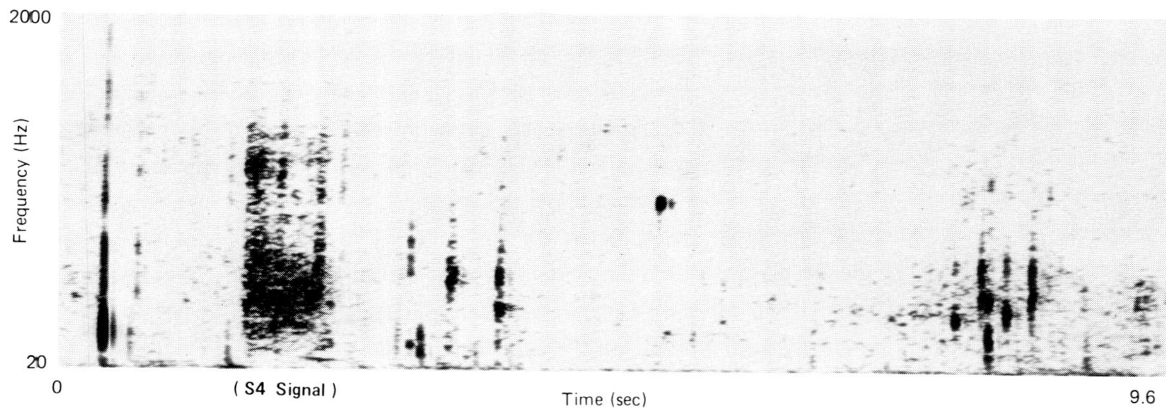

Fig. 11. *Spectrograph of a gray whale S4 signal.*

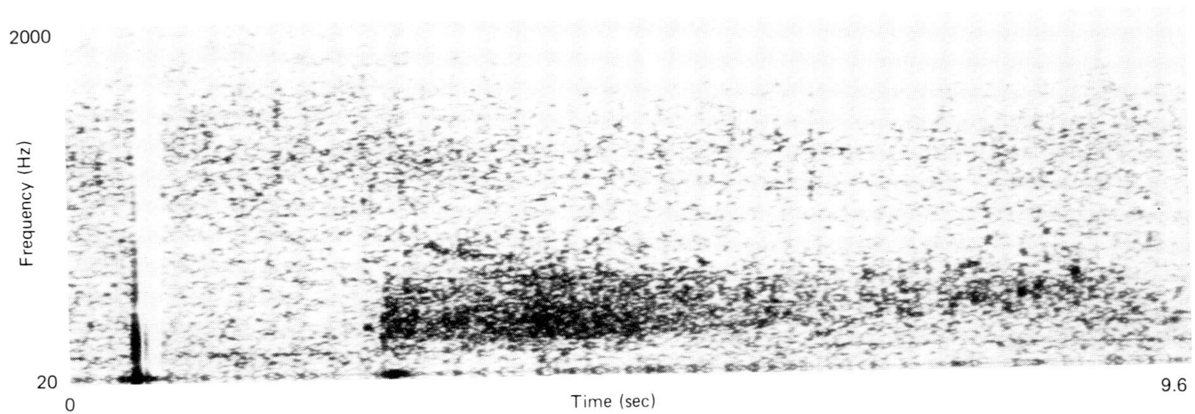

Fig. 12. *Spectrograph of a gray whale S5 signal (bubble blast).*

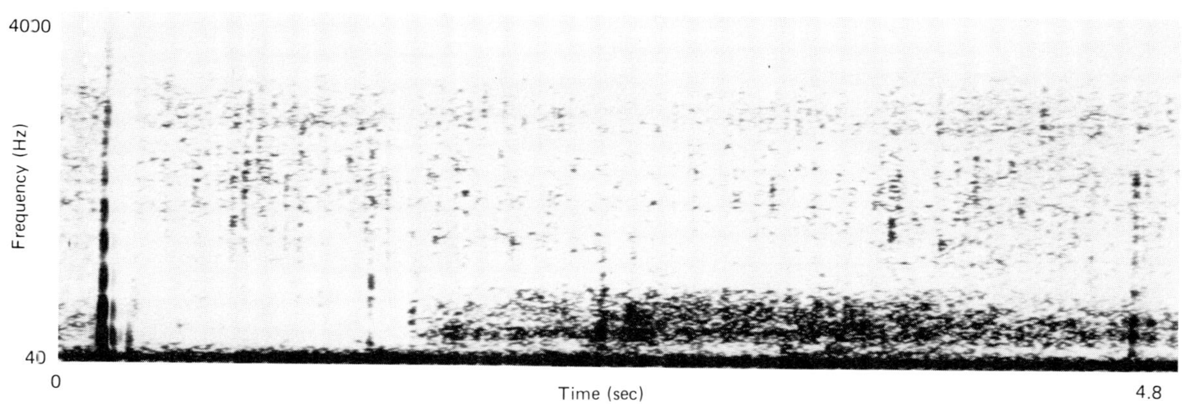

Fig. 13. *Spectrograph of a gray whale S6 signal (subsurface exhalation).*

The six major sound types were heard under a wide variety of behavioral activities and environmental conditions. We have not attempted to correlate sound production and behavior, with the exception of the bubble blast and the subsurface exhalation. Gray whales, however, appeared to be more soniferous under the following situations: (1) when numerous whales were together in a small area; (2) when whales were on collision courses with one another; (3) when single whales were observed chasing cow/calf pairs; (4) when whales were interacting with bottlenose dolphins; (5) when whales were on collision courses with vessels or skiffs; and (6) when whales were in the presence of increased nonbiological ambient noise conditions due to vessels or skiffs transiting the area.

Underwater recordings were made in the presence of breaching whales ($n = 20$) and in the presence of animals engaged in sexual activities ($n = 5$). No distinctive underwater sounds were heard or recorded in association with these two behaviors. In comparing types of signals produced and periodicity of phonations, no differences in gray whale sound production were found between day and night recordings.

Whales exhibiting "curious" behavior were encountered during our acoustical investigations. Engines kept in idle (engine running but propeller stopped) maintained curious whales in close proximity for periods up to 3 hr. Some whales terminated this activity when the engine was shut off. A spectrograph depicting outboard engine noise and a S1 signal we believed was produced by a curious whale is shown in Fig. 14. Upon inspection of this display, it becomes apparent that the main concentration of energy in the S1 signal falls within the frequency range emitted by the outboard engine noise.

To determine the frequency relationship of gray whale signals to the biological ambient noise, spectrographs were made expanding the frequency range to 4000 Hz. An interesting pattern emerged. A composite spectrograph representing a gray whale S1 series, outboard engine noise, and the naturally occurring biological ambient noise is shown in Fig. 15. Skiff noise predominates the frequencies below 500 Hz, with the biological noise dominating the frequencies above 2 kHz. Although gray whale signals do include energy up to 2 kHz, most energy of this pulsed call is below 1 kHz. All gray whale sounds analyzed emphasized the frequency bands below the main concentration of energy of the biological ambient noise. We analyzed 50 signals produced by the bottlenosed dolphin (*Tursiops truncatus*), the only other cetacean that occupied Laguna San Ignacio. This was accomplished to determine the relationship of the dolphin's signals to that of the high levels of natural occurring background noise (Fig. 16). In all cases, dolphin whistles occurred at frequencies above 7 kHz. The short duration whistles were frequency modulated, with the lowest frequency observed to rapidly rise when it encountered the ceiling of the biological ambient noise. Because dolphin signals were not treated in the same depth as those of the gray whale, these data should be interpreted with caution.

After comparing our spectrographs and recordings with those previously published, we attempted to synthesize the available data on calls produced by the gray whale (Table IV). Seven categories of sounds were established for the gray whale throughout its range.

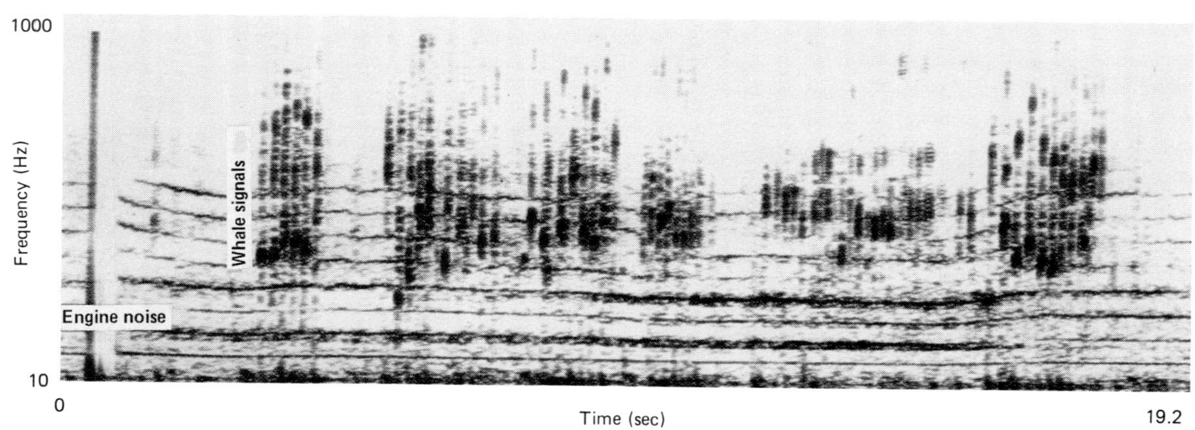

Fig. 14. *Spectrograph of a gray whale pulse series (S1) and outboard engine noise.*

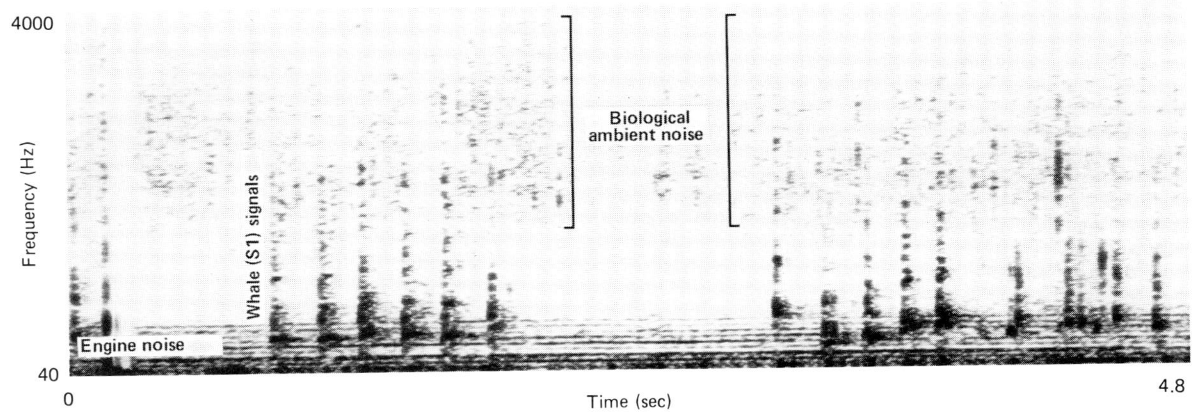

Fig. 15. *Composite spectrograph of a gray whale pulse series (S1), outboard engine noise, and biological ambient noise in Laguna San Ignacio.*

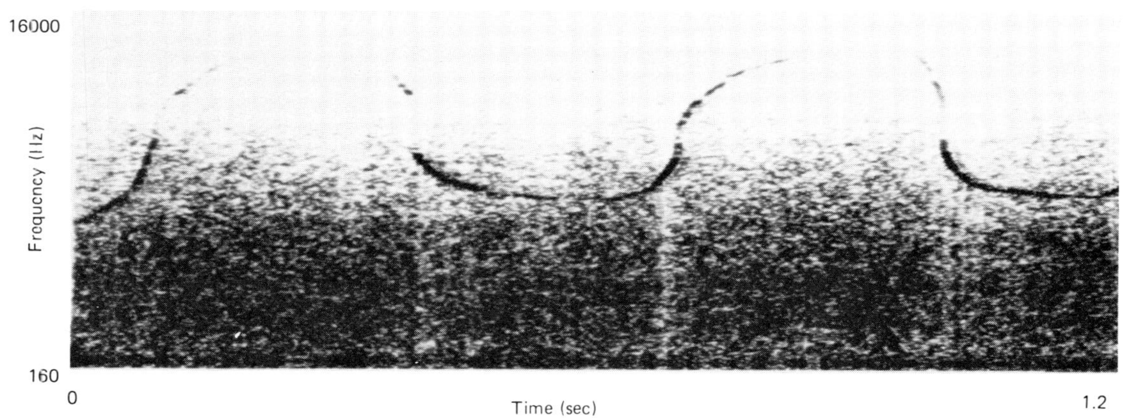

Fig. 16. *Spectrograph of a bottlenosed dolphin whistle and biological ambient noise.*

Table IV
An Integration of Sounds Produced by the Gray Whale Throughout Its Range Based on Acoustical Studies in Laguna San Ignacio and Published Accounts[a]

Sound type	Geographical area	Reference
S1[b]	S/N/M[c]	This chapter; Asa-Dorian and Perkins, 1967 (pulses); Fish *et al.*, 1974 (pulse trains); Swartz and Cummings, 1978 (knocks and pulses); Chapter 22, this volume (knocks).
S2	S	This chapter.
S3	S/N/M	This chapter; Cummings *et al.*, 1968 (moans); Fish *et al.*, 1974 (growls or moans); Chapter 22, this volume (moan).
S4	S/N/M	This chapter; Fish *et al.*, 1974 (grunt); Swartz and Cummings, 1978 (snort); Chapter 22, this volume (grunt).
S5	S/M	This chapter; Cummings *et al.*, 1968 (Bubble-type); Swartz and Cummings, 1978 (bubble bursts).
S6	S/M	This chapter; Cummings *et al.*, 1968 (blow sounds).
S7	M	Fish *et al.*, 1974 (clicks and click train).

[a]Only those accounts with sufficient data for comparison were included.

[b]An S1 signal produced on the northern range would be designated as an N1; the same type of signal produced during migration would be designated as M1.

[c]S, southern range; N, northern range; M, migration route.

Discussion

The high levels of biological ambient noise in Laguna San Ignacio are typical of coastal, shallow, warm-water lagoon environments (Johnson, 1943). Previous investigators attempting to record gray whale signals on the southern part of their range also noted high levels of background noise (Asa-Dorian and Perkins, 1967; Rasmussen and Head, 1965). Predominant frequencies of biological noise ranged from 2 to 5 kHz, with estimated one-third octave band levels of 94–110 dB re 1 μPa. We suspect that snapping (pistol) shrimp, either *Crangon dentipes* or *Synalpheus lockingtoni*, were responsible for most of the biological noise recorded during the present study. However, there are numerous other marine invertebrates and species of fish residing in the lagoon system that could be contributing to the overall noise.

Vessel and skiff traffic were the only sources of nonbiological noise measured for this report. These levels were higher than those observed for the naturally occurring ambient noise and emphasized different frequency ranges. Nonbiological noise predominated frequencies below 2 kHz. The received levels of noise reported do not truly represent the source level of the vessel, because these acoustical waves have undergone transmission losses and attenuation. Source levels would, of course, be greater than those received at some distance. Therefore, we must recognize that whales exist in a very dynamic acoustic environment.

We elected to use a numbering system to categorize gray whale sounds because onomatopoetic terms denote different things to people. Field observations assured us that these sounds were in fact produced by the gray whale. Signal strengths of all sounds increased as whales approached and decreased as they moved off. In some cases, whales apparently producing the sounds were within 3 m of the hydrophone, and associated high sound source levels were obtained.

Gray whales were more soniferous than expected from published accounts. The low-frequency sounds produced by these whales might easily have been masked by high levels of biological ambient noise characteristic of the southern range. Such high-level background noise may have prompted the reports that gray whales were silent. Recent advances in technology and the amount of time spent in the field compared to previous investigators may account for the numerous phonations we recorded. Gray whales appeared to be more soniferous under the six conditions listed in the results, suggesting that they produce sounds only when there is a need to be acoustically active.

By integrating all available sound types, a clarification of the gray whales' repertoire is obtained. Table IV suggests gray whales produce at least seven distinct sound types. Although variations exist within each sound category, each category was sufficiently distinct to be classified independently. Variations in these call types may reflect individual differences or responses to different environmental or behavioral situations.

With the exception of the S2 signal, all sounds listed in Table IV have been reported for gray whales on their southern range and during migration. Three sound types (S1, S3, S4) have also been recorded on their northern range (Chapter 23, this volume; recent recordings by the senior author). The most prevalent sound produced by the gray whale in Laguna San Ignacio and in the Bering Sea was the S1 signal. However, the S1 signal typically was frequency modulated in the breeding lagoon, which was not the case on the feeding grounds. The S1 call was also the most frequently heard signal recorded from the captive gray whale, Gigi (Fish *et al.*, 1974). The S7 call type was not recorded during investigations at Laguna San Ignacio; however, there appears to be sufficient information (Fish *et al.*, 1974) to warrant its categorization.

Behavioral correlations with sound production were not addressed for this report. Although several papers have attempted to ascribe a communicative or echolocative function to the calls of gray whales, it does not appear at present, that sufficient data exists to explain the nature of their calls. The S5 and S6 signals were associated with the bubble blast and the subsurface exhalation behavior, respectively. However, these two signals do not represent true phonations. No obvious differences in sound periodicity or types of signals produced by the gray whale were detected between day and night recordings. This agrees with observations by Swartz and Cummings (1978) in Laguna San Ignacio in which they reported similar behavioral activities by gray whales over 24-hr periods.

One explanation often given for whale breaching behavior is that the breach produces a low-frequency sound that might function in communication. We detected no distinctive underwater sound in the presence of breaching whales, although the airborne signal produced by a breaching whale falling on the water was readily audible and traveled a considerable distance (as far as 6 km, S. L. Swartz and M. L. Jones, personal communication). Watkins (1981) noted that the breach of a finback whale (*Balaenoptera physalus*) produced very little underwater sound even at low frequencies. The available data seem to indicate that breaching does not function in underwater acoustic communication.

No signals were recorded in the presence of courting whales. Right whales (*Balaena glacialis*) have been observed to produce many sounds during these activities (Chris Clark, personal communication), and the songs of humpback whales (*Megaptera novaeangliae*) during the breeding season have been well documented (Payne and McVay, 1971).

Curious gray whales were observed to actively seek out and physically contact slow-moving (2–4 knots) small skiffs. These whales also approached stationary skiffs and large vessels at idle and at anchor. Gilmore (1976), Swartz and Cummings (1978), and Swartz and Jones (1980, 1981) have documented "curious or friendly" whale behavior towards vessels and skiffs in Laguna San Ignacio over the past 7 years. This behavior has just recently been described in Laguna Guerrero Negro, Baja California Norte, Mexico (Chapter 15, this volume). Although we speculate the sound generated by the vessels or skiff engines are the initial stimulus in attracting the whales, other factors may be involved once the whale physically contacts the boat. It is noteworthy that these "curious" whales respond to underwater engine noise that occupies the same frequency ranges as their own signals.

The majority of gray whale signals were at frequencies below 2 kHz and occupied a different acoustical channel (20 Hz–2 kHz) than those of the biological noise (most energy occurred between 2 and 5 kHz). When gray whale signals were compared to the nonbiological sources, we obtained overlapping frequency ranges. Our two major sources of background ambient noise (biological and nonbiological) exhibited different frequency profiles. In addition, they also represented constant sources (biological) and temporary or intermittent sources (nonbiological). It appears reasonable that a cetacean would try to circumvent constant high levels of noise. Our gray whale signals were consistently below the biological ambient noise and our preliminary results on bottlenosed dolphin signals show that their frequencies are consistently above the biological ambient noise. These data imply that these cetaceans have different acoustical "niches" possibly dictated by the constant high levels of biological ambient noise in Laguna San Ignacio.

There is general agreement among scientists that the acoustical sense of marine mammals, especially that of the cetaceans, constitutes their most important sensory process. Although the sea provides a good medium for sound propagation and reception, ambient noise levels (both natural and manmade) may affect hearing abilities or act to inhibit sound production. Such reductions in acoustic reception or transmission could adversely affect the reproductive potential or even the survival of a given species. Calls that have these survival functions must overcome this noise to remain effective. The "acoustical niche" hypothesis would represent an adaptable strategy of these cetaceans to overcome or circumvent the high levels of biological noise prevalent in their environments, thereby insuring minimum interference and masking with the transmission and reception of their own signals. Future research is aimed at testing the acoustical niche hypothesis and determining whether or not nonbiological noise could mask the signals of the gray whale.

Summary

Underwater acoustical investigations were conducted in Laguna San Ignacio, Baja California Sur, Mexico, to determine levels of ambient noise prevalent in the lagoon system and to record underwater signals produced by the gray whale (*Eschrichtius robustus*). At 32 stations, estimated levels of biological noise (excluding cetaceans) ranged from 94–110 dB re 1 μPa, with most energy occurring at frequencies between 2 and 5 kHz. Received levels of noise from vessels ranging in length from 25 to 35 m occurred between 89 and 105 dB re 1 μPa, with most energy below 2 kHz. "Curious" whales were apparently attracted to underwater sounds generated by skiffs powered by outboard engines. This outboard engine noise occupied the same frequency ranges as signals produced by the gray whale ($<$2 kHz). Six distinct sound types (S1 to S6) were classified for the gray whale on its breeding grounds. An integration of our categorized sound types with published sounds resulted in a definition of a total of seven distinct sound types produced by this species throughout its range. Sounds produced by gray whales and bottlenosed dolphins (*Tursiops truncatus*) occurred at frequencies below or above the main concentration of energy of the biological noise, respectively. An "acoustical niche" hypothesis is proposed for the cetaceans occupying Laguna San Ignacio.

Acknowledgments

We are most indebted to the Mexican government for allowing us to conduct this work. We thank Jorge Carranza Frasier, Edith Polanco Jaime, Alfonso Yanez, Alonso Lopez, Walter O'Campo, Pedro Mercado Sanchez, and Luis Fleischer. Steven Swartz and Mary Lou Jones provided assistance and support during our two field seasons. Their knowledge and experience with the gray whales, and their understanding of the lagoon system, allowed for the success of our work. Robert F. Dahlheim, Jr., provided valuable support and encouragement during this study. Special thanks to several agencies and individuals who assisted at various stages of the research: Naval Acoustic Range, Bremerton, Washington (C. P. Henson); National Marine Mammal Laboratory, Seattle, Washington (Howard Braham, Mary Nerini, and Michael Tillman); M/V *Pacific Queen*, San Diego, California (Owner/Operator Ed McEwen); H & M Landing, San Diego, California (Catherine and Robert Miller); Margie Stinson; Larry Hobbs; James Sumich; Michael Symons; Michael Bursk; Francois Gohier; San Diego State University, San Diego, California (Frank Awbrey); and James Fish. William Watkins and Frank Awbrey critically reviewed this manuscript.

References

Aldrich, H. L. (1889). "Arctic Alaska and Siberia," pp. 32–35. Rand McNally, Chicago, Illinois.

Asa-Dorian, P. V., and Perkins, P. J. (1967). The controversial production of sound by the California gray whale, *Eschrichtius gibbosus*. *Nor. Hvalfangst-Tid.* **56**(4), 74–77.

Cummings, W. C., Thompson, P. O., and Cook, R. (1968). Underwater sounds of migrating gray whales, *Eschrichtius glaucus* (Cope). *J. Acoust. Soc. Am.* **44**(5), 1278–1281.

Eberhardt, R. L., and Evans, W. E. (1962). Sound activity of the California gray whale, *Eschrichtius glaucus*. *J. Audio Eng. Soc.* **10**(4), 324–328.

Fish, J. F., Sumich, J. L., and Lingle, G. L. (1974). Sounds produced by the gray whale, *Eschrichtius robustus*. *Mar. Fish. Rev.* **36**(4), 38–45.

Gilmore, R. M. (1976). The friendly whales of Laguna San Ignacio. *Terra* **15**(1), 24–28.

Hubbs, C. L. (1966). Comments. *In* "Whales, Dolphins, and Porpoises" (K. S. Norris, ed.), p. 444. Univ. of California Press, Berkeley and Los Angeles.

Johnson, M. W. (1943). Underwater sounds of biological origin. *OSRD* Sect. 6, No. 1270. 1-sr 30-412.

Norris, K. S., Goodman, R. M., Villa-Ramirez, B., and Hobbs, L. (1977). Behavior of the California gray whale, *Eschrichtius robustus*, in southern Baja California, Mexico. *Fish. Bull.* **75**(1), 159–172.

Painter, D. W., II (1963). Ambient noise in a coastal lagoon. *J. Acoust. Soc. Am.* **35**(9), 1458.

Payne, R., and McVay, S. (1971). Songs of humpback whales. *Science* **173,** 587–597.

Poulter, T. C. (1968). Vocalization of the gray whales in Laguna Ojo de Liebre (Scammon's Lagoon) Baja California, Mexico. *Nor. Hvalfangst-Tid.* **3,** 53–62.

Rasmussen, R. A., and Head, N. E. (1965). The quiet gray whale (*Eschrichtius glaucus*) *Deep-Sea Res.* **12,** 869–877.

Ray, G. C., and Schevill, W. E. (1974). Feeding of a captive gray whale. *Mar. Fish. Rev.* **36**(4), 31–38.

Swartz, S. L., and Cummings, W. C. (1978). Gray whales (*Eschrichtius robustus*), in Laguna San Ignacio, Baja California, Mexico. *U. S., NTIS, PB Rep.* **PB-276319,** 1–38.

Swartz, S. L., and Jones, M. L. (1980). Gray whales (*Eschrichtius robustus*) during the 1977–78 and 1978–79 winter seasons in Laguna San Ignacio, Baja California Sur, Mexico. *U. S., NTIS. PB Rep.* **PB80-202989,** 1–35.

Swartz, S. L., and Jones, M. L. (1981). Demographic studies and habitat assessment of gray whales (*Eschrichtius robustus*) in Laguna San Ignacio, Baja California Sur, Mexico. U. S., NTIS, PB Rep. **PB82-123373,** 1–56.

Tomilin, A. G. (1957). "Zveri SSSR i prilezhashchikh stran" (Mammals of the U.S.S.R. and Adjacent Countries), Kittobraznye, Vol. 9. Izd. Akad. Nauk SSSR, Moskva (Israel Program for Sci. Transl., Jerusalem, 1967).

Watkins, W. A. (1981). Activities and underwater sounds of fin whales. *Sci. Rep. Whales Res. Inst.* **33,** 83–117.

Wenz, G. M. (1964). Curious noises and the sonic environment in the Ocean. *In* "Marine Bio-Acoustics" (W. N. Tavolga, ed.), pp. 101–119. Macmillan, New York.

23

Gray Whales in the Beaufort, Chukchi, and Bering Seas: Distribution and Sound Production

Sue E. Moore and Donald K. Ljungblad

Introduction

Gray whales (*Eschrichtius robustus*) migrate into the Bering, Chukchi, and Beaufort Seas each summer, primarily to feed (Rice and Wolman, 1971). Major feeding areas have been documented in the north-central and western Bering Sea as well as in the southwestern Chukchi Sea (Votrogov and Bogoslovskaya, 1980; Bogoslovskaya *et al.*, 1981). Gray whales commonly range as far northeast as Point Barrow, Alaska (71°23′N, 156°25′W), with rarer sightings along the northern coast of Alaska to Barter Island (70°08′N, 143°35′W) (Maher, 1960). The easternmost records were those of three gray whales seen between Tuktoyaktuk and the Baille Islands, Canada, in the eastern Beaufort Sea (Rugh and Fraker, 1981).

Gray whale occurence, distribution, and behavior were studied during aerial surveys for whales, principally bowhead whales (*Balaena mysticetus*) in arctic waters during 1980 and 1981 (Ljungblad, 1981; Ljungblad *et al.*, 1982). These surveys primarily sought to corroborate existing data on gray whale distribution in the northeastern Bering Sea,

ISBN 0-12-389180-9

the eastern Chukchi Sea, and the Beaufort Sea. In addition, sounds were recorded near whales engaged in behavior interpreted as feeding. Although sounds believed to have been produced by gray whales have been quantitatively described for migrating, feeding, and captive whales (Eberhardt and Evans, 1962; Asa-Dorian and Perkins, 1967; Cummings *et al.*, 1968; Fish *et al.*, 1974), the only reports on gray whale sound production in arctic waters are anecdotal (Aldrich, 1889; Tomilin, 1957). Dahlheim *et al.* (Chapter 22, this volume) review published accounts of gray whale sound production throughout the range and add original data on sounds recorded in the lagoons of Baja California, Mexico. The recorded sounds described in the present paper are the first examined for gray whales in arctic waters.

Methods

STUDY AREA AND AERIAL SURVEYS

The study area was the northeast Bering Sea north of 62°N and east of the U.S./U.S.S.R. convention line, the southern and coastal Chukchi Sea, and the Beaufort Sea from shore to 72°N. Except for the northern Chukchi Sea, this area was divided into survey blocks suitable to the range of our aircraft (i.e., of a size such that each could be surveyed on a single flight) (Fig. 1).

Aerial surveys were chosen as the best means of sampling a large area over a short period of time. Surveys were conducted from April through October each year. The aircraft used was a Grumman Turbo Goose equipped with a Global Navigation System (GNS). The GNS has 0.37 km/hr precision and provides continuous position updating and transect turning point programming. Survey altitude was usually 153 m, but was adjusted relative to weather conditions as low as 100 m and as high as 350 m to maximize visibility. While circling to observe behavior and record whale sounds, altitude averaged 300 m. Air speed varied between 220 and 300 km/hr.

The three observers were positioned so that the person operating the GNS was in the right front seat and the other two observers were in opposite rear seats. Data were recorded by the left observer. Each observer was provided with a clinometer. The pilots and all observers were connected to a common communication system. The pilot acted as an observer only when other duties permitted.

Sightability of animals was reduced by such factors as percentage of ice coverage, rain and snow squalls, fog, haze, glare, and high sea states. Sea states ranged on the Beaufort scale from 01 (light ripple, wind 1–6 knots) to 06 (extensive whitecaps, wind 22–27 knots). Surveys were terminated as soon as possible when Beaufort 05–06 conditions were encountered. Surveys were conducted between the hours of 0830 and 1900.

At the time of sightings, the following were routinely recorded: aircraft altitude, time, latitude and longitude, Beaufort sea state, visibility range (subjective), number of animals, and behavior of the whales. Photographs were taken whenever possible using ASA 200 slide film and 80–230 mm zoom lens on a 35-mm Olympus OM-1 camera to catalog distinctive markings and record behavior patterns.

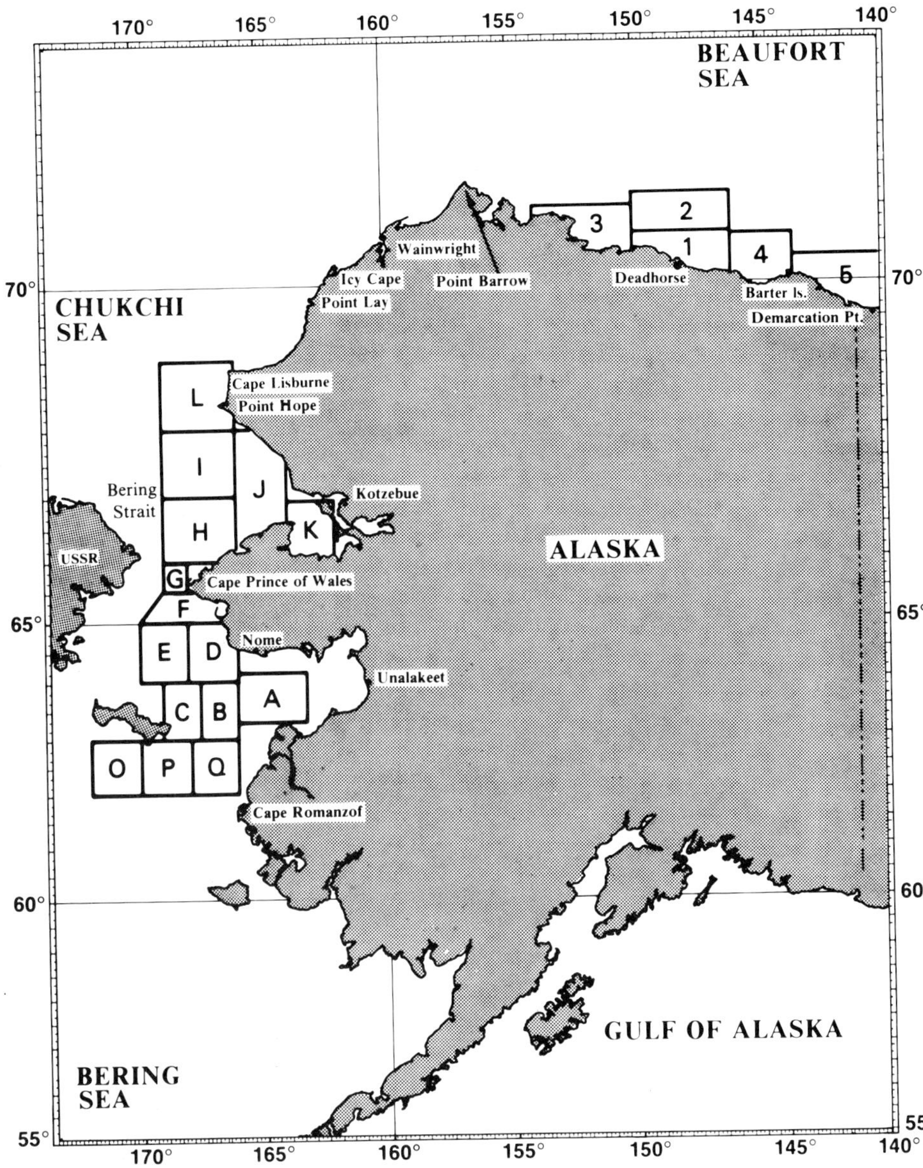

Fig. 1. *Survey blocks in the Beaufort, Chukchi, and northern Bering Seas.*

SOUND RECORDING AND ANALYSIS

Two types of sonobuoys, AN/SSQ 41A and AN/SSQ 41B, with frequency responses of 10 Hz to 6 kHz and 10 Hz to 20 kHz, respectively, were used to record gray whale sounds. These sonobuoys are passive listening systems that contain a hydrophone, signal processing electronics, and a VHF transmitter and are designed to be dropped from the aircraft by means of a rotochute or parachute. Once the sonobuoy is in contact with the water, a seawater-activated battery energizes the unit. The parachute assembly then jettisons and the hydrophone array drops to a preselected depth of 18.2 or 91.4 m. Sonobuoys were deployed in the northern Bering Sea (4 times), Norton Sound (1 time), and the Chukchi Sea (2 times) near gray whales thought to be feeding. Feeding was inferred from observations of frequent diving and the presence of mud plumes created when surfacing whales brought bottom sediments toward the surface. Sounds received by the hydrophones were amplified and transmitted to a Defense Electronics Instruments Model GPR-20 VHF broadband receiver aboard the aircraft. The receiver output was connected to a NAGRA IV-SJ tape recorder with a frequency response of 25 Hz–10 kHz ± 1.5 dB at a recording speed of 9.5 cm/sec. The response of the overall system was 25 Hz to 6 kHz ± 2 dB, or 25 Hz to 10 kHz ± 2 dB when recording from the 41A or 41B sonobuoy, respectively.

Sounds were first played back at recorded speed. All sounds of sufficient level (signal/noise ≥ 10 dB) to permit analysis were converted to continuous hard-copy spectograms using a Spectral Dynamics 350 real-time analyzer set at 10 Hz–5 kHz analyzing bandwidth. The response of the analyzing system was flat (±2 dB) from 50 Hz to 5 kHz. Sound frequency and duration parameters were measured from the spectrograms. A Kay Elemetrics 7029 A Sonagraph was used to make figures representative of each sound type.

Results

GRAY WHALE DISTRIBUTION AND BEHAVIOR

Gray whales were seen from Bristol Bay to Point Barrow, with total numbers of 326 whales in 1980 (Fig. 2A) and 546 whales in 1981 (Fig. 2B). The difference in counts may reflect different levels and distributions of aerial survey effort. Table I and Fig. 3 summarize survey efforts for both years. In 1980, 76% of all survey hours were flown in the Beaufort Sea, 11% in the Chukchi Sea, and 13% in the northern Bering Sea (Fig. 3A). In 1981, flight effort was more equally divided between the Beaufort (53%), Chukchi (14%), and northern Bering Seas (33%) (Fig. 3B). Although flight effort was concentrated in the Beaufort Sea (blocks 1–5) during both years, no gray whales were seen there.

Survey effort, sightings, and sighting rate in Beaufort 01–06 sea states are presented in Table II. When 1980 and 1981 survey effort was rounded to the nearest hour (n = 1055), 66% of all survey hours were spent in Beaufort 01 to 03 and 34% were in Beaufort 04 to 06 sea states. Eighty-three percent of all survey hours were spent in Beaufort 01 to 04 and 17% in Beaufort 05 to 06 conditions. Available data strongly suggest that effects of variable sea states on sighting frequency were not significantly different among the survey conditions encountered (Fig. 4). However, because it was not possible to test the relationship statistically or to determine effects of changing sea state on sightability as a function of distance, we attempted no density estimates from the data.

Gray whales were seen from 5 m to 150 km from shore in water from 3 to 40 m

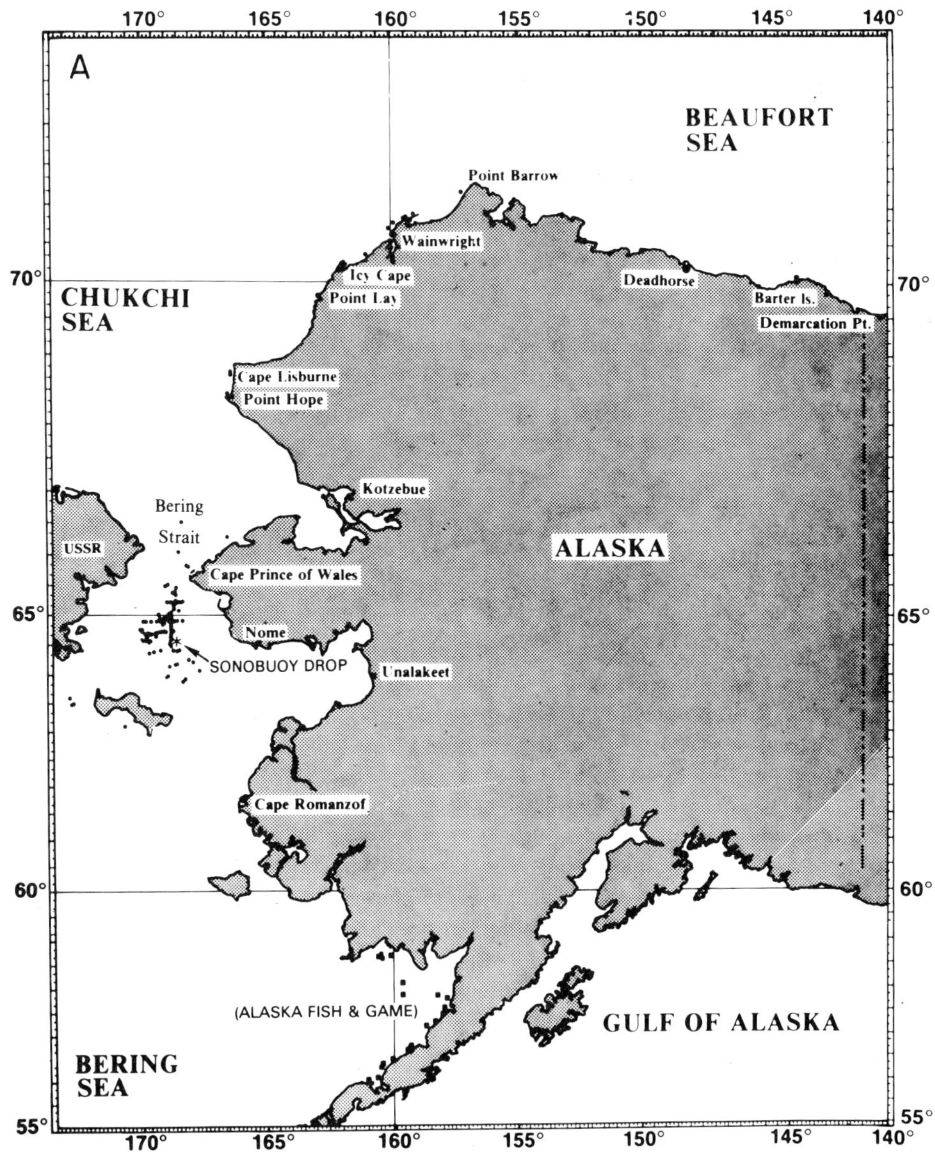

Fig. 2. *Locations of gray whale sightings in 1980 (A) and 1981 (B). Each point is a sighting of one or more whales.*

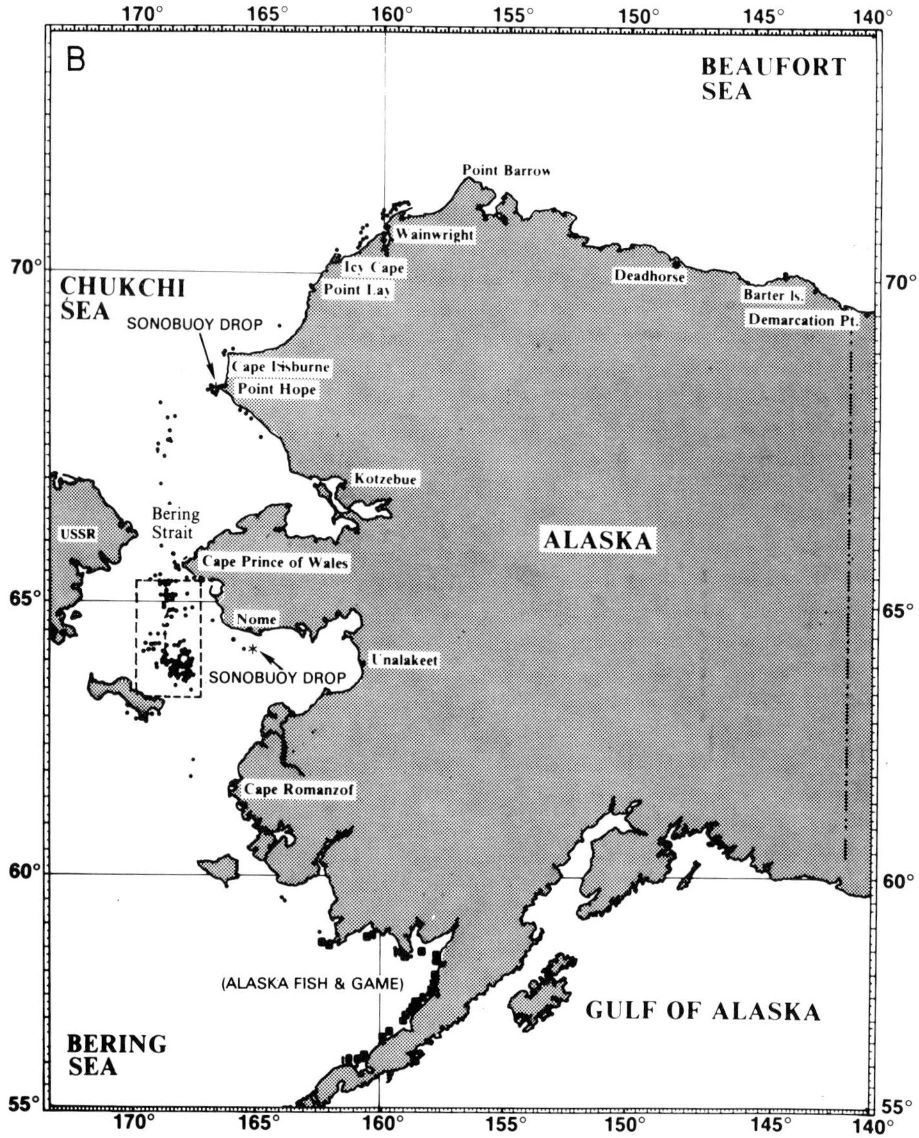

Fig. 2. (*Continued*)

deep. Some places were so shallow the whales appeared to be lying on the bottom, and sediment disturbance could be seen if the whales responded to the aircraft. Grays were occasionally seen in ice coverage as great as 30%.

In 1981, 63% ($n = 318$) of the gray whales seen in the northern Bering and southern Chukchi Seas (below 69°N latitude) were seen with brown mud plumes indicating feeding activity (Fig. 5). A concentration of gray whales, many with mud plumes, was

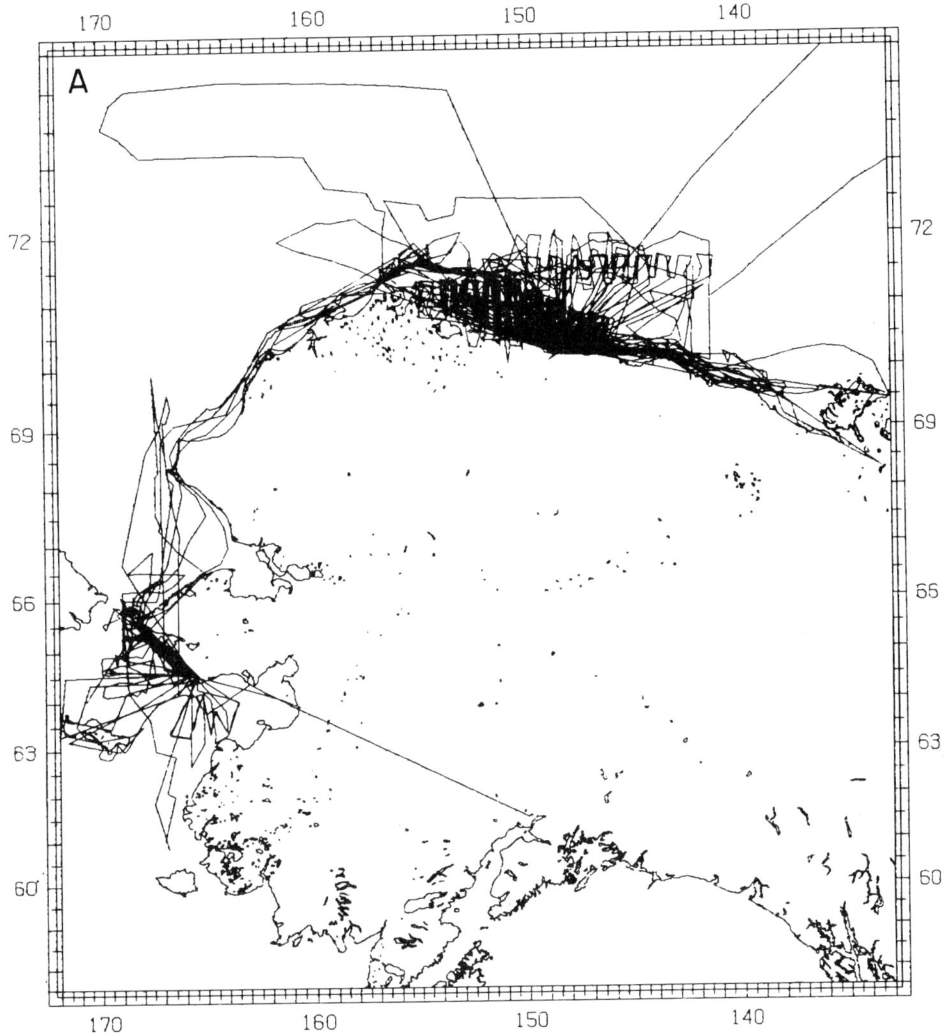

Fig. 3. *Aerial survey effort in 1980 (A) and 1981 (B).*

Table I
Aerial Survey Effort in 1980 and 1981

	1980		1981	
	Kilometers	Hours (%)	Kilometers	Hours (%)
Bering Sea	19,760	76 (13)	40,300	155 (33)
Chukchi Sea	16,640	64 (11)	17,160	66 (14)
Beaufort Sea	115,700	445 (76)	64,740	249 (53)

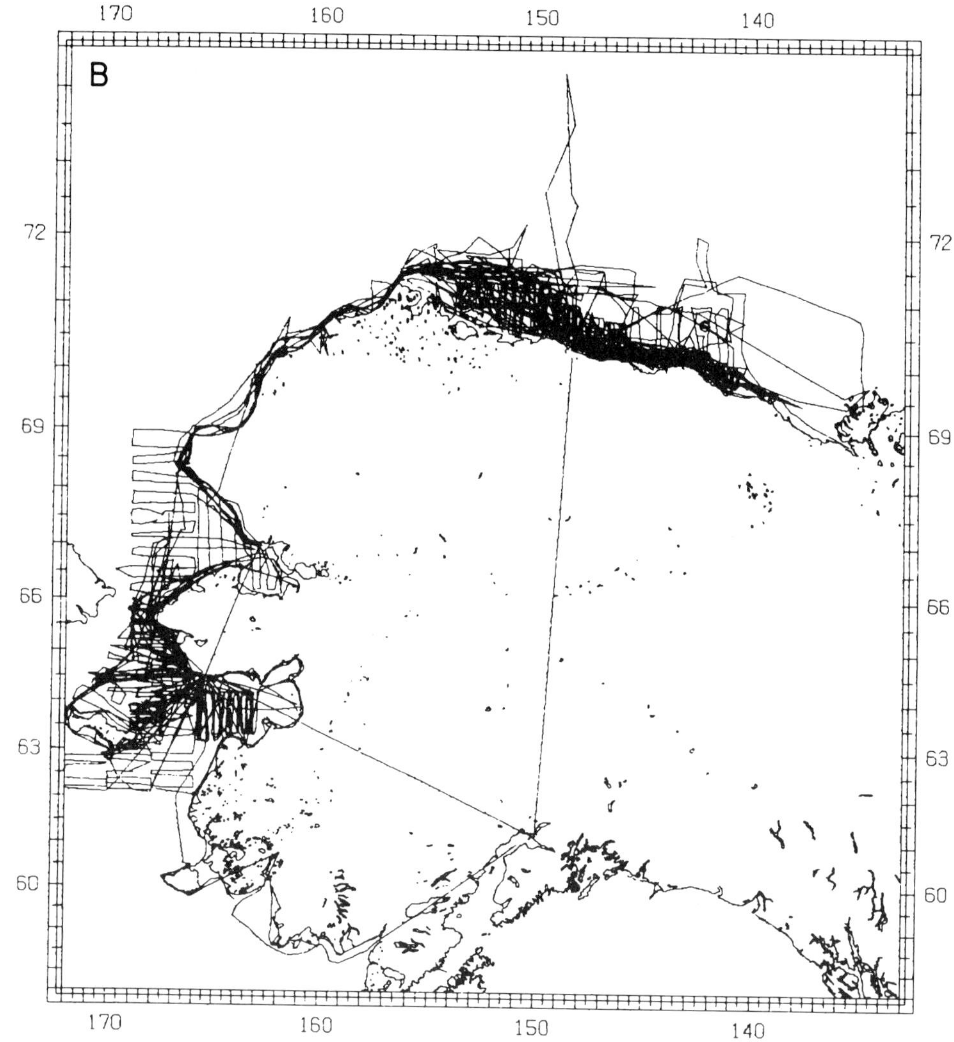

Fig. 3. *(Continued)*

Table II
Survey Effort, Sightings, and Sighting Rate in Beaufort 01–06 Sea States

Beaufort sea state	Hours (%)	Number of sightings (SI)	Number of SI/hour
0–01	137 (13)	119	0.87
02	243 (23)	307	1.26
03	317 (30)	135	0.43
04	179 (17)	87	0.49
05	116 (11)	171	1.47
06	63 (6)	53	0.84

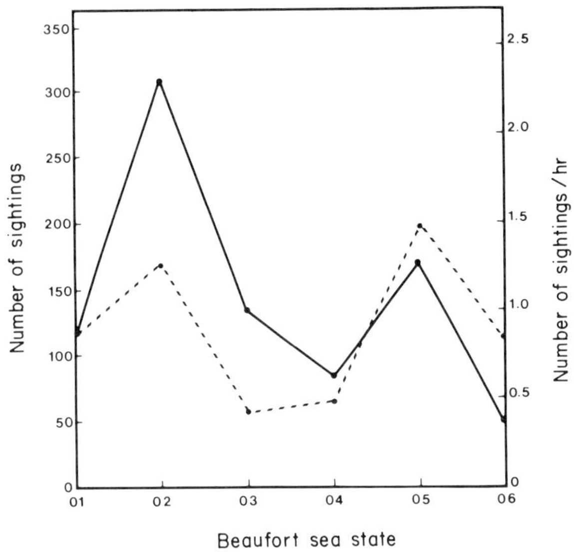

Fig. 4. *Effects of variable sea states on sighting frequency (solid line) and sighting rate (dashed line).*

Fig. 5. *Gray whales with mud plumes made during activities interpreted to be feeding.*

Fig. 6. *Gray whale with semicircular mud plume.*

seen in 1980 and 1981 in the Chirikov Basin, an area of the northern Bering Sea between St. Lawrence Island and the Bering Strait (see Fig. 2). Eighty-five percent (n = 299) of the gray whales seen in the Chirikov Basin between May 19 and June 22, 1981, were seen with plumes indicative of feeding. Most mud plumes (93%, n = 118) were straight or meandering trails behind the whales. Seven percent of the plumes seen were semicircular in shape (Fig. 6). Mud plumes, which attract feeding birds (Harrison, 1979), aid sightability and thus may positively bias the relative frequency of sightings of "feeding" gray whales. Plumes were timed to disperse within 5 min in a Beaufort 03 sea so are only a temporary sighting cue.

Gray whales seen in the northern Chukchi Sea (north of 69°N) in 1981 were never associated with mud plumes. Unfortunately specific behaviors of gray whales seen in the northern Chukchi Sea were seldom recorded. Because of small sample size, comparisons of behavior between whales north and south of 69°N latitude are limited to relative frequency of occurrence. Of sightings where behavior was recorded, 88% of the whales were traveling, 11% were resting, and none were feeding in the northern group (n = 9); in the southern group (n = 318), 34% were travelling, 3% were resting, and 63% were feeding.

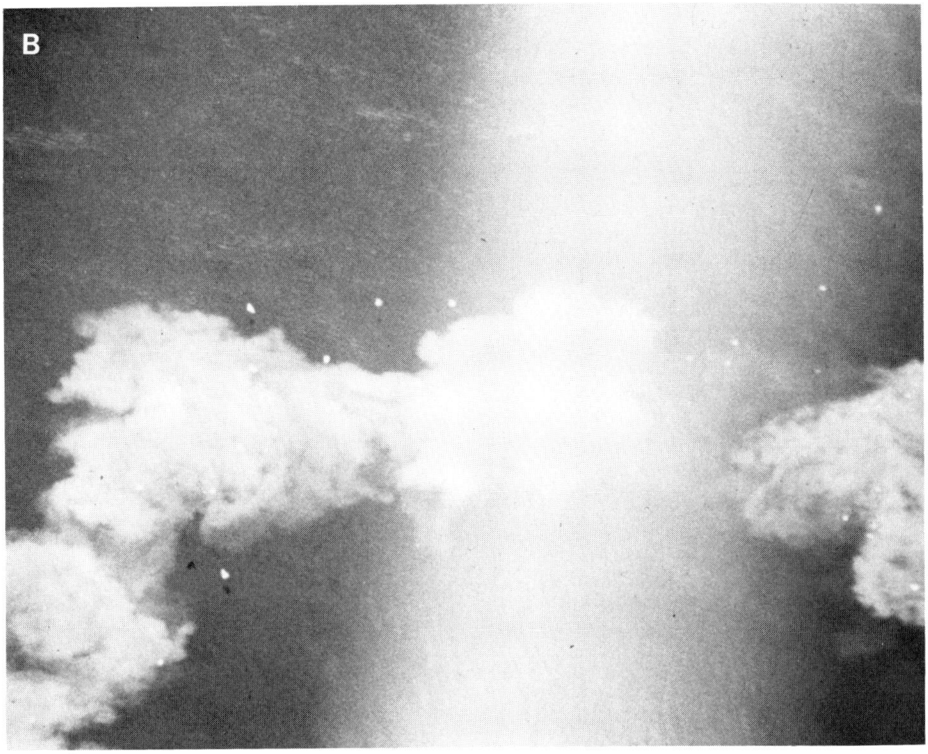

Fig. 6. *(Continued)*

SOUNDS RECORDED NEAR FEEDING GRAY WHALES

Three tapes totaling 4.5 hr contained approximately 2 hr of sounds of sufficient level to permit analysis. Sounds were recorded near whales assumed to be feeding off King Island in 1980, and in the Chukchi Sea (20 km SW of Point Hope) and Norton Sound in 1981 (Fig. 2A,B). We assumed the sounds recorded on tapes one and two were produced by gray whales because they were the only marine mammals seen that day. On the third tape (Norton Sound) calls of bearded seals (*Erignathus barbatus*) were prevalent, but other sounds similar to the gray whale sounds on tapes 1 and 2 were recorded and analyzed as gray whale sounds as well. Each sound was subjectively assigned to one of four available categories: N_1, N_2, N_3, and N_4 (no N_2 sounds were detected) following a classification scheme developed by Dahlheim *et al.* (Chapter 22, this volume). Several miscellaneous sounds could not be so classified. Table III presents range, mean, and standard deviation for measured parameters of each sound type.

Type N_1 sounds were pulsive signals that to the ear sounded like metallic knocks. They were the sounds most prevalent in all samples and were usually emitted in bursts. A burst was defined as a series of N_1 sounds in which silent intervals lasted less than 1

Table III
Frequency and Duration Measures on Three Gray Whale Sound Types

Sound type	Measured parameter	*N*	Range	$\bar{x} \pm$ SD
	Maximum energy (Hz)	330	238–2584	962.62 ± 652.66
	Duration (msec)	1530	45–227	106.38 ± 23.20
N_1	Interval (msec)	1379	14–955	139.64 ± 161.34
	Burst duration (sec)	127	0.25–19.05	2.76 ± 2.84
	Number pulses/burst (*n*)	127	2–69	11.86 ± 11.30
N_3	Initial (Hz)	14	102–578	308.4 ± 154.1
	Middle (Hz)	14	102–544	352.1 ± 121.2
	End (Hz)	14	170–544	325.6 ± 145.3
	Duration (msec)	14	621–1519	913.4 ± 259.8
N_3* Upper	Initial (Hz)	2	442–578	510 ± 96.2
	Middle (Hz)	2	374–476	425 ± 72.1
	End (Hz)	2	544	544 ± 0
	Duration (msec)	2	686	686 ± 0
N_3* Lower	Initial (Hz)	2	238–272	255.0 ± 24.0
	Middle (Hz)	2	308–340	289.0 ± 24.0
	End (Hz)	2	272–306	324.0 ± 22.6
	Duration (msec)	2	735–882	808.5 ± 103.9
N_4	Maximum energy (Hz)	19	136–748	388.3 ± 143.0
	Duration (msec)	19	245–588	344.4 ± 86.1

sec. Frequency of maximum energy, duration, interpulse interval, burst duration, and number of pulses per burst were measured whenever possible. (All values could not be determined for every N_1 burst due to masking by ambient noise). Type N_1 sounds were short ($\bar{x}$ = 106 msec) and broadband (to ~4 kHz) with maximum energy at a mean frequency of 963 Hz (Fig. 7). Interpulse intervals were quite variable (14–955 msec) but averaged 140 msec, resulting in a mean repetition rate of 7.14/sec. Bursts lasted as long as 19 sec, averaged less than 3 sec and were as short as 0.25 sec. Bursts contained from 2 to 69 pulses with an average of 12 pulses/series.

Of all sounds recorded from gray whales, type N_1 sounds were the most variable.

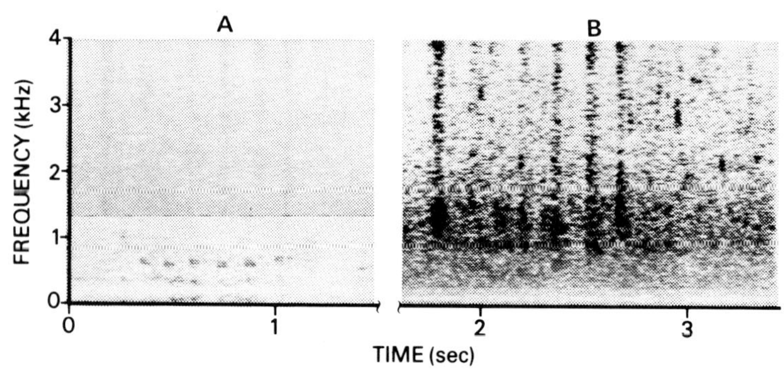

Fig. 7. *Sonogram of bursts of N_1-type sounds.*

Table IV
Comparison, using T Test[a] of N_1-Type Sounds Recorded near King Island and Others Recorded in Norton Sound

	King Island[b]		Norton Sound[c]		
	N	x̄ ± SD	*N*	x̄ ± SD	Comparison
Frequency of maximum energy (Hz)	299	550 ± 190	104[d]	1853 ± 333	<.001
Duration (msec)	229	109.5 ± 23	880	114 ± 25	<.100
Interval (msec)	229	240.4 ± 227	880	138 ± 146	<.001

[a]For T test, α = 0.001.
[b]Fig. 6A.
[c]Fig. 6B.
[d]This variable could not be measured on all knocks due to masking of sounds by calls of bearded seals (*Erignathus barbatus*).

When those recorded near King Island in 1980 were compared with those recorded in Norton Sound in 1981, significant differences were noted in frequency of maximum energy and interpulse interval but not in sound duration (t test, $p < .001$) (Table IV). The N_1 sounds recorded in Norton Sound were higher in frequency of maximum energy (1853 Hz, $n = 104$) and had a shorter interpulse interval (138 msec, $n = 880$) than those recorded near King Island (550 Hz, $n = 299$; 240 msec, $n = 229$). Reasons for these differences are unknown.

Type N_3 sounds were tonal, moanlike sounds. Initial, mid-sound, and end frequency and duration were measured. Type N_3 sounds were pulse and frequency modulated (FM) with mean fundamental energy at 325 Hz and an average duration of just under 1 sec ($\bar{x} = 913$ msec) (Fig. 8A). Two sounds classified as N_3* sounds sounded but did not look like other N_3 sounds. Type N_3* sounds consisted of two divergent fundamental bands with an approximate 100 msec delay in the onset of the lower frequency band (Fig. 8B). The lower frequency band averaged 300 Hz and 808 ms, while the higher

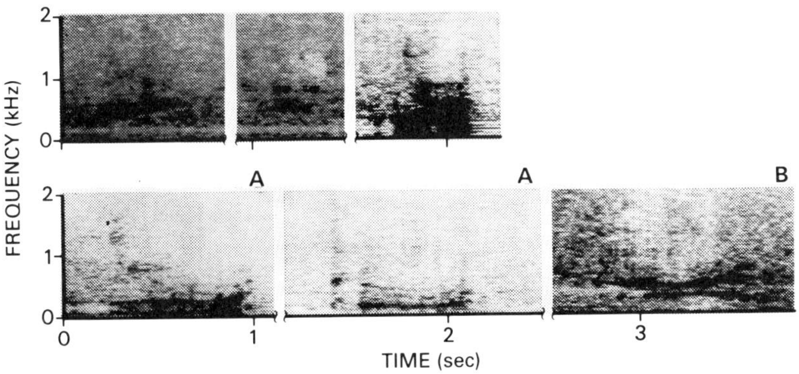

Fig. 8. *Top: Sonogram of three N_4-type sounds. Bottom: Sonogram of two N_3-type sounds (A) and one N_3*-type sound (B).*

frequency component averaged 525 Hz and 686 msec. Each FM band in the N_3*-type sounds appeared to be an independent component rather than a harmonic component of the sound. N_3*-type sounds comprised only 12% of the sample of total N_3-type sounds ($n = 16$).

Type N_4 sounds were pulse- and frequency-modulated signals that to the ear sounded belchlike. These sounds occurred singly or in pairs and sometimes at the end of an N_1 burst. Duration and frequency of maximum energy of N_4 sounds were measured. Compared to N_1 sounds, N_4 sounds were longer ($\bar{x} = 344$ msec) and of narrower frequency bandwidth (to 1.2 kHz), with maximum energy at a mean frequency of 388 Hz (Fig. 8, top).

Discussion

GRAY WHALE DISTRIBUTION AND BEHAVIOR

The distribution and behavior of gray whales in arctic waters may be dictated primarily by prey availability. We found gray whales ranging as far north as Point Barrow and as far south as northwestern Bristol Bay. Gray whales are reportedly widely seen in Bristol Bay in spring and fall (Chapters 10 and 11, this volume; Gill and Hall, 1983). During aerial surveys of Bristol Bay conducted by Alaska Fish and Game biologists (L. K. Lowry, personal communication) 32 gray whales were seen between April 16 and June 23, 1980, and 41 between April 7 and May 7, 1981 (Fig. 2A,B).

Notably the largest concentration of gray whales seen on our surveys was found in the northcentral Bering Sea. Up to 85% of the whales seen there were associated with plumes. Not surprisingly, dense beds of amphipods are reported to occur there (Nerini, 1982, Chapter 18, this volume; Oliver *et al.*, 1984), and stomach contents of whales taken from this area have contained primarily benthic amphipods (Rice and Wolman, 1971; Zimushko and Ivashin, 1979). Although additional nearshore feeding grounds have been documented in the western Bering and Chukchi Seas (Bogoslovskaya *et al.*, 1981), none have been reported in coastal areas on the eastern side of the Bering and Chukchi Seas. The fact that we saw no gray whales with mud plumes in the northern Chukchi Sea in 1980 or 1981 is consistent with the report by Stoker (1981) regarding the lack of benthic infauna in the eastern Bering/Chukchi continental shelf. A subsequent observation of gray whales feeding along the northern Chukchi coast in July 1982 lead us to speculate that gray whales will feed there when prey is available, however (Ljungblad *et al.*, 1983). Nerini (1982; Chapter 18, this volume) fully discusses gray whale feeding ecology.

On several occasions in the present study and in other observations (Chapter 11, this volume) feeding whales left semicircular, arcshaped mud plumes rather than the more common mud plume trails. The arcs appear to be formed by one whale but may be the result of several whales foraging together. We saw such arcshaped plumes with and without whales at the surface. These plumes may result from a whale pushing sediment from its baleen in several pulses as it swims an arc at or near the surface. This semicircular plume pattern may indicate that once a whale finds an area of dense benthic infauna, it will probably turn and make repeated dives in that area.

SOUNDS RECORDED NEAR FEEDING GRAY WHALES

The present chapter offers the first quantitative report of sound production by gray whales on their arctic feeding grounds. The three sound types reported are similar to those described for gray whales in other parts of their range (Chapter 22, this volume). Because our sound sample was small (2 hr), additional sound types will probably be described for gray whales in arctic waters as more recordings are made and analyzed.

The most prevalent and variable sounds recorded near feeding gray whales were bursts of metallic-sounding pulses termed N_1-type sounds. These were also the most prevalent sounds recorded by Dahlheim *et al.* (Chapter 22, this volume) in the breeding lagoons of Baja California, Mexico. Fish *et al.* (1974) described "metallic-sounding pulse train" recorded from a captive gray whale and from gray whales feeding off Vancouver Island, Canada. Gray whales feeding in Norton Sound emitted higher frequency N_1 sounds with a faster repetition rate than did whales feeding in the Chirikov Basin, although the duration of the sounds remained unchanged. Thus, whales can apparently alter the frequency and speed of type N_1 sound production. Analysis and comparison of N_1 sounds recorded in a variety of habitats and behavioral situations may further clarify this sound's variability and function.

The N_3 and N_4 sounds were rarely recorded from feeding whales. Both are of longer duration and lower frequency than N_1 sounds. Additionally, both are partially or wholly frequency-modulated signals. Cummings *et al.* (1968) report N_3-type sounds (moans) were the most common type recorded from migrating gray whales; N_1-type sounds were not heard at all. Both the N_3- and N_4-type sounds presented here fall within the range of (but have a shorter mean duration than) the S_3- and S_4-type sounds reported by Dahlheim *et al.* (Chapter 22, this volume) for gray whales in the breeding lagoons. The frequency range for both sound types recorded in arctic waters is comparable to those reported for similar sounds recorded in the lagoons. In the present study, too few N_3 and N_4 sounds were recorded to support speculation of their variability or function.

The N_3* sounds composed of two nearly simultaneous nonharmonic (divergent) tonal bands are a mystery. The physical appearance of the sound on a spectrogram or sonogram suggests either dual sound generation by a single whale (i.e., individual variation on N_3 type production), or nearly simultaneous N_3-type sound production by two whales. Until more sounds of this type are available we suggest they remain categorized as N_3 sounds, as they are aurally indistinct. No reliable associations could be made between any of the sounds we recorded and any observed surface behaviors of the whales.

Summary

Gray whales were seen from Bristol Bay to Point Barrow, Alaska on aerial surveys for endangered whales conducted in 1980 and 1981. In each year a concentration of gray whales was found in the northcentral Bering Sea between St. Lawrence Island and the Bering Strait. Eighty-five percent (n = 299) of the gray whales in this area in 1981 from

May 19 to June 22 were seen with mud plumes, indicating feeding activity. Sounds recorded in the presence of feeding gray whales were classified to three distinct types. We do not know what function these sounds serve.

Acknowledgments

This work was completed under the auspices of the Bureau of Land Management, interagency agreement #AA851-1A2-3. We thank Tim Sullivan, Bob Hansen, Jerry Imm, and Cleve Cowles for their support and assistance. Thanks also to D. R. Van Schoik and J. DeWald for their help in data collection and analysis, to the pilots and ground personnel from the Office of Aircraft Services, Anchorage, Alaska for their support of this project and to the personnel of the Naval Ocean Systems Center for their support and assistance. A special thanks to Frank T. Awbrey, John S. Oliver, and Marilyn E. Dahlheim for their comments and editorial review.

References

Aldrich, H. L. (1889). "Arctic Alaska and Siberia," pp. 32–35. Rand McNally, Chicago, Illinois.

Asa-Dorian, P. V., and Perkins, P. J. (1967). The controversial production of sound by the California gray whale, *Eschrichtius gibbosus*. *Nor. Hvalfangst-Tid.* **56,** 74–77.

Bogoslovskaya, L. S., Votrogov, L. M., and Semenova, T. N. (1981). Feeding habits of the gray whale off Chukotka. *Rep. Int. Whale. Commn.* **31,** 507–510.

Cummings, W. C., Thompson, P. O., and Cook, R. (1968). Underwater sounds of migrating gray whales, *Eschrichtius glaucus*. *J. Acoust. Soc. Am.* **44**(5), 1278–1281.

Eberhardt, R. L., and Evans, W. E. (1962). Sound activity of the California gray whale, *Eschrichtius glaucus*. *J. Audio Eng. Soc.* **10,** 324–328.

Fish, J. F., Sumich, J. L., and Lingle, G. L. (1974). Sounds produced by the gray whale, *Eschrichtius robustus*. *Mar. Fish. Rev.* **36**(4), 38–45.

Gill, R. E., and Hall, J. D. (1983). Use of nearshore and estuarine areas of the southeastern Bering Sea by gray whales (*Eschrichtius robustus*). *Arctic* **36**(3), 275–281.

Harrison, C. S. (1979). The association of marine birds and feeding gray whales. *Condor* **81,** 93–95.

Ljungblad, D. K. (1981). "Aerial Surveys of Endangered Whales in the Beaufort Sea, Chukchi Sea and Northern Bering Sea," Naval Ocean Systems Center Technical Document No. 449, 292 pp.

Ljungblad, D. K., Moore, S. E., Van Schoik, D. R., and Winchell, C. S. (1982). "Aerial surveys of Endangered Whales in the Beaufort, Chukchi and Northern Bering Seas," Naval Ocean Systems Center Technical Document No. 486, 306 pp.

Ljungblad, D. K., Moore, S. E., and Van Schoik, D. R. (1983). "Aerial Surveys of Endangered Whales in the Beaufort, Eastern Chukchi and Northern Bering Seas, 1982," Naval Ocean Systems Center Technical Document No. 605, 382 pp.

Maher, W. J. (1960). Recent records of the California gray whale (*Eschrichtius glaucus*) along the north coast of Alaska. *Arctic* **13**(4), 257–265.

Nerini, M. K., Jones, L., and Braham, H. (1982). "Gray Whale Feeding Ecology," Final report to NOAA, OCSEAP (RU 593), Contract No. R7120828. Natl. Mar. Mammal Lab., Washington, D.C.

Oliver, J. S., Slattery, P. N., Silberstein, M. A., and O'Connor, E. F. (1983). A comparison of gray whale feeding in the Bering Sea and Baja California. *Fish. Bull.* **81**(3), 513–522.

Rice, D. W., and Wolman, A. A. (1971). The life history and ecology of the gray whale (*Eschrichtius robustus*). Spec. Publ. Am. Soc. Mammal. **3,** 1–142.

Rugh, D. J., and Fraker, M. A. (1981). Gray whale (*Eschrichtius robustus)* sightings in the Eastern Beaufort Sea. *Arctic* **34**(2), 186–187.

Stoker, S. (1981). Benthic invertebrate macrofauna of the eastern Bering/Chukchi continental shelf. *In* "The Eastern Bering Sea Shelf—Oceanography and Resources" (D. W. Hood and J. A. Colder, eds.), pp. 1066–1091. Univ. of Washington Press, Seattle.

Tomilin, A. G. (1957). "Mammals of the U.S.S.R. and Adjacent Countries" (translated from Russian), pp. 295–326. Israel Program for Sci. Transl., Jerusalem.

Votrogov, L. M., and Bogoslovskaya, L. S. (1980). Gray whales off the Chukotka. Peninsula. *Rep. Int. Whal. Commn.* **30,** 435–437.

Zimushko, V. V., and Ivashin, M. V. (1979). Some results of Soviet investigations and whaling of gray whales (*Eschrichtius robustus*, Lilljeborg, 1861). *Rep. Int. Whal. Commn.* **30,** 237–246.

24

Dive Characteristics and Movements of Radio-Tagged Gray Whales in San Ignacio Lagoon, Baja California Sur, Mexico

James T. Harvey and Bruce R. Mate

Introduction

The study of baleen whale biology has traditionally relied on collected specimens or observations of individual whales for short periods of time. Continuous long-term observations of whale behavior and activities are infrequent because individuals are difficult to identify, can travel long distances underwater without detection, and may remain at the surface only for short periods. Recent radio miniaturization and attachment techniques have provided an alternative mechanism for continuous (24-hr) tracking of specific whales traveling great distances.

Gray whales, *Eschrichtius robustus*, are excellent subjects for the attachment of radio tags because they travel near shore and may occupy confined embayments during

ISBN 0-12-389180-9

winter. These animals tolerate large numbers of epizoic organisms and thus appear to be adapted to foreign objects in the skin. Additionally, in the breeding and calving lagoons of Baja California Sur, Mexico, some "inquisitive" gray whales can be closely approached (Gilmore, 1961). Gray whales, however, are less than ideal animals for radio-tagging studies because they rub against the seafloor while feeding and against each other during mating and raising of the young, thus increasing the chances of displacing an attached tag.

Recent studies of gray whales in Baja California have been conducted in Ojo de Liebre and Guerrero Negro lagoons (Rice *et al.*, 1981), San Ignacio Lagoon (Swartz and Cummings, 1978; Swartz and Jones, 1979, 1980, Chapter 14, this volume) and in Bahia de Magdalena (Norris *et al.*, 1977; Chapter 15, this volume). These studies did not concentrate on dive and surface patterns. Previous tagging efforts with gray whales involved the use of belly bands attached to restrained calves (Norris and Gentry, 1974) and a radio tag surgically sutured to a captive yearling (Evans, 1974; Sweeney and Mattsson, 1974).

The purpose of our study was to collect long-term information on the dive and surfacing characteristics and local movements of individual radio-tagged gray whales in San Ignacio Lagoon, Baja California Sur, Mexico.

Methods

San Ignacio Lagoon (26°50′N, 113°10′W) is a relatively small protected area and offers the possibility of closely approaching "inquisitive" whales, those which approach boats and often allow themselves to be touched. Two camps were established to monitor the activities of tagged whales. The base camp was located approximately halfway up the eastern shoreline of the lagoon, and the second camp was placed on Bronaugh Point on the north shore of the lagoon entrance (Fig. 1). The lagoon was arbitrarily demarcated from the ocean by a line between Bronaugh and Holcomb Points roughly approximating the position of the breaking surf.

The radio tag and its method of application have been described by Mate *et al.* (1983). The radio transmitters were 3.8 × 3.6 × 2.5 cm and transmitted pulsed signals between 148 and 149 MHz. Radio tags transmitted on one of six different frequencies with pulsed signals at one of six different rates. The tag was suspended from a tag applicator at the end of a 5-m pole; therefore, only whales within 3 m of the boat could be tagged. We only attempted to tag inquisitive whales because our previous experiences indicated it was difficult to closely approach other whales.

Tags were placed 1–2 m behind the blowhole on the middorsal line such that, at most surfacings, the tag's antenna was exposed and signals could be transmitted. Each tagged whale could be distinguished by its combination of signal frequency and transmission rate. A tagged whale could breathe without a signal being transmitted only when it raised its head vertically out of the water, which is a rare event.

From February 8 through March 20, 1980, radio-tagged whales were monitored

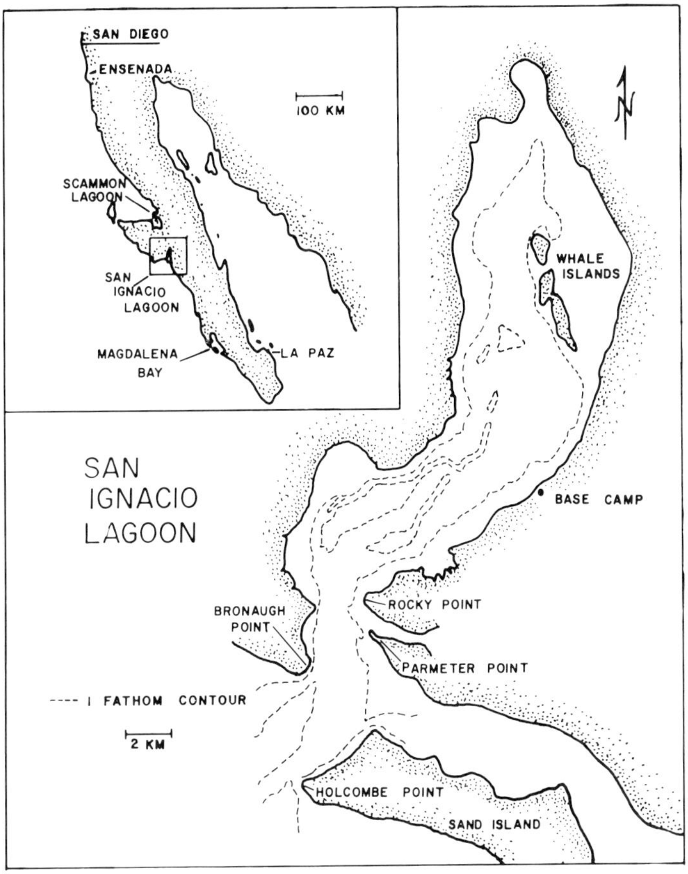

Fig. 1. *San Ignacio Lagoon, Baja California Sur, Mexico was the location of all tagging activities. A base camp and field camp at Bronaugh Camp were established to monitor radio-tagged whales. The lower lagoon was arbitrarily demarcated from the ocean by bearings originating from Bronaugh Point to Holcombe Point.*

from four platforms; base camp, Rocky Point, Bronaugh Point, and by boat. Rocky Point served as the monitoring location March 18–20 and Bronaugh Point was used from March 21 through the completion of the field season on April 29. At the base camp and Bronaugh Point, two 14-element yagi antennas were mounted perpendicularly to each other and separated vertically by 1 m on a 10-m mast. A switch box allowed reception from either antenna or both. Maximum reception range for these antennas was approximately 10 km. A direction finding (DF) antenna also was mounted on top of a 6-m mast and provided information on the bearing of the radio tags up to 10 km from camp. Handheld 2-element Yagi antennas and DF antennas were used for receiving signals while monitoring from boats. Each morning, afternoon, and evening, checks were made for tagged whales from the land camps. Weather permitting, daily boat surveys also were made to locate radio-tagged whales. During such times that a tagged whale was

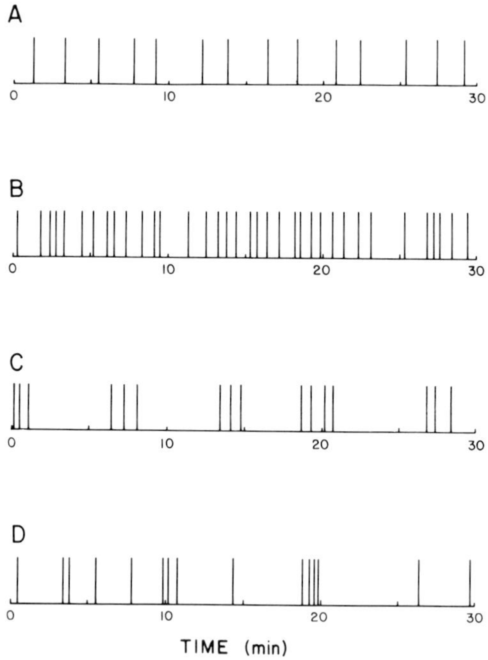

Fig. 2. *Examples of three dive patterns (from Whale 100R) which are typical of gray whales in San Ignacio Lagoon. (A) Regular–long; (B) regular–short; (C) clumped. The irregular or unpatterned dive series is shown in (D). Each vertical line represents a series of signals received when a whale surfaced.*

within monitoring range of the camps, nearly continuous data were collected on its activities.

As a whale surfaced, the times of the first and last signal received were recorded. Dive duration was defined as the time between the last audible signal during a surfacing and the first signal pulse upon resurfacing. Surface duration was defined as a period of continuous signal pulses. Relative signal strength was noted (strong, moderate, weak, or very weak). Records of very weak signals were not analyzed. The bearing of a tagged whale was determined approximately every 2 hr. Visual observations of tagged whales were made opportunistically from boats or 6-m observation towers at the shore camps.

Idealized diving and surfacing patterns were modeled (with the assistance of Fred Ramsey and Richard Hanlen) for each of three discernible patterns as a first-order Markov process, using transition matrices. The three patterns used were (1) "regular–long," regularly spaced dives greater than 1 min in duration; (2) "regular–short," dives less than 1 min in duration; and (3) "clumped," a long dive of greater than 1 min followed by two to six shorter dives of less than 1 min each (Fig. 2). W. Watkins (personal communication) has used the term "clustered" for a pattern similar to our "clumped." In addition, an irregular or unpatterned catch-all category was established for dives that did not fit the criteria of the other patterns. The dive pattern data were summarized from a series of seven dives at a time. A computer program calculated the probability that an observed series of dives fit a particular pattern by comparing the series of seven dives with the modeled, idealized dive patterns.

Results

DIVE AND SURFACE DURATIONS

Ten gray whales were radio tagged and tracked in San Ignacio Lagoon in 1980; three single adults (two females, one of undetermined sex) were tagged in early February, and the remaining seven whales were females with calves tagged between early March and mid-April (Table I). All remained in the lagoon at least 2 days. One female with calf (100R), tagged on March 19, lost her tag after 11 days (154.3 hr of monitoring). This animal was seen in the lagoon on 3 separate days subsequent to tag loss. Two other whales were monitored 4 and 5 days each during mid-March and early April. Seven of the tagged whales were later relocated along the Baja California and California coasts (Chapter 25, this volume), indicating that the tags had remained attached after the whales left the lagoon (Table I).

Duration of dives and surfacings were determined for radio-tagged gray whales monitored for a total 303.7 hr. Mean dive duration was 1.57 ± 0.02 min (SE) (n = 11,080 dives). Fifty-one percent (5,651) of these dives were less than 1 min duration, and 99% (10,969) were less than 6 min (Fig. 3). The mean dive duration for 5,455 dives longer than 1 min duration was 2.71 ± 0.03 min (Table II).

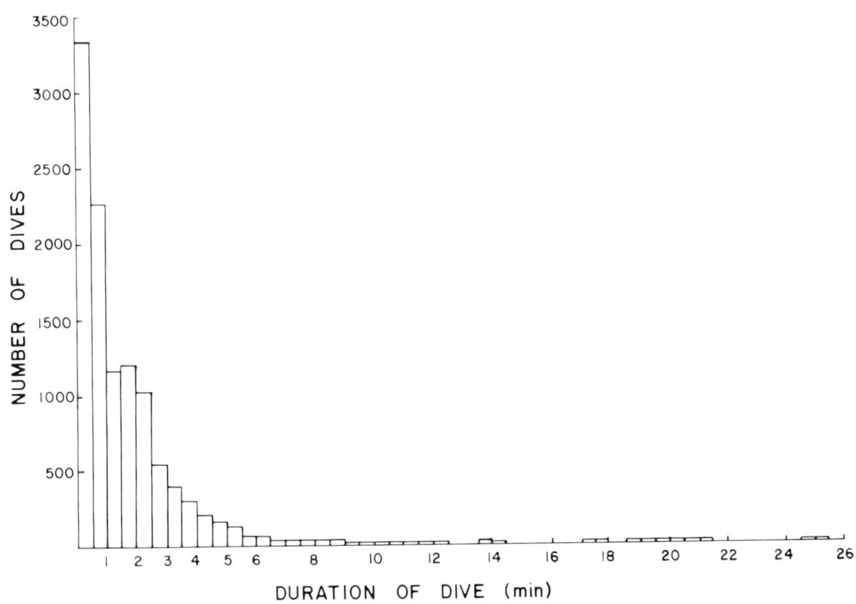

Fig. 3. *The frequency of distribution of 11,080 dive durations from 10 different radio-tagged whales monitored for 303.7 hr in San Ignacio Lagoon. The mean is 1.57 ± 0.02 (SE) min.*

Table I
Identification Number, Description, Radio Frequency, Tagging Date, Monitoring Time, and Duration of Tag Retention for 10 Gray Whales Radio Tagged in San Ignacio Lagoon, Baja California, Mexico (1980)

Whale I.D.	Whale description	Tag frequency (MHz)	Tagging date	Time monitored in lagoon vicinity (days)	Time monitored in lagoon vicinity (hr)	Known duration of tag retention (days)
50Y	Single adult	149.050	Feb. 8	2	8.0	7
80Y	Single female	149.080	Feb. 11	2	20.0	40
160Y	Single female	149.160	Feb. 11	1	1.6	50
50R	Female with calf	149.050	Mar. 18	5	29.6	19
100R	Female with calf	149.100	Mar. 19	11	154.3	13
120R	Female with calf	149.120	Apr. 5	4	47.2	4
140R	Female with calf	149.140	Apr. 9	1	3.1	1
50B	Female with calf	149.050	Apr. 11	2	7.1	16
160R	Female with calf	149.160	Apr. 11	2	10.9	20
80R	Female with calf	149.080	Apr. 14	2	21.6	13
				32	303.7	

Table II
Mean and Maximum Dive and Surface Durations for 10 Radio-Tagged Gray Whales in San Ignacio Lagoon[a]

	Dive duration (min)			Dives > Min		Surface durations (min)		
Whale I.D.	n	$\bar{x} \pm$ SE	Maximum	n	$\bar{x} \pm$ SE	n	$\bar{x} \pm$ SE	Maximum
50R	1,066	1.63 ± 0.04	8.63	665	2.33 ± 0.05	1,091	0.04 ± 0.01	0.18
50B	444	0.96 ± 0.05	7.68	102	2.28 ± 0.13	450	0.02 ± 0.01	0.35
80Y	388	2.58 ± 0.21	25.88	160	5.74 ± 0.40	392	0.50 ± 0.20	51.62
80R	1,043	1.21 ± 0.04	8.83	433	2.27 ± 0.06	1,053	0.03 ± 0.01	0.40
50Y	276	1.69 ± 0.16	12.28	94	4.24 ± 0.34	278	0.05 ± 0.01	0.15
100R	5,788	1.51 ± 0.02	16.27	2,893	2.59 ± 0.03	5,862	0.08 ± 0.01	12.13
120R	1,595	1.74 ± 0.04	11.90	835	2.86 ± 0.05	1,615	0.04 ± 0.01	0.75
140R	93	1.95 ± 0.14	6.22	65	2.54 ± 0.14	95	0.03 ± 0.01	0.07
160Y	45	2.04 ± 0.21	5.70	35	2.43 ± 0.23	46	0.06 ± 0.02	0.97
160R	342	1.87 ± 0.10	11.27	173	3.20 ± 0.13	347	0.03 ± 0.01	0.07
	11,080	1.57 ± 0.02	25.88	5,455	2.71 ± 0.03	11,229	0.07 ± 0.01	51.62

[a]Including separate analysis of dives longer than 1 min.

Table III
Summary of Dive Data Collected from Whale 100R

Date	Time monitored (hr)	Total surface time (hr)	Total dive duration (hr)	*n*	Mean dive duration (min)	Time at surface (%)
Mar. 19	2.26	0.06	2.21	94	1.14	2.7
Mar. 20	9.56	0.35	9.21	349	1.58	3.7
Mar. 23	14.7	0.77	13.93	529	1.58	5.2
Mar. 24	12.3	0.36	11.99	457	1.57	2.9
Mar. 25	15.45	1.38	14.07	509	1.66	8.9
Mar. 26	16.00	0.92	15.08	657	1.38	5.8
Mar. 27	20.38	0.96	19.42	778	1.50	4.7
Mar. 28	16.49	0.84	15.65	675	1.39	5.1
Mar. 29	14.30	0.31	13.99	527	1.59	2.2
Mar. 30	13.79	0.56	13.23	536	1.48	4.1
Mar. 31	12.01	0.64	11.37	469	1.45	5.3
	147.3[a]	7.15	140.15	5580	1.51	4.9

[a]Does not include 7 hr of data collected during periods of less than 10 min duration.

Mean dive durations for individual whales ranged from 0.96 to 2.58 min with a maximum dive duration of 25.88 min (Table II). Means for dives longer than 1 min ranged from 2.27 to 5.74 min. Whale 100R, for which we have the most data, was monitored for 154.3 hr and had a mean dive duration of 1.51 ± 0.02 min. Mean daily dive durations for 100R ranged from 1.14 to 1.66 min (Table III). The mean dive duration for single whales (50Y, 80Y, and 160Y) was generally longer than that of females with calves. No significance was attached to these data because a model is currently being generated to analyze the skewed distribution of duration of dives.

The mean surface duration for all 10 tagged whales was 0.07 ± 0.01 min (or 4.4 ± 0.6 sec). The maximum continuous surface duration was 51.62 min (whale 80Y). Mean duration per surfacing for individual whales ranged from 0.02 min (whale 50B) to 0.50 min (whale 80Y) (Table II). Overall, tagged whales spent 4.5% of the time at the surface (i.e., percentage of total time monitored that the antenna was above the water's surface). The percentage of time spent at the surface varied from 1.5 (whale 140R) to 16.3% (whale 80Y).

The mean rate of surfacing for tagged whales was 35.6 ± 0.8 surfacings/hr (Table IV). These were quite variable among individuals and ranged from 19.4 to 62.5 surfacings/hr. Gray whales surfaced significantly more often during the day ($x = 37.1 \pm 1.3$ surfacings/hr) than at night ($x = 30.3 \pm 1.5$ surfacings/hr) [$t(.05, 200) = 4.33$, $p < .05$]. The variation in surfacing rate was great among all 10 animals, and mean surfacing rate was not significantly different between morning, midday, and afternoon (weighted ANOVA, $p > .05$).

SURFACING PATTERNS

Surfacing patterns were analyzed for tagged whales 100R and 120R. During 30.07 hr of analyzed surfacing patterns, whale 100R surfaced 1108 times. Forty-five percent of

Table IV
Mean Rate of Surfacing (±SE) for 10 Radio-Tagged Gray Whales at Different Times of the Day[a]

	Mean surfacings/hr (n)					
Whale I.D.	(0600–1900) day	(1900–0600) night	(0600–1100) morning	(1100–1500) midday	(1500–1900) afternoon	Total
120R	33.7 ± 0.6 (20)	31.6 ± 0.5 (14)	33.25 ± 1.4 (8)	34.7 ± 0.8 (8)	32.5 ± 1.2 (4)	32.9 ± 0.5 (34)
100R	36.5 ± 0.6 (81)	37.0 ± 0.9 (27)	37.1 ± 1.0 (26)	36.5 ± 1.2 (27)	36.1 ± 1.3 (28)	36.7 ± 0.5 (108)
100Y	—[b]	34.0 ± 1.0 (7)	—	—	—	34.0 ± 0.6 (7)
80Y	22.0 ± 0.5 (7)	16.0 ± 1.4 (9)	25.0 ± 0.7 (4)	—	18.0 ± 0.8 (3)	18.6 ± 0.7 (16)
80R	51.4 ± 1.3 (10)	43.8 ± 0.3 (5)	57.8 ± 2.5 (5)	49.0 ± 2.0 (2)	42.3 ± 0.7 (3)	48.9 ± 0.9 (15)
50R	38.3 ± 0.6 (6)	30.2 ± 0.8 (6)	38.3 ± 1.2 (4)	38.5 ± 1.3 (2)	—	34.3 ± 0.3 (12)
160R	33.3 ± 0.5 (3)	32.3 ± 0.2 (3)	—	—	33.3 ± 0.9 (3)	32.8 ± 0.3 (6)
50B	67.0 ± 1.7 (3)	36.0 (1)	—	—	67.0 ± 3.0 (3)	59.3 ± 1.6 (4)
	37.1 ± 1.3 (130)	30.3 ± 1.5 (72)	36.9 ± 1.8 (47)	36.9 ± 7.4 (39)	36.6 ± 12.9 (44)	35.6 ± 0.8 (202)

[a]Number in parentheses is the number of hour intervals.
[b]No data.

Table V
Amount of Time and Number of Surfacings for Specific Surfacing Patterns from Two Radio-Tagged Gray Whales in San Ignacio Lagoon[a]

	Surfacing patterns				
	Clumped	Regular–Long	Regular–Short	Unpatterned	Total
Whale 100R					
Time in hr*	8.72	6.62	1.20	13.53	30.07*
Surfacings**	321	211	111	465	1108**
Whale 120R					
Time in hr*	7.94	8.34	1.59	21.85	39.72*
Surfacings**	368	284	134	869	1655**

[a]The total time and number of surfacings for each whale is in parentheses.
*Significant difference ($\chi^2 = 31.43$, $p < .05$).
**Not significant ($p > .05$).

the time (42% of the surfacings) the surfacings did not fit one of the previously described patterns (Table 5). The "clumped" pattern (Fig. 2) was most frequently observed for this whale; 29% (321) of the surfacings and 29% (8.72 hr) of the total time were represented by this pattern. The "regular–long" pattern was the second most common pattern, occurring 22% (6.62 hr) of the time and representing 19% (211) of the surfacings. The "regular–short" pattern occurred only 4% of the time (1.20 hr), or 111 dives.

Whale 120R surfaced 1655 times during the 39.72 hr analyzed for surfacing patterns. Forty-four percent of these surfacings fit into one of the three recognized patterns. The "regular–long" pattern of surfacings was recorded 21% (8.34 hr) of the time, and the "clumped" pattern occurred 20% of the time (7.94 hr). As with whale 100R, "regular–short" surfacing patterns were only observed 4% of the time (1.59 hr) (Table V). There was a significant difference between whales for the amount of time spent in each pattern ($\chi^2 = 31.43$, $p < .05$), but no difference was found for number of surfacings per pattern ($p > .05$).

LOCAL MOVEMENTS

Movements of three tagged females with their calves were monitored for the 4,5, and 11 days, respectively, that each pair remained in the lagoon vicinity after tagging. Female 120R and her calf were monitored for 4 days. They moved out of the lagoon during darkness (between 0100 and 0500) on two separate occasions, for 40 and 7 continuous hr, respectively, before returning to the lagoon. Whale 50R was monitored for 5 days in the vicinity of the lagoon; this whale moved out of the lagoon twice at 0100, for 2 and 8 hr, respectively.

The greatest amount of movement information was collected from whale 100R, who was monitored for 11 days. This whale and her calf went outside the lagoon on 7 occasions (Table VI). Six of these exits occurred in darkness between 1800 and 2300, and six of seven returns to the lagoon occurred during daylight. This whale moved out of

Table VI
Location and Activities of Whale 100R within the Vicinity of San Ignacio Lagoon[a]

Date	Time	Tidal current	Locations, activities, and comments
Mar. 19	1200	Ebbing	Tagged near Parmeter Point; stayed in area of Rocky Point until 2400
Mar. 20	0220	Ebbing	Animal near mouth of lagoon
	0300	Ebbing	Moving toward lagoon mouth
	0900	Flooding	Moving up lagoon from lagoon mouth
	1000	Ebbing	Sighted near Rocky Point
	1700	Low tide	0.5 m north of Rocky Point
Mar. 21	—	—	—
Mar. 22	—	—	—
Mar. 23	0730	Ebbing	Animal near mouth of lagoon
	1115	Flooding	Moving up lagoon toward Rocky Point
	2300[b]	Low tide	Moved out of lagoon
Mar. 24	0910	Ebbing	Offshore, in vicinity of lagoon mouth
	1140	Low tide	Offshore
	1600[c]	Flooding	Entered lagoon
	1800	High tide	Sighted off Bronaugh Point
Mar. 25	0500	High tide	Near Rocky Point
	1200	Low tide	Approached our boat near Rocky Point
	1800	High tide	Near Rocky Point
	2245	Ebbing	Near Rocky Point
Mar. 26	0525[b]	High tide	Offshore, near lagoon mouth
	0750[c]	Ebbing	Between Sand Island and Bronaugh Point
	1700	Flooding	Moved north, near Rocky Point
	1900[b]	High tide	Offshore
	2350	Low tide	Offshore
Mar. 27	0700	High tide	Offshore
	1330[c]	Low tide	Moved into lagoon
	1400	Flooding	1 km offshore Bronaugh Point in lagoon
	1700	Flooding	Near Rocky Point
Mar. 28	0500[b]	Flooding	Offshore
	0800[c]	Ebbing	Moved into lagoon
	1600	Flooding	Near Parmeter Point
	1700	Flooding	Near Rocky Point
	2100[b]	Ebbing	Moved offshore just after high tide
Mar. 29	0150	Low tide	Offshore
	0500[c]	Flooding	Moved into lagoon
	1400[b]	Low tide	Offshore
	1900	Flooding	Offshore
Mar. 30	0800[c]	High tide	Inside lagoon, near mouth
	1140	Ebbing	Between Bronaugh and Parmeter Points
	2040	High tide	Near Rocky Point
Mar. 31	0220[b]	Low tide	Offshore
	0650[c]	Flooding	Entered lagoon
	1230	Ebbing	Near Rocky Point
	1344	Ebbing	Animal lost transmitter

[a]Selected times are presented to depict movements of the whale, although monitoring was often continuous.

[b]Movement out of the lagoon.

[c]Movement into the lagoon.

the lagoon against the tide 44% (three times) of the time. In general, there was no significant relationship between whale movements out of the lagoon and ebbing tides ($\chi^2 = 1.2$, $p > .05$).

Discussion and Conclusions

DIVE AND SURFACE DURATIONS

Wyrick (1954) and Evans (1974) reported maximum dive durations for migrating gray whales of 12 and 16.5 min, respectively. The maximum recorded dive duration for a gray whale tagged in San Ignacio Lagoon was 25.9 min. All dives longer than 12 min in the lagoon were associated with resting animals, typified by a whale floating at or slightly below the water surface for periods up to 51.6 min and then submerging for 12–26 min. These dive and surface durations are 2 to 250 times greater than those recorded for active whales and are not expected to be the same for traveling animals.

The average surfacing rates for tagged adult whales indicated a breathing rate of approximately 36 breaths per hr. Norris *et al.* (1977) recorded a breathing rate of 50 breaths/hr for one female gray whale with a calf in Bahia de Magdalena. Although tagged whales made from 16.0 to 67.0 surfacings/hr (Table IV), our large sample size of surfacings and the certainty of detection of tagged whale signals suggests that a breathing rate of 36 breaths/hr is representative of gray whale activities in the lagoon.

Surfacing rates and surface durations are useful statistics in developing sightability correction factors for surveys. Many factors affect the sightability of a whale, such as weather, area covered, speed of sighting platform, and whale species and behavior. One important factor in sightability is the amount of time a whale is visible; radio tags can be used to estimate this time.

Surfacing rates of tagged whales were generally consistent throughout the day. Lower surfacing rates (15–20 surfacings/hr) appeared to be associated with resting animals and not with animals actively diving for long times. Swartz and Jones (1980) reported a midday reduction in whale activity, based on movements of whales in front of their observation tower. Tagged whale activity was not reduced in the midday using surfacing rates as an index.

Radio tags attached to whales can provide information on their behavior. Long dive durations, extremely long surface durations, and low surfacing rates were observed for whales resting at the surface. Whale 80Y, which had the longest average dive duration (2.58 min) and lowest surfacing rate (18.6 surfacings/hr) was found to be resting for 7 of the 20 hours that this whale was monitored. Surfacing patterns for other whales indicated resting periods up to 4 hr duration. Short average dive and surface durations and high surfacing rates were indicative of directed swimming. Whale 50B swam at a moderate speed (4 km/hr) for 2 of 7 hr monitored; he/she had an average dive duration of 0.96 min and a surfacing rate of 59.3 surfacings/hr. In the future, additional visual observations of behavior with data collected from radio tags attached to whales may allow further correlations between whale behavior and signal patterns from tags. The signal patterns from radio tags on whales then may allow continuous collection of behavioral information day and night, in bad weather, and when the whale is out of view.

SURFACING PATTERNS

The clumped surfacing pattern described in this study has been documented for migrating gray whales (Wyrick, 1954; Gilmore, 1961; J. Sumich, personal communication) and for whales occupying the breeding and calving lagoons (Norris *et al.*, 1977; Swartz and Jones, 1979). The regular pattern, however, has not been previously reported. This pattern occurred approximately 20% of the time and was observed almost as frequently as the clumped pattern. The regular surfacing pattern was probably a consequence of the behavior associated with lagoon residency. Whales in the lagoon, especially females with calves, spent many hours milling within the lagoon oriented into the current and remaining fairly stationary or resting. The regular surfacing pattern may therefore be a product of this inactive period.

LOCAL MOVEMENTS

Swartz and Jones (1980) observed gray whale movements from a tower positioned at Rocky Point in San Ignacio Lagoon. They reported a preponderance of whales moving northward into the upper lagoon areas in the morning; the predominant direction of travel in the afternoon was southward toward the lagoon entrance. These observations were made approximately 6 km from the lagoon entrance, so the observors could not determine if the whales actually moved out of the lagoon. Our study supports their observations; radio-tagged whales generally left the lagoon during the night and returned in the morning hours. Swartz and Jones (1980) also found that gray whales moved with the direction of the tide, especially in the shallow upper lagoon. The movement of radio-tagged whales out of the lagoon did not significantly coincide with tidal flow. They did, however, move with more often than against the tide.

The reason for nighttime movements of gray whales out of the lagoon is not known at this time. Swartz and Jones (1980) suggested that aggregations of shrimp, which occur near the lagoon mouth at night, may provide a food source for whales. Examination of two stranded yearling gray whales in the lagoon did not reveal evidence of shrimp predation, however (J. Harvey, unpublished data).

LAGOON RESIDENCY

Tagged whales in general did not remain in the lagoon for extended periods of time. Seven tagged whales were later located by radio signals along the Baja and California coasts, indicating the tags had remained attached beyond monitoring periods in the lagoon. The short residence time for single tagged whales was expected, because single animals appear to be somewhat transitory and use the upper lagoon only infrequently (Swartz and Jones, 1979). The peak abundance of females with calves in the lagoon occurs from mid-March to early April (Swartz and Jones, 1980). The three tagged females with calves, which remained in the lagoon for the longest time, were tagged during this period. By April, whales begin to leave the lagoon and move northward. The last four whales tagged (April 9–14) left the lagoon after 1–2 days, presumably commencing their northward migration.

The short residence time in the lagoon and daily movements for many of the tagged whales suggest that there is a considerable turnover in the lagoon population. Estimates of whale abundance in San Ignacio Lagoon fluctuated between consecutive daily counts (Swartz and Jones, 1980). This implies that some whales may be continuously moving out of the lagoon while others replace them. Estimates of the total number of whales using a particular lagoon, therefore, cannot be made simply from visual counts. Further tagging studies may help in developing turnover rates, which will allow the number of whales using a lagoon to be calculated based on visual surveys.

Radio tags attached to whales can provide a large amount of information on whale behavior that can be very difficult or impossible to collect using other techniques. The potential data which can be gathered is expanded to include 24 hr (continuous) observations during bad weather and when the whale is out of visual range. The coupling of visual observations of whale behavior with reception of radio-tag signals can be used to develop profiles of surfacing behavior for whales throughout the day. Additionally, radiotelemetry can collect physiological and hydrographic data as long as the tags remain attached.

Dive and surface durations and surfacing rates may be used to calculate sightability correction factors for cetaceans. Knowledge of time spent at the surface and the rate of surfacings may provide better estimates of whale abundance.

Summary

Ten gray whales, *Eschrichtius robustus*, were radio tagged and monitored in San Ignacio Lagoon, Baja California Sur, Mexico, from February 9 to April 15, 1980. Mean duration of dive for individual whales varied from 1.0 to 2.6 min ($\bar{x} = 1.6 \pm 0.02$ min). Ninety-nine percent of the 11,080 dives recorded were less than 6 min and 49% less than 1 min in duration. The longest dive was 25.9 min. Tagged whales averaged 4.4 ± 0.6 sec at the surface per surfacing. Eight of the tagged whales averaged less than 2.9% of the time at the surface (range, 1.56–16.3%). The tagged whales averaged 35.6 surfacings per hr. Three surfacing patterns were documented (regular–long, regular–short, and clumped) which accounted for approximately one-half of all dive sequences analyzed for two whales. Three radio-tagged whales were monitored for 4, 5, and 11 days, and moved into the ocean on 2, 2, and 7 occasions, respectively. Most oceanic movements were at night and 40% were against the tide. Seven of the tagged whales did not remain in the lagoon for more than 2 days.

Acknowledgments

We sincerely thank B. Blakesley, N. Dimeo, R. Gillette, M. L. Mate, R. Sanchez, R. Starr, and M. Yoklavich for their fine work and good humor in the field. We appreciate the talents and contributions of O. Batdorf, D. Beaty, and staff of Telonics, Inc., R. Maiefski, and R. Moeller. S. Swartz, M. L. Jones, M. Symons, J. and S. Sumich, T. Starr, Fisherman's Landing, and H & M Landings provided logistical help, equipment, and ideas. We thank R. Brown, J. Sumich, S. Swartz, W. Watkins, and M. Yoklavich for reviewing the manuscript; F. Ramsey, R. Hanlen, and S. Shiboski for statistical consultation; and N. Brown for assisting in manuscript preparation. This work was carried out under a Mexican scientific research permit issued by the Departamento de Pesca. Funding was provided by the Bureau of Land Management, Outer Continental Shelf Program (Contract #AA-730-79-4120-0109), and the Office of Naval Research (contract #700014-79-C-004).

References

Evans, W. E. (1974). Telemetering of temperature and depth from a free ranging yearling California gray whale, *Eschrichtius robustus. Mar. Fish. Rev.* **36,** 52–58.

Gilmore, R. M. (1961). "The Story of the Gray Whale," 2nd ed. Pioneer Printers, San Diego, California.

Mate, B. R., Harvey, J. T., Hobbs, L., and Maiefski, R. (1983). A new attachment device for radio-tagging large whales. *J. Wildl. Manage.* **47**(3), 868–872.

Norris, K. S., and Gentry, R. L. (1974). Capture and harnessing of young California gray whales, *Eschrichtius robustus. Mar. Fish. Rev.* **36**(4), 58–64.

Norris, K. S., Goodman, R. M., Villa-Ramirez, V., and Hobbs, L. (1977). Behavior of California gray whale, *Eschrichtius robustus*, in southern Baja California, Mexico. *Fish. Bull.* **75**(1), 159–172.

Rice, D. A., Wolman, A., Withrow, D., and Fleischer, L. (1981). Gray whales on the winter grounds in Baja California. *Rep. Int. Whal. Commn.* **31,** 477–493.

Swartz, S., and Cummings, W. C. (1978). Gray whales (*Eschrichtius robustus)* in Laguna San Ignacio, Baja California, Mexico. *U.S. NTIS, PB Rep.* **PB-276319,** 1–38.

Swartz, S., and Jones, M. L. (1979). The evaluation of human activities on gray whales, *Eschrichtius robustus*, in Laguna San Ignacio, Baja, California, Mexico. *U.S., NTIS, PB Rep.* **PB-289737,** 1–42.

Swartz, S., and Jones, M. L. (1980). Gray whales (*Eschrichtius robustus)* during the 1977–1978 and 1978–1979 winter seasons in Laguna San Ignacio, Baja California Sur, Mexico. *U.S., NTIS, PB Rep.* **PB80-202989,** 1–34.

Sweeney, J. C., and Mattsson, J. L. (1974). Surgical attachment of a telemetry device to the dorsal ridge of a yearling California gray whale, *Eschrichtius robustus. Mar. Fish. Rev.* **36,** 20–22.

Wyrick, R. F. (1954). Observations on the movements of the Pacific gray whale, *Eschrichtius glaucus* Cope. *J. Mammal.* **35,** 596–598.

25

Ocean Movements of Radio-Tagged Gray Whales

Bruce R. Mate and James T. Harvey

Introduction

Because of the propensity of gray whales (*Eschrichtius robustus*) to travel in near-shore waters, the annual migration of this species is better known than that of any other whale. Their movements have been observed by scientists and the general public for more than a century (Scammon, 1874), but have only recently become the subject of more detailed study (Rice and Wolman, 1971; Reilly *et al.*, 1980; Chapter 9, this volume). To date most research has concentrated on movements of the population rather than of individuals, because of the difficulty of tracking individual whales over long periods and distances. Some gray whales can be sufficiently identified by natural marks, such as scars, barnacles or pigment patterns, to make them recognizable year after year (Hatler and Darling, 1974; Leatherwood, 1974; Chapters 12 and 14, this volume). Attempts to conduct long-term, continuous tracking of individuals by visual means, however, have been unsuccessful.

The development during the last two decades of progressively smaller and more reliable radio packages for tracking and telemetering data from whales has made studies of cetacean movements possible (Leatherwood and Evans, 1979). Pioneering work in developing projectile radio tags was carried out by Schevill and Watkins (1966, and reported in Tillman and Johnson, 1976) and Ray *et al.* (1978). Recently these techniques have been applied to humpback (*Megaptera novaeangliae*) and fin whales (*Balaenoptera physalus*) (Watkins *et al.*, 1978) and Bryde's whales (*Balaenoptera edeni*) (Watkins *et al.*, 1979). Killer whales (Erickson, 1977), pilot whales (Evans, 1974), and

ISBN 0-12-389180-9

dolphins of several species (Evans, 1974; Leatherwood and Ljungblad, 1979; Wursig, 1979) have been captured, fitted with transmitters, released, and tracked in a number of locales.

Early attempts to track gray whales using acoustic techniques were unsuccessful (Schultz and Pyle, 1965). But Sweeney and Mattsson (1974) surgically sutured a 27-MHz radio transmitter to the dorsal ridge of a captive yearling gray whale before its release at sea and subsequent tracking by Evans (1974). Norris and Gentry (1974) captured suckling gray whale calves in Boca de Soledad, Baja, Mexico and attached similar transmitters with flexible harnesses and timed-release mechanisms (Norris and Gentry, 1974; Norris *et al.*, 1977). From each of these efforts, useful information was collected on dive duration, nearshore movements, and depths of dives of gray whales. This paper describes the first successful radio-tagging of unrestrained gray whales and examines their ocean movements along the west coast of North America.

Methods

Eighteen gray whales were tagged between February and April in 1979 and 1980 in Laguna San Ignacio, Baja, Mexico. Laguna San Ignacio (880 km south of San Diego, California) is one of three Pacific lagoons along the Baja coast where gray whales calve and breed (Fig. 1). We chose this area because of reports of inquisitive ("friendly" or "curious") behavior by a small proportion of the whales there (Gilmore, 1976; Swartz and Jones, 1980, Chapter 14, this volume). Inquisitive whales approach or tolerate boats and often allow themselves to be touched.

Full details of the transmitters and their attachment have been described by Mate *et al.* (1983) and are only summarized here. We used L2B5 transmitters (manufactured by Telonics) designed to transmit continuously for 4 months. The transmitters were 3.8 × 3.6 × 2.5 cm, weighed 70 g, and emitted a 10-mW, 30-msec pulse every 300–750 msec on discrete frequencies in the 148- to 150-MHz range. Two types of antennas were used; a 46-cm stainless steel $\frac{1}{4}$-wave whip encased in heat-shrink tubing, and a 22.9-cm truncated helix potted in polyurethane. Each transmitter had a 1 m × 7.5 cm colored plastic streamer and was attached to a stainless steel plate.

Two mounting and attachment techniques were used, umbrella tags and barnacle tags. Umbrella tags mounted to a plate (17.5 × 3.75 × 0.25 cm) were attached 1–2 m behind the whale's blowhole along the dorsal midline. In this position the transmitter's antenna cleared the water with each whale's breath. The attachment consisted of an umbrella anchor at each end of the plate. Each anchor consisted of a piston with six curved inconel legs (8 cm × 2 mm) each of which passed through a slot in the plate and spread laterally in the blubber layer to a depth of 6 cm (Fig. 2). The tag was applied by means of an applicator at the end of a hand-held 5-m pole. The applicator held the transmitter/plate assembly to align the umbrella anchors properly. Microswitches at both ends of the applicator assured the device was flat on the whale's back before electrically actuated pressure cartridges (Holex Mfg. Co.) deployed the anchors.

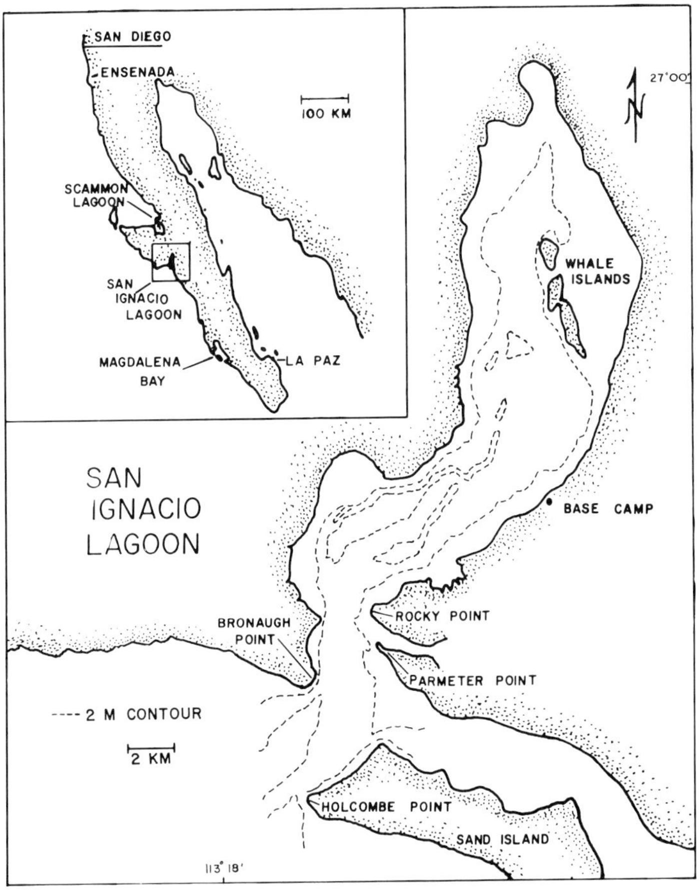

Fig. 1. *San Ignacio Lagoon, Baja California, Mexico was the location of all the tagging activities. A base camp along the eastern shore and field camp at Bronaugh Point were established to monitor radio-tagged whales.*

Barnacle tags used eight curved inconel legs to hold a 7.6-cm diameter circular base plate to the surface of the whales. On the upper surface of the plate the transmitter was embedded in a hemisphere of polyurethane resin (Fig. 3). A sliding ring held the distal ends of the legs in an 8-cm diameter circle prior to deployment. A truncated helix antenna was attached to the hemisphere of polyurethane. This projectile tag was shot by means of a compound bow at distances up to 8 m. A hollow arrow shaft covered the antenna and fitted into a molded recess at its base.

Whales were approached in the lagoon by inflatable boat, and tags were applied to approachable whales. We were extremely conservative in our tagging, taking precautions to assure good tag placement with both techniques. We monitored the surfacing and movements of tagged whales in the lagoon (Chapter 24, this volume) from inflatable boats and land-based camps (Fig. 1). Additional monitoring occurred from receiving stations at La Jolla, California; Newport, Oregon; Unimak Pass, Alaska; and during

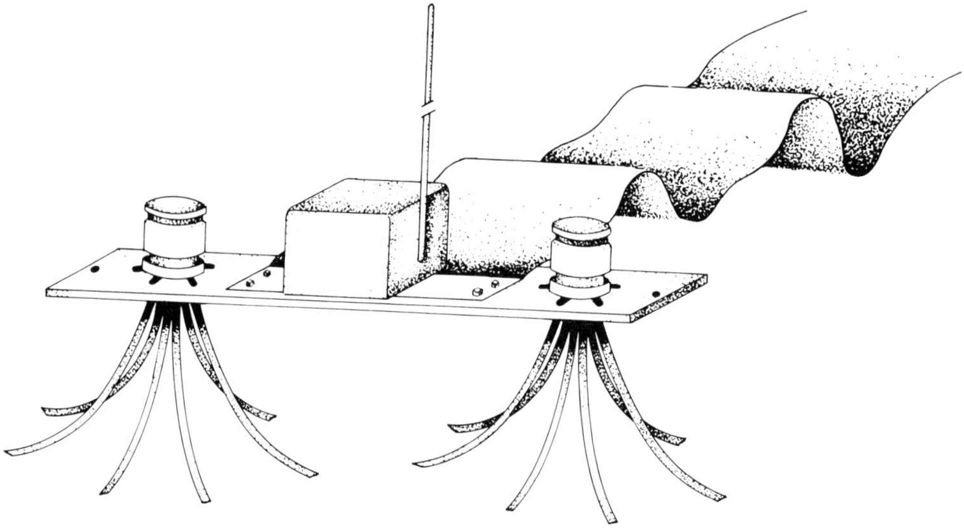

Fig. 2. *Umbrella tag showing transmitter (square box) with its antenna, visual identification streamer, and two umbrella attachments mounted on a base plate.*

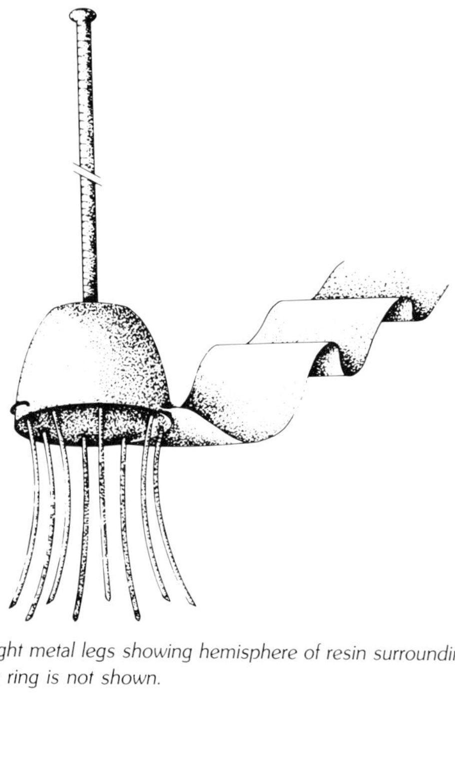

Fig. 3. *Barnacle tag with eight metal legs showing hemisphere of resin surrounding transmitter, antenna, and visual streamer. The retaining ring is not shown.*

aerial surveys in a Cessna 182. The plane was equipped with a Telonics TR-2 scanning receiver, two strut-mounted, two-element Yagi antennas, and a Telonics direction finder receiver (DF) using two belly-mounted DF antennas. During 1979, only one aerial survey (in early May) was conducted to locate tagged whales along the Washington, Oregon, and California coasts. During 1980, aerial surveys were flown principally along the Baja shoreline north and occasionally south of San Ignacio Lagoon in the process of supplying the camp and were the only source of whale movement information in Baja outside San Ignacio Lagoon.

Results and Discussion

DEPLOYMENT

Three (1979) and 11 (1980) umbrella tags were applied to gray whales in the 2 years: 3 adult females (without calves), 2 adults of unknown sex, and 9 females with calves. Whales did not react visibly to tagging with umbrella tags; instead, they continued their inquisitive behavior as before tagging. We tagged only one whale that did not exhibit inquisitive behavior. This female was sleeping at the surface with her calf nearby as we quickly approached. She awoke as the tag was attached and reacted with a vigorous fluke slap and rapid swimming. The calf was able to keep up with its mother, however, and both were observed slowly swimming in the lagoon the next day.

We noted some difference in the pressure needed to fully attach umbrella tags to adults with calves in comparison to those without calves. The amount of pressure that provided complete attachment on several adults without calves was insufficient to secure attachment on at least 8 of 9 adults with calves. We speculate that females with calves had used up some of their lipid reserves in the blubber layer during the development of a fetus and subsequent lactation, making the blubber a more compact layer of connective tissue and hence more resistant to penetration.

A barnacle tag was applied to each of four females with calves during 1980. Two of the barnacle tags hit the water before reaching the whales and did not fully deploy. The other two tags did attach fully. The very low angle from which the latter were applied resulted in a dorsolateral placement and a poor antenna angle. Vertical antenna orientation is preferred for the best signal output and is possible with a middorsal placement if barnacle tags are applied from a higher vantage point, such as from the bow of a larger vessel.

RETENTION

On four adults without calves, umbrella tags have remained attached with operative transmitters for at least 6, 40, 50, and 94 days. In the last of these cases when the animal was seen 810 days after tagging, the tag was still attached. Incompletely attached umbrella tags have remained attached on four females with calves for at least 3, 12.5, 18, and 20 days. Barnacle tags have remained attached for at least 20 days.

MOVEMENTS

Radio tracking has been possible for as long as 94 days (B. R. Mate and J. T. Harvey, unpublished data) with nearly continuous monitoring from within the lagoon for up to 12 days. The longest movement recorded was for a single adult (100W) tagged on February 27, 1979 (Fig. 4). Signals from this animal were subsequently received at the La Jolla receiver site 43 days after tagging, off Oregon by either aircraft or land-based receivers 61, 62, and 63 days after tagging, and at Unimak Pass, Alaska, 94 days after tagging (Table I). This animal had traveled at least 6680 km from the tagging site. The same animal was resighted by five researchers aboard a University of California research vessel 27 months after tagging.

In 1980, four female gray whales with calves (50R, 80R, 160R, 50B) and 3 single adults (50Y, 80Y, 160Y) were relocated a total of 16 times after they had left San Ignacio Lagoon (Fig. 5). One single female (80Y) was located 24 days after tagging near Magdalena Bay (192 km south of the lagoon). Ten days later the same animal was located >500 km to the north of San Ignacio Lagoon. Two tagged females with calves were located in the vicinity of Ojo de Liebre Lagoon, one on 2 consecutive days. The observations of whales moving both north and south to adjacent breeding and calving areas suggests that movement among lagoons may not be uncommon.

The longest radio tracks of whale movements in 1980 were for whales without calves (Table II). Tags may stay attached longer on single whales than on females with calves. For example, females with calves may have a higher risk of tag loss as a result of less complete initial attachment and the large amount of physical contact between mother and young. However, as single whales were tagged earlier in the season and aerial surveys terminated before most females with calves had moved north of San Francisco, the shorter tracks of females with calves may merely reflect a bias in efforts to locate tagged whales.

During an aerial survey in late April 1980, one radio-tagged whale was relocated near El Rosario along the northern Baja coast, where 60 other adults (most with calves) were found. These animals were consistently surfacing with mud streaming from their mouths, apparently feeding 0.2–1.2 km offshore.

SWIMMING SPEEDS

Whales tagged in 1980 travelled 208–1550 km in 6–50 days (Fig. 5). Minimum average swimming speeds, calculated for the shortest route between sightings, revealed considerable variability (Table II). Estimated speeds were generally higher for animals relocated after shorter periods of time. Longer intervals may include periods of wandering, resting, and feeding as well as directed travel, resulting in lower estimated swimming rates. Three of four females with calves had higher rates of travel (up to 96 km/day) than did single adult whales. Pike (1962) estimated an average northward speed of 91 km/day for gray whales, in northern waters, whereas Leatherwood (1974) estimated average speeds of 67 km/day off southern California. The single adult whale

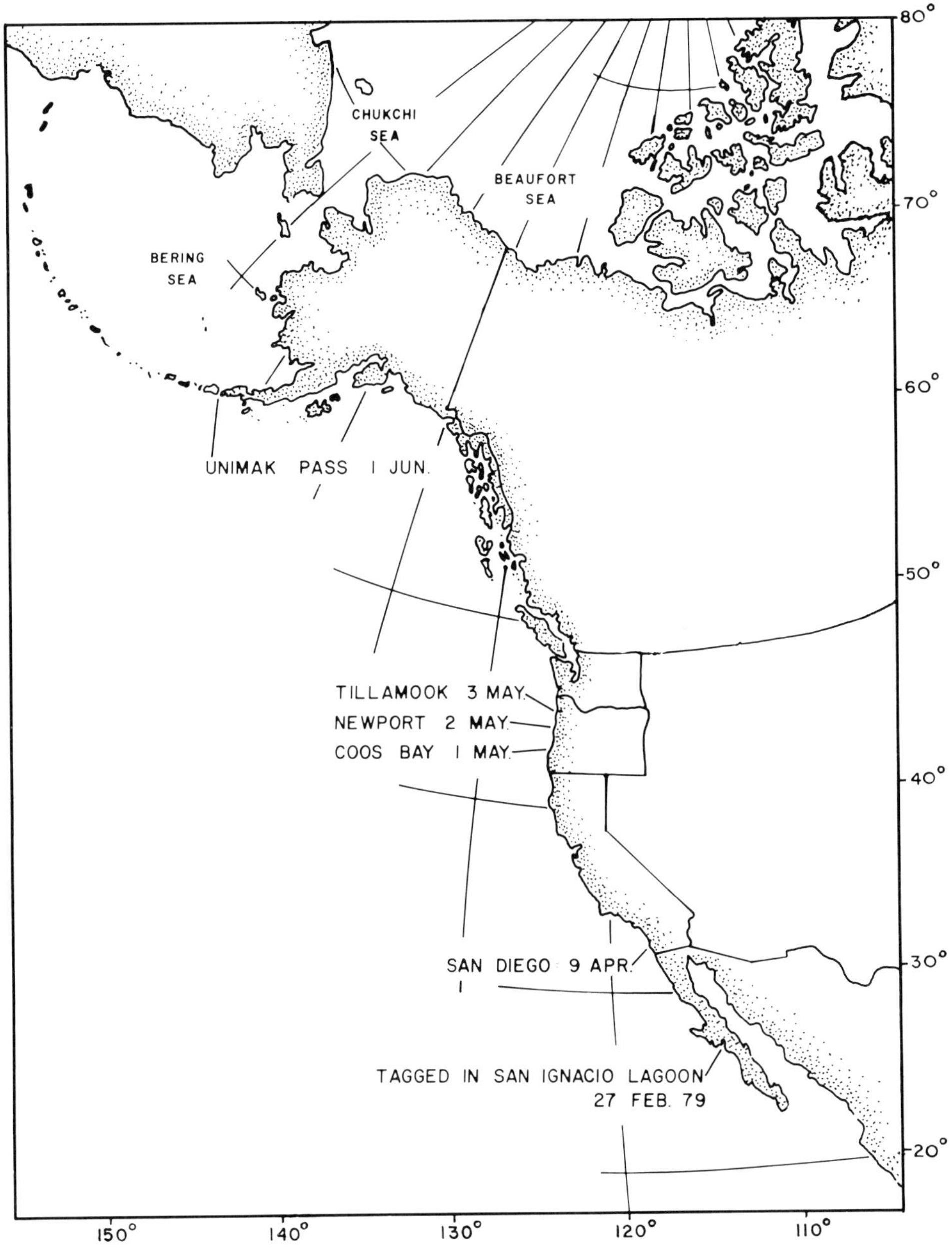

Fig. 4. *Locations and dates of signal receptions from a radio tag attached to an adult gray whale (100W) during the northward migration of 1979.*

Table I
Date, Location, Minimum Distance Traveled from Previous Location, Time from Previous Reception, Calculated Average Speed between Locations, and Number of Days after Tagging for Whale 100W[a]

Date	Location	Minimum distance (km)	Time (days)	Average speed between locations (km/day)	Number of days after tagging
February 27	San Ignacio Lagoon (26.8°N)	—	Date of tagging	—	—
March 14	San Ignacio Lagoon (26.8°N)	—	Last signals received in the lagoon	—	15
April 9	San Diego, California (32.5°N)	880	Departure date from lagoon unknown	>33	41
May 1	Coos Bay, Oregon (43.2°N)	1870	22	85	63
May 2	Newport, Oregon (44.4°N)	130	1.3	100	64
May 3	Tillamook, Oregon (45.3°N)	100	1	100	65
June 1	Unimak Pass, Alaska (54.3°N)	3700	29	127.6	94
	Total distance traveled	6680			

[a]Data from 1979.

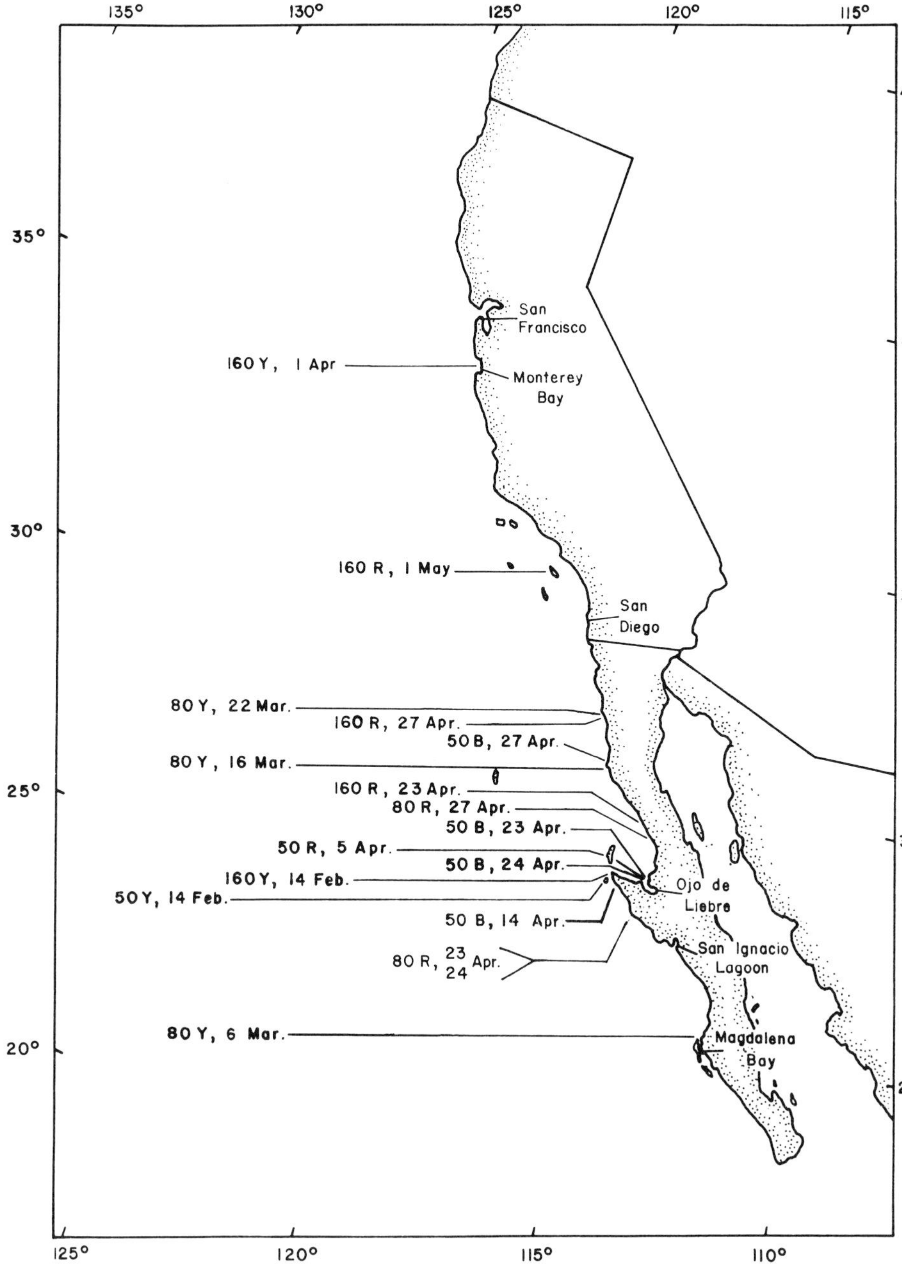

Fig. 5. *Ocean locations of gray whales tagged in San Ignacio Lagoon in 1980. The whale identification is followed by the date of signal reception.*

Table II
The Dates of Ocean Relocations, Minimum Distances between Locations, and Minimum Average Daily Swimming Speed for Radio-Tagged Whales

Whale I.D.	Dates observed in 1980	Distance (km) from prior sighting	Days between sightings	Minimum speed (km/day)
80Y (single)	Feb. 11	Tagged		
	Mar. 6	192 south	24	
	Mar. 16	706	10	70.6
	Mar. 22	176	6	29.3
		1074	40	$\bar{X}$ = 26.9
50Y (single)	Feb. 8	Tagged		
	Feb. 14	208	6	34.7
		208	6	
160Y (single)	Feb. 11	Tagged		
	Feb. 14	208	3	69.3
	Apr. 1	1342	47	28.6
		1550	50	$\bar{X}$ = 31.0
50R (female/calf)	Mar. 18	Tagged		
	Apr. 5	310	18	17.2
		310	18	
160R (female/calf)	Apr. 11	Tagged		
	Apr. 23	395	12	32.9
	Apr. 27	272	4	68
	Mar. 1	383	4	95.8
		1050	20	$\bar{X}$ = 52.5
80R (female/calf)	Apr. 14	Tagged		
	Apr. 23	144	8	18
	Apr. 24	0	1	0
	Apr. 27	226	3	75.3
		370	13	$\bar{X}$ = 28.5
50B (female/calf)	Apr. 11	Tagged		
	Apr. 14	184	3	61.3
	Apr. 23	128	9	9.0
	Apr. 24	0	1	0
	Apr. 27	280	3	93.3
		592	16	$\bar{X}$ = 37.0

(100W) from this study, which was relocated 5 times in 1979, averaged at least 33 km/day between San Ignacio Lagoon and San Diego; 81 km/day from San Diego, California to Coos Bay, Oregon; 100 km/day on 2 consecutive days of travel along the Oregon coast and 127 km/day for the subsequent 29 days of transit between Oregon and Unimak Pass, Alaska (Fig. 4). Although several days of high average daily speed were observed in Baja during 1980, the overall average was 31.4 km/whale-day. Slower speeds in Mexico and southern California may be due to activities such as mating and suckling, more common to the southern portion of the gray whales' range. Single whales appear to increase their rate of travel when they are north of these areas, and they perhaps swim at a more consistent rate as well.

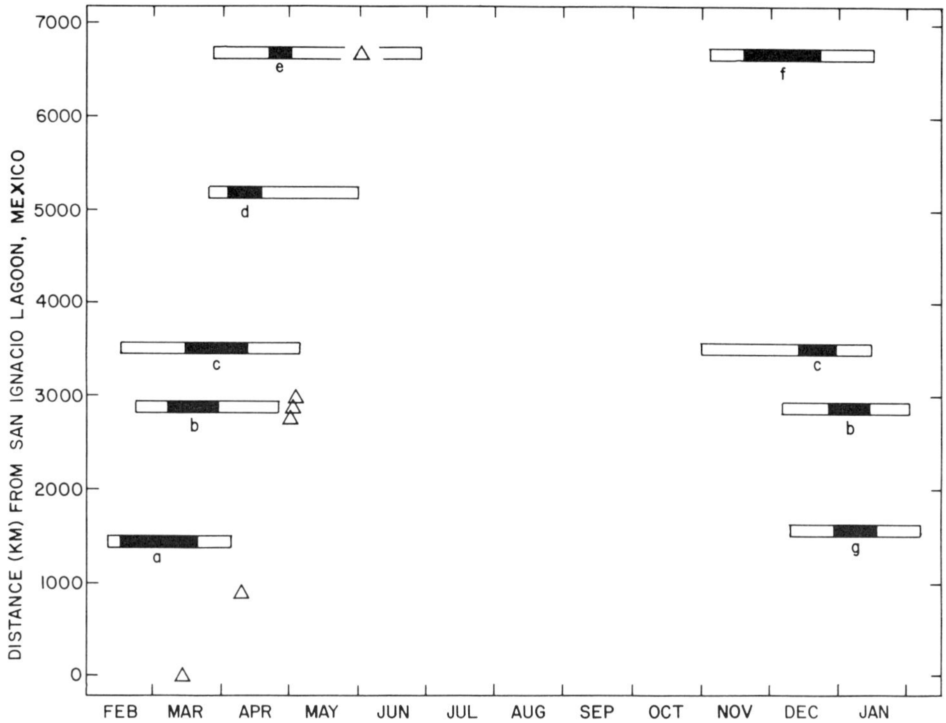

Fig. 6. *Times of northward and southward migrations of gray whales along western North America (open rectangle, with solid rectangle indicating period of peak passage rates; (A and G) Chapter 16, this volume; (B) Chapter 13, this volume; (C) Chapter 12, this volume; (D) Cunningham, no date, (E) P. Hessing, personal communication; (F) Chapters 9 and 10, this volume); and northward movements of one whale tagged in 1979 (△).*

These estimated rates of travel for individual whales can be compared to estimates for the bulk of the population by comparing the time at which maximum numbers of whales pass various locations in California, Oregon, and Alaska (Fig. 6). If the maximum passage rate occurring at each location is the result of a migratory "wave" of animals, then the estimated speed of the group is also an estimate of individual swimming speeds. The "wave" averaged 65 km/day between the central California and central Oregon coasts and 92 km/day from Oregon to Unimak Pass, Alaska (Chapters 9, 10, 13, and 16, this volume; Cunningham and Stanford, 1979; P. Hessing, personal communication). These figures compare favorably to the swimming speed estimates for the single whale tracked northward in 1979, supporting the idea that single animals move at a fairly consistent speed once they leave Mexico, and that there is a somewhat cohesive migratory wave. Note that the 1979 animal was one of the last single animals to leave San Ignacio Lagoon (still there March 14) and continued to be later than most other single whales in migrating north. We have included available data on the timing of peak numbers of whales passing study sites during the southbound migration in Fig. 6, which support Pike's (1962) contention that gray whales move faster during the southbound migration (approximately 125 km/day from Unimak Pass, Alaska to Monterey, California) than when they migrate north (approximately 88 km/day from Piedras Blancas to Unimak Pass in 1981).

Despite the advantages radio tags offer in migratory studies, they are still difficult logistically (ships, planes, people) and technically (receivers, ADFs, long-range transmitters). Our success with gray whales is attributable to their relatively well-known migration near shore and their accessibility in Baja. However, technological developments in radio transmitters, such as those by Jennings and Gandy (1980) for porpoises, suggest that satellite tracking of cetaceans in more pelagic situations may be possible in the future. It is predictable that future telemetered data from radio tags will not only better define migratory rates and routes, but also provide insight into cetacean behavior, navigation, physiology, and habitat preferences.

Summary

Eighteen gray whales were radio tagged using two different subdermal attachments: a projectile "barnacle" tag and an "umbrella" tag, applied at the end of a pole. Umbrella tags were easier to position than barnacle tags, and thus had better antenna orientation. Umbrella tags are known to have remained attached for up to 27 months, and signals from radio tags have been received up to 94 days after tagging (6680 km from the tagging site). Locations of radio-tagged whales have been determined using receivers on land and aboard aircraft. The average distance traveled per day during the northward migration was 85 km/day and was greater farther to the north, estimated at 127 km/day for the last 29 days. No differences were found between maximum calculated swimming speeds of single adults and those of females with calves. Some tagged animals moved both south and north from the tagging site (Laguna San Ignacio, Baja California Sur, Mexico) to adjacent calving/breeding areas. A few lingered around Laguna Ojo de Liebre and 1 whale was found apparently feeding with up to 60 other whales in an area along the northern Baja coast. Future technology may well allow collection of sensor data and location information telemetered by satellite.

Acknowledgments

We are extremely grateful to B. Blakesley, N. Dimeo, R. Gillette, M. L. Mate, R. Sanchez, R. Starr, and M. Yoklavich for their assistance in the field. We thank D. Beaty and Telonics, Inc., R. Maiesfski, O. Batdorff, R. Moeller, W. Watkins, and D. Wartzog for assistance in the development of the radio tags. S. Swartz, M. L. Jones, M. Symons, Fishermen's Landing, H & M Landing, J. and S. Sumich, and T. Starr provided logistic services. We thank J. Henderson, NMFS; M. Poole, U.S. Fish and Wildlife Service; B. Bartels, and J. Sarvis for monitoring shore receiver stations. We appreciate the critical review of the manuscript by R. Brown and W. Watkins. N. Brown assisted in preparation of the manuscript. The work was carried out under Mexican scientific research permits #1819 and #12541 issued by the Departamento de Pesca. Funding was provided by the Bureau of Land Management Outer Continental Shelf Program (Contract #AA-730-79-4120-0109), and the Office of Naval Research (Contract # N00014-79-C-0004). Thank you to J. Imm, C. Cowles, R. Tipper, and B. Zahuranec for patience and flexibility in contract monitoring.

References

Cunningham, W., and Stanford, S. (1979). Observations of migrating gray whales, (*Eschrichtius robustus*) at Cape St. Elias, Alaska. Unpublished manuscript.

Erickson, A. W. (1977). Killer whale population study. *In* "1976 Research in Fisheries, Annual Report of College of Fisheries," (D. Beall, ed.), pp. 66–67. University of Washington, Seattle.

Evans, W. E. (1974). Telemetering of temperature and depth information from a free-ranging yearling California gray whale, *Eschrichtius robustus*. *Mar. Fish. Rev.* **36**(4), 52–58.

Gilmore, R. M. (1976). The friendly whales of Laguna San Ignacio. *Terra* **15,** 24–28.

Hatler, D. G., and Darling, J. D. (1974). Recent observations of the gray whale in British Columbia. *Can. Field Nat.* **88,** 449–459.

Jennings, J. G., and Gandy, W. F. (1980). Tracking pelagic dolphins by satellite. *In* "A Handbook of Biotelemetry and Tracking" (C. J. Amlaner and D. W. MacDonald, eds.), pp. 753–755. Pergamon, Oxford.

Leatherwood, J. S. (1974). Aerial observations of migrating gray whales, *Eschrichtius robustus*, off southern California, (1969–1972). *Mar. Fish. Rev.* **36**(4), 45–49.

Leatherwood, J. S., and Evans, W. E. (1979). Some recent uses and potentials of radiotelemetry in field studies of cetaceans. *In* "Behavior of Marine Animals" (H. W. Winn and B. I. Olla, eds.), Vol. 3, pp. 1–31. Plenum, New York.

Leatherwood, J. S., and Ljungblad, D. K. (1979). Nighttime swimming and diving behavior of a radio tagged spotted dolphin, *Stenella attenuata*. *Cetology* **34,** 1–6.

Mate, B. R., Harvey, J. T., Hobbs, L., and Maiefski, R. (1983). A new attachment device for radio-tagging large whales. *J. Wildl. Manage.* **47**(3), 868–872.

Norris, K. S., and Gentry, R. L. (1974). Capture and harnessing of young California gray whales, *Eschrichtius robustus*. *Mar. Fish. Rev.* **36**(4), 58–64.

Norris, K. S., Goodman, R. M., Villa-Ramirez, V., and Hobbs, L. (1977). Behavior of California gray whale, *Eschrichtius robustus*, in southern Baja California, Mexico. *Fish. Bull.* **75**(1), 159–172.

Pike, B. G. (1962). Migration and feeding of the gray whale (*Eschrichtius gibbosus*). *J. Fish. Res. Board Can.* **19**(5), 815–838.

Ray, G. C., Mitchell, E. D., Wartzok, D., Kozicki, V. M., and Maiefski, R. (1978). Radio tracking of a fin whale, (*Balaenoptera physalus*). *Science* **202,** 521–524.

Reilly, S. B., Rice, D. W., and Wolman, A. A. (1980). Preliminary population estimate for the California gray whale based upon Monterey Shore Censuses, 1967/68 to 1978/79. *Rep. Int. Whal. Commn.* **30,** 359–368.

Rice, D. W., and Wolman, A. A. (1971). The life history and ecology of the gray whale (*Eschrichtius robustus*). *Spec. Publ.—Am. Soc. Mammal.* **3,** 1–142.

Scammon, C. M. (1874). "The Marine Mammals of the Northwestern Coast of North America." John Carmany & Co., San Francisco, California (reprinted by Dover, New York, 1968).

Schevill, W. E., and Watkins, W. A. (1966). "Radio-tagging of Whales," Ref. No. 66-17. Woods Hole Oceanogr. Inst., Woods Hole, Massachusetts (unpublished manuscript).

Schultz, J., and Pyle, C. (1965). Cat bites a whale. *Yachting* **118,** 48.

Swartz, S., and Jones, M. L. (1980). Gray whales (*Eschrichtius robustus*) during the 1977–1978 and 1978–1979 winter seasons in Laguna San Ignacio, Baja California Sur, Mexico. *U.S., NTIS PB Rep.* **PB80-202989,** 1–34.

Swartz, S., and Jones, M. L. (1983). Gray whale (*Eschrichtius robustus*) calf production and mortality in the winter range. *Rep. Int. Whal. Commn.* **33,** 503–507.

Sweeney, J. C., and Mattsson, J. L. (1974). Surgical attachment of a telemetry device to the dorsal ridge of a yearling California gray whale, *Eschrichtius robustus*. *Mar. Fish. Rev.* **36,** 20–22.

Tillman, M. F., and Johnson, J. H. (1976). "Radio Tagging of Humpback Whales," Report to the National Marine Fisheries Service (unpublished manuscript).

Watkins, W. A., Johnson, J. A., and Wartzok, D. (1978). "Radio Tagging Report of Finback and Humpback Whales," WHOI Ref. No. 78-51. Woods Hole Oceanogr. Inst., Woods Hole, Massachusetts.

Watkins, W. A., di Sciara, G. N., and Moore, K. E. (1979). "Observations and Radio Tagging of *Balaenoptera edeni* near Puerto La Cruz, Venezuela," WHOI Ref. No. 79-89. Woods Hole Oceanogr. Inst., Woods Hole, Massachusetts.

Wursig, B. (1979). Dolphins. *Sci. Am.* **240,** 136–148.

Index

C

D

R

S

T

U

V

W

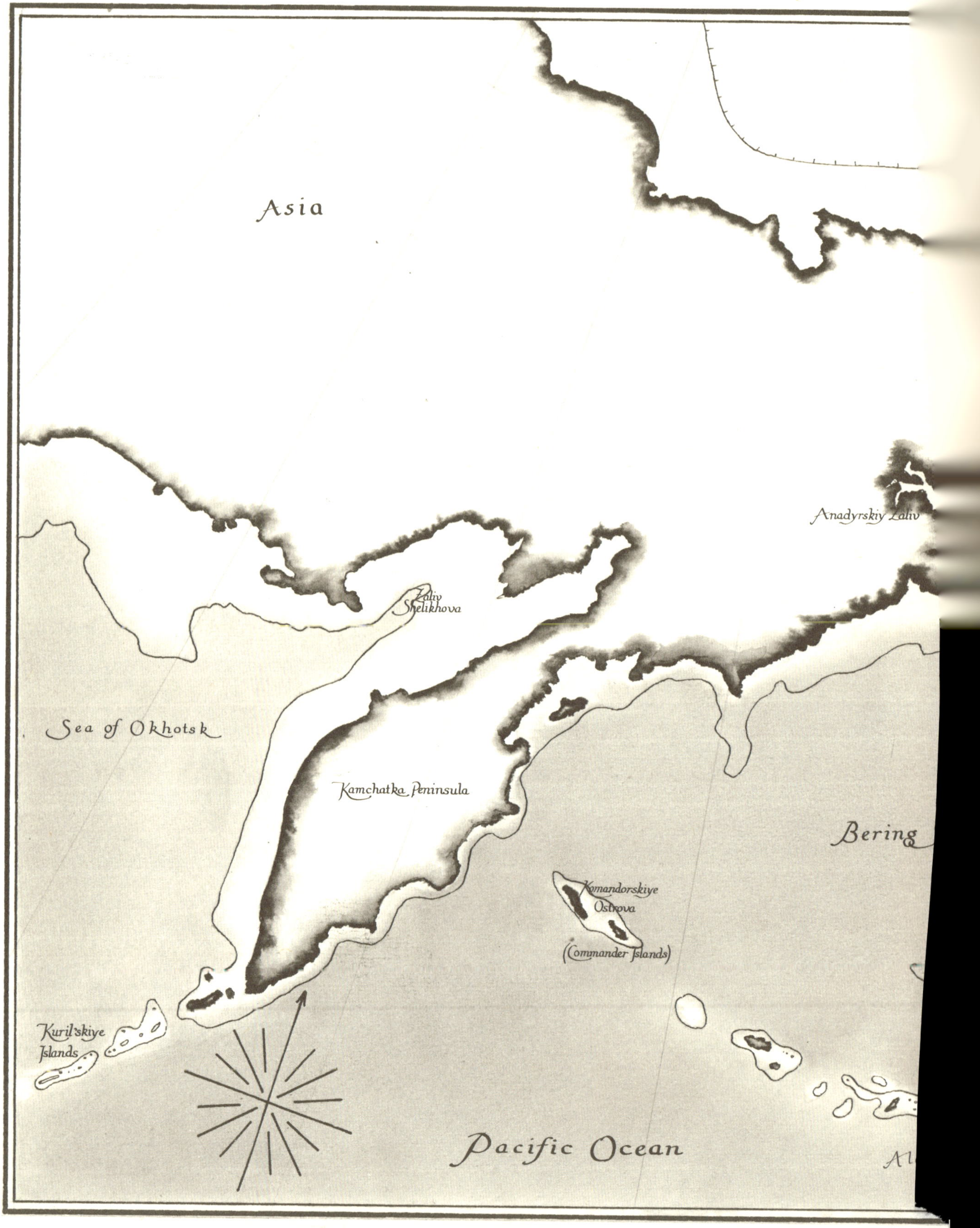

Asia
Anadyrskiy Zaliv
Zaliv Shelikhova
Sea of Okhotsk
Kamchatka Peninsula
Bering
Komandorskiye Ostrova
(Commander Islands)
Kuril'skiye Islands
Pacific Ocean